AF888817

Current Conveyors

Raj Senani • D.R. Bhaskar • A.K. Singh

Current Conveyors

Variants, Applications and Hardware Implementations

Springer

Raj Senani
Electronics and Communication
Engineering
Netaji Subhas Institute
of Technology
New Delhi, India

D.R. Bhaskar
Electronics and Communication
Engineering
Jamia Millia Islamia
New Delhi, India

A.K. Singh
Electronics and Communication
Engineering, Sharda University
Greater Noida, UP, India

ISBN 978-3-319-08683-5 ISBN 978-3-319-08684-2 (eBook)
DOI 10.1007/978-3-319-08684-2
Springer Cham Heidelberg New York Dordrecht London

Library of Congress Control Number: 2014948342

© Springer International Publishing Switzerland 2015
This work is subject to copyright. All rights are reserved by the Publisher, whether the whole or part of the material is concerned, specifically the rights of translation, reprinting, reuse of illustrations, recitation, broadcasting, reproduction on microfilms or in any other physical way, and transmission or information storage and retrieval, electronic adaptation, computer software, or by similar or dissimilar methodology now known or hereafter developed. Exempted from this legal reservation are brief excerpts in connection with reviews or scholarly analysis or material supplied specifically for the purpose of being entered and executed on a computer system, for exclusive use by the purchaser of the work. Duplication of this publication or parts thereof is permitted only under the provisions of the Copyright Law of the Publisher's location, in its current version, and permission for use must always be obtained from Springer. Permissions for use may be obtained through RightsLink at the Copyright Clearance Center. Violations are liable to prosecution under the respective Copyright Law.
The use of general descriptive names, registered names, trademarks, service marks, etc. in this publication does not imply, even in the absence of a specific statement, that such names are exempt from the relevant protective laws and regulations and therefore free for general use.
While the advice and information in this book are believed to be true and accurate at the date of publication, neither the authors nor the editors nor the publisher can accept any legal responsibility for any errors or omissions that may be made. The publisher makes no warranty, express or implied, with respect to the material contained herein.

Printed on acid-free paper

Springer is part of Springer Science+Business Media (www.springer.com)

Acknowledgements

The first author of this monograph (RS) started his academic career as a faculty member in the Electrical Engineering Department of Motilal Nehru Regional Engineering College (MNREC), Allahabad, in 1975 and started his independent research in the area of *Active Circuits* – an endeavor, in which he was very much inspired by the works of Professors R. W. Newcomb, S. C. Dutta Roy, A. S. Sedra, K. C. Smith, (Late) B. B. Bhattacharyya, and M. N. S. Swamy.

Besides undergraduate and postgraduate teaching, the first author (RS) was very keenly pursuing research on some problems related to inductance simulation, oscillator synthesis, and analog active filter design. *Current Conveyors* (CC) introduced by Sedra and Smith had already started attracting the attention and imagination of several researchers around the world (as an alternative to the more established IC op-amps) due to the promise which they appeared to hold in several applications even though no off-the-shelf IC CCs were available till then. As a teacher, he had the privilege of being granted ample academic freedom to introduce newly emerging ideas in a course on *Analog Integrated Circuits* in which he started teaching current-mode translinear circuits and current conveyors as early as in 1982 – a practice which he continued to follow even after shifting to Delhi Institute of Technology (DIT) in 1988. The first author (RS) since 1975 and several of his PhD students (including the second and third author of this monograph: DRB and AKS) since 1988 have thus been very intimately involved in the research on current-mode circuits in general and on current conveyors in particular.

After having written a monograph on *Current feedback Operational amplifiers and their Applications*, the authors of this monograph realized that since the CFOA is an offshoot of the more general building block – the *Current Conveyor (CC)*, which is a more fundamental concept than the CFOA and because of an extremely rich repertoire of literature available on CCs (spread over 1,500 research publications which have appeared in reputed journals during the past more than four decades!), it certainly deserves to be the topic of a full-fledged monograph in its own right. Having convinced ourselves about this, we then set out to write this monograph and proposed the same to Charles Glaser, the Executive Editor, Springer, who gave us the signal to go ahead.

The authors are thankful to the facilities provided by the Analog Signal Processing (ASP) Research Lab., Division of ECE, Netaji Subhas Institute of Technology (NSIT), New Delhi, where the first author (RS) works and where this entire project was carried out. The first author gratefully acknowledges the understanding and appreciation of Professor B. N. Misra, Chairman, Board of Governors of NSIT, who was also the founder Director of the erstwhile DIT – the Institute where the first research lab named *Linear Integrated Circuits Lab* was created way back in 1988 under his patronage and with his unflinching support.

The authors gratefully thank their respective family members for their continued encouragement, moral support, and understanding shown during the preparation of this monograph.

Thanks are due to Charles Glaser for his support, to Jessica Lauffer for her gentle reminders, and to Shashi Rawat in particular, who provided great support and help in the preparation of the manuscript.

The authors would also like to thank their colleagues from their research group, namely, V. K. Singh, S. S. Gupta, R. K. Sharma, and Pragati Kumar for their moral support and understanding.

The authors have also been involved in teaching a number of ideas contained in this monograph to their students in the popular course *Bipolar and MOS Analog Integrated Circuits,* during which a persistent query from our students has been *In which book the material taught to them could be found?* We thank our numerous students for this and do hope that this monograph, like its predecessor on *Current feedback Operational Amplifiers and their Applications,* provides answers to their queries in respect of *Current Conveyors.*

About the Authors

Raj Senani received B.Sc. from Lucknow University, B.Sc. Engg. from Harcourt Butler Technological Institute, Kanpur, M.E. (Honors) from Motilal Nehru National Institute of Technology (MNNIT), Allahabad and Ph.D. in Electrical Engg. from the University of Allahabad.

Dr. Senani held the positions of Lecturer (1975–1986) and Reader (1986–1988) at the EE Department of MNNIT, Allahabad. He joined the ECE Department of the Delhi Institute of Technology (now named as Netaji Subhas Institute of Technology) in 1988 and became a full Professor in 1990. Since then, he has served as Head, ECE Department, Head Applied Sciences, Head, Manufacturing Processes and Automation Engineering, Dean Research, Dean Academic, Dean Administration, Dean Post Graduate Studies and Director of the Institute, a number of times.

Professor Senani's teaching and research interests are in the areas of Bipolar and CMOS Analog Integrated Circuits, Electronic Instrumentation and Chaotic Nonlinear Circuits. He has authored/co-authored over 140 research papers in various international journals, four book chapters and one monograph '*Current feedback operational amplifiers and their Applications*' (Springer, 2013). He is currently serving as Editor-in-Chief for IETE Journal of Education and as an Associate Editor for the Journal on Circuits, Systems and Signal Processing, Birkhauser Boston (USA) since 2003, besides being on the editorial boards of several other journals and acting as an editorial reviewer for 30 international journals.

Professor Senani is a Senior Member of IEEE and was elected a Fellow of the National Academy of Sciences, India, in 2008. He is the recipient of Second Laureate of the 25th Khwarizmi International Award for the year 2012. Professor Senani's biography has been included in several editions of Marquis' Who's Who series (published from N.J., USA) and a number of other international biographical directories.

D. R. Bhaskar received B.Sc. degree from Agra University, B. Tech. degree from Indian Institute of Technology (IIT), Kanpur, M.Tech. from IIT, Delhi and Ph.D. from University of Delhi. Dr. Bhaskar held the positions of Assistant Engineer in DESU (1981–1984), Lecturer (1984–1990) and Senior Lecturer (1990–1995) at the EE Department of Delhi College of Engineering and Reader in ECE Department of Jamia Millia Islamia (1995–2002). He became a full Professor in January 2002 and has served as the Head of the Department of ECE during 2002–2005.

Professor Bhaskar's teaching and research interests are in the areas of Analog Integrated Circuits and Signal Processing, Communication Systems and Electronic Instrumentation. He has authored/co-authored over 75 research papers in various International journals, three book chapters and one monograph '*Current feedback operational amplifiers and their Applications*' (Springer, 2013). Professor Bhaskar is a Senior Member of IEEE. He has acted/has been acting as a Reviewer for several international journals. His biography is included in a number of international biographical directories.

Abdhesh Kumar Singh received M.Tech. in Electronics and Communication Engineering from IASED and Ph.D., in the area of Analog Integrated Circuits and Signal processing, from Netaji Subhas Institute of Technology (NSIT), University of Delhi. Dr. Singh held the positions of Lecturer and Senior Lecturer (June 2000-August 2001) at the ECE Department, AKG Engineering College, Ghaziabad. He joined ECE Department of Inderprastha Engineering College, Ghaziabad, India as a Senior Lecturer in August 2001 where he became Assistant Professor in April, 2002 and Associate Professor in 2006.

At present, he is a full Professor at the ECE Department of Sharda University, Greater Noida. His teaching and research interests are in the areas of Bipolar and MOS Analog Integrated Circuits and Signal Processing. Dr. Singh has authored/co-authored 40 research papers in various International journals, three book chapters and one monograph '*Current feedback operational amplifiers and their Applications*' (Springer, 2013).

Preface

It is well recognized that in spite of the dominance of digital circuits and techniques, analog circuits are indispensable since all natural signals are analog. Analog circuits and techniques are hence essentially required in realizing signal amplifiers, continuous-time filters, rectifiers, sinusoidal oscillators, analog-to-digital and digital-to-analog converters, data acquisition and signal conditioning, analog multipliers and dividers, and some types of artificial neural networks.

During the last four decades, there has been tremendous interest in current-mode techniques for the design of analog circuits, and these have given rise to a number of interesting configurations. In several cases, current-mode circuits provide attractive alternatives to their voltage-mode counterparts in terms of providing one or more of the several advantageous features such as better linearity, better accuracy, higher operational frequency range, larger dynamic range and realisability of the intended functions with the least possible number of components without requiring any component-matching conditions, etc.

It must be mentioned that although over two dozen new active building blocks have been introduced in the circuit theory literature for processing analog signals, however no other development has influenced and affected the field of analog electronic circuit design as prominently and as significantly as the new active elements named *current conveyors* introduced by Sedra and Smith in the late 1970s.

And yet, the huge amount of literature on current conveyors consisting of over 1,500 scholarly articles written by researchers from around the world has by and large remained confined to about two dozen professional journals only. The only exceptions are one book written on the *limited* topic of 'Low-voltage, Low-power CMOS Current Conveyors' published by Kluwer Academic Publishers in 2003 and two more books titled 'IC Analog filter design-a Current Conveyor approach' and 'CMOS Second generation Current Conveyors' published in 2011 and 2012 respectively by Lambert Academic Publishers Inc., both of which too are *limited* to only the topics of active filter design and some specific CMOS implementations of current conveyors.

Thus, to the best knowledge of the authors, no book has so far been written on current conveyors *in a comprehensive manner* in spite of the fact that the literature on the hardware implementation of current conveyors alone runs into several hundred research papers while their variants and applications are spread well over one thousand research publications!

During the past four decades, *current conveyors* have been employed in numerous applications such as precision rectifiers, universal voltage-mode and current-mode biquad filters, single-element-controlled sinusoidal oscillators, quadrature and multiphase oscillators, relaxation oscillators, multivibrators, VCOs, synthetic impedance realizations, chaos generators, the design of field-programmable analog arrays, etc. to name a few.

On the other hand, several IC-manufacturing companies have produced a number of current conveyor ICs such as CCII01 from LTP Electronics, PA630 from Phototronics Ltd. and AD844 from Analog Devices Inc. Also, several of the other ICs disguised as operational amplifiers or operational transconductance amplifiers such as OPA2662, OPA660, and OPA860 have in fact a current conveyor in their internal circuit architecture and therefore can be readily used as current conveyors too. Thus, it is now widely recognized and accepted by analog designers as well as the academic community that *current conveyors are the devices whose time has now come!*

In view of the above, the authors of this monograph thought that the time is now ripe for writing a comprehensive treatise on current conveyors, their variants, their discrete and integratable hardware implementations, and their numerous applications at a single place, and hence this monograph.

It is hoped that this monograph, which contains a discussion of over 500 current conveyor-based analog circuits with their relevant theory and design/performance details, should be useful for academicians, practicing engineers, and anybody interested in analog circuit design. Readers may also find a number of interesting and challenging problems worthy of further investigations from the suggestions given in the various chapters.

Lastly, we must acknowledge that in a monograph based upon over 1,500 published research papers, there might have been some inadvertent omissions of some references; however, the same is not intentional. Aggrieved authors whose works might have been omitted are most welcome to bring to our attention (using the e-mail ID senani@ieee.org) any missing reference(s), which we would surely like to include in the next edition of this monograph. Any other suggestions are also most welcome!

New Delhi, India — Raj Senani
New Delhi, India — D.R. Bhaskar
Greater Noida, India — A.K. Singh
July 01, 2014

Contents

Part I Evolution and Hardware Implementation of Current Conveyors

1 The Evolution and the History of *Current Conveyors* 3
1.1 Prologue 3
1.2 The Origin of the First Generation Current Conveyor 4
1.3 The Second Generation Current Conveyor 6
1.4 An Historical Overview of the Evolution of the Other Varieties of Current Conveyors 7
References 13

2 Hardware Implementations of CCs Using Off-the-Shelf ICs 17
2.1 Introduction 17
2.2 Hardware Implementations of CCs Using Off-the-Shelf ICs 17
2.2.1 Black-Friedmann-Sedra CC Implementation Using an Op-Amp with Uncommitted Leads 17
2.2.2 Bakhtiar-Aronhime's Entirely Op-Amp-Based Implementation 19
2.2.3 Senani's Op-Amp-OTA Based Implementation 20
2.2.4 Huertas's Entirely Op-Amp Based CC Implementation 21
2.2.5 Pookaiyaudom and Samootrut Implementation Using OTAs 22
2.2.6 Papazoglou-Karybakas' Modified Version of Senani's CC Implementation 24
2.2.7 Karybakas-Siskos-Laopoulos's Compensated, Tunable CC 24
2.2.8 Wilson's OMA-Based Implementations of CCII+/− 25
2.2.9 CCII Implementation Using Op-Amps and Only NPN Transistors 26

2.2.10 Current Conveyor Implementation Using New Mirror Formulation . . . 28
2.2.11 Conversion of CCII into CCI and Vice Versa . . . 28
2.2.12 OMA-Based Multiple-Output CCs . . . 28
References . . . 30

3 Integratable Bipolar CC Architectures and Commercially Available IC CCs . . . 33
3.1 Introduction . . . 33
3.2 Bipolar Circuit Architectures of Current Conveyors . . . 33
3.2.1 Fabre's Translinear CC . . . 34
3.2.2 Normand's Translinear CCs . . . 35
3.2.3 An Alternative CCII Implementation . . . 36
3.2.4 Two Simple CCII Implementations . . . 38
3.2.5 Surakampontorn and Thitimajshima Electronically-Controlled Conveyor (ECC) . . . 39
3.2.6 Filanovsky's Current Conveyor Modified from a Current Source . . . 40
3.2.7 Temperature-Compensated CCII . . . 40
3.2.8 CCII with Reduced Parasitic Resistance R_x . . . 42
3.2.9 CCII with Increased Input Impedance at Port-Y . . . 43
3.2.10 Bipolar CCII with Controllable Gain . . . 44
3.2.11 Bipolar Implementations of the CCI . . . 47
3.3 Commercially Available IC CCs . . . 49
3.3.1 CCII01 from LTP Electronics . . . 49
3.3.2 PA630 from Phototronics Limited . . . 50
3.3.3 AD844 from Analog Devices . . . 52
3.3.4 Using OPA-2662 as Current Conveyors . . . 53
3.3.5 CC from OPA 660/OPA 860 . . . 53
References . . . 56

4 CMOS Implementations of Current Conveyors . . . 59
4.1 Introduction . . . 59
4.2 Simple CMOS Realizations of CCII+ and CCII− . . . 61
4.3 Low-Voltage CMOS Current Conveyor . . . 61
4.4 Class AB First Generation Current Conveyors . . . 63
4.5 Wide Band CMOS Current Conveyors . . . 65
4.6 A 1.5 V CMOS Current Conveyor Based on Wide Range Transconductors . . . 67
4.7 High Speed High Precision Current Conveyors . . . 67
4.8 CMOS-Inverter-Based CCII . . . 68
4.9 High Accuracy CMOS Current Conveyors . . . 70
4.10 High Bandwidth Current Conveyor with Reduced R_X . . . 72
4.11 Current Conveyor with High Current Driving Capability, Operated from 1.5 V Power Supply . . . 73

4.12 CMOS Rail-to-Rail Current Conveyor 74
4.13 CMOS Rail-to-Rail Current Conveyor Operated from ± 0.75 V Supply 75
4.14 Low-Voltage Low-Power CCII Based on Folded Cascode Bulk-Driven OTA 75
4.15 Wide-band High Performance Current Conveyor 77
References 78

Part II The Early (First Generation) Applications of Basic CCI and CCII

5 Basic Analog Circuit Building Blocks Using CCs and Application of CCs in Impedance Synthesis 85
5.1 Introduction 85
5.2 The Basic Functional Circuits Using CCI and CCII 86
5.2.1 Variable-Gain Amplifiers: Constant-Bandwidth Structures 86
5.2.2 Constant-Bandwidth Instrumentation Amplifiers 89
5.2.3 Constant-Bandwidth Current-Mode Operational Amplifier 90
5.2.4 Integrators and Differentiators 92
5.2.5 Current-Mode and Voltage-Mode Summers 94
5.2.6 Grounded Negative Impedance Converters 95
5.2.7 Floating Negative Impedance Converters 97
5.2.8 Generalized Function Generator 100
5.3 Methods and Circuits for Simulating Inductors, FDNRs and Related Elements 101
5.3.1 CCII-Based Lossless Grounded Inductance Simulation Circuits 102
5.3.2 Active Gyrator Using a Single CCII 105
5.3.3 Single CCII-Based Low-Component-Count Grounded Impedance Simulators 106
5.3.4 Floating Impedance Realization Without any Component-Matching Constraints 108
5.3.5 Floating Generalized Impedance Converters/Inverters (GIC/GII) 111
5.3.6 Two-CC-Based FDNR and FGPIC/FGPII Implementations 116
5.3.7 A Family of Three-CC Floating Inductor/FDNR Simulators 119
5.3.8 Mixed-Source FIs Using CCIIs and Op-amps/OTAs 121
5.3.9 Novel FI Circuits Using CCII-Nullor Equivalence 124

5.3.10 Simulation of Higher Order Grounded/Floating Immittances Using CCs 127
5.3.11 Simulation of Mutually-Coupled Circuits 127
5.3.12 Grounded and Floating MOS VCRs and Transconductors 128
References 132

6 First, Second and Higher Order Filter Design Using Current Conveyors 139
6.1 Introduction 139
6.2 The First Order, the Second Order and the Higher Order Filter Realizations Using CCs 139
6.2.1 Single-CC First Order All Pass Filters 140
6.2.2 Single-CC Biquads 143
6.2.3 Multiple-CC Multifunction Biquads 145
6.2.4 Third Order Filters 175
6.2.5 MOSFET-C Integrators and Filters Using CCII 177
6.2.6 Higher Order Active Filter Design 178
References 184

7 Realization of Sinusoidal Oscillators Using CCs 193
7.1 Introduction 193
7.2 Single-CC SRCOs 194
7.3 SRCOs Employing Grounded Capacitors 198
7.4 SRCOs Employing All Grounded Passive Elements 202
7.5 Quadrature and Multi-phase Oscillators 205
7.6 Explicit Current Output (ECO) SRCOs 211
7.7 SRCOs with Grounded Capacitors and Reduced Effect of Parasitic Impedances of CCIIs 212
7.8 Fully-Uncoupled Oscillators 212
References 215

8 Nonlinear Applications of CCs 219
8.1 Introduction 219
8.2 Precision Rectifiers 219
8.3 Frequency Doubler and Full Wave Rectifier 225
8.4 Multipliers, Dividers, Squarers and Square Rooters 227
8.5 CCII-based Realization of Fuzzy Functions 232
8.6 Realization of Analog Switches 234
8.7 Pseudo-Exponential Circuit Realization 236
8.8 Built-in-Test Structures Using CCs 238
8.9 Schmitt Trigger and Waveform Generators Using CCs 239
8.10 Chaotic Oscillators Using CCs 246
8.11 Miscellaneous Other Applications 250
References 251

Part III Different Variants of Current Conveyors, Their Implementations and Applications

9 Second Generation Controlled Current Conveyors (CCCII) and Their Applications . . . 255
9.1 Introduction . . . 255
9.2 Bipolar/CMOS/BiCMOS CCCIIs . . . 256
9.3 Grounded and Floating Current-Controlled Positive/Negative Resistance Realization . . . 260
9.4 Current Controlled VM/CM Amplifiers . . . 264
9.5 Active-Only Summing/Difference Amplifiers . . . 264
9.6 Instrumentation Amplifiers . . . 265
9.7 Electronically-Tunable Grounded/Floating Synthetic Impedances and Related Circuits . . . 267
9.8 Electronically-Controllable Multifunction Voltage Mode Biquad . . . 274
9.9 Current-Mode Universal Biquad Filters . . . 276
9.10 Mixed-Mode Current-Controlled Multifunction Filters . . . 282
9.11 Tunable Ladder Filters Using Multiple-output CCCIIs . . . 285
9.12 Current-Controlled Sinusoidal Oscillators . . . 287
9.13 PID Controller Using CCCIIs . . . 294
9.14 CCCII-Based Precision Rectifiers . . . 295
9.15 Current-Mode Multiplier/Divider Using CCCIIs . . . 297
9.16 Squaring/Square Rooting Circuits . . . 299
9.17 ASK/FSK/PSK/QAM Wave Generator . . . 303
9.18 Advances in the Realization of Bipolar/ CMOS/Bi-CMOS CCCIIs . . . 303
References . . . 308

10 Varieties of Current Conveyors . . . 315
10.1 Introduction . . . 315
10.2 Different Variants of the Current Conveyors . . . 315
10.2.1 Current Voltage Conveyor . . . 316
10.2.2 Generalized Current Conveyor . . . 316
10.2.3 Operational Floating Conveyor . . . 317
10.2.4 Third Generation Current Conveyor . . . 319
10.2.5 Differential-Difference Current Conveyor . . . 320
10.2.6 Multiple-Output Current Conveyor . . . 323
10.2.7 Differential-Voltage Current Conveyor . . . 324
10.2.8 Inverting CCIIs . . . 324
10.2.9 Inverting Third Generation Current Conveyors . . . 326
10.2.10 Differential-Current Voltage Conveyor . . . 327
10.2.11 Fully-Differential CCII . . . 327
10.2.12 General Three-Port Conveyors . . . 328
10.2.13 Universal Current Conveyor (UCC) . . . 330

10.2.14 Modified Inverting CCII . . . 332
10.2.15 Dual-X Current Conveyor . . . 332
10.2.16 Fully-Balanced CCII . . . 333
10.2.17 Extended Current Conveyors . . . 333
10.2.18 Operational Conveyor . . . 336
10.2.19 Multiple-Input Differential CC (MIDCC) . . . 336
10.2.20 Multiplication-Mode Current Conveyor (MMCC) . . . 338
10.2.21 Balanced-Output Third Generation Inverting CC . . . 339
10.2.22 Voltage and Current Gain Second Generation Current Conveyor (VCG-CCII) . . . 339
10.2.23 TXTZ CCII . . . 341
10.2.24 Differential CCII . . . 342
10.2.25 Universal Voltage Conveyor . . . 342
10.2.26 Floating Current Conveyors . . . 343
References . . . 345

11 Other Building Blocks Having MTC or CC at Front-end and Their Applications . . . 349
11.1 Introduction . . . 349
11.2 CC-CFA . . . 350
11.3 Four-Terminal-Floating-Nullor (FTFN) . . . 351
11.4 Operational Trans-Resistance Amplifier (OTRA) . . . 353
11.5 Current-Differencing-Buffered-Amplifier (CDBA) and Its Variants . . . 354
11.6 Current Controlled Current-differencing Transconductance Amplifier (CC-CDTA) . . . 357
11.7 Current Controlled Current Conveyor Transconductance Amplifier (CCCC-TA) . . . 357
11.8 Current Follower Transconductance Amplifier (CFTA) . . . 360
11.9 Current Through Transconductance Amplifier (CTTA) . . . 360
References . . . 362

Part IV Second Generation Applications: Realization of Various Linear/Nonlinear Functions Using Other Types of Current Conveyors

12 Analog Filter Design Revisited: Circuit Configurations Using Newer Varieties of CCs . . . 371
12.1 Introduction . . . 371
12.2 Filter Design Using Different Varieties of CCs . . . 372
12.2.1 Filter Design Using DVCCs . . . 372
12.2.2 Filter Design Using DDCC . . . 391
12.2.3 Filter Design Using FDCCII . . . 398
12.2.4 Filter Design Using ICCII . . . 402
12.2.5 Filter Design Using DCVC or CDBA . . . 408

12.2.6 Filter Design Using CCIII 412
12.2.7 Filter Design Using DXCCII 414
12.2.8 Filter Design Using UVC 416
12.2.9 Filter Design Using CFCCII 418
12.2.10 Filter Design Using OFCC 419
12.2.11 Filter Design Using Balanced-dual-input Dual-output-CC (BDI-DOCC) 421
12.2.12 Filter Design Using Dual/Multi Output CCs (DOCC/MOCC) 422
References 438

13 Sinusoidal Oscillator Realizations Using Other Types of Current Conveyors 449
13.1 Introduction 449
13.2 A Dual-Mode Sinusoidal Oscillator Using a Single Operational Floating Current Conveyor 450
13.3 ICCII-Based Grounded-Capacitor (GC) SRCO 450
13.4 ICCII-Based All Grounded Passive Elements (AGPE) SRCO 451
13.5 Explicit Current Output (ECO) SRCO Using All Grounded Passive Components 453
13.6 Grounded-Capacitor Current-Mode SRCO Using a Single DVCCC 454
13.7 FDCCII-Based SRCOs 455
13.8 CM Quadrature Oscillator (QO) Using DVCCs 456
13.9 VM Quadrature Oscillator with AGPE Using DDCCs 458
13.10 MOCCII-Based VM/CM QO 459
13.11 VM/CM QO Using FDCCII 460
13.12 Electronically-Programmable Dual-Mode QO Using a DVCCCTA and Only Two GCs 461
References 465

14 *Second Generation* Applications of Other Types of Current Conveyors in Realizing Synthetic Impedances 469
14.1 Introduction 469
14.2 Simulated Lossless Floating Inductance Using Only Two CCs and Three Passive Components 470
14.3 DVCC-Based Floating Inductance/FDNR with *All Grounded Passive Elements* 470
14.4 Simulated Inductors Employing CCIII 471
14.5 Grounded R-L and C-D Immittances Using a Single DVCC 473
14.6 Electronically-Controllable Gyrator and Grounded Inductor Using DXCCII 474

14.7 Grounded Inductor Simulation Using the Modified Inverting CCII (MICCII) 476
14.8 DO-CCII-Based Synthetic Floating Immittances 478
14.9 A General Circuit for Converting a Grounded Immittance into Floating Immittance 479
14.10 Compensated Negative Impedance Converter 480
14.11 DDCC-Based FI with Improved Low Frequency Performance 481
14.12 Floating Simulator Employing DO-CCII and OTA 483
14.13 DO-CCCII Based Lossless Floating Inductance Simulator Employing a Grounded-Capacitor 483
14.14 Resistor-Less Simulated FI Using DXCCII 485
14.15 Tunable MOSFET-C FDNR Using a Single DXCCII 486
14.16 DXCCII-Based Grounded Inductance Simulation 487
14.17 FI Simulators with Only Two DVCCs 488
14.18 Lossless Grounded Inductor Using a Single FDCCII and Three Grounded Passive Elements 489
14.19 DX-CCII-Based Grounded Inductance Simulators 490
14.20 Grounded-Capacitor-Based Floating Capacitance Multiplier 491
14.21 Floating Lossy Inductance Simulators Using a Single DO-DDCC and a Grounded Capacitor 494
14.22 Grounded Inductance Simulator Using DCCII 495
References 497

15 *Second Generation* Miscellaneous Linear/Nonlinear Applications of Various Types of Current Conveyors 501
15.1 Introduction 501
15.2 PID Controllers 501
15.3 Wide-Band Controllable Low Noise Amplifiers 504
15.4 Single-Ended to Differential Converters 506
15.5 Precision Rectifiers Revisited 506
15.5.1 Precision Full Wave Rectifier Proposed by Koton, Herencsar and Vrba 506
15.5.2 Kumngern's Full Wave Rectifier 508
15.5.3 Precision Rectifier Proposed by Minaei and Yuce 509
15.6 Multivibrators and Relaxation Oscillators 510
15.6.1 Chien's Square/Triangular Wave Generator 510
15.6.2 Switch-Controllable Bi-stable Multivibrator 513
15.6.3 Single DVCC-Based Monostable Multivibrators 515
15.6.4 Chien's Relaxation Oscillators 517
15.6.5 Chien's DO-DVCC-Based Square/Triangular Wave Generator 519
15.7 Wide-Band Impedance Matching Circuits 521

15.8 Sample and Hold Circuits 521
15.9 CCII-Based Digital-to-Analog Converter 523
15.10 Chaos Generators: Revisited 523
15.11 Realization of Chua Family of Nonlinear Network Elements: Mutators, Rotators, Reflectors and Scalars 524
15.12 Memcapacitance and Meminductance Emulators 525
References 529

Part V Concluding Remarks and References for Further Reading

16 Recent Advances and Future Directions of Research 533
16.1 Introduction 533
16.2 Pathological Representations of Various Current Conveyors and Their Use in Systematic Circuit Synthesis 533
16.3 Recent Advances in the Hardware Implementation of Current Conveyors 534
16.3.1 New CCII Implementation Based Upon Modified Bipolar Translinear Cell 534
16.3.2 Bi-CMOS CCCII Realizations 536
16.3.3 FG-MOS Current Conveyors 538
16.3.4 Design of CCII Employing Bacterial Foraging Optimization 538
16.4 Current-Conveyor-Based Field Programmable Analog Arrays (FPAA) 539
16.5 Applications of the Current Conveyors in Realizing Logic Functions and Digital Circuits 540
16.6 Newer Varieties of Current Conveyors of More Recent Origin 541
16.7 Epilogue 541
References 542

Appendix: Additional References for Further Reading 545

Index 555

Abbreviations

ABB	Active Building Block
ADC	Analog to Digital Converter
A/D	Analog to Digital
AGPE	All Grounded Passive Elements
AM	Analog Multiplier
AP	All Pass
BDI-DOCC	Balanced Dual Input–Dual Output Current Conveyor
BE	Band Elimination
BiCMOS	Bipolar Complementary Metal Oxide Semiconductor
BIT	Built-in-Testing
BJT	Bipolar Junction Transistor
BP	Band Pass
BPF	Band Pass Filter
BR	Band Reject
BS	Band Stop
BSF	Band Stop Filter
CAB	Configurable Analog Block
CC	Current Conveyor
CCC	Composite Current Conveyor
CC-CBTA	Current Controlled Current Backward Transconductance Amplifier
CC-CCFOA	Current Controlled Current Feedback Operational Amplifier
CCCCTA	Current Controlled Current Conveyor Transconductance Amplifier
CC-CDBA	Current-Controlled Current Differencing Buffered Amplifier
CC-CDTA	Current-Controlled Current Differencing Transconductance Amplifier
CC-CFA	Current-Controlled Current Feedback Amplifier
CCCII	Controlled Current Conveyor (Second Generation)
CCCS	Current Controlled Current Source
CCDDCC	Current Controlled Differential Difference Current Conveyor
CCDDCCTA	Current Controlled Differential Difference Current Conveyor Transconductance Amplifier

CCI	Current Conveyor (First Generation)
CCII	Current Conveyor (Second Generation)
CCIII	Current Conveyor (Third Generation)
CCW	Counter Clock Wise
CDA	Complimentary Differential Amplifier
CDBA	Current Differencing Buffered Amplifier
CDTA	Current Differencing Transconductance Amplifier
CE	Characteristic Equation
CF	Current Follower
CFBCCII	Controlled Fully Balanced Current Conveyor (Second Generation)
CFOA	Current Feedback Operational Amplifier
CFTA	Current Follower Transconductance Amplifier
CM	Current Mode; *also*, Current Mirror
CMOS	Complementary Metal Oxide Semiconductor
CMRR	Common Mode Rejection Ratio
CO	Condition of Oscillation
COA	Current Mode Operational Amplifier
CR	Current Repeater
CTTA	Current Through Transconductance Amplifier
CVC	Current Voltage Conveyor
CW	Clock-Wise
D/A	Digital to Analog
DAC	Digital to Analog Converter
DC	Direct Current
DCC	Differential Current Conveyor
DCCCTA	Differential Current Controlled Conveyor Transconductance Amplifier
DCFDCCII	Digitally Controlled Fully Differential Second Generation Current Conveyor
DCVC	Differential Current Voltage Conveyor
DDCC	Differential Difference Current Conveyor
DDCCC	Differential Difference Complimentary Current Conveyor
DDCCTA	DDCC Transconductance Amplifiers
DIDO	Differential Input Differential Output
DOCC	Dual Output Current Conveyor
DPDT	Double-Pole Double-Throw
DVCC	Differential Voltage Current Conveyor
DVCCC	Differential Voltage Complimentary Current Conveyor
DVCCCTA	Differential Voltage Current-Controlled Conveyor Transconductance Amplifier
DVCCII	Second Generation Differential Voltage Current Conveyor
DVCCS	Differential Voltage Controlled Current Source

DVCCTA	Differential Voltage Current Conveyor Transconductance Amplifier
DXCCII	Dual-X Second Generation Current Conveyor
ECC	Electronically Controlled Current Conveyor
ECC	Extended Current Conveyor
ECCII	Electronically Tunable Second Generation Current Conveyor
ECO	Explicit Current Output
FAC	Floating Admittance Converter
FBCCII	Fully Balanced Second Generation Current Conveyor
FBDDA	Fully-Balanced Differential Difference Amplifier
FC	Floating Capacitance
FCC	Floating Current Conveyor
FCCNR	Floating Current Controlled Negative Resistance
FCCPR	Floating Current Controlled Positive Resistance
FDCCII	Fully Differential Second Generation Current Conveyor
FDNC	Frequency Dependent Negative Conductance
FDNR	Frequency Dependent Negative Resistance
FDPR	Frequency Dependent Positive Resistance
FET	Field Effect Transistor
FGPIC/FGPII	Floating Generalized Positive Immittance Converter/Inverter
FI	Floating Immittance
FO	Frequency of Oscillation
FPAA	Field Programmable Analog Array
FPGA	Field Programmable Gate Array
FTFN	Four-Terminal-Floating-Nullor
GBP	Gain Bandwidth Product
GC	Grounded Capacitor
GCC	Generalized Current Conveyor
GI	Grounded Impedance
GIC	Generalized Impedance Converter
GPIC	Generalized Positive Impedance Converter
GPII	Generalized Positive Impedance Inverter
GVC	Generalized Voltage Conveyor
HP	High Pass
HPF	High Pass Filter
IC	Integrated Circuit
ICCII	Inverting Second Generation Current Conveyor
ICCIII	Inverting Third Generation Current Conveyor
INIC	Current Inversion Negative Impedance Converter
KHN	Kerwin-Huelsman-Newcomb
LC	Inductance-Capacitance
LNA	Low Noise Amplifier
LP	Low Pass
LPF	Low Pass Filter

MCCCII	Multi Output Controlled Current Conveyor (Second Generation)
MCCIII	Modified Current Conveyor (Third Generation)
MDAC	Multiplying Digital-to-Analog Converter
MDO-DDCC	Modified Dual Output-Differential Difference Current Conveyor
MICCII	Modified Inverting Current Conveyor
MIDCC	Multiple Input Differential Current Conveyor
MIMO	Multiple-Input-Multiple-Output
MISO	Multiple-Input-Single-Output
MMCC	Multiplication-Mode Current Conveyor
MOCC	Multiple Output Current Conveyor
MO-CCCA	Multiple Output Current-Controlled Current Amplifier
MO-CC-CTTA	Multiple Output Current Controlled Current Through Transconductance Amplifier
MOCCII	Multiple Output Current Conveyor (Second Generation)
MOCF	Multiple Output Current Follower
MOSFET	Metal Oxide Semiconductor Field Effect Transistor
MRC	MOS Resistive Circuit
MTC	Mixed Translinear Cell
NAM	Nodal Admittance Matrix
NF	Notch Filter
NIC	Negative Impedance Converter
NMOS	N-type Metal Oxide Semiconductor
OC	Operational Conveyor
OCC	Operational Current Conveyor
OFA	Operational Floating Amplifier
OFC	Operational Floating Conveyor
OFCC	Operational Floating Current Conveyor
OMA	Operational Mirrored Amplifier
OTA	Operational Transconductance Amplifier
OTA-C	Operational-Transconductance-Amplifier-Capacitor
OTRA	Operational Trans-Resistance Amplifier
PIC	Positive Impedance Converter
PII	Positive Impedance Inverter
PMOS	P-type Metal Oxide Semiconductor
QO	Quadrature Oscillator
RC	Resistance-Capacitance
SCCO	Single-Capacitor-Controlled Oscillator
SCIC	Summing Current Immittance Converter
SECO	Single-Element-Controlled Oscillator
SFG	Signal Flow Graph
SIFO	Single Input Five Output
SIMO	Single Input Multiple Output
SISO	Single Input Single Output
SRCO	Single Resistance Controlled Oscillator

SVIC	Summing Voltage Immittance Convertor
TAM	Trans-Admittance–Mode
TCCII	Transconductance Current Conveyor (Second Generation)
THD	Total Harmonic Distortion
TI	Texas Instruments
TIM	Trans-Impedance-Mode
TL	Trans-Linear
TO-ICCII	Triple Output-Inverting Current Conveyor (Second Generation)
TX-TZ CCII	Two-X Two-Z Current Conveyor (Second Generation)
UCC	Universal Current Conveyor
UVC	Universal Voltage Conveyor
VC	Voltage Conveyor
VCG	Voltage and Current Gain
VCG-CCII	Voltage and Current Gain Current Conveyor (Second Generation)
VCO	Voltage-Controlled Oscillator
VCR	Voltage-Controlled-Resistance
VCVS	Voltage-Controlled-Voltage-Source
VDIBA	Voltage Differencing Inverting Buffered Amplifier
VDTA	Voltage Differencing Transconductance Amplifier
VF	Voltage Follower
VLSI	Very Large Scale Integrated Circuits
VM	Voltage Mirror; *also* Voltage-Mode
VMQO	VM Quadrature Oscillator
VNIC	Voltage Inversion Negative Impedance Converter
VOA	Voltage (mode) Operational Amplifier
WCDMA	Wide-band Code Division Multiple Access
ZC-CCCITA	Z-Copy Current Controlled Current Inverting Transconductance Amplifier

Part I
Evolution and Hardware Implementation of Current Conveyors

Chapter 1
The Evolution and the History of *Current Conveyors*

Abstract A historical account about the origin of Current Conveyors is given and important developments in the area, taken place during the past four decades have been highlighted.

1.1 Prologue

It is well recognized that in spite of the dominance of digital circuits and techniques, analog circuits are essentially required in many situations since all natural signals are analog. Analog circuits and techniques are hence, indispensable in realizing signal amplifiers, continuous-time filters, rectifiers, sinusoidal oscillators, analog-to-digital and digital-to-analog converters, analog multipliers and dividers and some types of artificial neural networks.

In the context of modern VLSI design, the *current mode approach* to analog circuit design has received dominant attention (see [1–73] and the references cited therein) due to the several advantages offered by it. Current-mode circuits provide attractive alternatives to their voltage-mode counterparts in terms of providing one or more of the following advantageous features such as realization of intended functions with the least possible number of external components without requiring any component-matching conditions or equality constraints, simpler circuit structures, better accuracy in realizing the intended characteristics, higher operational frequency range, larger dynamic range and better linearity etc.

The *Current Conveyors* (CC) invented by Sedra and Smith in the late seventies and early eighties are now well recognised to be the most prominent current-mode building blocks which have received worldwide attention of academicians, circuit designers and researchers as alternative building blocks in general, and in low-voltage, low-power circuit designs in particular. It has been now well established that the CCs can realize most of the above mentioned analog functions more efficiently and with several advantages than the corresponding circuits realized with the traditional op-amps. Indeed, CCs have had very significant effect on the evolution of a host of other building blocks in recent times and thus, have cast a profound influence on the development of modern analog circuit design.

© Springer International Publishing Switzerland 2015
R. Senani et al., *Current Conveyors*, DOI 10.1007/978-3-319-08684-2_1

However, the entire literature on CCs, which has been scattered over 1,000 journal papers, has still not been made available at a single source by any academic author/publisher so far. Although a few book chapters/books have been written on some limited aspects of CCs, any comprehensive treatise on CCs has still not been brought out.

This monograph aims to fill this void by bringing out a repertoire of important hardware implementations and applications of CCs and their numerous variants in modern analog circuit design and thereby bringing out a single-source reference on almost everything related to CCs at one place.

To put the contents of this book in the right perspective, in the following sections, we present a brief historical account of the story of the invention of the *CCs* as well as successive evolution of a large number of their variants which have been proposed by a number of researchers from time to time.

1.2 The Origin of the First Generation Current Conveyor

Although the story of how the concept of *current conveying* and its embodiment the *Current Conveyor* were originated has been mentioned by Sedra in [23] it is still interesting and inspirational to reiterate the same. It is told that when Sedra was working for his Master's thesis he was assigned the task of designing a good temperature-insensitive constant current source.

Sedra perceived that the common base arrangement shown in Fig. 1.1a was the natural choice to achieve this.

However, the output collector current in such a case is temperature-sensitive due to the presence of V_{EB1} which is equal to $V_T \ln\left(\frac{I_0}{I_{co}}\right)$ wherein V_T is directly proportional to the absolute temperature (being equal to KT/q where K is Boltzman constant, T is absolute temperature and q is electron charge). To get rid of this unwanted term, he thought of cancelling out V_{BE1} by ungrounding the base of transistor Q_1 and bringing another matched PNP transistor Q_2 in the loop, with the collector current of Q_2 made same as I_0. This modification clearly makes the voltage 'e' to be $e = (V_{EB1} - V_{EB2})$. Since Q1 and Q2 carry essentially the same amount of collector current (assuming the β of the transistors to be very high thereby considering the base currents to be negligibly small), it is clear that the corresponding V_{EBs} of the two transistors will cancel out thereby rendering the voltage $e = 0$ (a sort of virtual ground). Therefore, the resulting output current is now given by $I_0 = +V/R_0$ which is, as required, temperature- insensitive!

Now in order to ensure that Q_2 carries the same current as Q_1, a current copier comprised of three matched NPN transistors was added as shown in simplified version of Fig. 1.1c (the circuit devised by Smith and Sedra [1] contains three matched resistors added between the emitters of the transistors Q_3, Q4 and Q_5 and −V). This completed the design of the required temperature-insensitive current source.

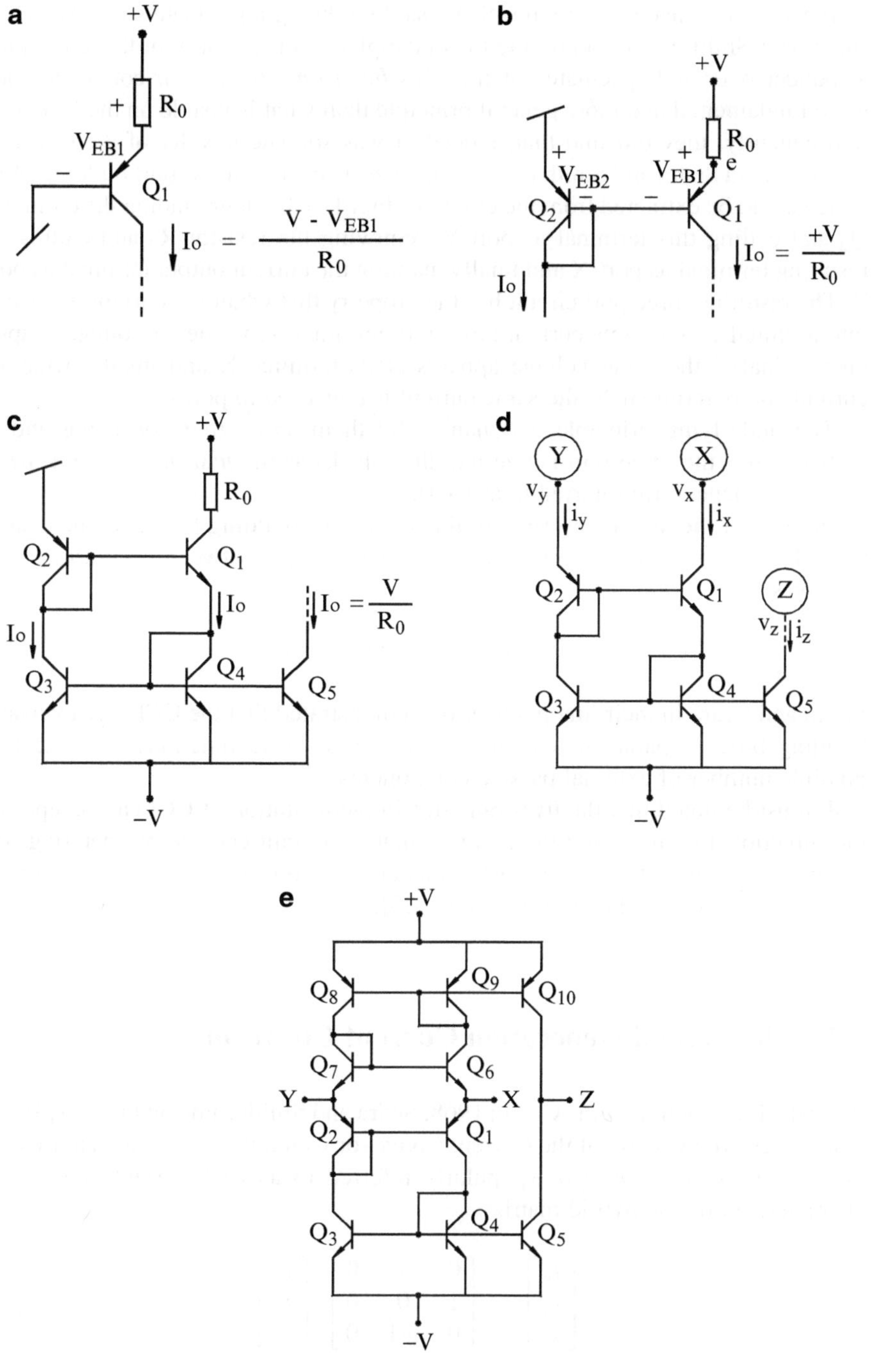

Fig. 1.1 The evolution of the first generation CC (CCI) (**a**) The common-base configuration as a constant current source (**b**) temperature compensation principle (**c**) temperature-compensated constant current source (**d**) The conceptualization of a generic three-port active building block-*The first generation Current Conveyor* (**e**) A class AB CCI

After having made this circuit, Sedra said he (being a graduate student) and his supervisor Smith had enough time to contemplate on the beauty of this circuit and to ponder over and speculate whether this *five transistor circuit* contained some more fundamental and more general principle than what is visible on the face of it? Subsequently, they did find that, indeed, it was so. The novelty of the circuit of Fig. 1.1c lead Sedra and Smith to conceptualize a versatile three-port building block which could be extracted from the circuit of Fig. 1.1c by ungrounding the emitter of Q_2 and calling this terminal as port-Y, removing the resistor R_0 and calling the resulting terminal as port-X and finally, naming the current output terminal as port-Z. The resulting three port circuit has the property that whatever current is injected into terminal-X, the same current flows into terminal-Y; whatever voltage is apply on terminal-Y the same voltage appears at its terminal-X and finally, whatever current enters terminal-X, the same current is *conveyed* to port -Z.

The underlying principle was named by them as *current conveying* and its embodiment (the three-port active building block) as the *first generation Current Conveyor* (henceforth referred to as CCI).

In view of the above description, this three-port building block can be characterized by the following instantaneous relationships between its various port variables:

$$\mathrm{i_y} = \mathrm{i_x}, \mathrm{v_x} = \mathrm{v_y} \text{ and } \mathrm{i_z} = \mathrm{i_x} \tag{1.1}$$

Smith and Sedra in their first publication demonstrated that the CCI was a versatile building block capable of realizing a number of basic functions with the least possible number of external passive components.

It must be noted that the five-transistor-implementation of CCI can accept only uni-directional currents and therefore, to make a circuit capable of operating with bi-directional currents, another complementary five-transistor circuit requires to be added on top of it as shown in Fig. 1.1e [23].

1.3 The Second Generation Current Conveyor

After the introduction of the CCI in 1968, Sedra and Smith came up in 1970 [3] with a more versatile version of the Current Conveyor which they chose to call a *second generation current conveyor* (popularly referred to as CCII). While the CCI is characterized by the hybrid matrix

$$\begin{bmatrix} i_y \\ v_x \\ i_z \end{bmatrix} = \begin{bmatrix} 0 & 1 & 0 \\ 1 & 0 & 0 \\ 0 & \pm 1 & 0 \end{bmatrix} \begin{bmatrix} v_y \\ i_x \\ v_z \end{bmatrix} \tag{1.2}$$

the CCII was proposed to be characterized by

$$\begin{bmatrix} i_y \\ v_x \\ i_z \end{bmatrix} = \begin{bmatrix} 0 & 0 & 0 \\ 1 & 0 & 0 \\ 0 & \pm 1 & 0 \end{bmatrix} \begin{bmatrix} v_y \\ i_x \\ v_z \end{bmatrix} \tag{1.3}$$

It was demonstrated in [3, 4] that CCII was, indeed, a versatile building block as it could realize all the four controlled sources, a negative impedance converter, a negative impedance inverter and can also be used to realize a number of other linear and non-linear functional circuits with a bare minimum number of external passive components quite easily and in a much simpler way than the way in which the same functional circuits were being realized with the traditional op-amps. Thus, the CCII was shown to be quite promising for realizing many basic functions needed for analog computation such as current amplifier, current differentiator, current integrator, current summer, weighted current summer and generalized nonlinear function generator- all employing only a bare minimum number of passive components. In [4], they demonstrated that CCIIs were also versatile building blocks for realizing the family of new non-linear elements namely, mutator, scalor, reflector and rotator, introduced by Chua in [74, 75].

1.4 An Historical Overview of the Evolution of the Other Varieties of Current Conveyors

Since the introduction of CCII by Sedra and Smith in 1970, the next decade witnessed a considerable interest by academicians and researchers in evolving a large number of their applications in realizing various analog signal processing and signal generation applications. However, on the other hand, very little work was done on the implementation of CCs themselves. It was only in 1989 that a fully integrated CC was proposed by Wadsworth which was produced by the Canadian company Phototronics Ltd. Although the now popular AD844 was already being manufactured by Analog Devices Inc., since it was disguised as a high slew rate op-amp. Although the internal circuit architecture of AD844 was based upon a translinear CCII+ and a unity gain voltage follower the CCII+ inherent in AD844 was not explicitly noticed till about 1990 or so. That AD844 did provide external access to the Z-terminal (as the compensation pin) makes it possible to use it as a CCII+ and also enables the realization of CCII− using two such devices.

A chronological listing of the important developments which have taken place on the realization and applications of CCs, starting from 1968 onwards, is as under:

- **1968:** Smith and Sedra introduced the basic concept of Current Conveying and its implementation as the *Current Conveyor* (later renamed as the first generation Current Conveyor).
- **1970:** Sedra and Smith introduced the more versatile second generation Current Conveyor (henceforth referred as CCII+). In another paper during the same year, Sedra and Smith demonstrated that CCIIs were useful building blocks for

realizing the Chua family of new non-linear network elements, namely, the *mutators, rotators, scalars and reflectors* with minimum possible number of passive elements.

- **1971:** Black et al. in [5] presented an integratable implementation of CCII using μA749 IC chip (containing two operational amplifiers with uncommitted output collectors) along with five additional transistors (obtained from an integrated transistor array CA3046) to realize CCII+, which, with some more transistors added, could also lead to a CCII− .
- **1978:** In [6], Bakhtiyar and Aronhime presented a CCII implementation using as many as five op-amps.
- **1979:** The first ever configuration for floating inductor simulation (parallel RL type) employing the least possible number of active and passive component namely, only one CCII−, a single capacitor and only two resistors, was introduced by Senani [9]. This result was important in view of the fact that no such circuit was known in the earlier literature using op-amps or any other kind of devices known till then. In the same year, a circuit for floating series-RL type lossy inductance simulation was also proposed by Singh in [10].
- **1980:** A novel implementation of the CCs was introduced by Senani in [7] who used an op-amp, an OTA and only one/two resistors to realize all the four varieties of CCs namely, CCI+, CCI−, CCII+ and CCII− from the same configuration. Huertas, in [8], came up with an entirely op-amp version of Senani's circuit.
- **1981:** In [11], Filanovsky and Stromsmoe introduced the *Current–voltage Conveyor* (CVC) with two versions called CVC1 and CVC2 which were shown to be useful elements for the realization of controlled sources, negative impedance converters and other active elements It was shown that the CVC could be realized by the OA-OTA based synthesis as done previously for CC realization by Senani in [7].
- **1982:** Dostal and Pospisil [13, 14] extended the idea of Current Conveyors to define three novel types of Voltage Conveyors called VC1, VC2 and VC3. They further demonstrated that the three-port immittance convertor is an useful transformation building block leading to two basic types of convertors which they chose to call summing voltage immittance convertor (SVIC) and summing current immittance convertor (SCIC) which were capable of leading to, as special cases, three varieties of Current Conveyors and three varieties of voltage conveyors.
- **1984:** In [15], Senani showed that the generalized current conveyor (characterized by the instantaneous terminal relations $i_y = h_{12}\, i_x$, $v_x = h_{21}\, v_y$ and $i_z = h_{32}\, i_x$) in conjunction with an active-RC network derived from a given grounded immittance simulation network i.e. one-port with $y_{11} = y(s)$, can be used to realize the same type of immittance function in floating form i.e. as a two-port characterized by $y_{11} = -y_{12} = -y_{21} = y_{2\ 2} = ky(s)$, k being a positive scale factor, with several advantages over earlier methods of floating immittance realization.

- **1984:** Whereas a number of ideal floating FDNR circuits were reported by a number of authors all of which employed three Current Conveyors two capacitors and their resistors, Senani demonstrated *for the first time* that the same element could be realized by using no more than two Current Conveyors (both CCII−) along with two capacitors and three resistors *without requiring any component-matching condition(s)* [16].
- **1986:** In [17], Senani proposed a family of three two-CC-five-passive-elements based floating generalized positive impedance convertors/invertors which could realize floating lossless inductor, floating frequency-dependent-negative-resistance (FDNR) and floating frequency-dependent negative conductance (FDNC)- all from the same configuration without requiring any component- matching conditions.
- **1989:** Two modified Current Conveyors were introduced by Pal in [21] out of which the first one had complementary Z outputs whereas the second one could accept differential Y input and was termed as the second generation differential voltage current conveyor (DVCCII). It was shown that the DVCCII was particularly attractive element for realizing a floating inductance circuit using all grounded passive components namely, only two grounded resistors along with a grounded capacitor.
- **1991:** Toumazou et al. [25] proposed the operational floating conveyor (OFC) characterized by the port relationships $V_x = V_y$, $i_y = 0$, $V_w = Z_t\ i_x$ and $i_z = -i_w$. A bipolar implementation of the OFC was given and it was demonstrated that the new building block exhibited the gain bandwidth independence of typical current mode circuits and was suitable for monolithic integration. It was also shown that the OFC is an attractive element for realizing accurate, closed loop, current amplifiers and current convertors including closed loop Current Conveyors.
- **1993:** Fabre and Saaid [27] presented a four-port translinear building block which was essentially a *CCI followed by a voltage buffer* and could implement a gyrator using transistor arrays such as ALA200 from ATT.
- **1995:** In [28], A new type of CC termed as the *third generation CC (CCIII)* was introduced by Fabre which is characterized by $i_y = -i_x$, $v_x = v_y$ and $i_z = \pm i_x$. The CCIII was shown to be a useful element for taking out current flowing through a floating branch of a circuit and it was envisaged that it may be advantageously employed as the input cell of the probes and current measuring devices.
- **1996:** A new versatile building block called *differential difference Current Conveyor (DDCC)* was introduced by Chiu et al. [29]. It was shown to be useful for realizing a number of nonlinear functional circuits such as multiplier, squarer and square rooter in conjunction with MOSFETs and was also shown to be convenient in realizing universal voltage mode and current mode biquads.
- **1996:** In [30], Wu and Al-Masry introduced a multiple output Current Conveyor (MOCC) containing both positive and negative outputs.
- **1997:** Elwan and Soliman [32] introduced the so-called *differential voltage Current Conveyor (DVCC)* as an extension of CCII. However, it went unnoticed

that this element had already been introduced by Pal in 1989 [21]. CMOS architecture of DVCC was proposed too, which however, was not attempted by Pal in his 1989 publication. Use of DVCCs in realizing grounded to floating positive/negative converters, floating gyrator, floating positive/negative generalized impedance inverters and MSOFET-C filters were shown which had not been conceived earlier.

- **1998**: In [35], Soliman introduced two new building blocks termed as Generalized Voltage Conveyor (GVC) and Generalized Current Conveyor (GCC).
- **1999**: In [37], Awad and Soliman introduced the so called inverting second generation current conveyor (ICCII) as the *missing* building block. However, it must be pointed out that these so called ICCII were already proposed earlier in [70] by Chong and Smith who termed them as new types of CCs denoted as CC_B whose terminals characteristic were defined as $i_y = 0$, $v_x = -v_y$ and $i_z = \pm i_x$ which is exactly the same characterization as described by Awad and Soliman in their paper [37]. Thus, the claim of Awad and Soliman of calling the ICCII as the *missing* building block is questionable, however, their proposition of the new pathological elements namely, the voltage mirror and current mirror, the nullor-mirror representations of the various types of current conveyors including ICCII, the proposed CMOS realization of the ICCII as well as the new voltage-mode and current-mode all pass filters using ICCII had not been known earlier and were, indeed, new.
- **2000**: In [40], Biolek et al. proposed a new general current conveyor definition from where they demonstrate that 12 different varieties of current conveyors can be defined. It was also shown that 8 of the 12 current conveyors can be realized by a five port current conveyor having differential Y input and complimentary Z outputs commonly known as DVCC, however, the conveyors that cannot be realized by the DVCC are CCI + ICCI+, CCI− and ICCIII (though these can be realized using an universal current conveyor (UCC) also) which was formerly proposed much later.

  In the same year, Adawy et al. [41] proposed a novel fully differential second generation current conveyor (FDCCII) which was projected to be a useful building block for mixed mode applications where fully-differential signal processing is required. A CMOS implementation of the FDCCII was suggested and its applications in realizing a differential input balanced output transconductor, a four quadrant analog multiplier, universal mixed-mode second order filter in voltage and current mode, fully differential amplifier and fully differential filters were outlined.

  In the same year Becvar et al. [42] introduced the so-called universal current conveyor (UCC) which is an eight-port building block having three high impedance input terminals (differential Y_1, Y_2 and summing Y_3), one low impedance input X and four current outputs ($Z_1, Z_2, \overline{Z}_1, \overline{Z}_2$) where the last named outputs are complimentary to the first two outputs. A CMOS implementation was suggested and a new type of inverting current conveyor namely, third generation inverting current conveyor (ICCIII) was shown to emerge as a consequence of the proposed UCC.

In [43], Beevao and Vrba proposed the so-called Universal Current Conveyor (UCC) as versatile building block which was capable of replacing any type of existing current conveyor. They particularly used UCC to derive novel generation of inverting current conveyors.

- **2001**: In [46], Gupta and Senani demonstrated that a modified form of DDCC called differential difference complimentary current conveyor (DDCCC) makes it possible to realise a novel single resistance controlled current oscillator employing two DDCCCs along with only five passive components (two capacitors and three resistors) all of which are grounded as advantageous from the viewpoint of integrated circuit implementation.
- **2002**: This year witnessed three important contributions related to Current Conveyors and their applications.

  Sato et al. [49] presented a novel variant of current conveyor called extended current conveyor (ECC) characterized by the following matrix equation:

$$\begin{bmatrix} V_x \\ I_y \\ I_z \end{bmatrix} = \begin{bmatrix} 0 & \alpha & 0 \\ \phi & 0 & 0 \\ \pm\beta & 0 & 0 \end{bmatrix} \begin{bmatrix} I_x \\ V_y \\ V_z \end{bmatrix} + \begin{bmatrix} V_{offsetx} \\ I_{offsety} \\ I_{offsetz} \end{bmatrix} \quad (1.4)$$

  Kuntman et al. [50] proposed a modified third generation current conveyor (MCCIII) with characterizing equation as $i_y = -i_x$, $v_x = v_y$, $i_{z2} = -2i_x$, $i_{z2} = i_x$. Through a number of applications it was shown that the four terminal MCCIII offers new design possibilities.

  In [51], Zeki and Toker presented another innovation which was called by them as the dual-X current conveyor (DXCCII) which was a five port building block characterized by as $i_y = 0$, $v_{xp} = v_y$, $v_{xn} = -v_y$, $i_{zp} = i_{xp}$, $i_{zn} = i_{xn}$. It was shown that DXCCII is particularly useful in MOSFET-C continuous-time filter design where the number of active devices and MOSFET resistors required are reduced considerably due to facilitating a novel opportunity of eliminating even order nonlinearities of the MOSFET characteristics.
- **2003**: Alzaher et al. [53] introduced another new variant of the current conveyor named fully balanced second generation current conveyor (FBCCII) which accepts all the signals in differential form. The FBCCII architecture proposed by them was obtained from fully-balanced version of CMOS differential difference amplifier by sensing its output currents. This innovation was shown to be useful for implementing CCII− based circuits to be used in practical IC applications.
- **2004**: Cajka and Vrba [58] formally defined an n-port voltage conveyor and introduced the terminology and the schematic symbols and their adjoints. It was hypothesized that the frequency characteristics of the three types of voltage conveyors discussed by them could be better than those of the classical op-amp.

- **2005**: In [59], Gift introduced a new circuit element referred to as an *operational conveyor* comprising of an op-amp and a current conveyor. This operational conveyor was demonstrated to be superior to a current conveyor in a number of situations.

  In the same year, Mahmood et al. [76] proposed a digitally-controlled current gain, fully-differential second generation CMOS current conveyor operated from a low voltage supply of the order of $\pm$ 1.5 V DC.
- **2008**: In [60], Soliman demonstrated that two types of inverting CCII namely, ICII+ and ICII− are universal building blocks.

  In the same year, Horng et al. [62] proposed an interesting variant of CCs called multi input differential current conveyor (MIDCC). MIDCC is basically an adjoint of MOCC and is, therefore, useful in deriving a universal voltage-mode biquad starting from universal current-mode biquad.
- **2009**: Hwang et al. [63] proposed the multiplication mode current conveyor (MMCC). The internal circuit structure of this was obtained by using a DVCC and a folded Gilbert cell. MMCC was shown to be useful element for designing four quadrant multiplier, squarer, summer and analog divider applications.

  Siripruchyanun et al. [64] introduced current controlled current conveyor transconductance amplifier (CCCCTA) whose parasitic resistance at the current input port is controllable by an external input bias current. This building block was shown to be useful for both current mode and voltage mode signal processing systems. An attractive feature of the CCCCTA is that a single unit is sufficient to design a variety of functional circuits such as VM and CM universal biquad or a grounded inductor simulator.

  In [65], Sobhy and Soliman presented balanced output inverting third generation current conveyors ICCIII$\pm$. The strength of these blocks was demonstrated by many applications in the design of integrators, filters and oscillators.

  Marcellis et al. [66] proposed a voltage and current gain second generation current conveyor which they chose to call VCG-CCII. CMOS architecture was obtained by replacing voltage and current buffers of the standard CCII by voltage and current amplifiers with tunable gains to obtain an extremely flexible and versatile building block.
- **2010**: Chunhua et al. [67] demonstrated that a current controlled fully balanced second generation current conveyor (CFBCCII) is particularly useful building block for designing multi loop feedback current-mode balanced filter.
- **2012**: In [68], Metin et al. came up with yet another variant of CCs called differential current conveyor (DCCII) which has a current difference capability. A DCCII is characterized by

$$\begin{bmatrix} v_{XN} \\ v_{XP} \\ i_Z \\ i_Y \end{bmatrix} = \begin{bmatrix} 0 & 0 & \beta_N \\ 0 & 0 & \beta_P \\ \alpha_P & -\alpha_N & 0 \\ 0 & 0 & 0 \end{bmatrix} \begin{bmatrix} i_{XP} \\ i_{XN} \\ v_Y \end{bmatrix} \tag{1.5}$$

where ideally $\beta_N = \beta_P = 1$ and $\alpha_N = \alpha_P = 1$ that represent the voltage and current transfer ratios of the DCCII, respectively.

- **2013**: Pandey et al. [70] proposed a current controlled differential difference current conveyor transconductance amplifier (CCDDCCTA) and demonstrated its suitability for designing a higher order wave active filter.

The various types of CCs have attracted so much of attention worldwide that the total number of references on CCs and their variants as on date runs into over a 1,000. It must also be mentioned that other than two IC Current Conveyors mentioned above, there have been two more attempts, one by LTP Electronics and another by Maxim who have also produced fully integrated Current Conveyor ICs.

More recently, Current Conveyors and their variants have inspired a large number of other current mode building blocks which constitute a dominant area of contemporary analog circuit research. This brief historical sketch gives a glimpse of only the origin and variety of Current Conveyors introduced in the literature. Interested readers may also refer to a number of review articles that have also appeared in literature on the hardware implementation and applications of Current Conveyors, for instance, see [12, 19, 20, 22, 23, 26, 33, 44, 45, 52, 55, 59, 61, 67, 69, 73].

Concluding Remarks

A brief historical account of the invention of the Current Conveyors by Sedra and Smith in 1968 and major developments taken place in the area of Current Conveyors since then was provided with particular reference to their numerous varieties.

The subsequent chapters of this monograph will gradually unfold details of the various discrete implementations of the CCs using commercially available ICs, the Bipolar/CMOS-BiCMOS implementations of CCs, a large number of the applications and numerous varieties of the CCs and their Bipolar and CMOS implementations.

It is hoped that this brief historical account puts the contents of the subsequent chapters in right perspective in the context of modern analog circuit design.

References

1. Smith KC, Sedra A (1968) The current conveyor-a new circuit building block. Proc IEEE 56:1368–1369
2. Smith KC, Sedra A (1969) A new simple wide-band current-measuring device. IEEE Trans Instrum Meas 18:125–128
3. Sedra A, Smith KC (1970) A second-generation current conveyor and its applications. IEEE Trans Circ Theor 17:132–134
4. Smith KC, Sedra A (1970) Realization of the Chua family of new nonlinear network elements using the current conveyor. IEEE Trans Circ Theor 17:137–139
5. Black GGA, Friedmann RT, Sedra AS (1971) Gyrator implementation with integrable current conveyors. IEEE J Solid State Circ 6:396–399

6. Bakhtiar MS, Aronhime P (1978) A current conveyor realization using operational amplifiers. Int J Electron 45:283–288
7. Senani R (1980) Novel circuit implementation of current conveyors using an O.A. and an O.T.A. Electron Lett 16:2–3
8. Huertas JL (1980) Circuit implementation of current conveyor. Electron Lett 16:225–226
9. Senani R (1979) Novel active RC circuit for floating-inductor simulation. Electron Lett 15:679–680
10. Singh V (1979) A new active—RC circuit realization of floating inductance. Proc IEEE 67:1659–1660
11. Filanovsky IM, Stromsmoe KA (1981) Current-voltage conveyor. Electron Lett 17:129–130
12. Kumar U (1981) Current conveyors: a review of the state of the art. IEEE Circ Syst Mag 3:10–14
13. Dostal T, Pospisil J (1982) Current and voltage conveyors-a family of three-port immittance converters. Int Symp Circ Syst:419–422
14. Dostal T, Pospisil J (1982) Hybrid models of 3-port immittance convertors and current and voltage conveyors. Electron Lett 18:887–888
15. Senani R (1984) Novel application of generalized current conveyor. Electron Lett 20:169–170; Errata (1984) Electron Lett 20: 356
16. Senani R (1984) Floating ideal FDNR using only two current conveyors. Electron Lett 20:205–206
17. Senani R (1986) On the realization of floating active elements. IEEE Trans Circ Syst 33:323–324
18. Wilson B (1986) Using current conveyors. Electron Wirel World 92:28–32
19. Wilson B (1989) Analogue current mode circuits. Int J Elect Eng Educ 26:206–223
20. Kumar U, Shukla SK (1989) The implementation and applications of current conveyors. Microelectron J 20:25–46
21. Pal K (1989) Modified current conveyors and their applications. Microelectron J 20:37–40
22. Wilson B (1990) Recent developments in current conveyors and current-mode circuits. IEE Proc 137:63–77
23. Sedra AS, Roberts GW, Gohh F (1990) The current conveyor: history, progress and new results. IEE Proc 137:78–87
24. Rathore TS (1991) Correspondence: some more published literature on current conveyors. IEE Proc Circ Devices Syst 138:432
25. Toumazou C, Payne A, Lidgey FJ (1991) Operational floating conveyor. Electron Lett 27:651–652
26. Wilson B (1992) Trends in current conveyor and current-mode amplifier design. Int J Electron 73:573–583
27. Fabre A, Saaid O (1993) Novel translinear impedance convertor and band pass filter applications. Electron Lett 29:746–747
28. Fabre A (1995) Third-generation current conveyor: a new helpful active element. Electron Lett 31:338–339
29. Chiu W, Liu SI, Tsao HW, Chen JJ (1996) CMOS differential difference current conveyors and their applications. IEE Proc Circ Devices Syst 143:91–96
30. Wu J, E-El M (1996) Current-mode ladder filters using multiple output current conveyors. IEE Proc Circ Devices Syst 143:218–222
31. Payne A, Toumazou C (1996) Analog amplifiers: classification and generalization. IEEE Trans Circ Syst-I 43:43–50
32. Elwan HO, Soliman AM (1997) Novel CMOS differential voltage current conveyor and its applications. IEE Proc Circ Devices Syst 144:195–200
33. Toumazou C, Barry G (1997) Intuitive analogue circuit design. Electron Commun Eng J 231–239
34. Cabeza R, Carlosena A, Arbel A (1997) Use of a CCII—as a universal building block. Microelectron J 28:543–550

35. Soliman AM (1998) Generalized voltage and current conveyors: practical realizations using CCII. IEICE Trans Fundament Electron Commun Comput Sci E81-A:973–975
36. Alami M (1999) Second generation current conveyors with enhanced input resistance. Int J Electron 86:405–412
37. Awad IA (1999) Inverting second generation current conveyors: the missing building blocks, CMOS realizations and applications. Int J Electron 86:413–432
38. Ridley R (2000) Current mode or voltage mode? Switching Power Mag 5:4–9
39. Schmid H (2000) Approximating the universal active element. IEEE Trans Circ Syst-II 47:1160–1169
40. Biolek D, Vrba K, Cajka J, Dostal T (2000) General three-port current conveyor: a useful tool for network design. J Electr Eng 51:36–39
41. El-Adawy AA, Soliman AM, Elwan HO (2000) A novel fully differential current conveyor and applications for analog VLSI. IEEE Trans Circ Syst-II 47:306–313
42. Becvar D, Vrba K, Zeman V, Musil V (2000) Novel universal active block: a universal current conveyor. ISCAS 3:471–474
43. Beevao D, Vrba K (2000) Novel generations of inverting current conveyor using universal current conveyor. Electron J Eng Tech 3:4
44. Gilbert B (2001) Analog at milepost 2000: a personal perspective. Proc IEEE 89:289–304
45. Takagi S (2001) Analog circuit designs in the last decade and their trends toward the 21st century. IEICE Trans Fundament E84-A:68–79
46. Gupta SS, Senani R (2001) Comment on ‘CMOS differential difference current conveyors and their applications’. IEE Proc Circ Devices Syst 148:335–336
47. Arbel AF (2002) Review of research on ASP by the author. IEEE Circ Syst-II 49:599–611
48. Mangelsdorf CW (2002) The changing face of analog IC design. IEICE Trans Fundament E85-A:282–285
49. Sato T, Wada K, Takagi S, Fujii N (2002) Extension of current conveyor concepts and its applications. IEICE Trans Fundament E85-A:414–421
50. Kuntman H, Cicekoglu O, Ozoguz S (2002) A modified third generation current conveyor, its characterization and applications. Frequenz 56:47–54
51. Zeki A, Toker A (2003) The dual-X current conveyor (DXCCII): a new active device for tunable continuous-time filters. Int J Electron 89:913–923
52. Schmid H (2003) Why ‘Current Mode’ does not guarantee good performance. Analog Integr Circ Sig Process 35:79–90
53. Alzaher HA, Elwan H, Ismail M (2003) A CMOS fully balanced second-generation current conveyors. IEEE Trans Circ Syst-II 50:278–287
54. Biolek D, Gubek T, Brno UFV (2004) New circuit elements for current-mode signal processing. Elektrorevue 28:12
55. Vidal E, Alarcon E, Gilbert B (2004) Up-to-date bibliography of current-mode design. Analog Integr Circ Sig Process 38:245–262
56. Cajka J, Dostal T, Vrba K (2004) General view on current conveyors. Int J Circ Theor Appl 32:133–138
57. Gilbert B (2004) Current mode, voltage mode or free mode? A few sage suggestions. Analog Integr Circ Sig Process 38:83–101
58. Cajka J, Vrba K (2004) The voltage conveyor may have in fact found its way into circuit theory. Int J Electron Commun 58:244–248
59. Gift SJG (2005) The operational conveyor and its application in an accurate current amplifier with gain-independent bandwidth. Int J Electron 92:33–47
60. Soliman AM (2008) The inverting second generation current conveyors as universal building blocks. Int J Electron Commun 62:114–121
61. Biolek D, Senani R, Biolkova V, Kolka Z (2008) Active elements for analog signal processing: classification, review, and new proposals. Radioengineering 17:15–32
62. Horng JW, Hou CL, Chang CM (2008) Multi-input differential current conveyor, CMOS realisation and application. IET Circ Devices Syst 2:469–475

63. Horng YS, Liu WH, Tu SH, Chen JJ (2009) New building block: multiplication-mode current conveyor. IET Circ Devices Syst 3:41–48
64. Siripruchyanun M, Silapan P, Jaikla W (2009) Realization of CMOS current controlled current conveyor transconductance amplifier (CCCCTA) and its applications. J Active Passive Electron Devices 4:35–53
65. Sobhy E, Soliman AM (2009) Novel CMOS realization of balanced-output third generation inverting current conveyor with applications. Circ Syst Sig Process 28:1037–1051
66. De Marcellis A, Ferri G, Guerrini NC, Scotti G, Stornelli V, Trifiletti A (2009) The VCG-CCII: a novel building block and its application to capacitance multiplication. Analog Integr Circ Sig Process 58:55–59
67. Chunhua W, Yang L, Qiujing Z, Yu F (2010) Systematic design of fully balanced differential current-mode multiple-loop feedback filters using CFBCCII. Radioengineering 19:185–193
68. Metin B, Herencsar N, Vrba K (2012) A CMOS DCCII with grounded capacitor based cascadable all-pass filter application. Radioengineering 21:718–724
69. Abdalla KK, Bhaskar DR, Senani R (2012) A review of the evolution of the current-mode circuits and techniques and various modern analog circuit building blocks. Nat Sci 10:1–13
70. Pandey N, Kumar P, Choudhary J (2013) Current controlled differential difference current conveyor transconductance amplifier and its application as wave active filter. ISRN Electron Article ID 968749:11
71. Chong CP, Smith KC (1986) Biquadratic filter sections employing a single current conveyor. Electron Lett 22:1162
72. Rathore TS (1976) Analogue computations using current conveyors. J Inst Electron Telecom Engr 22:510–511
73. Kuntman HH (2011) New advances and possibilities provided by alternative active elements in analogue circuit design. Elektrik-Elektronik ve Bilgisayar Sempozyumu 1:1–12
74. Chua LO (1967) The rotator- a new network component. Proc IEEE 55:1566–1577
75. Chua LO (1968) Synthesis of new non-linear network elements. Proc IEEE 56:1325–1340
76. Mahmoud SA, Hashiesh MA, Soliman AM (2005) Low-voltage digitally controlled fully differential current conveyor. IEEE Trans Circ Syst-I 52:2055–2064

Chapter 2
Hardware Implementations of CCs Using Off-the-Shelf ICs

Abstract In this chapter a variety of hardware implementations of CCs (CCI, CCII and their multiple output versions) using commercially available IC op-amps, IC OTAs, IC transistor arrays and IC current sources have been described and their limitations, advantages and performance ranges have been outlined. The circuits discussed herein can be advantageously employed to build many of the CC-based circuits described in this monograph.

2.1 Introduction

In this chapter, we present various hardware implementations of Current Conveyors based on commercially available ICs which have been proposed by various researchers [1–17] from time to time.

2.2 Hardware Implementations of CCs Using Off-the-Shelf ICs

Various off-the-shelf ICs have been employed in the past to construct CCs such as op-amps, operational transconductance amplifiers (OTA), combinations thereof and op-amps along with transistor arrays etc. In the following, various implementations proposed have been described and their merits and demerits have been highlighted.

2.2.1 Black-Friedmann-Sedra CC Implementation Using an Op-Amp with Uncommitted Leads

The simplified forms of Black, Friedmann and Sedra implementations of the CCII+ and CCII− are shown in Figs. 2.1 and 2.2 respectively (with trimming resistors used

© Springer International Publishing Switzerland 2015
R. Senani et al., *Current Conveyors*, DOI 10.1007/978-3-319-08684-2_2

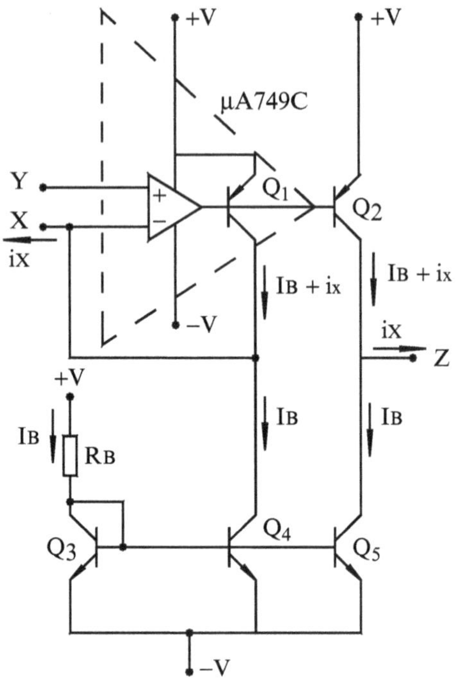

Fig. 2.1 Simplified form of integrable CCII+ proposed by Black et al. [1]

therein removed for the sake of simplicity which have been included in the circuits as presented in [1] for the purpose of reducing the errors in current followers due to finite β). In the circuit of Fig. 2.1, the uncommitted collector of Q_1 is connected to the inverting input terminal of the op-amp thereby creating the port-X which will therefore have $V_x = V_y$. From outside the chip four more transistors are added out of which the transistors Q_3-Q_4-Q_5 constitute a current repeater such that the DC bias current I_B generated in the resistor R_B through the diode connected Q_3 is mirrored in Q_4 and Q_5 as shown. As a consequence, the collector current of Q_1 which is $(I_B + i_x)$ is duplicated in the external transistor Q_2. Since the collectors of Q_2 and Q_5 are connected together and the junction of these two constitutes the Z-terminal of the CCII, it follows that current coming out of terminal-Z is exactly equal to the current applied at terminal-X (taken as coming out of terminal-X for simplicity). Thus $i_z = i_x$ and since $V_x = V_y$ and i_y is already zero, the circuit realizes a CCII+.

In the circuit schematic of Fig. 2.2 as many as seven external transistors have been used. It may be seen that the desired DC bias current is created in the resistor R_B through two diode connected transistors Q_3 and Q_6 connected to DC power supplies +V and −V respectively which is mirrored into the collector of Q_5 through the NPN current mirror Q_6-Q_5. Another replica of the DC bias current I_B is created in the collector of Q_4 through the PNP current mirror Q_3-Q_4. As in the earlier circuit, the open collector of Q_1 is connected to the inverting input of the op-amp and the collector of Q_5 thereby ensuring $V_x = V_y$ (note that i_y is already zero). It is

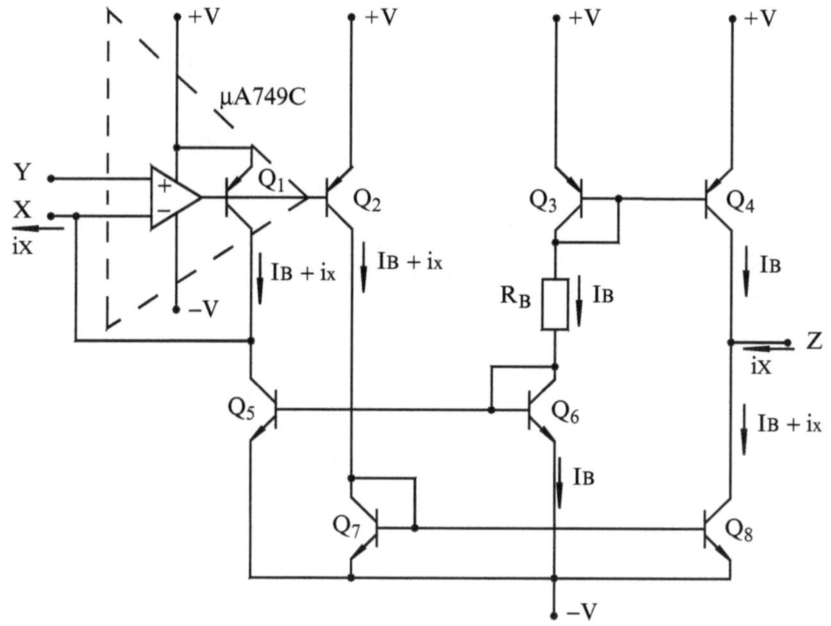

Fig. 2.2 Simplified form of integrable CCII− proposed by Black et al. [1]

now easy to see that the collector current of Q_1 which is $I_B + i_x$ is duplicated in Q_2 where from, through NPN current mirror Q_7-Q_8 it is replicated in collector of Q_8. Since the junction of the collectors Q_4 and Q_8 constitute the Z- port, from the various current components shown in the diagram it follows that current going into Z- terminal will be same as current coming out of X- terminal and therefore the circuit realizes a CCII−.

The CCII− made this way exhibited a voltage gain between ports Y and X of the order of 0.9997, an input impedance looking into port Y as 130 MΩ, output resistance looking into terminal X as 0.45 Ω and an offset of the order of 0.5 mV between ports Y and X. The output impedance looking into terminal Z could be raised by using Wilson current mirrors at the output end which was found to be of the order of 80 MΩ when the simple mirrors at the Z-output were replaced by an elaborate controlled positive feedback based mirror arrangement.

2.2.2 Bakhtiar-Aronhime's Entirely Op-Amp-Based Implementation

In reference [2], Bakhtiar and Aronhime demonstrated for the first time how a CCII (both + and −) can be realized exclusively using only op-amps and resistors. Their proposition is shown in Fig. 2.3. In this circuit, the first op-amp along with a resistor

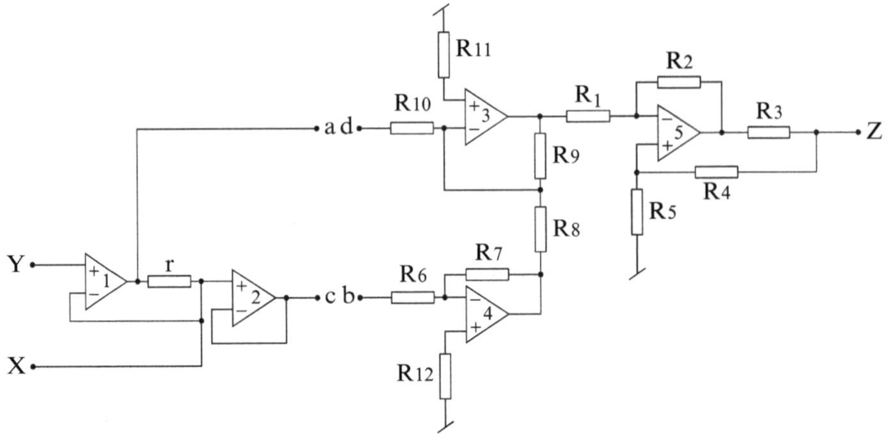

Fig. 2.3 CCII implementation proposed by Bakhtiar and Aronhime [2]

in negative feedback path ensures $I_y = 0$ and $V_x = V_y$. Among the remaining four amplifiers the two op-amps along with resistors R_6 to R_{12} ensure that the output voltage of the third op-amp is V_A-V_C when connections A-D and C-B are made. The fifth op-amp is configured as voltage to current converter. With $(R_6/R_{10}) = (R_7/R_8), R_7 = R_8, R_6 = R_9 = 10R_7, R_{11} = R_{12} = R_7 \| R_6$ and $R_2R_5 = R_1(R_3 + R_4)$, and $rR_2 = R_1R_3$, i_z equals $\pm i_x$, depending on the connections among terminals a, b, c and d. A major limitation of this circuit is the (employment of large number of op-amps along with a large number of resistors). In addition, the bandwidth of the current conveyor achieved is about 29 KHz only when the op-amp used (CA3130) has the gain bandwidth product $A_0\omega_p = 5.1 \times 10^7$.

2.2.3 *Senani's Op-Amp-OTA Based Implementation*

A novel three port active- R circuit capable of implementing all the four forms of current conveyors namely CCI+, CCI−, CCII+ and CCII− using an op-amp, an OTA and only one/two resistors was proposed by Senani in [3]. The circuit to realize CCII requires only one external resistor as shown in Fig. 2.4. A straight forward analysis of this circuit shows that $i_y = 0$, $V_x = V_y$ and $i_z = g_m R_{01} i_x$, thus $i_z = i_x$ is obtained if R_{01} is chosen as $1/g_m$. Alternatively, since the g_m of the OTA is electronically controllable through an external bias current I_B, the circuit realizes an electronically controllable current conveyor CCII+. If the polarities of the input terminals of the OTA are reversed, the same circuit realizes a CCII−.

To realize a CCI, one more resistor needs to be connected between the non-inverting input terminal and the output of the op-amp as shown in Fig. 2.5, if $R_{01} = R_{02} = 1/g$ the circuit realizes a CCI+. If the polarities of the input terminals of the OTA are reversed, the same circuit realizes a CCI−.

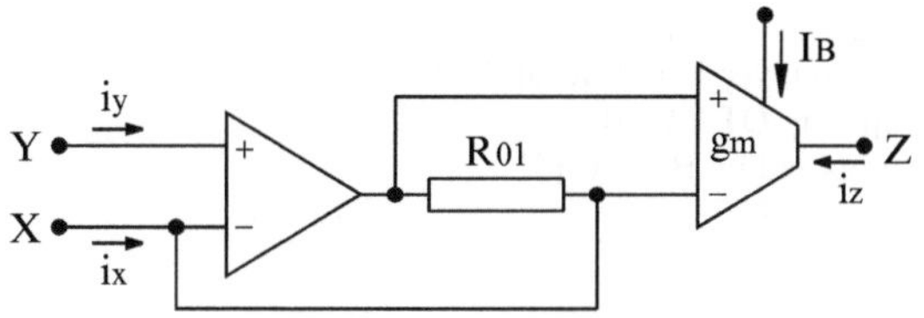

Fig. 2.4 Implementation of CCII+ and CCII− proposed by Senani [3]

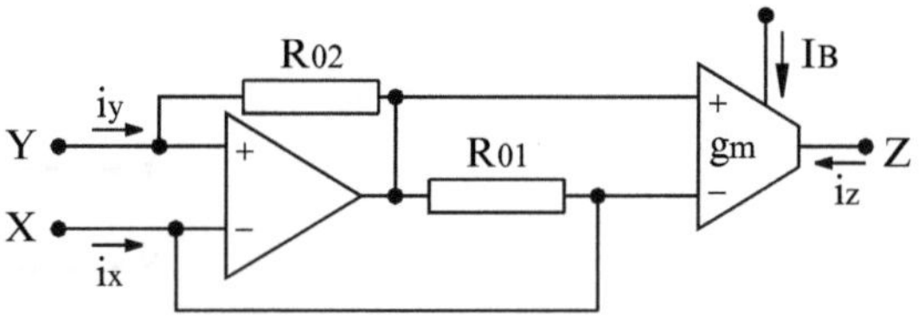

Fig. 2.5 Implementation of CCI+ and CCI− proposed by Senani [3]

It should be recalled that the CCII implementation by Black, Friedmann and Sedra requires an integrated circuit μA749 containing an op-amp with uncommitted output collectors, which when augmented by the introduction of several PNP and NPN transistors externally realizes a CCII. On the other hand the Bakhtiar-Aronhime circuit requires 5 op-amps along with 13 external resistors. In comparison with these approaches, the op-amp-OTA based implementation is simpler and more versatile because it can realize CCII as well CCI (positive and negative both).

2.2.4 Huertas's Entirely Op-Amp Based CC Implementation

Huertas [4] presented an entirely op-amp version of Senani's op-amp-OTA based circuit [3] which is shown here in Fig. 2.6. This circuit can be implemented from a single dual op-amp IC chip. The circuit as shown realizes a CCII; for realizing a CCI an additional resistor needs to be connected between the non-inverting input and the output of the first op-amp. The circuit in the dotted box acts as a differential voltage controlled current source (DVCCS) which is what an OTA does. An analysis of the circuit shows that it is not really necessary to take $r_1 = r_2$ and $r_3 = r_4$ as originally taken by Huertas. The same results can be obtained by selecting component values to satisfy the following condition.

$$\frac{r_1}{r_3} = \frac{r_2}{r_4} = \frac{r}{r'} \tag{2.1}$$

As compared to the op-amp-OTA-based circuit of [3] this circuit was claimed to provide the following advantages: (i) with appropriate choice of the op-amps this circuit can provide a better frequency response (ii) the differential input impedance of an OTA which is typically of the order of 20 KΩ, imposes an upper bound on the value of resistor r; using only op-amps relaxes this limitation (iii) it is possible to achieve offset correction in this circuit by appropriate choice of resistors r_1, r_2, r_3, r_4

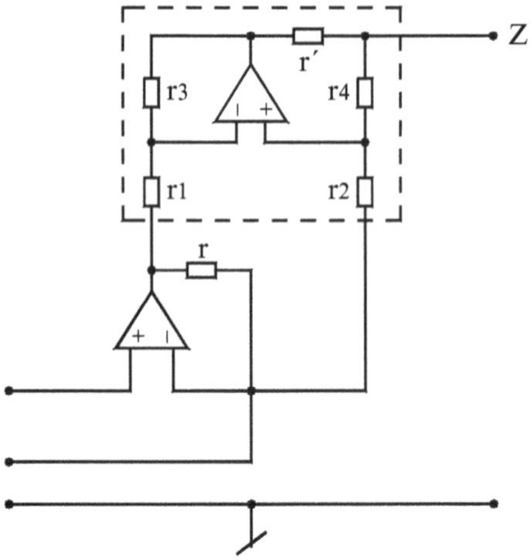

Fig. 2.6 An entirely op-amp version of Senani's op-amp-OTA CCII, proposed by Huertas [4]

and finally, (iv) in applications where several CCs are to be connected in parallel at port-Z, this implementation allows a practical sub-summing of many op-amps as shown in [4].

2.2.5 Pookaiyaudom and Samootrut Implementation Using OTAs

In [10] Pookaiyaudom and Samootrut demonstrated how a commercial dual operational transconductance amplifier (OTA) IC CA3280 together with a current source IC LM334 can be employed to realize both CCII+ and CCII−. These circuits are shown in Figs. 2.7 and 2.8 respectively.

The OTA CA3280 has the provision that the common junction of the emitters of the two transistors forming the differential pair is available as an external terminal outside the chip. In the CCII+ implementation, the basic idea is to keep the inverting input terminal of the OTA open circuited thereby forcing Q_1 to be off so that transistor Q_2 conducts the entire current ($I_{B\text{-}ix}$). Since Q_1 is off, the first current mirror (CM_1) is off and consequently, CM_3 is also off. The collector current of Q_2 is mirrored by CM_2 and reaches the output node Z where a constant current source LM334 is connected between terminal-Z and negative supply rail thereby ensuring $i_z = i_x$ as shown on the diagram.

It may be noted that the main departures from ideal CCII properties in this circuit are (i) the voltage drop of one V_{BE} between terminal Y and X and (ii) the nonzero input resistance at port-X. Since in most applications only AC signals are used therefore, the one V_{BE} voltage drop does not introduce serious application difficulty. Also, if one takes the maximum bias current of the order of 10 mA, the input resistance looking into port-X turns out to be the order of 2.5 Ω which is quite adequate for most practical applications.

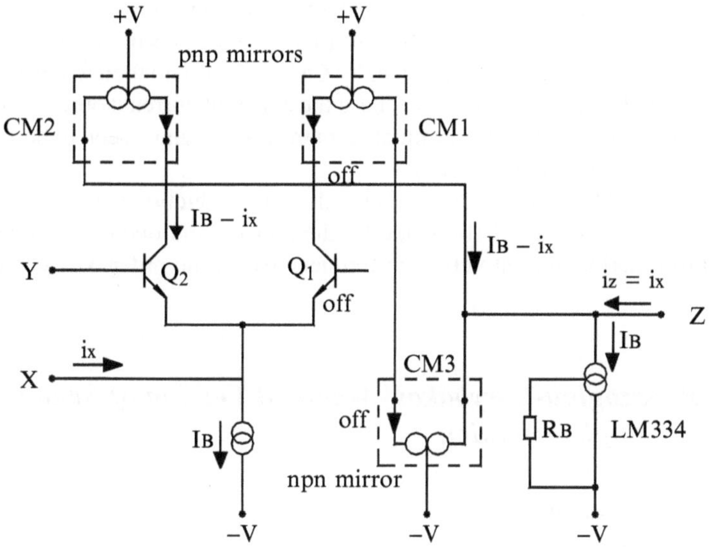

Fig. 2.7 Implementation of CCII+ proposed by Pookaiyaudom and Samootrut [10]

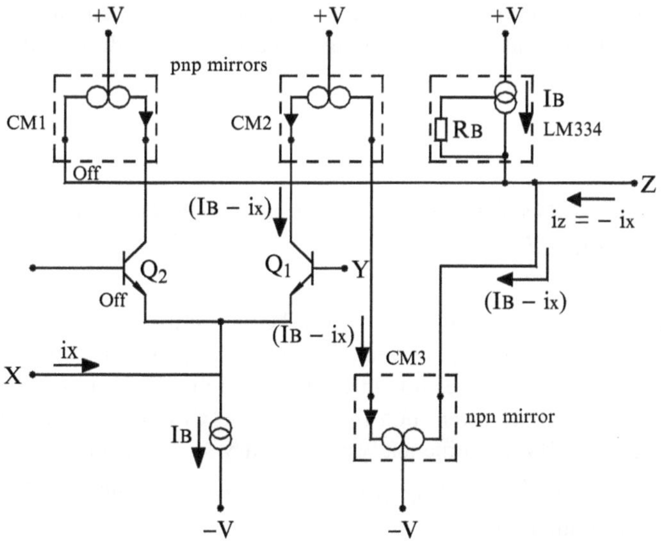

Fig. 2.8 Implementation of CCII− proposed by Pookaiyaudom and Samootrut [10]

For realizing CCII−, the non-inverting terminal of the OTA is left open; the base of Q_1 is treated as Q_y. In this case, Q_2 is off therefore, CM_2 is off while the collector current Q_1 is $(I_{B\text{-}ix})$ which is carried over by CM_1 to CM_3 and then to the Z- terminal of the circuit however, in this case, because the constant current source is connected from positive supply rail to the terminal- Z therefore, i_z can be seen to be $= -(I_{B\text{-}ix})$ and thus, the circuit realizes a CCII−.

Since CA3280 has reportedly a gain bandwidth product of around 10 MHz, it is expected that CCII+ and CCII− made by the above mentioned technique may be expected to function at relatively high frequencies (of the order of several MHz).

2.2.6 *Papazoglou-Karybakas' Modified Version of Senani's CC Implementation*

Papazoglou and Karybakas [15], while presenting a non-interacting electronically tunable CCII based current mode biquadratic filter, employed a slightly modified version of Senani's CC (see Fig. 1 therein) which is shown here in Fig. 2.9. Note that the DC bias current of the OTA which controls the transconductance gain g_m of the OTA is derived from an additional circuit consisting of an op-amp, two resistors and PNP transistor thereby arranging that I_B is equal to V_c/R_1. As a consequence the current gain between i_z and i_x which is symbolically denoted as h_{32} becomes electronically controllable by external control voltage V_c. Using this version of the CCII, they succeeded in implementing biquadratic filter circuit whose cut off frequency could be varied over a range of one decade through the external control voltage V_c.

2.2.7 *Karybakas-Siskos-Laopoulos's Compensated, Tunable CC*

Karybakas et al. in [12] presented a novel CC configuration based upon an op-amp and two OTAs which provides current gain tunability, has full compensation over temperature variation and does not suffer from the disadvantages of gain temperature dependence, input current non-linearities and CMRR effects. This circuit is shown in Fig. 2.10. By a routine analysis, it is easy to confirm that $V_x = V_y$, $I_y = 0$ and $I_z = \pm k\ i_x$. Since $\frac{i_z}{i_x} = \frac{G_2}{G_1} = \frac{I_{B2}/2V_T}{I_{B1}/2V_T} = \frac{I_{B2}}{I_{B1}} = k$, it is seen that the current gain is independent of temperature.

From the same circuit, a CCII− can also be realized by interchanging the input terminals of the OTA connected at Z- port. It has been shown in [12] that by changing I_{B2}, the current gain varied linearly over a range of three decades; the gain value remains temperature independent over a temperature from 0° to 100° and

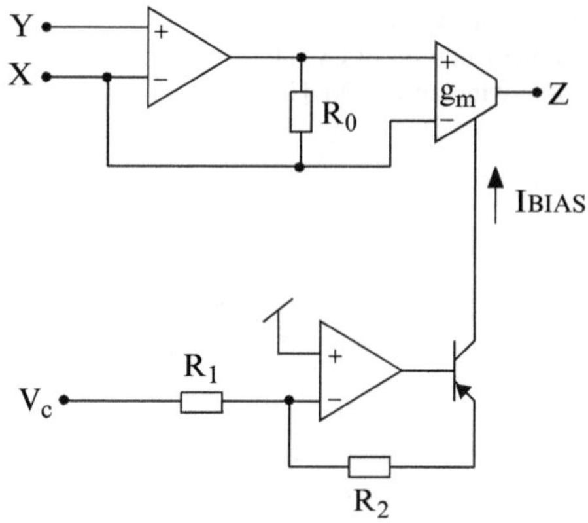

Fig. 2.9 Modified version of Senani's implementation of CC proposed by Papazoglou and Karybakas [15]

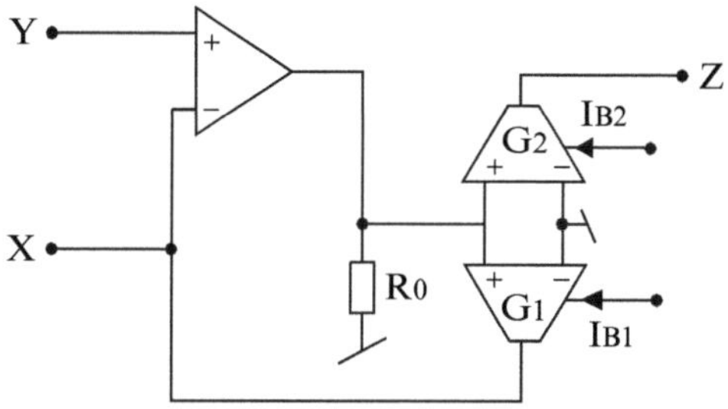

Fig. 2.10 CC realization proposed by Karybakas et al. [12]

non-linear phenomenon (owing to OTA input current and CMRR effects) do not affect the circuit performance.

2.2.8 Wilson's OMA-Based Implementations of CCII+/−

It was demonstrated by Wilson [7] that wide band precision current conveyor can be obtained by adopting the so called *power supply current sensing technique* whose embodiment is also popularly referred to as an Operational Mirrored Amplifier (OMA). An OMA-based CC implementation using two complementary modified Wilson current mirrors connected between the DC bias terminals and power supply rails to realize a CCII+ is shown in Fig. 2.11. It is well known that the output current of the op-amp in the internally compensated type op-amps such as LF356/μ A741 is equal to the difference of the two power supply currents I_p and I_N as shown in Fig. 2.11. Thus, x-terminal input current manifests itself as the difference between I_p and I_n each of which are copied by the current mirrors and their difference is made available at the high output impedance terminal Z thereby realizing $I_z = I_x$. Since the op-amp with a unity negative feedback ensures equality of the voltages at Y and X terminals leading to $V_x = V_y$ and the ideally infinite input impedance of the op-amp makes $I_y = 0$, it follows that the circuit realizes a CCII+.

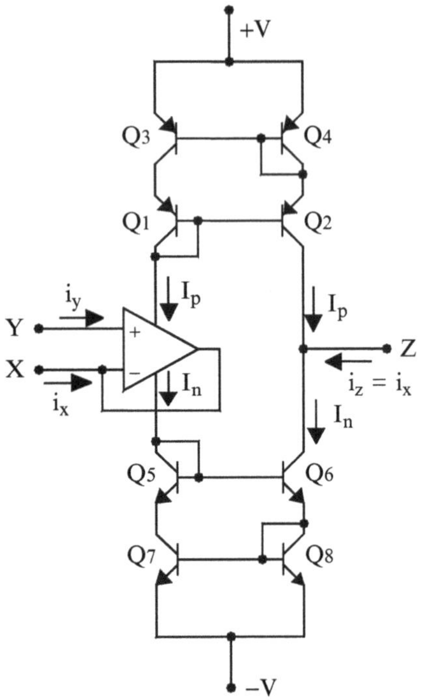

Fig. 2.11 Wilson's implementation of CCII+ employing an OMA+ [7]

By adding one more pair of current mirrors and a cross-coupling a CCII− is created as shown in Fig. 2.12.

A typical circuit made from CA3096AE mixed transistor arrays and LM301 op-amp with a compensating capacitor of 39 pF exhibits a small signal bandwidth of greater than 1 MHz which is only restricted by the PNP transistors' performance beyond this region. Transfer distortions around −68 to −70 dB can be expected depending upon the load. The X- terminal exhibits an effective parasitic capacitance of about 4 pF whose effect on the circuit performance can be reduced by good circuit layout techniques [8].

2.2.9 *CCII Implementation Using Op-Amps and Only NPN Transistors*

A circuit which requires only NPN transistors but two op-amps instead of one is shown in Fig. 2.13 which avoids use of PNP transistors and the associated frequency limitations which are now only limited by the performance of the op-amps employed. This circuit is based upon the operational mirrored amplifier proposed by Huijsing and Veelenturf [5].

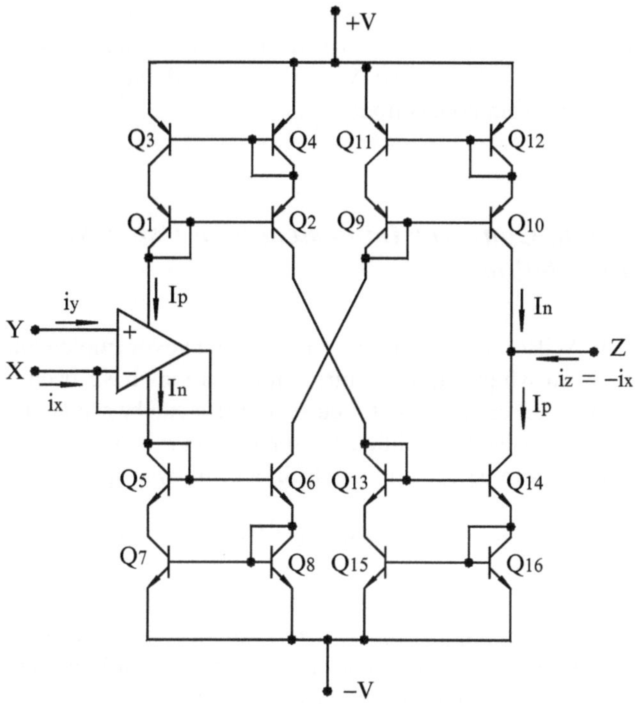

Fig. 2.12 Implementation of CCII− from an OMA- [8]

Fig. 2.13 A CCII implementation using op-amps and only NPN transistors based on the operational mirrored amplifier proposed by Huijsing and Veelenturf [5]

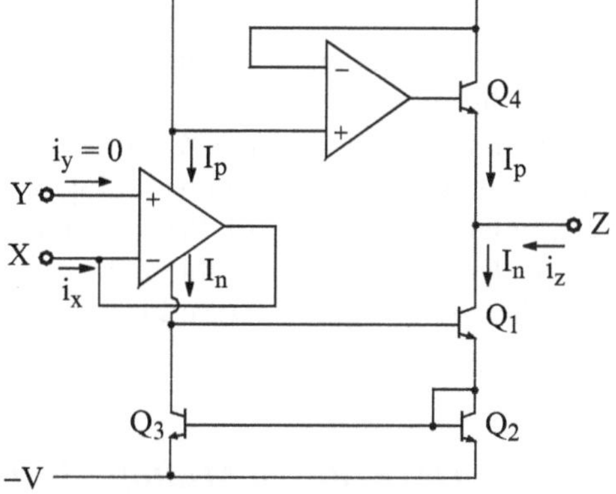

In this circuit, the PNP mirror employed in the Wilson's CC implementation to sense the current in the positive supply line is replaced by an elegant circuit using an op-amp configured as a voltage follower (VF) and three NPN transistors, two of which are diode-connected. With Q5 and Q6 matched, the VF ensures the same currents, each equal to Ip, flow in transistors Q5, Q6 and then Q4 so that at Z-terminal, the current going in is equal to (Ip-In) and hence, equal to i_x thereby realizing a CCII+ since $i_y = 0$ and $V_x = V_y$ are available by virtue of the first op-amp and its unity gain connection.

2.2.10 *Current Conveyor Implementation Using New Mirror Formulation*

In reference [9], Wilson proposed a new current conveyor implementation shown here in Fig. 2.14 in which the new mirror formulations have been shown in the dotted boxes. This new eight transistor current mirror can be seen to be obtained by stacking two four transistors modified Wilson current mirrors such that the output current I_0 can be seen to be equal to the difference of two input currents I+ and I− i.e., $I_0 = (I+ - I-)$.

This circuit was constructed using CA3096 AE mixed transistor arrays, LM301A op-amp with 27 pF compensation capacitor and a 4.7 KΩ bias resistor along with ± 15 V DC power supplies. The circuit exhibited a 3-dB bandwidth of more than 1 MHz for all loads less than 1 KΩ. The rise times of the order of 150 n sec irrespective of output of magnitude were achievable for low impedance of the order of 330 Ω or less. Maximum voltage slew rates greater than 15 V/μs were obtainable greatly exceeding slewing ability of the op-amps employed.

2.2.11 *Conversion of CCII into CCI and Vice Versa*

In [14] Bruun demonstrated that CCI can be realized from a multiple output CCII and on the other hand, a CCII can be similarly realized from a multiple output CCI. In the first case, one of the Z^+ outputs of the CCII is returned to its Y-terminal thereby making $i_y = i_x$ as shown in Fig. 2.15a.

Similarly, when one of the Z- output terminal of a multiple output CCI is returned to its Y-terminal this creates a net current entering the terminal-Y as zero thereby giving $i_y = 0$ (see Fig. 2.15b).

2.2.12 *OMA-Based Multiple-Output CCs*

Creating multiple output CCIIs (and therefrom multiple output CCIs using the conversion method outlined in the previous section) using op-amps in supply current sensing mode requires only additional current mirrors and is straight forward as shown in Fig. 2.16.

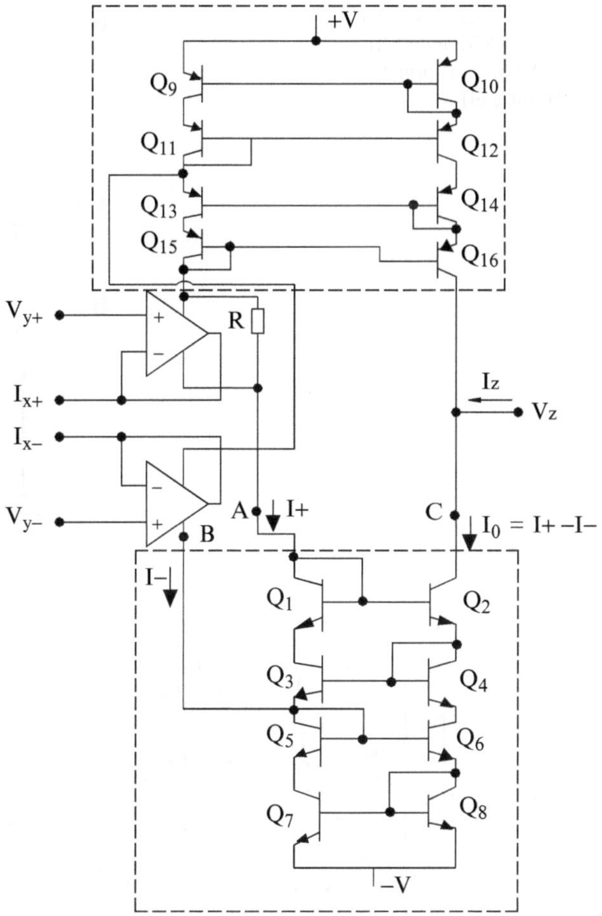

Fig. 2.14 CC implementation using new current mirror formulation [9]

Concluding Remarks

In this chapter we have discussed a variety of CC implementations using commercially available op-amps, OTAs and transistor arrays. Also discussed were methods for converting CCI into CCII and vice versa along with creation of multiple output CCs based on supply current sensing techniques. It may be mentioned that the op-amp-OTA implementation proposed by Senani, its entirely op-amp version proposed by Huertas and operational mirrored amplifier based versions proposed by Wilson for both CCII+ and CCII− have been extensively used in literature by a number of researchers to test and verify their circuit proposals employing current conveyors. Hardware implementation of CCs described here thus, can be readily used to build and test current conveyors based circuits when either IC current conveyors are not available or when multiple output CCs are needed.

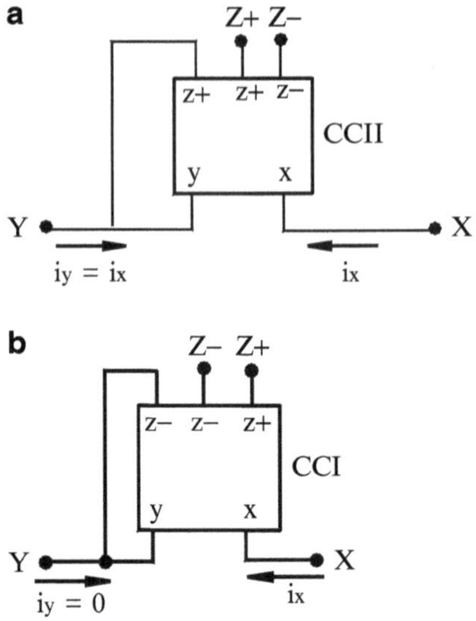

Fig. 2.15 Realizing CCI from MOCCII and CCII from MOCCI as proposed by Bruun [14]

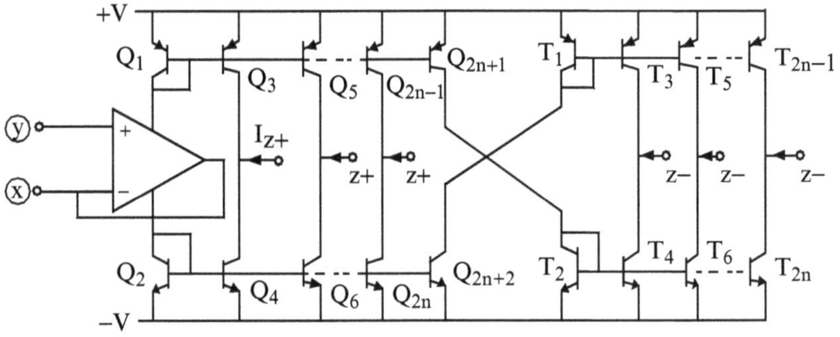

Fig. 2.16 Realization of multiple outputs CCII using power supply current sensing technique

References

1. Black GGA, Friedmann RT, Sedra AS (1971) Gyrator implementation with integrable current conveyors. IEEE J Solid State Circ 6:396–399
2. Bakhtiar MS, Aronhime P (1978) A current conveyor realization using operational amplifiers. Int J Electron 45:283–288
3. Senani R (1980) Novel circuit implementation of current conveyors using an O.A. and an O.T.A. Electron Lett 16:2–3
4. Huertas JL (1980) Circuit implementation of current conveyor. Electron Lett 16:225–226
5. Huijsing JH, Veelenturf CJ (1981) Monolithic class AB operational mirrored amplifier. Electron Lett 17:119–120
6. Wilson B (1981) A low-distortion bipolar feedback current amplifier technique. Proc IEEE 69:1514–1515

7. Wilson B (1984) High-performance current conveyor implementation. Electron Lett 20:990–991
8. Wilson B (1985) Floating FDNR employing a new CCII-conveyor implementation. Electron Lett 21:996–997
9. Wilson B (1986) Differential current-follower and conveyor using new mirror formulation. Electron Lett 22:708–710
10. Pookaiyaudom S, Samootrut K (1988) Efficient circuit implementations of current conveyors, negative impedance converters and nonlinear impedance converters using operational transconductance amplifiers. Int J Electron 64:941–945
11. Singh V (1989) An implementation of CCII- current conveyor, with application. IEEE Trans Circ Syst 36:1250–1251
12. Karybakas CA, Siskos S, Laopoulos THL (1992) A novel compensated, tunable current conveyor. J Circ Syst Comput 2:397–401
13. Carlosena A, Moschytz GS (1994) Design of variable-gain current conveyors. IEEE Trans Circ Syst-I 41:79–81
14. Bruun E (1995) A combined first- and second-generation current conveyor structure. Int J Electron 78:911–923
15. Papazoglou CA, Karybakas CA (1997) Noninteracting electronically tunable CCII-based current-mode biquadratic filters. IEE Proc Circ Devices Syst 144:178–183
16. Dual 9 MHz, Operational transconductance amplifier (OTA) (2002) Data Sheet: FN1174.6: CA3280, CA3280A:1–10
17. Rana S, Pal K (2007) Current conveyor simulation circuits using operational amplifiers. J Phys Sci 11:124–132

Chapter 3
Integratable Bipolar CC Architectures and Commercially Available IC CCs

Abstract Bipolar implementations of CCs based upon input stages consisting of translinear circuits as well as those having differential pair based input stages have been discussed. Circuit techniques for achieving temperature-compensated designs, x-port parasitic resistance reduction, increasing the input resistance at port-Y and obtaining current-controllable gains have also been elaborated. Together with these, the details of a number of commercially available IC CCs such as CCII01, PA 630, AD 844 or other IC building blocks which can also be used as CCs (such as OPA 2662, OPA 660/860, have also been provided.

3.1 Introduction

The object of this chapter is to describe the work done on the implementation of current conveyors in bipolar technology. Thus, this chapter presents the key developments and progress made on the implementation of CCs using bipolar junction transistors (BJT) as well as gives the details of the various integrated circuit current conveyors all of which have so far been implemented in bipolar technology only.

3.2 Bipolar Circuit Architectures of Current Conveyors

All current conveyors produced as off-the-shelf ICs have so far been implemented in bipolar technology. In view of this, in this section, we present some prominent topologies [1–24] and developments [25–40] made to realize Current Conveyors. In several circuit diagrams which follow, the various current mirrors and current repeaters are shown by their standard symbolic notations consisting of two or more circles touching each other with one connection to the positive/negative power supply rail with the input terminal shown with an arrow mark and the terminal (s) without arrow mark being the output terminal(s).

© Springer International Publishing Switzerland 2015

R. Senani et al., *Current Conveyors*, DOI 10.1007/978-3-319-08684-2_3

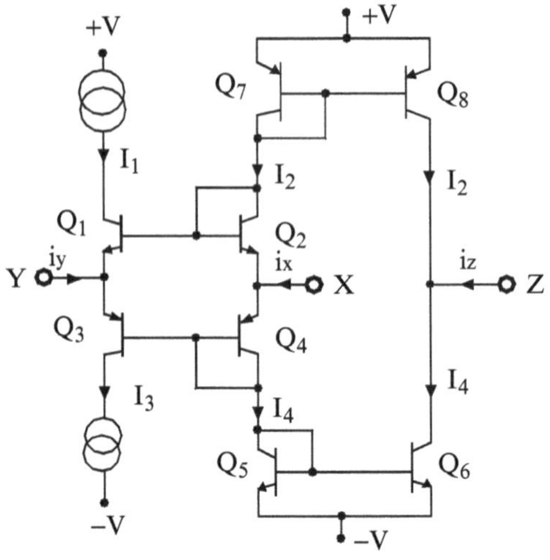

Fig. 3.1 Simplified form of the translinear CC proposed by Fabre [1]

3.2.1 Fabre's Translinear CC

The core of Fabre's CC [1] consists of a four transistor cell which is now commonly referred to as the *mixed translinear cell* as it consists of two PNP transistors Q_3 and Q_4 and two NPN transistors Q_1 and Q_2. Assuming matched transistors with large β (hence, negligible based currents), the application of the translinear principle on the translinear loop consisting of Q_1- Q_2-Q_3-Q_4 (Fig. 3.1) gives the equation: $I_1I_3 = I_2I_4$. Since $I_3 = (I_1 + i_y)$ and $I_4 = (I_2 + i_x) = (I_2 + i_z)$ and since I_1 and I_3 are identical, it follows that $i_y = 0$. Also, for peak-to-peak input currents i_x smaller than the bias current, it can be seen that the translinear loop ensures that $v_x = v_y$ and lastly, $i_z = i_x$ follows from the assumption that pnp and npn current mirrors have unity current transfer ratios.

It has been shown by Fabre [1] that the output port of his CCII has output impedance consisting of a resistance of R_0 in parallel with capacitance C_0. For this CCII+ implementation experimental measurements made at 100 Hz indicated that the equivalent resistance looking into the port-Y is dependent upon the DC bias current, for instance, its value is 100 KΩ for bias current of 330 μA and 40 KΩ for bias current of 1 mA.

With two identical Z outputs available, which requires only addition of one PNP and one NPN transistor, such dual output CCII+ s can be used to realize a CCI+, CCI− and CCII− as per schematics of Fig. 3.2.

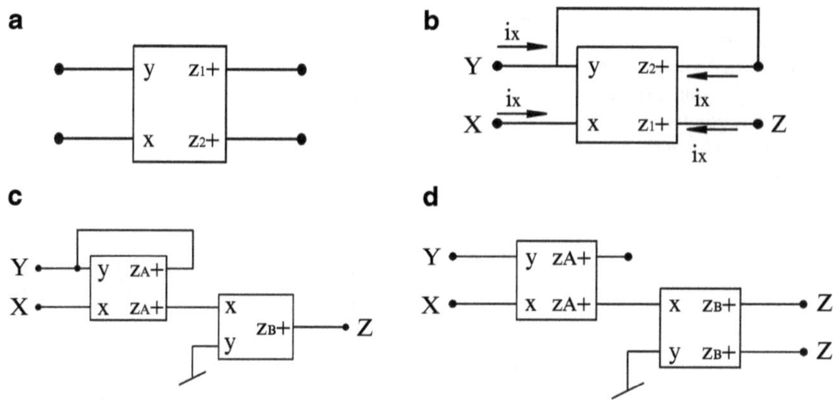

Fig. 3.2 Realization of various CCs from dual output CCII+ s (**a**) dual output CCII+ (**b**) CCI+ (**c**) CCI− (**d**) CCII−

3.2.2 Normand's Translinear CCs

Normand [2] presented two types of translinear CCs working either in class A mode or Class AB mode. In both the cases, a wide input range has been claimed. These architectures are based upon two types of basic cells both of which can be made either from NPN transistors or PNP transistors. The so-called type one cell is based upon Wilson current mirror and is shown in Fig. 3.3a. On the other hand, type two cell is based upon Pookaiyaudom-Quartet [2] as shown in Fig. 3.3b. Corresponding to these two basic cells, the two realizations of translinear CCs proposed by Normand [2] are as shown in Fig. 3.4a and b respectively.

A straight forward analysis of type- one cell with common-emitter current gain β reveals that it can be characterized by the following hybrid matrix in terms of dynamic currents and voltages.

$$\begin{bmatrix} i_y \\ v_x \\ i_z \end{bmatrix} = \begin{bmatrix} 0 & -\dfrac{1}{\beta^2} & 0 \\ 1 & \dfrac{\beta}{\beta+1} r_e & 0 \\ 0 & -1+\dfrac{1}{\beta^2} & 0 \end{bmatrix} \begin{bmatrix} v_y \\ i_x \\ v_z \end{bmatrix} \quad \text{with } r_e = kT/qI = V_T/I \tag{3.1}$$

It may be seen that for large β and very small r_e, the above matrix approximates to that applicable to a Current Conveyor. Similarly, the type two-cell is characterized by

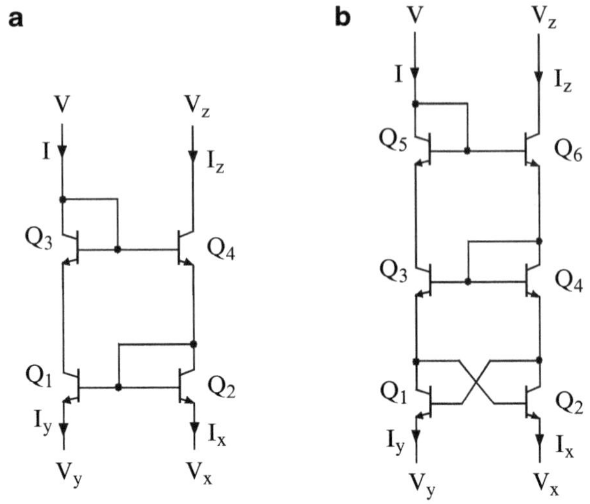

Fig. 3.3 The two types of cells considered by Normand [2] (**a**) type-one cell based on Wilson current mirror [3] (**b**) type-two cell based upon Pookaiyaudom-Quartet [4]

$$\begin{bmatrix} i_y \\ v_x \\ i_z \end{bmatrix} = \begin{bmatrix} 0 & -2/\beta & 0 \\ 1 & 2r_e/\beta & 0 \\ 0 & -\dfrac{\beta-1}{\beta+1} & 0 \end{bmatrix} \begin{bmatrix} v_y \\ i_x \\ v_z \end{bmatrix} \tag{3.2}$$

Like the type-one cell, the type- two cell also nearly works as a current conveyor.

An important property of Normand's translinear current conveyors of Fig. 3.4a, b is that they allow a very wide range input of I_x which is theoretically extendable to $\pm\beta I$. In fact, the circuits can operate in class A mode when the input current I_x is kept within the range of $\pm I$ or they can be operated in class AB mode where input current range can go even beyond these limits. The input resistance looking into X port has been found to be 130 Ω for the first configuration and equal to 5 Ω for the second configuration, for the DC bias current $I = 100\ \mu A$. Resistors of value 5 Ω are connected to reduce the effect of the offset voltages in the emitters of Q_6 and Q_8 (see Fig. 3.4b). The trimmer Tr1 cancels the offset current at Y input for Vin = 0 while Tr2 and Tr3 compensate the output offset current and reduces the even harmonics in the output for sine wave signal and Tr4 compensates the input offset voltage.

3.2.3 An Alternative CCII Implementation

Other than the four translinear mixed translinear loop employed in both Fabre [1] and Normand [2] implementations of the second generation current conveyor, there is an alternative four-transistor mixed-translinear-loop which is also used in quite a number of other CC implementations. While the mixed translinear cell employed in Fabre implementation has Y-input terminal on the junction of emitters of two

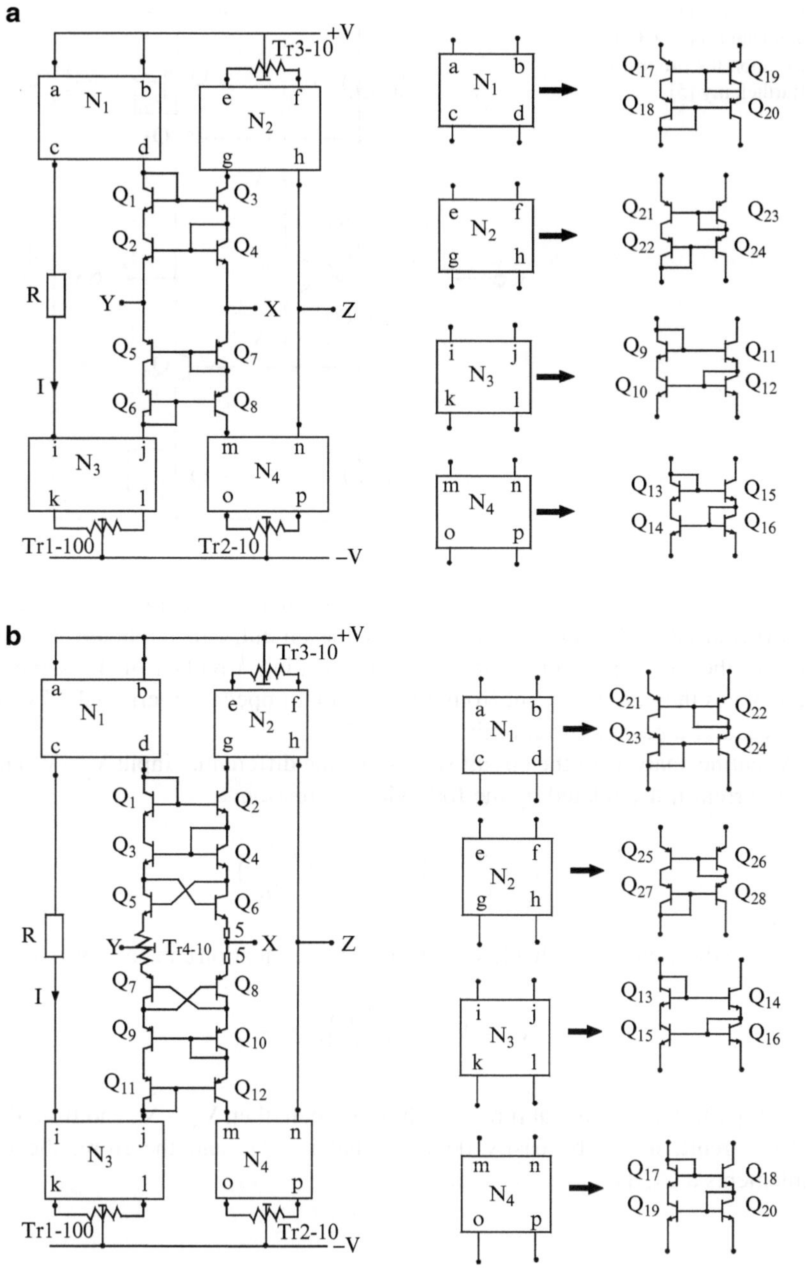

Fig. 3.4 The two CC implementation proposed by Normand [2] (**a**) Using type- one cells (**b**) Using type- two cells

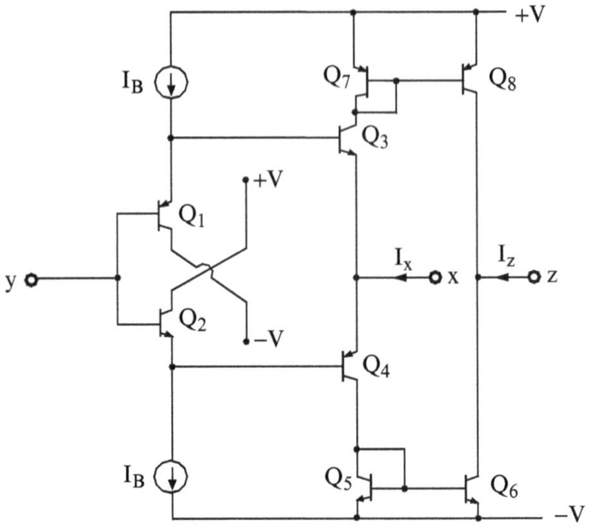

Fig. 3.5 An alternate implementation of CCII+ based upon the proposition of Barthelemy [5]

transistors, in this alternative structure [5] Y terminal is the common base terminal of two transistors Q_1 and Q_2. The rest of the circuitry to duplicate the current entering the low input impedance node and making a replica of the same made available as the output current from a high output impedance terminal- Z which is composed of a pair of current mirrors (Fig. 3.5).

A routine analysis of the circuit shows that the differential input $V_y - V_x$ and the input current i_x are related by the following expression:

$$\left(\frac{V_y - V_x}{V_T}\right) = Sin\,h^{-1}\left(\frac{i_x}{2I_0}\right) \tag{3.3}$$

Subject to the assumption that $i_x << 2I_0$, the above equation can be approximated as

$$(V_y - V_x) \cong \left(\frac{V_T}{2I_0}\right) i_x \cong r_x i_x \tag{3.4}$$

From Eq. (3.4) it is seen that if r_x is negligibly small, then $V_x \cong V_y$ and from the rest of the circuit, it can be easily deduced that $i_z = +i_x$ and therefore, the circuit implements a CCII+.

3.2.4 Two Simple CCII Implementations

In [6], Higashimura and Fukui proposed new Type 1 mutator circuit using CCs and to verify their workability, they employed simple CCII+ and CCII−

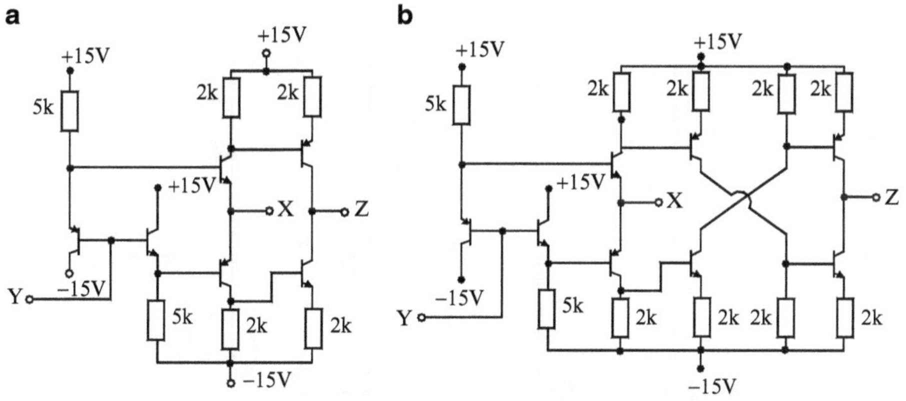

Fig. 3.6 CCII implementation using bipolar transistors (**a**) CCII+ (**b**) CCII− [6]

implementations proposed by Matsuura et al. [40] which are shown in Fig. 3.6. These circuits were realized using 2SA71 PNP transistors and 2SC1327 NPN transistors.

3.2.5 Surakampontorn and Thitimajshima Electronically-Controlled Conveyor (ECC)

Surakampontorn and Thitimajshima [8] presented a generalized current conveyor known as electronically tunable second generation current conveyor (ECCII) whose current transfer ratio between port currents i_z and i_x can be varied electronically. The circuit presented by them however provides two complementary Z outputs thereby realizing CCII+ and CCII− in the same circuit configuration which was claimed to be suitable for monolithic integrated circuit implementation. The basic principle of the circuit can be explained on the basis of the simplified schematic of Fig. 3.7. In this circuit, Q_1 and Q_2 constitute a differential pair with bases of transistor of Q_1 and Q_2 constituting port Y and X respectively. It is seen that the current source of value $I_1/2$ ensures that collector current of Q_1 and Q_2 would be equal where from, with identical transistors, it follows that their base emitter voltages which are also equal will cancel out thereby ensuring $v_x = v_y$. Through the current source I_2, the collector current of Q_3 becomes $I_{2+}i_x$. The differential current amplifier through the current mirror produces an output current which is $I_2 + A\ i_{x,}$ thus, yielding $i_z = A\ i_x$.

A fully integrable circuit configuration of the ECCII based upon this idea using NPN transistor array LM3046, matched discrete PNP transistors and integrated current sources LM334 to provide the DC bias currents has also been described by Surakampontorn and Thitimajshima in the same paper [8].

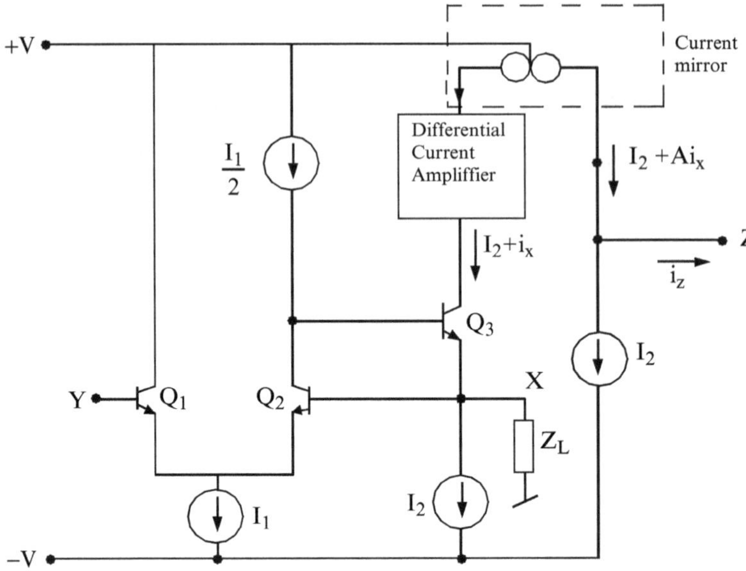

Fig. 3.7 The basic schematic of the electronically-controllable CC proposed by Surakampontorn and Thitimajshima [8]

3.2.6 *Filanovsky's Current Conveyor Modified from a Current Source*

The CCII structure proposed by Filanovsky [9] is actually obtained by appropriate modification of a current source and can be explained by the circuit shown in Fig. 3.8. The transistor circuit consisting of Q_7, Q_8, Q_9, Q_{10} ensures that two DC bias currents, each equal to I_B, are injected at nodes a and b in the circuit. If the β of transistors is very high then from the various total currents marked on the circuit of Fig. 3.8, it follows that the small signals at Y, X and Z ports are described by the equations; $i_y = 0$, $v_x = v_y$ and $i_z = i_x$ and the circuit, thus, realizes a CCII+.

3.2.7 *Temperature-Compensated CCII*

As in an OTA, the transconductance g_m is given by $g_m = I_0/2V_T$, likewise in a current conveyor the input resistance looking into port X is given by $R_x = V_T/2I_0$. Since V_T is directly proportion to absolute temperature, a method of temperature compensation applicable to both translinear CC and OTA was presented by Surakampontorn et al. [10]. The key idea of their preposition is the bias circuit for temperature compensation as shown in Fig. 3.9.

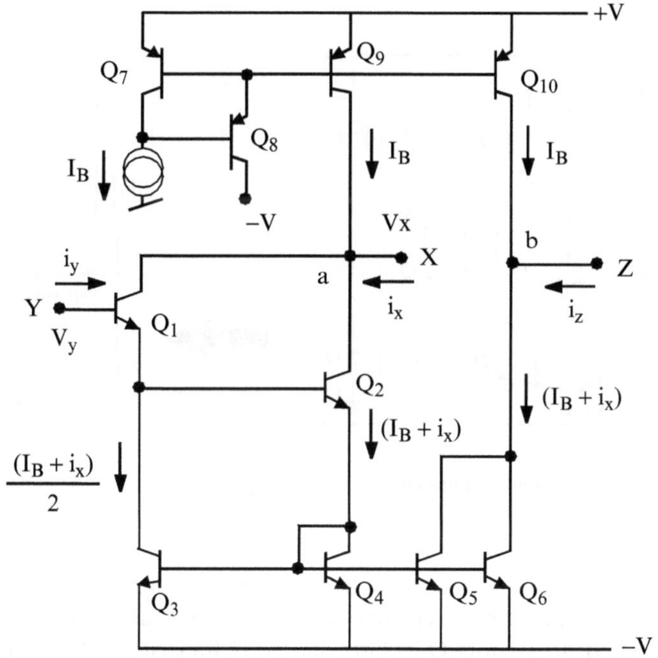

Fig. 3.8 A possible CCII implementation based upon Filanovsky's method of converting a current source into current conveyor [9]

This circuit generates a current which directly relates to the absolute temperature as well as is linearly variable. In this circuit, the transistor Q_1 along with Wilson current mirror realizes a reference current source which generates a current I_A of the form

$$I_A = (V_T \ln\alpha)/R \tag{3.5}$$

where α is the emitter ratio of the transistor Q_2 to Q_1 which in the present case is $\alpha = 4$. Now the current I_1 which is equal to $I_1 = 2I_A$ flows in the diode connected transistors Q_8 and Q_9. The transistors Q_4 to Q_{11} function as the Gilbert translinear cell which provides an output current I_0 given by

$$I_0 = I_1 I_B / I_2 = (2V_T \ln\alpha) I_B / I_2 R \tag{3.6}$$

where I_B is the external bias current and I_2 is a fixed constant current.

The temperature-compensated CC architecture, using this idea, is shown In Fig. 3.10.

In this circuit, using the above equations, it has been found that

$$R_x = V_T/2I_0 = V_T/\{4(V_T \ln\alpha) I_B / I_2 R\} = I_2 R / 4 \ln\alpha I_B \tag{3.7}$$

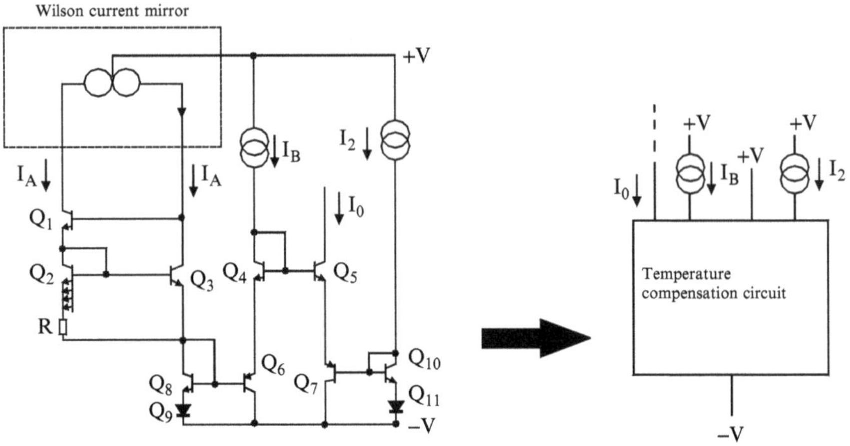

Fig. 3.9 Bias circuit for temperature compensation [10]

from which we see that R_x has now become temperature insensitive. The workability of this temperature-insensitive CC has been confirmed by realizing the circuit using PNP transistors Q2N2905 and NPN transistors Q2N3904 and it has been found that X port input resistance nearly remains invariant at a nominal value of 100 Ω for a temperature variation from 10° to 90°.

3.2.8 CCII with Reduced Parasitic Resistance R_x

Seguin and Fabre [11] presented a new architecture for the CC which is shown here in Fig. 3.11. As can be seen, this implementation has two supplementary transistors added as compared to the traditional implementation. Assuming the β of the transistors to be identical and assuming that $i_x << 2I_0$, the input current at port X is divided into two equal parts which respectively flow in the collectors of Q_5 and Q_7 so that the time-varying components of the currents are each equal to $i_x(t)/2$. This, in turn, induces time variant base currents for the transistors Q_5, Q_6, Q_7 and Q_8, each equal to $i_x(t)/2\beta$. As a consequence, a time-variant value of the collector currents of the transistors Q_2 and Q_4 are each approximately equal to $i_x(t)/\beta$. Therefore, at low frequencies, the parasitic equivalent input resistance looking into port-X is given by:

$$r_x^{/} = \frac{V_T}{2I_0} \cong \frac{2r_x}{\beta} \tag{3.8}$$

Through SPICE simulations, using the model parameters for the transistors from 2 μm HF2CMOS technology from SGS Thomson, with DC power supplies

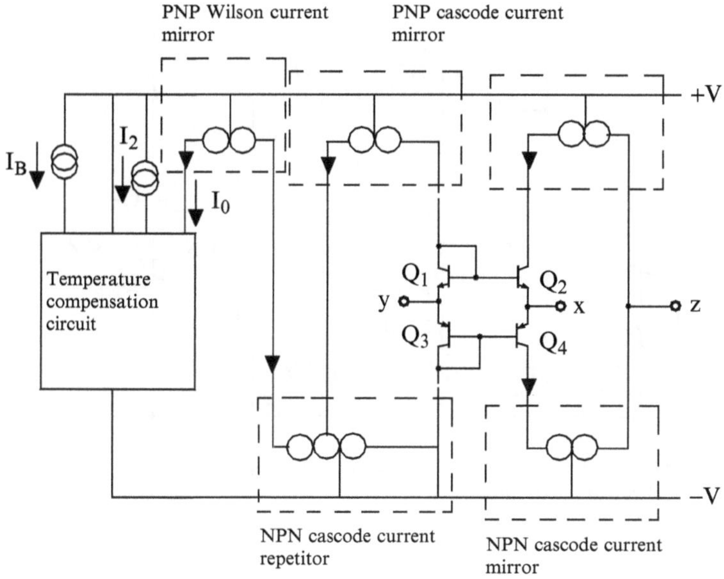

Fig. 3.10 An equivalent representation of the temperature-compensated translinear CC proposed by Surakampontorn et al. [10]

of ± 2.5 V and the DC bias current $I_0 = 300$ μA, it has been found that in this modified CC architecture, r_x is reduced to a value as low as 1.3 Ω (in contrast to a value of about 53 Ω for the conventional design).

3.2.9 CCII with Increased Input Impedance at Port-Y

The basic idea for increasing the Y-port input resistance of the CC was proposed by Alami et al. [12] and is shown in Fig. 3.12. It may be noted that the central core of the circuit consisting of transistors Q_1-Q_2-Q_3-Q_4-Q_5-Q_6-Q_7-Q_8 has been modified by adding two more transistor Q_9 and Q_{10} such that the sub-circuit consisting of Q_1-Q_2-Q_3-Q_4-Q_9-Q_{10} constitutes two translinear mixed loops in parallel which implies $V_y \cong V_x \cong V_0$. It may also be noted that employment of the modified Wilson current mirror consisting of transistors Q_{21}-Q_{22}-Q_{23}-Q_{24} connected to positive supply rail and a cascode mirror based current repeater consisting of Q_{15}-Q_{16}-Q_{17}-Q_{18}–Q_{19}-Q_{20} connected to negative power supply rail has been employed to ensure that the collector currents of transistors Q_1, Q_3, Q_5 and Q_7 are as near as possible to I_0. Similarly, to ensure that I_z is as nearly equal to I_x as possible and also to enhance the output impedance looking into terminal Z of modified Wilson current mirrors consisting of Q_{11}-Q_{12}-Q_{13}-Q_{14} and Q_{25}-Q_{26}-Q_{27}-Q_{28}.

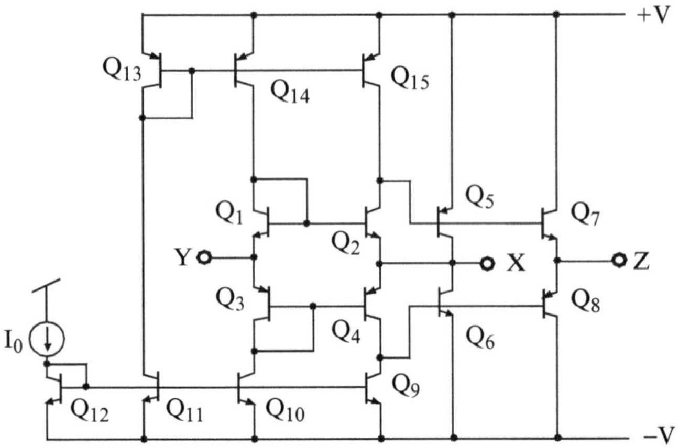

Fig. 3.11 A possible implementation of the CCII+ with reduced parasitic resistance, based upon the proposition of Seguin and Fabre [11]

By a straight forward analysis of the circuit assuming a load R_L connected at the X port, the equivalent input impedance looking into port Y is found to be:

$$R_Y = 2\beta^2(R_X + R_L) \tag{3.9}$$

which is β times larger than the Y-port input resistance of the conventional CCII, based upon the core circuit comprised of the four transistors *mixed translinear cell* (MTC).

The validity of this technique was confirmed by bread boarded versions of the circuit implemented from transistor arrays MPQ 2222 (for the NPN transistors) and MPQ 2907 for the (PNP transistors) which have demonstrated that as compared to $R_y < 100$ KΩ for the conventional design, the new architecture exhibits a value of R_y as high as 10 MΩ for a load resistance of 1 KΩ for the circuit architecture of Fig. 3.12.

3.2.10 Bipolar CCII with Controllable Gain

In many applications, it may be useful to have the current gain of the CC adjustable which could be useful for electronic control of the parameters of the functional circuits build from CCs of this type. Fabre and Mimeche [13] presented a new CCII architecture in which the current transfer from port X to port Z is controllable by an external bias current. The key idea of their proposition is a current mirror with adjustable gain which is shown in Fig. 3.13. If the input current $I_{in}(t) = I_0 + I(t)$ is

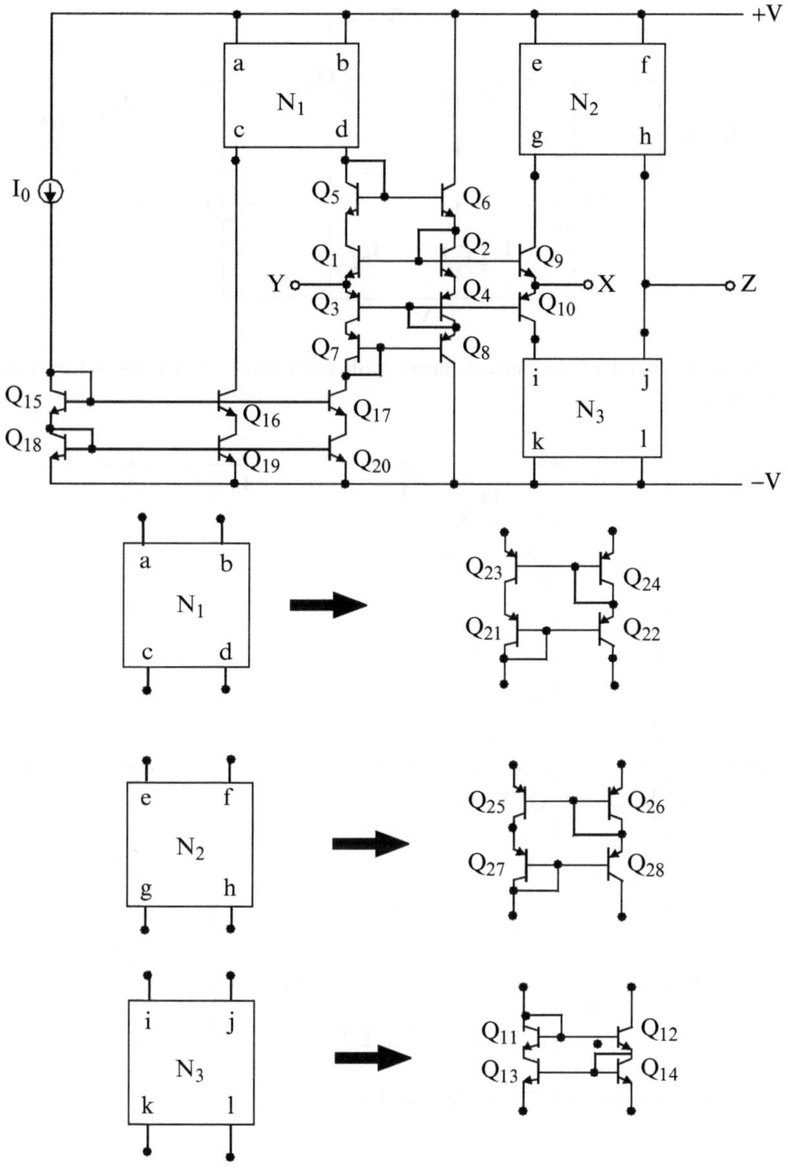

Fig. 3.12 CCII implementation with increased input impedance at port-Y [12]

composed of the DC bias current I_0 and a time varying input current signal I(t), application of translinear principle on the loop consisting of Q_3-Q_1-Q_2-Q_4 results in the following equation:

$$(I_0 + I(t))I_a = I_b I_{out} \tag{3.10}$$

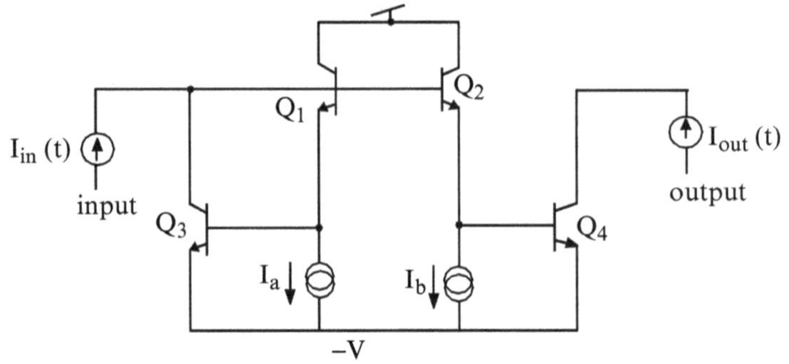

Fig. 3.13 Basic circuit of the current mirror with controllable gain proposed by Fabre and Mimeche [13]

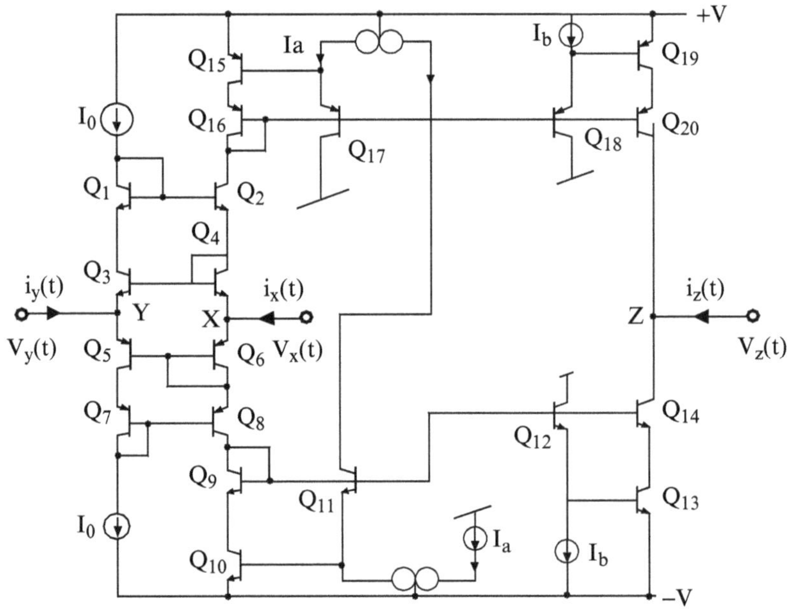

Fig. 3.14 Implementation of CCII+ with controllable gain [13]

$$I_{out} = G(I_0 + I(t)) \tag{3.11}$$

$$\text{where}\, G = \frac{I_a}{I_b} \tag{3.12}$$

It may be seen that the architecture of CCII+ with controllable gain (Fig. 3.14) essentially consists of eight transistors translinear cell between inputs Y and X which yields $I_y = 0$ and $V_x = V_y$ whereas the circuit employed to sense the current i_x and to convey the same current at Z port has been replaced by two modified circuits

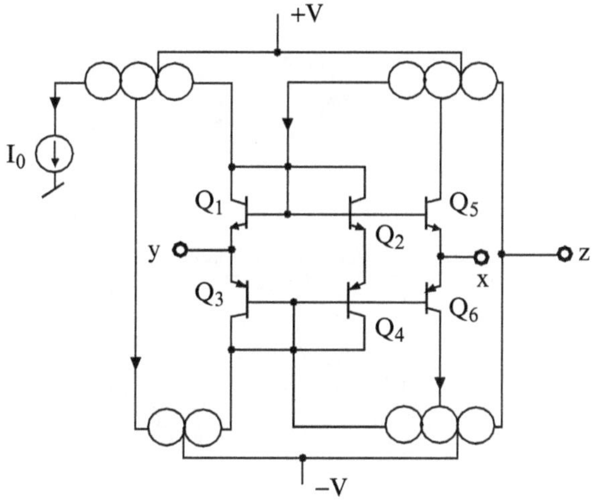

Fig. 3.15 A simplified schematic of the bipolar implementation of CCI+ proposed by Fabre et al. [14]

which essentially involve the mechanism of providing a current gain between i_z and i_x as explained above. As a consequence, the current transfer from X to Z port is, therefore, given by $i_z = G\ i_x$.

Implementation of the CCII+ with current gain described above using high performance ALA 200 transistor arrays from ATT with circuit biased from ±5 V DC power supplies and with DC bias current I_0 chosen as 50 μA, demonstrated that G could be varied from 0.5 to 5 with the resulting −3 dB bandwidth of the current gain remaining higher than 146 MHz. The voltage transfer between port Y and X was found to be 0.99 with −3 dB bandwidth of 182 MHz while the input impedance at port Y was found to be consisting of resistance $R_y = 160$ MΩ in parallel with capacitance $C_y = 1.1$ pF.

3.2.11 Bipolar Implementations of the CCI

A class AB CCI+ devised by Fabre et al. [14] is shown in Fig. 3.15 whose basic core consists of two translinear mixed loops in parallel. Due to the operation in class AB mode, the circuit can process current with magnitudes greater than the DC bias current I_0 and therefore, generate very low harmonic distortion.

The implementation of this CCI+ using transistor arrays ALA200 from AT&T which has NPN transistors with $f_T = 3.5$Ghz and $h_{FE} = 140$ and $f_T = 2$Ghz and $h_{FE} = 40$ for PNP transistors it has been found that with power supply as low as ±1.2 V and DC bias current I_0 as 100 μA, the circuit exhibits a bandwidth of 120 MHz, dissipates power less than 1 mW with the X-port parasitic input resistance of the order of 17 Ω. The circuit clearly is a good choice for low voltage and low power applications requiring CCI+ as a building block.

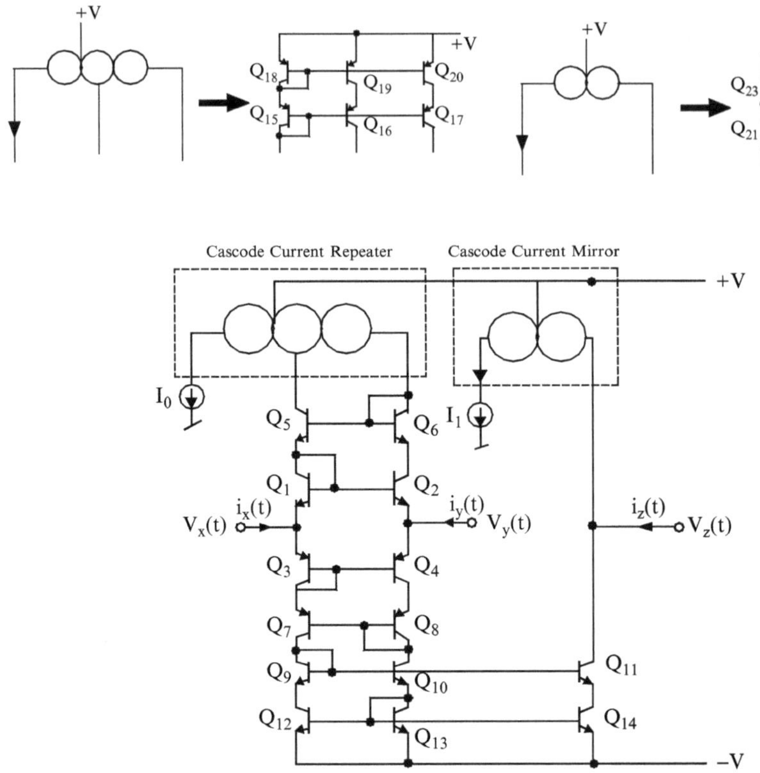

Fig. 3.16 CCI+ implementation proposed by Fabre et al. [15]

Another design of the CCI+ proposed by Fabre et al. [15] is shown in Fig. 3.16 which was derived from a translinear current convertor described previously by Fabre in [16]. In this circuit, the PNP current mirror consisting of transistors Q_{15} to Q_{20} forces the DC bias current I_0 to flow through the collectors of Q_1 and Q_2 thereby ensuring $V_x(t) = V_y(t)$. Now assuming that the β of the transistors are large enough, the application of the *translinear principle* on the mixed translinear loop consisting of Q_1-Q_2-Q_3-Q_4 implies that the collector currents of the transistors Q_3 and Q_4 would have the same value I_0 and consequently $i_y(t) = i_x(t)$. The mechanism to convey to port Z a current equal to the input current $i_x(t)$, consists of a modified Wilson current mirror based current repeater consisting of transistors Q_9-Q_{10}-Q_{11}-Q_{12}-Q_{13}-Q_{14} and a cascode current mirror consisting of PNP transistors Q_{21}-Q_{22}-Q_{23}-Q_{24} with a current source I_1 added which is to be adjusted to balance out the offsets resulting from non-identical values of β of NPN and PNP transistors in the ALA200 transistor array from AT&T.

With ±5 V DC power supplies and a DC bias current I_0 of 500 μA, the circuit dissipates about 20 mW, exhibits the current gains for Y and Z ports as 1.012 and 0.978 respectively with deviation in the current gains remaining less than 5 % for frequency ranges 0–70 MHz for output Y and 0–340 MHz for output Z. The output impedances ($R_0//C_0$) were found to be 220 KΩ//3.5 pF for port Y and 1.1 MΩ//1.1 pF for port Z. R_0 could be increased by reducing the DC bias current I_0.

3.3 Commercially Available IC CCs

In this section we describe various commercially available CCs as well as other devices which are although not labelled or termed as CCs in their manufacturer's data sheets and catalogues but nevertheless, can be either directly used as CCs or can be modified to realize CCs.

3.3.1 CCII01 from LTP Electronics

Figure 3.17 shows the simplified schematic of the LTP CCII01. The device was made available packaged as 8-pin DIL containing two CCIIs. It may be seen that the basic architecture is a voltage follower between ports Y and X with power supply current sensing mechanism implemented by the two current mirrors in the DC power supply rails in a typical operational mirrored amplifier formulation which senses the current input at port X and makes an equal current made available at the high output impedance terminal Z.

Each individual CCII has a high input impedance Y-terminal with input impedance as 80 kΩ at 1 kHz while X-terminal has low input impedance of the order of 10 Ω.

Two CCII01 devices could be combined to make a composite CC with enhanced characteristics with X- port input resistance reduced to a value as low as less than 200 mΩ at 1 MHz. The Z-port output impedance is of the order of 1 MΩ at 1 kHz.

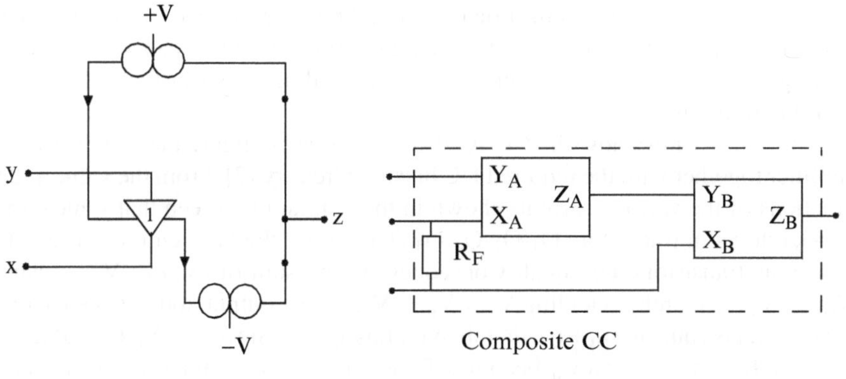

Fig. 3.17 The schematic of CCII01 IC current conveyor from LTP electronics [17]

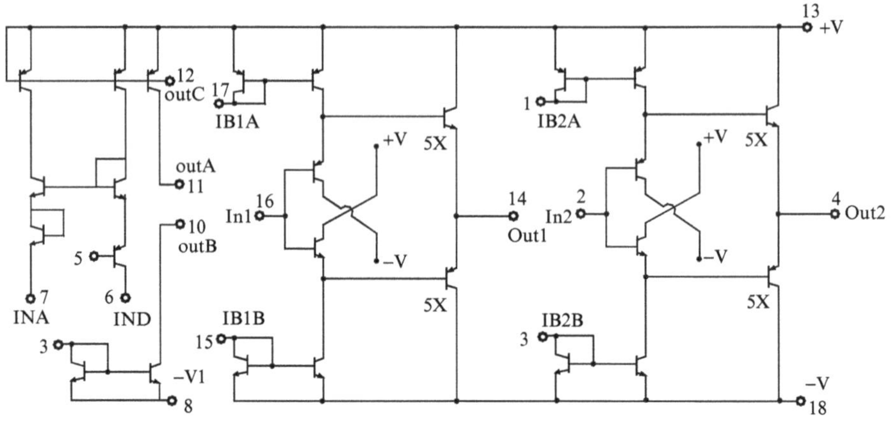

Fig. 3.18 Simplified diagram of the PA630 IC current conveyor [19]

3.3.2 *PA630 from Phototronics Limited*

In 1989, Wadsworth [18] reported an IC implementation of a novel CC topology whose simplified representation is shown in Fig. 3.18 [19]. This IC CC was manufactured by Phototronics Company; Ontario, Canada and took advantage of the true complementary bipolar process technology to produce a CC IC which offered significant improvement in the accuracy, bandwidth, transient response, output impedance and distortion.

As can be seen from Fig. 3.18, the basic architecture contains two base-input mixed translinear loops which implement two on-chip voltage buffers, a basic CC architecture and a number of current mirrors. The outside DC bias current can be set up in both the voltage buffers as well as the basic CC by connecting external resistors between the uncommitted terminals provided outside the chip for this purpose one such arrangement for one of the voltage follower is shown in Fig. 3.19 where the required DC bias current is set up by an external resistor R_5.

Figure 3.20 shows the connection of external resistors needed for biasing the CC part of the circuit which shows how an inverting amplifier can be realized by creating a virtual ground at pin number 7 and providing a resistor R_2 for DC biasing of all the transistors.

Figure 3.21 shows how the CC IC PA630 can be configured as a non-inverting amplifier together with the required DC biasing circuitry [7]. From the value and the directions of the various currents shown in the diagram it is seen that same current $(I_B + i_x)$ flows in transistors Q_1-Q_2-Q_4-Q_5. Therefore, the base emitter drops of all these four transistors are equal. Consequently, the sum of $(V_{BE4} + V_{BE5})$ cancels $(V_{BE2} + V_{BE1})$ thereby, yielding $V_y = V_{in} = V_x$. At the output node, it is easy to see that $(I_B + i_x)$ is equal to $(I_B + i_z)$, thereby yielding $i_z = i_x$. Since $i_z = V_0/R_2$ and $i_x = v_x/R_1 = Vin/R_1$, the gain v_0/v_{in} becomes R_2/R_1. Lastly, the remaining resistors are chosen as $R_4 = R_1$ and $R_5 = R_3$ to minimize the transistor output impedance error.

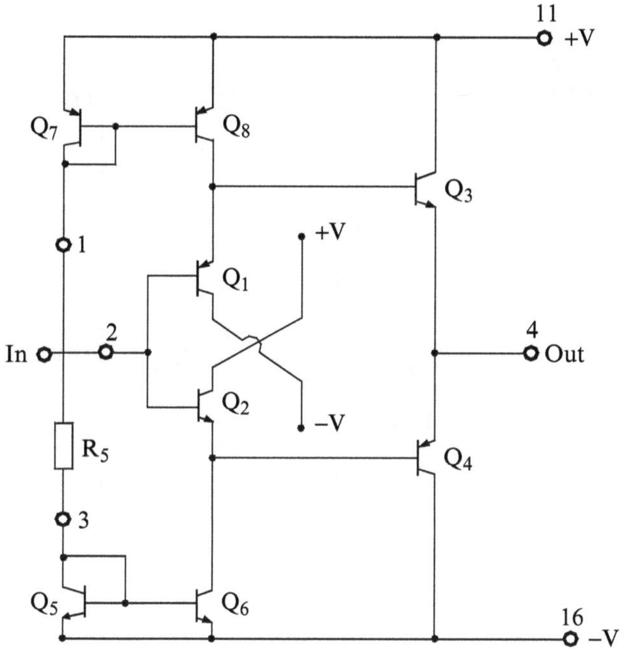

Fig. 3.19 DC bias current set up for the buffer of PA630 [19]

Fig. 3.20 PA 630 configured as CCI and arranged to realize an inverting amplifier [19]

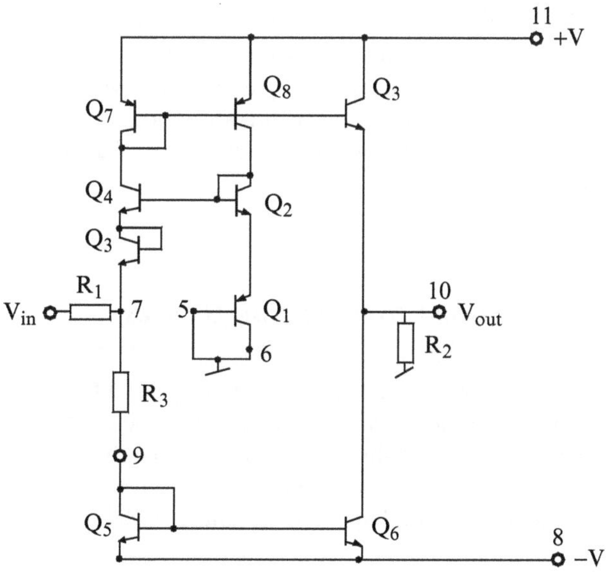

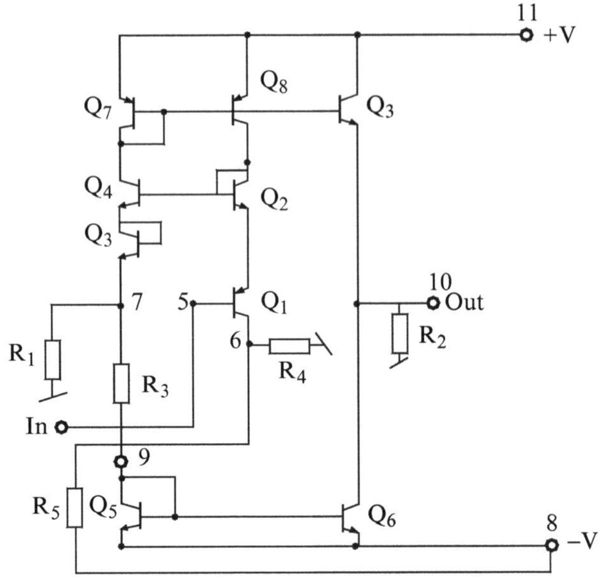

Fig. 3.21 PA 630 configured as CCII [19] and arranged to realize a noninverting amplifier

The main characteristic features of this IC are accuracy of 0.5 %, the bandwidth of CC as 80 MHz, CC distortion 0.02 % with buffer amplifier bandwidth of 50 MHz.

When PA630 was announced it was enthusiastically welcomed by the academic community particularly those involved in current conveyor research however, the practicing engineers were still hesitant in using these devices possibly due to their unfamiliarity with current mode circuits and techniques.

In retrospection, it appears that after the emergence of more versatile and easy to use AD844 employing a CCII+ along with a voltage buffer, PA630 perhaps did not receive so much attention in literature possibly due to the cumbersome DC bias arrangements needed and/or the difficulty of using it as standalone three-terminal CCI or CCII.

3.3.3 AD844 from Analog Devices

The current feedback operational amplifier (CFOA) AD844 from Analog Devices is composed of a translinear second generation CC (CCII+) based upon a four transistor mixed translinear cell followed by a translinear voltage buffer (see Fig. 3.22a) for the block diagram of the internal circuit and Fig. 3.22b for its transistor level implementation. Thus, AD844 is characterized by the following terminal equations: $i_y = 0$, $v_x = v_y$, $i_z = i_x$ and $v_w = v_z$.

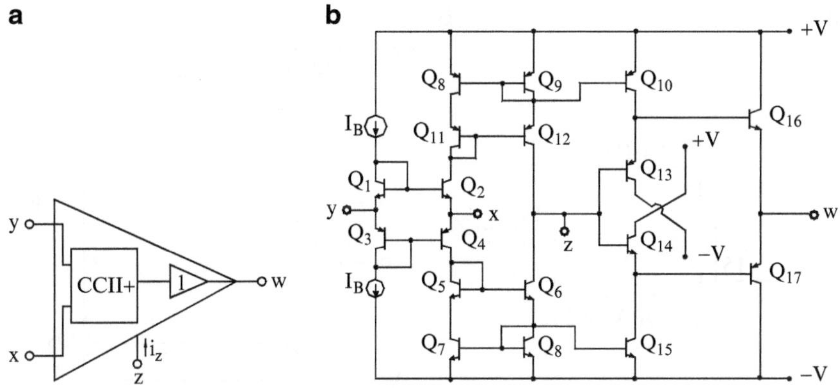

Fig. 3.22 Current feedback operational amplifier (**a**) Block diagram of the internal circuit (**b**) Bipolar implementation of AD844 type CFOA [20]

The AD844 is quite versatile as it can be used as a CCII+, can be used to realize CCII− (needing two AD844s), can be used as a direct replacement for the conventional op-amp (with its Z-terminal, the compensation pin, open) due to its pin-compatibility with the popular IC op-amps such as μA 741 or LF 356 and lastly, can be used as a four-port building block in its own right.

3.3.4 Using OPA-2662 as Current Conveyors

The operational transconductance amplifier OPA2662 [21] can be directly used as a CCII+ as it is (Fig. 3.23a). Its simplified circuit is shown in Fig. 3.23b. A CCII− can be implemented with the help of two such devices. The validity of such a realization has been demonstrated by Tomawski et al. [22] by using such CCII-s to realize a frequency dependent negative conductance.

3.3.5 CC from OPA 660/OPA 860

Although not mentioned anywhere in the producer's catalog of OPA660, the internal circuit architecture of the OPA 660 architecture (Fig. 3.24) is essentially a CCII+ made from transistors Q_1 to Q_{10} with terminals B, E and C respectively representing terminals Y, X and Z of the current conveyor CCII+. On the other hand, the part of the circuit composed of transistors Q_{11} to Q_{14} constitutes the translinear voltage follower (VF). In the entire circuit, the transistors Q_5 and Q_6 in the CCII+ part and transistor Q_{15} to Q_{20} in the VF part constitute the DC bias circuitry from which the bias current can be adjusted by an external resistor R_0

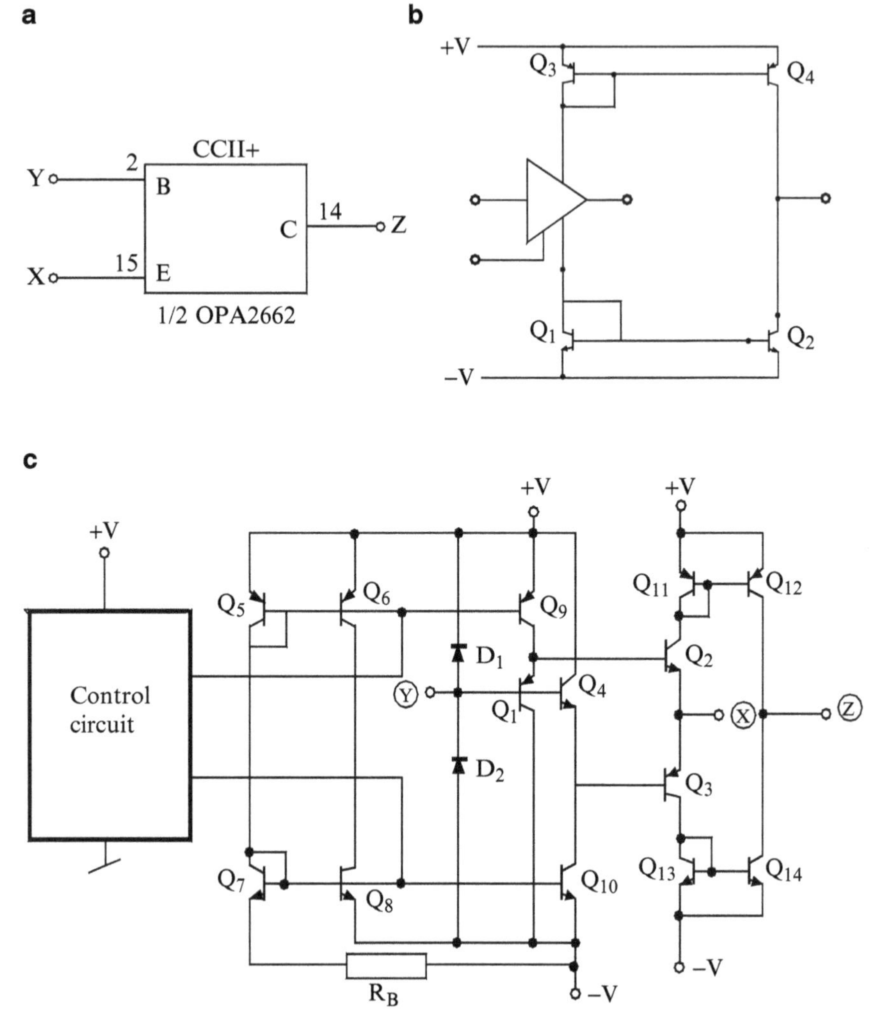

Fig. 3.23 (**a**) Realization of CCII+ from OPA2662 (**b**) Simplified circuit of CCII+ using OPA2662 (**c**) Simplified circuit diagram of OPA2662 showing the internal CC-architecture [21]

connected between terminals 1 and 4. The other terminal characterization are: terminal 7 for positive power supply, terminal 4 for negative power supply, terminal 3, 2 and 8 being terminals Y, X and Z of the current conveyor and terminals 5 and 6 being the input and output terminal of the VF. The nominal supply voltages specified are ± 5 V. The device offers a wide bandwidth of 850 MHz and exhibits a slew rate as high as 3,000 V/μs.

OPA860 [23] which is also functionally and architecturally similar to OPA660 has a bandwidth of 80 MHz and slew rate of 900 V/μs for the OTA (CCII+) part and has a close loop buffer structure providing 600 MHz bandwidth and 4,000 V/μs

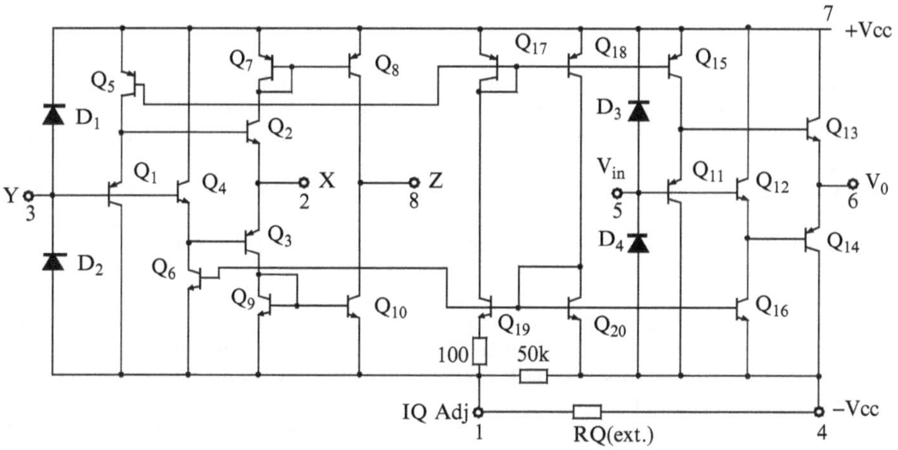

Fig. 3.24 Internal circuit architecture of OPA660 OTA which is essentially a CCII+ and a VF [23, 24]

slew rate. It is popularly referred to in the literature as a d*iamond transistor* since the CCII part acts like almost an ideal transistor and it is for this reason that the terminals X, Y and Z are respectively labeled as E (emitter), B (Base) and C (Collector). Of late, OPA860 has been extensively used in the literature as current conveyor to realize a variety of building blocks.

Concluding Remarks

This chapter was concerned with the bipolar implementation of current conveyors and has dealt with two types of bipolar CCs (i) a variety of bipolar CC architectures proposed in literature for implementing CCII as well as CCI having some specific features of interest such as electronic control of current gain, enhanced input impedance at terminal Y, reduced input impedance at terminals X etc. (ii) bipolar CCs which are available as of the shelf ICs such as PA630, LTPCC201, AD844, OPA2662, OPA660/860.

It is a matter of great curiosity to note that except the first two of this list all the remaining ones in spite of containing a CCII+ architecture internally and also having external access to the three terminals of the inherent CCII have neither been claimed as CCs in the nomenclature nor have been explicitly identified in the data sheet as a CC! Nevertheless, all of these have been used quite frequently in the literature by researchers and academicians as current conveyors. It is believed that after a clear cut identification of the above mentioned ICs as CCs, the readers and circuit designers would take advantage of this exposition and would be able to practically implement their CCI/CCII based functional circuits using any of the available variants discussed in this chapter.

References

1. Fabre A (1985) Translinear current conveyors implementation. Int J Electron 59:619–623
2. Normand G (1985) Translinear current conveyors. Int J Electron 59(6):771–777
3. Wilson B (1981) Low-distortion feedback voltage-to-current conversion technique. Electron Lett 17:157–158
4. Pookaiyaudom S, Kuhanont T (1977) High performance differential quartets. Proc Inst Elect Electron Engrs 65:1721–1723
5. Barthelemy H (1997) Low-output-impedance class AB bipolar voltage buffer. Electron Lett 33:1662–1664
6. Higashimura M, Fukui Y (1988) Type I mutator using current conveyor and its application to immittance simulation. Int J Electron 64:377–383
7. Wadsworth DC (1989) Professional audio current conveyor IC. Phototronics Co, Manotick
8. Surakampontorn W, Thitimajshima P (1988) Integrable electronically tunable current conveyors. IEE Proc 135:71–77
9. Filanovsky IM (1997) Current source is modified into a current conveyor. Circuits Devices 29
10. Surakampontorn W, Riewruja V, Kumwachara K, Fongsamut C (1998) Temperature compensation of translinear current conveyor and OTA. Electron Lett 34:707–709
11. Seguin F, Fabre A (2001) New second generation current conveyor with reduced parasitic resistance and band pass filter application. IEEE Trans Circ Syst-I 48:781–785
12. Alami M, Fabre A, Bouhdada A (1999) Second generation current conveyors with enhanced input resistance. Int J Electron 86:405–412
13. Fabre A, Mimeche N (1994) Class A/AB second-generation current conveyor with controlled current gain. Electron Lett 30:1267–1269
14. Fabre A, Amrani H, Barthelemy H (1999) A novel class AB first generation current conveyors. IEEE Trans Circ Syst-II 46:96–98
15. Fabre A, Siarry P, Lameche M (1991) Current-controlled translinear impedance converter. Int J Electron 70:795–801
16. Fabre A (1984) Wideband translinear current convertor. Electron Lett 20:241–242
17. Hickman I (1994) Current Conveyor ICs –a new building block. Electronics World + Wireless World 402–405
18. Wadsworth DC (1989) Accurate current conveyor integrated circuit. Electron Lett 25:1251–1253
19. Wadsworth DC (1990) Accurate current conveyor topology and monolithic implementation. IEE Proc 137:88–94
20. Devices A (1989) Linear products data book. Analog Devices Inc., Norwood
21. OPA 2662 (1991) Dual, wide bandwidth operational transconductance amplifier. Burr-Brown Corporation, Tucson, AZ 85734
22. Tomawski L, Gabrek R, Ole's S (2001) Inverting current conveyor from an OPA 1662: Electron World and Wireless World: Circuits Ideas. 792
23. OPA 660/OPA 860 (1990) Wide bandwidth operational transconductance amplifier and buffer. Burr-Brown Corporation
24. Henn C (1993) New ultra high-speed circuit techniques with analog ICs. Corporation, Burr-Brown, Tucson, AZ 85734
25. de Jager W, Smit J (1977) Application, design and symbolic analysis of a current follower. Electron Circ Syst 1:79–84
26. Jager W de, Smit J (1978) Design and symbolic analysis of current mode analog circuits. Technische Hogeschool Twente, Afdeling Der Elektrotechniek 4p
27. Hayashi Y, Goto N, Suzuki T (1984) Equivalent circuit of a gyrator using current conveyor II (CCII) and its active compensation. Trans IECE J 167:25–32
28. Goto N, Nishio M, Suzuki T (1986) A current conveyor II realized by mosfets and its application to switched capacitor filters. Electron Commun in Japan (Pt II: Electronics) 69:10–17

29. Wilson B (1986) Using current conveyors. Electron World and Wireless World 28–32
30. Wilson B (1990) Recent developments on current conveyors and current-mode circuits. IEE Proc Pt G Circ Devices Syst 137:63–77
31. Toumazou C, Lidgey FJ, Makris CA (1990) Extending voltage-mode op-amps to current-mode performance. IEE Proc 137:116–130
32. Lidgey J, Toumazou C (1990) Current amplifiers from voltage op-amps. Design: Electron World + Wireless World 568–573
33. Svoboda JA, McGory L, Webb S (1991) Applications of a commercially available current conveyor. Int J Electron 70:159–164
34. Fabre A, Saaid O (1993) Novel translinear impedance convertor and band pass filter applications. Electron Lett 29:746–747
35. Duncan RA, Chan KM, Sedra AS (1994) Design for a wideband current amplifier and a current conveyor. IEEE Trans Circ Syst-I 41:272–280
36. Matsumoto F, Miyake A, Noguchi Y (2000) A high-precision low-voltage bipolar current mirror circuit and its compensation for stability. Int J Electron 87:71–78
37. Robinson J (2008) New CCII current conveyor. Maxim Integr Appl Note 4198:1–6
38. Merz N, Kiranon W, Wongtachathum C, Pawarangkoon P, Narksarp W (2012) A modified bipolar translinear cell with improved linear range and its applications. Radioengineering 21:736–745
39. Abuelma'atti MT, Al-Zaher HA (1997) Nonlinear performance of the mixed translinear loop. Int J Electron 83:467–471
40. Matsumoto K, Horita M, Yamamoto G (1981) Analysis of a gyrator and a CR oscillator using current conveyor. Trans IECE Japan 12:891–892

Chapter 4
CMOS Implementations of Current Conveyors

Abstract In order to make CC-based circuits compatible with CMOS digital circuits, a lot of attention has been given in the literature over the past two decades to evolve CMOS-based Current Conveyors. In this chapter, we bring out some of the important developments in this area.

4.1 Introduction

In the previous chapters, the main focus has been on three different ways of realizing current conveyors (i) realization of CCs using commercially available integrated circuits (ii) realization of CCs using architectures based upon bipolar transistors (iii) commercially available IC implementation of current conveyors. The circuits in all of these categories, however, are based upon the use of bipolar junction transistors (BJT).

In view of the wide spread use of analog circuits in mixed-signal applications and requirement of analog circuits to be integrated along with the digital circuits on the same chip using the same technology, it is essential that there be CMOS implementable current conveyors and CMOS implementable analog functions using them so that they can be easily implemented in CMOS technology. Motivated by this requirement, therefore, a number of researchers have focused upon implementing the basic types of current conveyors namely CCI+/−and CCII+/− by CMOS hardware [1–69]. In this context, however, it is worth mentioning that simple CMOS realization of CCII+ and CCII− were proposed by Sedra et al. in [1] as early as in 1990 which are shown in Fig. 4.1a, b respectively. In this circuit diagram, as well as elsewhere subsequently, the various MOS current mirrors and MOS current repeaters are shown by their standard symbolic notations consisting of two or more circles touching each other with connection to positive/negative power supply rail(s), with input terminal shown with an arrow mark and the terminal (s) without arrow mark being the output terminal(s).

In this chapter, we consider, therefore, CMOS realizations of basic types of current conveyors only- a wide variety of which is scattered over dozens of publications which have appeared during the past three decades. Out of an

© Springer International Publishing Switzerland 2015
R. Senani et al., *Current Conveyors*, DOI 10.1007/978-3-319-08684-2_4

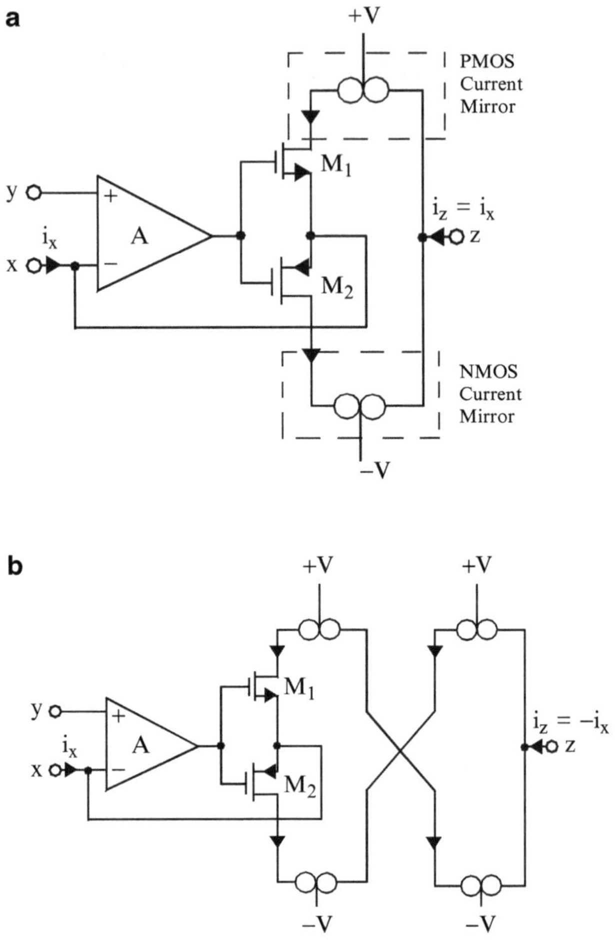

Fig. 4.1 Equivalent representations of the simple CMOS current conveyors proposed by Sedra et al. [1]

exhaustive repertoire of circuits available, we have chosen here only those architectures which have optimized one or more of the following important parameters or performance characteristics of the CCs such as: reduced DC power supply requirements, reduced power consumptions, high precision, reduced parasitic input resistance of the x-port, high current driving capability, high bandwidth, electronic tunability, high accuracy, capability of providing rail to rail operation etc.

We have also included some architecture which exhibits some novelty of design approach such as the use of new current mirror formulations or converting transconductors into current conveyors or realizing CCs using CMOS invertors etc. It may be mentioned that the CMOS implementations of other variants of the current conveyors are not included herein; they are dealt with in a separate chapter (Chap. 9) in part III of this monograph.

4.2 Simple CMOS Realizations of CCII+ and CCII−

Two simple CMOS realizations of CCIIs were proposed by Surakampontorn et al. in [5] which are shown in Fig. 4.2a, b. The circuit of Fig. 4.2a implements a CCII+. In this circuit, it can be observed that MOSFETs M_3-M_6 and M_7-M_8 represent, respectively, current repeater and current mirror. The basic core of the circuit consists of transistors M_1 and M_2 and the transistors M_1-M_2 as well M_3-M_4 are well-matched such that $(W/L)_1/(W/L)_2 = (W/L)_3/(W/L)_4$. It can be seen that the drain currents of M_1 and M_2 are identical so that V_{GS1} and V_{GS2} will cancel out each other and as a consequence, $V_x = V_y$ will be attainable. By simple inspection of the various currents shown on the circuit diagram, it can be observed that the source current of M_1 and the drain current of M_7 are equal which will force $I_{y=}0$. On the other hand, with $I_1 = I_2$, the node equation at node Z will yield $I_z = I_x$. Thus, the circuit of Fig. 4.2a implements a CCII+.

The circuit of Fig. 4.2b is obtained from that of Fig. 4.2a by adding an NMOS current mirror consisting of MOSFETs M_9-M_{10}. Since the rest of the circuit comprised of MOSFETs M_1-M_8 is exactly same as in the circuit of Fig. 4.2a, it follows that $I_y = 0$ and $V_x = V_y$. Furthermore, as the location of the DC bias current source I_2 has been changed, the node equation at node Z modifies to I_2-$i_z = I_1 + i_x$ and therefore, with $I_2 = I_1$, it follows that $i_z = -i_x$ and the circuit, thus, realizes a CCII−.

The authors of [5] have carried out a detailed non-ideal analysis of their proposed circuits from where they have estimated the values of the non-ideal parameters to be $R_x \approx 3.7\ \Omega$ and $R_y \approx 9.8\ \text{M}\Omega$ which are quite low and quite high respectively as desired. Both the circuits were constructed on proto-boards where all the devices were realized using IC CD4007 CMOS transistor arrays with DC bias current I_1 set to 200 μA. The circuits were found to exhibit an excellent linearity of the voltage transfer characteristics between V_y and V_x and current transfer characteristics between I_z and Ix with 3-dB bandwidth of the order of several MHz in both the cases.

4.3 Low-Voltage CMOS Current Conveyor

A MOS current conveyor design suitable for VLSI implementation was proposed by Cheng and Toumazou in [9]. The basic architecture proposed by them is based upon two generic circuits. The first of these circuits, composed of MOSFETs M_{p1}, M_{n3}, M_{p3} and M_{n7}, is essentially a voltage follower circuit which has an in-built DC offset cancellation scheme. On the other hand, the second generic circuit consisting of M_{n2} and M_{p2} is a basic N-channel and P-channel MOSFET V_{GS} matching circuit. These basic circuits along with two PMOS mirrors shown in the diagram symbolically as PMOS CM_1 and PMOS CM_2 and NMOS current repeater comprised of MOSFETs M_{n4}, M_{n5}, M_{n6} and M_{n1} constitute the final circuit as shown in Fig. 4.3.

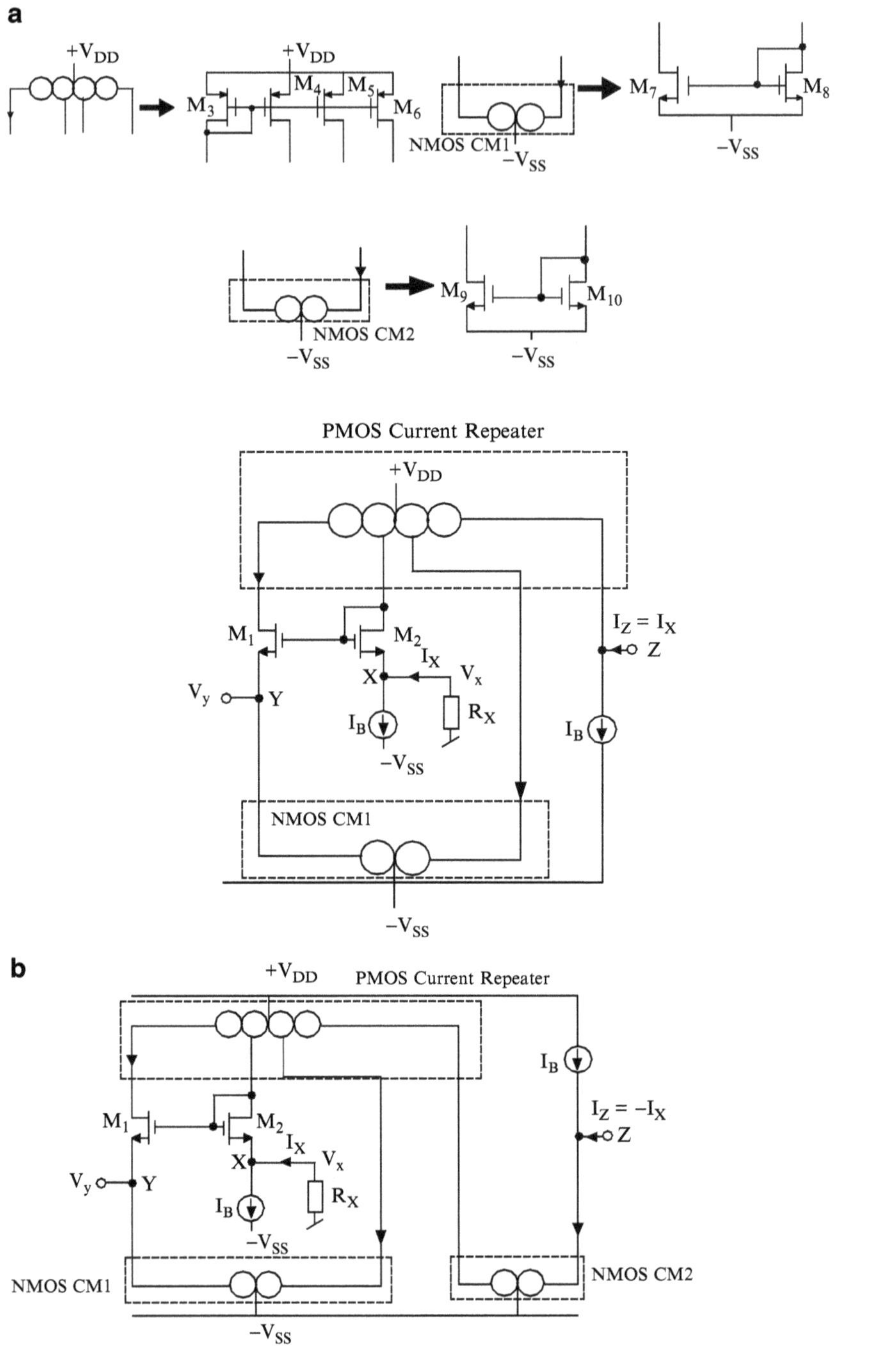

Fig. 4.2 CMOS based CCII+ and CCII− configurations proposed by Surakampontorn et al. [5]

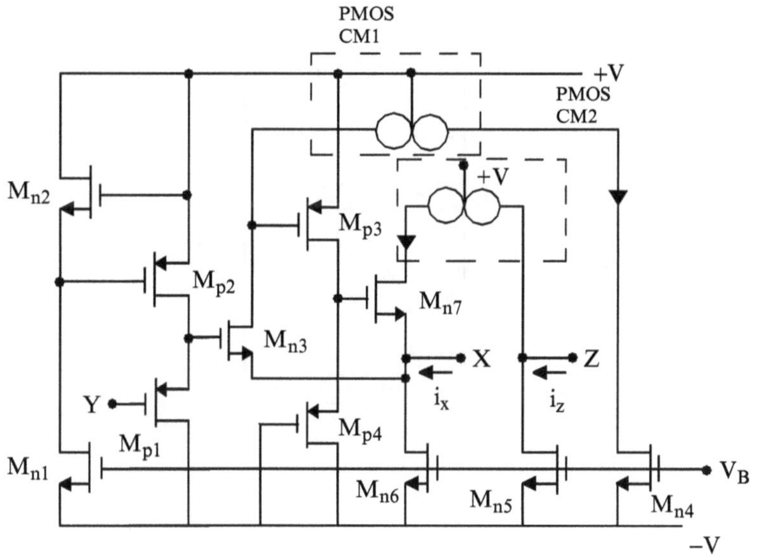

Fig. 4.3 An equivalent representation of the CMOS CCII+ circuit proposed by Cheng and Toumazou [9]

When this circuit was simulated in HSPICE biased with 3 V DC supply, at terminal-Y the circuit exhibited almost an infinite input resistance whereas terminal-X exhibited a very low and almost linear input resistance which was found to be of the order of 1.9 Ω (DC to 100 KHz). As the input signal frequency is increased R_x increases to 3.9 Ω at 1 MHz and 37.4 Ω at 10 MHz. The voltage gain between ports Y and X exhibited a bandwidth around 13 MHz whereas the bandwidth of the current gain between port Z and X was found to be around 10 MHz.

4.4 Class AB First Generation Current Conveyors

The traditional class AB CMOS current conveyor shown in Fig. 4.4a suffers from the drawback of poor control of quiescent current level as there is no mechanism for setting up the intended DC bias current in the circuit. To circumvent this problem, Bruun in [10] presented a CMOS architecture which contains an elegant bias stabilization scheme. This circuit is shown in Fig. 4.4b. This circuit incorporates a way to stabilize the quiescent current which consists of feeding back only a fraction $\lambda_{FB} < 1$ of the X-input branch current to the Y-input branch current and providing the remaining part of the Y-input branch current as a fixed current I_{ss} to the level shifting transistors M_1 and M_2. As a consequence, the quiescent current level in each branch is equal to

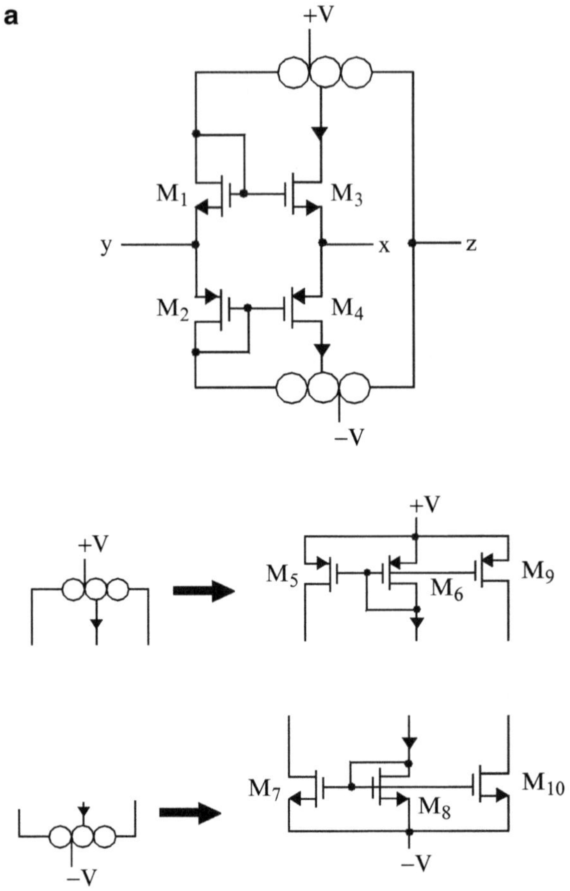

Fig. 4.4 Class AB first generation current conveyors proposed by Bruun (**a**) without bias stabilization (**b**) with bias stabilization [10]

$$I_{QY} = I_{D1} = I_{Qx} = I_{D3} = I_{SS}/(1 - \lambda_{FB}) \tag{4.1}$$

If $(1\text{-}\lambda_{FB})$ is reasonably large as compared to expected mismatch in the devices, a well-controlled quiescent current mechanism is, thus, available. Now, in order that as required for a CCI, i_y can be made equal to i_x, a fraction $(1\text{-}\lambda_{FB})$ of the X- branch current must be feedback directly to the Y-input. This function is performed by the MOSFETs M_{5b} and M_{7b}.

Subject to the assumption of matched transistors having same g_m and neglecting the bulk effect, the input resistance looking into the port-X is given by:

$$R_x = \frac{1}{2_{gm}}(1 - \lambda_{FB}) \tag{4.2}$$

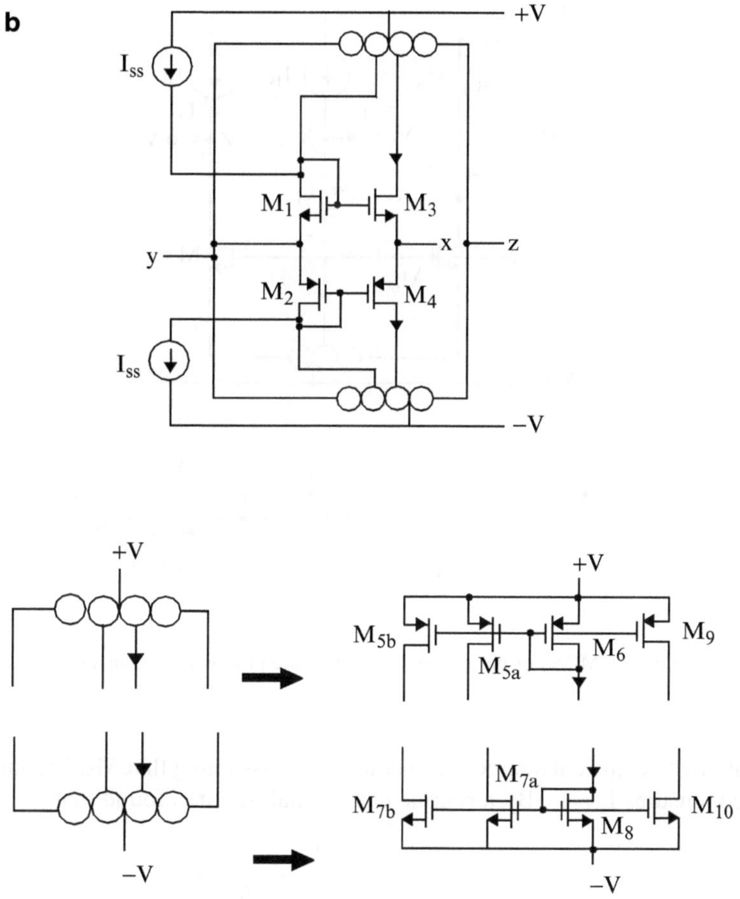

Fig. 4.4 (continued)

From SPICE simulations using device parameters from a commercially available CMOS process it has been verified in [10] that this topology offers improvements in quiescent current control over earlier design and that it has a well-defined input impedance.

4.5 Wide Band CMOS Current Conveyors

A CMOS CCII+ suitable for high frequency current mode signal processing was proposed by Cha and Watanabe in [13] which is shown here in Fig. 4.5. In this circuit, the transistors M_2, M_3, M_4, and M_6 form the regulated current cell for current input. The transistor M_1 is configured as a source follower. To accept

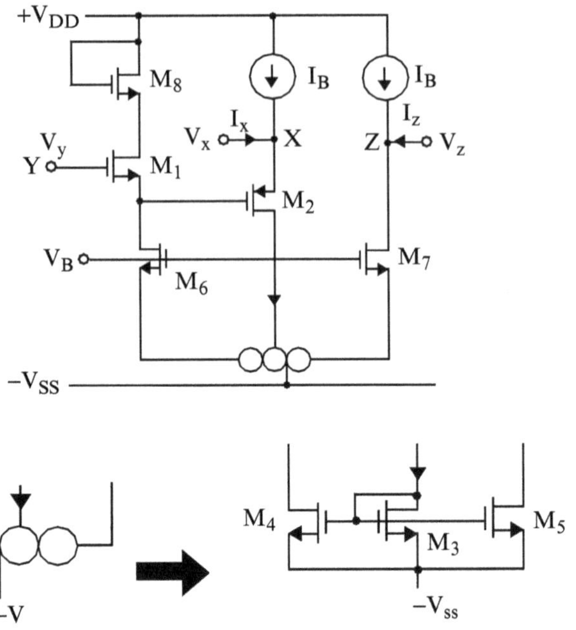

Fig. 4.5 Wide band CMOS current conveyor architecture proposed by Cha and Watanabe [13]

Y-input, the two current sources each equal to I_B, assuming that M_3, M_4 and M_5 are matched, ensures $I_z = I_x$. By a routine circuit analysis, it is found that

$$v_X = \frac{g_{m1} g_{m2} R_X}{g_{m1} - g_{m2} + g_{m1} g_{m2} R_X} v_Y \tag{4.3}$$

If $g_{m1} = g_{m2}$ or $g_{m1} g_{m2} R_x >> (g_{m1} - g_{m2})$ then v_x exactly follows v_y. Since no current flows into node Y we have $i_Y = 0$. The input impedance looking into the node X is given by

$$R_X = \frac{1}{g_{m2}} - \frac{1}{g_{m1}} + \frac{1}{g_{m2} g_{m3} r_{ds2}} \tag{4.4}$$

The input resistance R_x can be made negligibly small by selecting the aspect ratios of the MOSFETs M_1 and M_2 appropriately.

Using MOS transistor model for a 1.2 μm standard CMOS process with dimensions of M_1, M_2 as 160/1.2 μm and 480/1.2 μm with all other NMOS transistors having dimensions 120/2 μm, the current source design for $I_B = 300$ μA with supply voltages of ±5 V along with $V_B = -2.5$ V, the current transfer characteristic and voltage transfer characteristic as well as the voltage gain between Z and Y port (with suitable termination at ports X and Z) demonstrated a 3-dB bandwidth of more than 700 MHz.

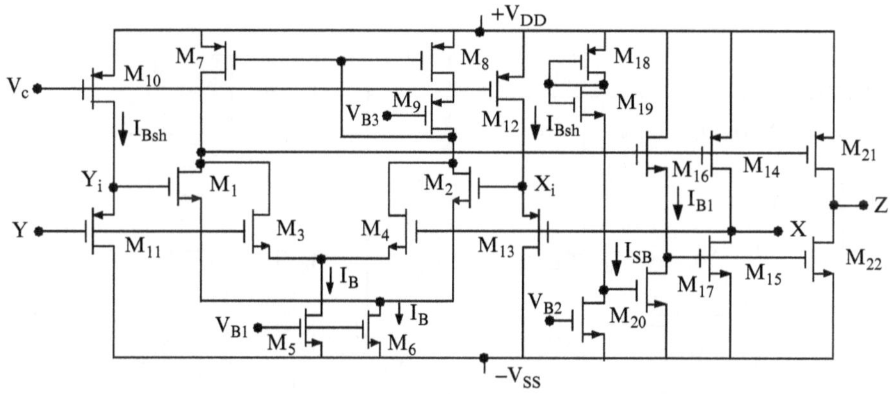

Fig. 4.6 CMOS realization of CCII+ (Adapted from [48] ©2006 Springer)

4.6 A 1.5 V CMOS Current Conveyor Based on Wide Range Transconductors

Madian et al. [48] presented a CCII+ architecture which uses two N-channel differential pair instead of complimentary differential pairs for making the input stage. This results in the availability of almost a rail-to-rail input and output operation. Another advantage of this structure is that the number of current mirrors required in the input stage is reduced. This circuit from [48] is shown in Fig. 4.6. Part of the circuit consisting of transistors M_1-M_{13} is essentially a wide band transconductance circuit which has been converted into a voltage follower circuit by adding the MOSFETs M_{14}-$M_{20.}$ Finally, the complete CCII circuit is obtained with the addition of two more transistors namely, M_{21} and M_{22}.

It has been shown that this architecture can be operated from a DC power supply of ±0.75 V with a total standby current of 133 μA.

4.7 High Speed High Precision Current Conveyors

Calvo et al. in [43] presented the design of a high speed high performance CMOS current conveyor which was based upon a new high performance voltage follower topology. The circuit proposed by them is shown in Fig. 4.7. In this circuit, MOSFET pair $M_{3,4}(M_{7,8})$ constitutes a complementary follower cascade whereas the MOSFETs $M_{1,2}(M_{5,6})$ are two complementary level shifters. The positive feedback path sampling the output current of M_7 (M_8) and adjusting the bias current of M_4 (M_3) is implemented by transistors $M_{9,\ 10,\ 17}$ ($M_{11,\ 12,\ 18}$). The current flowing from node X is reproduced at node Z by the transistors $M_{13,\ 14}$ ($M_{15,\ 16}$). Thus, a high output impedance node and low voltage input node are created simultaneously.

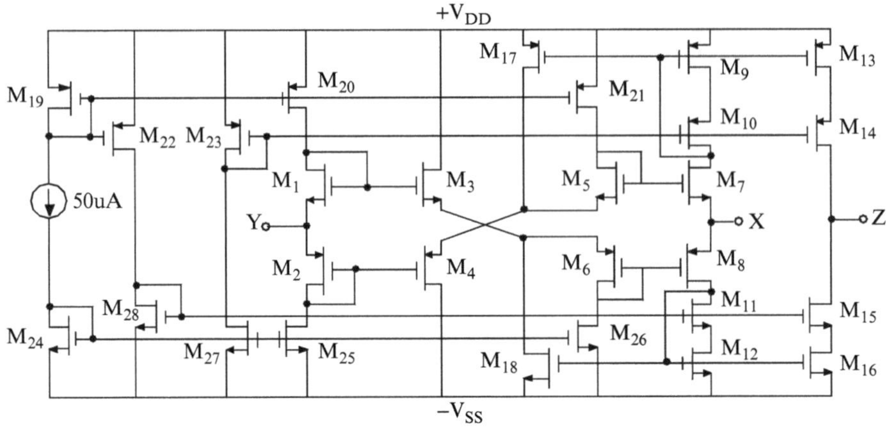

Fig. 4.7 Class-AB CCII+ realization proposed by Calvo et al. (Adapted from [43] © 2003 Springer)

The simulation results using 0.8 μm CMOS technology of this CC with ±3.3 V power supplies show that the resistance at X-terminal is 42 Ω, and impedance at Y-terminal is 174 kΩ||1.23 pF while the Z-terminal parasitic output impedance consists of 2.24 MΩ||0.21 pF. A very low input offset voltage of value 0.2 mV and input offset current of 50.2 nA are achievable.

4.8 CMOS-Inverter-Based CCII

Barthelemy et al. [51] presented a CCII+ implementation based upon classical CMOS inverters operating in transconductance mode. The aspect ratios of the NMOS and PMOS transistors are assumed to be set such that a voltage transition is obtained around $V_{IN} = V_{OUT} = V_{DD}/2$. This condition leads to a common-mode voltage $V_{CM} = V_{DD}/2$. Around this voltage, the inverter exhibits a transconductance mode between the input voltage V_{in} and the output current i_{out}. In the transconductance mode of operation, the small signal output current of the inverter is given by $i_{out} = -g_m v_{in}$ (see Fig. 4.8a), where $g_m = g_{mN+}g_{mP}$, where g_{mN} and g_{mP} are the NMOS and PMOS transconductance in saturated mode of operation. The proposed CCII+ topology based upon six such inverters is shown in Fig. 4.8b. By routine analysis of this circuit, assuming load R_L to be connected at node X, the expression of V_X in terms of V_Y turns out to be:

$$V_X = \frac{g_m R_L}{1 + g_m R_L} V_Y \text{ for } g_{m1} = g_{m2} \text{ and } g_{m3} = g_{m4} = g_m \quad (4.5)$$

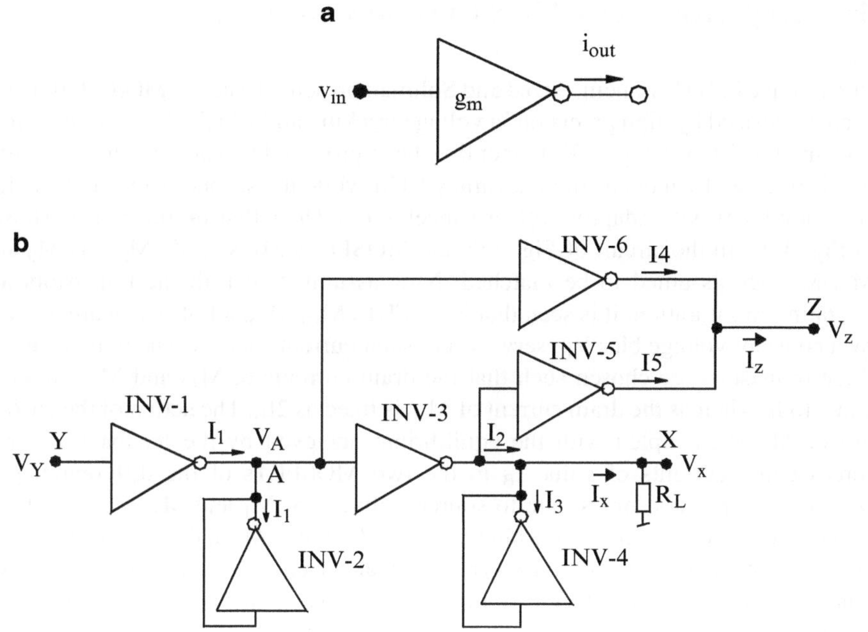

Fig. 4.8 (**a**) Symbolic notation of CMOS transconductor (**b**) CMOS inverter based CCII+ proposed by Barthelemy et al. (Adapted from [51] ©2007 Springer)

$$\cong V_Y \,\text{for}\, g_m R_L >> 1 \tag{4.6}$$

$$I_X = g_{m3}\frac{g_{m1}}{g_{m2}}V_Y - g_{m4}V_X \tag{4.7}$$

$$I_Z = g_{m6}\frac{g_{m1}}{g_{m2}}V_Y - g_{m5}V_X \tag{4.8}$$

$$\text{Now if we take}\, g_{m6} = g_{m3} \,\text{and}\, g_{m5} = g_{m4},\, I_Z = I_X \tag{4.9}$$

Simulation of this circuit using BSIM3 version 3 model of a typical 0.35 μm CMOS process parameters shows that for a voltage supply of 2.5 V DC, the circuit exhibits a current gain of 0.986 with 3 dB bandwidth of 200 MHz and a voltage gain of 0.972 with 3 dB bandwidth of 103 MHz. On the other hand with six CMOS inverters HEF 4069 UBP from Philips Semiconductors, the circuit exhibits current gain of 1.045 with 3 dB bandwidth of 1.17 MHz and a voltage gain of 0.9 with 3 dB bandwidth of 1.4 MHz with parasitic resistance $R_X = 540\ \Omega$ when all the inverters were biased by 5 V DC supply. The circuit exhibited wide current and voltage swings. When CMOS inverters are used to implement digital cells, this CCII+ implementation is particularly useful since it makes possible to implement numerous analog functions also using the same basic cells.

4.9 High Accuracy CMOS Current Conveyors

In reference [44] Hassanein, Awad and Soliman presented a new CMOS CCII which is characterized by high precision in voltage tracking and which also exhibits a very low input resistance at port X. In fact they have proposed two architectures: the first one is referred by them as high accuracy CCII while the second is referred as high accuracy CCII with adaptive offset cancellation. Their first proposition is shown in Fig. 4.9a. In the circuit of Fig. 4.9a the MOSFETs M_1-M_2, M_3-M_5, M_7-M_9 and M_{11}-M_{13} are assumed to be matched. Now assuming that all the transistors are operating in saturation, it is seen that MOSFETs M_{10}-M_{13}, all of which are derived by a constant voltage bias V_B, serve as constant current sources. The dimensions of these transistors are chosen such that the drain currents of M_{12} and M_{13} are each equal to I_B whereas the drain current of M_{10} is fixed as $2I_B$. The action of the current mirror M_3-M_4, coupled with the equilibrium achieved by the feedback control, forces equal currents of value I_B in the two MOSFETs of the differential pair M_1-M_2. In turn, this forces gate-to-source voltages of M_1 and M_2 to be identical, as a consequence of which, one obtains $V_x = V_y$. Since M_8 and M_9 carry the same current as M_7 and since M_{12} and M_{13} carry equal currents each equal to I_B, it follows that i_z will be same as i_x. Since I_y is already equal to zero, the circuit, thus, realizes a CCII+.

The offset cancelling mechanism of this circuit can be explained as follows. The voltage offset of this circuit is given by the following equation:

$$\Delta V = V_x - V_y \approx [\lambda_n(V_{D1} - V_{D2})]\sqrt{\frac{I_B}{\mu_n C_{ox} W_1/L_1}} \tag{4.10}$$

where λ_n is the channel length modulation parameter; V_{D1} and V_{D2} being the drain voltages of MOSFETs M_1 and M_2. It is, therefore, clear that offset can be cancelled by making V_{D1} and V_{D2} identical. It may be noted that I_{D10} is taken equal to $2I_B$. Now to ensure that $I_{D4} = I_{D5} = I_B$, it is required that I_{D11} should be made (1/2)/ $I_{D10} = I_B$.

$$(V_{D1} - V_{D2}) = \sqrt{\frac{I_{D4}}{\mu_p C_{ox}\frac{W_4}{L_4}}} - \sqrt{\frac{I_{D5}}{\mu_p C_{ox}\frac{W_5}{L_5}}} \tag{4.11}$$

Although this mechanism of offset cancellation is independent of input current and input voltage, still the required doubling of current cannot be achieved precisely because of drains of M_{10} and M_{11} are at different voltage levels.

To circumvent this problem, Hassanein et al. proposed [44] a modified version of their circuit which is shown here in Fig. 4.9b.

In the circuit of Fig. 4.9 b the added mosfets M_{14} is matched to the mosfets M_4 and the added M_{15} is matched to M_{11}. Thus the drain current of M_4 is recreated in M_{14} which through the mirror M_{15}_M11 is delivered to M_5 thus ensuring $I_{d5} = I_{d4}$. Thus, the additional circuit implements an adaptive offset cancellation technique.

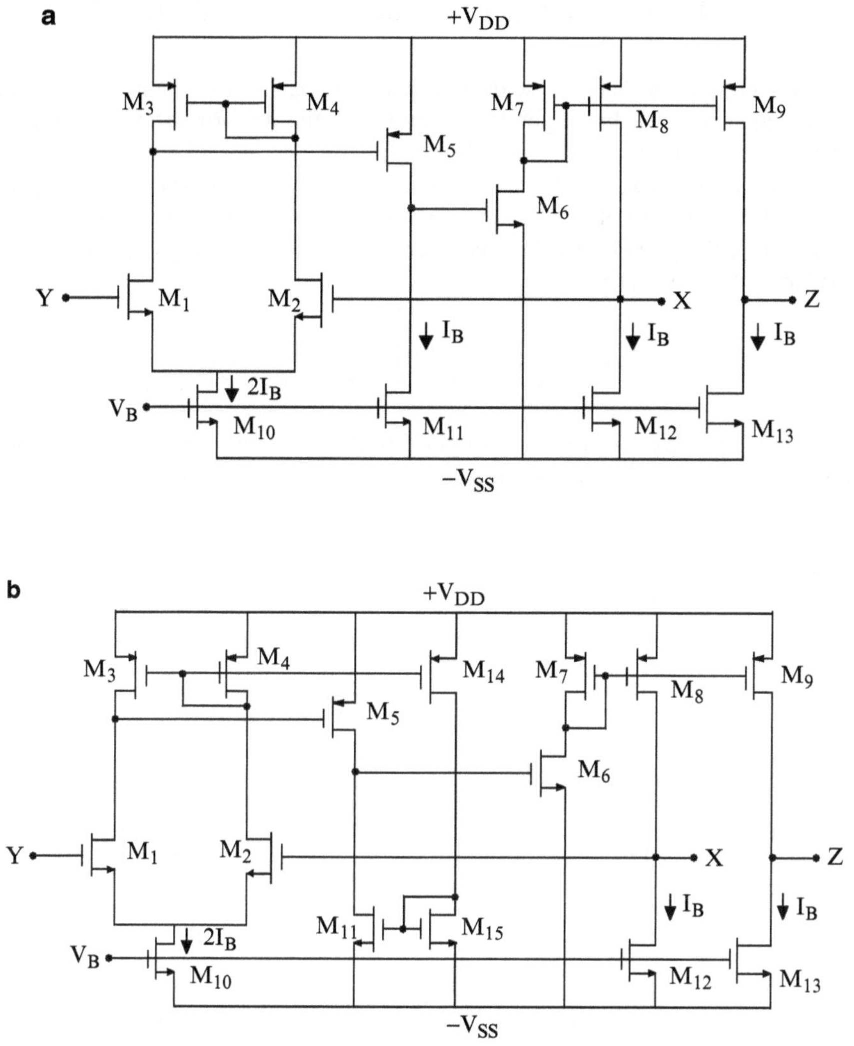

Fig. 4.9 High accuracy CMOS current conveyors proposed by Hassanein et al. (**a**) The first proposed high accuracy CCII+ (**b**) High accuracy CCII+ with adaptive offset cancellation (Adapted from [44] © 2004 Elsevier GmbH)

SPICE simulation results have demonstrated that the topology of Fig. 4.9a has a voltage gain of 1.00003, current gain of 1.000001 with their 3-dB bandwidth being 1.15 and 15 MHz respectively; R_x is as small as 0.037 Ω and input voltage range is −0.4–0.8 V, with voltage offset variation from 0.00027 to 0.071 mV only. On the other hand, the CCII with adaptive offset cancellation has both voltage gain and current gain of unity with their 3-dB bandwidths being 0.48 and 2.16 MHz with voltage offset variation from 0 to 0.00026 and $R_x = 0.06$ Ω.

4.10 High Bandwidth Current Conveyor with Reduced R_X

An improved CMOS CCII whose design was optimized using a heuristic algorithm for optimal sizing to achieve good static and dynamic performances was proposed by Salem et al. [46] and is shown in Fig. 4.10.

In this configuration, the main modification is the introduction of a new type of signal path for the current transfer between ports X and Z which ensures very low impedance looking into port X. It has been deduced [46] that the parasitic input impedance R_X is given by:

$$R_X = \frac{1}{g_{m8}(1 + g_{m2}r_{02})} \Bigg|\Bigg| \frac{1}{g_{m5}(1 + g_{m4}r_{04})} \tag{4.12}$$

As compared to the conventional CMOS CCII, the above expression leads to a quite reduced value of R_X. The algorithm used by the authors was set up for minimizing X-port input resistance, maximizing Y and Z port resistances, maximizing the 3 dB bandwidths of the current and voltage transfers and minimizing the noise effects and silicon area. With DC bias current of 100 μA, the 3 dB bandwidth of the voltage gain (0.96) was found to 3.9 GHz, that of the current gain (0.976) as 2.6 GHz while the parasitic impedance levels were found to be $R_X = 18.47$ Ω; $R_Z||C_Z = 34.5$ kΩ|| 9.45 fF and $R_Y||C_Y = 25.25$ kΩ||49.7 fF.

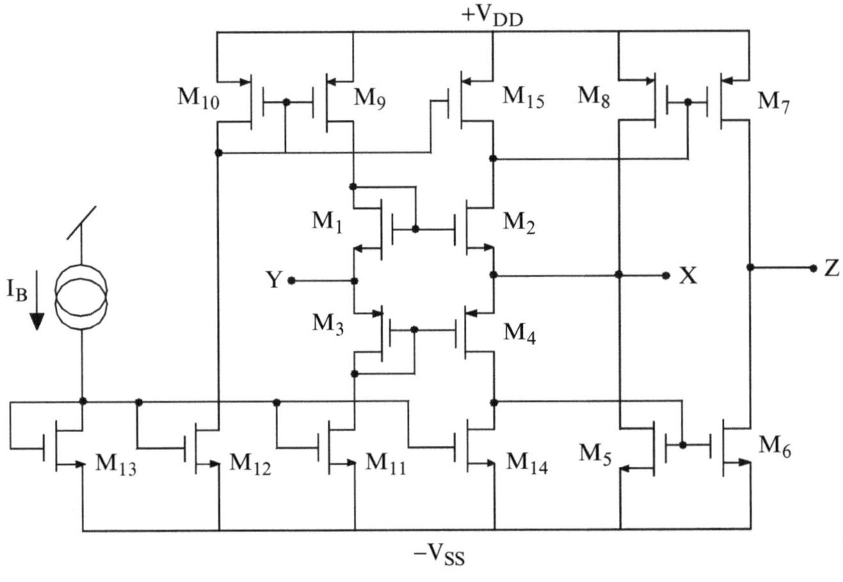

Fig. 4.10 The improved translinear based CCII structure implementation advanced by Salem et al. (Adapted from [46] © 2006 Springer)

The utility of this improved CMOS CCII architecture was demonstrated by successfully devising a band pass filter and a current controlled oscillator, tunable in high frequency ranges: of 1.22–1.56 GHz and 290–475 MHz respectively.

4.11 Current Conveyor with High Current Driving Capability, Operated from 1.5 V Power Supply

A CMOS CCII+ implementation with rail-to-rail input–output operation and high deriving capability was proposed by Mita et al. in [38] and is shown here in Fig. 4.11. This circuit is based upon three main blocks which are: a transconductance amplifier with a differential input and two matched inverting transconductance output stages with their inputs connected in parallel. The output of one of the inverting transconductance is fed back to non-inverting terminal of the input transconductance thereby creating a unity voltage gain between terminals Y and X of the current conveyor. In the schematic of Fig. 4.11, the first transconductance block is constructed from two complimentary differential amplifiers made from the MOSFETs M_1-M_2 and a PMOS current mirror (shown symbolically) and MOSFET pair M_5-M_6 and two NMOS mirrors (shown symbolically) along with the DC bias current sources I_{B1} and I_{B2}. It is this part of the circuit which ensures rail-to-rail capability. The MOSFET M_5 is configured as a common source amplifier and the transistors M_6, M_7, M_8, M_{11} and M_{16} setup the quiescent current and help in properly deriving the two output transconductance stages which are based upon a topology proposed by Palumbo [67]. The MOSFETs M_6-M_7 and M_{11}, forming the current mirror, provide the signal inversion required for the Miller compensation which is implemented through a compensating resistor R_c and compensating

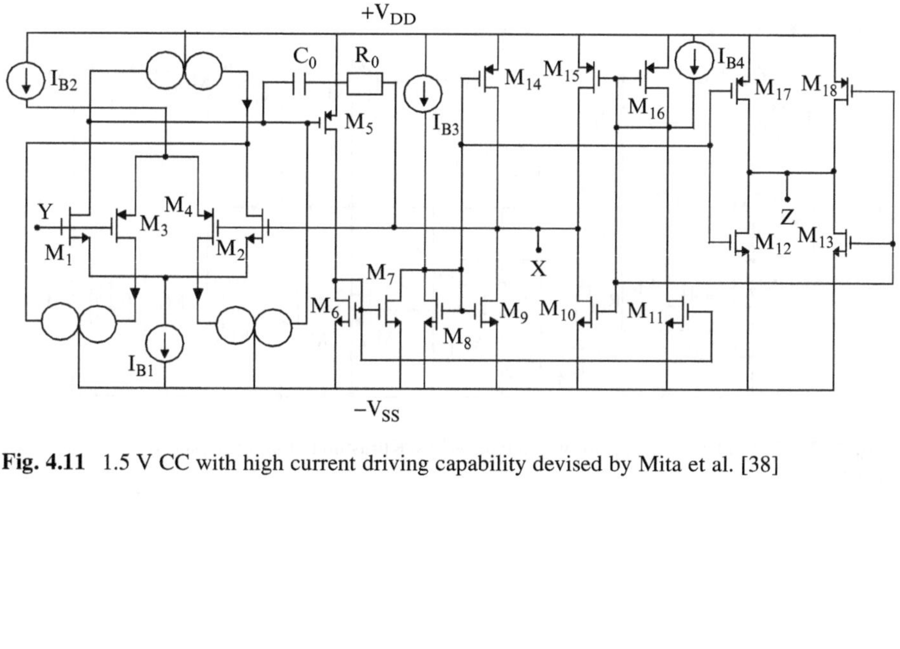

Fig. 4.11 1.5 V CC with high current driving capability devised by Mita et al. [38]

capacitor C_c. Finally, the group of MOSFETs M_9, M_{14}, M_{10} and M_{15} constitute the second inverting transconductance stage while the M_{12}, M_{13}, M_{17} and M_{18} constitute the third inverting transconductance stage.

Using 1.5 V supply, the test on a prototype, fabricated in a 0.35 μm technology, exhibits an input resistance lower than 100 Ω at the X-terminal of the CCII. The current drive capability achieved was better than ± 0.9 mA which is about 45 times of the quiescent current in the output branch. The frequency response demonstrated that the 3-dB bandwidth for the current gain between Z and X port was about 1.2 MHz and that for the voltage gain between ports Y and X as 2.4 MHz.

4.12 CMOS Rail-to-Rail Current Conveyor

To meet the requirements of wide dynamic range and wide bandwidth operations under low supply voltage environment, the circuits employing CCs which themselves should have rail-to-rail operation capability are of particular interest. A number of authors have focused on the design of such CMOS CC architectures.

One such CMOS CC architecture was advanced by Kurashina et al. in [32] and is shown here in Fig. 4.12. This circuit consists of a rail-to-rail complementary differential input stage for the voltage input V_Y formed by the MOSFETs M_1-M_8 where R_{C1}, R_{C2}, C_{C1}, C_{C2} have been used for phase compensation whereas the transistors M_9-M_{12} constitute the level shifting stage. The MOSFETs M_{13} and M_{14} constitute the rail-to-rail class AB push-pull stage which provides an output terminal which acts either as a voltage output or current input for input current I_x. Finally, the MOSFETs M_{15} to M_{22} create appropriate current mirrors and the two complementary Z-output terminals Z+ and Z−.

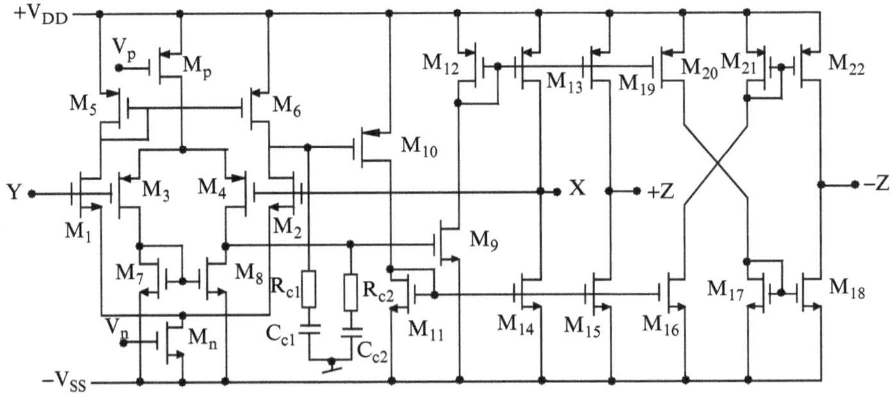

Fig. 4.12 CMOS CC architecture was advanced by Kurashina et al. (Adapted from [32] © 2002 IEICE)

The measured performance of a prototype, fabricated using a double-poly triple-metal 0.6 μm n-well CMOS technology, demonstrated that this configuration achieves a rail-to-rail swing of ±2.3 V under ±2.5 V DC biasing power supply with current transfer ratios as $\frac{I_{Z+}}{I_X} = 1.00$ and $\frac{I_{Z-}}{I_X} = -1.03$ respectively. The measured input impedance was found to be $R_X = 41.7\ \Omega$ and the 3-dB bandwidth of the voltage and current ratios around 120 MHz.

4.13 CMOS Rail-to-Rail Current Conveyor Operated from ± 0.75 V Supply

Another rail-to-rail CMOS CC architecture operable from ±0.75 V supply was proposed by Kasemsuwan and Nakhlo [52] which is shown in Fig. 4.13. This circuit employs a complementary source follower with a common source output in parallel with a pseudo source follower.

Simulation results using 0.13 μm CMOS technology demonstrate the voltage follower and current follower bandwidths in excess of 300 MHz. The input output voltage swing was found to be 1.2 V peak-to-peak and the impedance levels at the X, Y and Z terminals were 7.2 Ω, 61 MΩ and 22 kΩ respectively whereas the power dissipation was of the order of 532 μw.

4.14 Low-Voltage Low-Power CCII Based on Folded Cascode Bulk-Driven OTA

As seen from the previous illustrations, the design of low voltage and low power CCs has attracted considerable attention due to the increasing demand of analog circuits which can be integrated along with their digital counterparts in the same technology particularly in case of portable systems demanding least possible power consumption and least possible DC power supply voltages. A technique which has been found to be attractive, particularly for this purpose, is the class of circuits known as bulk-driven MOSFET circuits.

The conventional MOSFET is a four terminal device having these four terminals as *gate, drain, source and bulk* (substrate). In most common applications, the bulk terminal is connected to positive or negative supply rail or simply tied to the source terminal and then is forgotten/ignored. However, if this bulk terminal is used as the input terminal, rather than the gate terminal, the threshold voltage obstruction does not appear in the signal path, as a consequence of which, the resulting analog cell can be applied to very low supply voltage. Furthermore, to enable to use the bulk terminal as an input terminal, the gate terminal needs to be connected to an appropriate DC bias voltage which is adequate enough to create a conduction

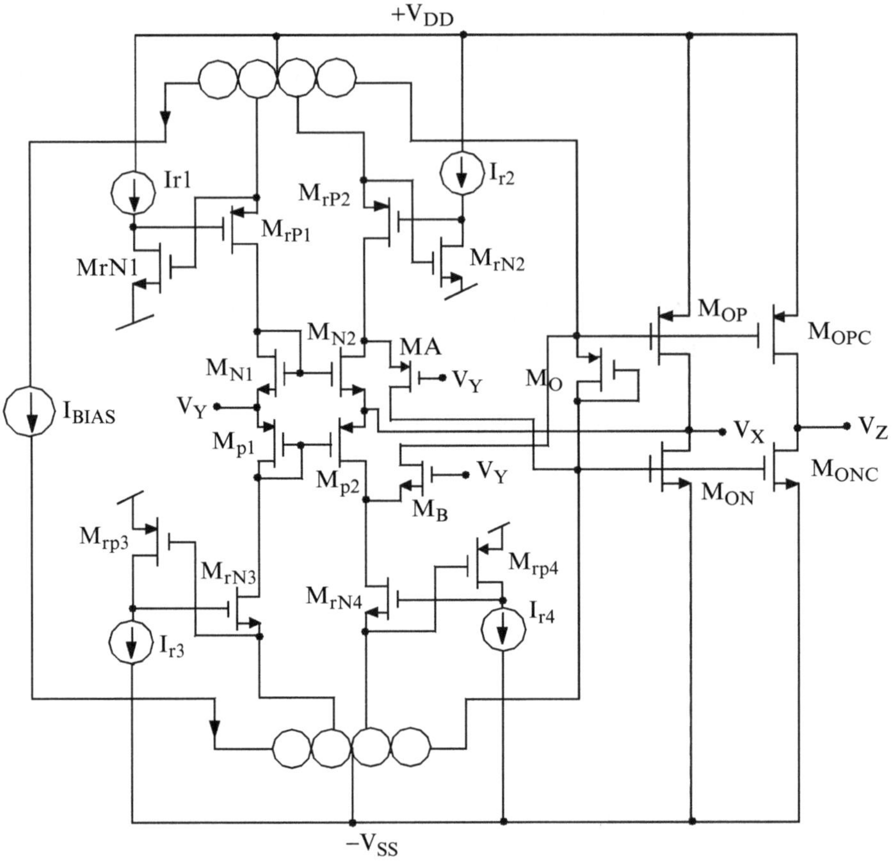

Fig. 4.13 CC proposed by Kasemsuwan and Nakhlo [52]

channel inversion layer. Thus, a bulk-driven MOSFET can effectively work under negative, zero or even slightly positive biasing conditions.

Khateb et al. [60] proposed a CCII architecture using a folded-cascode bulk-driven OTA structure consisting of transistors M_1-M_{11} where M_1-M_2 constitute the input stage, M_6-M_7 provide a level shift stage and the MOSFETs M_8-M_{8c}, M_9-M_{9c} constitute a cascode current mirror which performs the operation of differential to single-ended conversion (Fig. 4.14). The MOSFETs M_3-M_4 provide the current necessary for the input and output branches. The transistors M_{10}-M_{11} constitute the second stage of the OTA. Transistors M_{12}-M_{13} and M_{14}-M_{15} are identical to transistors M_{10} and M_{11} respectively. Since the gates of M_{11}, M_{13} and M_{15}, M_{10} and M_{12}, M_{14} are tied together, the currents in the branches M_{10}-M_{11}, M_{12}-M_{13} an M_{14}-M_{15} are necessarily same.

The simulation results of this low-voltage, low-power bulk-driven CCII demonstrate that both the current and voltage ratios are *exactly unity* with the 3 dB bandwidth in excess of 10 MHz, with DC biasing voltage as low as 0.38 V the

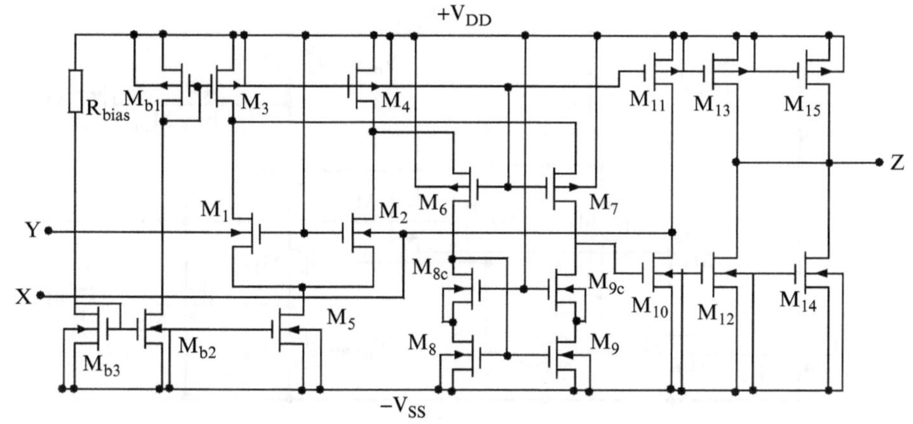

Fig. 4.14 A bulk-driven CCII proposed by Khateb et al. (Adapted from [60] © 2011 Elsevier)

circuit exhibited the power consumption of 64 μW, $R_X = 27\ \Omega$ and $R_Z \| C_Z$ of the order of 0.89 MΩ‖40 fF.

4.15 Wide-band High Performance Current Conveyor

Arslan et al. [63] proposed a wide band high performance architecture which consisted of an improved voltage follower stage which employs two source followers along with a local feedback to obtain reduced equivalent impedance (see Fig. 4.15).

The expression for the equivalent resistance looking into port-X is given by:

$$R_X = \frac{1}{g_{m11}g_{m20}r_{o11}(1 + g_{m10}r_o/2)} \Big\| \frac{1}{g_{m12}g_{m21}r_{o12}(1 + g_{m9}r_o/2)} \tag{4.13}$$

The output of the improved voltage follower stage is put to a current follower consisting M_6 and M_{19} which senses the current flowing through X-port and copies it to the high impedance Z-port. Between ports X and Z, the current gain α is given by:

$$\alpha = \frac{i_Z}{i_X} \cong \frac{g_{m6,19}}{g_{m20,21}} \cong \frac{(W/L)_{6,19}}{(W/L)_{20,21}} \tag{4.14}$$

With the matching of the MOSFET M_6 with M_{20} and that of M_{19} with M_{21}, α is very close to unity.

On the other hand, the equivalent resistance seen at the output port is $R_Z = r_{o6} \| r_{o19}$. SPICE simulation results, with CMOS circuit biased with ± 1.65 V, exhibit

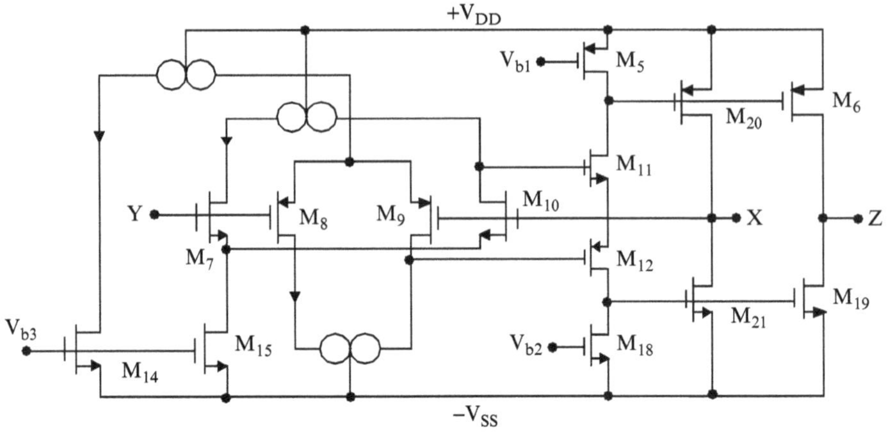

Fig. 4.15 CMOS CCII proposed by Arslan et al. [63]

that the voltage gain of the circuit is 1.01, current gain is 0.99 and the corresponding 3-dB bandwidths are about 810 MHz and 915 MHz respectively. The values of the parasitic impedances were found to be $R_x = 9.6\ \Omega$ and $R_z \| C_z = 147\ \mathrm{k}\Omega \| 33.5$ fF.

Concluding Remarks

In this chapter, we have presented a variety of CMOS CC architectures chosen from a large number of circuits [1–67, 69] reported in the literature. It must be pointed out that a very good coverage of a number of CMOS CCs till 2003 has already been dealt with in the book [68] by Ferri and Guerrini hence; we have mainly focused upon developments not included in [68] or those taken place after 2003. Furthermore, only a selected works have been included and the interested readers are referred to the references of [1–69] given in the end of this chapter as well as those in the supplementary list of references given in Appendix-I. CMOS/BiCMOS implementation of the CCs is an area which has continued to attract the attention of researchers over the past two decades and continues to do so even now.

References

1. Sedra AS, Roberts GW, Gohh F (1990) The current conveyor: history, progress and new results. IEE Proc Circ Devices Syst 137:78–87
2. Liu SI, Tsao HW, Wu J, Lin TK (1990) MOSFET capacitor filters using unity gain CMOS current conveyors. Electron Lett 26:1430–1431
3. Liu SI, Tsao HW, Wu J (1991) CCII-based continuous-time filters with reduced gain-bandwidth sensitivity. IEE Proc Circ Devices Syst 138:210–216
4. Chen JJ, Chen CC, Tsao HW, Liu SI (1991) Circuit-mode oscillators using single current follower. Electron Lett 27:2056–2059

5. Surakampontorn W, Riewruja V, Cheevasuvit F (1991) Integrable CMOS-base realization of current conveyors. Int J Electron 71:793–798
6. Surakampontorn W, Kumwachara K (1992) CMOS-based electronically tunable current conveyor. Electron Lett 28:1316–1317
7. Laopoulos TH, Siskos S, Bafleur M, Givelin PH (1992) CMOS current conveyor. Electron Lett 28:2261–2262
8. Bruun E (1993) CMOS high speed, high precision current conveyor and current feedback amplifier structures. Int J Electron 74:93–100
9. Cheng MCH, Toumazou C (1993) 3 V MOS current conveyor cell for VLSI technology. Electron Lett 29:317–318
10. Brunn E (1995) Class AB CMOS first –generation current conveyor. Electron Lett 31:422–423
11. Bruun E (1995) A combined first-and second-generation current conveyor structure. Int J Electron 78:911–923
12. Piovaccari A (1995) CMOS integrated third-generation current conveyor. Electron Lett 31:1228–1229
13. Cha HW, Watanabe K (1996) Wideband CMOS current conveyor. Electron Lett 32:1245–1246
14. Oliaei O, Loumeau P (1996) Current-mode class AB design using floating voltage-source. Electron Lett 32:1526–1528
15. Bruun E (1996) Analysis of the noise properties for CMOS current conveyors. Analog Integr Circ Sig Process 12:71–78
16. Hassan O, Elwan HO, Soliman AM (1996) A novel CMOS current conveyor realization with an electronically tunable current mode filter suitable for VLSI. IEEE Trans Circ Syst-II 43:663–670
17. Oliaei O, Porte J (1997) Compound current conveyor (CCII+ and CCII−. Electron Lett 33:253–254
18. Elwan HO, Soliman AM (1997) Low-voltage low-power CMOS current conveyors. IEEE Trans Circ Syst-I 44:828–835
19. Arbel A (1997) Towards a perfect CMOC CCII. Analog Integr Circ Sig Process 12:119–132
20. Cha HW, Ogawa S, Watanabe K (1998) Class A CMOS current conveyors. IEICE Trans Fundament E81-A:1164–1167
21. Tarim N, Yenen B, Kuntman H (1998) Simple and accurate non-linear current conveyor macromodel suitable for simulation of active filters using CCIIs. Int J Circ Theor Appl 26:27–38
22. Ismail AM, Soliman AM (1998) Wideband CMOS current conveyor. Electron Lett 34:2368–2369
23. Awad IA, Soliman AM (1999) New CMOS realization of the CCII-. IEEE Trans Circ Syst-II 46:460–463
24. Fabre A, Amrani H, Barthelemy H (1999) A novel class AB first generation current conveyor. IEEE Trans Circ Syst-II 46:96–98
25. Premont C, Abouchi N, Grisel R, Chante JP (1999) A BiCMOS current conveyor based four—quadrant analog multiplier. Analog Integr Circ Sig Process 19:159–162
26. Palmisano G, Palumbo G, Pennisi S (1999) Design strategies for class A CMOS CCIIs. Analog Integr Circ Sig Process 19:75–85
27. Ismail AM, Soliman AM (2000) Low-power CMOS current conveyors. Electron Lett 36:7–8
28. Yodprasit U (2000) High-precision CMOS current conveyor. Electron Lett 36:609–610
29. Palumbo G, Pennisi S (2001) A high-performance CMOS CCII. Int J Circ Theor Appl 29:331–336
30. Emami S, Wada K, Takagi S, Fujii N (2001) A novel design strategy for class A CMOS second generation current conveyors. IEICE Trans Fundament 84-A:552–558
31. Ferri G, Laurentiis PD, Stochino G (2001) Current conveyors II. Analogue design. Electron World 300–302

32. Kurashina T, Ogawa S, Watanabe K (2002) A CMOS rail-to-rail current conveyor. IEICE Trans Fundament 85-A:2894–2900
33. Nero AL, Aguiar Ruil L, Santos DM (2002) Bandwidth aspects in second generation current conveyors. Analog Integr Circ Sig Process 33:127–136
34. Cicekoglu O, Tarim N, Kuntman H (2002) Wide dynamic range high output impedance current-mode multifunction filters with dual-output current conveyors. Int J Electron Commun (AEU) 56:55–60
35. El-Adawy AA, Soliman AM, Elwan HO (2002) Low voltage digitally controlled CMOS current conveyor. Int J Electron Commun (AEU) 56:137–144
36. Calvo B, Celma S, Martinez PA, Sanz MT (2003) Novel high performance CMOS current conveyor. Microelectron Reliabil 43:955–961
37. Kurashina T, Ogawa S, Watanabe K (2003) A CMOS rail-to-rail current conveyor and its applications to current-mode filters. IEICE Trans Fundament E86-A(6):1445–1450
38. Mita R, Palumbo G, Pennisi S (2003) 1.5-V CMOS CCII+ with high current-driving capability. IEEE Trans Circ Syst-II 50:187–190
39. Awad IA, Soliman AM (2004) High accuracy class AB CCII-. Int J Electron Commun (AEU) 58:237–243
40. Rajput SS, Jamuar SS (2004) Low voltage, low power and high performance current conveyors for low-voltage analog and mixed mode signal processing applications. Analog Integr Circ Sig Process 41:21–34
41. Hassanein WS, Awad IA, Soliman AM (2004) New wide band low power CMOS current conveyors. Analog Integr Circ Sign Process 40:91–97
42. Ferri G, Guerrini NC (2004) Noise determination of differential pair-based second generation current conveyors. Analog Integr Circ Sig Process 41:35–46
43. Calvo B, Celma S, Martinez PA, Sanz MT (2003) High-speed high-precision CMOS current conveyor. Analog Integr Circ Sig Process 36:235–238
44. Hassanein WS, Awad IA, Soliman AM (2005) New high accuracy CMOS current conveyors. Int J Electron Commun 59:384–391
45. Hassanein WS, Awad IA, Soliman AM (2005) Long tail pair based positive CMOS current conveyors: a review. Frequenz 59:186–194
46. Salem SB, Fakhfakh M, Masmoudi DS, Loulou M, Loumeau P, Masmoudi N (2006) A high performance CMOS CCII and high frequency applications. Analog Integ Circ Sig Process 49:71–78
47. Hassan HM, Soliman AM (2006) Novel accurate wideband CMOS current conveyor. Frequenz 60:234–236
48. Madian AH, Mahmoud SA, Soliman AM (2006) New 1.5-V CMOS second generation current conveyor based on wide range transconductor. Analog Integr Circ Sig Process 49:267–279
49. Minaei S, Sayin OK, Kuntaman H (2006) A new CMOS electronically tunable current conveyor and its application to current-mode filters. IEEE Trans Circ Syst-I 53:1448–1457
50. Ferri G, Stornelli V, Fragnoli M (2006) An integrated improved CCII topology for resistive sensor application. Analog Integr Circ Sig Process 48:247–250
51. Barthelemy H, Fillaud M, Bourdel S, Gaubert J (2007) CMOS inverters based positive type second generation current conveyor. Analog Integr Circ Sig Process 50:141–146
52. Kasemsuwan V, Nakhlo W (2007) A simple 1.5 V rail-to-rail CMOS current conveyor. J Circ Syst Comput 16:627–639
53. Arcamone J, Misischi B, Graells FS, Boogaart MAFVD, Brugger J, Torres F, Abadal G, Barniol N, Murano FP (2008) Compact CMOS current conveyor for integrated NEMS resonators. IET Circ Devices Syst 2:317–323
54. Motlak HJ, Ahmed SN (2008) CMOS CCII-based on modified dual output-OTA for high frequency applications. Int J Electron 95:879–889
55. Al-Absi MA (2009) A novel highly accurate current mirror. Int J Electron 96:781–786
56. Thankachan S, Pattanaik M, Rajput SS (2009) A ± 0.5 V BiCMOS class-A current conveyor. Int J Electr Electron Eng 3:607–610

57. Mahmoodi A, Abrishamifar A (2010) A novel current conveyor with high functionality and optimized ports. IEICE Electron Express 7:1480–1485
58. Chatterjee A, Fakhfakh M, Siarry P (2010) Design of second-generation current conveyor employing bacterial foraging optimization. Microelectron J 41:616–626
59. Motlak HJ, Ahmed SN (2010) Wide bandwidth CMOS CCII+ using resistive-compensation technique. J Active Passive Electron Devices 5:163–180
60. Khateb F, Khatib N, Kubanek D (2011) Novel low-voltage low-power high-precision CC± based on bulk-driven folded cascode OTA. Microelectron J 42:622–631
61. Kapur G, Mittal S, Markan CM, Pyara VP (2012) Design of analog field programmable cmos current conveyor. Sci J Circ Syst Sig Process 1:9–21
62. Ahmadpoor N, Mohamadzade S, Ahmadzadeh M (2012) A novel linear low voltage rail to rail second generation current conveyor for RF applications. J Basic Sci Res 2:12306–12310
63. Arslan E, Minaei S, Morgul A (2013) On the realization of high performance current conveyors and their applications. J Circ Syst Comput 22:23 pages
64. Hwang YS, Ku YT, Chen JJ, Wang SF (2013) A low-voltage current conveyor using inverter-based error amplifier and its oscillator application. IEICE Electron Express 10:1–7
65. Alzaher H, Tasadduq N, Al-Ees O, Al-Ammari (2013) A complementary metal-oxide semi-conductor digitally programmable current conveyor. Int J Circ Theor Appl 41:69–81
66. Hwang YS, Ku YT, Chen JJ, Yu CC (2013) Inverter-based low-voltage CCII-design and its filter application. Radioengineering 22:1026–1033
67. Palumbo G (1999) 1.2 V CMOS output stage with improved drive capability. Electron Lett 35:358–359
68. Ferri G, Guerrini NC (2003) Low-voltage low-power CMOS current conveyors. Kluwer, Dordrecht
69. Manhas PS, Pal K, Sharma S, Mangotra LK, Jamwal KKS (2009) New low-voltage class AB current conveyor II for analog applications. Ind J Pure Appl Phys 47:306–309

Part II
The Early (First Generation) Applications of Basic CCI and CCII

Chapter 5
Basic Analog Circuit Building Blocks Using CCs and Application of CCs in Impedance Synthesis

Abstract This chapter discusses basic functional circuits realizable using CCI and CCII of the normal kind. This includes constant bandwidth variable gain voltage amplifiers, current-mode op-amp, integrators, differentiators, instrumentation amplifiers, summers, impedance converters/inverters, simulated inductors and FDNRs (in both grounded and floating forms) and a generalized function generator.

5.1 Introduction

This chapter discusses a variety of analog circuits which can be made using the basic types of Current Conveyors, namely, the CCI and CCII. The circuits considered here include, variable gain voltage amplifiers, instrumentation amplifiers, current-mode op-amp, integrators, differentiators, summers, impedance converters/inverters and simulated inductors, FDNR (frequency dependent negative resistance; an element having impedance $Z(s) = 1/Ds^2$) and a generalized function generator. Indeed, a vast amount of literature exists which deal with the above mentioned functional circuits using CCs, see [1–148, 149, 150–164, 165].

As is well known, the four controlled sources namely VCVS, VCCS, CCVS and CCCS are the fundamental two-port building blocks which can be used to realize other building blocks among which impedance converters and inverters (both positive and negative types) are the most fundamental ones. Impedance converters and inverters are useful as building blocks in their own right but are more commonly employed to simulate inductors, FDNR, FDNC (frequency dependent negative conductance; an element having $Z(s) = Ms^2$) and several forms of negative elements such as negative resistances, negative capacitances and negative inductances. Such synthetic impedances find interesting applications in the realization of filters, oscillators (which include linear oscillators, relaxation oscillators as well as chaotic oscillators), loss-cancellation in inductors/capacitors and in enhancing the frequency range of oscillators, to name a few. Other basic blocks include integrators, differentiators, instrumentation amplifiers (or finite variable-gain amplifiers).

The Current Conveyor literature has an abundance of circuits of varying complexities and notable properties, which have been developed during the period of

© Springer International Publishing Switzerland 2015
R. Senani et al., *Current Conveyors*, DOI 10.1007/978-3-319-08684-2_5

over four decades, to create a variety of synthetic impedances in both grounded and floating form, for instance, see [4, 5, 9, 10, 12–22, 24–45, 47–59, 61–65, 68–72, 74, 77, 78, 80–98, 100–110, 113–141, 150–153, 166, 165].

In this chapter, we describe some of the most significant developments on the various functional circuits mentioned above. For the interested reader, who wishes to explore other available options than those chosen for inclusion here, a comprehensive list of additional references has been provided in Appendix A, at the end of this monograph.

5.2 The Basic Functional Circuits Using CCI and CCII

Traditionally, the IC op-amp is regarded to be a versatile element and it is said that in analog circuits area, almost anything can be realized with op-amps. However, over the years, it has been proved that Current Conveyors are not only as versatile as the op-amps but can outperform op-amps in many situations and applications where the CC-based circuits require a bare minimum number of external passive components, do not require any component-matching, do not suffer from the effects of gain bandwidth conflict and do not have severe finite slew rate based effects exhibited by their op-amp counterparts. It is, therefore, widely believed now that CCs are building blocks whose time has now come. This belief has been further strengthened by the commercial availability of CCs as off-the-shelf ICs and excellent workability and versatility of several CC-based functional circuits which have been evolved during the last four decades. In this section, we elaborate some basic circuits which can be made from the traditional three-terminal CCs since they are commercially available as ICs as well as can be built from other available ICs.

5.2.1 Variable-Gain Amplifiers: Constant-Bandwidth Structures

It is well known that the traditional voltage mode op-amp (VOA) based inverting and non-inverting amplifiers (see Fig. 5.1) both exhibit a conflict between the maximum gain and 3-dB bandwidth.

The finite gain bandwidth product (GBP) of the VOA is incorporated by representing the open-loop gain as

$$V_0 = A(V_1 - V_2) \tag{5.1}$$

where the open loop gain A of the op-amp is expressed as

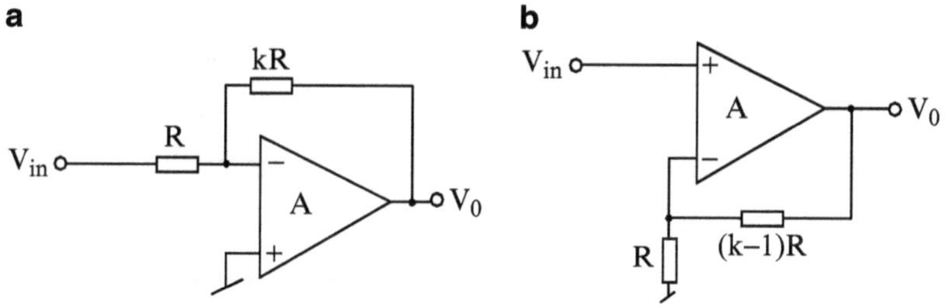

Fig. 5.1 VOA-based K-gain amplifiers (**a**) Non-inverting (**b**) Inverting

$$A = \frac{A_0\omega_p}{s+\omega_p} \cong \frac{A_0\omega_p}{s} \cong \frac{\omega_t}{s} \text{ for } \omega > \omega_p \tag{5.2}$$

where A_0 is the maximum gain (at DC) and ω_p is the pole frequency in radians/sec with $\omega_t = A_0\,\omega_p$ being the gain-bandwidth-product (GBP) of the op-amp.

By straight forward analysis, it can be shown that the inverting amplifier exhibits the maximum gain = −K and 3-dB bandwidth $= \frac{\omega_t}{k+1}$ while the non-inverting amplifier exhibits the maximum gain = +K and 3-dB bandwidth $= \frac{\omega_t}{k}$.

Thus, in both the cases, the moment the gain 'K' is fixed, the 3-dB BW also gets fixed. In other words, increasing the closed loop gain 'K' reduces the available bandwidth and a larger bandwidth can be accomplished only for smaller values of 'K'. There is clearly a conflict between the requirement of a larger gain and a larger bandwidth as the two are interdependent and cannot be met simultaneously.

In the following, we show that when the same configurations are realized with CCs (as shown in Fig. 5.2a, b), the gain bandwidth conflict is eliminated and it becomes possible to design variable gain amplifiers with nearly constant bandwidth independent of the gain [162].

A straight forward analysis of the first circuit taking into account the parasitic Z-port impedance Z_p of the CCs which is equal to a parallel RL ($Z_p = R_p/(1/sC_p)$) reveals its non-ideal transfer function as:

$$\frac{V_o}{V_{in}} = -\frac{\frac{1}{C_pR_1}}{s+\frac{1}{C_p}\left(\frac{2}{R_2}+\frac{1}{R_p}\right)} \tag{5.3}$$

Similarly, an analysis of the non-inverting configuration reveals:

$$\frac{V_o}{V_{in}} = +\frac{\frac{1}{C_p}\left(\frac{1}{R_1}+\frac{2}{R_2}\right)}{s+\frac{1}{C_p}\left(\frac{2}{R_2}+\frac{1}{R_p}\right)} \tag{5.4}$$

From the above, it can be easily found that the maximum gains in case of the two circuits are found to be:

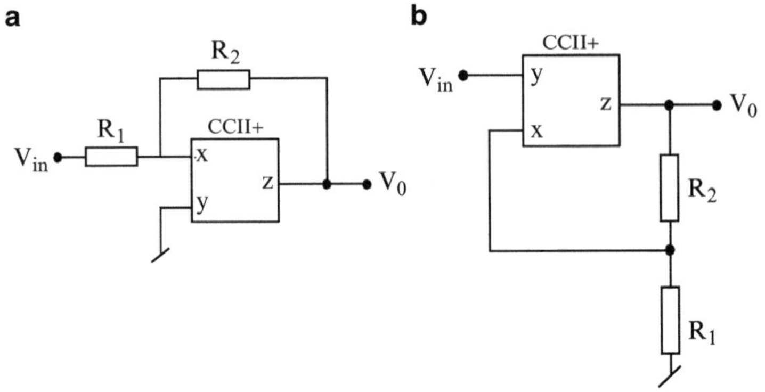

Fig. 5.2 CCII-based amplifiers (**a**) Inverting (**b**) Non-inverting

For the inverting amplifier:

$$-\frac{\frac{R_2}{2R_1}}{\left(1+\frac{R_2}{2R_p}\right)} \approx -\frac{R_2}{2R_1} \text{for } 2\mathrm{R_p} >> \mathrm{R_2} \tag{5.5}$$

For the non-inverting amplifier:

$$+\frac{\left(1+\frac{R_2}{2R_1}\right)}{\left(1+\frac{R_2}{2R_p}\right)} \approx \left(1+\frac{R_2}{2R_1}\right) \tag{5.6}$$

whereas both the circuits have the same −3 dB frequency given by

$$\omega_{3-\mathrm{dB}} = \frac{1}{C_p}\left(\frac{2}{R_2}+\frac{1}{R_p}\right) \approx \frac{2}{C_p R_2} \tag{5.7}$$

It is, therefore, interesting that in both the cases, the 3-dB bandwidth can be set by the feedback resistor R_2 after which the variable gain can be obtained by varying the resistance R_1. Thus, there is no gain-bandwidth conflict in case of the inverting and non-inverting amplifiers made from the CCIIs.

It has been experimentally verified that setting R_2 at 4.7 Ω and varying R_1 produced voltage gain varying from unity to over 100 with −3 dB bandwidth of 2.7 MHz at all values of the gains when the current conveyor was realized by BiFET op-amp LF351 and CA3096AE transistor arrays using the operational mirrored amplifier based formulation. It was also noticed that because in the inverting configuration the Y-terminal has been grounded this forces the output terminal of the internal op-amp to be virtually grounded thereby suppressing the slewing mechanism of the internal op-amp due to which maximum slew rate recorded for the inverting amplifier has been found to be as high as 100 V/us!

In this context, it is worth pointing out that a translinear CCII, based upon the input stage composed of a four-transistor mixed translinear cell is characterized by $i_z = i_x = 2I_B$ *Sinh* $[(V_y\text{-}V_x)/V_T]$ and therefore, exhibits a very large (theoretically unlimited) current for large differential inputs. This leads to theoretically unlimited charging current in the z-port parasitic capacitance thereby resulting in a very high value of the slew rate. By contrast, the CCII architectures which are based upon differential-pair-based input stages, would have an i-v equation of type $i_z = i_x = I_B$ *tanh* $[(V_y\text{-}V_x)/2V_T]$ thereby leading to the maximum charging current as $\pm I_B$ only (due to the saturation of the differential pair) for large differential inputs $(V_y - V_x)$ and hence, such CCII structures are likely to have comparatively much lower slew rates.

Interestingly, the property of gain bandwidth decoupling is available even if the inverting and non-inverting amplifiers are realized from CCII using the configurations of Fig. 5.3. The expressions for the non-ideal gains for these circuits have been found to be:

$$\frac{V_o}{V_{in}} = \mp\left(\frac{R_2}{R_1}\right)\frac{\frac{1}{C_pR_2}}{s + \frac{1}{C_p}\left(\frac{1}{R_2} + \frac{1}{R_p}\right)} \tag{5.8}$$

from where the maximum gain and the 3-dB frequency are found to be:

$$\text{Maximum gain} = \frac{R_2}{R_1}/\left(1 + \frac{R_2}{R_p}\right);\ 3 - \text{dB bandwidth} = \frac{1}{C_p}\left(\frac{1}{R_2} + \frac{1}{R_p}\right) \tag{5.9}$$

From the above, it can be seen that the bandwidth can be held constant by fixing R_2 while the gain can be continuously adjusted by varying the other resistor R_1.

5.2.2 *Constant-Bandwidth Instrumentation Amplifiers*

Wilson [67] demonstrated that an excellent instrumentation amplifier can be made using only two CCII+ and no more than two resistors which is a very significant improvement over the conventional VOA-based topology which typically employs as many as three VOAs along with as many as seven resistors along with the requirement of precise matching of at least four of them.

A simplified form of the circuit proposed by Wilson is shown in Fig. 5.4 which has the voltage gain of $V_0/(V_1 - V_2) = \left(\frac{R_2}{R_1}\right)$. By taking the Z-port parasitic impedances $R_p/(1/sC_p)$ into account, it can be easily verified that this circuit exhibits a constant bandwidth set by R_2 and a variable gain adjustable with R_1.

Figure 5.5 shows an improved instrumentation amplifier proposed by Khan et al. [142]. This modified circuit features high CMRR without requiring matched

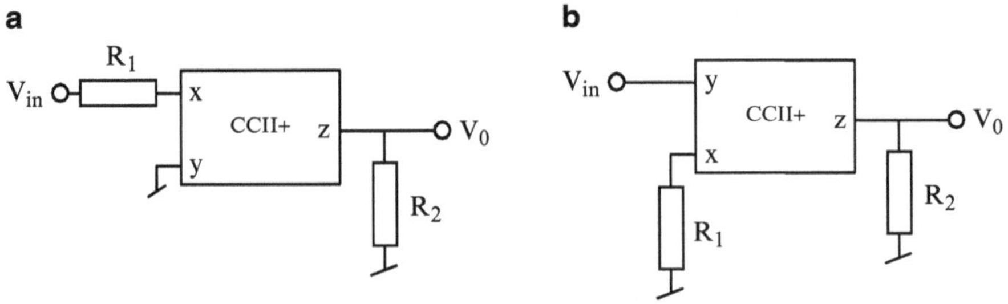

Fig. 5.3 Alternative realizations of CCII-based (**a**) Non-inverting (**b**) Inverting amplifier

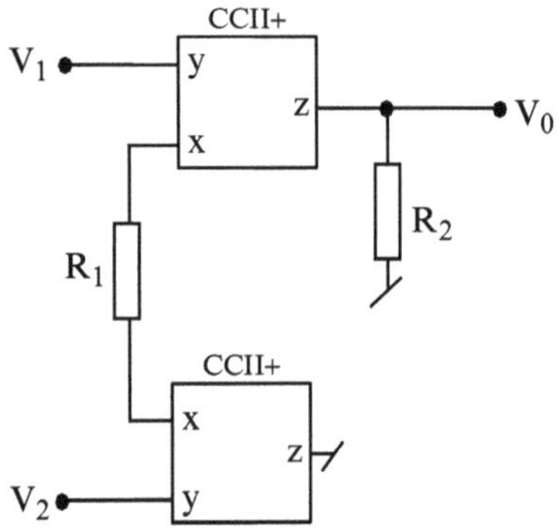

Fig. 5.4 A simplified form of the instrumentation amplifier proposed by Wilson [67]

resistors. Moreover, it has been claimed in [142] that high differential gain and bandwidth can be achieved simultaneously.

For other works which have dealt with the design of constant bandwidth instrumentation amplifiers using CCs see [66, 99, 111, 112, 160, 161].

5.2.3 Constant-Bandwidth Current-Mode Operational Amplifier

A constant BW current mode op-amp (COA) was proposed by Bruun [143] as shown in Fig. 5.6 which requires three CCIIs.

When such a COA is configured as a finite gain current amplifier as shown in Fig. 5.6b, from a straight forward analysis, assuming Z_T to be comprised of a parasitic resistance R_T in parallel with parasitic capacitance C_T, it is found that the maximum current gain is given by:

$$A_I = \frac{I_{out}}{I_{in}} \cong -\left(1 + \frac{R_1}{R_2}\right) \tag{5.10}$$

Fig. 5.5 An alternative instrumentation amplifier proposed by Khan et al. [142]

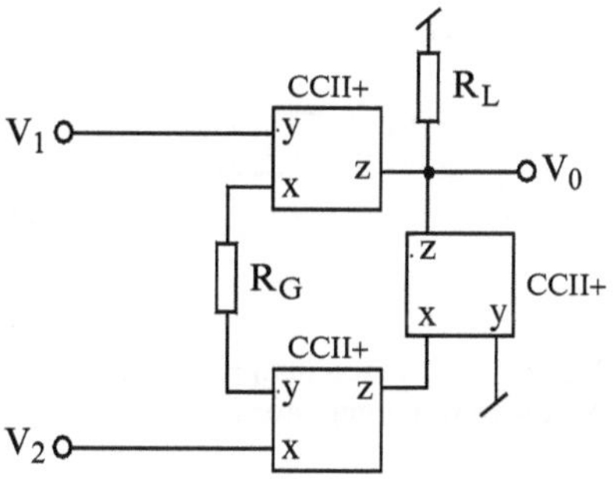

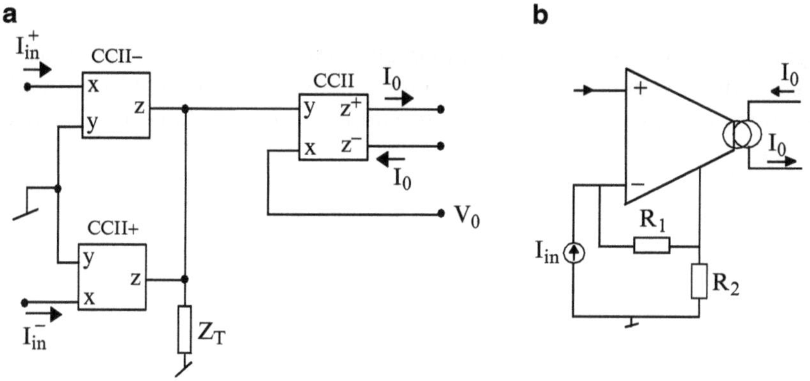

Fig. 5.6 Current-mode op-amp proposed by Bruun [143] (**a**) COA topology (**b**) Finite gain variable current amplifier

while the 3-dB bandwidth is given by:

$$BW \cong \frac{1}{2\pi C_T R_1} \tag{5.11}$$

It is, thus, seen that in contrast to VOA-based circuits, the COA-based circuit exhibits a decoupling of the gain and BW in that having fixed the BW by R_1 gain can be still varied by R_2.

For an alternative formulation of the current mode op-amp and a number of other related building blocks such as translinear operational current amplifier, Operational floating Conveyor and differential output current conveyor-all of which can be made from CCII+ and CCII−, the reader is invited to see [73].

Experimental results of the configuration of Fig. 5.6 with the CCIIs realized by LM307 op-amps (in power supply current sensing mode) and transistor arrays CA3096 used to implement current mirrors, showed that the circuit provided nearly gain independent bandwidth of around 500 KHz (since the LM307 op-amp employed had its unity gain bandwidth of approximately 700 KHz only) for several different value of R_1/R_2.

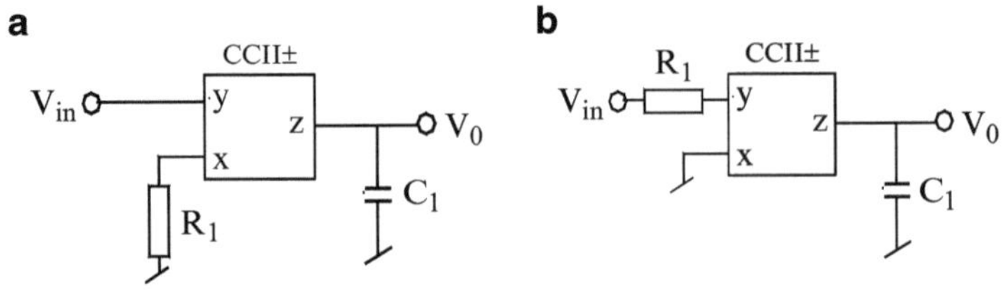

Fig. 5.7 CC-based integrators (**a**) Non-inverting/Inverting integrator from CCII± (**b**) Inverting/Non-inverting integrator from CCII±

5.2.4 Integrators and Differentiators

Integrators and differentiators are important building blocks with numerous applications in the design of waveform shaping circuits, filters, oscillators etc. and their realizations using CCs have been dealt with by several authors, for instance, see [75, 154–159].

The realization of non-inverting integrator is attained by connecting a resistor at port-X and terminating port-Z into a capacitor connected to ground (see Fig. 5.7a). The transfer function of both the circuits can be expressed as:

$$\frac{V_0}{V_{in}} = \pm\frac{1}{sC_1R_1} \tag{5.12}$$

where the + sign is applicable for CCII+ and negative sign corresponds to CCII−. Alternatively, a CCII+ can also be used to realize an inverting integrator by the circuit configuration shown in Fig. 5.7b.

A differential input integrator can be realized with the circuit arrangement shown in Fig. 5.8 which also uses only two CCIIs+ and a voltage follower but can be realized using only two AD844. The circuit is characterized by the following equation:

$$V_0 = \frac{1}{sC_1R_1}(V_1 - V_2) \tag{5.13}$$

In principle, from all the three circuits described above, the corresponding differentiator circuits can be obtained by replacing the resistor by a capacitor and vice-versa.

In current mode filter design, current mode oscillators and other current mode signal processing applications, there may be the requirement of a current mode integrator/differentiator. Three configurations to realize such integrators are shown in Fig. 5.9a–c.

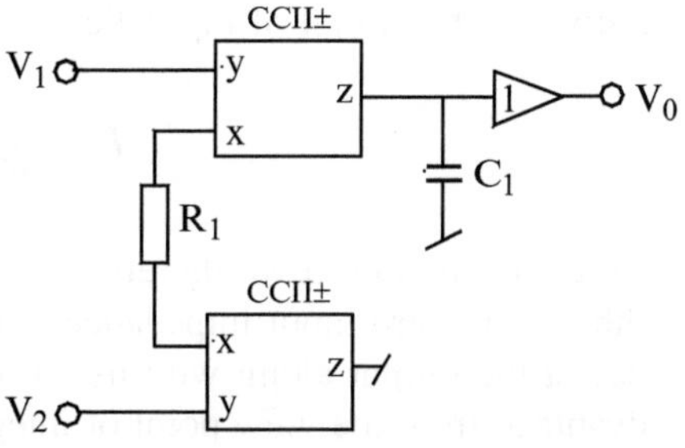

Fig. 5.8 Differential integrator

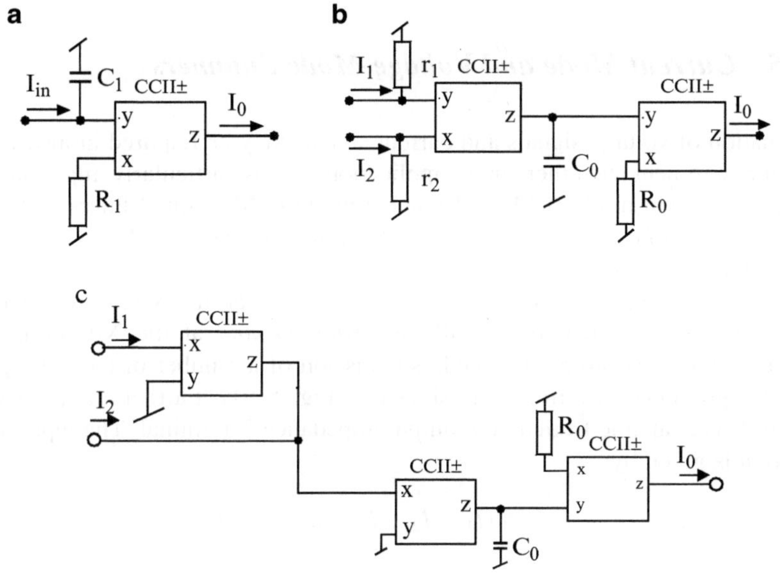

Fig. 5.9 Current-mode integrators (**a**) basic configuration (**b**) integrator with tunable time constants (**c**) differential integrator with zero input impedances at both inputs

By a straight forward analysis for the circuit of Fig. 5.9a, the current ratio transfer function is given by:

$$\frac{I_o}{I_{in}} = \frac{1}{sC_1R_1} \tag{5.14}$$

whereas for the circuit of Fig. 5.9b the output current is related to the two input currents by the following equation:

$$I_O = \frac{1}{sC_OR_O}\left(-I_2 + \frac{r_1}{r_2}I_1\right) \tag{5.15}$$

Lastly, for the circuit of Fig. 5.9c,

$$I_0 = \frac{1}{sC_0R_0}(I_1 - I_2) \tag{5.16}$$

The circuit of Fig. 5.9c, therefore, realizes a differential current mode integrator with ideally zero input impedance at its both input terminals, high output impedance at the output along with use of only grounded RC components which is an advantage from the view point of integrated circuit implementation.

5.2.5 Current-Mode and Voltage-Mode Summers

Summation of voltage signals and current signals may be required in instrumentation, measurement and other circuit applications, but is particularly important in the design of current-mode (CM) and voltage-mode (VM) biquad filters and also in many methods of oscillator synthesis. In the following, we describe a variety of CM and VM summer circuits.

Since CCII has ideally zero input impedance at its X-terminal when its Y-terminal is grounded, there will be *virtual ground* at the X-terminal. This makes it extremely simple to enable submission of a number of current inputs at virtually grounded X-terminal as shown in Fig. 5.10a. Furthermore, the output current I_0 is available from a high output impedance Z-terminal. The input -output equation is given by:

$$I_0 = \mp(I_1 + I_2 + I_3 + \ldots\ldots\ldots + I_n) \tag{5.17}$$

By adding one more CC it is possible to add **n** number of current inputs and subtract **m** number of current signals (see Fig. 5.10b) leading to the equation of kind

$$I_0 = \pm(I_1 + I_2 + I_3 + \ldots\ldots\ldots + I_n) \mp (i_1 + i_2 + i_3 + \ldots\ldots. + i_m) \tag{5.18}$$

A variety of other current mode and voltage mode summers are shown in Fig. 5.11a–c. The current output of the circuit Fig. 5.11a is given by:

$$I_0 = \pm\left(\frac{R_1}{R_{01}}\right)I_1 \pm \left(\frac{R_2}{R_{02}}\right)I_2 \pm \left(\frac{R_3}{R_{03}}\right)I_3 \pm \ldots\ldots\ldots. \pm \left(\frac{R_n}{R_{0n}}\right)I_n \tag{5.19}$$

For the circuit of Fig. 5.11b, the output current is given by

$$I_0 = \mp\left(\frac{R_{01}}{R_{02}}\right)\left\{\frac{V_1}{R_1} + \frac{V_2}{R_2} + \frac{V_3}{R_3} + \ldots. + \frac{V_n}{R_n}\right\} \tag{5.20}$$

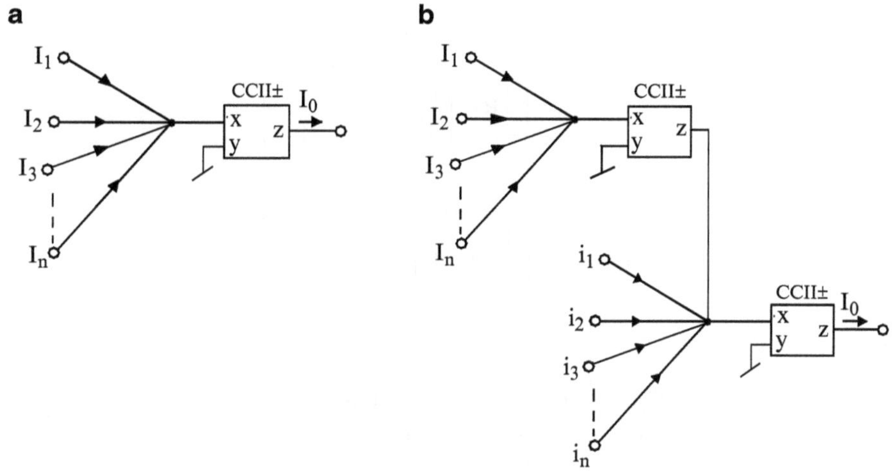

Fig. 5.10 Current-mode summers (**a**) single-CC adder (**b**) adder/subtractor

Lastly, for the circuit of Fig. 5.11c, the output voltage is given by

$$V_0 = \pm R_0 \left\{ \frac{V_1}{R_1} + \frac{V_2}{R_2} + \frac{V_3}{R_3} + \ldots + \frac{V_n}{R_n} \right\} \tag{5.21}$$

The circuit of Fig. 5.1c has the attractive feature of offering ideally infinite input impedance at all its input terminals, providing voltage output from a low impedance terminal and employing all grounded resistors.

5.2.6 Grounded Negative Impedance Converters

The negative impedance converter (NIC) is the basic 2-port active element which may be used to create negative circuit impedances. Although the use of negative impedances as employed in filters is handicapped by their associated notorious stability problems, there are several applications such as in oscillators, loss-compensation or frequency range extension where negative impedances can still be put to use effectively by judicious design.

In view of the above, a study of techniques and circuits to create negative elements using CCs is interesting and important. An NIC can be implemented by a CCI by using only its x and y ports; whereas it is realizable from a CCII+ by shorting its y and z terminals thereby leading to four different circuits to create a grounded negative impedance equal to $-Z_L$ (see Fig. 5.12).

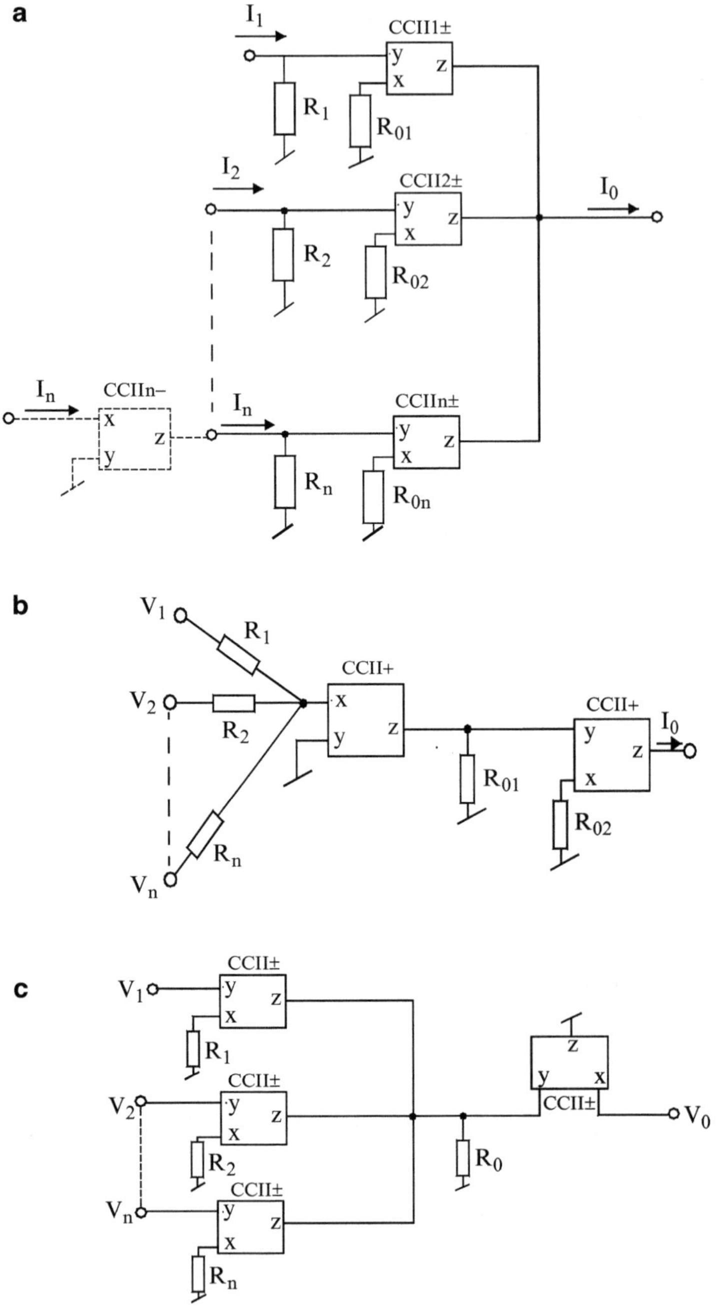

Fig. 5.11 Other varieties of current mode/voltage-mode summers

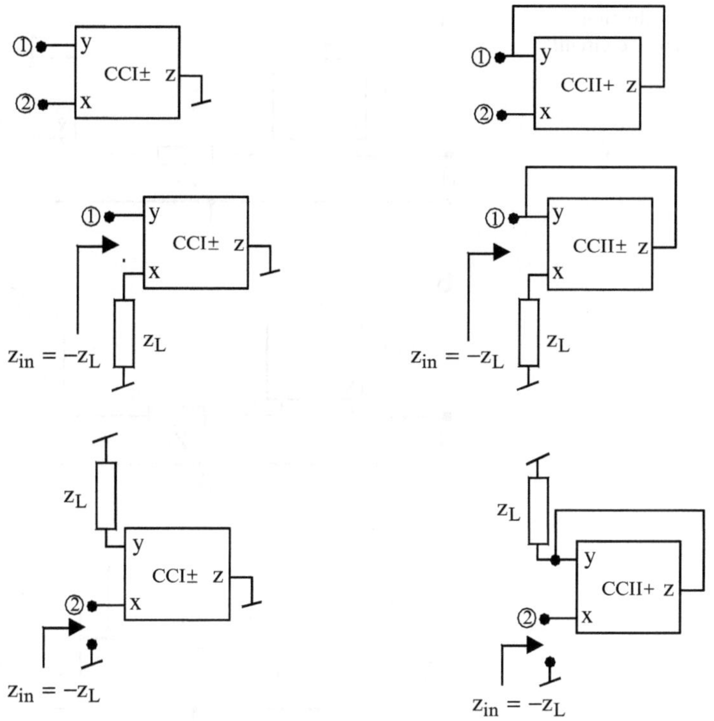

Fig. 5.12 Two ways of realizing an NIC leading to four different ways of creating negative impedance

For more elaborate designs of negative impedance converters employing more than one CC which exhibit a reduced effect of the parasitic impedances of the CCs, see [153].

5.2.7 Floating Negative Impedance Converters

The CCII-based circuits to realize floating negative impedances have been proposed in the literature by Paul, Dey and Patranabis in [38] and by Nandi in [50, 51, 53]. Here, we show how such circuits can be evolved systematically from first principles.

If floating negative impedance is required, the same can be obtained by the well-known method of putting the load impedance in between two NICs; all three elements being connected in cascade. Thus, in principle, four such circuits, each realizable with only two CCs, are possible which are shown in Fig. 5.13 out of which that of Fig. 5.13c has been reported by Nandi in [50] whereas that of Fig. 5.13d has been discussed by Nandi in [51].

Fig. 5.13 Four floating negative impedance circuits

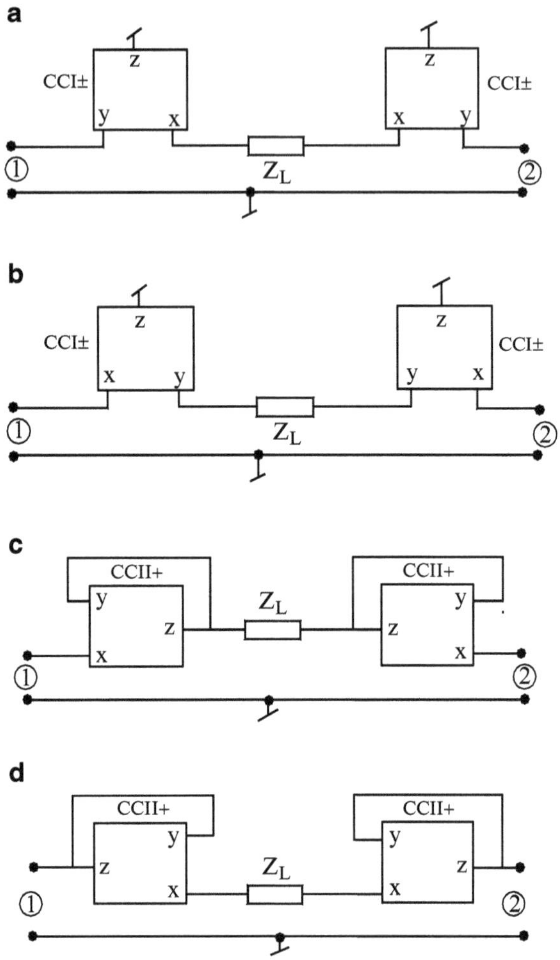

An interesting question which arises is whether two CCII-s can be used to realize floating negative impedance? The answer to this question is in the affirmative and two interesting ideas lead to the creation of such circuits.

Firstly, there are two classical nullor-based lattice structures which realize floating negative impedance. These are shown in Fig. 5.14. Note that the circuit of Fig. 5.14a is the current inversion type floating NIC (INIC) while that of Fig. 5.14b is the voltage inversion type NIC (VNIC). Secondly, if one observes the terminal equations of a CCII- i.e. $i_y = 0$, $v_x = v_y$ and $i_z = -i_x$ (see Fig. 5.15a) and the terminal equations of a three terminal nullor, it is seen that these are exactly equivalent to a CCII− (see Fig. 5.15b).

At this point, it must be emphasized that after the introduction of CCs in 1968 and 1970, it was not until 1978 onwards when first De Jager and Smit in [144] and

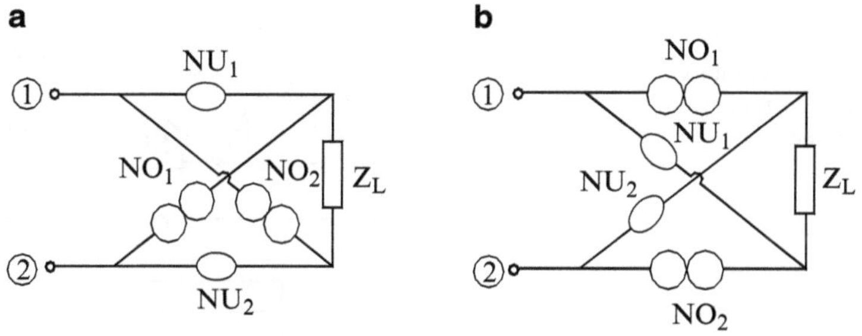

Fig. 5.14 Floating INIC and VNIC circuits using nullors

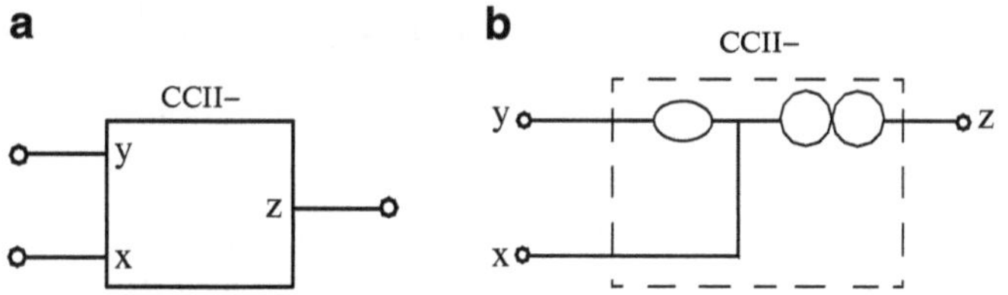

Fig. 5.15 The equivalence between a CCII- and three-terminal-floating-nullor

later Senani[1] in [56] and Higashimura and Fukui in [52] *explicitly* employed the 'CCII−: nullor' equivalence in the analysis and synthesis of CCII-based circuits. From the floating negative impedance structures of Fig. 5.14 and the 'CCII−: nullor' equivalence of Fig. 5.15, it, therefore, appears that four different floating negative impedance circuits should be possible to be evolved as follows: (i) from the nullor model of Fig. 5.14a, the first two-CCII− based circuit follows by combining nullator-norator pair NU_1-NO_1 to form the first CCII− and combining NU_2-NO_2 to form the second CCII− (ii) an alternative circuit is then made by grouping NU_1-NO_2 to form the first CCII− and combining Nu_2-NO_1 to form the second CCII−. In a similar way, two other floating negative impedance structures, each using two CCII-s, can also be conceived from the nullor model of Fig. 5.14b by proceeding exactly in the same way as outlined in steps (i) and (ii) above. A close examination, however, reveals that, in this set of four circuits, the last two turn out to be exactly same as the first two! Thus, only two distinctly different circuits emerge from the above procedure and these are shown in Fig. 5.16. The circuit of Fig. 6.16 has been described by Paul et al. in [38].

[1] The first author (RS) cannot resist commenting that the facts that a composite connection of two CCs can be used to realize a fully floating four-terminal-floating-nullor (FTFN) and that FTFNs make it possible to convert an op-amp-based grounded impedance network into a floating impedance simulation network systematically were *first proposed explicitly* in references [145] and [146]. Both of these ideas, now looking backwards in time, might appear simple and obvious, but were not well-understood or well-recognized in the open technical literature till then.

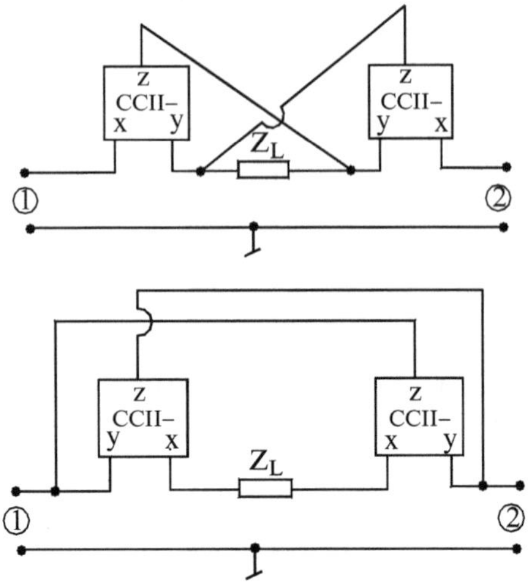

Fig. 5.16 Two distinctly different floating negative impedance configurations using CCII− emerging from the nullor models of Fig. 5.14

It may be mentioned that the four CCII+ based structures of Fig. 5.13 as well as the two CCII− based structures of Fig. 5.16, all are characterized by the terminal equations: $i_1 = - i_2 = (-1/Z_L)$ $[V_1\text{-}V_2]$ and represent floating impedance $-Z_L$, between ports 1 and 2.

In the last, it is also worthwhile to mention that four electronically-controllable negative resistance converters, each employing two electronically- tunable CCIIs, have been advanced by Surakampontorn and Thitimajshima in [55]. Another scheme of realizing floating current controlled resistance converter using electronically controlled conveyor (ECC) has been reported by Riewruja and Petchmaneelumka in [126].

5.2.8 Generalized Function Generator

In their seminal 1970 paper where the second generation current conveyor was introduced (see reference [3] of Chap. 1) Sedra-Smith demonstrated how CCIIs along with ideal diodes can be employed to realize a generalized function generator. Their proposition is shown here in Fig. 5.17

Assuming the diodes to be characterized by the equation

$$i_d = I_s e^{v_d/v_T} \tag{5.22}$$

a straight forward analysis of the circuit, assuming all matched diodes, leads to the following output equation:

$$I_0 = i_1^{m_1} i_2^{m_2} \ldots\ldots i_n^{m_n} \tag{5.23}$$

$$\text{where } m_j = \pm\left(R_0/R_j\right) \text{ with } m_1 + m_2 + \ldots. + m_n = 1 \tag{5.24}$$

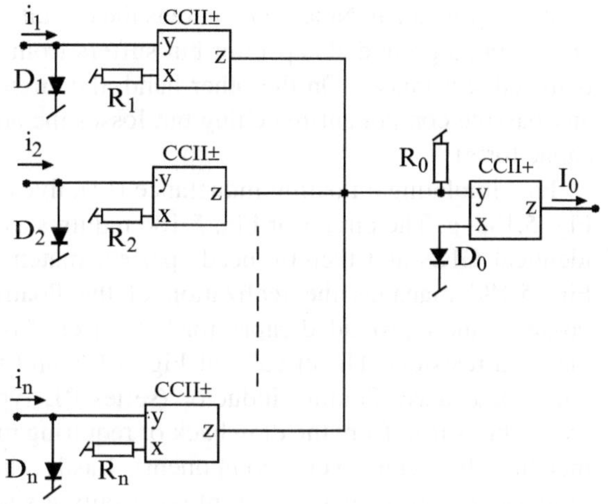

Fig. 5.17 The generalized non-linear function generator proposed by Sedra and Smith (See reference [3] of Chap. 1)

From this generalized configuration, a number of interesting functions can be obtained by judicious choice of the number and polarity of the current conveyors, for instance, choosing $n = 3$, $m_1 = +1$, $m_2 = +1$ and $m_3 = -1$, the resulting circuit, consisting of four current conveyors, will yield the following equation:

$$I_0 = \frac{I_1 I_2}{I_3} \tag{5.25}$$

and thus, the realization of a current mode multiplier/divider has been achieved.

5.3 Methods and Circuits for Simulating Inductors, FDNRs and Related Elements

Among many other applications of CCs evolved so far, their use in simulating inductors, FDNRs and related elements has been one of the most prominent and significant one due to the fact that the type of circuits for realizing these elements as possible with CCs have never been possible to be made from other prevalent building blocks most notably the ubiquitous IC op-amp. In view of this, before embarking upon CC-based impedance simulation networks, it appears necessary to take a quick view of a number of popular op-amp-based impedance simulation networks such as those in [1–3, 6–8, 11, 23, 46, 60] and in the references cited therein to be able to appreciate the advantages of CC-based circuits in the right perspective.

In Fig. 5.18 we display some popular op-amp-based inductance/general impedance simulation networks. The circuit of Fig. 5.18a shows how a lossless grounded inductor (GI) based upon the gyrator method can be simulated from two op-amps whereas the circuit of Fig. 5.18b shows how the Antoniou's generalized impedance convertor (GIC) made from two op-amps and five impedances can realize a lossless floating inductance. Note that whereas the circuit of Fig. 5.18a has the advantage of employing a grounded capacitor but suffers from the drawback of requiring seven matched resistances. On the other hand, the circuit of Fig. 5.18b does not require any passive component-matching but losses the advantage of employing grounded capacitor(s).

For simulating a floating inductance (FI), five exemplary solutions are shown in Fig. 5.18c–g. The circuit of Fig. 5.18c requires as many as four op-amps with two identical GICs and thereby needs perfect matching between them; the circuit of Fig. 5.18d manages the realization of the floating inductance using only three op-amps and a grounded capacitor but suffers from the drawback of requiring ten matched resistors. The circuits of Fig. 5.18e and f each require only two op-amps but realize lossy floating inductor (series-RL type and parallel-RL type respectively) but suffer from the drawback of requiring more than one capacitor as well as matching between several components. Lastly, the circuit of Fig. 5.18g realizes a lossless floating inductance employing only a single capacitor but still requires a number of conditions in terms of various resistances to be fulfilled to enable the required realization.

From the exhaustive literature survey of op-amp based inductance simulation circuits, it has been well established that using op-amps no circuit exists which can realize a lossless grounded inductor using a bare minimum number of passive components (namely only two resistors and a grounded capacitor) and that using op-amps, it is impossible to simulate a floating impedance of any kind (lossy or lossless) without requiring any passive component-matching.

In the following sub-section, we will show that using CCs it is possible to devise synthetic impedance networks which can overcome both the difficulties mentioned above.

5.3.1 CCII-Based Lossless Grounded Inductance Simulation Circuits

Sedra and Smith (see reference [3] of Chap. 1) while introducing the second generation current conveyor had outlined a number of basic applications of CCII± including the interesting application of CCII in realizing PII/gyrator. It was shown that a gyrator can be made from a parallel back-to-back connection of two voltage controlled current sources of opposite polarity, each realizable from a single CCII+ or single CCII− with X-port terminated into a grounded resistor or a grounded impedance. With port 2 of such a circuit terminated into third impedance (as in Fig. 5.19a) the input impedance of the circuit is given by:

$$Z_{in} = \frac{Z_1 Z_2}{Z_3} \tag{5.26}$$

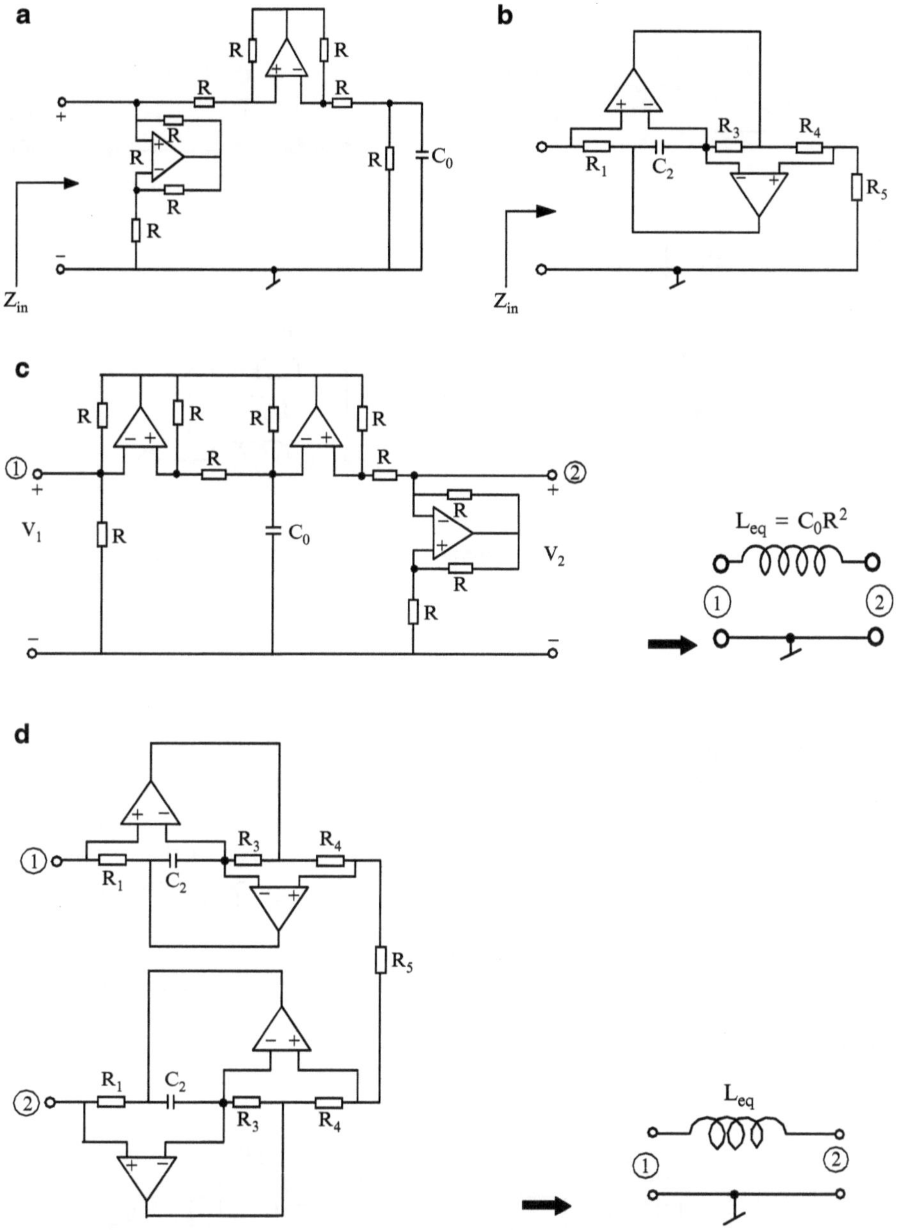

Fig. 5.18 Some inductance simulation networks employing IC op-amps (**a**) Gyrator-based lossless GI, $Z_{in} = sC_0R^2$ [1] (**b**) Antoniou's [1] GIC-based grounded inductance simulator, $Z_{in} = sC_2R_1R_3R_5/R_4$ (**c**) Three-op-amp lossless FI due to Glover, $Z_{1\text{-}2} = sC_0R^2$ [6] (**d**) Two-GIC-based FI simulator, $Z_{1\text{-}2} = sC_2R_1R_3R_5/R_4$ (**e**) Dutta Roy [2] and Wise [3] method of FI simulation using unity gain amplifiers, $Z(s) = 2(R_1 + R_2) + 2sC_1R_1R_2$ (**f**) A lossy tunable parallel-RL type floating inductance simulator derived from lossless FI of Sudo and Teramoto [8] (**g**) A lossless FI from *The* and *Yanagisawa's* proposition [7] with $R_6 = R_7 = R_8 = R$; $R_1 = R_2 = R_4 = 2R$; $R_3 = 2R/3$; $R_5 = 10R/7$ and $R_9 = R/5$; the simulated floating inductance has a value $L_{eq} = C_0 R^2$

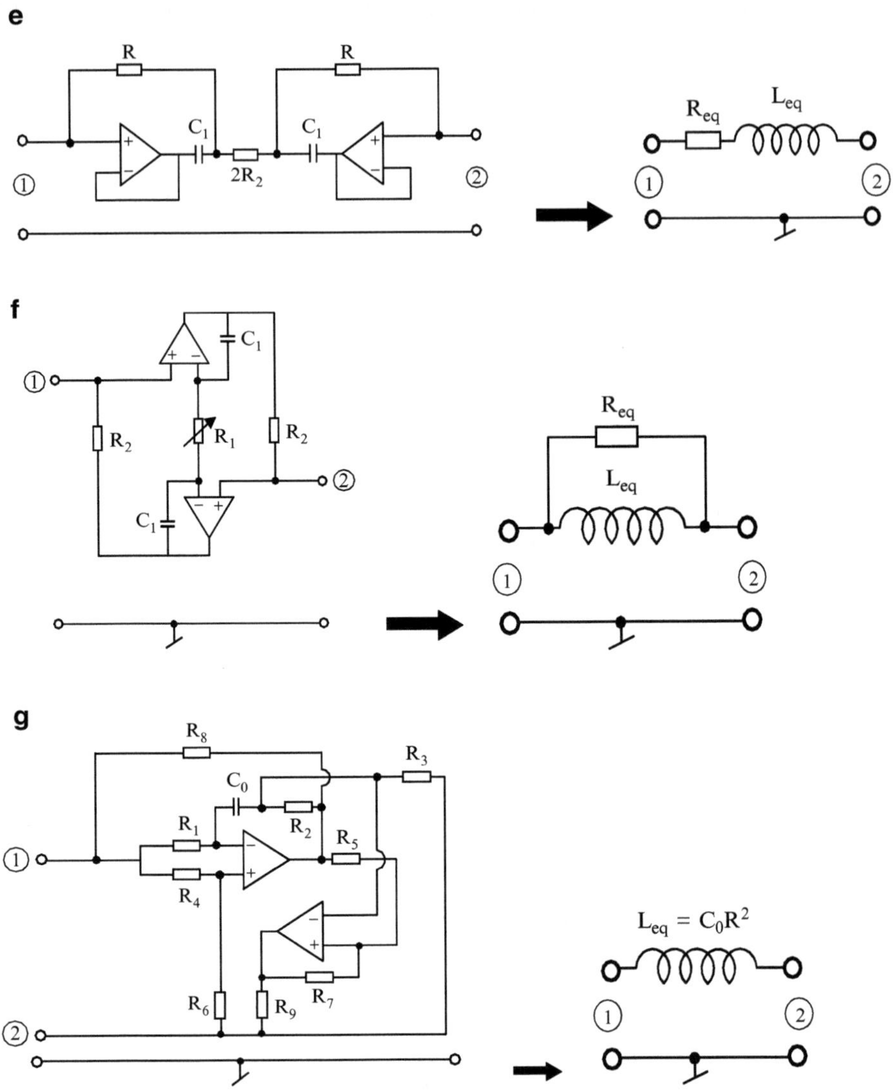

Fig. 5.18 (continued)

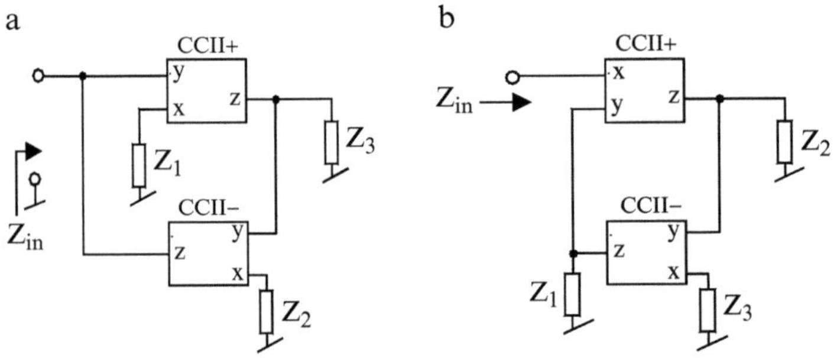

Fig. 5.19 Circuits for simulating a lossless grounded inductor using CC-based gyrators

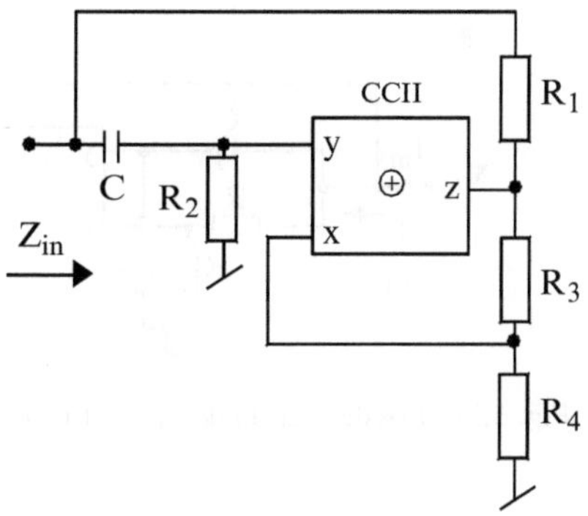

Fig. 5.20 Active gyrator proposed by Soliman [9] (Also by Paul et al. [33])

It is, thus, clear that with Z_3 as a capacitor C_0 and the remaining impedances being resistors R_1 and R_2, the input impedance would be purely inductive with equivalent simulated inductance being given by $L_{eq} = C_0R_1R_2$. An alternative circuit which realizes exactly the same type of input impedance while using all the three grounded components only is shown in Fig. 5.19b. This alternative form of the active gyrator was recently studied by Cicekoglu [89] considering the voltage and current tracking errors of the CCIIs. It has been found that this alternative version does offer better performance than the Sedra-Smith gyrator of Fig. 5.19a. Several other forms of lossless grounded inductors using current conveyors have been advanced from time to time such as [24, 34, 93, 105, 106], however, these invariable require more than two CCs.

It is worth pointing out that grounded ideal FDNRs can be realized from either of the circuits of Fig. 5.19 by choosing Z_1 and Z_2 as capacitors and Z_3 as a resistor. Furthermore, if both the CCII are taken as CCII+, then a frequency dependent positive resistance (FDPR: an element having $Z(s) = -1/Ds^2$) is realizable as in [18].

Lastly, it may be pointed out that instead of one CCI+ and one CCII−, an entirely CCII + based circuit (thereby employing three CCII+ s) can be easily obtained by realizing a CCII− by two CCII+ s. In addition, there are a number of other ways of deriving such three CCII+ based lossless GI circuits from those of Fig. 5.19 as elaborated in [102].

5.3.2 *Active Gyrator Using a Single CCII*

The *active gyrators* using CCs have been proposed by many authors [63, 65, 69, 70, 85, 114, 116, 121, 122, 125, 128, 129], however, the first circuit to realize an active gyrator and thereby a lossless inductance was proposed by Soliman in [9] (and also *independently* by Paul-Dey-Patranabis in [33]) and is shown in Fig. 5.20.

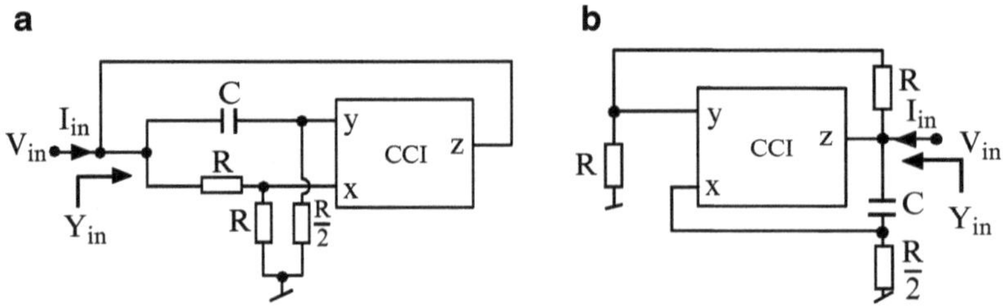

Fig. 5.21 Lossless grounded inductors using CCI

The circuit can realize a lossless grounded inductor or a lossy inductance in the form of series RL or parallel RL subject to the fulfillment of appropriate conditions. The input impedance of the circuit is given by:

$$Z_{in} = \frac{R_1R_2 + R_2(R_3 + R_4) - R_1R_4 - sC_1R_1R_2R_4}{sC\{(2R_2R_3) - R_1R_4\} - R_4} \tag{5.27}$$

Subject to fulfillment to the following conditions:

$$2R_2R_3 = R_1R_4 \; and \; R_2(R_1 + R_3 + R_4) = R_1R_4 \tag{5.28}$$

the input impedance becomes purely inductive and is given by:

$$Z_{in} = \frac{sCR_1R_2R_4}{R_4} = sCR_1R_2 \tag{5.29}$$

Recently, Arslan, Cam and Cicekoglu in [105] have presented two new lossless grounded inductance simulators using a single first generation current conveyor (CCI). The circuits although employ a floating capacitor, like that in Soliman's circuit [9], but require one less resistor. These circuits are shown in Fig. 5.21.

With the component values shown, both the circuits realize $Z_{in} = sCR^2/3$. The practical workability of these circuits has been successfully demonstrated [105] in the realization of a third order high pass Butterworth filter with a CMOS CCI for cut off frequency of the order of 106 KHz.

5.3.3 *Single CCII-Based Low-Component-Count Grounded Impedance Simulators*

As compared to lossless grounded inductance, a lossy inductor can often be simulated with a canonical number of components typically using only one CC, two resistors and a capacitor. A number of such circuits have been proposed in

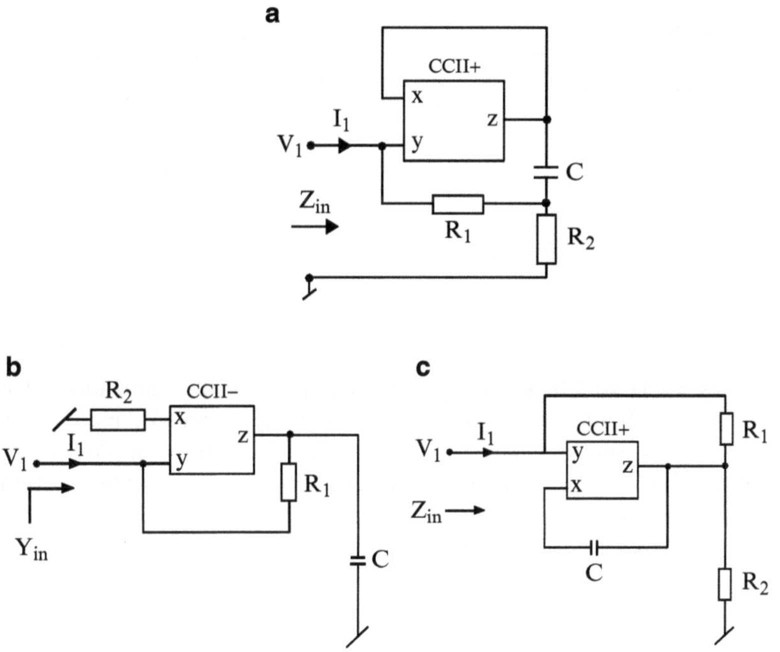

Fig. 5.22 Low-component-count lossy inductance simulation circuits proposed by Nandi (**a**) and (**b**) from [5] (**c**) from [12]

technical literature by a number of researchers. Some of them are almost replicas of the well-known op-amp-based circuits. Some such circuits are shown in Fig. 5.22 which were proposed by Nandi in [5] and [12].

In the first circuit (Fig. 5.22a), the CCII+ can be readily identified to be configured as a unity gain voltage follower thereby making it clear that the circuit is, in fact, a CC-version of the well-known Prestcott circuit based on an op-amp [147]. The circuit simulates a series-RL type grounded inductor with $L_{eq} = CR_1R_2$ and equivalent resistance $R_{eq} = (R_1 + R_2)$.

On the other hand, the circuit Fig. 5.22 b although simulates bilinear inductor with a series combination of a resistor R_a and a parallel-RL consisting of resistive part R_b and inductive part L_{eq} with their values given by $L_{eq} = CR_2(R_1 - R_2)$, $R_a = R_2$ and $R_b = (R_1 - R_2)$, subject to $R_1 > R_2$. This circuit can be considered to be a CC-analog of Berndt-Dutta Roy circuit [167] but has the advantage of employing a grounded capacitor over Berndt-Dutta Roy circuit [167] which employs a floating capacitance.

In the circuit of Fig. 5.22 c exactly the same number of components have been employed as in the earlier two circuits while this circuit also simulates series-RL type grounded inductor with $R_{eq} = (R_1 + R_2)$ but equivalent inductance is given by $L_{eq} = 2CR_1R_2$.

Figure 5.23 shows the generalized form of another single-CC configuration which was proposed by Soliman [10] to realize the Ford-Girling [148] equivalent

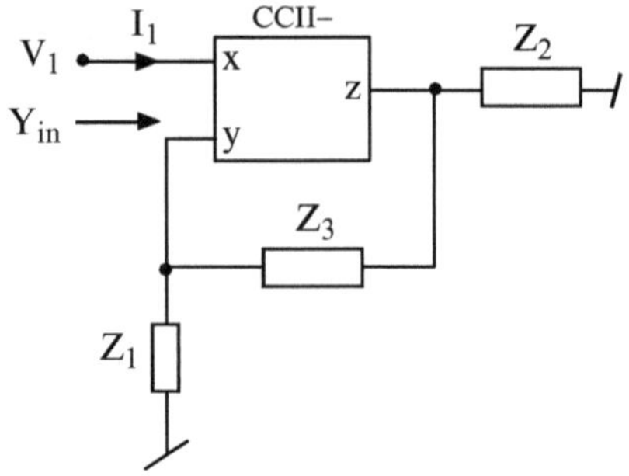

Fig. 5.23 Generalized form of Soliman's CCII-based Ford-Girling equivalent FDNR simulator [10]

circuit using CCII which, in the special case, permits the realization of a lossy FDNR employing both grounded capacitors if the impedances Z_1 and Z_2 are chosen as capacitors with the impedance Z_3 being a resistor. This non-ideal FDNR was shown [10] to be a useful implementation for realizing low pass filter sections with transmission zero in the open left half s-plane.

For other CC-based lossy grounded inductance simulators or non-ideal FDNR/FDNC simulators, employing one or two CCs, the readers may refer to other works [17, 21, 22, 31, 58, 59, 78, 94, 95, 96–98, 102].

5.3.4 Floating Impedance Realization Without any Component-Matching Constraints

There have been numerous investigations and studies on simulating floating lossless inductors using CCs such as [16, 19, 26, 36, 68, 92, 101, 103, 104, 107, 119, 127, 131, 132, 136, 137, 150] as well as on realizing V-I scalars/floating capacitances/floating capacitance multipliers, such as [90, 109, 134, 135].

In this subsection, however, we deal with some exemplary circuits [16, 19] and [36] from this repertoire which are of specific significance, as will be demonstrated in the following.

In this context, it must be mentioned that for simulating a lossless floating inductance either one can use CCII-based gyrators or CCII-based mutators (such as those in 4, 64, 100, 139, 141]); however, in both the cases, four or more CCs will be required. Furthermore, since these methods employ two identical gyrators/mutators, the lossless floating inductance is realizable only subject to fulfillment of some prescribed component-matching conditions. Obviously, any mismatch in the component values would, therefore, lead to inevitable degradation of the simulating inductance resulting in appearance of undesirable parasitic components in its equivalent circuit, besides the intended inductance.

The most desirable circuit for FI simulation will be the one which does not require any component matching. In this context however, it must be mentioned that it had been conclusively proved that using op-amps such a circuit was impossible to be devised.

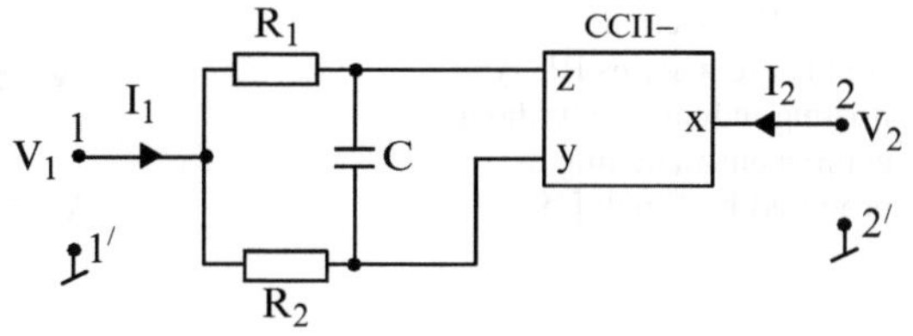

Fig. 5.24 Senani's lossy floating inductance simulator *without any component-matching* [16]

In [16], Senani demonstrated, *for the first time ever*, as to how a lossy FI (parallel RL-type) can be realized by using only one CCII- along with only three passive components (only two resistors and a capacitor) *without requiring any component matching condition*. This circuit is shown in Fig. 5.24.

By straight forward analysis, assuming to CCII− to be characterized by $i_y = 0$, $v_x = h_{21}\, v_y$ and $i_z = -h_{32}\, i_x$ where nominally $h_{21} = 1 = h_{32}$, the non-ideal Y-matrix of the circuit is found to be:

$$\left[Y'\right] = \left(\frac{1}{R_1} + \frac{1}{R_2}\right) + \frac{1}{sCR_1R_2}\begin{bmatrix} 1 & -\dfrac{1}{h_{21}} \\ -\dfrac{1}{h_{32}} & \dfrac{1}{h_{21}h_{32}} \end{bmatrix} \tag{5.30}$$

From the above, it can be seen that for $h_{21} = 1 = h_{32}$, the circuit simulates parallel RL floating impedance with R_{eq} and L_{eq} given by:

$$R_{eq} = R_1R_2/(R_1 + R_2),\ L_{eq} = CR_1R_2 \tag{5.31}$$

From the above [Y] matrix, it is easy to deduce that the circuit enjoys very low active and passive sensitivities all of which can be found to be belonging to the range

$$0 \le \left\| S_{xj}^{F} \right\| \le 1 \tag{5.32}$$

It is interesting to point out that from the above circuit, a grounded inductor circuit can be obtained in two different ways.

(a) If terminals 2-2′are short-circuited a grounded inductor is realized between 1-1′with equivalent inductance and resistance given by the same values as shown above. However, since the non-ideal driving point admittance of the circuit is given by:

$$Y_{11}'(s) = \left(\frac{1}{R_1} + \frac{1}{R_2}\right) + \frac{1}{sCR_1R_2} = \mathrm{Y_{11}(s)ideal} \tag{5.33}$$

The above equation implies that

$$S_{h_{21}}^{L_{eq}} = S_{h_{32}}^{L_{eq}} = S_{h_{21}}^{R_{eq}} = S_{h_{32}}^{R_{eq}} = 0 \tag{5.34}$$

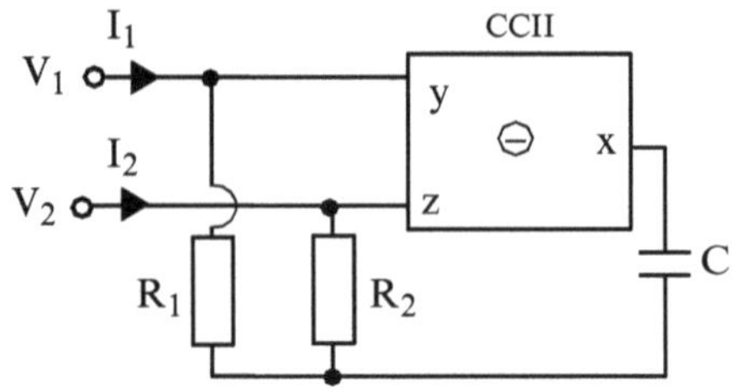

Fig. 5.25 A circuit for simulating a series RL type floating inductance without component-matching as proposed by Singh [19]

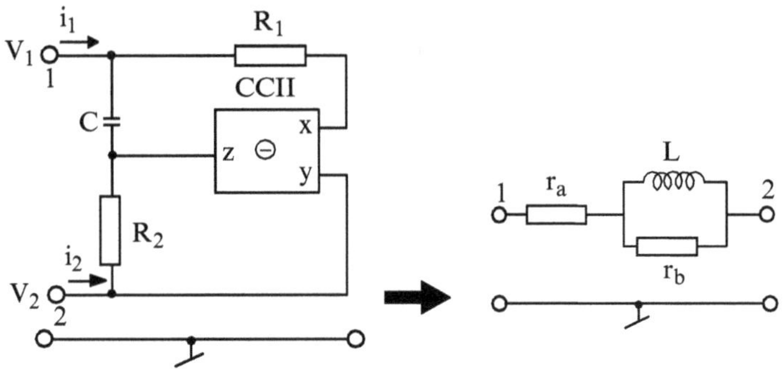

Fig. 5.26 Bilinear floating inductor proposed by Nandi and Nandi [36]

Thus, this grounded inductor circuit has the attractive feature that it is completely insensitive to active parameter variations.

(b) If instead, terminals 1-1′are short-circuited one obtains Soliman's Ford-Girling equivalent circuit realization [10] with equivalent inductance and resistance still given by the same values. Soliman's circuit [10] is more advantageous in realizing an FDNR in parallel with a capacitance, owing to the use of both grounded capacitors which is a preferable feature from the point of view of integrated circuit implementation.

At about the same time, Singh [19] reported a single-CC circuit independently which employs exactly the same number of active and passive components but simulates a floating series RL impedance without requiring any component-matching condition. This circuit from [19] is shown in Fig. 5.25.

The equivalent resistance (R_{eq}) and inductance (L_{eq}) for this circuit are given by $R_{eq} = (R_1 + R_2)$, $L_{eq} = CR_1R_2$.

A non-ideal analysis of this circuit, however, shows that the non-ideal parameters of this circuit are quite complex in contrast to the circuit of Fig. 5.24 and as a consequence, all the sensitivity coefficients of this circuit *are not low*.

In 1983 (i.e. exactly 4 years after the introduction of the single CCs FI by Senani in [16] and Singh in [19]) Nandi and Nandi in [36] reported another single CCII-based circuit with similar property but their circuit could realize only a bilinear floating inductor with $r_a = R_1$, $r_b = (R_2 - R_1)$ and $L = CR_1R_2$. This circuit is shown in Fig. 5.26.

5.3.5 Floating Generalized Impedance Converters/Inverters (GIC/GII)

Floating GIC/GII constitute a more general class of active elements which are useful for creating a variety of floating impedances and a number of CC-based configurations to realize them have so far been known, for instance, see [35, 43, 52, 87, 104, 108, 113].

Higashimura and Fukui [52] presented a number of circuits for floating immittance simulation using as many as four CCs and as many as four admittances out of which two were required to be identical.

Their circuits are shown here in Fig. 5.27a–c. All the circuits realize a floating admittance given by:

$$Y_{in} = \frac{Y_1 Y_3}{Y_2} \tag{5.35}$$

By the judicious choice of the components (resistive/capacitive), the circuits can be used to realize either lossless FIs or lossless FDNR but the drawback of requiring two matched admittances might be a deterrent in many practical applications.

Another circuit, which although employs four CCs but has the advantage of employing all CCII+ only, was proposed by Yuce et al. [113] and is shown in Fig. 5.28. This circuit is characterized by

$$[Y] = \left(\frac{y_1 y_2}{y_3} + \frac{y_1 y_2}{y_4}\right)\begin{bmatrix} 1 & -1 \\ -1 & 1 \end{bmatrix} = Y_{eq}\begin{bmatrix} 1 & -1 \\ -1 & 1 \end{bmatrix} \tag{5.36}$$

Although the circuit realization does not depend upon matching of component values but two capacitors would be needed to simulate a lossless FI although both the capacitors can be grounded. On the other hand, in the case of floating FDNR simulation, the two capacitors employed will be both floating.

A practical advantage of this circuit is that it is readily implementable using AD844 due to the employment of only CCII+ s.

Another four-CC-based generalized floating impedance convertor was advanced by Toumazou and Lidgey [43] which is based upon an appropriate connection of two differential voltage controlled current sources (Fig. 5.29).

By straight forward analysis, it can be easily verified that the floating impedance realized by this circuit is characterized by:

$$i_1 = -i_2 = \frac{V_1 - V_2}{Z_{eq}}, Z_{eq} = \frac{Z_1 Z_4}{(Z_2 + Z_3)} \tag{5.37}$$

Yet another four-CC-based circuit which does not require two matched impedances like the circuits of Higashimura and Fukui [52] and does not need four impedances like the circuit of Yuce et al. [113] as well as the circuit of Toumazou and Lidgey [43], is shown in Fig. 5.30 and was proposed by Senani in [35].

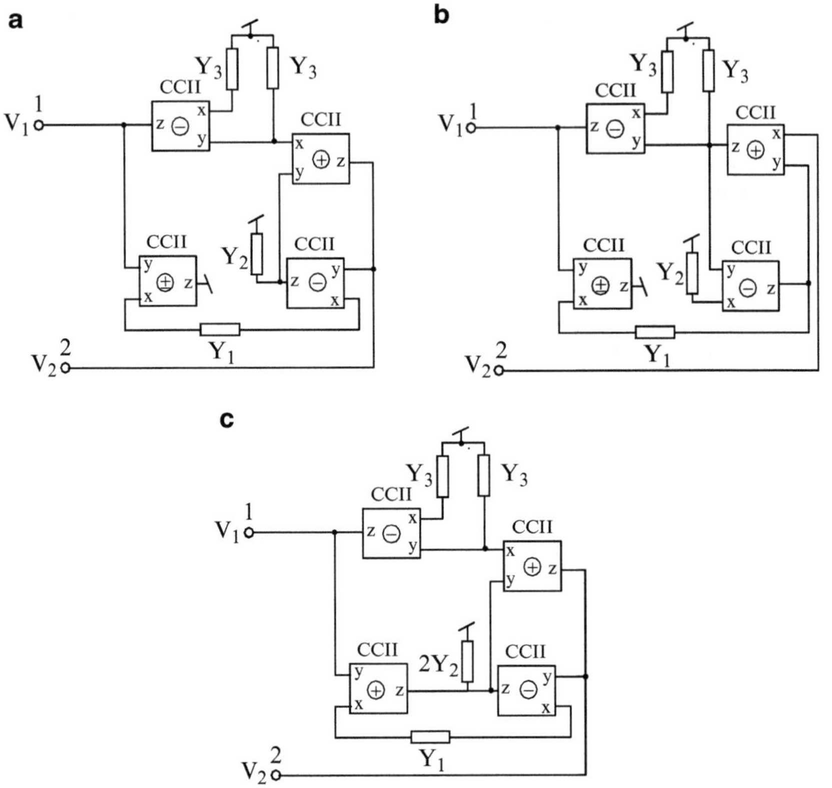

Fig. 5.27 General floating impedance simulation circuits presented by Higashimura and Fukui [52]

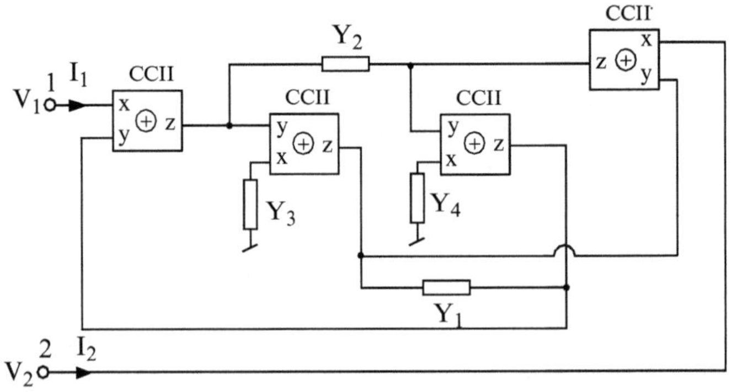

Fig. 5.28 A general floating impedance simulation circuit proposed by Yuce et al. [113]

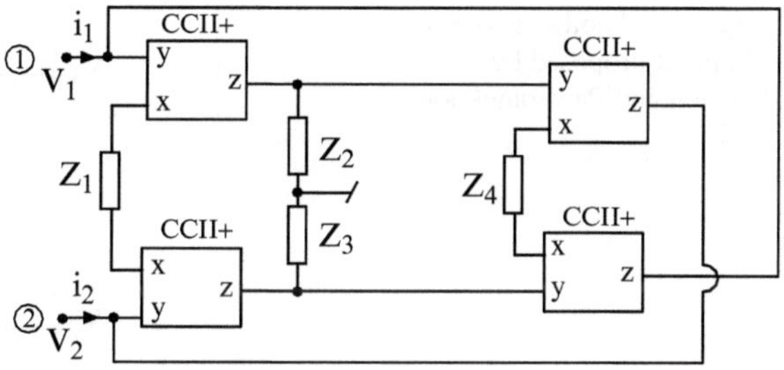

Fig. 5.29 A generalized floating impedance realization proposed by Toumazou and Lidgey [43]

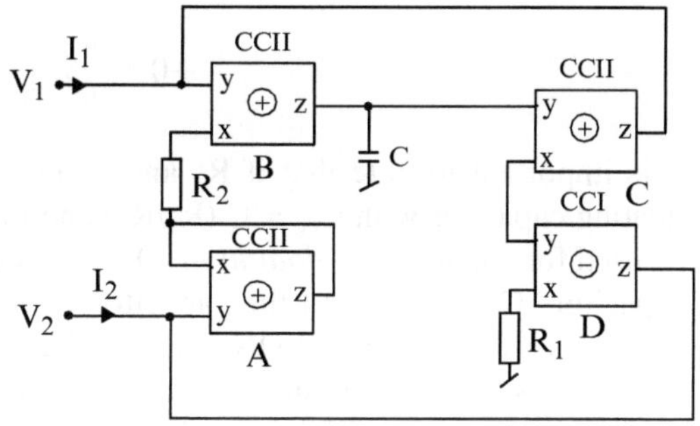

Fig. 5.30 The lossless floating inductor proposed by Senani [35]

Assuming CCII and CCI to be characterized by the following hybrid matrices:

$$\begin{bmatrix} i_y \\ v_x \\ i_z \end{bmatrix} = \begin{bmatrix} 0 & 0 & 0 \\ h_{21} & 0 & 0 \\ 0 & h_{32} & 0 \end{bmatrix} \begin{bmatrix} v_y \\ i_x \\ v_z \end{bmatrix} \tag{5.38}$$

$$\begin{bmatrix} i_y \\ v_x \\ i_z \end{bmatrix} = \begin{bmatrix} 0 & h_{12} & 0 \\ h_{21} & 0 & 0 \\ 0 & h_{32} & 0 \end{bmatrix} \begin{bmatrix} v_y \\ i_x \\ v_z \end{bmatrix} \tag{5.39}$$

the non-ideal Y- matrix of the circuit is found to be:

$$[Y] = \frac{h_{32C}h_{12D}h_{21D}h_{21C}h_{32B}}{sCR_1R_2} \begin{bmatrix} +h_{21B} & -h_{21A} \\ \dfrac{-h_{21B}h_{32D}}{h_{32C}h_{12D}} & \dfrac{+h_{21A}h_{32D}}{h_{32C}h_{12D}} \end{bmatrix} \tag{5.40}$$

From the above matrix, it can easily be seen that even non-ideally; all the y-parameters of the circuit still represent lossless inductance i.e. the basic nature of the Y parameters of the FI remains unchanged even non-ideally. Furthermore, from the non-ideal matrix above, it can be easily confirmed that the circuit also enjoys very low active and passive sensitivities all of which are confined to the range

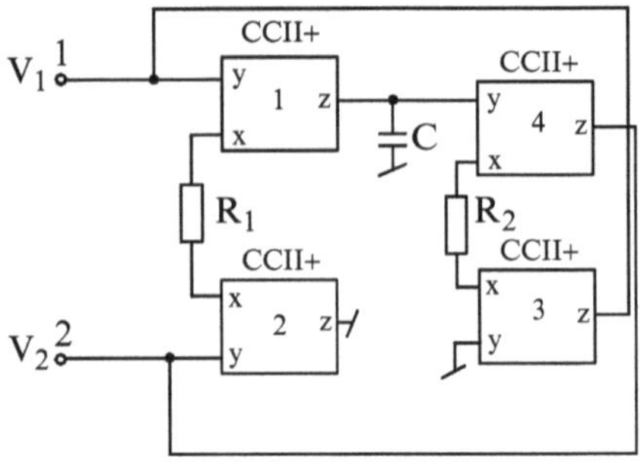

Fig. 5.31 Another lossless FI circuit proposed by Kiranon and Pawarangkoon [87]

$$-0 \leq \left\| S_{xj}^{F} \right\| \leq 1 \tag{5.41}$$

It is important to note that if R_1 and C are interchanged, the circuit simulates a floating capacitor with $C_{eq} = C\ (R_1/R_2)$ and therefore, in this mode, the circuit can be used for *capacitance-floatation.* On the other hand, if R_1 and R_2 are replaced by capacitors C_1 and C_2 and the capacitor C is replaced by a resistor R, the circuit simulates a ideal floating FDNR ($Z(s) = 1/D_{eq}s^2$) with D_{eq} given by $D_{eq} = C_1C_2R$.

Whereas Senani's circuit [35] employs three CCII+ as well as one CCI+, it was shown by Kiranon and Pawarangkoon [87] that the same kind of lossless floating inductance can also be realized by a circuit comprised of all CCII+ only along with only two resistors and a grounded capacitor as shown here the Fig. 5.31.

This circuit between ports 1 and 2, simulates a lossless floating inductance $L_{eq} = CR_1R_2$. For a systematic derivation of number of other current-mode lossless floating inductance circuits using CCs, the reader is referred to [84, 86, 88, 117].

A more versatile floating generalized impedance convertor employing as many as five CCII+ s along with five general impedances was proposed by Pal in [108] which simulates a floating impedance given by $Z_{1-2} = Z_1Z_3Z_5/Z_2Z_4$ and is shown in Fig. 5.32. Note that the nature of this expression is identical to the one encountered in the classical Antoniou's GIC [1] (which, however, realizes a grounded impedance of the same kind). It is obvious that by judicious choice (resistive/capacitive) of the five impedances, the circuit can be used to realize a floating inductance, floating capacitance and floating FDNR, as follows:

Floating inductor is obtained by choosing Z_4 as a capacitor and all the remaining impedances as resistors thereby leading to $Z_{1-2} = sC_4\ R_1R_3R_5/R_2$. Similarly, choosing Z_3 as capacitance, the circuit can be seen to be a grounded capacitor to floating capacitance convertor with the value of the floating capacitance given by $C_{1-2} = C_3\ R_1R_5/R_2R_4$. Finally, with Z_2 and Z_4 chosen as capacitors, the input impedance becomes $Z_{1-2} = s^2C_2C_4R_1R_3R_5$ and the circuit, therefore, simulates a floating frequency dependent negative conductance (FDNC).

A method for compensating for the series losses of the simulated floating inductor employing four CCs was presented by Ferri et al. in [104] which involved addition of two more current conveyors along with two resistors to the structure. The scheme proposed by them is shown here in Fig. 5.33.

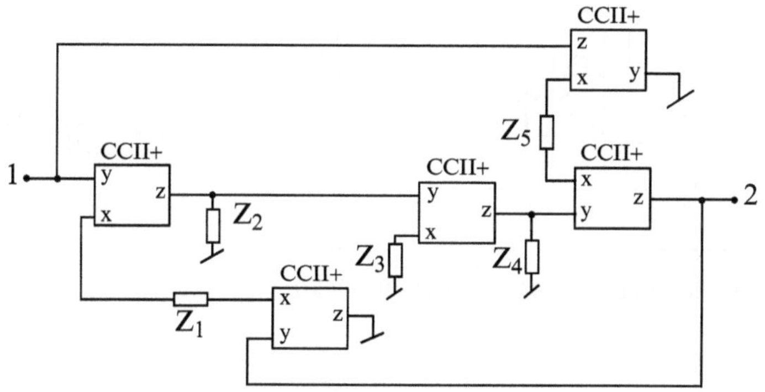

Fig. 5.32 A floating generalized impedance simulator due to Pal [108]

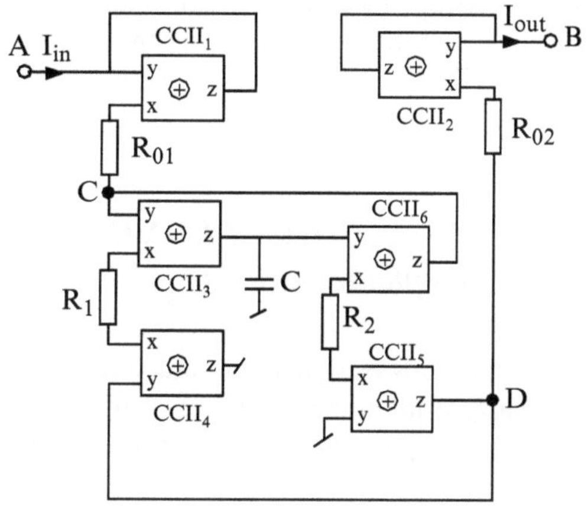

Fig. 5.33 A method for compensating for the series losses of the simulated floating inductor proposed by Ferri et al. [104]

Considering the various parasitic impedances of the CCIIs particularly modeling the Z-port parasitics by $R_z/(1/sC_z)$ and Y-port input impedance as C_y, analysis has shown that the equivalent floating impedance realized by the circuit is given by

$$Z_{eq} = \left(\frac{R_1 R_2}{R_Z} - 2R_x\right) + 2C_T R_1 R_2 \tag{5.42}$$

where C_T represents the effective total capacitance at the node connecting Z-port of CCII+ 3 and Y-port CCII+ 6 i.e. $C_T = (C + C_{Z3} + C_{Y6})$. From the analysis, it follows that the condition for theoretical zeroing of the series resistance R_S is:

$$\frac{R_1 R_2}{R_Z} = 2R_x \tag{5.43}$$

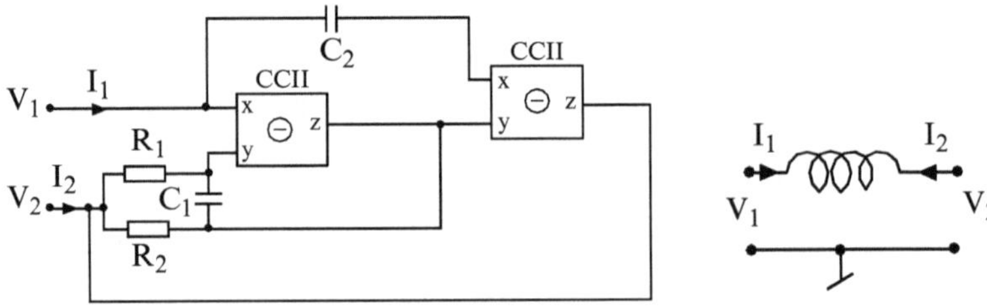

Fig. 5.34 A lossless FI proposed by Yuce [120] in 2007

The effectiveness of the proposed compensation method has been confirmed by realizing CCII+ with AD844 and choosing $R_1 = 100$ KΩ, $R_2 = 3.162$ KΩ and $C = 307$ nf (with the compensation condition satisfied with $R_x = 50$ Ω and $R_Z = 3$ MΩ). It has been found that the series resistance reduces from 80 Ω to about 11 μΩ thereby confirming the validity of the proposed compensation which also exhibits a considerable operating frequency range extension towards low frequencies from two or three decades to about six decades.

5.3.6 *Two-CC-Based FDNR and FGPIC/FGPII Implementations*

In the previous section, FI circuits were represented which require four or more CCs to realize a lossless inductance or more generalized impedance convertor. In this section, we present a collection of a number of interesting circuits which can perform the same task with no more than two CCs.

The circuit of Fig. 5.34 was proposed by Yuce [120] in 2007 and is characterized by:

$$\begin{bmatrix} I_1 \\ I_2 \end{bmatrix} = \frac{1}{sL_{eq}} \begin{bmatrix} 1 & -1 \\ -1 & 1 \end{bmatrix} \begin{bmatrix} V_1 \\ V_2 \end{bmatrix} \tag{5.44}$$

In the above expression $L_{eq} = C_1R_1R_2$ is obtainable subject to the fulfilment of a single matching condition $C_2 = C_1\ (R_1 + R_2)/R_2$.

It is interesting to point out that a two-CC-based floating FDNR as well as a number of two-CC-based floating GIC circuits were first proposed more than three decades ago by Senani in [40] and [47] respectively, which are shown in Figs. 5.35 and 5.36 respectively.

The circuit of Fig. 5.35 simulates an ideal FDNR having value $D_{eq} = C_1C_2R_1R_2/R_3$. Subsequent to the publication of this configuration, Wilson [44] used this configuration to verify his CCII− implementation based upon an operational mirrored amplifier formulation. The experimental results of [44], therefore, confirmed the practical validity of this floating FDNR.

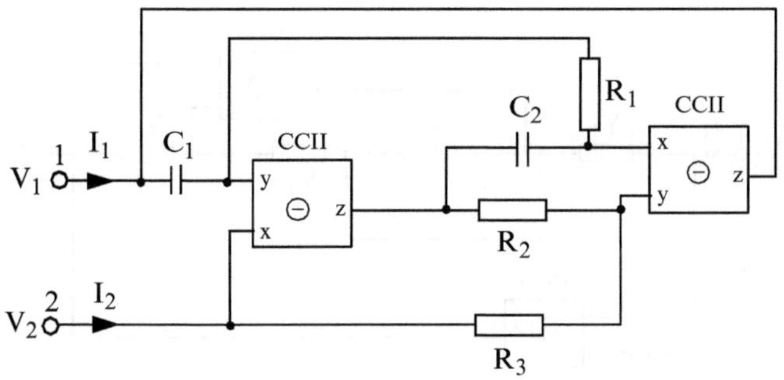

Fig. 5.35 The first Ideal floating FDNR using only two CCs proposed by Senani [40] in 1984

Three distinctly different floating generalized positive immittance converter/inverter (FGPIC/FGPII) circuits were proposed in [47] which are shown here in Fig. 5.36a–c.

A straight forward analysis of these circuits assuming ideal CCII-s shows that the impedance realized between ports one and two is given by ($Z_1Z_4Z_L/Z_2Z_3$). From the generalized expression, it is obvious that all the three circuits can be used to realize a variety of floating impedances using different choice of circuit impedances.

When the circuits of Fig. 5.36a–c are re-analyzed by taking into account the non-ideal h parameters of the CCII-s, it is found that the resulting Y-parameters of the circuits of Fig. 5.36a, b are more complex and unsymmetrical than those of circuit of Fig. 5.36c which after considering only the non-ideal current gains between ports X and Z of the two CCII-s, namely, h_{32A} and h_{32B} of the two CCII-s are found to be:

$$Y_{11} = \frac{Y_1Y_4Y_L}{Y_2Y_3}h_{32A} + Y_1(1 - h_{32A}) = -Y_{12} \tag{5.45}$$

$$-Y_{21} = \frac{Y_1Y_4Y_L}{Y_2Y_3}\left(\frac{1}{h_{32B}}\right) + Y_1\left(\frac{Y_4}{Y_2}\right)\left(\frac{1}{h_{32B}} - 1\right) = Y_{22} \tag{5.46}$$

The above expressions clearly show that, other than the desirable first part, the equivalent circuit of the realized impedance will contain unwanted parasitic elements. Nevertheless, the fact that these circuits can be successfully used in the design of filters has been demonstrated in [61] even though the CCII− was realized therein by Huertas' op-amp-based CC-implementation (see reference [4] of Chap. 2).

For other two CC-based lossy FI simulators, the interested readers may refer to [18, 84].

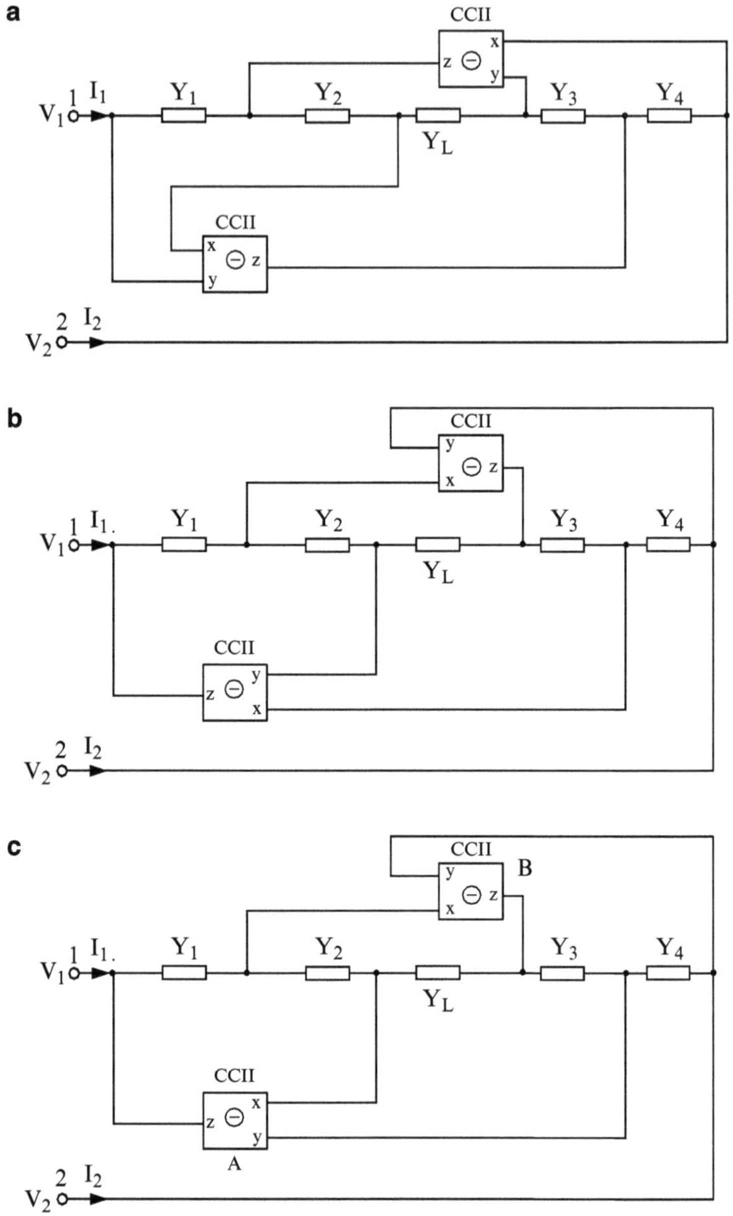

Fig. 5.36 Floating generalized positive immittance converter/inverter (FGPIC/FGPII) circuits proposed by Senani [47] in 1986

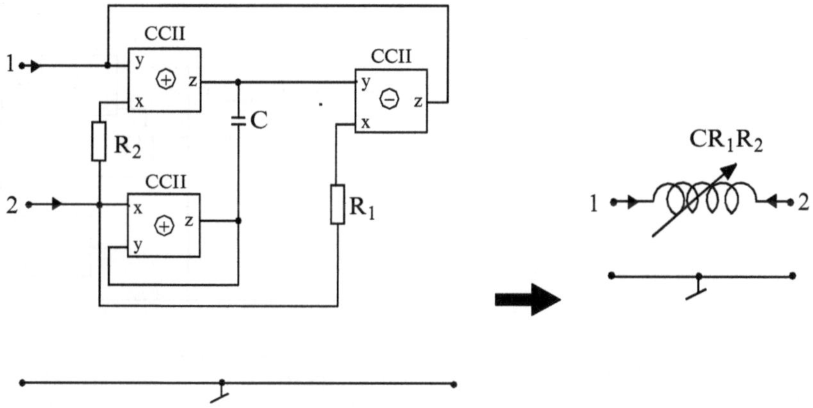

Fig. 5.37 Lossless FI circuit proposed by Senani [28]

5.3.7 A Family of Three-CC Floating Inductor/FDNR Simulators

There is an important result in circuit theory [163, 164] related to gyrators realizable using nullors and resistors, which states that if only two nullors are permitted then the realization of a gyrator requires at least five resistors (impedances) and if only three resistors are allowed then at least three nullors will be required for the same purpose.

There is another old but less well known theory [168] as per which an impedance converter, with respect to some port, can be looked upon as an impedance converter with respect to some other port and vice versa.

In view of the above mentioned results, the FGPIC/FGPII circuits of the previous section can be looked upon as two nullor five impedance circuits recalling that each CCII− is, in fact, a three-terminal nullor. It is, therefore, apparent that there should be a class of three-nullor three impedances, converters and inverters. More specifically, this means that a floating GPIC/GPII network should also be realizable by three CCs and no more than three passive elements. This section presents a variety of such three-CC based FGPIC/FGPII configurations or special cases thereof, which have been presented in the literature as lossless floating inductors or lossless floating FDNRs.

A floating lossless inductance using only three CCs and only three passive elements was first proposed by Senani in 1980 [28] which is shown in Fig. 5.37. The circuit realizes a lossless floating inductance of value $L_{eq} = CR_1R_2$ between ports 1 and 2.

A floating ideal FDNR was reported by Nandi et al. [37] in 1983. Assuming CCIIs to be characterized by $i_y = 0$, $v_x = v_y$ and $i_z = \pm h_k\, i_x$ (k = 1, 2, 3) where, ideally, h = 1 and the sign associated with **h** determines the polarity of the CCII

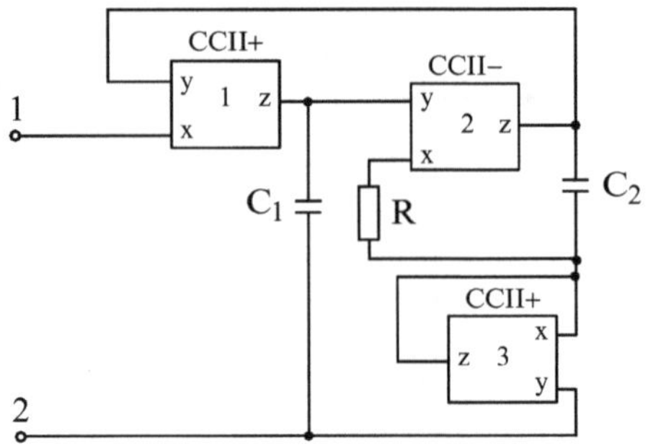

Fig. 5.38 Floating FDNR circuit realization proposed by Nandi et al. [37]

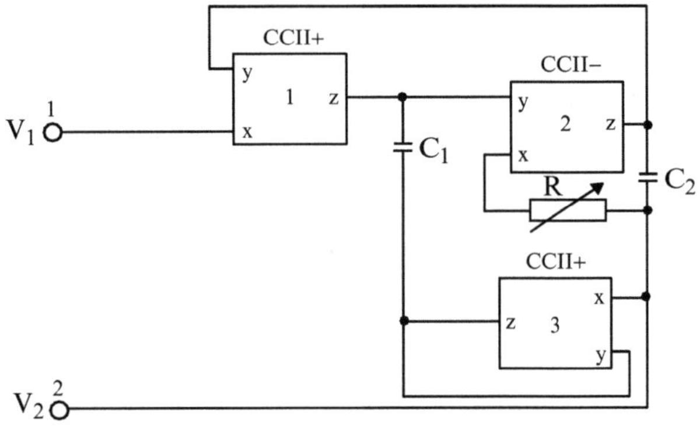

Fig. 5.39 Lossless floating FDNR proposed by Nandi et al. [41]

concerned. Analysis leads to the following admittance matrix of this circuit (Fig. 5.38):

$$[Y] = s^2 C_1 C_2 R \left(\frac{h_2}{h_1}\right) \begin{bmatrix} 1 & -1 \\ -h_1 & h_1 \end{bmatrix} \tag{5.47}$$

From the above matrix it can be deduced that all active and passive and sensitivities of this circuit are very low. It may be noted that if resistors are replaced by capacitors and the capacitor is replaced by a resistor, the same circuit was shown in [28] to realize an ideal floating FDNR.

Another circuit (Fig. 5.39) employing exactly the same number of CCs and RC elements to realize an ideal floating FDNR was subsequently published in 1984 (4 years after the publication of the first ideal floating FDNR in 1980 in [28]) by Nandi et al. [41] for simulating ideal tunable FDNR having value $D_{eq} = C_1 C_2 R$ where the D_{eq} can be tuned by the only resistor R used in the circuit.

In 1985 Abdalla [42] commented that by a small modification in the circuit of Nandi et al. [41], it becomes possible to employ only one type of CCs i.e. either all the three CCII− or all the three CCII+.[2] This proposed formulation is shown in Fig. 5.40.

[2] A careful re-check of the circuit, however, reveals that this assertion is not correct.

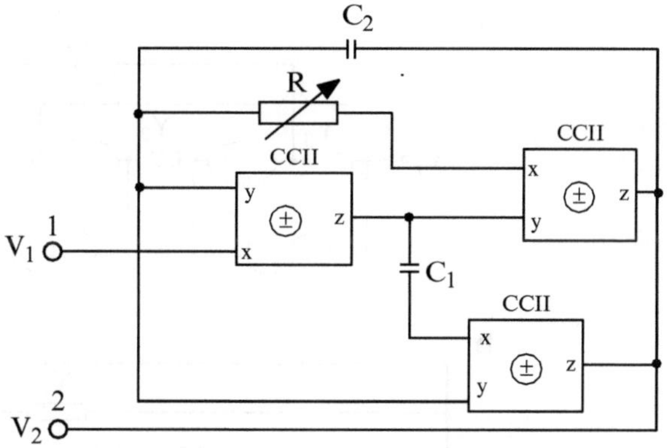

Fig. 5.40 Floating ideal tunable FDNR simulation scheme proposed by Abdalla [42]

In 1987, Higashimura and Fukui [53] derived two new three-CC-based generalized floating immittance simulators based upon the nullor model shown in Fig. 5.41. Recalling that a CCII- is equivalent to a three terminal nullor, this model leads to two possible realizations of the general floating immittances as shown in Fig. 5.41b, c respectively. Both the circuits simulate a floating impedance between ports 1 and 2 having value of equivalent admittance as $Y_{1\text{–}2} = Y_1Y_3/Y_2$.

Higashimura and Fukui in another publication [54] during 1987, came up with four new lossless tunable floating FDNR simulation circuits each using two CCIIs and a current inversion type negative immittance convertor (NIC) as active elements. Needless to say, like all previously discussed circuits, these circuits also have the advantage of not requiring any component matching conditions and providing tunability of the FDNR value through a single variable resistor. These circuits are shown in Fig. 5.42a–d.

All the circuits of Fig. 5.42a–d realize the admittance matrix:

$$[Y] = s^2D_0\begin{bmatrix} 1 & -1 \\ -1 & 1 \end{bmatrix}, D_0 = C_1C_2R, \tag{5.48}$$

It is interesting to point out that the same authors in [48] presented two more ideal floating FDNR circuits, each employing two CCIIs, one inverting/non-inverting buffer, two capacitors and a resistor. The circuit based upon inverting buffer, therein however, will require two more equal-valued resistors if the inverting buffer is implemented from a CCII or an op-amp.

5.3.8 Mixed-Source FIs Using CCIIs and Op-amps/OTAs

Recently, some researchers have also attempted to create floating inductance configurations using CCs along with other building blocks like op-amps and

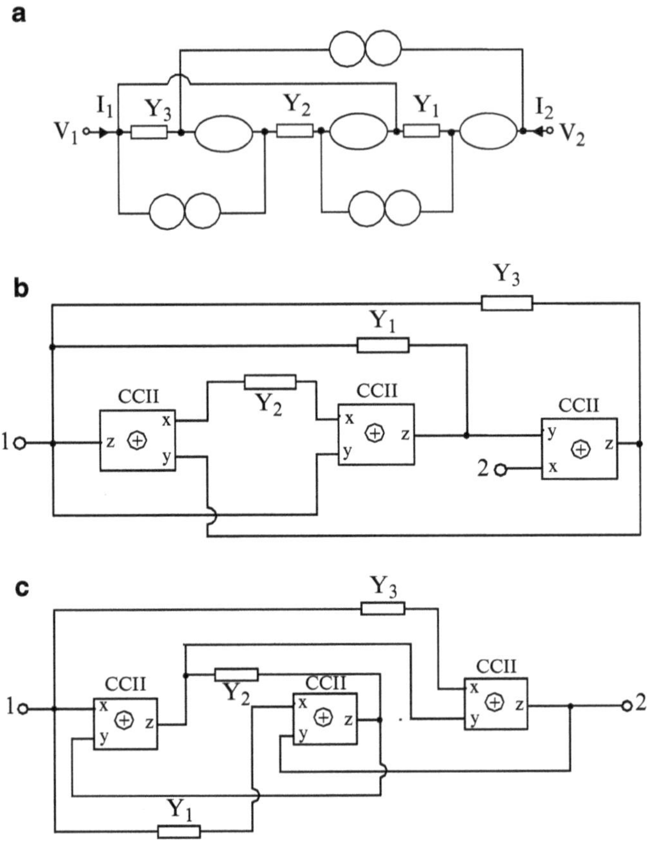

Fig. 5.41 CC based floating impedance realization derived from a nullor models proposed by Higashimura and Fukui [53] (**a**) Nullor model of FI (**b**), (**c**) CCII– based FIs derived from the nullor model

OTAs to achieve properties which are not feasible with circuits containing only CCs. Two such examples are outlined in the following.

In the circuit of Fig. 5.43, Maundy et al. [122] have presented a mixed source arrangement for realizing a floating inductance using two CCs and two op-amps. In their proposition, the two CCs along with a resistor have been employed to create a differential voltage controlled current source. By standard analysis, it can be easily found out that between ports 1 and 2, the circuit realizes a floating impedance of value $Z_{1-2} = sC_1R_1R_2$.

In the second case proposed by Higashimura and Fukui [62], (Fig. 5.44), an OTA is used in conjunction with two CCs, a resistor and a grounded capacitor, to create an inductance which is electronically tunable through the external bias current of the OTA. By straight forward analysis, it has been found that to realize a floating inductance of value $L_{eq} = CR_1/G_m$ since for bipolar OTA, such as CA3080, its transconductance G_m is given by $G_m = I_{bias}/2V_T$. It is, thus, clear that simulated inductance value is linearly controllable through the external bias current I_{bias}.

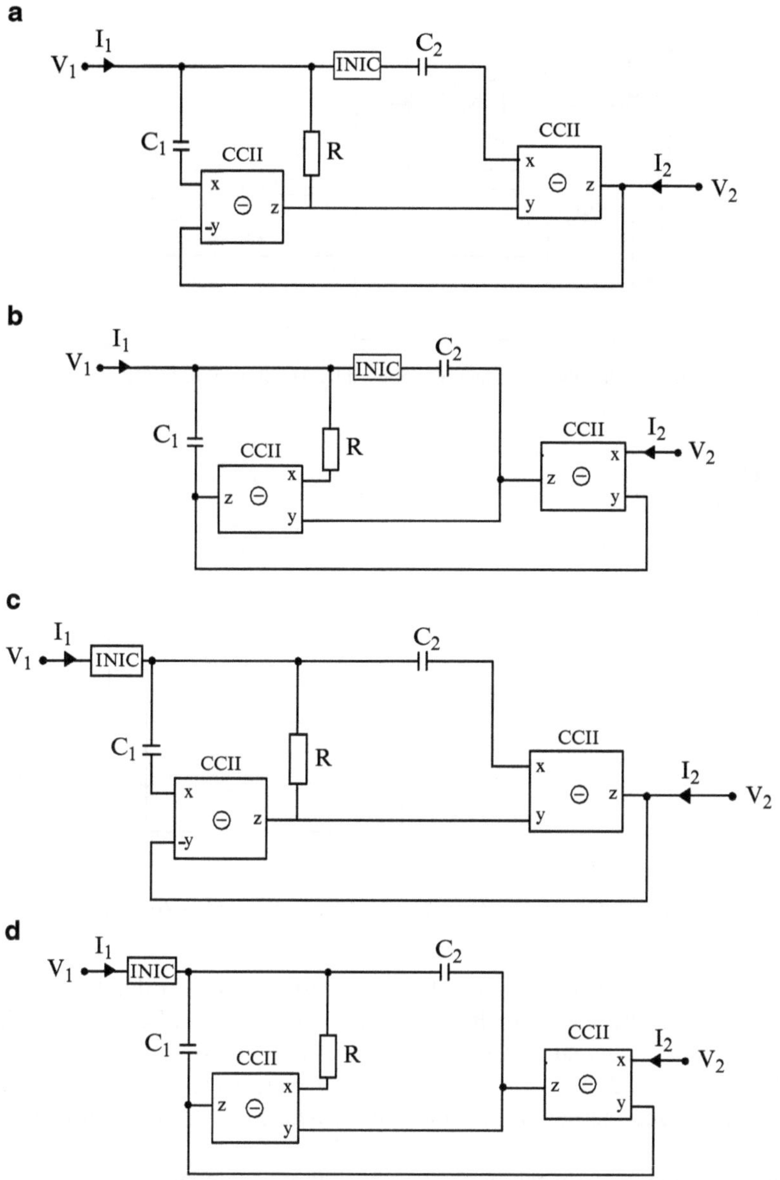

Fig. 5.42 Circuit configurations for lossless tunable floating FDNR simulations proposed by Higashimura and Fukui [52]

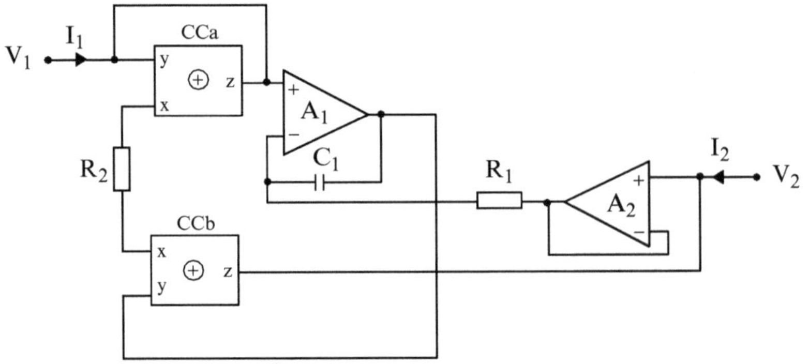

Fig. 5.43 A floating bidirectional inductor proposed by Maundy et al. [122]

Fig. 5.44 An electronically-controllable floating lossless inductance using combination of CCII+ and OTA proposed by Higashimura and Fukui [62]

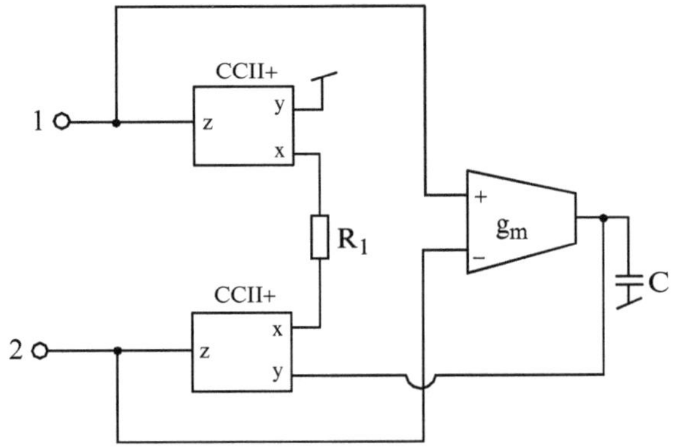

5.3.9 *Novel FI Circuits Using CCII-Nullor Equivalence*

In view of the methods and circuits discussed earlier it is clear that lossy/non-ideal inductors and FDNRs can be realized with a smaller number of CCs than their ideal counterparts and many such circuits are known in literature, for instance, see [13, 14, 20, 24, 29, 30, 32, 39, 45, 49, 115, 138], however, in the following we present a novel nullor-based methodology which can generate such circuits in a systematic manner.

The equivalence between a CCII− and three-terminal floating nullor was employed by Senani in [56] to generate a class of new FI structures providing single resistance tunability of the realized floating inductance value while employing only two CCII-s along with a bare minimum of only three passive components, namely, two resistors and a capacitor. Subsequently, the nullor approach has also been used by other researchers as in [57, 74, 76].

The starting point of the derivation of the circuits presented by Senani [56] are the two nullor models shown in Fig. 5.45 which simulate a parallel RL and series

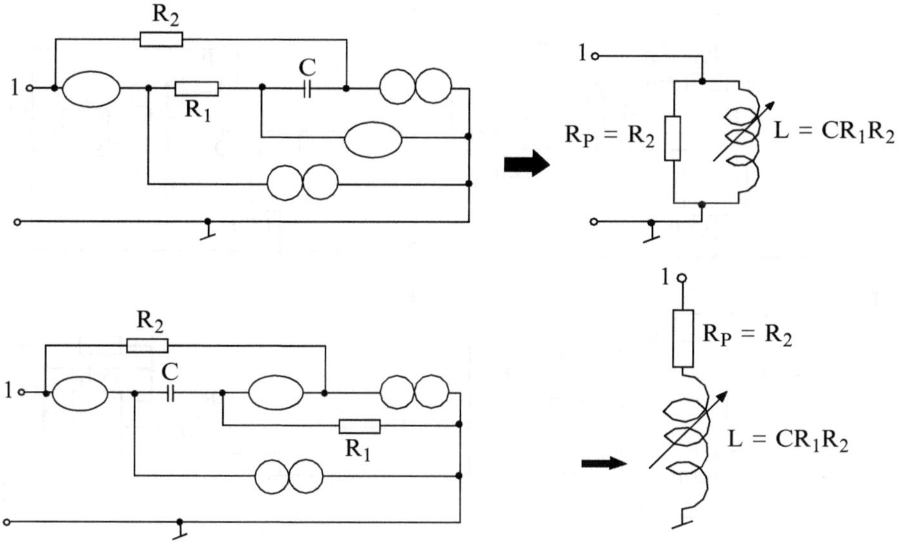

Fig. 5.45 Nullors model employed for synthesizing CCII− based floating impedance configurations [56]

RL impedance respectively. From these models, the nullor models of floating impedances can be obtained by ungrounding all the passive elements, nullators and norators. From these FI models, two additional FI models can be obtained by swapping nullators by norators and vice versa. Lastly, four additional nullor models can be evolved by carrying out the operation of *port-transposition* on the preceding four models. This operation, however, will change the nature of impedances simulated in the sense that if the original model represented a floating series RL the transposed model will represent a floating parallel RL and vice versa.

The actual CCII− based circuits emanating from the above described eight nullor models turn out to be as shown in Fig. 5.46. It can be verified that the circuits of Fig. 5.46a, c, f, and g have the Y matrix given by:

$$[Y] = \left(\frac{1}{R_2} + \frac{1}{sCR_1R_2}\right)\begin{bmatrix} 1 & -1 \\ -1 & 1 \end{bmatrix} \tag{5.49}$$

Hence, all these circuits realize a floating parallel RL impedance with

$$\mathrm{L} = \mathrm{CR_1R_2};\ \mathrm{R_p} = \mathrm{R_2} \tag{5.50}$$

On the other hand, the circuit configurations of Fig. 5.46b, d, e and h are characterized by the Y- matrix:

$$[Y] = \frac{1}{(R_2 + sCR_1R_2)}\begin{bmatrix} 1 & -1 \\ -1 & 1 \end{bmatrix} \tag{5.51}$$

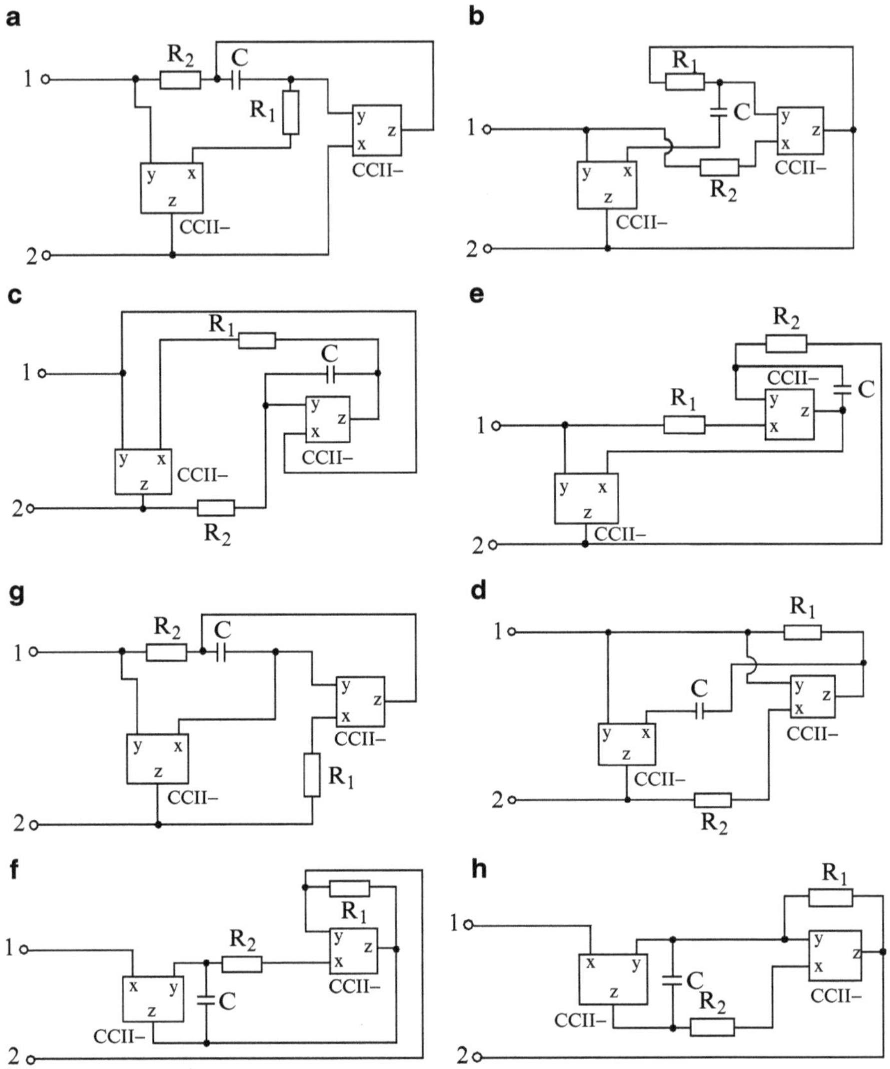

Fig. 5.46 A family of two CC-two- resistor-one capacitor FI simulators derived by Senani [56]

which represents floating series RL impedance with

$$L = CR_1R_2;\ R_s = R_2 \tag{5.52}$$

From all the eight circuits of Fig. 5.46a–h, non-ideal floating FDNRs can be realized by applying RC: CR transformation.

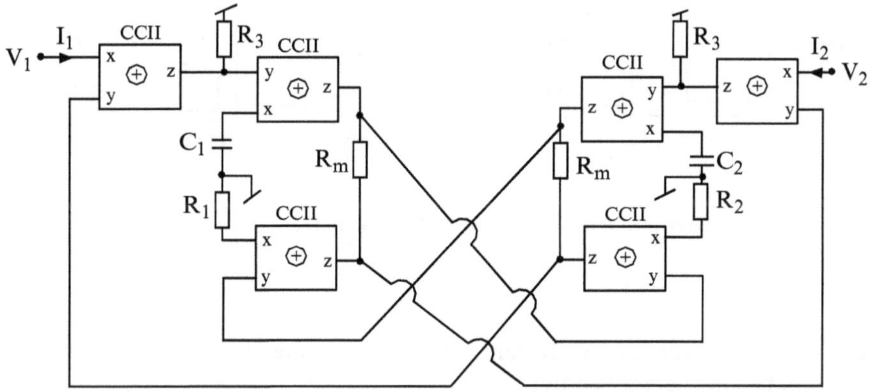

Fig. 5.47 Simulated coupled circuit proposed by Abuelma'atti et al. [110]

5.3.10 Simulation of Higher Order Grounded/Floating Immittances Using CCs

While a majority of works on CC-based impedance simulation have dealt with inductors (first order impedance) or FDNR/FDNC elements (both second order immittances), a number of authors have devised CC-based circuits for simulating higher order immittances, in both grounded and floating forms, for instance, see [80, 81, 83, 140, 151].

5.3.11 Simulation of Mutually-Coupled Circuits

There are numerous applications in Communication, Instrumentation and Control which call for the use of mutually-coupled circuits. A mutually-coupled circuit essentially has a primary self-inductance, a secondary self-inductance and mutual inductance. Apart from the simulation of inductors, the simulation of mutually coupled circuits using CCs has also received attention of researchers [110, 123, 124].

A circuit for mutually-coupled circuit using only CCII+ s was proposed by Abuelma'atti et al. in [110] and is shown here in Fig. 5.47.

An analysis of this circuit shows that the voltages V_1 and V_2 are given by:

$$V_1 = sC_1R_1R_3I_1 + sC_1R_3R_mI_2 \tag{5.53}$$

$$V_2 = sC_1R_3R_mI_1 + sC_1R_2R_3I_2 \tag{5.54}$$

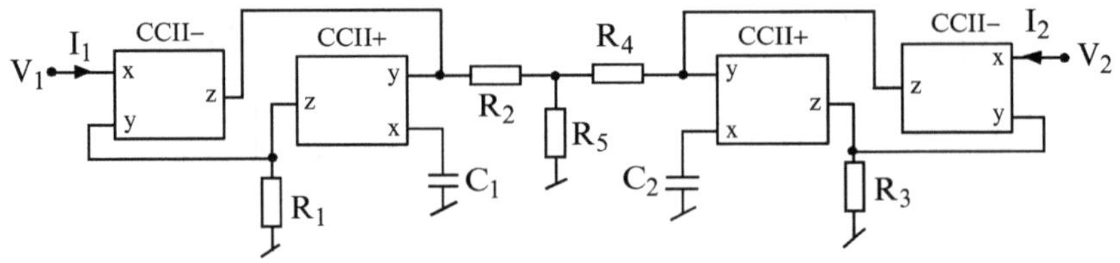

Fig. 5.48 Mutually-coupled circuit proposed by Yuce et al. [124]

from where it is seen that equivalent parameters i.e. the two self- inductances and mutual inductance are given by:

$$L_1 = C_1R_1R_3, L_2 = C_1R_2R_3, M = C_1R_mR_3 \tag{5.55}$$

Another circuit for simulating a mutual-coupled circuit employing only four CCs two of which are CCII− with complete circuit requiring as many as six CCIIs was proposed by Yuce et al. [124] and is shown in Fig. 5.48. This mutually-coupled circuit is characterized by the following matrix equation.

$$\begin{bmatrix} V_1 \\ V_2 \end{bmatrix} = \begin{bmatrix} s(L_1 + M_{11}) & sM_{12} \\ sM_{21} & s(L_2 + M_{22}) \end{bmatrix} \begin{bmatrix} I_1 \\ I_2 \end{bmatrix} \tag{5.56}$$

where

$$L_1 = C_1R_1R_2, L_2 = C_2R_3R_4 \tag{5.57}$$

$$M_{11} = M_{12} = M_1 = C_1R_1R_5; M_{21} = M_{22} = M_2 = C_2R_3R_5 \tag{5.58}$$

If one takes $C_1R_1 = C_2R_3$, $M_1 = M_2 = M$ is obtained.

It may be mentioned that both the circuits suffer from the drawback of requiring a large number of CCIIs (six in the circuit of Fig. 5.47 and four in the circuit of Fig. 5.48) as well as using two capacitors at the X-terminal of the current conveyor which is often associated with stability problems [116]. Thus, there is enough scope for devising improved alternative realizations of the mutually-coupled circuits, free from these difficulties.

5.3.12 Grounded and Floating MOS VCRs and Transconductors

In many applications, electronically-variable resistors and other impedances are required for achieving electronic control of the parameters of the functional circuits incorporating such impedances. Several authors have presented circuits to realize such functions. In this section, we present some representative circuits dealing with electronically-controllable impedances using CCs.

In the circuit of Fig. 5.49a, the basic idea of which was proposed by Tabei et al. [72], a CCII-based NIC is employed as shown by Soliman in [79], to cancel out the square non-linearities of the MOSFETs M_1 with that of M_2 thereby leading to the realization of equivalent input impedance given by:

$$R_{in} = 1/[2K(V_{G1} - V_{G2})] \tag{5.59}$$

where $K = \frac{\mu_s}{2} C_{ox} \left(\frac{W}{L}\right)$ and symbols have their usual meaning.

The circuit of Fig. 5.49b realizes a voltage-controlled resistor of same value in floating form. It must be mentioned that in the grounded and floating VCRs of Fig. 5.49, a negative grounded VCR and a negative floating VCR can be obtained simply by selecting $V_{G2} > V_{G1}$.

An extension of the above idea, when applied to a known two-NIC-based gyrator leads to a MOSFET-based gyrator shown in Fig. 5.50. When this gyrator is terminated into a capacitance C_L the input impedance represents pure voltage-controlled inductance controllable by the voltage ($V_{G1} - V_{G2}$).

By a routine analysis, the input impedance of the circuit, with output port terminated into a capacitance C_L, is given by

$$Z_{in} = \frac{sC_L}{4K^2 (V_{G1} - V_{G2})^2} \tag{5.60}$$

and the circuit, thus, realizes a voltage controlled inductance value.

An extension of the ideas incorporated in the circuits of Fig. 5.49 to realize a linear transconductor was presented by Soliman in [79]. His proposition is shown in Fig. 5.51. By a straight forward analysis (assuming the triode region operation of both the MOSFETs), the output current (I_o) can be obtained as:

$$I_0 = 2K[V_{G1} - V_{G2}] V_{in} \tag{5.61}$$

A CC-based circuit for realizing a voltage/current controlled grounded resistor with wide dynamic range extension was given by Maundy et al. in [130]. This circuit incorporates the use of bisection of the drain to source voltage of a JFET to produce the linear resistor where CCII+ along with three resistors has been employed to extend the linearity of the variable resistor whose value can be controlled by a variety of methods finally culminating into the circuit shown in Fig. 5.52.

With $R_a = R_b = R_1/2$, $R_1 = \left(\frac{a}{1-a}\right) R_2$ and assuming that $V_c > \frac{aV_{in}}{2} + V_{TH}$ to ensure that the MOSFET is operating in triode region, the input resistance of the circuit is found to be:

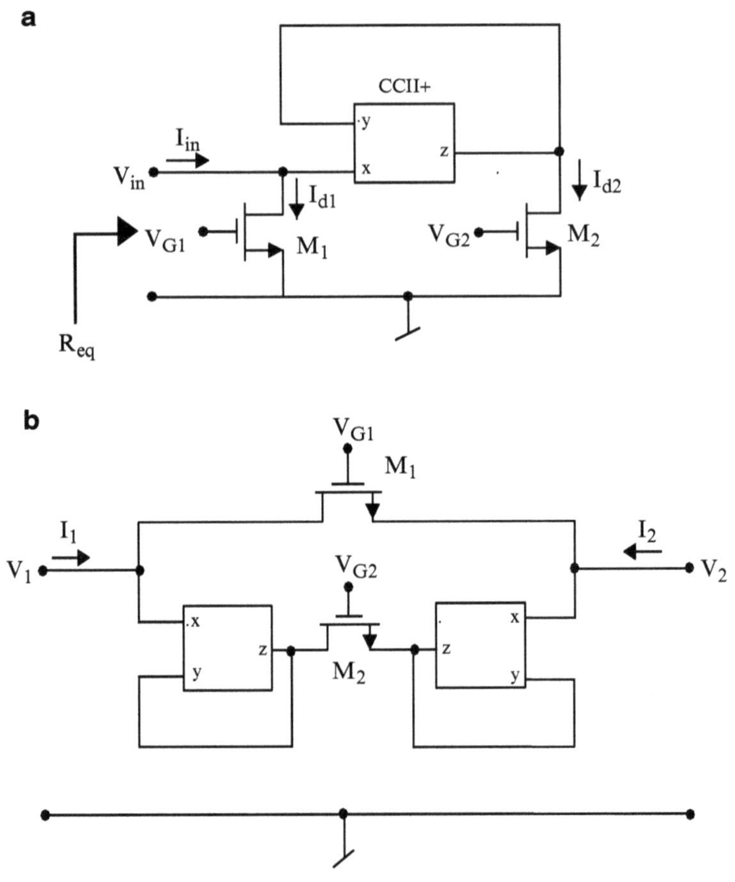

Fig. 5.49 Linear voltage controlled resistance realization using current conveyors based upon the NIC-based propositions of Tabei et al. [72]; also discussed by Soliman [79] (**a**) grounded VCR (**b**) floating VCR

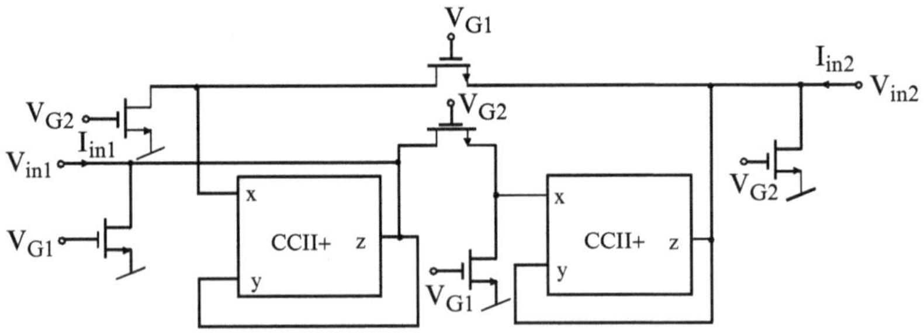

Fig. 5.50 Voltage controlled gyrator realization (**a**) the prototype gyrator (**b**) the MOS version using CCIIs, based upon the proposition of Tabei et al. [72]

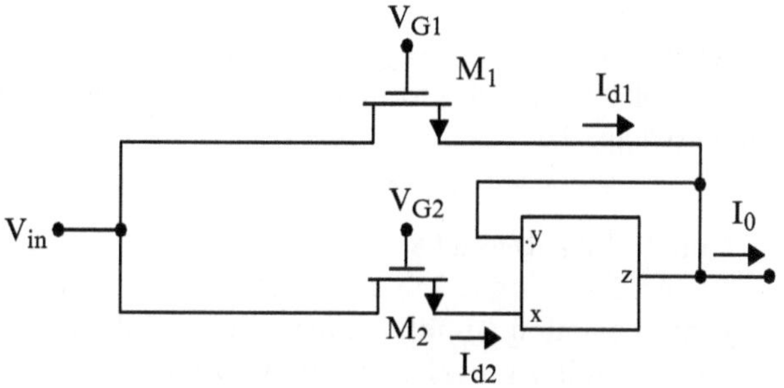

Fig. 5.51 A MOS transconductor obtained by a rearrangement of the floating VCR of Fig. 5.49b as proposed by Soliman [79]

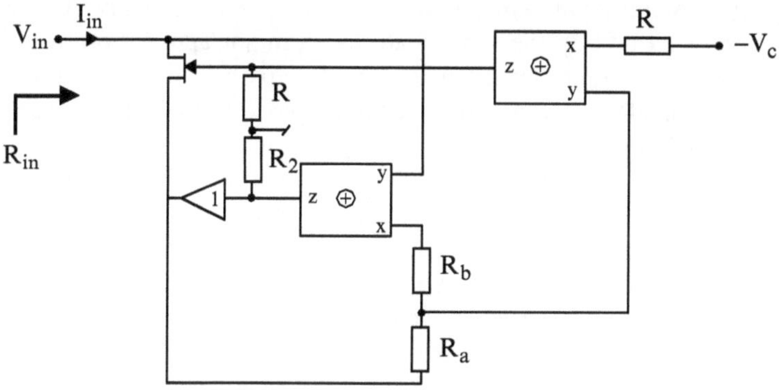

Fig. 5.52 The liner VCR proposed by Maundy et al. [130]

$$R_{in} = \frac{2}{aK(V_c - V_{TH})} \tag{5.62}$$

where $K = \mu_n C_{ox}(W/L)$ and the symbols have their usual meaning. It may be noted that the operation of the MOSFET in deep triode region requires that $V_c >> V_{TH}$. Thus, it is seen that the circuit realizes a linear voltage-controlled resistance with square nonlinearity of the MOSFET cancelled out.

The workability of this circuit has been verified by using AD844 to realize CCIIs and voltage follower.

In the context of the above circuit, it may be pointed out that [169] many similar ideas leading to a variety of linear VCR circuits as well as more versatile voltage-controlled impedances in positive as well as negative, grounded as well as floating forms were advanced by Senani and Bhaskar in [165, 170, 171] using op-amps; see references cited in [169]. Surely, several of them can be converted into circuits employing CCs in which case, a reduction in the number of the external passive components can be envisaged together with the extension of the frequency range of the operation of the resulting circuits. This, however, requires further research and is, therefore, open to investigation.

Lastly, it is worth pointing out that Current-controllable resistances can also be realized by Current Conveyors as shown by Saaid and Fabre in [82], however, the circuits of this kind are formally dealt with in a subsequent chapter of this monograph (Chap. 9).

Concluding Remarks

It was demonstrated that CC-based realisations of inverting/non-inverting/instrumentation amplifiers, make it possible to eliminate the gain and bandwidth conflict prevalent in op-amp based circuits. Furthermore, inverting/non-inverting/differential integrators/differentiators all can be realized with the least possible number of passive components in both VM and CM. Furthermore, positive/negative impedance converters/inverters can be implemented exhibiting the same features. A major application of CCs has been in the realization of impedance simulation particularly in the realization of floating impedances wherein it becomes possible to realize a variety of floating impedances *without requiring any component matching condition(s)* – *a* feat which is impossible to be achieved with op-amp-based circuits. From a vast amount of literature available on this topic, some notable grounded impedance and floating impedance simulators using CCs, including the voltage controlled impedances, were outlined and some ideas for further work in these areas have been suggested.

References

1. Antoniou A (1969) Realisation of gyrators using operational amplifiers and their use in RC-active network synthesis. Proc IEE 116:1838–1850
2. Dutta Roy SC (1974) A circuit for floating inductance simulation. Proc IEEE 64:521–523
3. Wise DR (1974) Active simulation of floating lossy inductances. Proc IEE 121:85–87
4. Soliman AM (1976) Generalized immittance inverters and their realizations. Int J Electron 41:59–64
5. Nandi R (1976) Inductor simulation using a current conveyor. Proc Lett 65:1511–1512
6. Glover J (1976) Lossless ungrounded inductor realisation. Electron Lett 12:171–173
7. The LQ, Yanagisawa T (1977) Some new lossless floating inductance circuits. Proc IEEE 65:1071–1072
8. Sudo S, Teramoto M (1977) Constitution of floating inductance using operational amplifiers. IEICE Trans E60:185–186
9. Soliman AM (1978) New active-gyrator circuit using a single current conveyor. Proc IEEE 66:1580–1581
10. Soliman AM (1978) Ford-girling equivalent circuit using CCII. Electron Lett 14:721–722
11. Senani R, Tiwari RN (1978) New canonic active RC realizations of grounded and floating inductors. Proc IEEE 66:803–804
12. Nandi R (1978) Active inductances using current conveyors and their application in a simple band pass filter realization. Electron Lett 14:373–375
13. Senani R (1978) Active simulation of inductors using current conveyor. Electron Lett 14:483–484

14. Senani R (1979) (Comment) Active simulation of inductors using current conveyor. Electron Lett 15:112–114
15. Nandi R (1979) A new equal-valued grounded-capacitor resonator realization using current conveyor. Proc IEEE 67:370–371
16. Senani R (1979) Novel active RC circuit for floating-inductor simulation. Electron Lett 15:679–680
17. Nandi R (1979) Insensitive grounded-capacitor simulation of grounded inductors using current conveyors. Electron Lett 15:693–694
18. Pookaiyaudom S, Srisarakham W (1979) Realization of stable current controllable frequency-dependent positive resistance. Proc IEEE 67:1660–1662
19. Singh V (1979) A new active–RC circuit realization of floating inductance. Proc IEEE 67:1659–1660
20. Singh V (1979) Realization of floating inductance using current conveyor. Proc IEEE 67:1573–1574
21. Rathore TS, Singh BM (1979) Simulation of RL impedances suitable for micro-miniaturization. Microelectron Reliab 19:233–235
22. Rathore TS (1980) Corrections to simulation of RL impedances suitable for micro-miniaturization. Microelectron Reliab 20:767–768
23. Patranabis D, Tripathi MP, Roy SB (1979) A new approach for lossless floating inductor. IEEE Trans Circ Syst 26:892–893
24. Patranabis D, Paul AN (1980) Floating inductor with two current conveyors. Int J Circ Theor Appl 8:457–468
25. Pal K (1980) Realization of ideal grounded inductances and frequency dependent negative resistances using current conveyors and without matched components. Int J Circ Theor Appl 9:242–245
26. Singh V (1980) Novel active RC realisations of tunable floating inductors. Electron Lett 16:758
27. Pal K (1980) Novel F.D.N.C Simulation using current conveyors. Electron Lett 16:639–641
28. Senani R (1980) New tunable synthetic floating inductors. Electron Lett 16:382–383
29. Senani R (1980) Novel active RC realisations of tunable floating inductors. Electron Lett 16:154–155
30. Rathore TS, Singh BM (1980) Active RC synthesis of floating immittances. Int J Circ Theor Appl 8:184–188
31. Rathore TS (1980) Grounded capacitor synthesis of RL impedances based on lagging current concept. Int J Circ Theor Appl 8:461–465
32. Salawu RI (1980) Realization of frequency dependent negative resistance. Microelectron Reliab 20:853–857
33. Paul AN, Patranabis D (1981) Active simulation of grounded inductors using a single current conveyor. IEEE Trans Circ Syst 28:164–165
34. Singh V (1981) Active RC single–resistance controlled lossless floating inductance simulation using single grounded capacitor. Electron Lett 17:920–921
35. Senani R (1982) Novel Lossless synthetic floating inductor employing a grounded capacitor. Electron Lett 18:413–414
36. Nandi R, Nandi S (1983) Novel insensitive floating inductor simulation using single current conveyor. Int J Electron Commun (AEU) 37:393–396
37. Nandi S, Jana PB, Nandi R (1983) Floating ideal FDNR using current conveyors. Electron Lett 19:251
38. Paul AN, Dey AN, Patranabis D (1983) A floating NIC without constraint and its application. IEEE Trans Circ Syst 30:181–183
39. Senani R (1984) Novel application of generalised current conveyor. Electron Lett 20:169–170; also see *ibid* (1984) 20: 356
40. Senani R (1984) Floating ideal FDNR using only two current conveyors. Electron Lett 20:205–206

41. Nandi S, Jana PB, Nandi R (1984) Novel floating ideal tunable FDNR simulation using current conveyors. IEEE Trans Circ Syst 31:402–403
42. Abdalla YAN (1985) Comment on Novel floating ideal tunable FDNR simulation using current conveyors. IEEE Trans Circ Syst 32:303
43. Toumazou C, Lidgey FJ (1985) Floating-impedance convertors using current conveyors. Electron Lett 21:640–642
44. Wilson B (1985) Floating FDNR employing a new CCII-conveyor implementation. Electron Lett 21:996–997
45. Senani R (1985) Novel higher-order active filter design using current conveyors. Electron Lett 21:1055–1057
46. Von Grunigen DC, Ramseier D, Moschytz GS (1986) Simulation of floating impedances for low-frequency active filter design. Proc IEEE 74:366–367
47. Senani R (1986) On the realization of floating active elements. IEEE Trans Circ Syst 33:323–324
48. Higashimura M, Fukui Y (1986) Novel lossless tunable floating FDNR simulation using two current conveyors and a buffer. Electron Lett 22:938–939
49. Senani R (1987) Network transformations for incorporating nonideal simulated immittances in the design of active filters and oscillators. IEE Proc 134:158–166
50. Nandi R (1987) Precise realisation of an insensitive floating negative admittance convertor. Electron Lett 23:775–777
51. Nandi R (1987) Novel floating negative immittance convertor. IEE Proc 134:236–238
52. Higashimura M, Fukui Y (1987) Novel method for realising lossless floating immittance using current conveyors. Electron Lett 23:498–499
53. Higashimura M, Fukui Y (1987) Realization of floating immittance using three current conveyors. IEICE Trans JPN J70-A:1203–1204
54. Higashimura M, Fukui Y (1987) New lossless tunable floating FDNR simulation using two current conveyors and an INIC. Electron Lett 23:529–531
55. Surakampontorn W, Thitimajshima P (1988) Integrable electronically tunable current conveyors. IEE Proc 135:71–77
56. Senani R (1988) Floating immittance realisation: Nullor approach. Electron Lett 24:403–405
57. Higashimura M, Fukui Y (1988) Type 1 mutator using current conveyor and its application to immittance simulation. Int J Electron 64:377–383
58. Ishida M, Okazaki Y, Fukui Y, Ebisutani K (1988) Realization of immittance function using current conveyors. Trans IEICE J71-A:1205–1207
59. Higashimura M, Fukui Y (1988) Realization of impedance function using current conveyors. Int J Electron 65:223–231
60. Senani R (1989) Three op-amp floating immittance simulators: a retrospection. IEEE Trans Circ Syst 36:1463–1465
61. Singh V (1989) An implementation of CCII-current conveyor, with application. IEEE Trans Circ Syst 36:1250–1251
62. Higashimura M, Fukui Y (1989) Simulation of lossless floating inductance using two current conveyors and an operational transconductance amplifier. Int J Electron 66:633–638
63. Himura A, Fukui Y, Ishida M, Higashimura M (1989) Immittance function simulator using a single current conveyor. IEICE Trans E72:1279–1284
64. Kumar U, Shukla SK (1989) The implementation and applications of current conveyors. Microelectron J 20:25–46
65. Himura A, Fukui Y, Ishida M, Higashimura M (1989) Immittance function simulator using a single current conveyor. Trans IEICE 72:1279–1284
66. Toumazou C, Lidgey FJ (1989) Novel current-mode instrumentation amplifier. Electron Lett 25:228–230
67. Wilson B (1989) Universal conveyor instrumentation amplifier. Electron Lett 25:470–471
68. Wilson B (1990) Recent developments in current conveyors and current-mode circuits. IEE Proc 137:63–77

69. Himura A, Fukui Y, Ishida M, Higashimura M (1990) Series impedance simulators using one CCII. Electron Lett 26:269–270
70. Fabre A, Alami M (1992) Insensitive current-mode band pass implementation-based non-ideal gyrators. IEEE Trans Circ Syst 39:152–155
71. Fabre A, Siarry P, Lameche M (1991) Current controlled translinear impedance converter. Int J Electron 70:795–801
72. Tabei K, Czarnul Z, Takagi S (1991) Realization of highly linear MOS circuits using negative impedance convertors. Electron Lett 27:1416–1417
73. Bruun E, Olesen OH (1992) Conveyor implementations of generic current mode circuits. Int J Electron 73(1):129–140
74. Higashimura M, Fukui Y (1992) Realization of immittance floatator using nullor. IEICE Trans Fundament E-75:644–649
75. Lee JY, Tsao HW (1992) True RC integrators based on current conveyors with tunable time constants using active control and modified loop technique. IEEE Trans Instrum Measure 41 (5):709–714
76. Hou Chun-Li WY-P, Lu F-C (1993) Synthetic methods for floating immittances of one-ports and z and y parameters of multiports using CCII. Int J Electron 74:577–586
77. Hou CL, Chen RD, Wu YP, Hu PC (1993) Realization of grounded and floating immittance function simulators using current conveyors. Int J Electron 74:917–923
78. Chang C-M, Wang H-Y, Chien C-C (1994) Realization of series impedance functions using one CCII+. Int J Electron 76:83–85
79. Soliman AM (1996) A linear trans conductor-multiplier using a matched pair of MOS transistors and a current conveyor. Frequenz 50:292–293
80. Vrba K, Cajka J, Zeman V (1996) Floating RC networks using current conveyors. Radioengineering 5:8–11
81. Liu S-I, Yang C-Y (1996) Higher-order immittance function synthesis using CCIIIs. Electron Lett 32:2295–2296
82. Saaid O, Fabre A (1996) Class AB current-controlled resistor of high performance current-mode applications. Electron Lett 32:4–5
83. Vrba K, Cajka J (1997) High-order one-port elements for low pass realization. J Electr Eng 48:31–34
84. Layos MC, Haritantis I (1997) On the derivation of current-mode floating inductors. Int J Circ Theor Appl 25:29–36
85. Wang HY, Lee CT (1997) Immittance function simulator using a single current conveyor. Electron Lett 33:574–576
86. Al-Walaie Soliman A, Alturaigi MA (1997) Current mode simulation of lossless floating inductance. Int J Electron 83:825–829
87. Kiranon W, Pawarangkoon P (1997) Floating inductance simulation based on current conveyors. Electron Lett 33:21
88. Abuelma'atti MT (1998) (Comment) Floating inductances simulation based on current conveyors. Electron Lett 34:1037–1038
89. Cicekoglu O (1998) New current conveyor based active-gyrator implementation. Microelectron J 29:525–528
90. Karybakas CA, Horopanitis EE (1998) V-I scalar circuits based on CCIIs. Electron Lett 34:317–318
91. Ozoguz S, Acar C (1998) On the realization of floating immittance function simulators using current conveyors. Int J Electron 85:463–475
92. Cabeza R, Carlosena A (1998) Computational synthesis of arbitrary floating impedances. Int J Circ Theor Appl 26:463–475
93. Cicekoglu MO (1998) Active simulation of grounded inductors with CCII+ s and grounded passive elements. Int J Electron 85:455–462
94. Abuelma'atti MT, Tasadduq NA (1999) New negative immittance function simulators using current conveyors. Microelectron J 30:911–915

95. Abuelma'atti MT (2000) Comment on active simulation of grounded inductors with CCII+ s and grounded passive elements. Int J Electron 87:177–181
96. Cicekoglu O (2000) Reply to comment on 'Active simulation of grounded inductors with CCII+ s and grounded passive elements'. Int J Electron 87(2):183–184
97. Wang HY, Lee CT (2000) Systematic synthesis of R-L and C-D immittances using single CCIII. Int J Electron 87(3):293–301
98. Fabre A, Saaid O (2000) Phase compensation of ideal inductances based second-generation current conveyors. Analog Integr Circ Sig Syst 24:153–162
99. Gift SGJ (2001) An enhanced current-mode instrumentation amplifier. IEEE Trans Instrum Meas 50:85–88
100. Tomita Y, Agu M (2001) A multi-port D-R mutator using CCIIs with current followers. Int J Electron 88:31–39
101. Ferri G, Guerrini N (2001) High-valued passive elements simulation using low-voltage low-power current conveyors for fully integrated applications. IEEE Trans Circ Syst-II 48:405–408
102. Cicekoglu O, Toker A, Kuntman H (2001) Universal immittance function simulators using current conveyors. Comput Electr Eng 27:227–238
103. Minaei S, Cicekoglu O, Kuntman H, Turkoz S (2002) Electronically tunable, active only floating inductance simulation. Int J Electron 89:905–912
104. Ferri G, Guerrini NC, Diqual M (2003) CCII-based floating inductance simulator with compensated series resistance. Electron Lett 39:22
105. Arslan E, Cam U, Cicekoglu MO (2003) Novel lossless grounded inductance simulators employing only a single first generation current conveyors. Frequenz 57:204–206
106. Khan IA, Zaidi MH (2003) A novel ideal floating inductor using translinear conveyors. Active Passive Electron Comput 26:87–89
107. Gift SJG (2004) New simulated inductor using operational conveyors. Int J Electron 91:477–483
108. Pal K (2004) Floating inductance and FDNR using positive polarity current conveyors. Active Passive Electron Comput 27:81–83
109. Anand Mohan PV (2005) Floating capacitance simulation using current conveyors. J Circ Syst Comput 14:123–128
110. Abuelma'atti MT, Al-Shahrani SM, Al-Absi MK (2005) Simulation of a mutually coupled circuit using plus-type CCIIs. Int J Electron 92:49–54
111. Ghallab YH, Badawy W, Kaler VIS, Maundy BJ (2005) A novel current-mode instrumentation amplifier based on operational floating current conveyor. IEEE Trans Instrum Meas 54:1941–1949
112. Galanis C, Haritantis I (1996) International conference on electronics, circuits and systems 1:65–68
113. Yuce E, Cicekoglu O, Minaei S (2006) Novel floating inductance and FDNR simulators employing CCII+ s. J Circ Syst Comput 15:75–81
114. Yuce E, Minaei S, Cicekoglu O (2006) Limitations of the simulated inductors based on a single current conveyors. IEEE Trans Circ Syst-I 53:2860–2867
115. Metin B, Cicekoglu O (2006) A novel floating lossy inductance realization topology with NICs using current conveyors. IEEE Trans Circ Syst-II 53:483–486
116. Yuce E, Cicekoglu O (2006) The effects of non-idealities and current limitations on the simulated inductances employing current conveyors. Analog Integr Circ Sig Process 46:103–110
117. Fakhfakh M, Loulou M, Tlelo-Cuautle E (2007) Synthesis of CCIIs and design of simulated CCII-based floating inductance. IEEE international conference on electronics, circuits and systems, Marrakech, pp 379–382
118. Yuce E, Cicekoglu O, Minaei S (2006) CCII-based grounded to floating immittance converter and a floating inductance simulator. Analog Integr Circ Sig Process 46:287–291
119. Yuce E (2006) On the realization of the floating simulators using only grounded passive components. Analog Integr Circ Sig Process 49:161–166

120. Yuce E (2007) Inductor implementation using a canonical number of active and passive elements. Int J Electron 94:317–326
121. Yuce E (2007) Comment and reply 'The effects of non-idealities and current limitations on the simulated inductances employing current conveyors'. Analog Integr Circ Sig Process 51:55
122. Maundy B, Gift S, Aronhime P (2007) A novel hybrid active inductor. IEEE Trans Circ Syst-II 54:663–667
123. Yuce E, Minaei S (2007) A new active network suitable for realizing ladder filters and transformer simulator. J Circ Syst Comput 16:29–41
124. Yuce E, Minaei S, Ibrahim MA (2007) A new simulation of mutually coupled circuit based on CCIIs. Int J Electron 94:367–372
125. Saad RA, Soliman AM (2008) Generation, modelling and analysis of CCII-based gyrators using the generalized symbolic frame work for linear active circuits. Int J Circ Theor Appl 36:289–309
126. Riewruja V, Petchmaneelumka W (2008) Floating current-controlled resistance converters using OTAs. Int J Electron Commun (AEU) 62:725–731
127. Minaei S, Yuce E (2008) Realization of tunable active floating inductance simulators. Int J Electron 95:27–37
128. Yuce E (2008) Grounded inductor simulators with improved low-frequency performances. IEEE Trans Instrum Meas 57:1079–1084
129. Ferri G, Guerrini N, Silverii E, Tatone A (2008) Vibration damping using CCII-based inductance simulators. IEEE Trans Instrum Meas 57:907–914
130. Maundy B, Gift S, Aronhime P (2008) Practical voltage/current-controlled grounded resistor with dynamic range extension. IET Circ Devices Syst 2:201–206
131. Sagbas M, Ayten UE, Sedef H, Koksal M (2009) Floating immittance function simulator and its applications. Circ Syst Sig Process 28:55–63
132. Sagbas M, Ayten UE, Sedef H, Koksal M (2009) Electronically tunable floating inductance simulator. Int J Electron Commun (AEU) 63:423–427
133. Yuce E, Minaei S (2009) On the realization of simulated inductors with reduced parasitic impedance effects. Circ Syst Sig Process 28:451–465
134. Marcellis AD, Ferri G, Guerrini NC, Scotti G, Stornelli V, Trifiletti A (2009) The VCG-CCII: a novel building block and its application to capacitance multiplication. Analog Integr Circ Sig Process 58:55–59
135. Khan IA, Afzal N, Khan MR (2009) Digitally programmable impedance multiplier using CCIIs with high resolution capability. J Active Passive Electron Devices 8:247–257
136. Lahiri A (2009) Comment on electronically tunable floating inductance simulator. Int J Electron Commun (AEU) 63:878
137. Saad RA, Soliman AM (2010) On the systematic synthesis of CCII-based floating simulators. Int J Circ Theor Appl 38:935–967
138. Metin B, Minaei S (2010) Parasitic compensation in CCI-based circuits for reduced power consumption. Analog Integr Circ Sig Process 65:157–162
139. Soliman AM (2010) New CCII and ICCII based realizations of L-C and L-R mutators. Circ Syst Sig Process 29:1181–1191
140. Horng JW, Hou CL, Chun L, Chang CM, Yang H, Shyu WT (2009) Higher order immittance functions using current conveyors. Analog Integr Circ Sig Process 61:205–209
141. Swamy MNS (2011) Mutators, generalized impedance converters and inverters and their realization using generalized current conveyors. Circ Syst Sig Process 30:209–232
142. Khan AA, Al-Turaigi MA, El-Ela MA (1995) An improved current-mode instrumentation amplifier with bandwidth independent of gain. IEEE Trans Instrum Meas 44
143. Bruun E (1991) Constant bandwidth current-mode operational amplifier. Electron Lett 27:1673–1674
144. De Jager W, Smit J (1978) Design and symbolic analysis of current-mode analog circuits. Alfdelingder Electrotechneik

145. Senani R (1987) Generation of new two-amplifier synthetic floating inductors. Electron Lett 23:1202–1203
146. Senani R (1987) A novel application of four terminal floating nullors. Proc IEEE 75:1544–1546
147. Prestcott AJ (1966) Loss compensated active gyrator using differential input operational amplifiers. Electron Lett 2:283–284
148. Ford RL, Girling FEJ (1966) Active filters and oscillators using simulated inductance. Electron Lett 2:52
149. Cabeza R, Carlosena A, Serrano L (1994) Unified approach to the implementations of universal active devices. Electron Lett 30:618–620
150. Pal K (1981) Novel floating inductance using current conveyors. Electron Lett 17:638
151. Horng JW (2012) General high-order grounded and floating immittance structures using current conveyors. Analog Integr Circ Sig Process 71:265–274
152. Cabeza R, Carlosena A (1997) Analog universal active device: theory, design and applications. Analog Integ Circ Sig Process 12:153–168
153. Yuce E (2008) Negative impedance converter with reduced nonideal gain and parasitic impedance effects. IEEE Trans Circ Syst 55:276–283
154. Patranabis D, Ghosh DK (1984) Integrators and differentiators with current conveyor. IEEE Trans Circ Syst 31:567–569
155. Moreira JP, Silva MM (2001) Limits to the dynamic range of low-power continuous-time integrators. IEEE Trans Circ Syst 48:805–817
156. Abuelma'atti MT (1994) On the realization of current-mode integrator using current conveyors. Active Passive Elec Comp 17:79–82
157. Nandi R, Ray SB (1993) Precise realisation of current mode integrators using current conveyor. Electron Lett 29:1152–1153
158. Liu S-I, Kuo J-H, Tsao H-W, Wu J, Tsay J-H (1991) New CCII-based differentiator and its applications. Int J Electron 71:645–652
159. Liu S-I, Hwang Y-S (1994) Dual-input differentiators and integrators with tunable time constants using current conveyors. IEEE Trans Instrum Meas 43:650–654
160. Bruun E (1992) High speed, current conveyor based voltage mode operational amplifier. Electron Lett 28:742–744
161. Gift SJ, Maundy B, Muddeen F (2007) High-performance current-mode instrumentation amplifier circuit. Int J Electron 94:1015–1024
162. Wilson B (1988) Constant bandwidth voltage amplification using current conveyors. Int J Electron 65:983–988
163. Deprettere E (1975) On the minimal realisation of the gyrator by means of nullors and resistors I. Int J Circ Theor 3:383–390
164. Deprettere E (1976) On the minimal realisation of the gyrator by means of nullors and resistors II. Int J Circ Theor 4:285–297
165. Senani R, Bhaskar DR (1994) Versatile voltage-controlled impedance configuration. IEE Proc Circ Devices Syst 141:414–416
166. Hou CL, Wang WY (1998) Realization of floating immittance function simulators using CCII +. Microelectron J 29:59–63
167. Berndt DF, Dutta Roy SC (1969) Inductor simulation using a single unity gain amplifier. IEEE J Solid State Circuits 6:161–162
168. Daniels RW (1969) Gyrators, negative impedance converters and related circuits. IEEE Trans Circ Theor 16:261–262
169. Senani R, Bhaskar DR (2008) Comment on Practical voltage/current-controlled grounded resistor with dynamic range extension. IEE Proc 2:465–466
170. Senani R, Bhaskar DR (1991) Realization of voltage-controlled impedances. IEEE Trans Circ Syst 38:1081–1086; also see ibid, 1991;39:162
171. Senani R, Bhaskar DR (1992) A simple configuration for realizing voltage-controlled impedances. IEEE Trans Circ Syst 39:52–59

Chapter 6
First, Second and Higher Order Filter Design Using Current Conveyors

Abstract A number of first order, second-order and higher-order active filter configurations, chosen from the vast literature available on this topic, using classical types of CCI and/or CCII, have been discussed and their salient features have been highlighted.

6.1 Introduction

Soon after their introduction by Sedra and Smith, CCI and CCII started attracting prominent attention of the researchers. Early works on CCs mainly dealt with the explorations of their use in the realization of filters, oscillators and immittance simulation. Out of these applications, this chapter focuses on the realization of filters using CCs which can be classified into various types such as first, second and higher order types; single-CC and multiple-CC types; fixed-topology and variable-topology types; voltage-mode (VM), current-mode (CM) and mixed-mode types and finally, single-input-single-output (SISO), single input-multiple-output (SIMO), multiple-input-single-output (MISO) and multiple-input-multiple-output (MIMO) types. Selecting from an exhaustive repertoire of filter circuits available in the literature [1–166], published during the last four decades, this chapter is confined to only a limited number of configurations in each of the categories mentioned above which, in the opinion of the authors, have some novelty in their characteristic features.

6.2 The First Order, the Second Order and the Higher Order Filter Realizations Using CCs

In the following, we present some novel first order, second order and higher order active filter configurations and methodologies, realizable from CCI/CCII of the normal kind. In choosing the circuits from a large number of available alternatives, the main considerations have been to include only those circuits which provide as

© Springer International Publishing Switzerland 2015
R. Senani et al., *Current Conveyors*, DOI 10.1007/978-3-319-08684-2_6

many as possible of the following desirable features: (i) employment of a minimum possible number of capacitors preferably grounded as desirable for integrated circuit implementation (ii) availability of sequential/independent tunability of as many characteristic parameters (namely, radian frequency ω_0, quality factor Q_0 and gain H_0) as possible (iii) ease of cascadability-which can be attained by infinite input impedance and/or zero output impedance in case of the voltage-mode configurations and by infinite output impedance and/or zero input impedance in case of the current-mode configurations (iv) employment of a canonic number of resistors for the realization of the intended functions and (v) availability of multifunction capability wherever feasible.

6.2.1 Single-CC First Order All Pass Filters

For an economic and efficient design, it is often necessary to keep the number of passive as well as active elements as small as possible to reduce the power consumption, noise, space and cost. For this purpose, therefore, active filter circuits realizable with a single CC are of interest. In the literature on CCs, there exist a wide variety of single-CC filters, for instance, see [1–6, 11, 30, 42–46, 53, 56–60, 62, 64, 68, 69, 73, 74, 139, 151, 153, 154, 166].

A lot of attention has been given in the CC literature on realizing all pass filters for instance, see [2, 3, 5, 12, 16, 22, 52, 60, 83, 87, 100, 129, 134, 148, 149, 156] because of their utility as phase shifters and their use in the design of oscillators.

Higashimura and Fukui [1] demonstrated that a first order all pass network can be realized by a single CCII− along with only three resistors and a single capacitor. Two of their prepositions are shown in Fig. 6.1. Assuming CCII− to be characterized by $i_y = 0$, $v_x = v_y$ and $i_z = -i_x$, the open circuit voltage ratio transfer function for these circuits is given by

$$T(s) = \frac{V_{out}}{V_{in}} = \pm \frac{1 - sCR}{1 + sCR} \tag{6.1}$$

where the + sign corresponds to the circuit of Fig. 6.1a and the −sign corresponds to the circuit of Fig. 6.1b. The circuit of Fig. 6.1a can provide a phase shift from 0° to −180° whereas that of Fig. 6.1b can provide a phase shift from +180° to 0°. Both the circuits have ideally infinite input impedance which is helpful for easy cascading without requiring any impedance-matching device.

The same authors presented in 1990 [2] two equally simple configurations for realizing first order all pass functions in current-mode with the feature of the availability of ideally infinite output impedance and thereby, facilitating easy cascadability. These circuits from [2] are shown in Fig. 6.2.

Assuming CCII− to be characterized by $i_y = 0$, $v_x = v_y$ and $i_z = -i_x$, the current transfer functions for the two circuits are given by

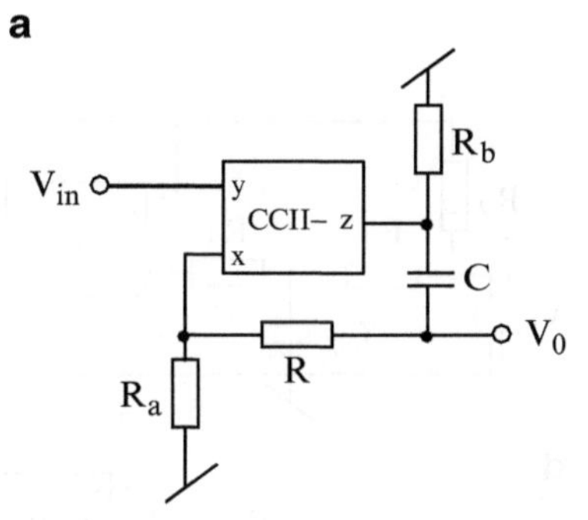

Fig. 6.1 First order voltage-mode all pass filters proposed by Higashimura and Fukui [1]

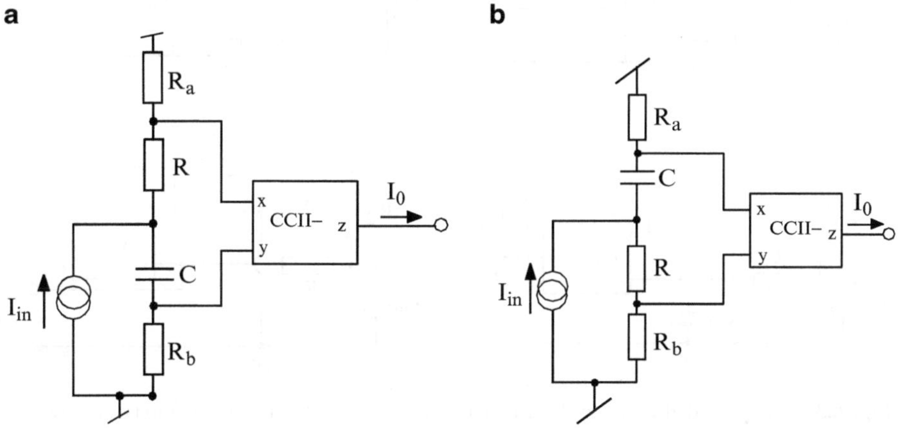

Fig. 6.2 First order current-mode all pass filters proposed by Higashimura and Fukui [2]

$$i_{out} = \frac{I_{out}}{I_{in}} = \pm \frac{1 - sCR}{1 + sCR} \tag{6.2}$$

where the + sign is applicable to the circuit of Fig. 6.2a and the – sign is applicable to the circuit of Fig. 6.2b. The SPICE simulation results of the circuit of Fig. 6.2b using a CCII−, based upon *power supply current sensing technique* employing LF356 type op-amps together with the current mirrors made from pnp transistors 2SC733 and npn transistors 2SC945, have shown that these circuits perform as predicted by the theory.

Cicekoglu et al. in [3] have conducted a systematic generation of CCII+ based first order all pass sections and derived a set of 22 circuits employing single capacitor and 2–3 resistors. Out of these 22 circuits, as many as 16 employ a floating capacitor while the remaining 6 employ a grounded capacitor. In view of the importance of grounded-capacitor structures, as preferred from IC implementation view point, these 6 circuits are shown here in Fig. 6.3.

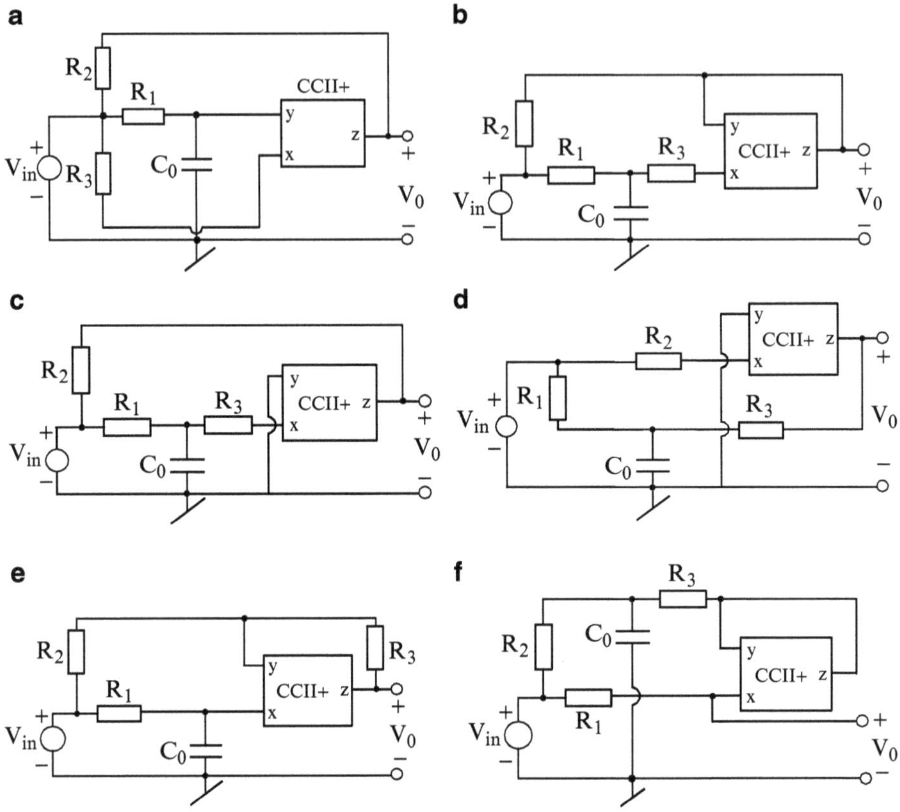

Fig. 6.3 The grounded-capacitor-based first order all pass filter configurations proposed by Cicekoglu et al. [3]

The voltage ratio transfer function of the circuits given in Fig. 6.3 can be expressed in the following general form

$$\frac{V_0(s)}{V_i(s)} = K\,\frac{1 - sT}{1 + sT} \tag{6.3}$$

By a straight forward analysis, the specific values of the transfer voltage ratio, gain K, radian frequency ω_0 and the design constraints for the various circuits are shown in Table 6.1.

It is worthwhile to point out that the circuits of Fig. 6.3b–d have the advantage that the X-port parasitic input impedance r_x can be readily absorbed in the external resistors employed, whereas the circuit of Fig. 6.3f has the advantage of providing a voltage output from a low output impedance terminal since the same is being taken from the X-terminal of the CCII+.

With CCII implemented from AD844, there is flexibility of taking out a low impedance output in circuits of Fig. 6.3a–e by exploiting the buffer between

Table 6.1 The various Transfer functions and the realizability conditions for the circuits of Fig. 6.3

Transfer function	K	$\omega_0 = 1/T$	Design constraints
$\frac{\frac{1}{R_1C_0}-s}{\frac{1}{R_1C_0}+s}$	1	$\left(\frac{1}{R_1C_0}\right)$	$R_2 = 2R_3$
$\frac{\left(\frac{1}{R_1C_0}-\frac{2}{R_2C_0}\right)-s}{\left(\frac{1}{R_1C_0}-\frac{2}{R_2C_0}\right)+s}$	1	$\left(\frac{1}{R_1C_0}-\frac{2}{R_2C_0}\right)$	$R_2 = 2R_3$ $R_2 > 2R_1$
$\frac{\left(\frac{1}{R_3C_0}+\frac{1}{R_1C_0}\right)-s}{\left(\frac{1}{R_3C_0}+\frac{1}{R_1C_0}\right)+s}$	−1	$\left(\frac{1}{R_3C_0}+\frac{1}{R_1C_0}\right)$	$R_2 = 2(R_3 + R_1)$
$\frac{\left(\frac{1}{R_1}-\frac{1}{R_2}\right)\left(\frac{1}{R_1C_0}-s\right)}{\frac{2}{R_1}\left(\frac{1}{R_1C_0}+s\right)}$	$\left(\frac{R_2-R_1}{2R_2}\right)$	$\left(\frac{1}{R_1C_0}\right)$	$R_3 = \left(\frac{R_2-R_1}{2}\right)$ $R_2 > R_1$
$\frac{\left(\frac{1}{R_1C_0}-\frac{1}{R_2C_0}\right)-s}{\left(\frac{1}{R_1C_0}-\frac{1}{R_2C_0}\right)+s}$	1	$\left(\frac{1}{R_1C_0}-\frac{1}{R_2C_0}\right)$	$R_3 = R_1$ $R_2 > R_1$
$\frac{\left(\frac{1}{R_2C_0}-\frac{2}{R_1C_0}\right)-s}{\left(\frac{1}{R_2C_0}-\frac{2}{R_1C_0}\right)+s}$	1	$\left(\frac{1}{R_2C_0}-\frac{2}{R_1C_0}\right)$	$R_1 = 2R_3$ $R_1 > 2R_2$

the Z- terminal of the CCII+ and the voltage output terminal-W of AD844. With AD844 employed to realize the CCII+, the circuits are found to operate well up to a frequency in the MHz range.

Another circuit, which uses a CCII− and also employs only two equal-valued resistors and a single capacitor, was advanced by Khan and Maheshwari [4] and is shown in Fig. 6.4 the transfer function of this circuit is given by (for $R_1 = R_2 = R$)

$$\frac{V_{out}}{V_{in}} = \frac{sCR-1}{sCR+1} \quad \text{with } R_1 = R_2 = R \tag{6.4}$$

6.2.2 Single-CC Biquads

A four or five passive elements-based configuration employing two different types of CCIIs was proposed by Chong and Smith in [5] which can realize BPF, LPF and HPF filter responses by appropriate choice(s) of the type of CCIIs and passive elements. The first circuit configuration as shown in Fig. 6.5a employs a CC_A (CCII− which is characterized by $i_y = 0$, $v_x = v_y$ and $i_z = -i_x$,) and can realize a BPF response whose transfer function is given by

$$\frac{V_0}{V_{in}} = \frac{sC_2R_4}{k\left[s^2C_1C_2R_3R_4 + sR_4(C_1+C_2)+1\right]} \tag{6.5}$$

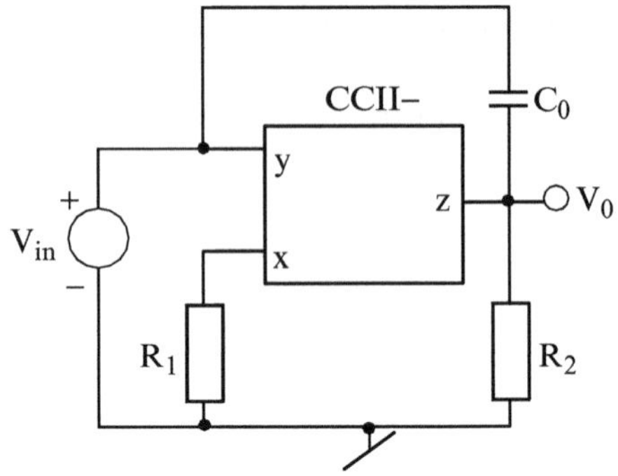

Fig. 6.4 All pass filter proposed by Khan and Maheshwari [4]

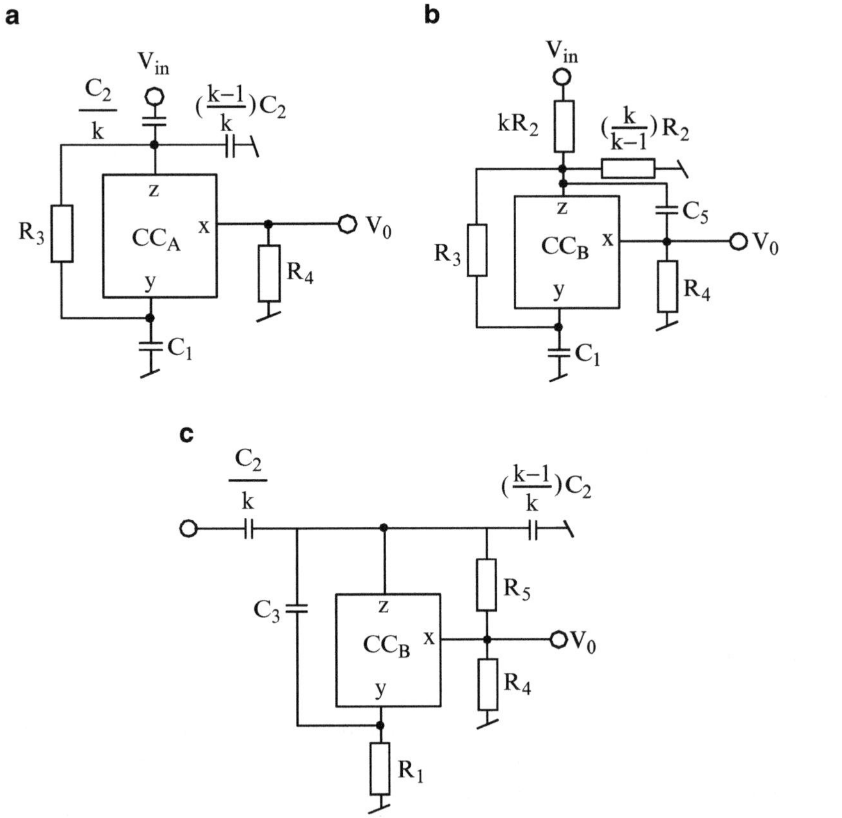

Fig. 6.5 Single-CC-based biquad configurations proposed by Chong and Smith [5] (**a**) BPF realization (**b**) LPF realization (**c**) HPF realization

The circuit of Fig. 6.5b employs a new type[1] of CC (termed in [5] as CC_B) whose terminal characteristics are $i_y = 0$, $v_x = -v_y$ and $i_z = i_x$. This circuit yields the transfer function of an inverting type LPF given by

[1] Note that this is nothing but the so-called 'Inverting Current Conveyor' (ICCII) as later re-introduced by Awad and Soliman; see Sect. 10.2.9 of Chap. 10.

$$\frac{V_0}{V_{in}} = \frac{-R_4}{k\left[2s^2C_1C_5R_2R_3R_4+s\{4C_5R_2R_4+C_1R_4(R_2+R_3)\}+(R_2+R_4)\right]} \tag{6.6}$$

Figure 6.5c also uses a CC_B and can provide an inverting HPF response with the transfer function given by

$$\frac{V_0}{V_{in}} = \frac{-s^2C_2C3R_1R_4R_5}{k\left[s^2C_2C_3R_1R_4R_5\ +s\{C_3(4R_1R_4+R_1R_5+R_4R_5)+C_2R_4R_5\}+2R_4)\right]} \tag{6.7}$$

A generalized single-CC configuration was proposed by Metin et al. in [6] from where a LPF/BPF, a HPF/BPF and Notch/APF/HPF were realizable as special cases. Two specific configurations from [6] are shown in Fig. 6.6.

For the circuit of Fig. 6.6a, a straight forward analysis assuming CCI+ to be characterized by $i_y = i_x$, $v_x = v_y$ and $i_z = i_x$, gives the two transfer functions realizable by the circuit as

$$\frac{V_{01}}{V_{in}} = \frac{s^2C_3C_4R_1R_4+sC_3(R_1-R_4)}{s^2C_3C_4R_1R_4+sC_3(R_1-R_4)+2sC_4R_4+2} \tag{6.8}$$

$$\frac{V_{02}}{V_{in}} = \frac{-sC_3R_4}{s^2C_3C_4R_1R_4+sC_3(R_1-R_4)+2sC_4R_4+2} \tag{6.9}$$

Thus, for $R_1 = R_4$, Eqs. (6.8) and (6.9) will provide HPF and BPF responses respectively.

Similarly, for the circuit of Fig. 6.6b, the two transfer functions are given as

$$\frac{V_{01}}{V_{in}} = \frac{2s^2C_1C_2R_3R_4+s(C_2-C_1)R_4+1}{2s^2C_1C_2R_3R_4+s(C_2-C_1)R_4+2sC_1R_3+1} \tag{6.10}$$

$$\frac{V_{02}}{V_{in}} = \frac{2s^2C_1C_2R_3R_4+s(C_2-C_1)R_4}{2s^2C_1C_2R_3R_4+s(C_2-C_1)R_4+2sC_1R_3+1} \tag{6.11}$$

Thus, for $C_1 = C_2$, the transfer functions in Eqs. (6.10) and (6.11) can realize notch and HPF responses, respectively. Furthermore, if we select $C_1 = 2C_2$ and $R_4 = 2R_3$, Eq. (6.10) can provide an APF response also.

Using a CMOS CCI, pole frequencies of the order of 1.8 and 3.98 MHz have been easily realised in practice.

6.2.3 Multiple-CC Multifunction Biquads

In this section, we present a number of multiple-CC biquads capable of realizing three or more filter functions simultaneously under different conditions and/or at

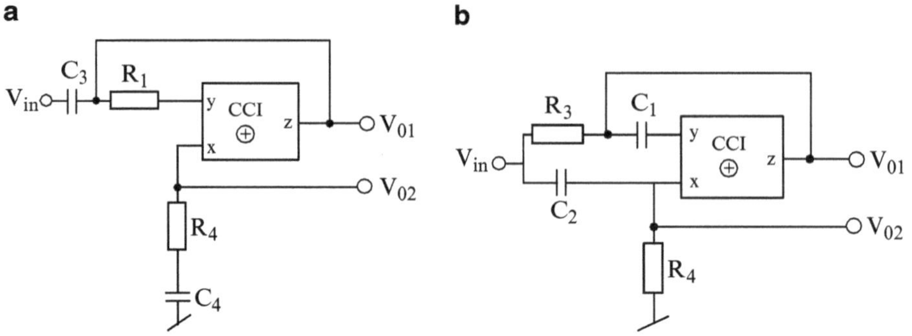

Fig. 6.6 Special cases of Metin et al. [6] single-CC-biquad configuration (**a**) HPF/BPF realization (with RC: CR transformation the same circuit would realize a LPF/BPF) (**b**) Realization of Notch/APF/HPF

different output terminals. Such multifunction biquads can be broadly categorized in the following categories; single-input-single-output (SISO) type, single-input-multi-output (SIMO) type, multiple-input-single-output (MISO) type and multi-input-multi-output (MIMO) type. Furthermore, depending upon whether the input and output variables are voltages or currents or both, they can also be classified as voltage-mode (VM), current-mode (CM), trans-admittance–mode (TAM) and trans-impedance-mode (TIM). Those biquads which are capable of realizing all the five second-order filter functions namely, low pass (LP), high pass (HP), band pass (BP), notch and all pass (AP) from the same configuration are popularly referred to as the *universal* biquads. Finally, the term *mixed-mode universal biquad* is used for those configurations which can realize all the five standard filter functions in both VM and CM (or in two or more of the four possible modes namely VM, CM, TAM and TIM).

6.2.3.1 Single-Input-Single-Output (SISO)-Type VM Biquads

The circuits in this class are usually of variable-topology type in which certain impedances need to be open-circuited (or short-circuited) for realizing a specific type of transfer function. A number of such circuits are reported in technical literature, for instance, [7, 8, 47, 63, 66, 71, 72, 79, 84, 88, 98, 102, 105, 106, 109, 111, 132, 137, 144, 153, 159].

In the following, we present two representative circuits which were advanced by Toumazou and Lidgey [7] and Higashimura and Fukui [8] respectively.

The universal active filter using seven CCIIs and a voltage buffer introduced by Toumazou and Lidgey [7] can realize all the five standard second-order filter responses with high input impedance and independently controllable voltage gain

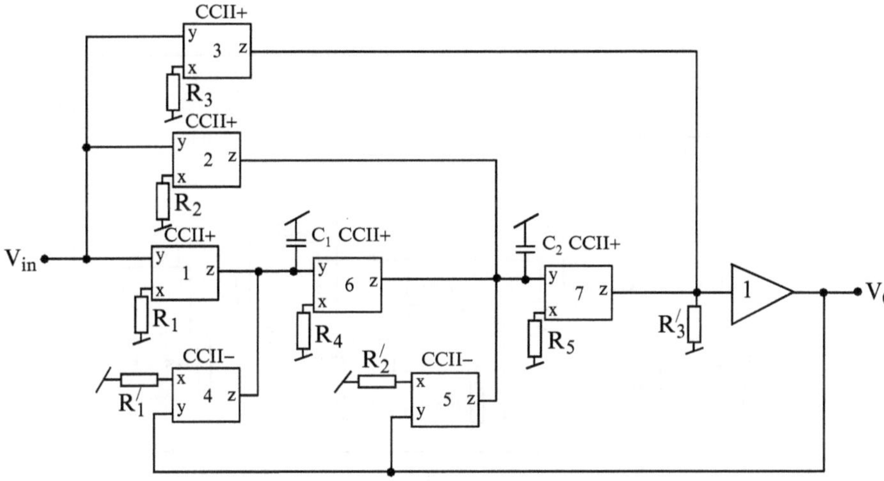

Fig. 6.7 Universal active filter proposed by Toumazou and Lidgey [7]

and is shown in Fig. 6.7. Using ideal port relations of the CCIIs, the transfer function realized by this configuration is given by

$$\frac{V_0}{V_{in}} = \frac{s^2C_1C_2R_4R_5/R_3 + sC_1R_4/R_2 + 1/R_1}{s^2C_1C_2R_4R_5/R_3' + sC_1R_4/R_2' + 1/R_1'} \tag{6.12}$$

from where the various filter responses can be derived as follows: HPF by making $R_1 = R_2 = \infty$; LPF by making $R_3 = R_2 = \infty$; BPF by making $R_1 = R_3 = \infty$; Notch by making $R_2 = \infty, R_3 = R_1$ and finally, APF by making $R_1 = R_2 = R_3, R_1' = R_2' = R_3'$ with CCII2 taken as CCII−.

Using high performance CCII+ and CCII− implemented from Wilsons' OMA-based CCII implementations (see references [7, 8] of Chap. 2) using 741-type op-amps and CA3096E mixed transistor arrays, the circuit was found to give extremely good performance with band pass Q_0 of the order of 200 with very close performance between experimental and theoretical results.

Another high input impedance universal active filter proposed by Higashimura [8] is shown in Fig. 6.8. This circuit uses only plus type CCIIs and grounded capacitors and can yield all the five types of standard filters under appropriate conditions.

In this circuit, the filter parameters namely, ω_0, Q_0 and voltage gain H_0 can be controlled by separate resistors. The transfer function of this circuit can be expressed as:

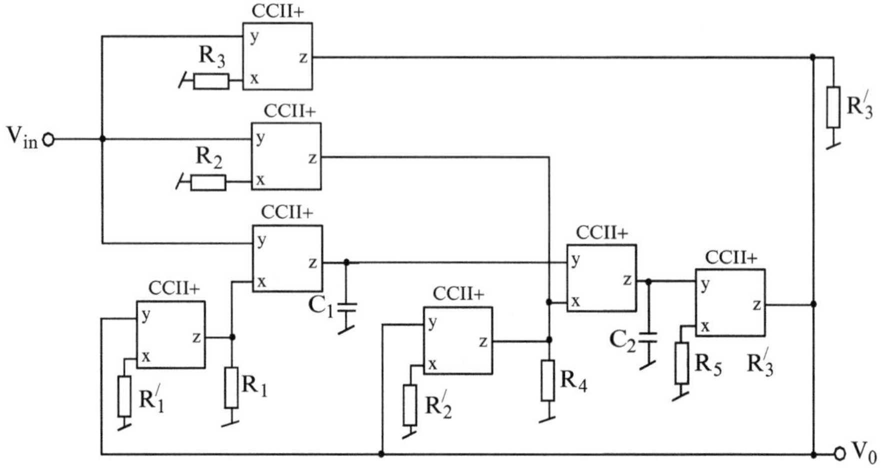

Fig. 6.8 VM universal active filter proposed by Higashimura and Fukui [8]

$$\frac{V_2}{V_1} = \frac{s^2 C_1 C_2 R_4 R_5 / R_3 - s C_1 R_4 / R_2 + 1/R_1}{s^2 C_1 C_2 R_4 R_5 / R_3' - s C_1 R_4 / R_2' + 1/R_1'} \tag{6.13}$$

The conditions required to realize various filters are found to be LPF: $R_3 = R_2 = \infty$, BPF: $R_1 = R_3 = \infty$, HPF: $R_1 = R_2 = \infty$, Notch: $R_2 = \infty$, and APF: $R_3/R_3' = R_2/R_2' = R_1/R_2'$.

As compared to its predecessor of the circuit by Toumazou and Lidgey [7], the important advantage of this circuit is that it does not require the type of CCII to be changed (hence change in topology) to realize the AP function. It has been demonstrated in [8] that using supply current sensing based CCII+ realized with LF356 op-amps and current mirrors constructed from 25A945 and 25C733 bipolar transistors, the circuit can realize various filters with f_0 of the order of 10 KHz (restricted only due to the bandwidth of the op-amps employed).

6.2.3.2 Single-Input-Multiple-Output (SIMO)-Type Voltage Mode (VM) Biquads

One of the common categories for VM biquads has been undoubtedly the so-called SIMO-type biquads which have been in vogue since the times of op-amp-based biquads such as the well-known Kerwin-Huelsman-Newcomb (KHN) biquad [167], Tow-Thomas biquad [168, 169] Akerberg-Mossberg biquad [170], to name a few. It was, therefore, natural that with the advent of CCs, the possibility of realizing SIMO-type VM biquads using CCs caught immediate attention of the researchers, as a consequence, a large number of such biquads are now known, for instance, see [9, 10–16, 17, 77, 78, 83, 85, 92, 94, 95, 110, 118, 119, 130, 131, 148,

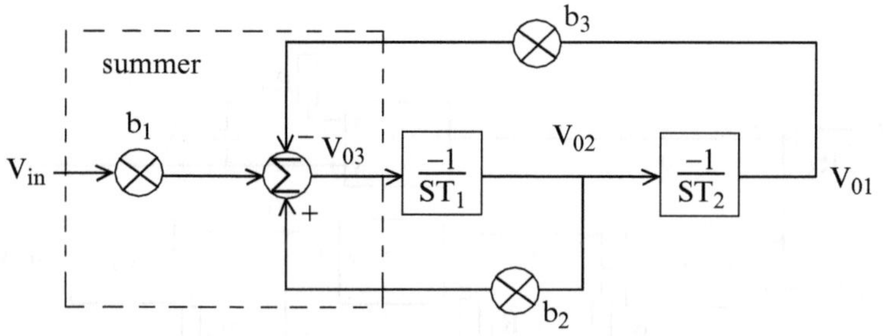

Fig. 6.9 State variable representation of the classical KHN-biquad

152, 157, 163]. In this section, we present some prominent configurations belonging to this class.

The first KHN-equivalent biquad using CCIIs was presented by Soliman in [78] which was capable of realizing exactly the same transfer functions as the op-amp-based KHN-biquad. Several alternative KHN equivalent biquads were subsequently reported independently by Soliman in [83] and one by Senani and Singh in [9]. Since the latter circuit offers a number of advantages over the circuit proposed by Soliman in [78], in the following, we describe the genesis of the quoted circuit from [167].

The original KHN-biquad was derived from the state variable block diagram shown in Fig. 6.9.

The three transfer functions realized by this block diagram are given by:

$$\begin{aligned} T_{LP}(s) &= (b_1/T_1T_2)/D(s);\ T_{BP}(s) = -(sb_1/T_1)/D(s) \text{ and } T_{HP}(s) \\ &= s^2b_1/D(s) \\ &\text{where } D(s) = s^2 + s\,(b_2/T_1) + (b_3/T_1T_2) \end{aligned} \tag{6.14}$$

The various transfer functions of the Senani-Singh circuit [9] of Fig. 6.10, with CCIIs characterized by $i_y = 0$, $v_x = v_y$, $i_z = \pm\, i_x$, are given by

$$T_1(s) = \frac{V_{01}}{V_{in}} = \left[\frac{R_4}{R_5}\right]\left(\frac{R_6/R_4}{R_1R_2C_1C_2}\right)/D(s) = H_0{\omega_0}^2/D(s) \tag{6.15}$$

$$T_2(s) = \frac{V_{02}}{V_{in}} = -\left[\frac{R_3}{R_5}\right]\left(\frac{R_6}{R_1R_3C_1}\right)s/D(s) = -H_0\left(\frac{\omega_0}{Q_0}\right)/D(s) \tag{6.16}$$

and

$$T_3(s) = \frac{V_{03}}{V_{in}} = \left[\frac{R_6}{R_5}\right]s^2/D(s) = H_0s^2/D(s) \tag{6.17}$$

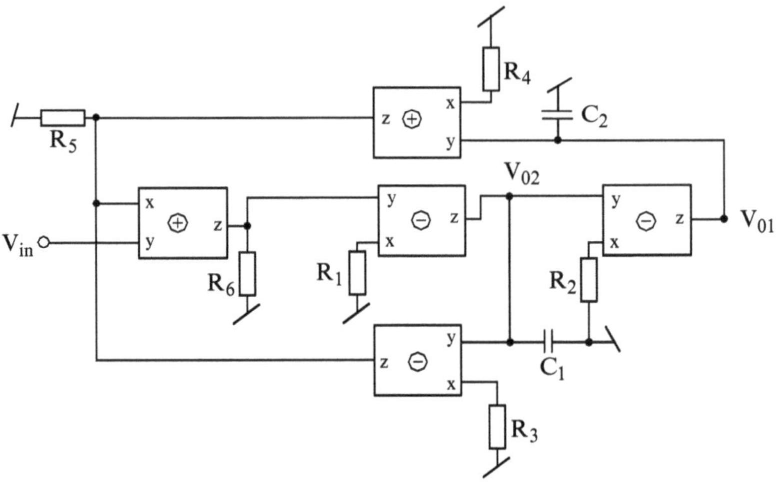

Fig. 6.10 KHN-equivalent CC biquad proposed by Senani and Singh [9]

where

$$D(s) = s^2 + s\left(\frac{\omega_0}{Q_0}\right) + \omega_0^2 = s^2 + \left(\frac{R_6}{R_1 R_3 C_1}\right)s + \left(\frac{R_6}{R_1 R_2 R_4 C_1 C_2}\right) \tag{6.18}$$

The advantages offered by this KHN-equivalent biquad are: (i) Infinite input impedance in contrast to the original KHN biquad as well as the CCII-based KHN equivalent biquad of [78] (ii) employment of both the grounded capacitors as preferred for IC implementation, as well as, the use of all grounded resistors which is particularly attractive in facilitating electronic control of the various parameters of the realized filters by replacing the resistors by FET/MOSFET-based linear voltage-controlled-resistances (VCR) and (iii) independent control of H_0 and Q_0 in case of LP and HP filters by R_5 and R_3 respectively; independent tunability of the center frequency ω_0 by R_2 along with independent tunability of the BW by resistor R_3 in case of BP filter and finally, independent tunability of the gain H_0 by the resistor R_5.

A number of other biquads of varying complexities and features were advanced by a number of authors, for instance, see [10–16] and the references cited therein. However, the circuits advanced by Chang [17] are particularly appealing due to the employment of only four CCs. Out of the various circuits presented in [17], three SIMO-type VM biquads are shown here in Figs. 6.11, 6.12, and 6.13.

Using standard notation and usual characterization of the CCIIs, the following transfer functions are obtained for the circuit of Fig. 6.11.

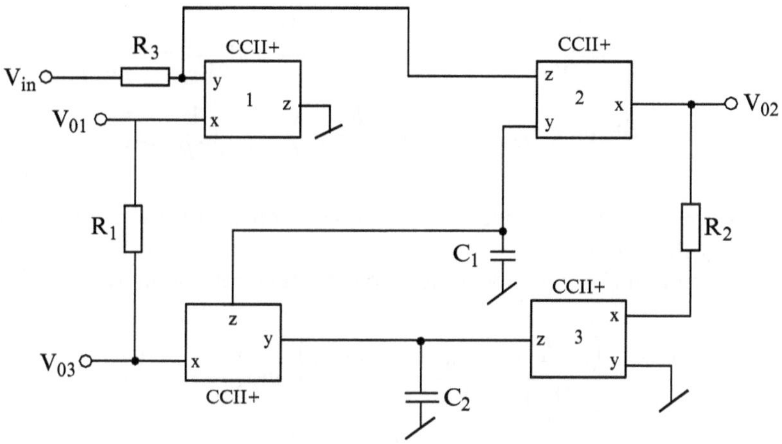

Fig. 6.11 Voltage-mode notch, LPF and BPF proposed by Chang [17]

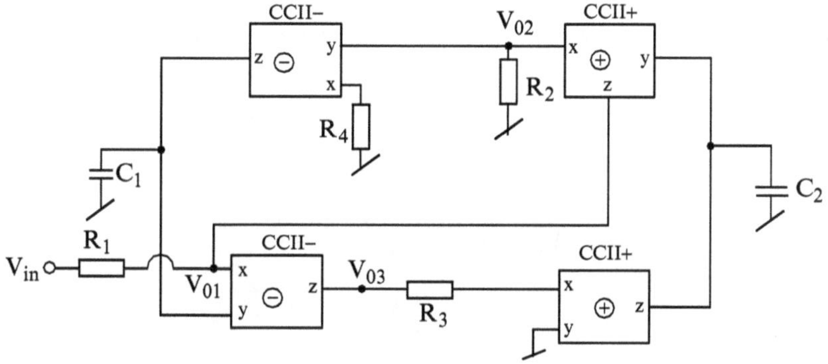

Fig. 6.12 Voltage-mode HPF, BPF and LPF proposed by Chang [17]

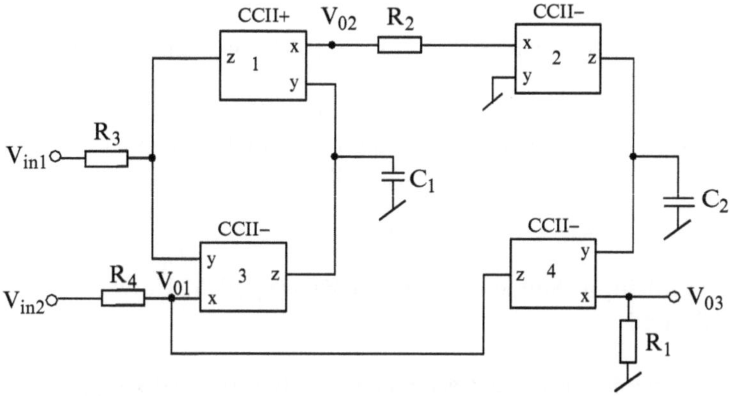

Fig. 6.13 Voltage-mode APF, notch, BPF and LPF proposed by Chang [17]

$$\frac{V_{01}}{V_{in}} = \frac{s^2 + \left(\frac{1}{C_1C_2R_1R_2}\right)}{D(s)}, \frac{V_{02}}{V_{in}} = \frac{-\left(\frac{1}{R_1C_2}\right)s}{D(s)} \text{ and } \frac{V_{03}}{V_{in}} = \frac{\frac{1}{C_1C_2R_1R_2}}{D(s)}$$

$$\text{where } D(s) = s^2 + s\left(\frac{R_3}{C_2R_1R_2}\right) + \frac{1}{C_1C2R_1R2} \tag{6.19}$$

Thus, it is seen that an inverting BPF and non-inverting LPF and notch filter are realizable by this circuit.

For the circuit of Fig. 6.12, the various transfer functions of interest are given by:

$$\frac{V_{01}}{V_{in}} = \frac{\left(\frac{1}{C_1C_2R_1R_4}\right)}{D(s)}, \frac{V_{02}}{V_{in}} = \frac{-s\left(\frac{1}{C_2R_1}\right)}{D(s)} \text{ and } \frac{V_{03}}{V_{in}} = \frac{s^2\left(\frac{R_3}{R_1}\right)}{D(s)}$$

$$\text{where } D(s) = s^2 + s\left(\frac{R_3}{C_2R_1R_2}\right) + \frac{1}{C_1C_2R_1R_4} \tag{6.20}$$

Thus, LPF, BPF and HPF are realizable from this circuit.

Finally, the various transfer functions for the circuit of Fig. 6.13 are given by

$$V_{01} = \frac{V_{in1}\left(s^2 + \frac{1}{C_1C_2R_1R_2}\right) + V_{in2}\left(\frac{sR_3}{C_1R_2R_4}\right)}{D(s)} \tag{6.21}$$

$$V_{02} = \frac{\left(\frac{s}{C_1R_4}\right)(V_{in2} - V_{in1})}{D(s)} \tag{6.22}$$

$$V_{03} = \frac{\left(\frac{1}{C_1C_2R_2R_4}\right)(V_{in2} - V_{in1})}{D(s)} \tag{6.23}$$

$$\text{where } D(s) = s^2 + s\left(\frac{R_3}{C_1R_2R_4}\right) + \frac{1}{C_1C_2R_1R_2} \tag{6.24}$$

Thus, the filters realizable by this circuit are BPF, notch and all pass (if $V_{in1} = V_{in}$ and $V_{in2} = -V_{in}$).

Good agreement has been obtained between the theory and the experimental results of the circuits when they were experimentally tested with CCII+ realized by one AD844 and CCII− realized by two AD844s.

6.2.3.3 Multi-Input-Single-Output (MISO)-Type VM Universal Biquads

In this sub-section, we present a number of interesting MISO-type VM biquads from amongst those known in literature [12–20, 133, 138].

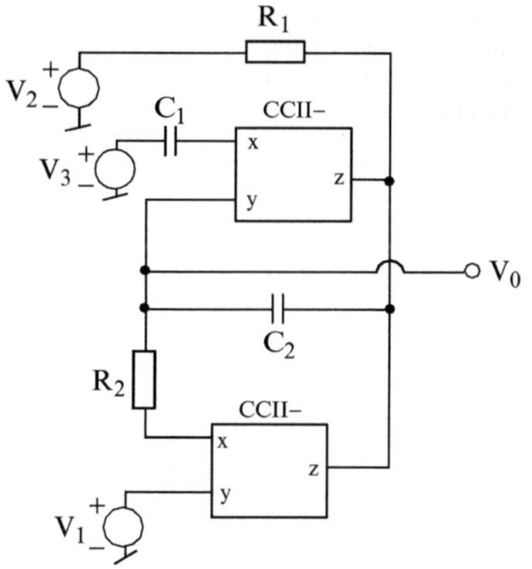

Fig. 6.14 VM MISO-type universal biquadratic filters proposed by Horng [12]

Horng's Universal Biquad Two MISO-types VM universal biquadratic filters were introduced by Horng [12] employing two minus type CCIIs, two capacitors and two resistors.

Both the topologies are capable of realizing LPF, BPF, HPF and notch functions whereas the APF could also be obtained by using one more inverting amplifier to invert one of the voltage signals. One of these circuits is shown in Fig. 6.14.

Assuming ideal CCIIs, the expression for V_0 for the circuit of Fig. 6.14 is given by

$$V_0 = \frac{(s^2C_1C2R_1R_2)V_3 + (sC_2R_2)V_2 + V_1}{s^2C_1C_2R_1R_2 + sC_2R_2 + 1} \tag{6.25}$$

The various filter responses can be realized as follows: HPF: if $V_1 = V_2 = 0$ (grounded) and $V_3 = V_{in}$; BPF: if $V_1 = V_3 = 0$ and $V_2 = V_{in}$; LPF: if $V_2 = V_3 = 0$ and $V_1 = V_{in}$; Notch: if $V_2 = 0$ and $V_1 = V_3 = V_{in}$ and finally, an APF: if $V_1 = -V_2 = V_3 = V_{in}$.

This circuit successfully realizes all the five filter responses using AD844 with the all-pass also realizable by implementing the required inverting amplifier using an AD844 and two additional resistors.

Chang and Tu Universal Biquad The circuit of Fig. 6.15 proposed by Chang and Tu [13] employs two CCII+ s, two capacitors and three resistors with four inputs and one output.

The circuit is capable of realizing all standard second-order filter responses where ω_0 and Q_0 can be controlled orthogonally. Circuit analysis yields the expression for the output voltage V_{out} (assuming ideal CCIIs) as:

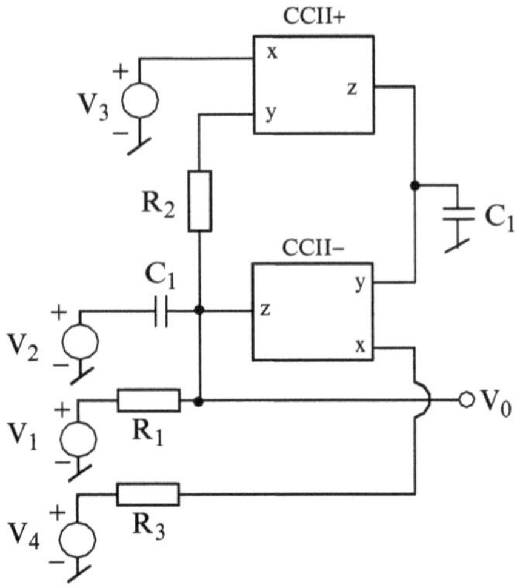

Fig. 6.15 VM universal filter using two CCII+ proposed by Chang and Tu [13]

$$V_{out} = \frac{\left(s^2 C_1 C2 R_1 R_2 R_3\right) V_1 + (sC_1 R_2 R_3) V_2 + (sC_1 R_1 R_3 + R_1) V_3 - \left(sC_1 R1 R_2\right) V_4}{s^2 C_1 C_2 R_1 R_2 R_3 + sC_1 R_3 (R_1 + R_2) + R_1} \tag{6.26}$$

The various filters can be obtained by the appropriate choice(s) of the input signals as follows: HPF: if $V_2 = V_3 = V_4 = 0$ (grounded) and $V_1 = V_{in}$; BPF: if $V_1 = V_3 = 0$, $V_2R_3 \neq V_4R_1$ and V_{in} is either V_2 or V_4; LPF: if $V_1 = 0$, $(V_2/R_1 + V_3/R_2) = V_4/R_3$ and $V_{in} = V_3$; Notch: if $(V_2/R_1 + V_3/R_2) = V_4/R_3$, and $V_{in} = V_1 = V_3$; APF: if $\{V_2/R_1 + V_3\ (2/R_2 + 1/R_1)\} = V_4/R_3$ and $V_{in} = V_1 = V_3$.

The experimental verification of the operation of this circuit in various modes has been successfully demonstrated by building the circuit with $R_1 = R_2 = 4$ KΩ, $R_3 = 1$ KΩ $C_1 = 250$ pf, $C_2 = 25$ nF and CCII+ s implemented with AD844.

Liu and Lee Universal Biquads Three VM universal filter configurations using two CCIIs, three resistors and two capacitors were presented by Liu and Lee [14] and are shown in Fig. 6.16a–c.

By a routine circuit analysis (assuming ideal CCIIs), the output voltage of the circuits of Fig. 6.16a, b is given by:

$$V_1 = \frac{\left(s^2 C_1 C2 R_2 R_3\right) V_{in1} - (sC_2 R_3) V_{in2} + V_{in3}}{s^2 C_1 C_2 R_2 R_3 + s\left(\frac{C_2 R_2 R_3}{R_1}\right) + 1} \tag{6.27}$$

The various filter responses can be realized as follows: (i) LPF: if $V_{in1} = V_{in2} = 0$ (grounded) and $V_{in} = V_{in3}$ (ii) BPF: if $V_{in1} = V_{in3} = 0$ and $V_{in} = V_{in2}$ (iii) HPF: if $V_{in2} = V_{in3} = 0$ and $V_{in} = V_{in1}$ and BPF can also be obtained with V_2/V_{in1} (iv) Notch: if $V_{in2} = 0$, $V_{in1} = V_{in3} = V_{in}$ and (v) APF: if $V_{in1} = V_{in2} = V_{in3} = V_{in}$ and $R_1 = R_2$

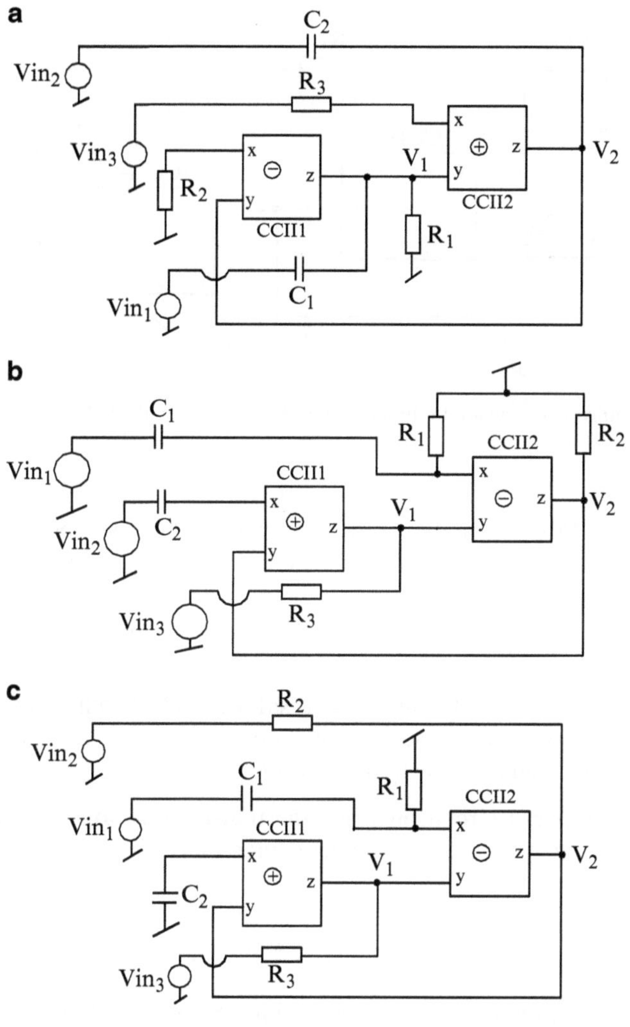

Fig. 6.16 VM universal filters using CCIIs proposed by Liu and Lee [14]

If CCIIs of Fig. 6.16b are exchanged, another universal filter can also be achieved.

The output voltage of the circuit of Fig. 6.17c is found to be:

$$V_1 = \frac{\left(s^2C_1C2R_2R_3\right)V_{in1} + (sC_2R_3)V_{in2} + V_{in3}}{s^2C_1C_2R_2R_3 + s\left(\frac{C_2R_2R_3}{R_1}\right) + 1} \tag{6.28}$$

This filter structure is particularly attractive because grounded-capacitor realization of LPF and BPF is available which is suitable for monolithic integrated circuit

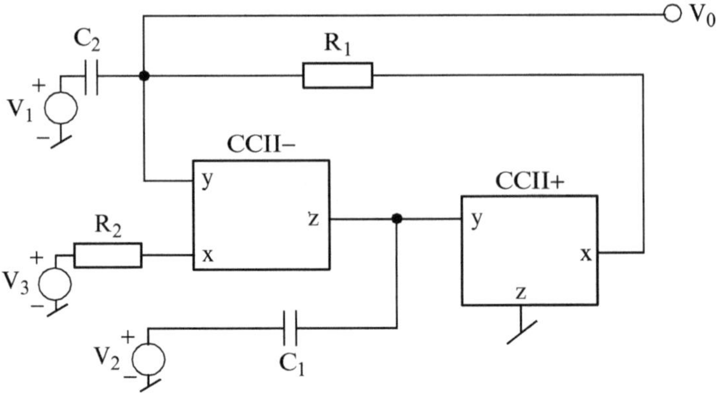

Fig. 6.17 VM universal biquadratic Filter proposed by Horng et al. [15]

implementation. In addition, these circuits provide orthogonal control of ω_0 and Q_0 by grounded resistors, are easily convertible into a voltage-controlled filter and have low total-component- count.

The practical feasibility of all the three circuits has been verified using AD844 to realize CCIIs.

Horng, Lee, Cheng and Chang Universal Biquad Another low-component-count MISO-type two CCII-based VM universal biquadratic filter was introduced by Horng et al. [15] and is shown in Fig. 6.17.

By straight forward circuit analysis, the transfer functions realizable by this circuit can be expressed by the following expression:

$$V_0 = \frac{(s^2 C_1 C2 R_1 R_2)V_1 + (sC_1 R_2)V_2 + V_3}{s^2 C_1 C_2 R_1 R_2 + s(C_1 R_2) + 1} \tag{6.29}$$

From the above equation, one can see that all the five generic filter functions can be realized as follows: (i) LPF: if $V_1 = V_2 = 0$ (grounded) and $V_3 = V_{in}$ (ii) BPF: if $V_1 = V_3 = 0$ (grounded) and $V_2 = V_{in}$ (iii) HPF: $V_2 = V_3 = 0$ (grounded) and $V_1 = V_{in}$ (iv) Notch: $V_2 = 0$ (grounded) and $V_3 = V_{in} = V_1$ and (v) APF: if $V_1 = -V_2 = V_3 = V_{in}$. Thus, the penalty to be paid for realizing an APF is that one more CCII, along with two resistors will be needed to realize an inverting amplifier to negate the sign of the input V_2.

The circuit has been found to work as predicted by theory when CCIIs are realized with AD844. The Q-factors of the order of 10 and f_0 of the order of 40 KHz were found to be easily realizable.

Horng, Tsai and Lee Universal Biquad A two CCII-based MISO-type VM filter using three resistors and two capacitors was proposed by Horng et al. in [16] (Fig. 6.18).

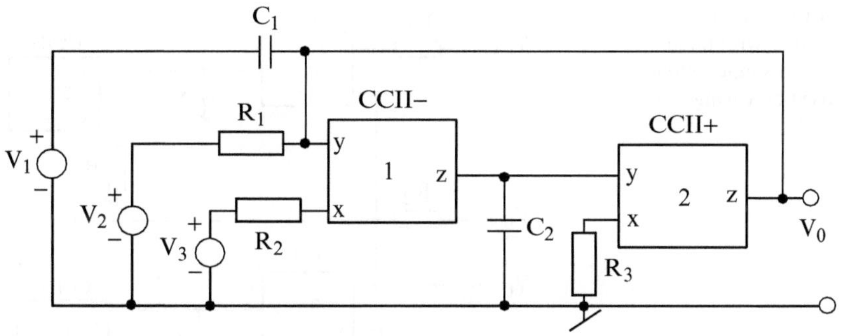

Fig. 6.18 Universal active VM filter proposed by Horng et al. 16]

Assuming ideal CCIIs, a straight forward circuit analysis yields the following relation between the output voltage in terms of the various input voltages:

$$V_{out} = \frac{\left(s^2C_1C2R_1R_2R_3\right)V_1 + (sC_2R_2R_3)V_2 + (R_1)V_3}{s^2C_1C_2R_1R_2R_3 + sC_2R_2R_3 + R_1} \tag{6.30}$$

The various filter functions can be obtained are (i) HPF: if $V_2 = V_3 = 0$ (grounded) and $V_1 = V_{in}$ (ii) BPF: if $V_1 = V_3 = 0$ (grounded) and $V_2 = V_{in}$ (iii) LPF: $V_1 = V_2 = 0$(grounded) and $V_3 = V_{in}$ (iv) Notch: $V_2 = 0$ (grounded) and $V_1 = V_3 = V_{in}$ and (v) APF: if $V_1 = -V_2 = V_3 = V_{in}$. Note that this circuit also requires an additional CCII and two resistors for inverting one of the voltage inputs to enable the realization of the APF.

With CCII+ implemented from one AD844 and CCII− from two AD844s, and components chosen as $R_1 = R_2 = R_3 = 1\ K\Omega$ and $C_1 = C_2 = 0.1\ \mu F$, the circuit has been found to realize all the five filters as predicted by theory.

Chang Universal Biquad Another MISO-type VM biquad was proposed by Chang [17] as shown in Fig. 6.19.

By straight forward analysis, the output voltage of this circuit as a function of the various inputs is given by

$$V_{out} = \frac{\left(s^2C_1C2R_2R_3R_4R_5\right)V_a + (sC_1R_1R_3R_5)V_b + (R_1R_2)V_c}{s^2C_1C_2R_2R_3R_4R_5 + sC_1R_1R_3R_5 + R_1R_2} \tag{6.31}$$

Like the previous cases, it can be easily seen that by appropriate selection of various input signal(s), all the types of fitter responses can be derived.

The attractive features of this circuit are: realisability of all the five functions, lack of component-matching requirements, employment of both grounded capacitors as preformed for IC implementation, orthogonal tunability of ω_0 and Q_0 and the low-sensitivity performance.

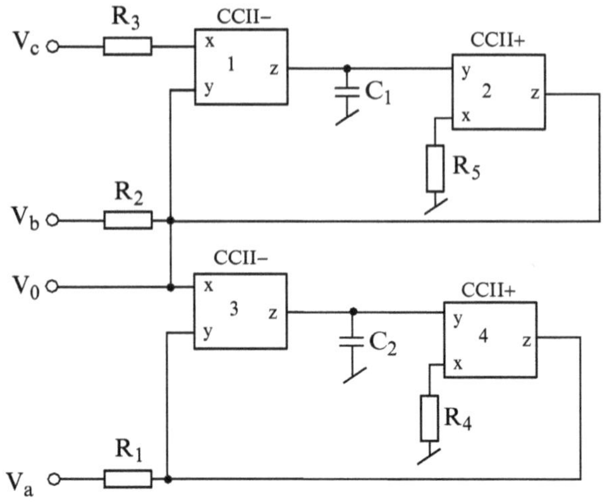

Fig. 6.19 VM universal active filter with three inputs and single output proposed by Chang [17]

Ozoguz and Gunes Universal Biquad The universal filter with three inputs and single output using plus type CCIIs was presented by Ozoguz and Gunes [18] as shown in Fig. 6.20.

The configuration is capable of realizing all type of second-order filter responses with proper choice(s) of input signal(s). Assuming ideal CCIIs, routine circuit analysis yields the following expression of the output voltage in terms of the various input voltages:

$$V_{out} = \frac{V_{in3}\left(s^2 C_1 C2R_1R_2R_3 + sC_1R_1R_3\right) - V_{in2}sC_1R_1R_2 + V_{in1}R_2}{s^2C_1C_2R_1R_2R_3 + sC_1R_1R_3 + R_2} \quad (6.32)$$

The practical workability of this circuit had been demonstrated by realizing various filters with $Q_0 = 1$ and $f_0 = 1.59$ KHz which are typical of those needed when a second order all pass filter is used to linearize the phase of an nth order low pass filter.

Chang and Lee Universal Biquads The filter circuits proposed by Chang and Lee [19] employ three CCIIs and one voltage follower and realizes a VM biquad having three inputs and one output (Fig. 6.21)

Assuming ideal CCIIs, circuit analysis yields the following expression of the output voltage:

$$V_{out} = \frac{\left(s^2 C_1 C2R_1R_2R_3\right)V_1 + (sC_2R_2R_3)V_2 + (R_1)V_3}{s^2C_1C_2R_1R_2R_3 + sC_2R_2R_3 + R_1} \quad (6.33)$$

From the inspection of the Eq. (6.33), it is clear that all the five filtering functions can be realized by appropriate choice(s) of input(s) which are: $V_1 = 0$, $V_2 = 0$, $V_3 = V_{in}$, for LPF; $V_1 = 0$, $V_3 = 0$, $V_2 = V_{in}$ for BPF; $V_2 = 0$, $V_3 = 0$ for HPF; $V_2 = 0$, $V_1 = V_3 = V_{in}$ for notch and with V_2 inverted, $V_1 = V_2 = V_3 = V_{in}$ for all pass.

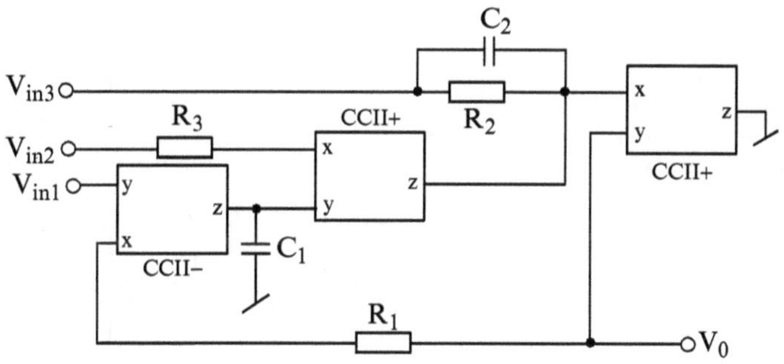

Fig. 6.20 VM universal filter with three inputs single output using CCIIs+ proposed by Ozoguz and Gunes [18]

Fig. 6.21 Universal active VM filter with three inputs and one output proposed by Chang and Lee [19]

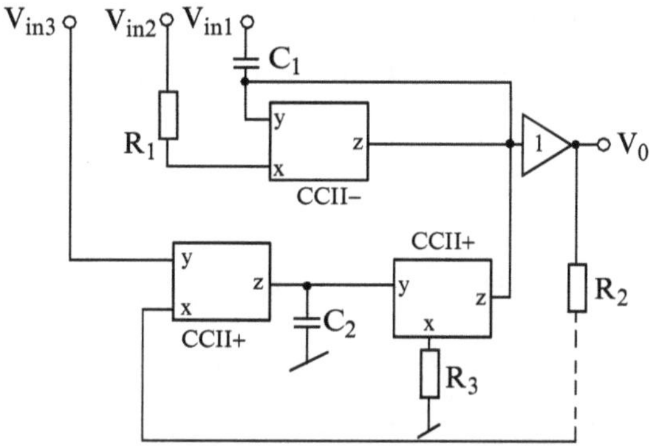

The circuit shown in Fig. 6.22 has been proposed by the same authors [20] using only two CCIIs and one voltage follower as a simplification to their proposition of Fig. 6.21. A little reflection into the role of the CCII− in the circuit of Fig. 6.21 shows that it is indeed redundant and can be deleted.

The expression of the output for the simplified circuit of Fig. 6.22, however, remains unchanged by this simplification and is, therefore, still given by Eq. (6.33).

With the CCIIs realized from AD844, good correspondence between theory and experiments has been observed [19]. The circuit of Fig. 6.22 [20] has the advantage of requiring one less CCII while providing all the features of its predecessor of Fig. 6.21 [19]. Both the circuits are completely implementable with AD844 ICs.

6.2.3.4 Multi-Input-Multi-Output (MIMO)-Type VM Biquads

In this sub-section, we present a number of interesting MIMO-type VM biquads.

Chen Universal Biquad A MIMO-type universal filter configuration composed of only one CCII−, two capacitors and three resistors was proposed by Chen [11],

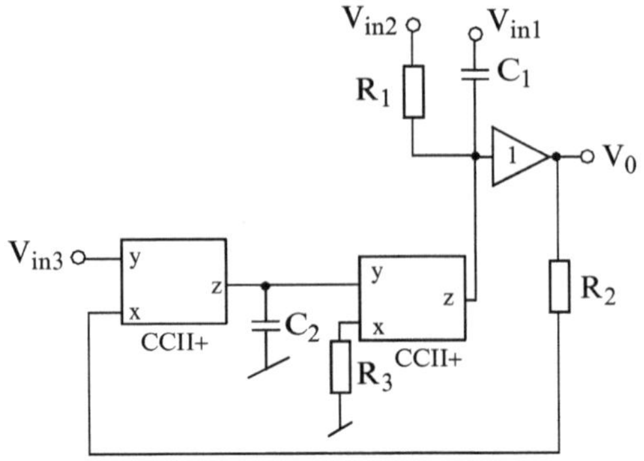

Fig. 6.22 Another VM filter with three inputs and one output proposed by Chang and Lee [20] which is obtained by removing one redundant CCII-from the circuit of Fig. 6.21

which can realize all the five standard filter functions. This circuit is shown in Fig. 6.23.

Assuming ideal CCII−, the expressions for V_{01} and V_{02} in terms of its inputs can be derived as:

$$V_{01}=\frac{1}{\nabla}\left[\left(s^2C_1C_2+sC_1/R_3\right)V_{i1}+\left(sC_2/R_1\right)V_{i2}+\left\{\left(1/R_2R_3\right)-sC_2/R_1\right)\right\}V_{i3}\right]$$

$$V_{02}=\frac{1}{\nabla}\left[\left(s^2C_1C_2+sC_1/R_2\right)V_{i1}-\left(sC_1/R_1+sC_1/R_2+sC_2/R_1+1/R_1R_2\right)V_{i3}\right.$$
$$\left.+\left\{\left(1/R_1R_2\right)+sC_2/R_1+sC_1/R_1\right)\right\}V_{i2}\right]$$
$$\text{where } \nabla = s^2C_1C_2+s\left(C_1C2\right)/R_3+1/R_2R_3 \qquad (6.34)$$

Thus, the various filters can be realized as follows: (i) for $V_{i1}=0$ (grounded) and $V_{i2}=V_{i3}=V_{in}$, LPF can be realized at V_{01} and BPF at V_{02}; (ii) for $V_{i1}=V_{i3}=0$ and $V_{i2}=V_{in}$, a BPF can be obtained at V_{01}; (iii) for $V_{i1}=V_{i3}=V_{i2}=V_{in}$, a HPF can be realized at V_{02}; (iv) for $V_{i2}=0$, $V_{i1}=V_{i3}=V_{in}$ and $C_2R_3=C_1R_1$, a notch can be obtained at V_{01}; (v) for $V_{i2}=0$, $V_{i1}=V_{i3}=V_{in}$ and $C_2R_3=R_1\ (2C_1+C_2)$, an APF can be obtained at V_{01}.

Using CCII-implementable in 0.35 μm CMOS process, biased with ± 1.65 V DC power supplies, the circuit has been observed to realize all the five filters with f_0 of the order of 1 MHz. The simulation results were found to be in close conformity to the theoretical ones. However, in actual hardware, the circuit can also be realized using two AD844 ICs to realize the CCII−

Chang Universal Biquad Figure 6.24 shows a two input three output multi-function biquad presented by Chang [17].

The expressions for the three outputs in terms of the two inputs (assuming ideal CCIIs) are found to be:

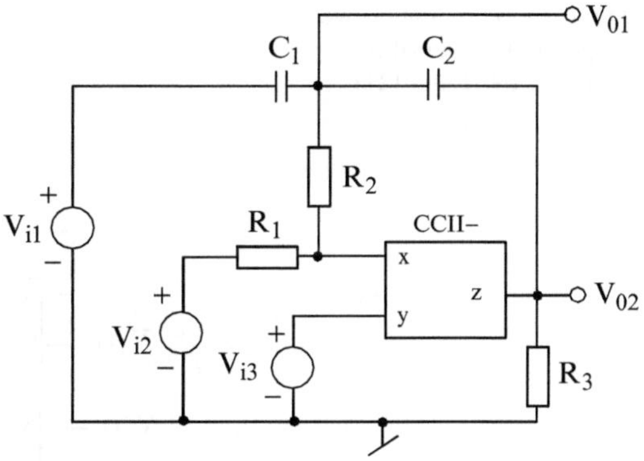

Fig. 6.23 CCII-based MIMO-type VM universal filter proposed by Chen (Adapted from [11] © 2009 Springer)

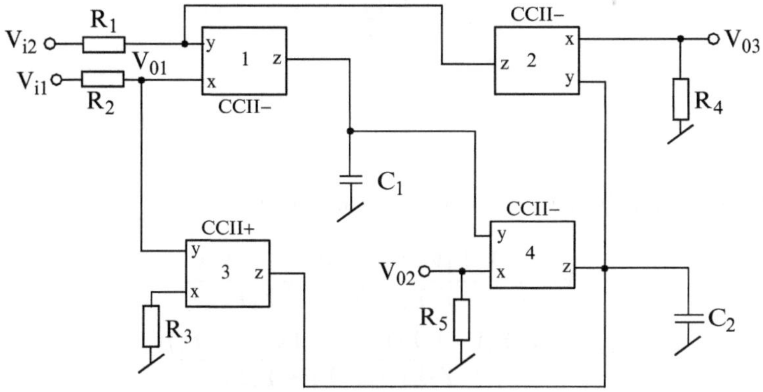

Fig. 6.24 VM Notch, HPF, LPF and BPF proposed by Chang [17]

$$V_{o1} = \frac{\left(s^2C_1C2R_2R_3R_4R_5\right)V_{i1} + (R_1R_3)V_{i2}}{s^2C_1C_2R_2R_3R_4R_5 + sC_1R_1R_2R_5 + R_1R_3} \tag{6.35}$$

$$V_{o2} = \frac{\left(-sC_2C2R_3R_4R_5\right)V_{i1} + \left(sC_2R3R_4R5 + R_1R_5\right)V_{i2}}{s^2C_1C_2R_2R_3R_4R_5 + sC_1R_1R_2R_5 + R_1R_3} \tag{6.36}$$

$$V_{o3} = \frac{\left(sC_1C2R_2R_4R_5 + R_3R_4\right)V_{i1} + (-R_3R_4)V_{i2}}{s^2C_1C_2R_2R_3R_4R_5 + sC_1R_1R_2R_5 + R_1R_3} \tag{6.37}$$

From the above, it may be seen that by appropriate choice of input signal(s), VM Notch, HPF, LPF and BPF responses can be obtained.

Horng Universal Biquad Two VM filter configurations for possessing high input impedance employing three plus-type CCIIs and having three inputs were presented by Horng [21]. The first filter circuit is shown in Fig. 6.25.

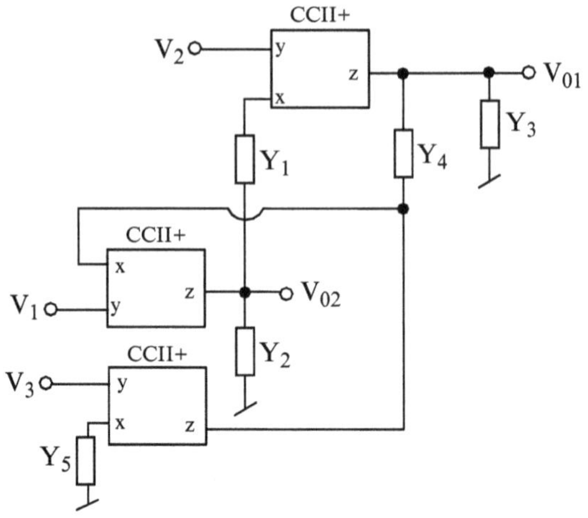

Fig. 6.25 Three-input-two-output VM universal filter proposed by Horng [21]

The expressions for the output voltages are given by:

$$V_{01} = \frac{Y_2Y_4V_1 + Y_1Y_2V_2 + Y_1Y_5V_3}{Y_2Y_4 + Y_2Y_3 + Y_1Y_3} \tag{6.38}$$

$$V_{02} = \frac{Y_3Y_4V_1 + Y_1Y_3V_2 - Y_5(Y_3 + Y_4)V_3}{Y_2Y_4 + Y_2Y_3 + Y_1Y_3} \tag{6.39}$$

Two choices of the circuit admittances leading to two different biquad filters as special cases have been elaborated. However, one of these requires three capacitors leading to a non-canonic structure and is hence omitted. Likewise, the second configuration proposed in [21] has three capacitors in both the choices of components therein; hence, this particular configuration is also omitted.

In the configuration shown in Fig. 6.25 only the following choice of components leads to a canonic structure: $Y_1 = 1/R_1$, $Y_2 = sC_2$, $Y_3 = 1/R_3$, $Y_4 = sC_4$ and $Y_5 = 1/R_5$. Under these conditions from Eqs. (6.38) and (6.39), the various filter responses can be obtained as follows:

(i) For $V_2 = V_3 = 0$ (grounded) and $V_1 = V_{in}$, HPF can be realized at V_{01} and BPF at V_{02} (ii) for $V_1 = V_3 = 0$ (grounded) and $V_2 = V_{in}$, BPF can be realized at V_{01} and LPF at V_{02} (iii) for $V_1 = V_2 = 0$ (grounded) and $V_3 = V_{in}$, a LPF can be realized at V_{01} (iv) for $V_2 = 0$ (grounded), $R_3 = R_5$ and $V_1 = V_3 = V_{in}$ a notch can be obtained at V_{01} (v) for $V_1 = -V_2 = V_3 = Vin$, $R_1 = R_3 = R_5$, an APF can be realized.

The workability of this circuit has been confirmed with circuits implemented from AD844 ICs. It was observed that input levels of with 17.3 V p-p are possible without significant distortion when the AD844 were biased from ± 10 V Dc power supplies.

6.2.3.5 Single-Input-Multi-Output (SIMO)-Type CM Biquads

In the following sub-section, we present a selected number of current-mode biquads chosen from a large number of available ones [22–27, 70, 71, 75, 76, 80, 87, 97, 101, 104, 106, 115, 118, 136, 140, 141, 143, 146, 147, 148, 162].

Ozoguz and Acar CM Universal Biquad The circuit shown in Fig. 6.26 was proposed by Ozoguz and Acar [22] and is based on an approach involving RLC shunt circuit (presented in [24] and to be spelt out explicitly later in this subsection) which realizes inverting type LPF, HPF and BPF simultaneously employing a minimum number of passive components and without requiring any component matching conditions. Although this filter circuit does not offer independent control of Q_0 and ω_0, nevertheless, Q_0 can be made controllable by connecting a grounded resistor in parallel with the input signal $I_{in.}$

The various transfer functions realized by this configuration are given by:

$$\frac{I_{01}}{I_{in}} = \frac{-s^2}{s^2 + s\left(\frac{1}{C_1R_1}\right) + \frac{1}{C_1C2R_1R_2}};\ \frac{I_{02}}{I_{in}} = \frac{-s\left(\frac{1}{C_1R_1}\right)}{s^2 + s\left(\frac{1}{C_1R_1}\right) + \frac{1}{C_1C2R_1R_2}}$$

$$\frac{I_{03}}{I_{in}} = \frac{-\left(\frac{1}{C_1R_1C_1R_2}\right)}{s^2 + s\left(\frac{1}{C_1R_1}\right) + \frac{1}{C_1C2R_1R_2}}. \tag{6.40}$$

The notch filter can be obtained by adding I_{01} and I_{03} while by replacing $CCII_2+$ by $CCII_2-$ and then adding I_{01}, I_{02} and I_{03}, an APF can also be realized.

Abuelma' atti and Shabra CM Universal Biquad: A novel CC-based universal CM filter with one input and four outputs employing all grounded passive components, presented by Abuelma' atti and Shabra [23], is shown in Fig. 6.27.

The proposed configuration can yield all type of responses simultaneously. In all the filter responses, ω_0 can be controlled by a grounded resistor without disturbing $\frac{\omega_0}{Q_0}$ which can also be tuned by another resistor without disturbing ω_0. The circuit, when implemented with AD844s has shown results which are in good conformity with the theoretical analysis.

A simple approach of deriving SIMO-type CM biquad filters A simple approach of deriving SIMO-type CM biquad filters was presented by Senani in [24]. The basic idea of this methodology is the simple (but not so obviously recognized before its disclosure in [24]) observation that the currents through the inductor, resistor and the capacitor represent respectively a second-order LPF, BPF and HPF responses and hence, if these currents can be sensed out by using either CCII− or CCI- as shown in Fig. 6.28.

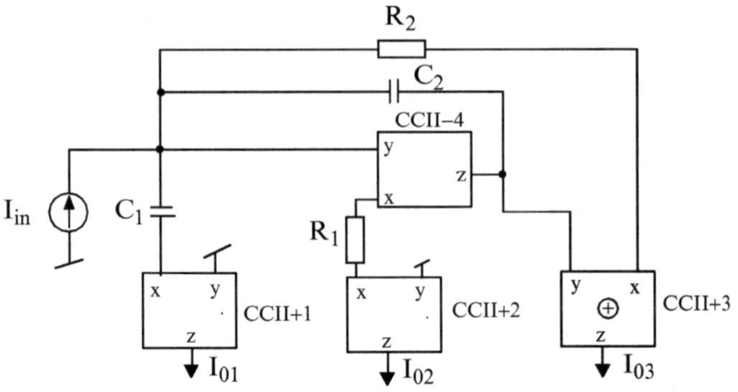

Fig. 6.26 CM universal biquadratic filter proposed by Ozoguz and Acar [22]

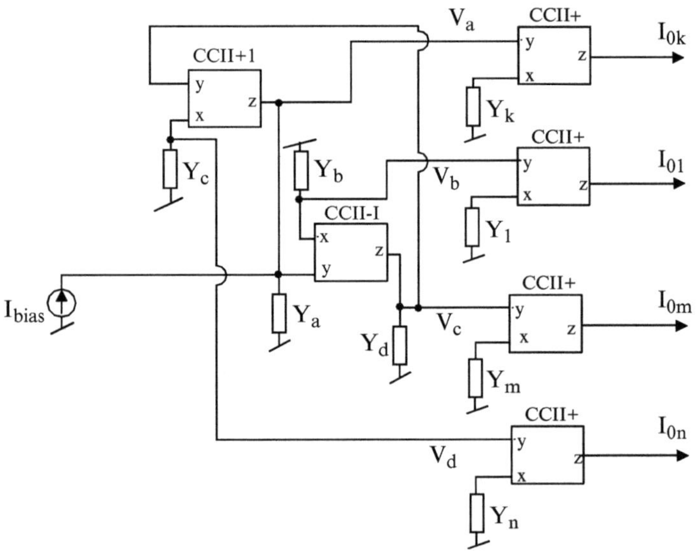

Fig. 6.27 CC-based CM universal filter with single input and four outputs proposed by Abuelma'atti and Shabra (Adapted from [23] © 1996 Elsevier Science Limited)

This circuit, therefore, is an excellent vehicle for generating CM universal biquads by replacing the inductor by CC-simulated inductor and then realizing the remaining two functions namely, notch and APF by adding LPF and HPF currents in the former case and by adding LPF, inverting BPF and HPF currents in the latter case.

The five transfer functions, thus, realizable, are given by:

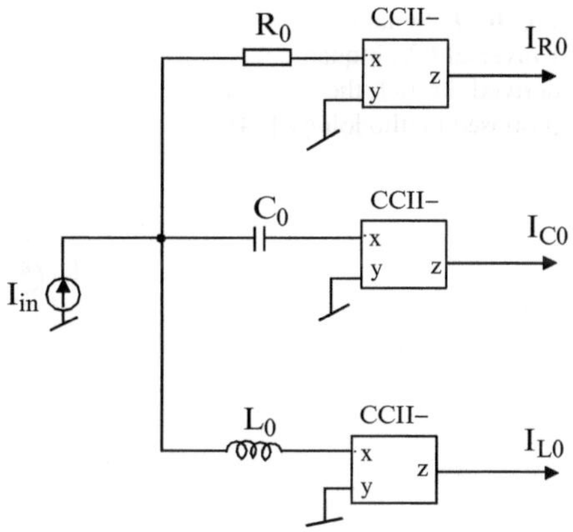

Fig. 6.28 The method of derivation of a CM biquad filter from a passive RLC prototype, proposed by Senani [24]

$$\frac{I_{L0}}{I_{IN}} = \frac{\frac{1}{L_0C_0}}{D(s)};\ \frac{I_{R0}}{I_{IN}} = \frac{s\frac{1}{R_0C_0}}{D(s)};\ \frac{I_{L0}}{I_{IN}} = \frac{s^2}{D(s)};\ \frac{I_{01}}{I_{IN}} = \frac{s^2+\frac{1}{L_0C_0}}{D(s)} \text{ and}$$

$$\frac{I_{02}}{I_{IN}} = \frac{s^2 - s\frac{1}{C_0R_0}+\frac{1}{L_0C_0}}{D(s)} \quad \text{where } D(s) = s^2 + s\frac{1}{C_0R_0}+\frac{1}{L_0C_0} \tag{6.41}$$

It is interesting to note that depending upon the type of simulated inductor employed, this basic scheme can give rise to a variety if CC-based universal biquads. An exemplary circuit is shown in Fig. 6.29 which realizes all the five functions mentioned above with $L_o = C_1R_1R_2$.

Another exemplary CM biquad realizable with only four CCIIs is shown in Fig. 6.30.

The various transfer functions realized by this circuit are given by:

$$\frac{I_{01}}{I_{IN}} = \frac{s^2}{D(s)};\ \frac{I_{03}}{I_{IN}} = \frac{-s\frac{1}{R_2C_1}}{D(s)};\ \frac{I_{02}}{I_{IN}} = \frac{s+\frac{1}{R_1C_1}}{D(s)}$$
$$\text{where } D(s) = s^2 + s\frac{1}{C_1R_1}+\frac{1}{C_1R_1C_2R_2} \tag{6.42}$$

A LPF can be realized by adding currents I_{02} and I_{03} (with $R_1 = R_2$); Notch can be realized by obtaining $I_{notch} = I_{01} + I_{02} + I_{03}$ (with $R_1 = R_2$) and finally, an APF can be realized by adding I_{01}, I_{02}, and I_{03} (with $R_1 = 2R_2$)

Senani's CM Universal Biquad The CM filter structure shown in Fig. 6.31 was proposed by Senani [25] and offers the realization of all generic filter functions without requiring critical matching conditions/ cancellation constraints using all grounded passive elements.

This configuration also provides (i) availability of non-interacting controllability of the parameters of the realized filters through grounded resistors (advantageous for obtaining voltage control through incorporation of FET-based voltage

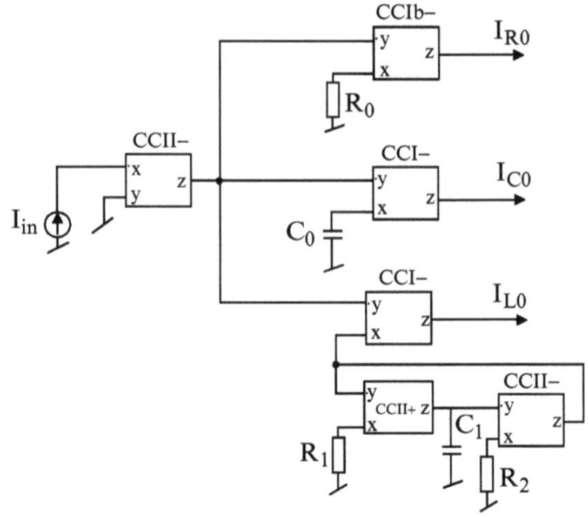

Fig. 6.29 Senani's Universal CM biquad derived through the proposed methodology [24]

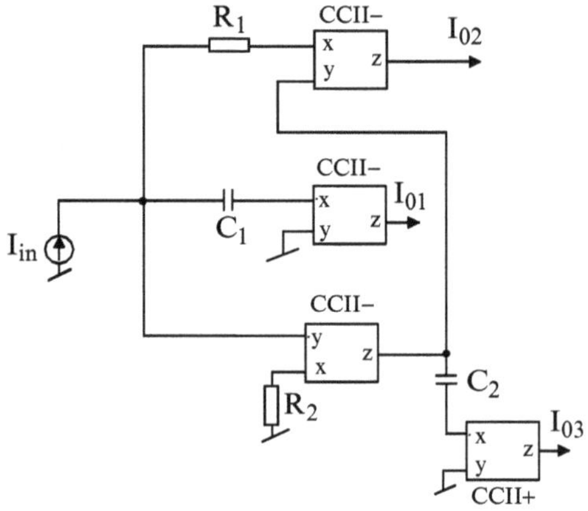

Fig. 6.30 Senani's CM biquad realizable with only four CCIIs [24]

controlled resistors) (ii) easy conversion into a VM biquad and (iii) suitability for IC implementation. Assuming ideal CCs, a routine circuit analysis yields the following current transfer functions:

$$\frac{I_{01}}{I_{in}} = \frac{s^2}{D(s)}; \frac{I_{02}}{I_{in}} = \frac{-s\left(\frac{1}{R_1C_1}\right)}{D(s)}; \frac{I_{03}}{I_{in}} = \frac{\frac{1}{C_1C_2R_1R_2}}{D(s)}$$

$$\text{where } D(s) = s^2 + s\frac{R_3}{C_1R_1R_4} + \frac{1}{C_1C_2R_1R_2} \tag{6.43}$$

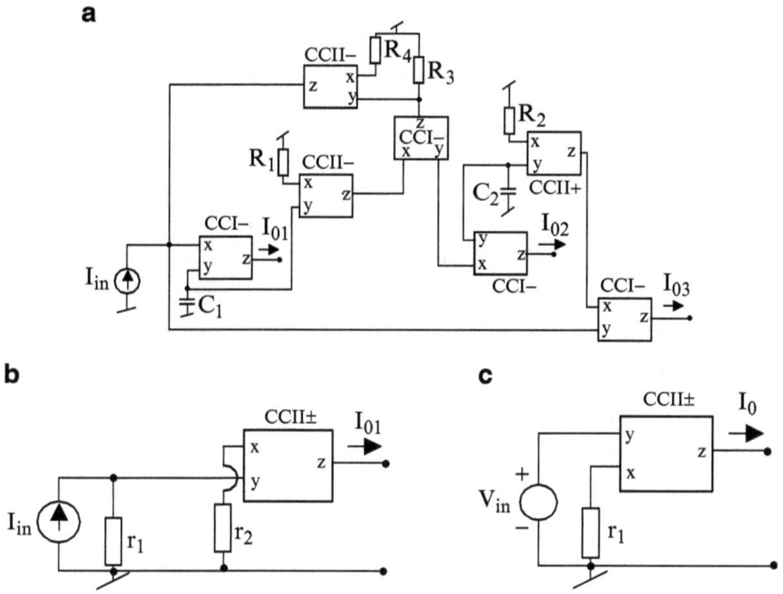

Fig. 6.31 CM biquad structure employing CCs proposed by Senani [25]

A CM notch filter can be obtained by joining the outputs providing $I_{04} = I_{01} + I_{03}$ and APF by adding I_{01}, I_{02} and I_{03} which gives $I_{05} = I_{01} + I_{02} + I_{03}$. The respective transfer functions are given by:

$$\frac{I_{04}}{I_{in}} = \frac{s^2 + \frac{1}{C_1C_2R_1R_2}}{D(s)} \quad \text{and} \quad \frac{I_{04}}{I_{in}} = \frac{s^2 - s\left(\frac{1}{R_1C_1}\right) + \frac{1}{C_1C_2R_1R_2}}{D(s)} \tag{6.44}$$

provided that $R_3 = R_4$.

Chang's Universal CM Biquad Figure 6.32 shows a universal CM biquad presented by Chang [26].

This filter configuration employs seven CCIIs, eight grounded resistors and two grounded capacitors and offers independent control of ω_0, Q_0 and current gain through separate grounded resistors. Assuming ideal CCIIs, the various transfer functions are given by:

$$\frac{I_{high}}{I_i} = \frac{s^2\left(\frac{R_1}{R_8}\right)}{D(s)};\ \frac{I_{band}}{I_i} = \frac{-s\left(\frac{R_1}{R_4R_7C_2}\right)}{D(s)};\ \frac{I_{low}}{I_i} = \frac{\frac{R_1}{R_4R_5R_8C_2C_1}}{D(s)}$$

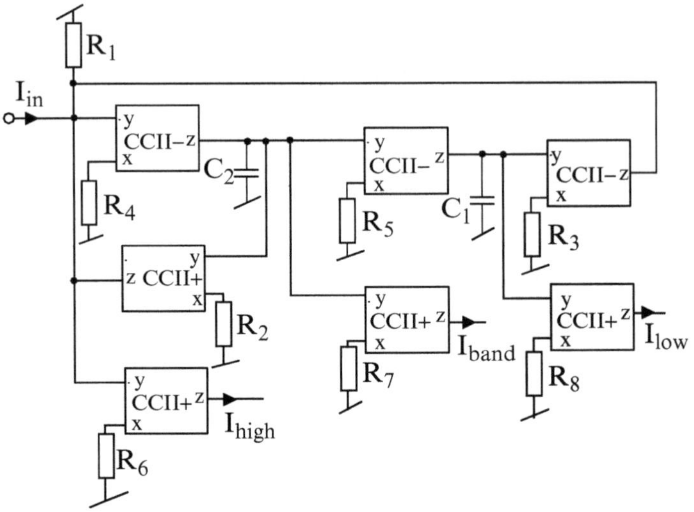

Fig. 6.32 CCII-based CM universal biquad proposed by Chang [26]

$$\frac{I_{notch}}{I_i} = \frac{s^2\left(\frac{R_1}{R_6}\right) + \frac{R_1}{R_4R_5R_3C_1C_2}}{D(s)} \text{ for } R_8 = R_3. \tag{6.45}$$

$$\frac{I_{all}}{I_i} = \frac{s^2 - s\left(\frac{R_1}{C_2R_2R_4}\right) + \frac{R_1}{R_4R_5R_3C_1C_2}}{D(s)} \text{ for } R_6 = R_1,\ R_7 = R_2 \text{ and } R_8 = R_3$$

$$\text{where } D(s) = s^2 + s\frac{R_1}{C_2R_2R_4} + \frac{R_1}{C_1C_2R_3R_4R_5} \tag{6.46}$$

This circuit provides all the advantages of the earlier circuit by Senani [25] and in addition has the advantages of (i) employing active elements of the same type: only CCIIs (ii) tunable current gain and (iii) independent control of ω_0, Q_0 and current gain through separate resistors in all the five responses.

The above mentioned aspects were verified by bread boarded implementations of the circuit where CCIIs were realized by LF 356 N type op-amps and CA 3096AE type mixed-transistor arrays used for implementing current mirrors for op-amps engaged in supply current sensing mode.

Chang's five CC Universal CM Biquad Another CM biquad was presented by Chang in [27] which had single input and three outputs and employed five CCs, three grounded resistors and two grounded capacitors as shown in Fig. 6.33.

This filter structure too realizes all the five second-order filter functions without the requirement of critical component matching/cancellation constraints and offers

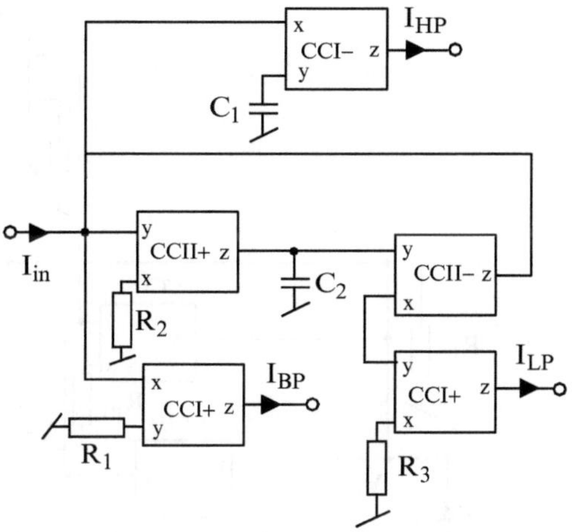

Fig. 6.33 Universal active current filter proposed by Chang [27]

orthogonal control of ω_0 and Q_0 through grounded resistors. The filter responses realized by the circuit are given by:

$$\frac{I_{HP}}{I_i} = \frac{s^2}{D(s)}\ ;\ \frac{I_{BP}}{I_i} = \frac{-s\left(\frac{1}{R_1C_1}\right)}{D(s)}\ ;\ \frac{I_{LP}}{I_i} = \frac{\frac{1}{C_1C_2R_2R_3}}{D(s)} \tag{6.47}$$

The notch (NH) and all pass (AP) responses can be obtained by the following equations:

$I_{NH} = (I_{HP} + I_{LP})$ and $I_{AP} = (I_{HP} + I_{BP} + I_{LP})$ i.e.,

$$\frac{I_{NH}}{I_i} = \frac{s^2 + \frac{1}{C_1C_2R_2R_3}}{D(s)} \text{ and } \frac{I_{AP}}{I_i} = \frac{s^2 - s\frac{1}{C_1R_1} + \frac{1}{C_1C_2R_2R_3}}{D(s)} \tag{6.48}$$

where $D(s) = s^2 + s\frac{1}{C_1R_1} + \frac{1}{C_1C_2R_2R_3}$

The advantages of this configuration are (i) requirement of only five CCs (ii) realisations of APF without any realization constraints and (iii) orthogonal tunability of ω_0 and Q_0 through grounded resistors.

With CCs realized by OMA based CC formulations based upon LF356 op-amps and CA3096E mixed transistor arrays, the practical implementations of the circuit have been reported to yield measured frequency responses of the notch and all pass functions in good agreement with the theory.

6.2.3.6 Single-Input-Single-Output (SISO)-Type CM Biquad

A number of SISO Cm biquads have been reported by various researchers in literature, for instance, [28, 61, 67, 86, 88, 98, 99, 103, 109, 112, 114, 116, 117, 118, 154, 158, 165].

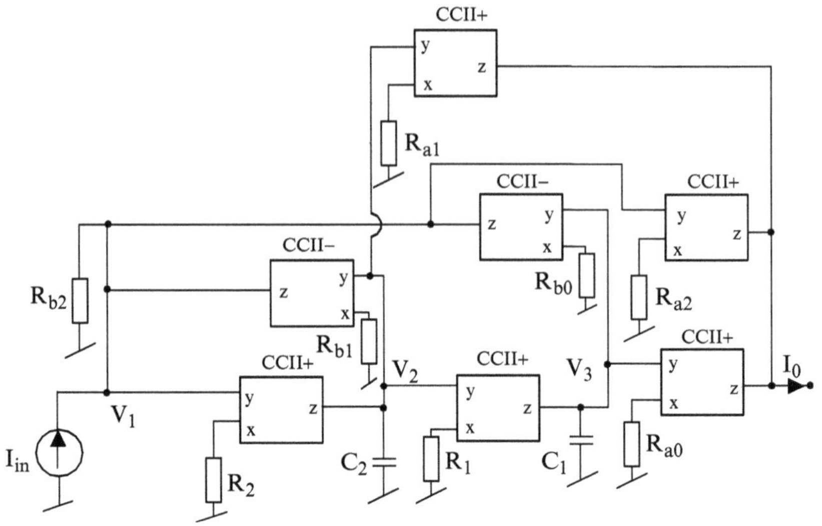

Fig. 6.34 A multipurpose active biquad based on CCIIs proposed by Sun and Fidler [28]

Figure 6.34 shows an exemplary multi-purpose CCII-RC biquad filter homologous to the traditional KHN-op-amp biquad proposed by Sun and Fidler in [28].

Assuming ideal CCs, a routine circuit analysis yields the following transimpedance and current transfer functions:

$$\frac{V_1}{I_{in}} = \frac{s^2(C_1R_1C_2R_2)}{D(s)};\ \frac{V_2}{I_{in}} = \frac{s(C_1R_1)}{D(s)};\ \frac{V_3}{I_{in}} = \frac{1}{D(s)} \tag{6.49}$$

$$\text{and } \frac{I_{out}}{I_{in}} = \frac{s^2\left(C_1R_1C_2R_2\frac{1}{R_{a2}}\right) + s\left(C_1R_1\frac{1}{R_{a1}}\right) + \frac{1}{R_{a0}}}{D(s)} \tag{6.50}$$

$$\text{where } D(s) = s^2\left(C_1C_2R_1R_2\frac{1}{R_{b2}}\right) + s\left(R_1C_1\frac{1}{R_{b1}}\right) + \frac{1}{R_{b0}} \tag{6.51}$$

The various CM filter functions can be realized as follows.

$$\frac{I_{LP}}{I_i} = \frac{\frac{1}{R_{a0}}}{D(s)} \text{ for} R_{a1} = \infty = R_{a2};\ \frac{I_{BP}}{I_{in}} = \frac{s\left(C_1R_1\frac{1}{R_{a1}}\right)}{D(s)};\text{for} R_{a0} = \infty$$

$$= R_{a2}\frac{I_{HP}}{I_{in}} = \frac{s^2\left(C_1R_1C_2R_2\frac{1}{R_{a2}}\right)}{D(s)} \text{ for } R_{a0} = \infty = R_{a1} \tag{6.52}$$

$$\frac{I_{Notch}}{I_{in}} = \frac{s^2\left(C_1R_1C_2R_2\frac{1}{R_{a2}}\right) + \frac{1}{R_{a0}}}{D(s)}\text{; provided that } \frac{R_{b2}}{R_{a2}} = \frac{R_{b0}}{R_{a0}} \tag{6.53}$$

$$\frac{I_{AP}}{I_{in}} = \frac{s^2\left(C_1R_1C_2R_2\frac{1}{R_{a2}}\right) - s\left(\frac{1}{R_{a1}}C_1R_1\right) + \frac{1}{R_{a0}}}{D(s)} \tag{6.54}$$

provided that $\frac{R_{b2}}{R_{a2}} = \frac{R_{b0}}{R_{a0}} = \frac{R_{b1}}{R_{a1}}$ and CCII+ connected with R_{a1} should be replaced by CCII−.

he filter configuration of Fig. 6.34 offers the controllability of its parameters by independent grounded resistors and the CCII tracking errors were found [28] to be having insignificant effect on the performance of the realized filters and can be precisely compensated [28].

The significant advantages of this configuration are the tunability of filter parameters through separate grounded resistors and employment of grounded capacitors as preferable from the view point of IC implementation.

6.2.3.7 MISO-Type Universal CM Biquads

We now present some exemplary MISO-type CM universal biquads from amongst those existing in literature [30–34, 133].

Ozcan, Cicekoglu and Kuntman Biquad A MISO-type CM filter employing a single CCII+ along with only four passive components was presented by Ozcan et al. in [30] and is shown in Fig. 6.35a.

This configuration can realize the following filters:

$\frac{i_{out}}{i_2} = \frac{\frac{1}{C_1C_2R_2R_1}}{D(s)}$ for $i_1 = i_3 = 0$; $\frac{i_{out}}{i_3} = \frac{\frac{-s}{C_2R_1}}{D(s)}$ for $i_1 = i_2 = 0$ and

$\frac{i_{out}}{i_1} = \frac{s^2}{D(s)}$; provided that $i_2 = -i_1$ and $i_3 = i_1\left(\frac{R_1}{R_2}\right)\left(1 + \frac{C_2}{C_1}\right)$

$$\text{where } D(s) = s^2 + s\left(\frac{1}{R_2}\right)\left(\frac{1}{C_1} + \frac{1}{C_2}\right) + \frac{1}{C_1C_2R_1R_2} \tag{6.55}$$

Thus, the circuit realises LPF, BPF and HPF. Through RC:CR transformation HPF can also be realized without any element matching condition.

The experimental results of this circuit designed for Butterworth characteristics with gain = 1, and pole frequency of 225 kHz with element values $C_1 = C_2 = 1$ nF, $R_1 = 1$ kΩ, $R_2 = 500\Omega$ using AD844 to realize CCII+ biased with ±12 V supply are shown in Fig. 6.35b, whereas a typical output waveform is shown in Fig. 6.35c. Very good correspondence between theoretical and experimental result has been reported in [30]

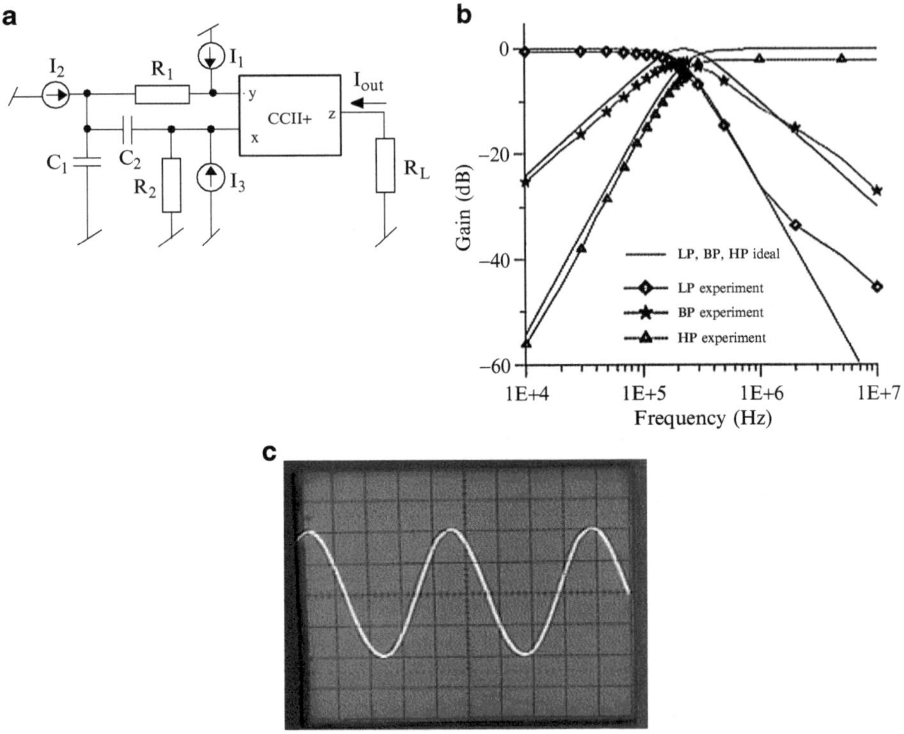

Fig. 6.35 Filter configuration proposed by Ozcan et al. (Adapted from [30] © 2003 Elsevier) (**a**) The circuit configuration (**b**) The theoretical and measured LP, BP and HP responses (**c**) A typical output waveform of the filter (input current 1 mA, 205 kHz; load resistance 10 kΩ: vertical axis- 5 V/div., horizontal axis- 1 μs/div)

Abuelma'atti and Tasadduq Biquad Figure 6.36 shows a three input and single output universal CM biquad proposed by Abuelma'atti and Tasadduq [31]. The presented filter configuration employs six CCIIs+, two grounded capacitors and six grounded resistors. The filter configuration enjoys independent control of ω_0 and ω_0/Q_0 through separate grounded resistance and independent gain control through separate grounded resistor of all the filter responses.

The various CM transfer functions for this circuit can be deduced from the following input output equation:

$$I_0 = -\left(\frac{1}{R_8}\right)\frac{s^2 I_3 - s\left(\frac{1}{C_3 R_4}\right)I_2 + \left(\frac{1}{C_1 C_3 R_2 R_4}\right)I_1}{s^2\left(\frac{1}{R_5}\right) + s\left(\frac{1}{C_3 R_4 R_7}\right) + \frac{1}{C_1 C_3 R_2 R_4 R_6}} \tag{6.56}$$

From Eq. (6.56), the following five generic filter responses can be obtained: LPF:

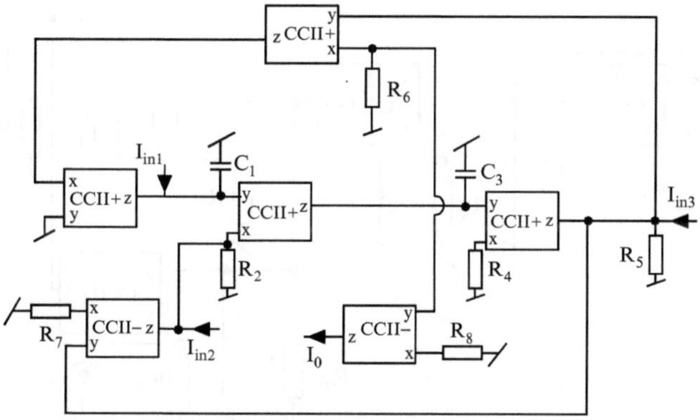

Fig. 6.36 MISO-type universal CM biquad using CCII+ s proposed by Abuelma'atti and Tasadduq (Adapted from [31] © 1997 Elsevier)

$I_3 = I_2 = 0$, $I_1 = I_{in}$; BPF: $I_1 = I_3 = 0$, $I_2 = I_{in}$; HPF: $I_1 = I_2 = 0$ and $I_3 = I_{in}$; Notch: $I_2 = 0$, $I_1 = I_3 = I_{in}$ and $R_5 = R_6$; and APF: $I_1 = I_2 = I_3 = I_{in}$, and $R_5 = R_6 = R_7$.

This circuit was experimentally tested using AD844 as CCII+. The results obtained had excellent agreement with theoretical results.

Abuelma'atti and Al-Zaher Biquad Another three-input-single-output CM biquad employing four CCIIs+, five grounded resistors and three grounded capacitors, which was proposed by Abuelma'atti and Al-Zaher in [32], is shown in Fig. 6.37.

This filter configuration offers independent control of filter parameters ω_0 and $\left(\frac{\omega_0}{Q_0}\right)$ through grounded resistors and has low active and passive sensitivities.

Assuming ideal CCIIs+ the expression of the output current in terms of the various input currents is found to be:

$$I_0 = \left(\frac{R_4}{R_7}\right) \frac{s^2\left(\frac{C_3}{C_6}\right)I_3 - s\left(\frac{1}{C_6R_4}\right)I_2 + \left(\frac{1}{C_1C_6R_2R_4}\right)I_1}{s^2 + s\left(\frac{1}{C_6R_6}\right) + \frac{1}{C_1C_6R_2R_5}} \tag{6.57}$$

From Eq. (6.57), the following filter responses can be derived: LPF: $I_3 = I_2 = 0$, $I_1 = I_{in}$; BPF: $I_1 = I_3 = 0$, $I_2 = I_{in}$; HPF: $I_1 = I_2 = 0$ and $I_3 = I_{in}$; Notch: $I_2 = 0$, $I_1 = I_3 = I_{in}$ with $R_5 = R_6$; and APF: $I_1 = I_2 = I_3 = I_{in}$ along with $R_5 = R_4$ and $C_3 = C_7$.

The circuit had been practically implementable with AD844 and provides the following advantages: use of only one type of CCs namely, CCII+, all resistors and capacitors being grounded, low active and passive sensitivities, independent control of ω_0 and band width using grounded resistors and the feasibility of realizing all the five standard filter functions without any component matching.

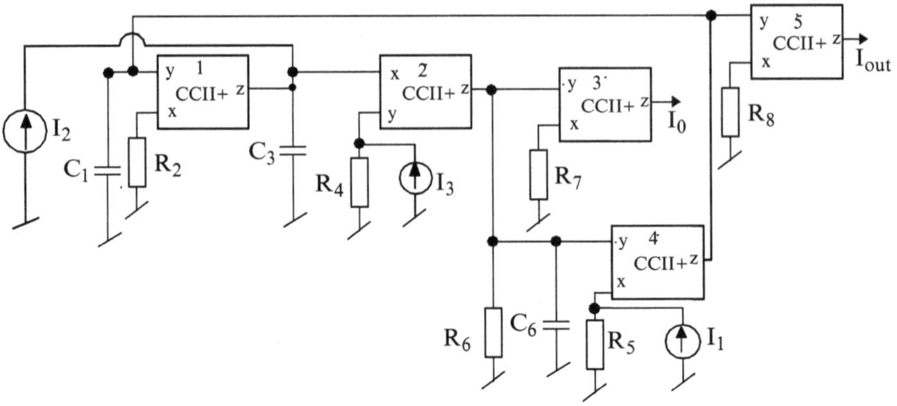

Fig. 6.37 CM universal filter proposed by Abuelma'atti and Al-Zaher [32]

Fig. 6.38 Universal CM active filter proposed by Chang [33]

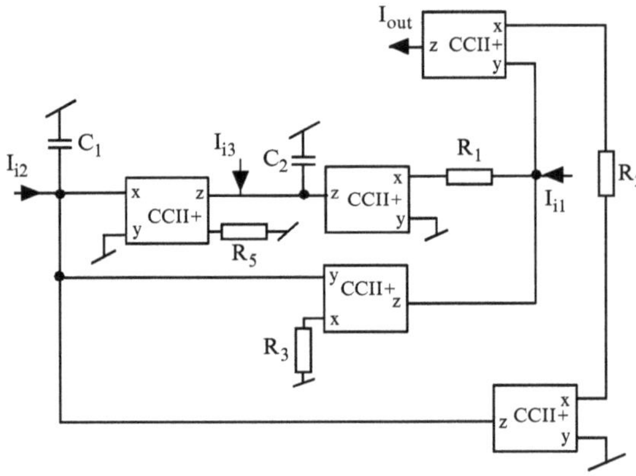

Chang's CM Biquad The filter circuit shown in Fig. 6.38 is a three inputs and one output CM biquad employing five CCIIs+, four resistors and two grounded capacitors proposed by Chang [33].

Although the filter parameters ω_0 and bandwidth are orthogonally controllable but the circuit suffers from the drawback of component matching and sign inversion of one of the currents in case of the realization of an APF.

A routine circuit analysis using ideal CCIIs+ yields the following expression for the output current in terms of the input currents:

$$I_{out} = \frac{s^2\left(\frac{1}{R_1R_2}\right)I_{i1} + s\left(\frac{R_1}{C_1R_2R_3}\right)I_{i2} + \left(\frac{R_1}{C_1C_2R_2R_3R_5}\right)I_{i3}}{s^2 + s\left(\frac{R_1}{C_1R2R_3}\right) + \frac{1}{C_1C_2R_3R_5}} \tag{6.58}$$

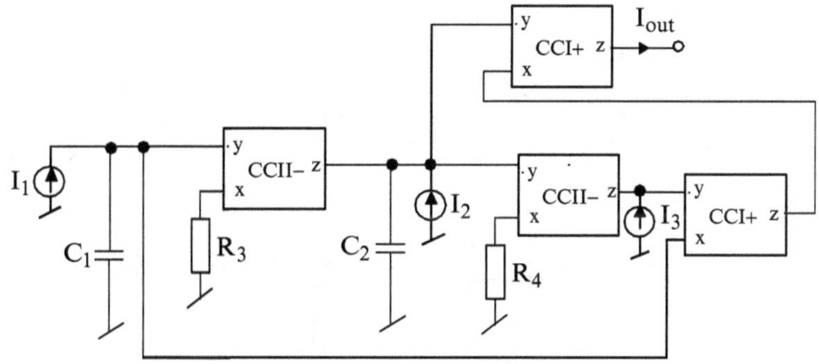

Fig. 6.39 Universal active CM filter proposed by Chang et al. [34]

From the above equation, the various filter responses can be obtained as follows:

HPF : $I_{i3} = I_{i2} = 0$, $I_{i1} = I_{in}$; BPF : $I_{i1} = I_{i3} = 0$, $I_{i2} = I_{in}$; LPF : $I_{i1} = I_{i2} = 0$ and $I_{i3} = I_{in}$; Notch : $I_{i2} = 0$, $I_{i1} = I_{i3} = I_{in}$and $R_5 = R_6$; and APF : $I_{i1} = -I_{i2} = I_{i3} = I_{in}$, and $R_2 = R_1$.

Chang, Chien and Wang CM Biquad The filter configuration shown in Fig. 6.39 was proposed by Chang et al. in [34] which uses two CCI+ and two CCII− with two grounded resistors and two grounded capacitors. The circuit is capable of yielding all the five generic filter responses without any component matching or sign inversion of any input current(s).

Assuming ideal CCs, analysis of the circuit gives the following expression for the output current in terms of the various input currents:

$$I_{out} = \frac{s^2 I_3 - s\left(\frac{1}{C_2 R_4}\right) I_2 + \left(\frac{1}{C_1 C_2 R_3 R_4}\right) I_1}{s^2 + s\left(\frac{1}{C_2 R_6}\right) + \frac{1}{C_1 C_2 R_3 R_4}} \tag{6.59}$$

From Eq. (6.59), the following filter functions can be obtained: LPF: $I_3 = I_2 = 0$, $I_1 = I_{in}$; BPF: $I_1 = I_3 = 0$, $I_2 = I_{in}$; HPF: $I_1 = I_2 = 0$ and $I_3 = I_{in}$; Notch: $I_2 = 0$, $I_1 = I_3 = I_{in}$ and $R_5 = R_6$; and APF: $I_1 = I_2 = I_3 = I_{in}$.

This circuit offers several advantages such as: low sensitivities, use of grounded capacitors and versatility to realize any type of active filter transfer function.

6.2.4 *Third Order Filters*

Several authors have considered realization of third order Butterworth filters using CCs [35, 36, 48–50, 81, 96, 113]. Figure 6.40 shows two third order CM Butterworth filter structures employing equal-valued capacitors and resistors with

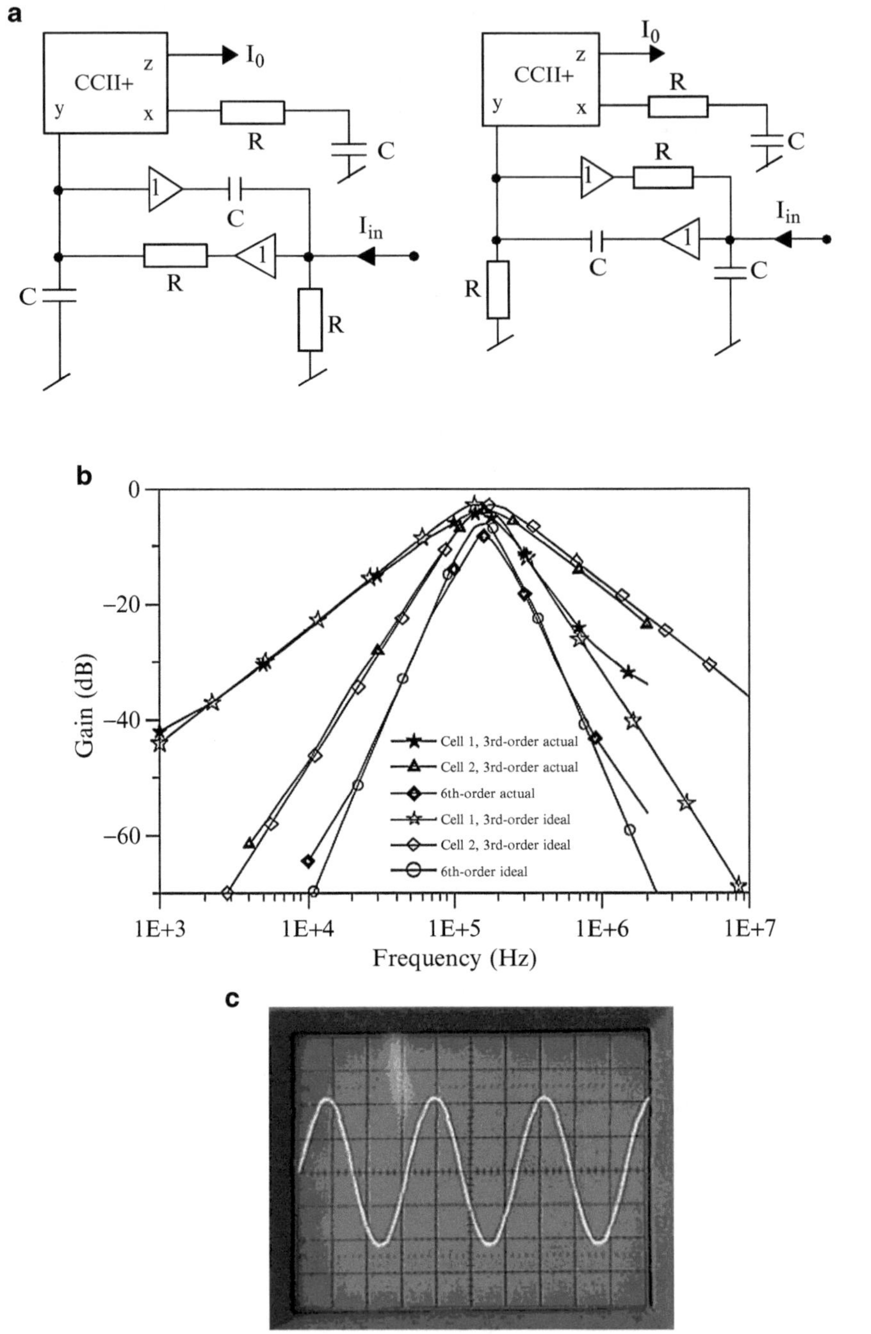

Fig. 6.40 Third order Butterworth filters with equal valued capacitors and resistors proposed by Kuntman et al. (Adapted from [35] © 2002 Kluwer) (**a**) The circuit configuration (**b**) The experimental and theoretical frequency responses of third order and six order filters obtained by

unity gain active elements proposed by Kuntman et al. [35]. Assuming ideal CCII+ s and voltage followers, the current transfer function for the circuit of Fig. 6.40a is given by:

$$\frac{i_O}{i_{in}} = \frac{s}{1 + 2s + 2s^2 + s^3} \text{ (for R } = 1\Omega \text{ and C } = 1 \text{ F)} \tag{6.60}$$

Using RC-CR transformation, another BPF can be obtained whose current transfer function is given by:

$$\frac{i_O}{i_{in}} = \frac{s^2}{1 + 2s + 2s^2 + s^3} \text{ (for R } = 1\Omega \text{ and C } = 1 \text{ F)} \tag{6.61}$$

The circuits, when implemented with AD844 ICs used to realize CCII+ as well as voltage followers, showed excellent correspondence with theoretical responses when designed for $f_O = 150$ KHz with AD844 biased from ± 12 V DC power supplies with load resistance $R_L = 13$ kΩ, equal valued R-C components taken as 1 kΩ and 1 nF.

6.2.5 *MOSFET-C Integrators and Filters Using CCII*

It has been well known that MOS resistive circuit (MRC) can cancel completely the even and odd MOS nonlinearities of the four matched MOSFETs employed therein. Assuming ideal CCII+ and the devices in MRC to be identical (matched) and operated in triode region, the expression for its output voltage of the circuit of Fig. 6.41 is given by:

$$V_{out} = -\frac{1}{RC}\int V_{in} d\tau, \text{ where } R = \pm\frac{1}{\mu C_{ox}\left(\frac{W}{L}\right)\{|V_{GA}| - |V_{GB}|\}} \tag{6.62}$$

where $V_{in} = (V_1 - V_2)$ and the positive or negative R is used to represent an inverting or non-inverting integrator.

This integrator can be easily employed to design higher order filters. In [37] it has been demonstrated in the design of a third order low pass filter based upon doubly-terminated ladder filter with 0.01 dB pass band ripple up to 1 MHz. The functionality of this integrator has also been verified by designing a voltage-controlled oscillator realized by a bipolar CC, the MRC and a Schmitt Trigger (both implemented from CD 4007). The structure of Fig. 6.41 is suitable for realizing high frequency MOSFET-C filters using CCs.

Fig. 6.40 (continued) cascading the two circuits of Fig. 6.40 (**a**) referred to as cell-one and cell-two respectively. (**c**) A typical output waveform (input current 1 mA, 150 kHz sinusoidal wave: vertical axis 5 V/div., horizontal axis 2 μs/div.)

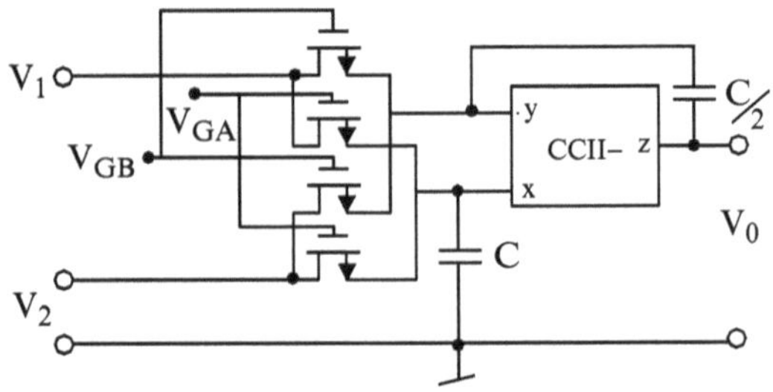

Fig. 6.41 MOSFET-C integrator proposed by Liu et al. [37]

For the use of an alternative MOSFET-C integrator using CCII+ (along with voltage buffer) in the design of a sixth order ladder filter, the reader is referred to [65].

6.2.6 Higher Order Active Filter Design

In this section we outline some exemplary methods of designing higher order active filters using current conveyors from amongst a number of such techniques and circuits reported in literature such as those in [29, 35, 38–40, 41, 51, 54, 55, 72, 82, 89–91, 93, 107, 108, 113, 145, 164].

Senani's method of incorporating non-ideal simulated inductors and FDNRs into higher order filter design: A number of CC-based grounded and floating inductance simulation circuits have been described in Chap. 5 of this monograph. Of course, such circuits can be easily used to design active filters from passive LC prototypes by replacing passive inductor of the prototype LC ladders by CC-simulated inductors and FDNRs. It has been recognized that lossy inductors/FDNRs can be realized more economically than their ideal counter parts. Four new transformations were proposed by Senani in [35, 38] which make it possible to incorporate even such non-ideal simulated immittances as direct elements in the design of active filters based upon LC ladder filters.

In this sub-section, we consider one of these transformations which make it possible to incorporate even non-ideal (lossy) simulated inductance and FDNR elements directly into the design of active filters. Table 6.2 shows one of the transformations (T2: $F(s) = s/(1 + s)$) from [35, 38] wherein it has been shown how R, L and C will be transformed if their impedances are multiplied by this particular frequency dependent scaling factor F(s).

The main advantage of this method is that it requires only **n** number of CCII− s for converting an **nth** order LC filter into an **nth** order CCII− based active filter. The practical workability of this method had been demonstrated in [35, 38] using CCII− implementations based on op-amps and OTAs discussed in Chap. 2 of this monograph (see Fig. 2.3a and reference [3] therein).

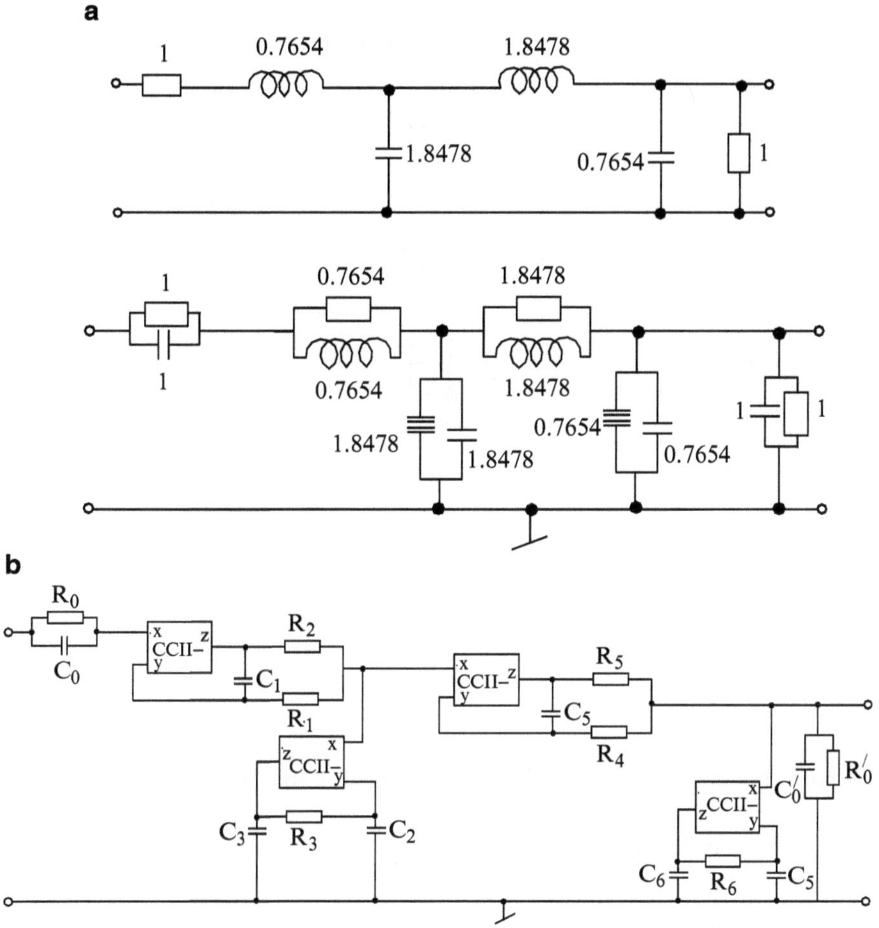

Fig. 6.42 Application of Senani's transformation T-2 in the design of a fourth order LPF (**a**) Normalized fourth-order Butterworth LPF and transformed filter obtained by applying T-2 (with F (s) = s/(s + 1) to the LC prototype (**b**) Active implementation of the transformed filter of Fig. 6.42 (**b**) using CCs; cut-off frequency = 1 KHz (**c**) The final circuit implementation

Hwang, Liu, Wu and Wu method Another method for systematic design of doubly terminated lossless ladder filters using CCs which is suitable for higher-order filters and is based upon linear transformations, was proposed by Hwang et al. in [39]. The proposed method was demonstrated by designing a third-order Chebychev LPF with 50 kHz bandwidth and 1 dB ripple, employing a minimum number of CCIIs and grounded capacitors.

The basic building blocks required to implement the design methodology proposed in [39], the following four sub-circuits need to be considered:

$$\text{Input termination}: \; y_{21} = \frac{E - x_{21}}{as + 1} \tag{6.63}$$

Table 6.2 Application of Senani's transformation T-2 [35, 38]

Element in the LC prototype	Transformed elements	Component values
R_0	R_0', C_0'	$R_0' = \frac{R_0}{b_0}$; $C_0' = \frac{b_1}{R_0}$
L_0	R_0', L_0'	$R_0' = \frac{L_0}{b_1}$; $L_0' = \frac{L_0}{b_0}$
C_0	C_0', D_0'	$C_0' = b_0 C_0$; $D_0' = b_1 C_0$

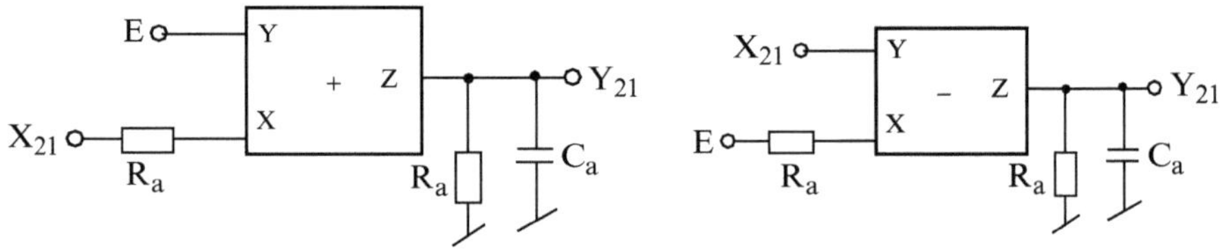

Fig. 6.43 Integrators

The corresponding CCII-based circuits to realize Eq. (6.63) are shown in Fig. 6.43.

$$\text{Middle terminations } : y_{1i} = y_{2i} = \frac{x_{1i} - x_{2i}}{as}, i = 2, 3, \ldots, (n-1) \tag{6.64}$$

The corresponding CCII-based circuits to realize Eq. (6.64) are shown in Fig. 6.44.

$$\text{Output termination} : y_{1n} = \frac{x_{1n}}{as + 1} = V_0 \text{ for } \mathrm{R_s} = \mathrm{R_L} \tag{6.65}$$

The corresponding CCII-based circuits to realize Eq. (6.65) are shown in Fig. 6.45.

$$\text{If, } R_s \neq R_L,\ y_{1n} = \frac{x_{1n}}{as + 1} = V_0 \tag{6.66}$$

The corresponding CCII-based circuits are shown in Fig. 6.46 with $R_s = R$, $a = RC$ or L/R, and $b = R/R_L$ or R_L/R.

The design equations are given by:

$$a = R_a C_a \text{ and } b = \frac{R_a}{R_b} \tag{6.67}$$

Using the methodology of [39], the final circuit for realizing the third-order Chebychev LPF is shown in Fig. 6.47.

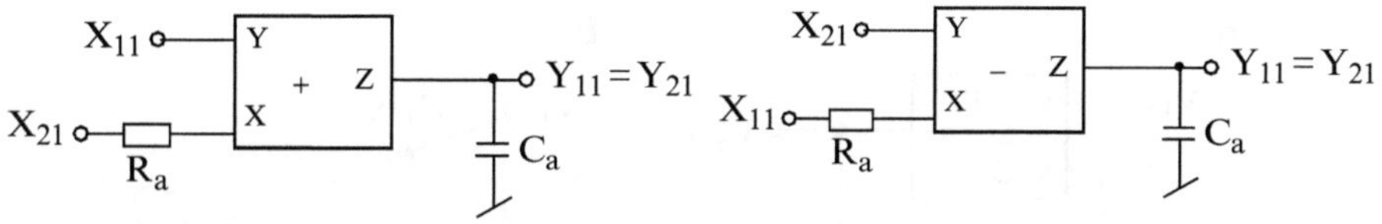

Fig. 6.44 CC-based realisations of Eq (6.64)

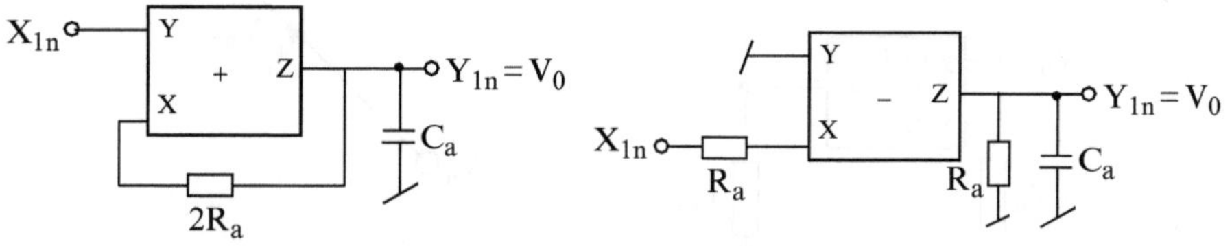

Fig. 6.45 CC-based realisations of Eq (6.65)

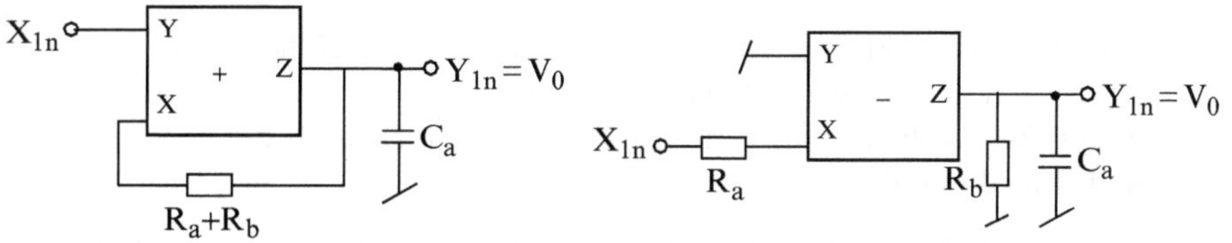

Fig. 6.46 CC-based realisations of Eq. (6.66)

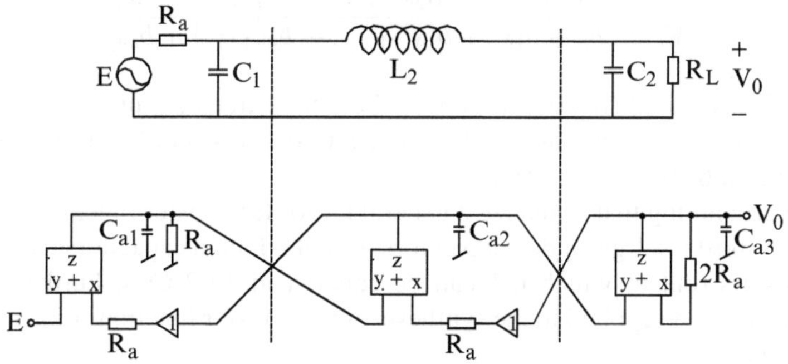

Fig. 6.47 The 3rd-order LC prototype filter and the corresponding CCII-based filter [39]

The practical workability of the filter designed through the approach briefly presented above has been verified by hardware implementations using AD844 to realize CCIIs.

Anday and Gunes method based upon signal flow graph: Another method of realization of nth-order transfer functions employing CCs was proposed by Anday and Gunes in [40] using signal flow graph technique. Figure 6.48 (a) shows the

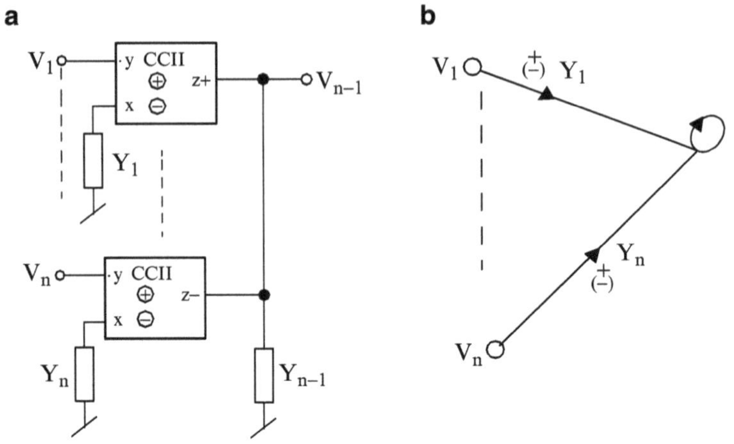

Fig. 6.48 (**a**) CCII-based sub-network (**b**) its associated graph (Adapted from [40] © 1992 Wiley)

CCII-based sub-network and (b) shows its sub flow graph. If the nth-order transfer function is represented by a suitable signal flow graph composed of sub-graphs of the form of Fig. 6.48b, then the corresponding circuit realization can be found easily as demonstrated in [40].

If we consider the general form of the nth-order voltage transfer function given by:

$$\frac{V_0}{V_i} = \frac{a_0 s^0 + a_1 s^1 + a_2 s^2 + \ldots + a_{n-1} s^{n-1} + a_n s^n}{b_0 + b_1 s^1 + b_2 s^2 + \ldots + b_{n-1} s^{n-1} + b_n s^n} \tag{6.67}$$

the signal flow graph representation of the above transfer function and its final CC-based implementation as per the procedure outlined in [40] turns out to be as shown in Fig. 6.49a, b respectively.

It is worth noting that to realize a general nth-order transfer function this method requires at most n + 1 grounded capacitors (out of which n capacitors can be equal), 3n-1 resistors (out of which n-1 can be equal) and 3n-2 CCs. Furthermore, this method can also be applied to the synthesis of the transfer functions having negative coefficients.

Concluding Remarks

From the references [1–166] given at the end of this chapter, it is apparent that a vast amount of literature exists on the realization of analog continuous time filters using CCs. From this repertoire of CC-based filter circuits, in this chapter, we have presented a selected number of configurations and have highlighted their salient features. Thus, we presented some representative

(continued)

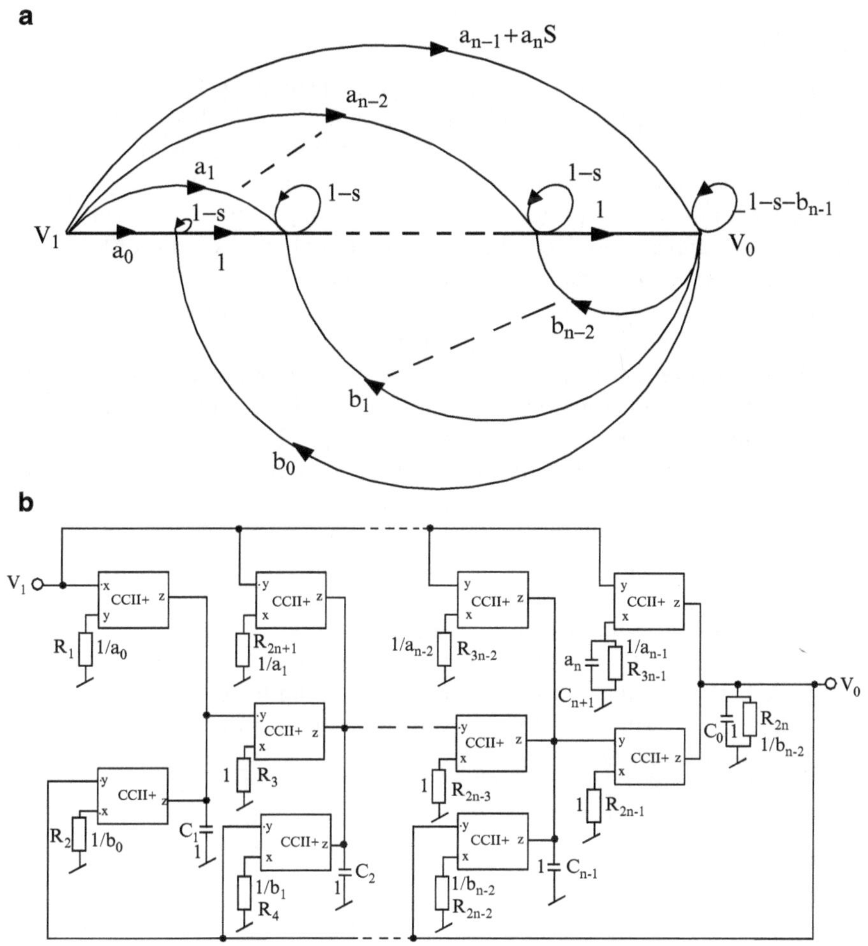

Fig. 6.49 (**a**) Signal flow graph corresponding to Eq. (6.67) and (**b**) CCII-based realization of nth-order voltage transfer function (Adapted from [40] © 1992 Wiley)

(continued)

structures for realizing first order all pass filters and second order and higher order filters. We have included a number of novel configurations for realizing VM and CM filters which have been classified in categories like SISO, SIMO, MISO and MIMO universal biquads. Pertinent equations and design equations have been spelt out for all chosen circuits. The circuits described in this chapter present a sample of various options available for analog filter design using CCs.

(continued)

(continued)

It may be noted that all the circuits presented in this chapter can be practically realized using commercially available IC AD844 (with CCII+ realized by one IC and CCII− realized with two ICs) or using CC-implementations based upon op-amps and current mirrors in operational-mirrored amplifier formulations or using PA630 IC CCs although the AD844 appears to have been the preferred choice of most of the researchers whose propositions we have included in this chapter.

It must also be pointed out that in this chapter we have limited ourselves to only those configurations which use the basic CCI and CCII as per their original definitions; the discussion of a variety of other filter configurations realizable from numerous other variants of CCs is the subject matter of a subsequent chapter of this monograph (see Chap. 12).

Lastly, from a critical study of various configurations described here, as well as those omitted from inclusion, it can be readily inferred that there is still enough scope for deriving new circuits having properties superior to those included herein. For example, any circuit providing ideally zero input impedance, very high output impedance and capable of realizing all the five filtering functions without any realization constraints and with independent tunability of all the three parameters H0, ω_0 and Q_0 having all passive grounded elements and yet not requiring more than three CCs, has still not been discovered! This, therefore, appears to be an interesting problem which is open to investigation.

References

1. Higashimura M, Fukui Y (1988) Realization of all-pass networks using a current conveyor. Int J Electron 65:249–250
2. Higashimura M, Fukui Y (1990) Realization of current mode all-pass networks using a current conveyor. IEEE Trans Circ Syst 37:660–661
3. Cicekoglu O, Kuntman H, Berk S (1999) All-pass filters using a single current conveyor. Int J Electron 86:947–955
4. Khan IA, Maheshwari S (2000) Simple first order all-pass section using a single CCII. Int J Electron 87:303–306
5. Chong CP, Smith KC (1986) Biquadratic filter sections employing a single current conveyor. Electron Lett 22:1162–1164
6. Metin B, Yuce E, Cicekoglu O (2007) A novel dual output universal filter topology using a single current conveyor. Electr Eng 89:563–567
7. Toumazou C, Lidgey FJ (1986) Universal active filter using current conveyors. Electron Lett 22:662–664
8. Higashimura M, Fukui Y (1996) Universal filter using plus-type CCIIs. Electron Lett 32:810–811
9. Senani R, Singh VK (1995) KHN equivalent biquad using current conveyors. Electron Lett 31:626–628

10. Horng JW, Hou CL, Chang CM, Chou HP, Lin CT (2006) High input impedance voltage mode universal biquadratic filter with one input and five outputs using current conveyors. Circ Syst Sig Process 25:767–777
11. Chen HP (2010) Single CCII-based voltage mode universal filter. Analog Integr Circ Sig Process 62:259–262
12. Horng JW (2004) Voltage-mode universal biquadratic filters using CCIIs. IEICE Trans Fundament 87:406–409
13. Chang CM, Tu SH (1999) Universal voltage-mode filter with four inputs and one output using two CCII+ s. Int J Electron 86:305–309
14. Liu SI, Lee JL (1997) Voltage mode universal filters using two current conveyors. Int J Electron 82:145–149
15. Horng JW, Lee MH, Cheng HC, Chang CW (1997) New CCII based voltage mode universal biquadratic filter. Int J Electron 82:151–155
16. Horng JW, Tsai CC, Lee MH (1996) Novel universal voltage mode biquad filter with three inputs and one output using only two current conveyors. Int J Electron 80:543–546
17. Chang CM (1997) Multifunction biquadratic filters using current conveyors. IEEE Trans Circ Syst-II 44:956–958
18. Ozoguz S, Gunes EO (1996) Universal filter with three input using CCII+. Electron Lett 32:2134–2135
19. Chang C, Lee MS (1994) Universal voltage mode filter with three inputs and one output using three current conveyors and one voltage follower. Electron Lett 30:2112–2113
20. Chang CM, Lee MS (1995) Comment: universal voltage mode filter with three inputs and one output using three current conveyors and one voltage follower. Electron Lett 31:353
21. Horng JW (2004) High input impedance voltage mode universal biquadratic filters with three input using plus-type CCIIs. Int J Electron 91:465–475
22. Ozoguz S, Acar C (1997) Universal current-mode filter with reduced number of active and passive components. Electron Lett 33:948–949
23. Abuelma'atti MT, Shabra AM (1996) A novel current conveyor-based universal current-mode filter. Microelectron J 27:471–475
24. Senani R (1996) A simple approach of deriving single input multiple output current mode biquad filters. Frequenz 50:124–127
25. Senani R (1992) New current-mode biquad filter. Int J Electron 73:735–742
26. Chang CM (1993) Universal active current filter with single input and three outputs using CCIIs. Electron Lett 29:1932–1933
27. Chang CM (1993) Novel universal current-mode filter with single input and three outputs using only five current conveyors. Electron Lett 29:2005–2007
28. Sun Y, Fidler JK (1994) Versatile active biquad based on second generation current conveyors. Int J Electron 76:91–98
29. Senani R (1985) Novel higher order active filter design using current conveyors. Electron Lett 21:1055–1057
30. Ozcan S, Cicekoglu O, Kuntman H (2003) Multi-input single output filter with reduced number of passive elements employing single current conveyor. Comput Electr Eng 29:45–53
31. Abuelma'atti MT, Tasadduq NA (1999) A novel three inputs and one outputs universal current mode filter using plus type CCIIs. Microelectron J 30:287–292
32. Abuelma'atti MT, Al-Zaher HA (1998) Novel current conveyor based universal current mode biquad filter with three inputs and one output. Active Passive Elec Comp 20:235–240
33. Chang CM (1997) Universal active current filter with three inputs and one output using plus type CCIIs. Electron Lett 33:1207–1208
34. Chang CM, Chien CC, Wang HY (1994) Universal active current filter with three input using current conveyors Part 2. Int J Electron 76:87–89
35. Kuntman H, Cicekoglu O, Ozcan S (2002) Realization of current mode third order Butterworth filters employing equal valued passive elements and unity gain buffers. Analog Integr Circ Sig Process 30:253–256

36. Hwang YS, Chen JJ, Li JP (2007) New current mode all pole and elliptic filters employing current conveyors. Electr Eng 89:457–459
37. Liu SI, Tsao HW, Wu J, Lin TK (1990) MOSFET capacitor filters using unity gain CMOS current conveyors. Electron Lett 26:1430–1431
38. Senani R (1987) Network transformations for incorporating nonideal simulated immittances in the design of active filters and oscillators. IEE Proc 134:158–166
39. Hwang YS, Liu SI, Wu DS, Wu YP (1995) Linear transformation all-pole filters based on current conveyors. Int J Electron 79:439–445
40. Anday F, Gunes O (1992) Realization of nth-order transfer functions using current conveyors. Int J Circ Theor Appl 20:693–696
41. Soliman AM (1972) Active RC realization of current transfer function using voltage generalized-immittance converters. Int J Electron 33:273–280
42. Soliman AM (1973) Inductorless realization of an all-pass transistor function using the current conveyor. IEEE Trans Circ Theory 20:80–81
43. Soliman AM (1973) Another realization of an all-pass or a notch filter using a current conveyor. Int J Electron 35:135–136
44. Aronhime P (1974) Transfer-function synthesis using a current conveyor. IEEE Trans Circ Syst 21:312–313
45. Gopal K (1974) Comment on 'Inductorless realization of an All pass transfer function using the current conveyor'. IEEE Trans Circ Syst 21:704–705
46. Naimpally SV (1975) Comments on "Transfer-function synthesis using a current conveyor". IEEE Trans Circ Syst 22:960
47. Soliman AM (1977) Two novel active RC canonic band pass networks using the current conveyor. Int J Electron 42:49–54
48. Nandi R (1978) Equal valued earthed-capacitor realisation of a third-order low pass Butterworth characteristic using current conveyors. Electron Lett 14:699–700
49. Stephenson FW, Davies JD (1979) Simplified design procedures for a third-order system using current conveyors. Electron Lett 15:215–216
50. Davies JD, Stephenson FW (1980) Sensitivity optimization of active filters containing current conveyors and controlled sources. Int J Electron 48:283–289
51. Rathore TS (1980) Active complementary networks. IEEE Trans Circ Syst 27:1278–1279
52. Pal K (1981) Realization of current conveyor all-pass networks. Int J Electron 50:165–168
53. Higashimura M, Ishida M, Hara M, Fukui Y (1988) Realization of biquadratic transfer function using single current conveyor. Trans IEICE J71-A: 228–234
54. Tek H, Anday F (1989) Voltage transfer function synthesis using current conveyors. Electron Lett 25:1552–1553
55. Roberts GW, Sedra AS (1989) All current-mode frequency selective circuits. Electron Lett 25:759–761
56. Fabre A, Martin F, Hanafi M (1990) Current mode all pass/notch and band pass filters with reduced sensitivities. Electron Lett 26:1495–1496
57. Aronhime P, Nelson D, Adams C (1990) Applications of a first-generation current conveyor in current-mode circuits. Electron Lett 26:1456–1457
58. Liu SI, Tsao HW, Wu J (1990) Cascadable current-mode single CCII biquads. Electron Lett 26:2005–2006
59. Sagbas M, Koksal M (2007) Voltage-mode three-input single-output multifunction filters employing minimum number of components. Frequenz 61:87–93
60. Aronhime P, Dinwiddie A (1991) Biquadratic current-mode filters using a single CCI. Int J Electron 70:1063–1071
61. Higashimura M (1991) Current-mode transfer function using CC IIs with grounded passive elements. IEICE Trans 74:1017–1019
62. Chang CM (1991) Current mode all pass/notch and band pass filter using single CCII. Electron Lett 27:1812–1813
63. Higashimura M (1991) Realization of voltage-mode biquads using CCIIs. Electron Lett 27:1345–1346

64. Liu SI, Tsao HW (1991) New configurations for single CCII biquads. Int J Electron 70:609–622
65. Liu SI, Tsao HW, Wu J (1991) CCII-based continuous-time filters with reduced gain-bandwidth sensitivity. IEE Proc Pt-G 138:210–216
66. Chang CM, Chen PC (1991) Realization of current-mode transfer function using second-generation current conveyors. Int J Electron 71:809–815
67. Alami M, Fabre A (1991) Insensitive current-mode band pass filter implemented from two current conveyors. Electron Lett 27:897–899
68. Chang CM (1991) Universal active current filters using single second-generation current conveyor. Electron Lett 27:1614–1417
69. Liu SI, Tsao HW (1991) The single CCII biquads with high-input impedance. IEEE Trans Circ Syst 38:456–461
70. Liu SI, Kuo JH, Tsay JH (1992) New CCII-based current-mode biquadratic filters. Int J Electron 72:243–252
71. Fabre A, Houle JL (1992) Voltage-mode and current-mode Sallen-Key implementation based on translinear conveyors. IEE Proc Pt-G 139:491–497
72. Hou CL, Wu YP (1992) New methods of synthesizing transfer functions by applying CCIIs. Int J Electron 72:119–128
73. Higashimura M (1992) A novel configuration for voltage-mode biquads using a single current conveyor. Microelectron J 23:359–362
74. Chang CM, Chien CC, Wang HY (1993) Universal active current filters using single second-generation current conveyor. Electron Lett 29:1159–1160
75. Chang CM (1993) Current-mode low pass, band pass and high pass biquads using two CCIIs. Electron Lett 29:2020–2021
76. Abuelma'atti MT (1993) New current-mode-active filters employing current conveyors. Int J Circ Theor Appl 21:93–99
77. Svoboda JA (1994) Comparison of RC op-amp and RC current conveyor filters. Int J Electron 76:615–626
78. Soliman AM (1994) Kerwin-Huelsman-Newcomb circuit using current conveyors. Electron Lett 30:2019–2020
79. Fabre A, Dayoub F, Duruisseau L, Kamoun M (1994) High input impedance insensitive second-order filters implemented from current conveyors. IEEE Trans Circ Syst-I 41:918–921
80. Wu DS, Hwang YS, Liu SI, Wu YP (1994) New multifunction filter using an inverting CCII and a voltage follower. Electron Lett 30:551–552
81. Nandi R (1994) Insensitive current mode realization of third-order Butterworth characteristic using current conveyors. IEEE Trans Circ Syst-I 41:925–927
82. Gunes EO, Anday F (1995) Realization of nth-order voltage transfer function using CCII+. Electron Lett 31:1022–1023
83. Soliman AM (1995) Current conveyors steer universal filter novel arrangement delivers low pass, band pass, and high pass response, inverting and non-inverting. IEEE Circ Devices Mag 11:45–46
84. Anandamohan PV (1995) New current mode biquad on Friend-Deliyannis active RC biquad. IEEE Trans Circ Syst-II 42:225–228
85. Abuelma'atti MT, Quddus A (1995) Programmable current-conveyor-based voltage mode filters with single input and five outputs. Active Passive Electron Comp 18:273–278
86. Liu SI, Chen JJ, Hwang YS (1995) New current mode biquad filters using current followers. IEEE Trans Circ Syst-I 42:380–383
87. Fabre A, Alami M (1995) Universal current mode biquad implemented from two second generation current conveyors. IEEE Trans Circ Syst-I 42:383–385
88. Celma S, Sabadell J, Martinez P (1995) Universal filter using unity-gain cells. Electron Lett 31(21):1817–1818
89. Acar C, Ozoguz S (1996) High-order voltage transfer function synthesis using CCII+ based unity gain current amplifiers. Electron Lett 32:2030–2031

90. Acar C (1996) Nth-order all pass voltage transfer function synthesis using CCII+ s: signal-flow graph approach. Electron Lett 32:727–729
91. Acar C (1996) Nth-order low pass voltage transfer function synthesis using CCII+ s: signal-flow graph approach. Electron Lett 32:159–160
92. Soliman AM (1996) New inverting-non-inverting band pass and low pass biquad circuit using current conveyors. Int J Electron 81:577–583
93. Vrba K, Cajka J, Zeman V (1996) RC-active N-th order low pass filter. J Electr Eng 47:257–259
94. Nakai K, Yamamoto G, Nakamura T (1996) A configuration of state variable biquad filters using current conveyors. IEICE Trans Fundament E-79:639–641
95. Abuelma'atti MT, Quddus A (1996) Programmable voltage-mode multifunction filter using two current conveyors and one operational transconductance amplifier. Active Passive Electron Comp 19:133–138
96. Nandi R (1996) Precise insensitive current-mode third-order low pass Butterworth characteristics. IEE Proc Circ Devices Syst 143:223–224
97. Abuelma'atti MT, Al- Qahtani MA (1996) Current mode universal filters using unity-gain cells. Electron Lett 32:1077–1078
98. Celma S, Martinez PA, Sabadell J (1996) A transformation method for equivalent infinite-gain Op-amp to unity-gain CCII networks. IEEE Trans Circ Syst-I 43:61–63
99. Acar C, Kuntman HK (1996) Limitations on input signal level in current-mode active-RC filters using CCIIs. Electron Lett 32:1461–1462
100. Soliman AM (1996) Comment on 'The single CC II biquads with high-input impedance'. IEEE Trans Circ Syst-I 43(1):65
101. Fabre A, Amrani H, Saaid O (1996) Current-mode band-pass filters with Q-magnification. IEEE Trans Circ Syst-II 43:839–842
102. Cajka J, Lindovsky D (1997) Universal RC-active networks using CCII+. J Electr Eng 48:98–100
103. Soliman AM (1997) Theorems relating to port interchange in current mode CCII circuits. Int J Electron 82:585–604
104. Acar C (1997) On the realization of current-mode filters using second-generation current conveyors. Int J Circ Theor Appl 25:229–233
105. Zhang X, Kambayashi N, Shinada Y (1997) A realization of active current-mode resonator with complex coefficients using CCIIs. IEICE Trans Fundament 2:413–415
106. Lata ZJ, Aronhime PB (1997) Cascadable current-mode biquads. Analog Integr Circ Sig Process 13:275–284
107. Cajka J, Dostal T, Vrba K (1997) Realization of Nth-order voltage transfer function using current conveyors CCII. Radioengineering 6:22–25
108. Vrba K, Cajka J (1997) High-order one–port elements for low pass filter realization. J Electr Eng 48:31–34
109. Soliman AM (1997) Generation of current conveyor-based all-pass filter from op-amp based circuits. IEEE Trans Circ Syst-II 44:324–330
110. Cajka J, Vrba K, Lindovsky D (1997) Network function synthesis using current conveyors. J Electr Eng 48:209–211
111. Hou CL, Wu JS (1997) Universal cascadable current-mode biquad using only four CCIIs. Int J Electron 82:125–129
112. Gunes ECEO, Anday F (1997) Realization of voltage and current-mode transfer functions using unity-gain cells. Int J Electron 83:209–213
113. Papazoglou CA, Karybakas CA (1997) Noninteracting electronically tunable CCII-based current mode biquadratic filters. IEE Proc Circ Devices Syst 144:178–184
114. Abuelma'atti MT, Farooqui AA (1998) Universal current-conveyors-based current-mode filters with single input and three outputs. Active Passive Electron Comp 20:195–200
115. Abuelma'atti MT (1998) Programmable current-mode universal active filters employing current conveyors. Active Passive Electron Comp 21:221–230
116. Cabeza R, Carlosena A (1998) A cautionary note on stability of current conveyor based circuits. Int J Circ Theor Appl 26:215–218

117. Soliman AM (1998) Current conveyor filters: classification and review. Microelectron J 29:133–149
118. Soliman AM (1998) Generation of CCII and CFOA filters from passive RLC filters. Int J Electron 85:293–312
119. Horng JW (1999) Inverting and/or non-inverting biquad circuit using second-generation current conveyors. Int J Electron 86:297–303
120. Karybakas CA, Papazoglou CA (1999) Low-sensitive CCII-based biquadratic filters offering electronic frequency shifting. IEEE Trans Circ Syst-II 46:527–539
121. Cicekoglu O (1999) Comments on 'Multifunction biquadratic filters using current conveyors'. IEEE Trans Circ Syst-II 46:658–659
122. Bhaskar DR, Sharma VK, Monis M, Rizvi SMI (1999) New current-mode universal biquad filter. Microelectron J 30:837–839
123. Toker A, Ozoguz S (2000) Insensitive current-mode universal filter using dual output current conveyors. Int J Electron 87:667–674
124. Abuelma'atti MT, Tasadduq NA (2000) Current-mode low pass, band pass and high pass filter using CCII+ s. Frequenz 54:162–164
125. Toker A, Cicekoglu O, Ozcan S, Kuntman H (2001) High-output-impedance transadmittance-type continuous-time multifunction filter with minimum active elements. Int J Electron 88:1085–1091
126. Horng JW (2003) High input impedance voltage-mode universal biquadratic filter using two OTAs and one CCII. Int J Electron 90:185–191
127. Saraswat R, Pal K, Rana S (2003) Novel grounded capacitor all-pass and notch filters using current conveyors and differential amplifier. Active Passive Electron Comp 26:167–170
128. Aksoy M, Ozcan S, Cicekoglu O, Kuntman H (2003) High output impedance current-mode third-order Butterworth filter topologies employing unity gain voltage buffers and equal-valued passive components. Int J Electron 90:589–598
129. Pal K, Rana S (2004) Some new first-order all-pass realizations using CCII. Active Passive Electron Comp 27:91–94
130. Horng JW, Tang HW, Wen YH (2005) Voltage-mode high input impedance inverting and/or non-inverting high pass, band pass and low pass filers using three CCIIs. J Active Passive Electron Devices 1:145–158
131. Hayat M, Kumar U (2005) Novel variable current gain active filter using current conveyors. J Active Passive Electron Devices 1:91–96
132. Kumar P, Pal K (2005) Variable Q all-pass, notch and band-pass filters using single CCUU. Frequenz 59:9–10
133. Pandey N, Paul SK, Bhattacharyya A, Jain SB (2006) A new mixed mode biquad using reduced number of active and passive elements. IEICE Electron Express 6:115–121
134. Minaei S, Goknar IC, Cicekoglu O (2006) A new differential configuration suitable for realization of high CMRR all-pass/notch filters. Electr Eng 88:317–326
135. Kumar P, Pal K (2006) High input impedance band pass, all pass and notch filters using two CCIIs. HAIT J Sci Eng A 3:2–13
136. Khan IQ, Khan MR, Afzal N (2006) Digitally programmable multifunctional current mode filter using CCIIs. J Active Passive Electron Devices 1:213–220
137. Dostal T, Axman V (2007) Biquads based on single negative impedance converter implemented by classical current conveyor. Radioengineering 16:96–102
138. Kumar P, Pal K, Rana S (2008) High input impedance universal biquadratic filters using current conveyors. J Active Passive Electron Devices 3:17–27
139. Kumar P, Pal K (2008) Universal biquadratic filter using a single current conveyor. J Active Passive Electron Devices 3:7–16
140. Yuce E, Minaei S (2008) Signal limitations of the current-mode filters employing current conveyors. Int J Electron Commun (AEU) 62:193–198
141. Soliman AM (2009) Current mode universal filters with grounded passive elements filters with grounded passive elements and using single outputs current conveyors. J Active Passive Electron Devices 4:55–62

142. Soliman AM (1998) Equal-R, Equal-C current mode Butterworth lowpass filters. IEICE Trans Fundament 2:340–342
143. Weng RM, Lai JR, Lee MH (2000) New universal biquad filters using only two unity-gain cells. Int J Electron 87:57–61
144. Metin B, Pal K, Cicekoglu O (2012) A new approach for high-input impedance in voltage mode filters using first-generation current conveyor in place of second-generation current conveyor. Int J Electron 99:131–139
145. Hwang YS, Hung PT, Chen W, Liu SI (2002) CCII-based linear transformation elliptic filters. Int J Electron 89:123–133
146. Ozoguz S, Toker A, Cicekoglu O (1998) High output impedance current mode multifunction filter with minimum number of active and reduced number of passive elements. Electron Lett 34:1807–1809
147. Soliman AM (2008) Current mode universal filters using current conveyors: classification and review. Circ Syst Sig Process 27:405–427
148. Soliman AM (2008) History and progress of the Tow-Thomas Bi-quadratic filter. Part I. Generation and op-amp realizations. J Circ Syst Comput 17:33–54
149. Soliman AM (2008) History and progress of the Tow-Thomas biquadratic filter Part II: OTRA, CCII, and DVCC realizations. J Circ Syst Comput 17:797–826
150. Soliman AM (2008) History and progress of the Kerwin-Huelsman Newcomb filter: generation and op-amp realizations. J Circ Syst Comput 17:637–658
151. Metin B, Toker A, Terzioglu H, Cicekoglu O (2003) A new all-pass section for high performance signal processing with a single CCII-. Frequenz 57:11–12
152. Singh VK, Senani R (1990) New multifunction active filter configuration employing current. Electron Lett 26:1814–1815
153. Higashimura M, Fukui Y (1988) Realization of all-pass and notch filter using a single current conveyor. Int J Electron 65:823–828
154. Hou CL, Wu YP, Liu SI (1991) New configuration for single-CCII first order and biquadratic current-mode filters. Int J Electron 71:637–644
155. Khan IA, Zaid MH (2000) Multifunctional translinear-C current mode filter. Int J Electron 87:1047–1051
156. Horng JW (2005) Current conveyors based all pass filters and quadrature oscillators employing grounded capacitors and resistors. Comput Electr Eng 31:81–92
157. Senani R (1994) Two ICs make biquad filter. Circuit ideas, electronics and wireless world. Reed Business Publishing Ltd. UK, p 922
158. Fabre A, Alam M (1993) Insensitive current mode notch filter implemented form translinear conveyors. Int J Electron 74:735–739
159. Roberts GW, Sedra AS (1992) A general class of current amplifier-based biquadratic filter circuits. IEEE Trans Circ Syst-I 39:257–263
160. Hassan TM, Mahmoud SA (2009) Fully-programmable universal filer with independent gain $-\omega_0$-Q control based on new digitally programmable CMOS CCII. J Circ Syst Comput 18:875–897
161. Cicekoglu O, Ozcan S, Kuntman H (1999) Insensitive multifunction filter implemented with current conveyors and only grounded passive elements. Frequenz 53:158–160
162. Ozoguz S, Toker A, Cicekoglu O (1999) New current mode universal filters using only four (CCII+)s. Microelectron J 30:255–258
163. Horng JW, Lay JR, Chang CW, Lee MH (1997) High input impedance voltage mode multifunction filters using plus-type CCIIs. Electron Lett 33:472–473
164. Svoboda JA (1994) Transfer function synthesis using current conveyors. Int J Electron 76:611–614
165. Horng JW, Hou CL, Chang CM, Chung WY, Liu HL, Lin CT (2006) High output impedance current-mode first order all pass networks with four grounded components and two CCIIs. Int J Electron 93:613–621

166. Horng JW (2010) Voltage/current mode universal biquadratic filter using single CCII+. Indian J Pure Appl Phys 48:149–756
167. Kerwin W, Huelsman L, Newcomb R (1967) State variable synthesis for insensitive integrated circuit transfer functions. IEEE J Solid State Circ 2:87–92
168. Thomas L (1971) The biquad Part-I – some practical design considerations. IEEE Trans Circ Theor 18:350–357
169. Tow J (1969) A step by step active filter design. IEEE Spectrum 6:64–68
170. Akerberg D, Mossberg K (1974) A versatile active RC building block with inherent compensation for the finite bandwidth of the amplifier. IEEE Trans Circ Syst 21:75–78

Chapter 7
Realization of Sinusoidal Oscillators Using CCs

Abstract Single capacitance/resistance controlled sinusoidal oscillators using CCs have been discussed which include both single CC-based as well as two or more CC-based oscillators including quadrature and multiphase oscillators. Also included are fully-uncoupled oscillators and oscillators employing all grounded passive elements as well as those providing explicit current output.

7.1 Introduction

This chapter is concerned with the realization of single-capacitance/resistance controlled sinusoidal oscillators using CCs a wide variety of which have been advanced during the last 35 years or so [1–56].

Because of their numerous applications in electronics, communication, instrumentation, measurement and control, sinusoidal oscillators have been a major topic of research since long. In communication systems, sinusoidal oscillators are used to generate a carrier signal in modulators, in instrumentation and measurement systems they may be employed as test oscillators employed in measurement and sensor applications as well as transducer oscillators in conjunction with resistive/capacitive/ inductive transducers.

With the advent of the integrated circuit (IC) op-amps, it was a logical consequence to have op-amp-based configurations for realizing a number of classical sinusoidal oscillator topologies such as Wien bridge oscillator, RC-phase shift oscillator and twin-T oscillators etc. for low to medium frequency applications. A common difficulty with the mentioned topologies was recognized to be their inability to provide variable frequency oscillations through only a single variable element, preferably a variable resistor. Hribsek and Newcomb were the first to propose a single-resistance-controlled oscillator using two IC op-amps [62]. Since then, efforts of the researchers were soon directed towards devising new topologies for such single-element-controlled oscillators (SECO)/SRCOs employing a single op-amp along with a canonical number of passive components namely, only two capacitors and at most five resistors. A number of such single-op-amp SRCOs have since been discovered, for instance, see [63–69] and those cited therein.

© Springer International Publishing Switzerland 2015

R. Senani et al., *Current Conveyors*, DOI 10.1007/978-3-319-08684-2_7

Since the current conveyors were already introduced in the literature, a number of research groups around the world also started looking for the possibility of realizing sinusoidal oscillators using CCs in general and SECOs/SRCOs in particular. After a considerable research it was found that a canonic SECO/SRCO using a single CC can be realized with only five passive components namely, either three resistors or two capacitors or with three capacitors and two resistors. With the introduction of a large variety of extensions of the basic current conveyor types, the research on the realization of SRCOs including quadrature and multi-phase oscillators has still continuing. Because of the enormity of the literature on SRCOs using CCs, this chapter focuses only on SECOs/SRCOs realizable with the basic types of current conveyors namely CCI± and CCII± only with the circuits realizable from other modified variants of current conveyors have been reserved for a separate treatment in subsequent chapter (Chap. 13) of this monograph.

In the subsequent sections of this chapter, we focus on the single-CC-based SECOs/SRCOs and a number of other varieties of sinusoidal oscillator configurations evolved over the past 35 years [1–56] using the basic three-terminal types of CCs.

7.2 Single-CC SRCOs

It is interesting to note that although both CCI and CCII had distinctly different terminal characterizations as compared to the ubiquitous op-amp, the equality of the voltages between Y and X terminals was quite similar to that of an infinite gain op-amp (when connected in a negative feedback arrangement) which was responsible for a number of circuits which were derived in the initial phase of research on current conveyors through analogy of corresponding op-amp-based circuits. Thus, a number of early current conveyor based oscillators which were based upon classical Wien bridge oscillator and its variants using op-amp were obvious, for instance see [1, 24, 29, 36] and therefore, they also suffered from the same limitations or difficulties as their op-amp-based counterparts. Besides this there have been a number of other single-CC based single frequency [11, 12, 15, 20] as well as variable frequency [3–10, 13, 16, 19, 20][1] oscillators reported in literature.

In the following, however, we focus on only variable frequency oscillators capable of providing frequency control through a single variable element (a capacitance or resistance).

In fact, the first SECO using a single CCII+ was proposed by Soliman in [3] which is shown here Fig. 7.1. This circuit is characterized by the condition of oscillation (CO) as

[1] Apart from a single CC, this circuit, however, also incorporates an OTA.

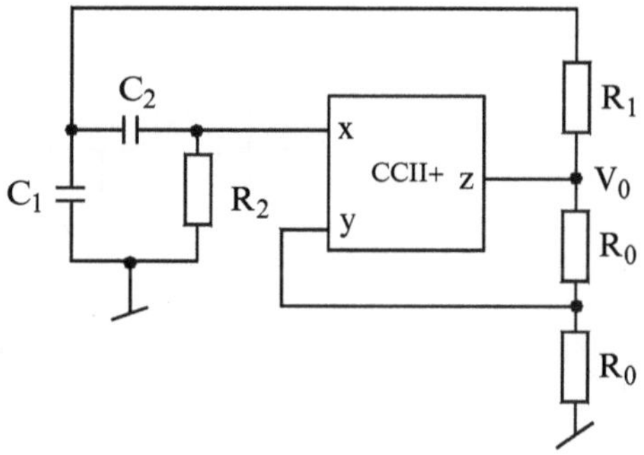

Fig. 7.1 Soliman's single-capacitor-controlled oscillator [3]

$$\text{CO}: \frac{1}{R_2} = \frac{1}{R_0} + \frac{2}{R_1} \tag{7.1}$$

and the frequency of oscillation (FO) as

$$\text{FO}: \omega_0 = \frac{1}{R_1}\sqrt{\frac{2}{C_1 C_2}} \tag{7.2}$$

From the above equations, it is seen that the condition of oscillation can be controlled by the grounded resistor R_2 whereas the frequency of oscillation is adjustable through a single variable capacitor C_2 or C_1. It is worth pointing out that because none of the two capacitors appears in the CO, with one of the external capacitor namely C_1 taken out, the remaining circuit must be a lossless grounded inductor subject to the fulfillment of the same condition as given in Eq. (7.1). It is, indeed, true as was subsequently shown by Soliman himself (in [9] of Chap. 5, see Fig. 5.20 of Sect. 5.3.2) that with C_1 omitted, the remaining circuit is, indeed, an *active gyrator.*

In the same year, Soliman came up with another single CCII+ based circuit which could realize an ideal FDNR, a lossy FDNR, a parallel tuned circuit or an oscillator from the same structure subject to the fulfillment of appropriate condition in terms of RC elements used in the circuit. This circuit is shown in Fig. 7.2

The CO and FO for this circuit in oscillator mode are given by:

$$\text{CO}: C_2R_2\left(1+\frac{R_1}{R_4}\right) - C_1R_1\left(1+\frac{2R_3}{R_1}+\frac{R_2+R_3}{R_4}\right) = 0 \tag{7.3}$$

$$\text{FO}: f_0 = \frac{1}{2\pi C_1}\sqrt{\left\{\frac{\frac{1}{R_4}-\frac{1}{R_3}}{R_1\left(2+\frac{R_1}{R_3}\right)}\right\}} \tag{7.4}$$

provided that the following conditions should be satisfied:

$$C_2 = C_1\left(1+\frac{R_3}{R_2}\right), R_2 = (R_1+R_3) \text{and} R_3 > R_4 \tag{7.5}$$

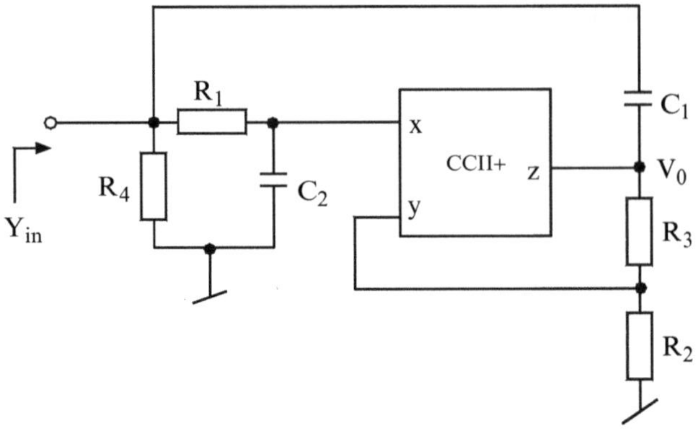

Fig. 7.2 A canonic active-RC circuit capable of realizing an ideal FDNR, a lossy FDNR, parallel tuned circuit and an oscilltor proposed by Soliman [4]

If the component values are selected such that $R_1 = R_3$, $R_2 = 2R_1$, $R_1 > R_4$ and $C_2 = (3/2)C_1$, then

$$f_0 = \frac{1}{2\pi C_1 R_1}\sqrt{\left\{\frac{1}{3}\left(\frac{R_1}{R_4} - 1\right)\right\}} \tag{7.6}$$

Thus, it is seen that f_o can be controlled by a single grounded resistance R_4 and can also be controlled by external voltage by replacing this resistor by an FET working in the triode region, however, the circuit lacks the provisions of providing independent control of CO through a single variable resistance.

Another single-CC SRCO which does not have the above mentioned limitation was proposed by Senani in [5]. This circuit employs one more resistor than Soliman's circuit and is shown here as Fig. 7.3. Subject to the fulfillment of the condition

$$\frac{C_1}{C_2} = 1 + R_3\left(\frac{1}{R_1} + \frac{1}{R_2}\right) \tag{7.7}$$

the characterizations of this circuit are:

$$\text{CO} : \text{R}_4 = \text{R}_2 \tag{7.8}$$

$$\text{FO} : \omega_0 = \left(\frac{R_2}{2C_1C_2R_1R_3R_0}\right)^{1/2} \tag{7.9}$$

It is, thus, seen that in this circuit, CO can be satisfied by adjusting the grounded resistor R_4 without effecting FO which is also independently controllable by

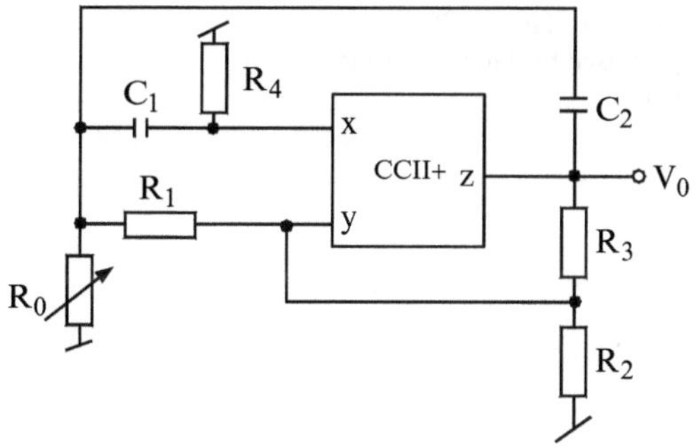

Fig. 7.3 Senani's single-resistance-controlled oscillator [5]

another single variable grounded resistance R_0 which does not appear either in Eqs. (7.7) or (7.8).

In integrated circuit realization, it is desirable to have all the capacitors grounded as this not only helps in the fabrication process simplified it also helps in absorbing the parasitic capacitances of the active device into the external capacitors. The previously described single CC SECOs/ SRCOs do not have both the capacitors grounded. A single-CC SRCO, with all grounded capacitors (although three in numbers), was introduced by Jana and Nandi in [6] and is shown here in Fig. 7.4.

The CO and FO for this circuit are given by:

$$C_x = C_2\left(1 + \frac{R_2}{R_1}\right) + C_1 \tag{7.10}$$

$$\omega_0 = \sqrt{\frac{R_x - R_1}{R_1 R_2 R_x C_1 C_2}} \tag{7.11}$$

In this circuit, the capacitance C_x can be used to set the circuit to produce sinusoidal oscillations whereas the oscillation frequency can be varied by varying R_x subject to ensuring that $R_x > R_1$. Chong and Smith in [7] carried out a detailed study on the evaluation of the effects of non-zero offset current at terminal-Z of the CCs on the operation of CC-based sinusoidal oscillators. This study revealed that to reduce the effect of non-zero output offset current, the oscillator circuit proposed by Jana and Nandi [6] needs to be modified by removing the resistor R_1 connected across C_1 and instead of this, putting another resistance R_0 in parallel with the capacitor C_2. With this modification, CO and FO are modified to:

$$\text{CO}: C_x = C_1\left(1 + \frac{R_2}{R_0}\right) + C_2 \tag{7.12}$$

$$\text{FO}: \omega_0 = \sqrt{\frac{R_x - R_0}{R_0 R_2 R_x C_1 C_2}} \tag{7.13}$$

It may be mentioned that one still requires R_x to be larger than R_0 for low frequency oscillations, it is possible to choose R_2 to be very large without affecting the DC

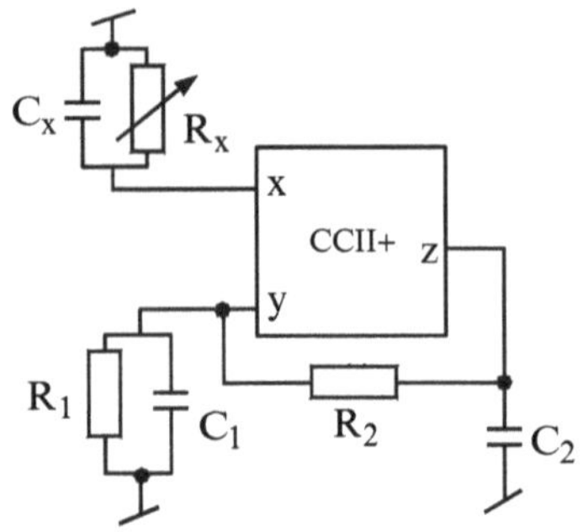

Fig. 7.4 The SRCO proposed by Jana and Nandi [6]

offset voltage at terminal-Z. This is an advantage of the suggested modification in the oscillator of Jana and Nandi [6].

After the introduction of the above mentioned circuits by different researchers, a number of systematic studies were conducted by several authors such as by Abuelma'atti and Humood in [8–10] and Celma et al. in [16] to derive single-CC-based cananoic SECOs/SRCOs and it was eventually established that only five passive components are really necessary alongwith a single CC, to make either a single-capacitor-controlled oscillator (SCCO) or SRCO; two resistors and three capacitors for the former and two capacitors and three resistors in case of the latter. An exemplary SCCO/SRCO (both derivable from each other through RC:CR transformation) is shown in Fig. 7.5. It is interesting to note that the SRCO of Fig. 7.5 was independently proposed by Abuelma'atti and Humood in [10] as well as by Celma et al. in [16].

The CO and FO for this circuit are given by

$$\frac{R}{R_2} = 1 + \frac{C_2}{C_1} \tag{7.14}$$

$$\text{and } \omega_0 = \sqrt{\frac{2}{C_1 C_2 R R_1}} \tag{7.15}$$

from where it is readily seen that CO is controllable by R_2 whereas FO is independently adjustable by R_1.

7.3 SRCOs Employing Grounded Capacitors

It is well known that the circuit configurations employing all grounded capacitors are attractive not only from the view point of IC implementation but also due to the fact that they can easily accommodate the parasitic capacitances of the active devices employed which can be absorbed in them. It is worth pointing out that using the traditional op-amps, a number of circuits are known in literature which employ

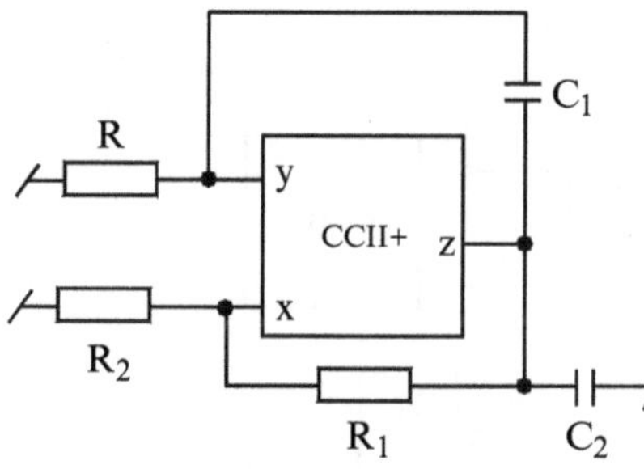

Fig. 7.5 Canonic SRCO using a CCII + proposed by Abuelma'amatti and Humood [10] and also by Celma et al. [16]

three/two op-amps along with two grounded capacitors to realize SRCOs however, similar circuit with a single op-amp is although in principle possible but the required design constraints are very complex and so are the resulting tuning laws.

In this section, we show that the current conveyors outperform the op-amps in this respect. In particular, not only such circuits can be realized using current conveyors while avoiding complex design equations but also the resulting circuits employ invariably the least possible number of passive components. We highlight some exemplary circuits exhibiting these features from amongst many such circuits advanced by various researchers in the literature, for instance, [2, 14, 18, 31, 41, 43, 47, 51, 54, 62].

We first present two minimum-component oscillators which employ only two CCIIs, only two resistors and both grounded capacitors. Such circuits were proposed by Horng et al. in [31] and are shown in Fig. 7.6. Both the circuits are characterized by the same characteristic equation and have the same CO and FO given respectively by:

$$s^2C_1C_2R_1R_2 + sR_1(C_2 - C_1) + 1 = 0 \tag{7.16}$$

$$\mathrm{CO} : C_1 = C_2 \tag{7.17}$$

$$\text{and FO} : \omega_0 = \sqrt{\frac{1}{C_1C_2R_1R_2}} \tag{7.18}$$

In both these circuits, the only point of inconvenience is the adjustment of the CO which will require in practice a variable capacitor which is not a very convenient option.

Two other circuits, with exactly same number of active and passive components and also governed by exactly the same equations, are shown in Fig. 7.7a, b and were proposed by Horng [41] and Fongsamut et al. [47] respectively.

From a non-ideal analysis of both the circuits assuming the current gains of the CCIIs as α_1 and α_2 and the voltage gains as β_1 and β_2. The non-ideal expressions for the FO of the two circuits are given by:

$$\mathrm{FO} : \omega_0 = \sqrt{\frac{\beta_1\alpha_1\alpha_2}{C_1'C_2'R_1'R_2'}} \quad \text{for the Fig.7.7a} \tag{7.19}$$

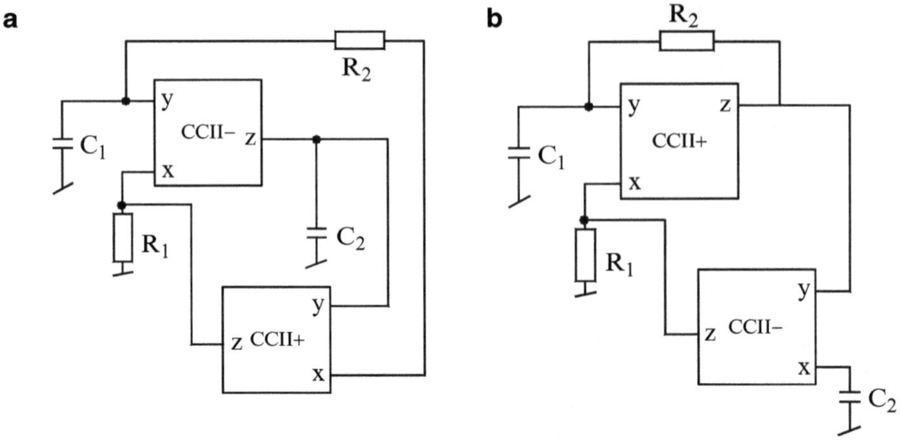

Fig. 7.6 Minimum-component oscillators proposed by Horng et al. [31]

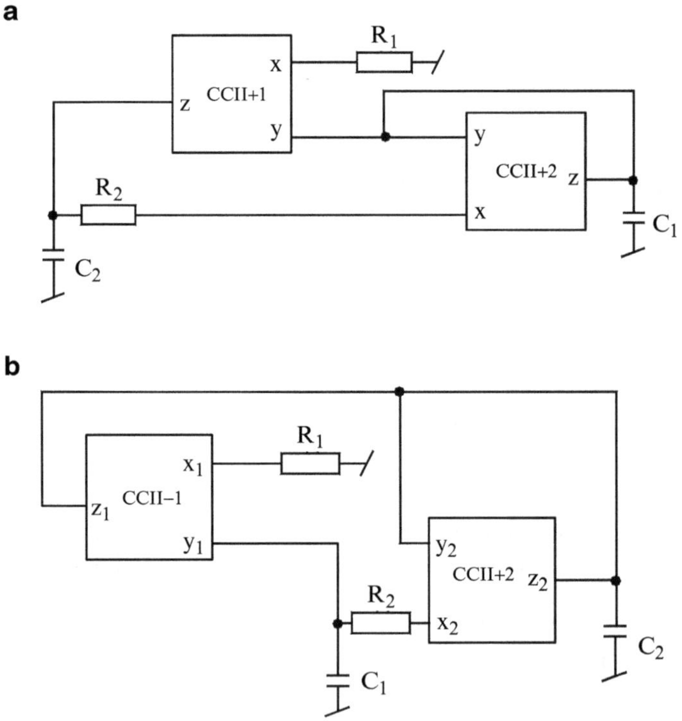

Fig. 7.7 Grounded capacitors oscillators (**a**) Proposed by Horng [41] (**b**) Proposed by Fongsamut et al. [47]

$$\text{and } \omega_0 = \sqrt{\frac{\beta_1\alpha_1\beta_2}{C_1^{'}C_2^{'}R_1^{'}R_2^{'}}} \quad \text{for the Fig.7.7b} \tag{7.20}$$

Where

$$C_1^{/} = C_1 + C_{y1} + C_{y2} + C_{z2},\ C_2^{/} = C_2 + C_{y2} + C_{z1} + C_{z2},\ R_1^{/} = R_1 + R_{x1} \text{ and } R_2^{/} = R_2 + R_{x2} \text{ for the Fig7.7a} \tag{7.21}$$

$$C_1^{/} = C_1 + C_{y1} + C_{y2} + C_{z2},\ C_2^{/} = C_2 + C_{z1},\ R_1^{/} = R_1 + R_{x1} \text{ and} R_2^{/} = R_2 + R_{x2} \text{ for the Fig.7.7b} \tag{7.22}$$

Fongsamut et al. have demonstrated [47] that since the value of voltage gain β is much closer to unity than the current gain α, therefore by comparing the above equations, it is obvious that the FO of the circuit of Fig. 7.7b would be in a closer agreement of the theoretical value than that of the circuit of Fig. 7.7a. Using the translinear current conveyor as well as using CCII+ from AD844 and CCII− implemented from two AD844s, the validity of this assertion has been confirmed in [47].

The difficulty of adjusting the CO through a variable capacitance is eliminated if an SRCO could be synthesized with at least three resistors such that one of the two resistors can control the CO and the remaining third resistor controls the FO. Such a circuit using a CCII− and a voltage follower was first proposed by Bhaskar and Senani [18] and is shown in Fig. 7.8.

The CO and FO for this circuit are given by

$$\text{CO}: \frac{R_1}{R_2} = \frac{C_2}{C_1} \tag{7.23}$$

$$\text{and FO}: f_o = \frac{1}{2\pi}\sqrt{\frac{1}{R_3R_2C_1C_2}} \tag{7.24}$$

Slightly different version of this circuit is also possible with the same functionalities but slightly altered expression for CO when the resistor R_2, instead of being connected to V_o, is returned to the Z- terminal of the CCII−. The resulting CO is then given by

$$\text{CO}: \frac{R_1}{R_2} = \left(\frac{C_2}{C_1} + 1\right) \tag{7.25}$$

whereas the expression for the FO remains the same as earlier.

A remarkable property of this circuit is that with $R_1 = R_2 = R$; $C_1 = C_2 = C$ and $R_3 = R/n$, the frequency stability factor S^F, for the resistor ratio n much larger than

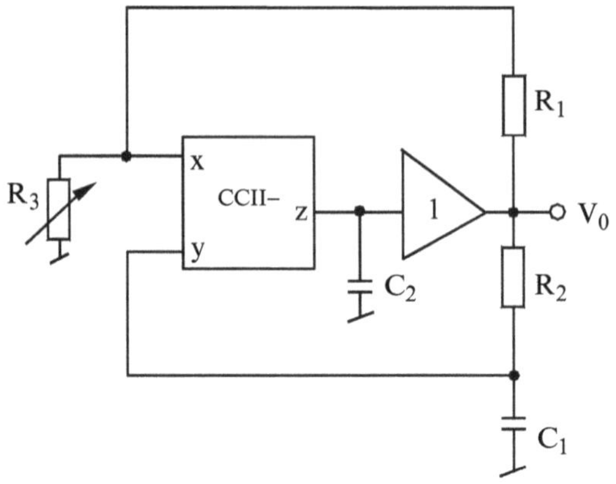

Fig. 7.8 SRCO proposed by Bhaskar and Senani [18]

unity can be approximated as $S^F \cong 2\sqrt{n}$ which can be made quite large (in fact this value is almost the largest stability factor attainable by any SRCO).

Another noticeable feature of this circuit is that because of control of FO through a grounded resistor R_3 the FO can be easily made voltage- controllable by replacing the frequency controlling resistor R_3 by an FET used as a voltage controlled resistance.

Using CCII− made from LF356 type op-amp in supply current sensing mode implemented through current mirrors made from CA3096E mixed transistor arrays and the voltage buffer made from LM310N, the various claims of the circuit have been verified and the capability to realize voltage-controlled-oscillator have been verified by using the simplified single buffer based linearized VCR of [62].

It is also worth mentioning that this configuration is also implementable by the IC CC, PA630A which however requires additional current mirrors to enable the realization of a CCII−. In this case it is interesting to note that the VCO version can be implemented rather easily without requiring any additional voltage followers because of the availability of two-on chip buffers in the PA630A CC chip.

7.4 SRCOs Employing All Grounded Passive Elements

Oscillator circuits employing all grounded passive components are interesting because of three basic reasons. Firstly, having all resistors and capacitors grounded is attractive from the point of view of integrated circuit implementation of an oscillator having all elements grounded. Secondly, all resistors being grounded is useful from the point of view of electronic control of the oscillation frequency since the frequency controlling grounded resistor(s) can be easily replaced by FET/MOSFET-based linear voltage-controlled-resistances (VCR) [61] thereby making the oscillation frequency voltage-controllable thereby converting the oscillator into a VCO. Thirdly, if the resistor controlling the condition of oscillation is also grounded, this is useful in incorporating of an automatic feedback system for stabilizing/varying

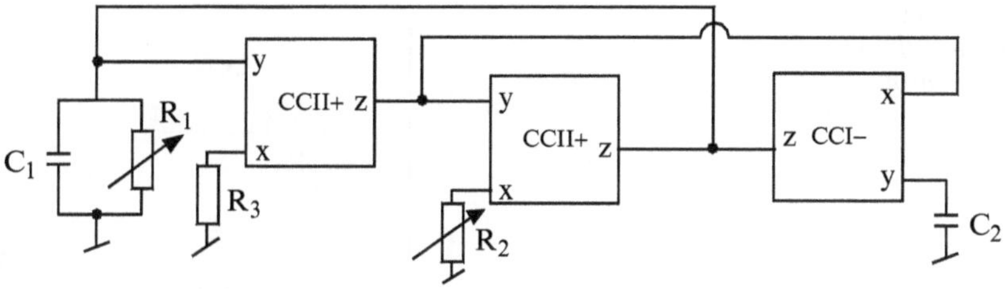

Fig. 7.9 SRCO employing AGPE proposed by Chang [17]

the amplitude of such oscillators wherein this condition setting grounded resistor can also be replace by a grounded VCR as part of the feedback control circuit.

In the literature several CC-based oscillators have been reported which employ all grounded passive elements (AGPE), for instance, [17, 22, 25, 26, 28–30, 32, 37, 46, 50, 53, 57]. In the following we discuss some exemplary circuits out of these.

A circuit which although requires three CCs and does not have the advantage of providing a buffered outlet but has the advantage of employing all grounded passive elements (AGPE) was proposed by Chang [17] and is shown in Fig. 7.9.

The circuit is characterized by the following CO and FO:

$$\mathrm{CO}: R_1 = R_3 \tag{7.26}$$

$$\mathrm{FO}: \omega_0 = \sqrt{\frac{1}{C_1 C_2 R_2 R_3}} \tag{7.27}$$

One of the drawbacks of the circuit of Chang (use of three CCs) described above was overcome by Liu in [22] who came up with a novel configuration which could achieve all the features of Chang's circuit while employing only two CCs. This circuit is shown in Fig. 7.10 and is characterized by

$$\mathrm{CO}: R_1 = R_3 \tag{7.28}$$

$$\mathrm{FO}: \omega_0 = \sqrt{\frac{1}{C_1 C_2 R_1 R_2}} \tag{7.29}$$

This circuit lets the control of CO through R_3 and that of FO by R_2. Also, the circuit enjoys very good frequency stability properties since the frequency stability factor is found to be as $S^F \cong 2\sqrt{n}$ which can be made quite large by making the resistor ratio **n** large as in the circuit of [18].

Another similar circuit with five grounded passive elements and using both CCI- was advanced by Abuelma'atti and Al-Ghumaiz in [28] and is shown in Fig. 7.11. This circuit is characterized by CO and FO which are as follows:-

$$\mathrm{CO}: R_1 = R_2;\ \mathrm{FO}: \omega_0 = \sqrt{\frac{1}{C_1 C_2 R_2 R_3}} \tag{7.30}$$

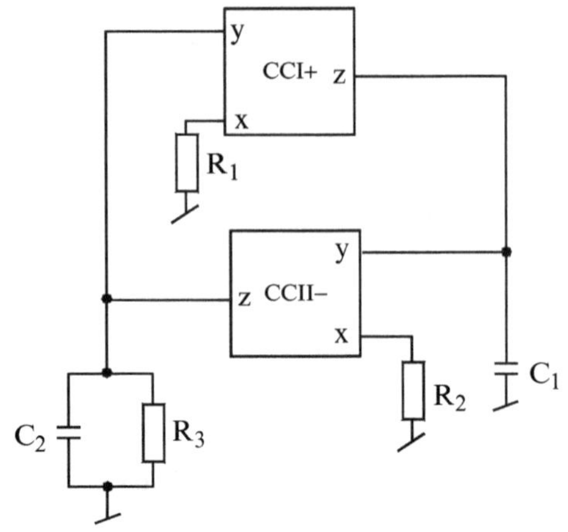

Fig 7.10 Liu's SRCO employing *all grounded passive elements* [22]

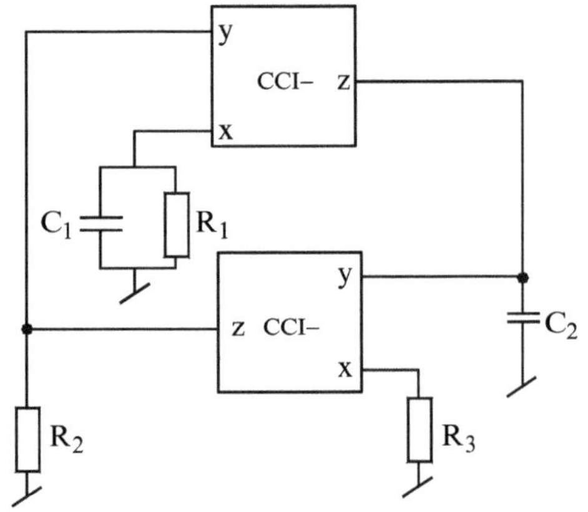

Fig. 7.11 SRCOs proposed by Abuelma'atti and Al-Ghumaiz [28]

It may be mentioned that the authors of [28] had started from a general configuration containing two CCI- and four grounded admittances from which eight special cases were derived using three/two resistors and two/three capacitors. However, there are only four distinctly different cases therein (see Table 1 of [28]) i.e. circuit numbered 1 and 3 are same and so are circuit numbered 2 and 4. Likewise, special cases five and six are the same as well as seven and eight are the same. This can be easily checked by swapping the 'first CCI- alongwith Y_1 and Y_4' by 'second CCI- alongwith admittances Y_3 and Y_2'. Equation (7.30) can be obtained by selecting Y_1 as parallel combination of R_1 and C_1, Y_2 as capacitor, Y_3 as resistor and Y_4 as resistor respectively.

An interesting circuit which also employs two CCII+ and only five grounded external passive components and yet is able to take cognizance of the X-port

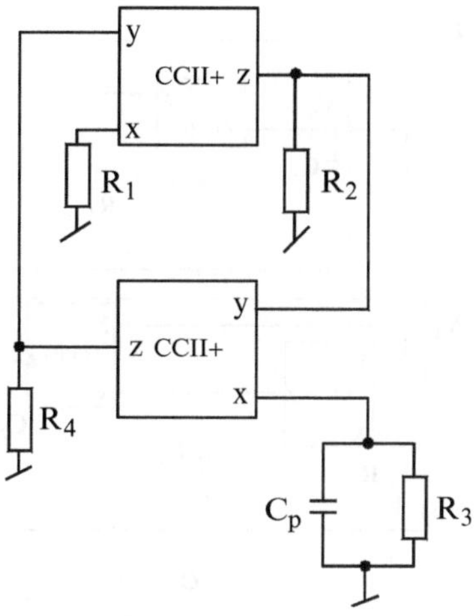

Fig. 7.12 Sinusoidal oscillator proposed by Abuelma'atti and Khan [30]

parasitic input resistance R_x and Z-port parasitic impedance consisting of R_p in parallel with C_p for both the current conveyors and assuming them identical, was proposed by Abuelma'atti and Khan in [30] and is shown here in Fig. 7.12.

A straight forward analysis shows that the condition of oscillation and the frequency of oscillation of this circuit are given by:

$$\mathrm{CO}: R_2R_4 \cong R_1R_x,\ \mathrm{FO}: \omega_0^2 = \frac{R_1(R_3+R_x) - R_2R_4}{C_p(C_z+C_y)R_1R_3R_x(R_2+R_4)} \tag{7.31}$$

Thus, two interesting features of this circuit are: requirement of only a single grounded capacitor and incorporation of all the parasitic impedances of the CCs into the design thereby incorporating their effect, *a priori,* in the design. The latter feature means that this circuit can be designed to generate frequencies much higher than those possible with other two-CC-based circuits described earlier. That this is indeed so has been demonstrated by the authors of [30] realizing CCII+ s from AD844 ICs from which oscillation frequencies up to 11.11 MHz have been attained easily.

7.5 Quadrature and Multi-phase Oscillators

A special type of oscillators known as quadrature oscillators which are capable of generating two sinusoidal signals having a phase difference of 90° (in quadrature) which are useful in quadrature mixers and single sideband modulators. In measurement applications, quadrature oscillators are useful in vector generators and selective voltmeters. In addition four-phase quadrature oscillators find applications in sub-harmonic mixers to reduce noise and intermodulation distortion. Lastly, multiphase oscillators may find applications in power electronic systems and

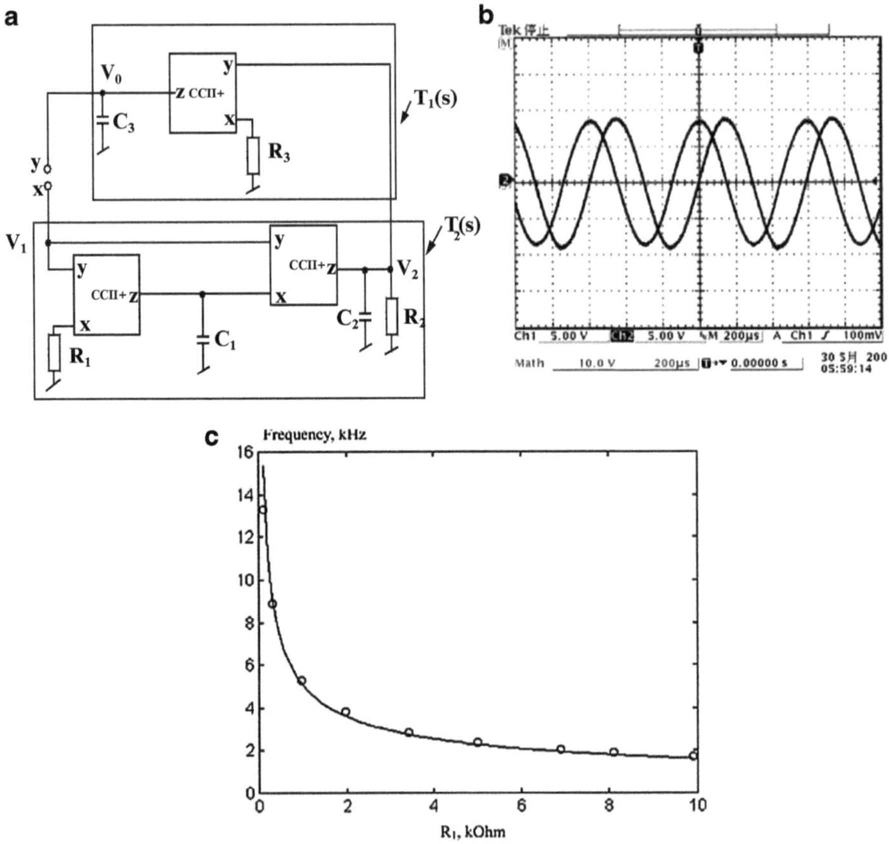

Fig. 7.13 Quadrature oscillator proposed by Horng (Adapted from [44] © 2005 Elsevier) (**a**) The circuit configuration (**b**) Experimental result (**c**) Variation of oscillation frequency with respect to R_1

sub-harmonically pumped frequency conversion circuits. A number of quadrature and multi-phase oscillator topologies using CCs have been reported in literature, for example, [21, 23, 27, 33, 34, 44, 45, 48, 49, 52]. In the following, we discuss a number of quadrature and multi-phase oscillators.

Quadrature oscillators are circuits which provide two sinusoidal signals having 90° phase difference. In the literature there has been lot of efforts on making quadrature oscillators out of which the designs based upon all pass filters are quite common. One such circuit proposed by Horng [44] is shown in Fig. 7.13.

The condition of oscillation and oscillation frequency are given by:

$$C_1R_2 = C_3R_3;\ \omega_o = \sqrt{\frac{1}{C_2C_3R_1R_3}} \tag{7.32}$$

The phase between the two generated signals V_2 and V_1 is given by:

$$\varphi = -\tan^{-1}\omega C_1 R_1 - \tan^{-1}\omega C_2 R_2 \tag{7.33}$$

Using the value of C_1 from Eq. (7.32) i.e. $C_1 = \frac{C_3 R_3}{R_2}$, the phase different ϕ can be then represented as:

$$\varphi = -\tan^{-1}\left(\sqrt{\frac{C_3 R_1 R_3}{C_2 R_2}}\right) - \tan^{-1}\left(\sqrt{\frac{C_2 R_2}{C_3 R_1 R_3}}\right) = -\frac{\pi}{2} \tag{7.34}$$

which shows that V_2 and V_1 are, indeed, in quadrature.

The experimental results of this circuit were taken with CCII+ realized by AD844 IC biased with power supply ± 12 V and component values taken as $C_1 = C_2 = C_3 = 10$ nF, $R = 9.9$ KΩ, $R_2 = 9.88$ KΩ, $R_3 = 9.78$ KΩ. A typical waveform obtained from this circuit is shown in Fig. 7.13 b while the variation of the oscillation frequency with the resistance R_1 is shown in Fig. 7.13c.

A voltage-mode (VM) two-phase quadrature oscillator using only CCII+ s proposed by Minhaj [49] is shown in Fig. 7.14.

The CO and FO for this circuit are given by:

$$\text{CO} : R_1 \geq R_4 \, \text{and FO} : \omega_o = \sqrt{\frac{1}{C_1 C_2 R_2 R_3}} \tag{7.35}$$

It is easy to see that because of the second CCII+ configured as a non-inverting integrator, its input and output i.e. the two outputs V_2 and V_4 of the circuit will be in quadrature.

A generalized circuit for realizing a multiphase oscillator which consists of 'n' numbers of identical CCII-RC sections was presented by Wu et al. [27] and is shown in Fig. 7.15. The condition for which the proposed circuit will provide the sustained oscillations of frequency (ω_0) is given by:

$$(j\omega_0 + \omega_2)^n - (-\omega_1)^n = 0 \tag{7.36}$$

The Eq. (7.36) would have a solution only if the value of n is odd ($n \geq 3$). By equating the real and imaginary parts of Eq. (7.36) to zero, the CO and FO can be then expressed as:

$$\text{CO} : R_2 \geq R_1 \sec\frac{\pi}{n} \, \text{and} \, \text{FO} : \omega_0 = \omega_1 \sin\frac{\pi}{n} \tag{7.37}$$

The circuit was experimentally tested using AD844AN CFOAs for a three phase oscillator. Since FO of each phase is independently controllable by a grounded resistor, this makes it possible for the circuit to be converted into a voltage controlled multiphase oscillator. Another notable feature of the circuit is its suitability for IC fabrication because of the use of AGPEs.

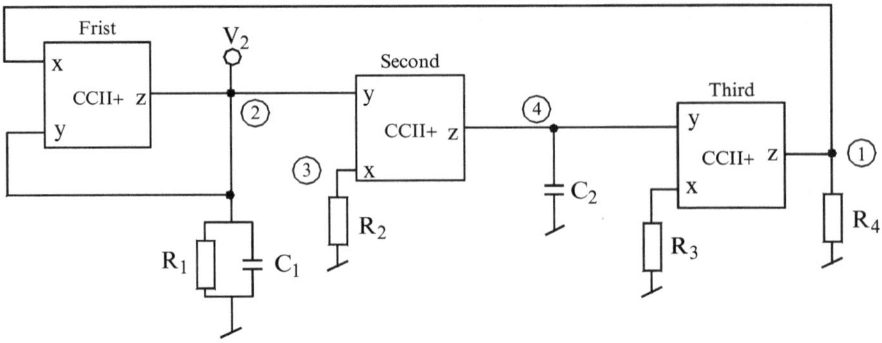

Fig. 7.14 VM two-phase quadrature oscillator proposed by Minhaj [49]

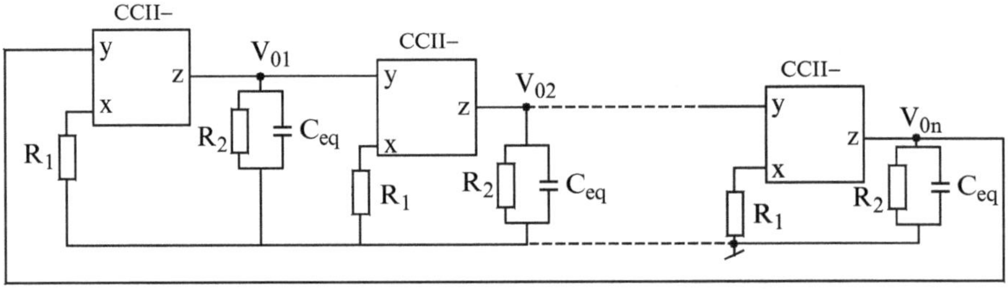

Fig. 7.15 Generalized structure for realizing a multiphase oscillator proposed by Wu et al. [27]

Another three phase sinusoidal oscillator using only CCII+ s advanced by Skotis and Psychalinos [52] is shown in Fig. 7.16 which is having in its internal structure a cascade connection of two lossy non-inverting integrators, each made from a single CCII+ and a lossy inverting integrator made from two CCII+ s.

A sample experimental result of the circuit from [54] obtained by using AD844 for realizing CCs and using component values as $R_1 = 10$ K, $R_2 = 20$ K, $C = 270$ pF with DC bias supply used for biasing AD844 as ± 10 V is shown in Fig. 7.16b. While the measured value of the frequency has been found to be 45 KHz. That of the phase difference was found to be $-60°$ between V_{02} and V_{01} and $-114°$ between V_{03} and V_{01}.

By addition of two inverting amplifies, each implemented from two resistors and a CCII+, the circuit is readily converted into a six-phase quadrature oscillator as shown in Fig. 7.17.

The experimental results of the circuit from [54], as shown in Fig. 7.17b, demonstrate that by using AD844 for realizing CCs and using component values as $R_1 = 10$ K, $R_2 = 20$ K, $C = 270$ pF with DC bias supply used for biasing AD844 as ± 10 V, the measured value of the frequency has been found to be 46.9 KHz whereas the measured phase difference was found to be 160° between $V_{01}{}'$ and V_{01}, $-60°$ between V_{02} and V_{01}, 102° between $V_{02}{}'$ and V_{01}, $-115°$ between V_{03} and V_{01} and 56° between $V_{03}{}'$ and V_{01}.

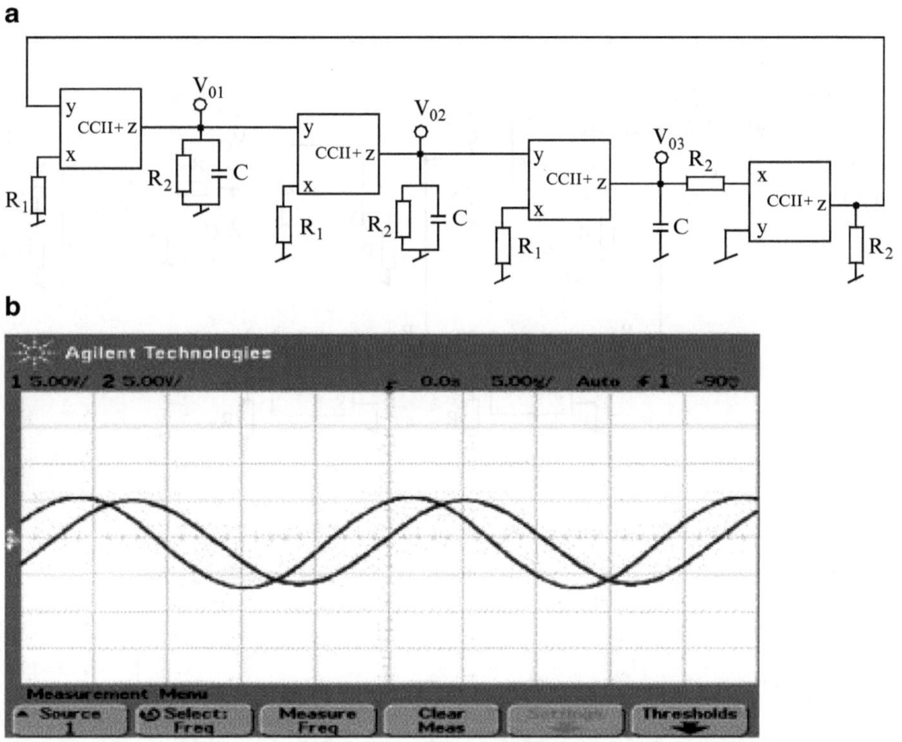

Fig. 7.16 Realization of three-phase oscillator employing CCII + s proposed by Skotis and Psychalinos (Adapted from [52] © 2009 Elsevier GmbH) (**a**) The circuit (**b**) Typical waveforms

Both the circuits follow essentially from the general block diagram of Fig. 7.18 which shows the basic principle of synthesizing an n-phase oscillator for 'n' being odd or even. The second circuit then follows from this schematic by adding two inverting amplifiers.

By a straight forward analysis, the open loop transfer function assuming 'n' identical stages is given by:

$$L(s) = -\left(\frac{K}{1 + s\tau}\right)^n \tag{7.38}$$

Denoting the maximum gain (R_2/R_1) as K of non-inverting lossy integrators and the time constant $CR_2 = \tau$, the criterion of oscillation at a specific frequency ω_0 requires the following conditions to be fulfilled:

$$\tau = \frac{1}{\omega_0}\tan\left(\frac{\pi}{n}\right) \tag{7.39}$$

a

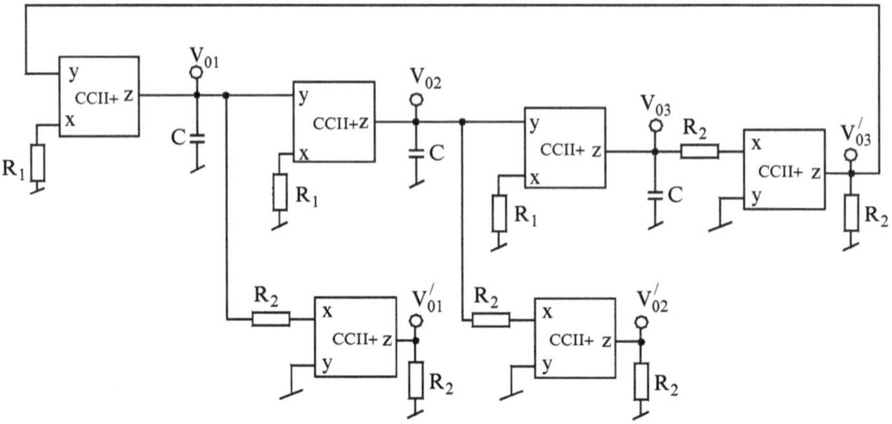

b

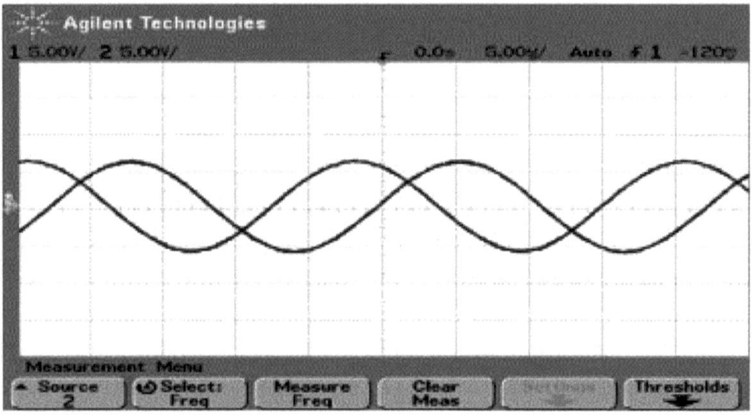

Fig. 7.17 Realization of six-phase oscillator employing CCII+s proposed by Skotis and Psychalinos (Adapted from [52] © 2009 Elsevier GmbH) (**a**) The circuit (**b**) Typical waveforms

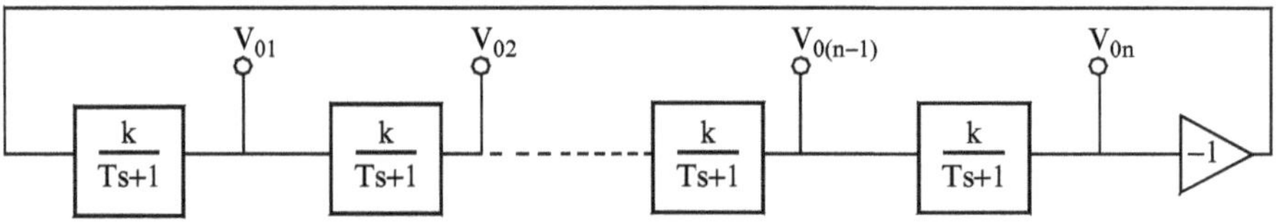

Fig. 7.18 General block diagram of an even and odd phase sinusoidal oscillator proposed by Skotis and Psychalinos (Adapted from [52] © 2009 Elsevier GmbH)

$$K = \left(1 + \left[\tan\left(\frac{\pi}{n}\right)^2\right)^{1/2} \tag{7.40}$$

Both the circuits are practically realizable with CCII+ from AD844 ICs with reasonably reduced total components than those required in other design methodologies using CCs [21, 27, 33, 34].

7.6 Explicit Current Output (ECO) SRCOs

With the advent of current mode signal processing particularly current mode filters, rectifiers, instrumentation amplifiers etc., it is imperative that for testing of such circuits one should have signal generators capable of providing an *explicit* current output from a high output impedance terminal. In this context, therefore, sinusoidal oscillators having *explicit current output* from high output impedance node are obviously potential candidates for such applications. Several such oscillators have been reported in literature such as those in [29, 32, 38, 42, 46, 55, 56].

Senani and Gupta [38] proposed two methods for devising such oscillators, resulting thereby in different configurations which are shown here in Fig. 7.19.

The circuit of Fig. 7.19a has been synthesized by simulating a negative resistor with a CCI in parallel with series RLC branch, wherein the series RL part is created by two unity gain voltage followers B_1 and B_2, resistors R_1 and R_2 and capacitance C_2 thereby leading to an equivalent RL impedance with $R_{eq} = R_1$ and equivalent inductance $L_{eq} = C_2R_1R_2$. The resulting oscillator, therefore, has the following characterization:

$$\mathrm{CO} : \mathrm{R}_1 = \mathrm{R}_3; \mathrm{FO} : \omega_0 = \sqrt{\frac{1}{C_1C_2R_1R_2}} \tag{7.41}$$

On the other hand, the second circuit (Fig. 7.19b) is derived through the application of the state variable synthesis technique proposed by Senani and Gupta in [58–60]. In this case, assuming $C_1 = C_2 = C$, CO is same as in Eq. (7.41) while the frequency of oscillation is given by:

$$\omega_0 = \frac{1}{C\sqrt{R_2R_3}} \tag{7.42}$$

A comparison with the other alternative AGPE oscillators mentioned earlier is of interest. In this context, it may be noted that the remaining oscillators quoted earlier suffer from one or more of the following drawbacks: use of more than three resistors and/or more than two capacitors as in [32, 39]; non-availability of single element controls for CO and FO as in [29, 55] and requirement of more active elements than those in the circuit of Fig. 7.19 considered here [56].

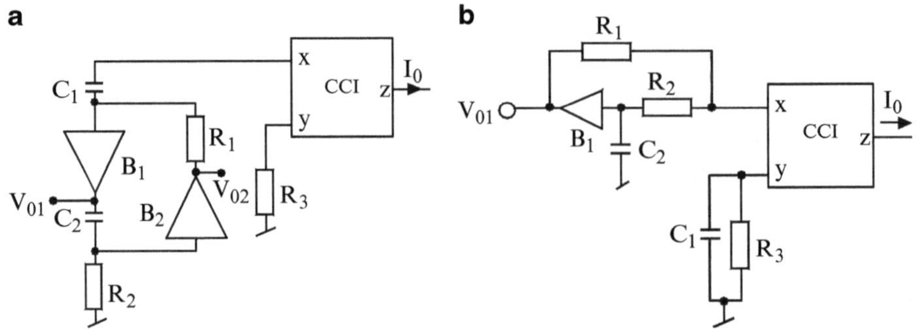

Fig. 7.19 SRCOs proposed by Senani and Gupta [38] (The circuit of Fig. 7.19b has also been derived *independently* in [42] by the approach adopted therein is quite different than the one employed in [38])

7.7 SRCOs with Grounded Capacitors and Reduced Effect of Parasitic Impedances of CCIIs

Out of the three circuits derived systematically by Martinez et al. [35], through a formulation based upon two integrator loop biquadratic general structure, the two circuits which appear to be the better than the remaining ones are shown here in Fig. 7.20.

A notable feature of these circuits is that the parasitic X-terminal input resistance R_x of all the current conveyors can be absorbed in the external resistances connected to terminal-X of the CCs namely, R_A and R_2 in the circuit of Fig. 7.20a and R_A, R_1 and R_2 in case of circuit of Fig. 7.20b. On the other hand, the Z-port parasitic capacitance of the CCs can be merged in the external capacitors C_1 and C_2 as they are connected from the Z-terminals of the CCs only. Thus, it is expected that the circuits would exhibit good high frequency performance.

Experimental results using CCIIs implemented from AD844 confirm that these circuits can be satisfactorily employed to generate oscillations up to 10 MHz. It has been possible to obtain signals from 40 to 400 KHz with average peak-to-peak amplitude of 15 V and THD of less than 1 %.

It is also interesting to note that the circuit of Fig. 7.20a has an additional advantage of providing an explicit current mode output from the Z-terminal of the first CCII+ while the voltage output can be taped from any appropriate node.

7.8 Fully-Uncoupled Oscillators

Fully-uncoupled oscillators are those in which the components which control the condition of oscillation do not appear in the expression for frequency of oscillation and on the other hand, the components appearing in FO do not appear in CO.

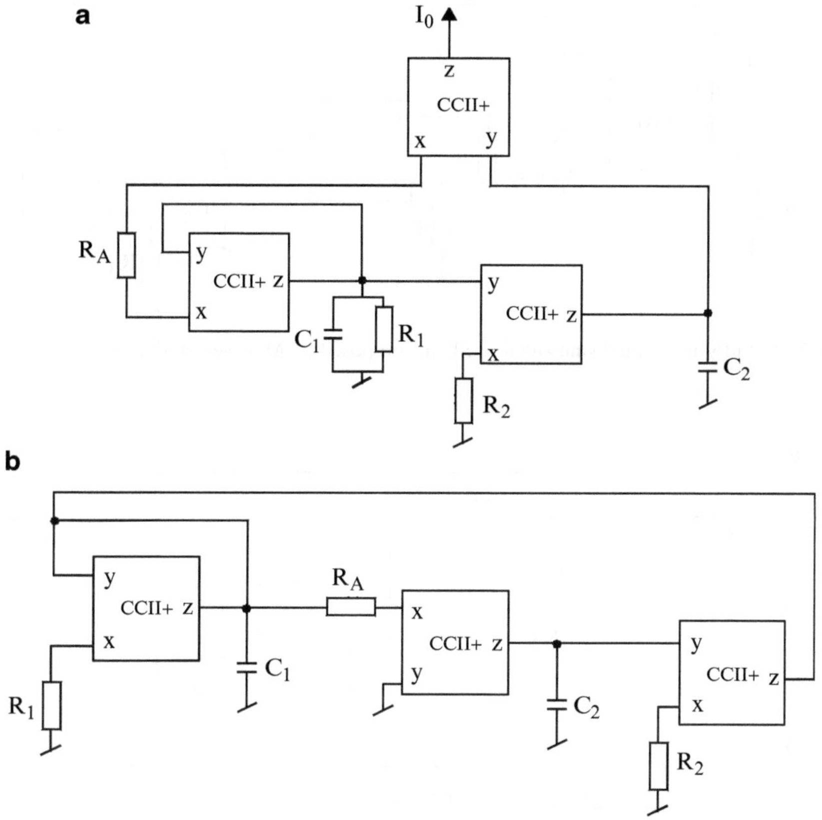

Fig. 7.20 Variable frequency sinusoidal oscillators employing CCII + s proposed by Martinez et al. [35]

Such oscillators can neither be made from a single CC nor with two CCs and therefore, call for the use of a minimum number of three CCs.

One such circuit was proposed by Martinez et al. in [35] and is shown here in Fig. 7.21.

The circuit is characterized by the following equations:

$$\mathrm{CO}: \mathrm{R}_1 = \mathrm{R}_\mathrm{A};\ \mathrm{FO}{:}\,\omega_0 = \sqrt{\frac{1}{C_1 C_2 R_2 R_3}} \tag{7.43}$$

Abuelma'atti in [40] derived a number of three CCII+ based circuits each employing exactly the same number of active and passive components namely four resistors and two capacitors all of which are grounded. Two exemplary circuits from [40] are shown in Fig. 7.22.

Both the circuits are characterized by same CO and FO which are given by:

$$\mathrm{CO}: \mathrm{R}_4 = \mathrm{R}_5;\ \mathrm{FO}: \omega_o^2 = \sqrt{\frac{1}{C_1 C_4 R_2 R_3}} \tag{7.44}$$

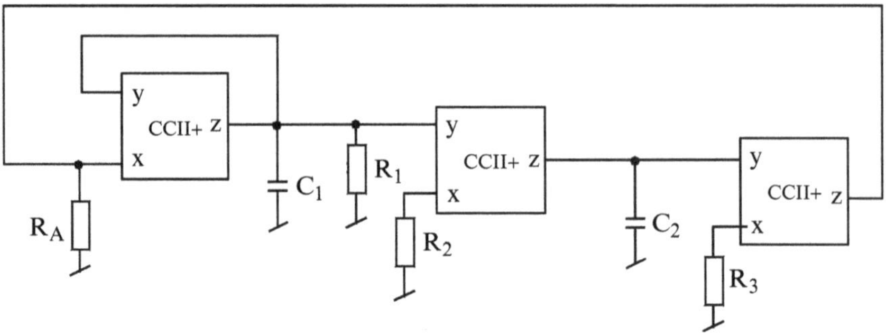

Fig. 7.21 Fully-uncoupled sinusoidal oscillator proposed by Martinez et al. [35]

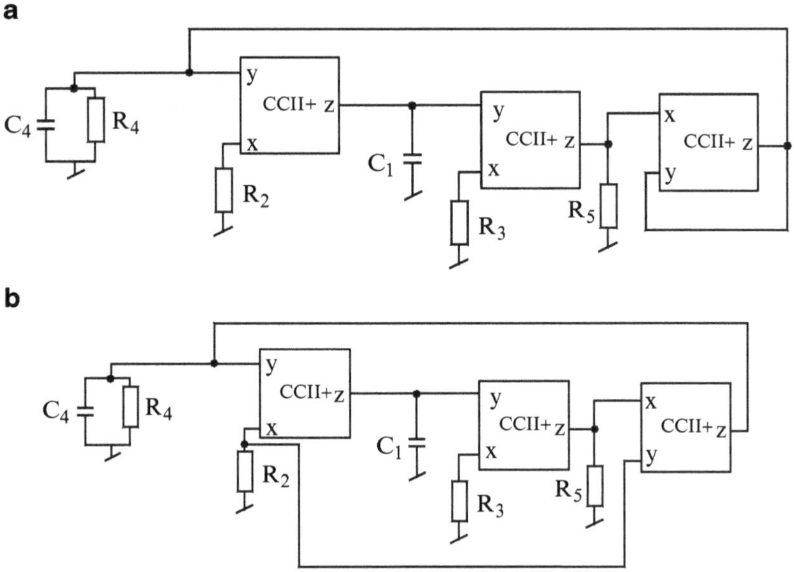

Fig. 7.22 Fully uncoupled oscillators proposed by Abuelma'atti (Adapted from [40] © 2000 Springer)

In non-ideal analysis of these circuits assuming the voltage gains of the three CCII+ as α_i, $i = 1-3$ and current gains as β_i , it is found that the non-ideal expressions for the FO of these circuits is found to be exactly same i.e. $\frac{\alpha_1\alpha_2\beta_1\beta_2\beta_3}{C_1C_4R_2R_3}$ whereas the CO in both the circuits only slightly different as can be seen from the following:

$$\frac{R_5}{\alpha_3\beta_3} = R_4;\ \text{for the circuit of Fig.7.22a} \tag{7.45}$$

$$\frac{R_5}{\alpha_1\alpha_3\beta_3} = R_4;\ \text{for the circuit of Fig.7.22b} \tag{7.46}$$

From the above expressions it can be seen that various active and passive sensitivities of these circuits would be very low and that the non-ideal, non-unity gains of the CCs may not result in larger errors between the theoretical and experimental frequencies. This has indeed been substantiated by the experimental results of these circuits by using AD844 to implement CCII+ s.

Concluding Remarks

In this chapter we have presented a variety of variable frequency oscillators with particular emphasis on the so-called SRCOs offering a variety of features such as employment of grounded capacitors, employment of all grounded passive elements, availability of explicit current output, fully un-coupled tuning laws for CO and FO or structures offering least possible effect of the CC parasitic impedances on the high frequency performance of the oscillators etc. Quadrature and multiphase oscillator circuits have also been included which find numerous applications in telecommunication and measuring systems.

From a much larger variety and number of circuits available in the vast literature on this topic we have attempted to include only those which, in our opinion, are better than those omitted from inclusion. Besides this, a major factor in selecting the circuits has been that all of them use only the basic types of current conveyors and therefore, can be readily implemented practically using AD844 CCII+ elements or CCs constructed from IC op-amps like μA741/ LF356 type op-amps and CA3096 type mixed transistor arrays or OPA660/OPA860 (which could be easily used as CCs).

It is expected that the catalogue of SRCO circuits provided here serves as a good source for circuit designers to choose an appropriate circuit for specific application. Of course, readers interested in knowing more can look into other alternatives through the references provided at the end of this chapter and additional references given in the Appendix, at the end of this monograph.

References

1. Nandi R (1977) Wien bridge oscillators using current conveyors. Proc IEEE 65:1608–1609
2. Nandi R (1977) New RC oscillators using current conveyors. Int J Electron 42:309–311
3. Soliman AM (1978) A novel variable frequency sinusoidal oscillator using a single current conveyor. Proc IEEE 66:800
4. Soliman AM (1978) Realization of frequency dependent negative resistance circuits using two capacitors and a single current conveyor. Proc IEE 125:1336–1337

5. Senani R (1979) New canonic single-resistance controlled sinusoidal oscillator using a single current conveyor. Electron Lett 15:568–569
6. Jana PB, Nandi R (1984) Single current conveyor tunable sinewave RC oscillator. Electron Lett 20:44–45
7. Chong CP, Smith KC (1987) Sinusoidal oscillators employing current conveyors. Int J Electron 62:515–520
8. Abuelma'atti MT, Humood NA (1987) Two new minimum-component Wien-Bridge oscillators using current-conveyor. Int J Electron 63:669–672
9. Abuelma'atti MT, Humood NA (1987) Current-conveyor RC oscillator. Electron Wireless World 93:1618
10. Abuelma'atti MT, Humood NA (1988) Current-conveyor sine-wave oscillators. Electron Wireless World 94:282–284
11. Chen JJ, Chen CC, Tsao AW, Liu SI (1991) Current-mode oscillators using single current follower. Electron Lett 27:2056–2059
12. Svoboda JA, McGory L, Webb S (1991) Applications of a commercially available current conveyor. Int J Electron 70:159–164
13. Abuelma'atti MT (1992) Minimal realization for single resistor controlled sinusoidal oscillator using single CCII. Electron Lett 28:1264–1265
14. Senani R, Singh VK (1992) Single-element controlled sinusoidal oscillator employing single current conveyor IC. Electron Lett 28:414–415; Errata Electron Lett 28: 895
15. Abuelma'atti MT (1992) Grounded capacitor current-mode oscillator using single current follower. IEEE Trans Circ Syst-I 39:1018–1020
16. Celma S, Martinez PA, Carlosena A (1992) Minimal realization for single resistor controlled sinusoidal oscillator using single CCII. Electron Lett 28:443–444
17. Chang CM (1994) Novel current-conveyor-based single-resistance controlled/voltage-controlled oscillator employing grounded resistors and capacitors. Electron Lett 30:181–183
18. Bhaskar DR, Senani R (1993) New current-conveyor-based single-resistance-controlled/voltage controlled oscillator employing grounded capacitors. Electron Lett 29:612–614
19. Abuelma'atti MT (1993) New current-controlled CCII based RC oscillators with grounded capacitors and resistors. Int J Circ Theor Appl 21:487–491
20. Celma S, Martinez PA, Carlosena A (1994) Approach to the synthesis of canonic RC-active oscillators using CCII. IEE Circ Devices Syst 141:493–497
21. Hou CL, Shen B (1995) Second–generation current conveyor-based multiphase sinusoidal oscillators. Int J Electron 78:317–325
22. Liu SI (1995) Single-resistance controlled/voltage controlled oscillator using current conveyors and grounded capacitors. Electron Lett 31:337–338
23. Vosper JV, Heima M (1996) Comparison of single-and dual element frequency control in a CCII-based sinusoidal oscillator. Electron Lett 32:2293–2294
24. Martinez PA, Celma S, Gutierrez I (1995) Wien-type oscillators using $CCII^{+}$. Analog Integr Circ Sig Process 7:139–147
25. Abuelma'atti MT, Al-Ali AK, Ahsan M (1995) Programmable current-conveyor-based oscillator employing grounded resistors and capacitors. Active Passive Electron Comp 18:259–265
26. Abuelma'atti MT, Al-Ghumaiz AA, Khan MH (1995) Novel CCII-based single-element controlled oscillator employing grounded resistors and capacitors. In J Electron 78:1107–1112
27. Wu DS, Liu SI, Hwang YS, Wu YP (1995) Multiphase sinusoidal oscillator using second generation current conveyors. Int J Electron 78:645–651
28. Abuelma'atti MT, Al-Ghumaiz AA (1996) Novel CCI-based single-element-controlled oscillators employing grounded resistors and capacitors. IEEE Trans Circ Syst-I 43:153–155
29. Vrba K, Cajka J, Zeman V (1997) New RC-active networks using current conveyors. Radioengineering 6:18–21
30. Abuelma'atti MT, Khan MH (1997) New sinusoidal oscillator employing the CCII internal pole. Int J Electron 83:817–823

31. Horng JW, Chang CW, Lee MH (1997) Single-element controlled sinusoidal oscillators using CCIIs. Int J Electron 83:831–836
32. Soliman AM (1998) Current mode CCII oscillators using grounded capacitors and resistors. Int J Circ Theor Appl 26:431–438
33. Abuelma'atti MT, Al-Qahtani MA (1998) A grounded-resistor current conveyor-based active-R multiphase sinusoidal oscillator. Analog Integr Circ Sig Process 16:29–34
34. Abuelma'atti MT, Al-Qahtani M (1998) Low component count second-generation current conveyor-based multiphase sinusoidal oscillators. Int J Electron 84:45–52
35. Martinez PA, Sabadell J, Aldea C, Celma S (1999) Variable frequency sinusoidal oscillators based on CCII+. IEEE Trans Circ Syst-I 46:1386–1390
36. Soliman AM, Elwakil AS (1999) Wien oscillators using current conveyors. Comput Electr Eng 25:45–55
37. Soliman AM (1999) Synthesis of grounded capacitor and grounded resistor oscillators. J Franklin Inst 336:735–746
38. Senani R, Gupta SS (2000) Novel SRCOs using first generation current conveyor. Int J Electron 87:1187–1192
39. Papazoglou CA, Karybakas CA (2000) An electronically tunable sinusoidal oscillator suitable for high frequencies operation based on single dual-output variable-gain CCII. Analog Integr Circ Sig Process 23:31–44
40. Abuelma'atti MT (2000) New sinusoidal oscillators with fully uncoupled control of oscillation frequency and condition using three CCII+ s. Analog Integr Circ Sig Process 24:253–261
41. Horng JW (2001) A sinusoidal oscillator using current-controlled current conveyors. Int J Electron 88:659–664
42. Bhaskar DR, Prasad D, Imam SA (2004) Grounded-capacitor SRCOs realized through a simple general scheme. Frequenz 58:175–177
43. Khan AA, Bimal S, Dey KK, Roy SS (2005) Novel RC sinusoidal oscillator using second-generation current conveyor. IEEE Instrum Meas 54:2402–2406
44. Horng JW (2005) Current conveyor based allpass filters and quadrature oscillators employing grounded capacitors and resistors. Comput Electr Eng 31:81–92
45. Horng JW, Hou CL, Chang CM, Chung WY, Tang HW, Wen YH (2005) Quadrature oscillators using CCIIs. Int J of Electron 92:21–31
46. Martinez PA, Sanz BMM (2005) Generation of two integrator loop variable frequency sinusoidal oscillator. Int J Electron 92(10):619–629
47. Fongsamut C, Anuntahirunrat K, Kumwachara K, Surakampontorn W (2006) Current-conveyor-based single-element-controlled and current-controlled sinusoidal oscillators. Int J Electron 93:467–478
48. Khan IA, Hasan S (2006) Current mode four phase quadrature oscillator using CCIIs. J Active Passive Electron Devices 1:273–279
49. Minhaj N (2007) Current conveyor-based voltage-mode two-phase and four-phase quadrature oscillators. Int J Electron 94:663–669
50. Kumar A, Vyas AL (2007) A single resistance controlled eight possible oscillator circuits for analog signal processing applications. Acad Open Internet J 21:1–8, ISSN 1311–4360
51. Kumar V, Pal K (2009) Single element controlled sinusoidal oscillators using current conveyors. J Active Passive Electron Devices 8:197–202
52. Skotis GD, Psychalinos C (2010) Multiphase sinusoidal oscillators using second generation current conveyors. Int J Electron Commun 64:1178–1181
53. Abdalla KK, Bhaskar DR, Senani R (2010) Configuration for realizing a current-mode universal filter and dual-mode quadrature single resistor controlled oscillator. IET Circ Devices Syst 6:159–167
54. Lahiri A (2011) New canonic active RC sinusoidal oscillator circuits using second-generation current conveyors with application as a wide-frequency digitally controlled sinusoid generator. Active Passive Electron Comp 2011, 274394, 8p

55. Lahiri A (2012) Current-mode variable frequency quadrature sinusoidal oscillators using two CCs and four passive components including grounded capacitors. Analog Integr Circ Sig Process 71:303–311
56. Sotner R, Lahiri A, Kartci A, Herencsar N, Jerabek J, Vrba K (2013) Design of novel precise quadrature oscillators employing ECCIIs with electronic control. Adv Electr Comput Eng 13:65–72
57. Abuelma'atti MT, Alsuhaibani ES, Obadi ASB, Khalifa ZJ (2013) Independent control of the frequency and condition of oscillation: a caution. Int J Electron 100:384–392
58. Senani R, Gupta SS (1997) Synthesis of single-resistance-controlled oscillators using CFOAs: simple state-variable approach. IEE Proc Circ Devices Syst 144:104–106
59. Gupta SS, Senani R (1998) State variable synthesis of single resistance controlled grounded capacitor oscillators using only two CFOAs. IEE Proc Circ Devices Syst 145:135–138
60. Gupta SS, Senani R (1998) State variable synthesis of single resistance controlled grounded capacitor oscillators using only two CFOAs: additional new realizations. IEE Proc Circ Devices Syst 145:415–418
61. Senani R, Bhaskar DR (1992) Simple configuration for realizing voltage-controlled impedances. IEEE Trans Circ Syst 39:52–59
62. Hribsek M, Newcomb RW (1976) VCO controlled by one variable resistor. IEEE Trans Circ Syst 23:166–169
63. Soliman AM, Awad SS (1978) A canonical voltage controlled oscillator realized using a single operational amplifier. Frequenz 32:153–154
64. Senani R (1979) New canonic sinusoidal oscillator with independent frequency control through a single grounded resistor. Proc IEEE 67:691–692
65. Dutta Roy SC, Pyara VP, Jamuar S (1983) Identification and design of single amplifier single resistance controlled oscillators. IEE Trans Circ Syst 30:176–181
66. Senani R (1994) On equivalent forms of single op-amp sinusoidal oscillators. IEEE Trans Circ Syst I41:617–624
67. Darkani MT, Bhattacharyya BB (1984) A unified approach to the realization of RC-active, single as well as variable, frequency, oscillators using operational amplifiers. J Franklin Inst 317:413–439
68. Kaliyugvaradan S (1981) Single resistance controlled RC-oscillator using a single operational amplifier. Int J Electron 50:153–155
69. Singh V (2001) Realization of operational floating amplifier based 1-op-amp based sinusoidal oscillators. IEEE Trans Circ Syst 48:377–381

Chapter 8
Nonlinear Applications of CCs

Abstract This chapter discusses the applications of CCs in realizing a number of non-linear functions such as multipliers, dividers, Squarers, square-rooters, fuzzy functions, analog switches, pseudo-exponential circuits and built-in-test structures. Also discussed are a variety of Schmitt triggers, relaxation oscillators, wave form generators and chaotic oscillators.

8.1 Introduction

Admittedly, a major chunk of work on the CCs has dominantly been concerned with applications of the CCs in realizing linear circuits like analog filters, oscillators and impedance simulators etc. Nevertheless, several researchers have investigated the use of CCs in the realization of non-linear functions as well and it has been found that CCs could be used to advantage even in many non-linear function applications [1–44]. This chapter and subsequently Chap. 13 of this monograph deal with various nonlinear applications of the CCs.

8.2 Precision Rectifiers

In many signal processing, signal conditioning and instrumentation applications when signal levels are low particularly lower than the cutin voltage of the diodes, precision rectifiers are required which are normally made from op-amps and diodes such that the cutin voltage of the diode is made ineffective because of very large gain of the op-amps employed. Some of the applications areas where such precision rectifiers are useful are: RMS to DC convertor, peak detectors and floor detectors in ultrasonic systems.

Several authors have demonstrated [3–7, 21, 31, 35, 37, 40] that current conveyors can be advantageously employed to design precision full wave rectifiers with operating frequencies in MHz range quite easily with a reduced number of resistors as compared to precision rectifiers made from IC op-amps.

© Springer International Publishing Switzerland 2015
R. Senani et al., *Current Conveyors*, DOI 10.1007/978-3-319-08684-2_8

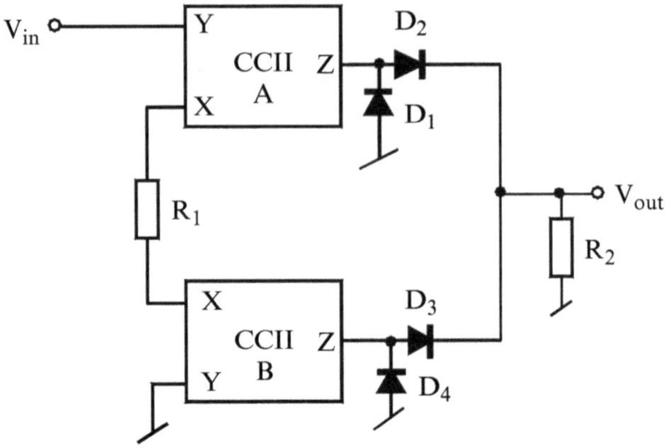

Fig. 8.1 Precision full wave rectifier using CCs proposed by Toumazou et al. [3]

Toumazou et al. [3] demonstrated the construction of a precision full wave rectifier using current conveyors CCII01 from LTP Electronics which has bandwidth f_T of the order of 100 MHz. The circuit from [3] is shown in Fig. 8.1.

The operation of the circuit can be explained as follows. The two CCs are configured to act as voltage to current convertors. During the positive input cycle, a current V_{in}/R_1 flows out of the Z-terminal of CCIIA and the same amount of current flows into the Z-terminal of CCIIB which makes diodes D_2 and D_4 conducting thereby making a voltage V_{out} across R_2 available and ensuring $V_{out} = V_{in}$ provided $R_1 = R_2$. During the negative cycle of the input voltage, diodes D_3 and D_1 conduct in which case also the output current of CCIIB creates a voltage across R_2 thereby again making $V_{ou} = V_{in}$. Unfortunately, for low level input voltages, the slewing characteristic of the input stage of the current conveyor architecture provides a small current derive as opposed to virtually unlimited current drive for large input signals. As a consequence of this, the output wave form shows significant distortion because during the point at which the inputs signal changes polarity, all the diodes are off. Thus, the output voltage of the rectifier remains at zero volt until the voltage at the appropriate Z-terminal reaches the level to turn the diodes on, at which point the output voltage changes sharply. Because of the reduced value of the current conveyor's small signal dV/dT the performance of the precision rectifier deteriorates at frequency much below the f_T.

From the above discussion, it is clear that since the distortion in the output waveform of the precision rectifier of Fig. 8.1 is due to small signal dV/dT limitations, it can be corrected by a suitable modification of the circuit by appropriately biasing all the diodes. One such modification is shown in Fig. 8.2 where a low impedance voltage difference circuit producing a voltage of the order of 0.6 V has been appropriately added. By experiments it has been found that this modification improves the performance of the rectifier up to the frequency of about 30 MHz.

In [4], Hayatleh et al. subsequently demonstrated that although the addition of voltage bias (V_B) as shown in Fig. 8.2 results in improvement of the performance of the precision rectifier, this design exhibits rather a poor temperature characteristics

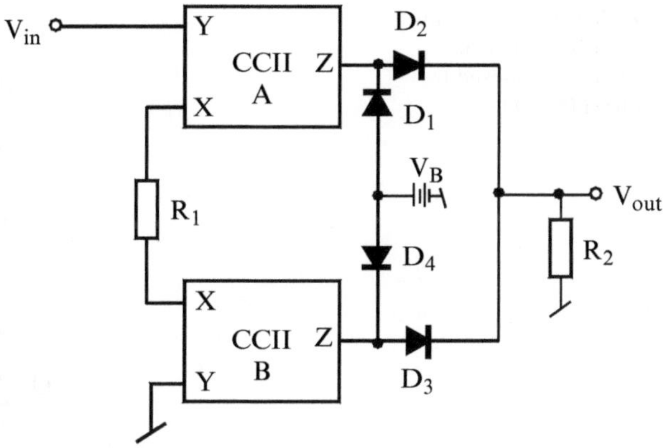

Fig. 8.2 Modified high frequency precision full wave rectifier proposed by Toumazou et al. [3]

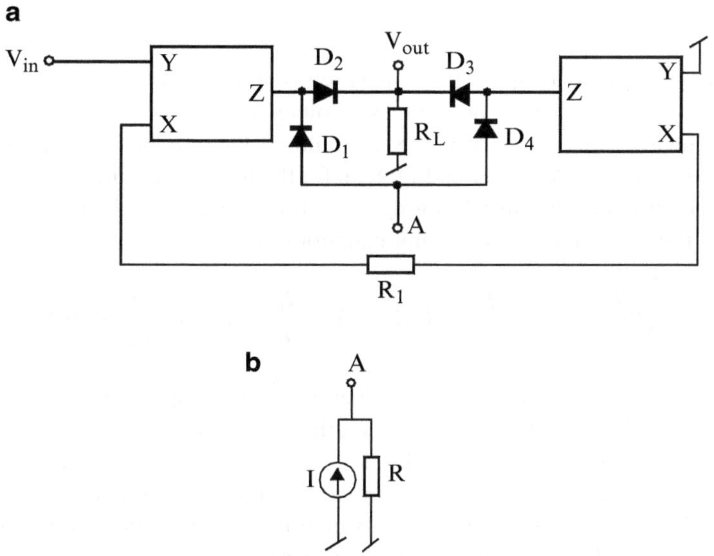

Fig. 8.3 (**a**) Precision full wave reciter employing CCs (**b**) Current biasing technique proposed by Hayatleh et al. [4]

with the additional draw back that any excess of the bias voltage will also add directly to the DC offset voltage in the final output of the circuit. A better alternative is to employ a current biasing technique as shown in Fig. 8.3b. It has been demonstrated that far better temperature stability can be achieved by biasing the diode with a constant DC current source. Fig.8.3a is a CC-based precision full wave rectifier from [4].

In [5], Stiurca demonstrated that the use of current mirror based biasing arrangement to bias the rectifying diodes provides higher temperature stability than

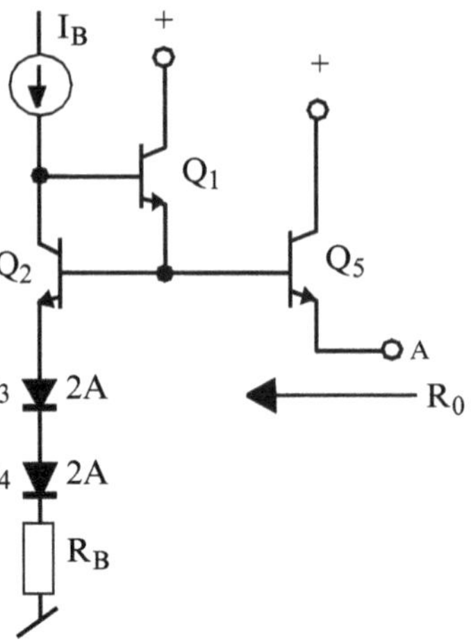

Fig. 8.4 Current control-mirror based biasing technique proposed by Stiurca [5] 1995

previously described designs. His proposition is shown in Fig. 8.4. Another advantage of this design is that the output offset voltage is easily adjustable to the desired optimum level. The scheme of Fig. 8.4 should, in fact, be looked as an alternative arrangement to those described in Fig. 8.3b for the rectifier architecture of Fig. 8.3a.

If we assume that the rectifying diodes are matched with the transistors in the circuit of Fig. 8.4, then the following equations apply:

$$V_T \ln\frac{I_B}{I_S} + 2V_T \ln\frac{I_B}{2I_S} + I_B R_B = V_T \ln\frac{I_D}{I_S} + 2V_T \ln\frac{I_D/2}{I_S} + I_D R_L \tag{8.1}$$

If $R_L = R_B$, it follows that $I_D = I_B$ and therefore, the temperature independent DC output offset voltage is obtained. The operation of this scheme in the rectifier circuit of Fig. 8.3a was tested [5] through SPICE using full transistor level schematic of the IC CC PA630. Simulations have confirmed that the use of this circuit results in a rectifier with considerable improvements in terms of current consumption, temperature stability, reduced output offset control as compared to schemes described earlier

Khan et al. in [7] proposed the full wave rectifier circuit of Fig. 8.5 which employs two CCII+ and is implementable with AD844. The circuit operation can be explained as follows:

The current in the resistance R_1 is given by $i_{xB} = V_{in}/ (R_1 + 2R_x)$ where R_x is the non-ideal input resistance looking into the terminal X of each CCII+. Now, using the characterization of a CCII+, it follows that $i_{zA} = -i_{zB} = -i_{xB} = i_{xA}$. The diodes D_1-D_2 and D_3-D_4 are so arranged that they would allow the Z-port output current to flow either into the load resistance R_L or to ground depending upon the polarity of the input voltage V_{in}.

WhenV_{in} is positive the Z-output current of CCII+ A puts D_1 ON and D_2 OFF thereby allowing current to flow in load R_2, during the same condition, CCII+ B

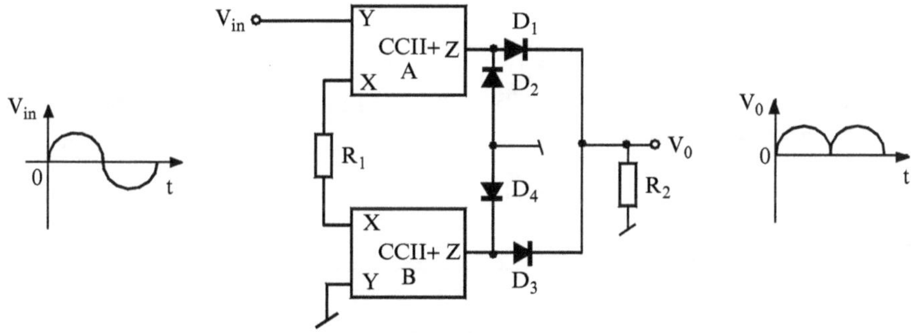

Fig. 8.5 Proposed precision full-wave rectifier scheme using CCII+ proposed by Khan et al. [7]

z-output current makes D_4 ON and D_3 OFF. Thus, for $V_{in} > 0$, it follows that V_o too will be positive and would be given by:

$$V_0 = \frac{R_2}{(R_1 + 2R_x)} V_{in} \tag{8.2}$$

In the next half cycle, when V_{in} becomes negative, the roles of the diode pairs D_1-D_2 and D_3-D_4 are interchanged but the current in the load resistor still flows in the positive direction only whose value is still given by the above equation. Thus, a full wave rectification is attained.

It is easy to visualize that by reversing the terminals of all the four diodes, an inverting full wave rectifier can be realized. This circuit with CCII+ realized using AD844 is found to work well till a frequency of the sinusoidal input of 1 MHz, exhibits very good linearity in its transfer characteristics. However, V_{in} and R_1 need to be selected such that the maximum X-terminal current does not exceed 6 mA [7].

Wilson and Mannama demonstrated in [6], that the basic bridge-type full wave rectifier topology of Fig. 8.2 can be modified as shown in Fig. 8.6 which results in enhanced performance. The bias voltage V_B is created on the junction of the diodes D_1 and D_3 through a current source I_B, a separate bias bridge of similar diodes D_5–D_8 and a voltage follower. The current I can be created in the simplest case by a resistor returned to positive DC power supply. The output section of the modified circuit consists of the current summation node and a transimpedance amplifier. It may be noted that in the circuits of Figs. 8.1 and 8.2, deriving the load directly from the bridge results in full output voltage being reflected back through the bridge detracting from accurate operation through voltage modulation of the CCII+ output terminals. Moreover, due to the biasing current flowing into the load there would be an unwanted offset voltage. By contrast, in the modified circuit of Fig. 8.6, due to the formation of the current summing node and a virtual ground at the junction of the diodes D_2 and D_4, the effect of the reflected voltage modulation is eliminated thereby providing a much greater linear range while still providing a voltage output for interfacing through subsequent stages through the transimpedance amplifier output stage.

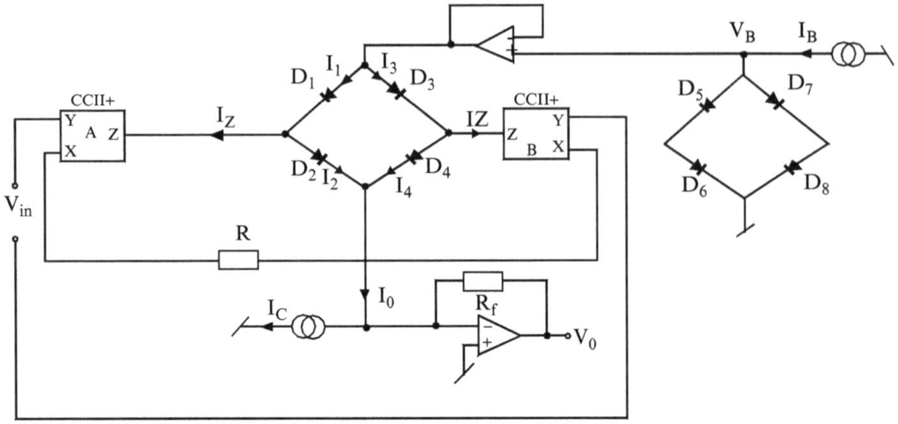

Fig. 8.6 Precision full-wave rectifier proposed by Wilson and Mannama [6]

A routine analysis of the circuit using translinear (TL) principle applied to the TL loop consisting of the four diodes D1-D4, the following equation results for the output voltage of the circuit:

$$V_o = -R_f\left\{(V_{in}/R)^2 + I_B^2\right\}^{1/2} \tag{8.3}$$

The effect of bridge bias current has been found to be negligible for low bias currents. The workability of this circuit has been tested [6] using CCII+ from AD844, transimpedance amplifier realized with AD847 and a bridge bias current of 40 μA with ICs biased with ±15 V, voltage swings greater than 10 V were attainable.

Monpapassorn [21] has recently introduced another two-CC full wave rectifier which is shown here in Fig. 8.7. Assuming that the DC bias currents I_1, I_2 and I_3 are zero, the operation of the circuit can be explained as follows:

When V_{in} is positive diode D_1 is ON and current $+V_{in}/R_1$ flows out of the Z-terminal of CCII+ 1. On the other hand, when V_{in} is negative, D_2 conducts and a current $-V_{in}/R_2$, even the current coming out of the terminal-Z is still positive and equal to $+V_{in}/R_1$. As a consequence, full wave rectified output is available with V_0 given by

$$V_{out} = +\left(\frac{R_L}{R_{in}+R_x}\right)V_{in};\ \text{for } V_{in} > 0 \text{ as well as for } V_{in} < 0 \tag{8.4}$$

where $R_1 = R_2 = R_{in}$

If one chooses $R_L = (R_1 + R_x) = (R_2 + R_x)$ then the magnitudes of the input and output waveforms would be equal. Also, the output impedance is given by $R_z//R_L$ which is smaller than other earlier full wave rectifiers, for smaller values of R_L.

The workability of this circuit has been demonstrated [21] by realizing the circuit with AD844 CCII+, $R_1 = R_2 = 50\,\Omega$ $R_L = 100\,\Omega$ so that with $R_x = 50\,\Omega$ for the AD844

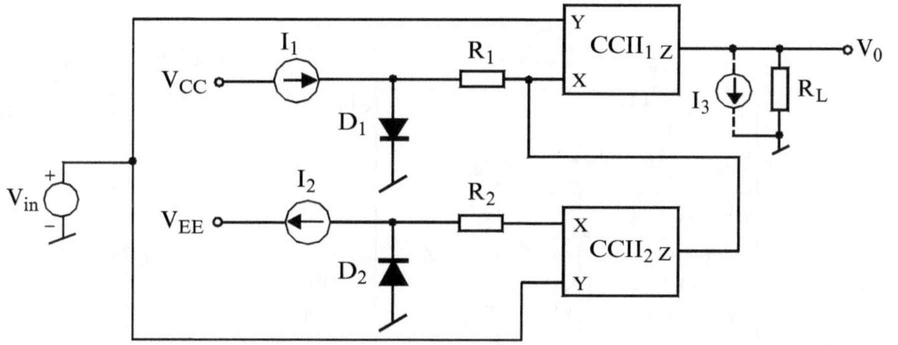

Fig. 8.7 Low output impedance dual-CCII full wave rectifier proposed by Monpapassorn [21]

CCII+, the gain of the circuit becomes unity. The AD844 was biased with ±12 V DC supply and DC bias currents were taken as $I_1 = I_2 = 5$ mA and $I_3 = -9.71$ mA. With these values, for an input sine wave of 1 V p-p, 1 MHz, the circuit functioned as a full wave rectifier quite well with the output offset voltage reduced to as low as −132 μV while the output impedance was found to be as low as 99.9 Ω.

8.3 Frequency Doubler and Full Wave Rectifier

A number of authors have reported full wave rectifiers and frequency doublers made from CCs such as [12, 34, 36, 38, 41]. We present here some exemplary circuits out of these.

The frequency doubler and full wave rectifier circuit devised by Surakampontorn et al. in [12] is shown here in Fig. 8.8.

From the analysis of the translinear loop consisting of the transistors Q_1–Q_2–Q_3–Q_4, it can be deduced that

$$I_1 = \left[\left\{I_{in}^2 + 4I_0^2\right\}^{1/2} - I_{in}\right]/2 \tag{8.5}$$

$$I_2 = \left[\left\{I_{in}^2 + 4I_0^2\right\}^{1/2} + I_{in}\right]/2 \tag{8.6}$$

Also, by noting that $I_A = (2I_0 + 2I_1)$ and $I_B = (3I_0 + 2I_2)$, the output current of the circuit can be written as:

$$I_{out} = (I_A + I_B) - I_C = 5I_0 + 2(I_1 + I_2) - I_C \tag{8.7}$$

If we choose $I_C = 5I_0$, we get

$$I_{out} = 2\left\{I_{in}^2 + 4I_0^2\right\}^{1/2} \tag{8.8}$$

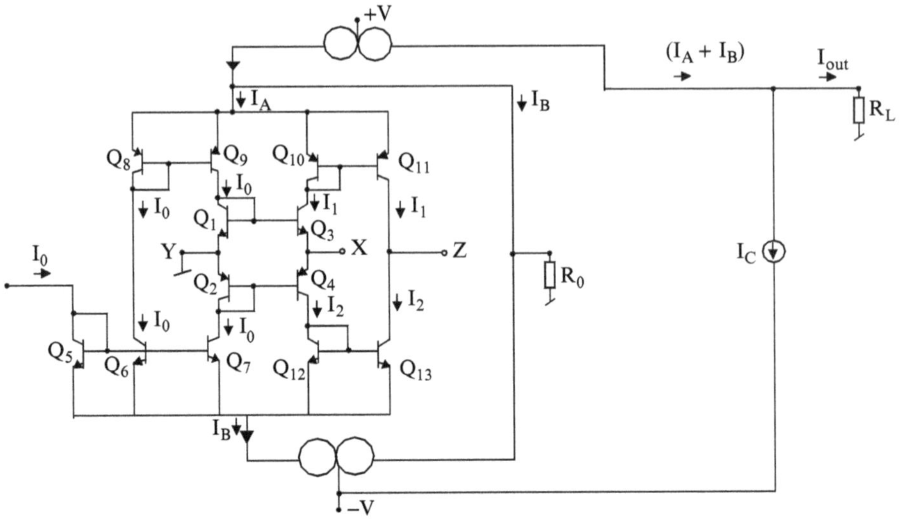

Fig. 8.8 Full wave rectifier circuit and frequency doubler proposed by Surakampontorn et al. [12]

which can be re-written as

$$I_{out} = 4I_0\left(1 + \frac{I_{in}^2}{4I_0^2}\right)^{1/2} \tag{8.9}$$

Consider now the following cases:

(a) Sinusoidal frequency doubler: Choosing $I_{in} = I_m$ Sin ωt if $|I_{in}| << 2I_0$, then using the power series expansion-

$$\sqrt{1+x} = 1 + \frac{1}{2}x - \frac{1}{8}x^3 + \ldots. \tag{8.10}$$

Now defining $k = \frac{I_{in}^2}{4I_0^2}$ and considering k <1, the output current can be written as

$$I_{out} \cong I_{DC} + I_{2w}\cos 2\omega t \tag{8.11}$$

$$I_{DC} \cong \left\{1 + \frac{1}{4}k - \frac{3}{64}(k)^2 + \frac{5}{256}(k)^3\right\}4I_O \tag{8.12}$$

$$I_{2w} \cong \left\{-\frac{1}{4}k + \frac{1}{16}(k)^2 - \frac{15}{512}(k)^3\right\}4I_O \tag{8.13}$$

It is, thus, seen that the output current contains a signal whose frequency is double of the frequency of the input signal.

(b) Full wave rectifier: If we choose the input signal current such that $|I_{in}| >> 2I_0$, then from equations for I_1 and I_2 it turns outs that

$$I_1 \cong 0 \text{ and} I_2 \cong 2I_{in}, \text{for} I_{in} > 0 \tag{8.14}$$

$$I_1 \cong 2I_{in} \text{ and} I_2 \cong 0 \text{for} I_{in} < 0 \tag{8.15}$$

From the above, it follows that the output current is given by $I_{out} = 2|I_{in}|$ and the circuit, therefore, functions as a full wave rectifier.

8.4 Multipliers, Dividers, Squarers and Square Rooters

Liu et al. demonstrated in [1] that using current conveyors in conjunction with MOSFETs interesting circuits can be obtained for realizing fully integrable non-linear building blocks such as multipliers, dividers, Squarers, square rooters and vector submission circuits. In all such circuits, the MOSFET is generally operated in triode region.

In the circuit of the Fig. 8.9a if the two MOSFETs are assumed to be operating in the non- saturated region then the current output of the multiplier can be expressed as:

$$I_0 = 2KV_1V_2 \text{ where} K = \frac{W}{2L}\mu_s C_{ox} \tag{8.16}$$

The operational range of this circuit is governed by the following constraint:

$$V_1 \leq \min[V_G + V_2 - V_{TH}, V_G - V_{TH}] \tag{8.17}$$

It is worth mentioning that if we exchange the gate voltages of the two MOSFETs, the resulting circuit will still be a multiplier with $I_0 = -2KV_1V_2$.

Using the multiplier proposed above an analog divider is obtained from the configuration of Fig. 8.10. A routine analysis of this circuit reveals that the output voltage of this circuit is given by:

$$V_0 = -\frac{1}{2KR}\left(\frac{V_1}{V_2}\right) \tag{8.18}$$

An alternative CC-based voltage-to-current converter using a MOS resistive circuit, containing four MOSFETs, is shown in Fig. 8.11.

Assuming the MOSFETs to be matched and operating in triode region, it turns out that the output current I_0 of this circuit is given by:

$$I_0 = 2K(V_{GA} - V_{GB})(V_{in1} - V_{in2}) \tag{8.19}$$

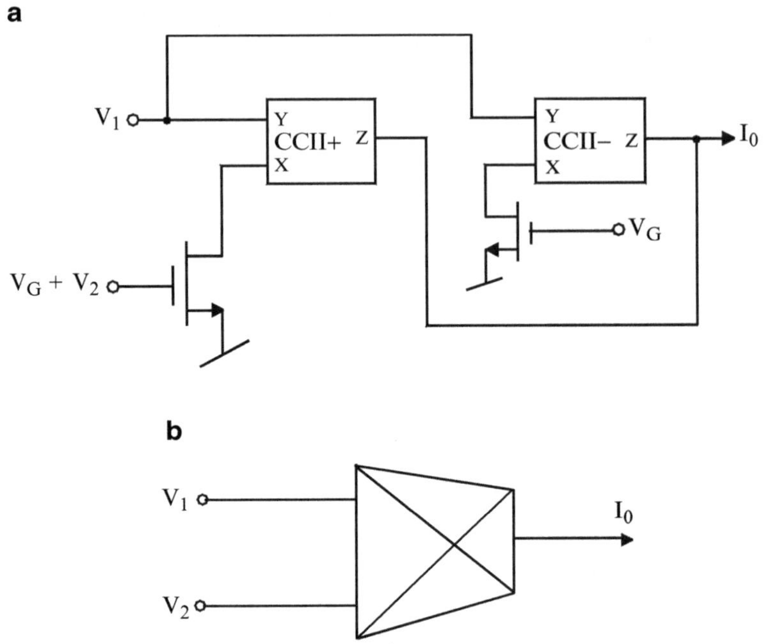

Fig. 8.9 (**a**) A transconductance multiplier circuit and (**b**) its symbolic notation proposed by Liu et al. [1]

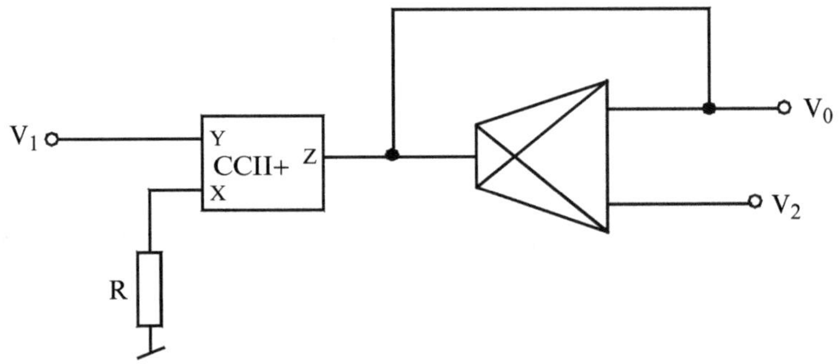

Fig. 8.10 Analog divider [1]

Using this multiplier circuit, another analog divider similar to that of Fig. 8.10 can be obtained which is shown in Fig. 8.12.

By straight forward analysis assuming the aspect ratios of all the MOSFETs to be identical, the output voltage of this circuit is given by:

$$V_0 = \frac{(V_{GA} - V_{GB})(V_{in1} - V_{in2})}{V_2} \tag{8.19}$$

A circuit to carry out the function of a square rooter is shown in Fig. 8.13.

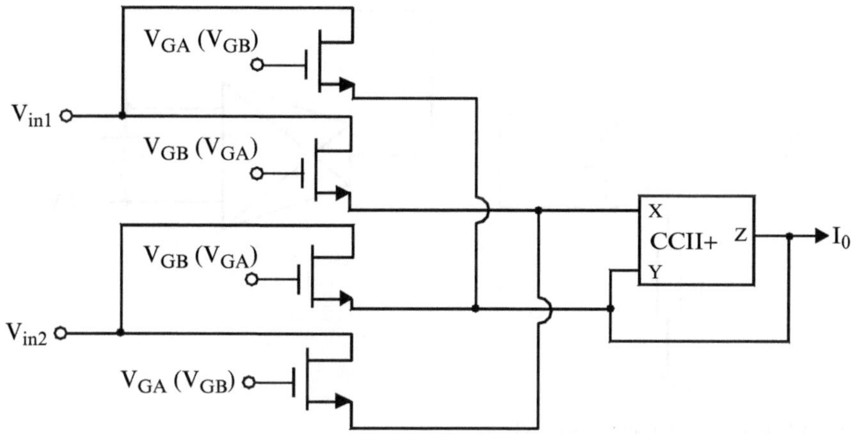

Fig. 8.11 Voltage- to- current converter using MOS resistive circuit [1]

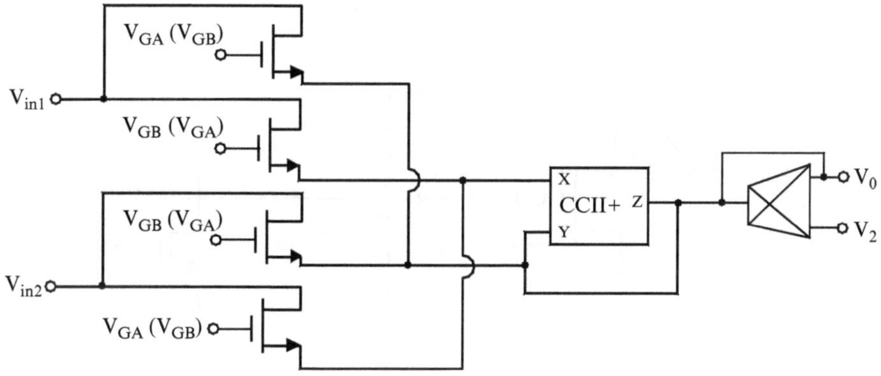

Fig. 8.12 An alternative analog divider circuit [1]

The workability of all the mentioned circuits has been verified in [1] by implementing the CCII+ using the supply current sensing technique based upon LF356 IC op-amps and CA3096 type transistor arrays. However, it is obvious that all the circuits can be implemented in entirely CMOS technology by employing CMOS CCII+ architectures instead of the CCII+ obtained by power supply sensing technique.

In Fig. 8.14, assuming that the MOSFETs are operating in triode region and are characterized by the following relationship between the drain current and gate-to-source and drain-to-source voltages.

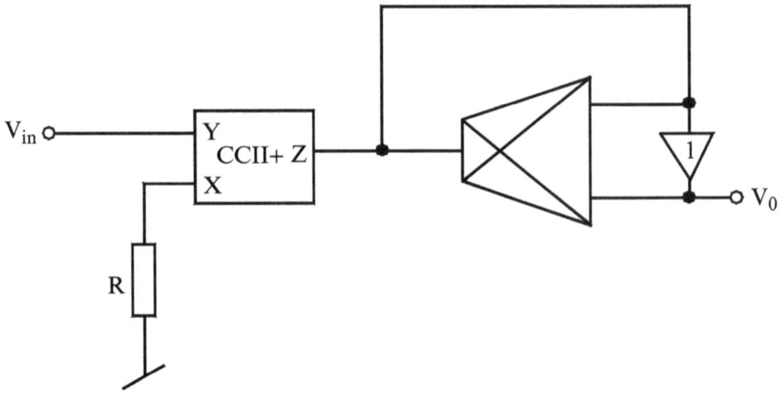

Fig. 8.13 Circuit of a root squarer [1]

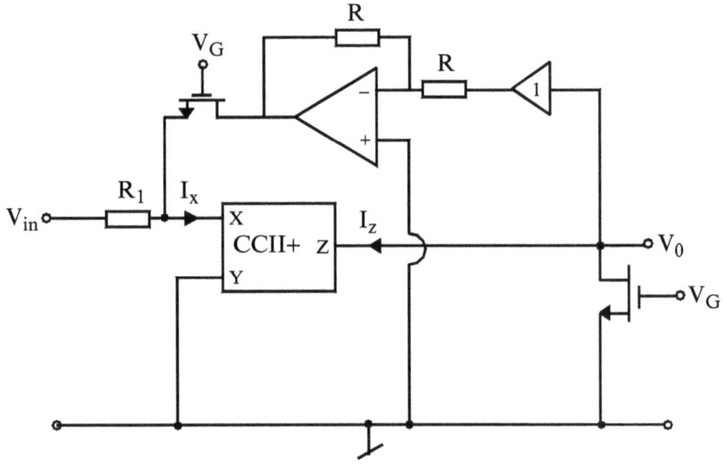

Fig. 8.14 Alternative square-rooting circuit proposed by Liu [8]

$$I_D = \mu_s C_{ox} \frac{W}{L} \left[(V_{GS} - V_{TH}) V_{DS} - \frac{V_{DS}^2}{2} \right] \tag{8.20}$$

it can be easily determined that the output voltage of this circuit is given by:

$$V_0 = -\left(\frac{V_{in}}{R_1 \mu_s C_{ox} \frac{W_1}{L_1}} \right)^{1/2} \text{ for } V_{in} > 0 \tag{8.21}$$

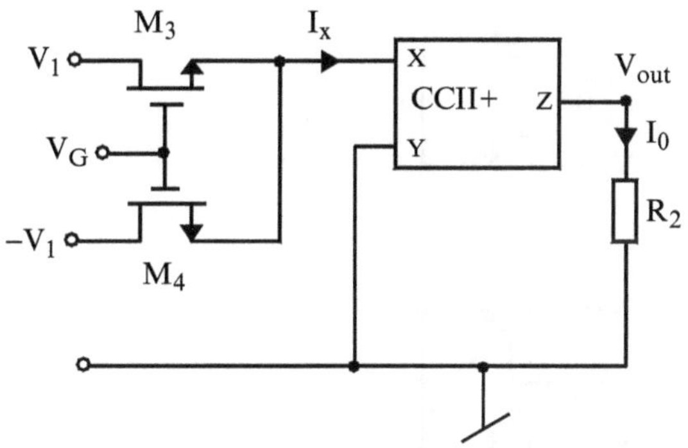

Fig. 8.15 Squaring circuit proposed by Liu [8]

To ensure the operation of the circuit as predicted by the analysis, the following condition should be satisfied:

$$V_G > |V_0| + V_{TH} \tag{8.22}$$

Thus, the circuit of Fig. 8.14 functions as a square-rooting circuit.

If complementary inputs are easily available and can be applied then the circuit of Fig. 8.15 performs the operation of squaring. The output current of this squaring circuit, assuming the same aspect ratios of the two MOSFETs, is given by:

$$I_0 = \mu_s C_{ox} \frac{W_3}{L_3} V_1^2 \tag{8.23}$$

The basic schematic of the squaring circuit can be generalized by adding more input variables along with additional the modified expression of the output current as pairs of matched MOSFETs thereby leading to

$$I_0 = \mu_s C_{ox} \left(\frac{W}{L}\right) \left(V_1^2 + V_2^2 + \ldots.. + V_n^2\right) \tag{8.24}$$

Finally, if the circuit is further modified as shown in Fig. 8.16 it performs the function of vector summation for which the relation between the output voltage and the various input signals is found to be

$$V_0 = -\left(\frac{R_2}{R_1}\left(V_1^2 + V_2^2 + \ldots + V_n^2\right)\right) \tag{8.25}$$

The workability of the above mentioned circuits has been confirmed [8] by realizing CCII+ using AD844, the unity gain inverting amplifier using LF356 and the MOSFETs from the CMOS transistor arrays CD40007. The bandwidth of a two input vector summation circuit has been found to be around 400 KHz.

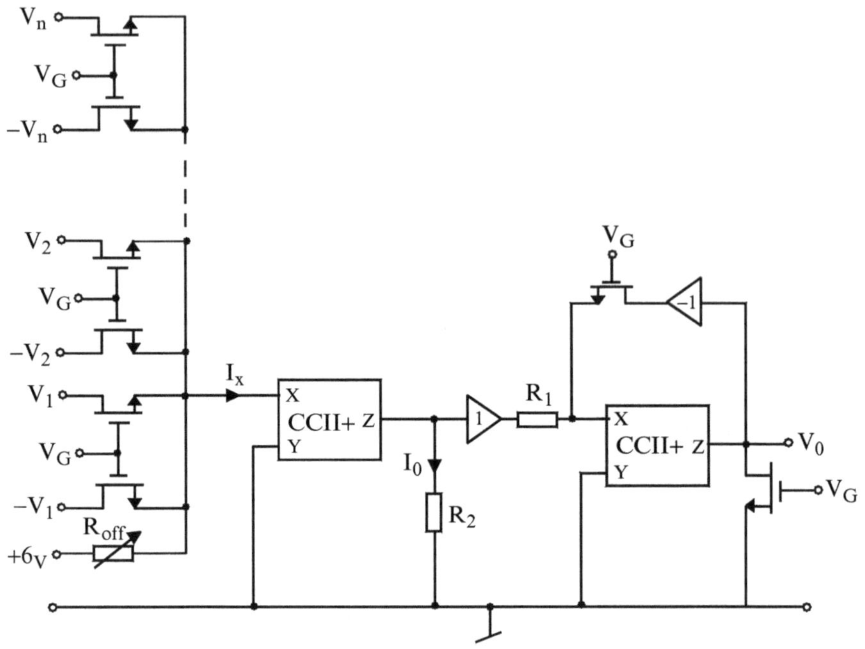

Fig. 8.16 Vector summation circuit with multiple inputs proposed by Liu [8]

8.5 CCII-based Realization of Fuzzy Functions

There has been lot of interest on applying the ideas of fuzzy set theory and fuzzy logic in various areas like process control, robotics, expert systems etc. Consequently, attention has also been focused by a number of researchers in implementing the basic building blocks of fuzzy systems by analog current mode electronic circuits. Thus, CCs have been the prominent building blocks whose applications in realizing fuzzy member functions and Max/Min circuits and a number of realizations have been advanced in literature. In this section, we outline some important contributions made in this area. Analog implementations of fuzzy operators are expected to provide the advantages of lower power supply voltages, high frequency operation and smaller chip area as compared to fuzzy information processing through the use of digital computers.

Liu et al. in [2] propose CCII+ based fuzzy membership function and Max/Min circuits which are shown here in Figs. 8.17 and 8.18 respectively. The operation of circuit of Fig. 8.17 can be explained as follows:

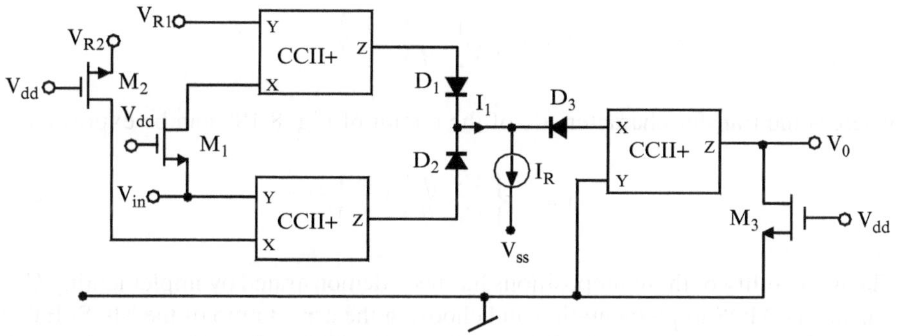

Fig. 8.17 CCII based fuzzy membership function proposed by Liu et al. [2]

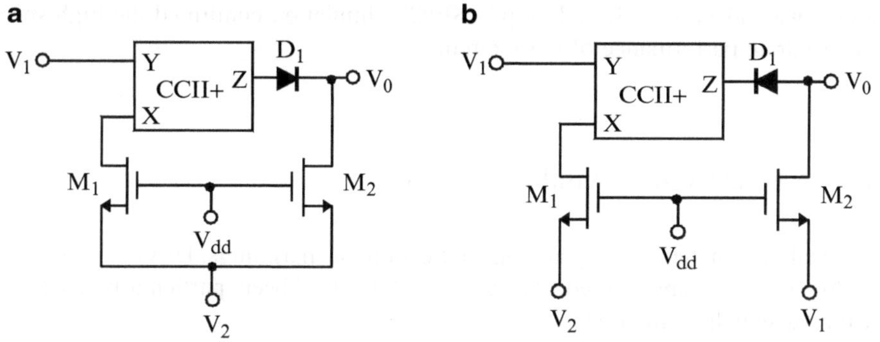

Fig. 8.18 Two-input Max and Min circuits proposed by Liu, Hwang and Tsay (**a**) Max (**b**) Min [2]

$$I_1 = \begin{cases} g_{m1}(V_{R1} - V_{in}) & \textit{if } V_{in} < V_{R1} \\ 0 & \textit{if } V_{R1} \leq V_{in} \leq V_{R2} \\ g_{m1}(V_{in} - V_{R2}) & \textit{if } V_{in} > V_{R2} \end{cases} \tag{8.26}$$

for equal transconductance of M_1 and M_2.

If the conductance of the MOSFET M_3 is taken as g_{m3} the output voltage of the circuit can be expressed as:

$$V_0 = \begin{cases} \{g_{m1}(V_{R1} - V_{in})/g_{m3} \ \textit{if } V_{R1} - I_R/g_{m1} \leq V_{in} < V_{R1} \\ I_R/g_{m3} \ \textit{if } V_{R1} \leq V_{in} \leq V_{R2} \\ g_{m1}(V_{in} - V_{R2})/g_{m3} \ \textit{if } V_{R2} < V_{in} \leq V_{R2} + I_R/g_{m1} \end{cases} \tag{8.27}$$

The circuit thus, realizes a fuzzy membership function consisting of three piecewise-linear sections. The circuits which can function as two-input Max or Min circuits are shown in Fig. 8.18.

In these circuits, the MOS transistors are acting as resistors. The transfer characteristic of the circuit of Fig. 8.18 a can be expressed as:

$$V_0 = \begin{cases} V_1 & if\, V_1 > V_2 \\ V_2 & if\, V_2 \geq V_1 \end{cases} \tag{8.28}$$

whereas the transfer characteristic of the circuit of Fig. 8.18b can be expresses as:

$$V_0 = \begin{cases} V_2 & if\ V_1 \geq V_2 \\ V_1 & if\ V_2 > V_1 \end{cases} \tag{8.29}$$

The workability of these prepositions has been demonstrated by implementing CCII + using a CMOS implementation and choosing the aspect ratio of the MOSFETs as W/L = 2 μm/2 μm for the MOSFETs, M_1 and M_2; W/L = 3 μm/2 μm for the diode-connected MOSFETs D_1, D_2, D_3; W/L = 2 μm/8 μm for the MOSFET M_3. The CMOS CCII was biased from ±5 V DC power supply and the reference current source was taken to be $I_R = 34.4$ μA. SPICE simulation confirmed the high speed and accurate performance of these circuits.

8.6 Realization of Analog Switches

It is well known that in many analog applications such as filters, D/A converters etc. analog switches are needed. MOS technology has been particularly useful in realizing switching methods.

Premont et al. in [11] demonstrated a new approach of designing of analog switches using two current conveyor cell as switching element for continuous time application which was claimed to offer high linearity, low noise and high frequency characteristics. In this context, it may be noted that a CCII can be easily used to realize either a voltage follower (between ports Y and X; with terminal-Z grounded) or a current follower (between ports X and Z; with terminal-Y grounded). The switching element proposed by Premont et al. [11] is based upon two current conveyor cell as shown in Fig. 8.19.

This switching element contains two ways signal path. By tuning the biasing of two current conveyors, the circuit simulates two ways analog switch. This behavior is depicted in Fig. 8.20. The input resistance R_y is more than 10 MΩ for OFF state of the switch. On the other hand, R_x equals around 350 Ω (for I_{bias} of 20 mA) below 5 MHz for the ON state of the switch. For the OFF state, R_x is considered to be greater than 10 MΩ below 5 MHz. This circuit when implemented with a CMOS current conveyor, exhibits characteristics which is comparable to the traditional commercial circuit of the analog switch such as CD4066.

Figure 8.21 shows an analog switch employing a CCII− [15]. Using CCII− characterization, it is clear that $V_A = V_B$ also $i_B = i_{RL} = (V_B\text{-}V_C)/R_L = (V_A - V_C)/R_L = i_A$. These equations mean that the proposed configuration can operate as an analog switch in both cases when $V_A > V_C$ and $V_A < V_C$ by supplying DC bias

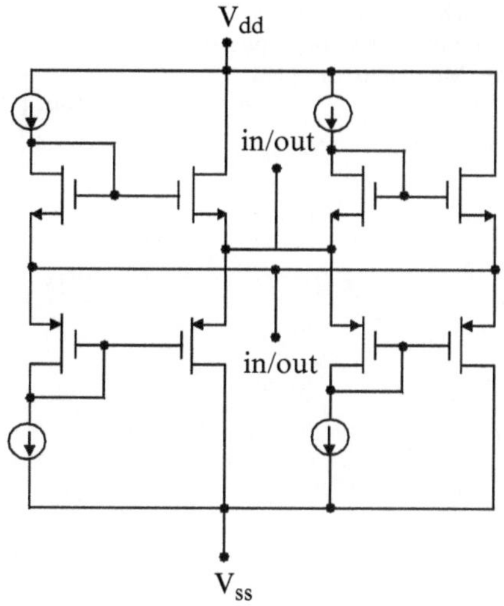

Fig. 8.19 Two-current-conveyor cell proposed by Premont et al. [11]

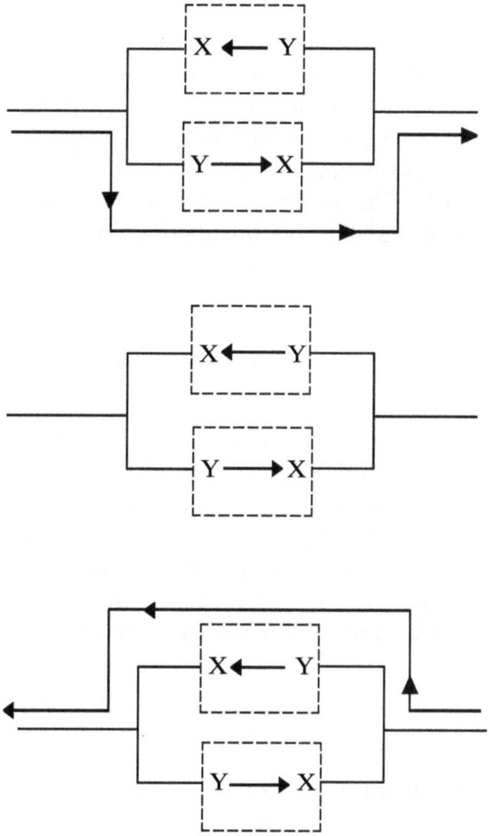

Fig. 8.20 The behavior of the analog switch [11]

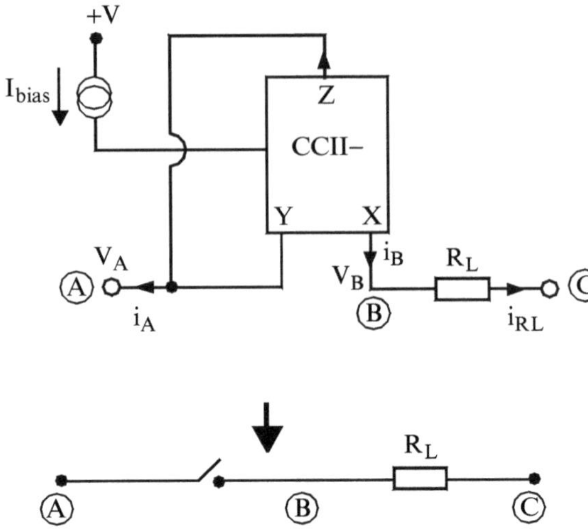

Fig. 8.21 An analog switch employing CCII- proposed by Monpapassorn [15]

current to the CCII−, when it is proposed to be turned ON and by not supplying I_{bias} current when one wants to turn the switch OFF.

A SPICE simulation check [15] for the workability of this switch using a CMOS CCII− implementation in 0.5 μm CMOS technology has demonstrated that this analog switch can operate satisfactorily with DC bias voltage of ±1.2 V and exhibits an operational range from −300 to +300 mV up to a frequency of 1 MHz. This simple switch appears to be a suitable choice for mixed analog and digital signal processing applications.

8.7 Pseudo-Exponential Circuit Realization

Pseudo-exponential function circuits are useful in a number of communication and signal processing systems. Such circuits have an important role in applications which require allowing wide gain control changes with a control parameter. In principle, the exponential v-i characteristics of a BJT or a MOSFET operating in week inversion mode can be exploited to design exponential circuits.

A pseudo-exponential function can alternatively be realized by analog circuits using its rational function approximation such as:

$$e^{2\beta x} \approx \frac{1+\beta x}{1-\beta x} \tag{8.30}$$

Two circuits to realize the above functions were proposed by Maundy and Gift [16] and are shown in Fig. 8.22.

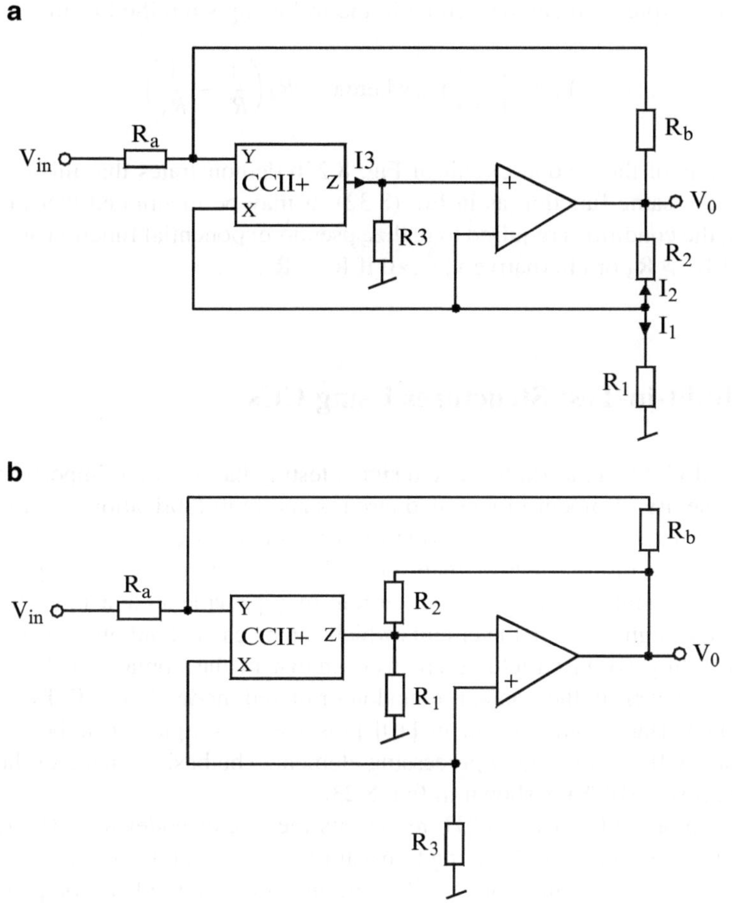

Fig. 8.22 Pseudo-exponential function circuits proposed by Maundy and Gift [16]

In the circuit of Fig. 8.22a, the voltage at node-Y (assuming identical resistors $R_a = R_b$) is given by:

$$V_y = \frac{V_{in} + V_o}{2} \tag{8.31}$$

Due to the properties of the CCII+, the voltage applied on terminal-Y is transferred to terminal-X. By a routine analysis of the circuit, it can be easily found that:

$$V_o = \frac{1 + R_2\left(\frac{1}{R_1} - \frac{1}{R_3}\right)}{1 - R_2\left(\frac{1}{R_1} - \frac{1}{R_3}\right)} V_{in} \tag{8.32}$$

The above expression can be written in the following simplified form:

$$V_o = \frac{1+x}{1-x} V_{in} \text{ where } x = R_2 \left(\frac{1}{R_1} - \frac{1}{R_3} \right) \tag{8.33}$$

An analysis of the second circuit of Fig. 8.22b demonstrates that this circuit also realizes the same function as in Eq. (8.32). It may be mentioned that in both the circuits, the conditions required to realize pseudo-exponential function are given by $x < 0$, if $R_1 > R_3$ or alternatively, $x > 0$ if $R_1 < R_3$.

8.8 Built-in-Test Structures Using CCs

The use of CM techniques for mixed signal testing has acquired importance due to the complexity of modern integrated circuits and high fabrication densities. It has been recognized that in the detection of certain classes of faults, current sensing techniques are more effective than the techniques based upon voltage measurements. The monitoring of quiescent current may provide a clear indication of the faults and design errors in the circuits which can be carried out on- or off-chip. The built-in-testing (BIT) structures are well-known in the domain of digital circuit design. However, in the domain of analog or mixed mode circuit, BITs are not yet established. Hatzopoulos et al. in [10] presented a simple linear built-in-sensor based on CCII+ as the signal processing element. The basic scheme of the built in current sensor (BICS) is shown in Fig. 8.23.

In the proposed scheme, CUT represents the circuit under test. The quiescent current I_{PS} close into the detecting transistor Q_1 which is mirrored and scaled by a constant k ($k \leq 1$) to transistor Q_2. The current flowing into Q_2 is compared with a reference current I_{ref}. It may be noted that the CCII+ ensures that the collector-emitter voltages of the transistor Q_1 and Q_2 are forced to be equal and the collector–base voltages are equal to zero. During normal operation, CCII+ provides an output current at node-Z equal to the input current at node-X. In the proposed scheme, if I_{ref} is set equal to the quiescent current, the CCII+ detects only the deviations in I_{ps} and provides an output current and hence, an output voltage proportional to the deviation in the quiescent current i.e., $V_{out} = \Delta I_{ps} R_L$.

Hatzopoulos et al. have also demonstrated in [10] how built-in- test-structures can be implemented for voltage input data as well as current input data using a number of voltage controlled current conveyors in which the current gain between the ports X and Z of the CCII+ s can be made either zero or one depending upon whether an external control voltage is kept high or low. The interested readers are referred to [10] for further details.

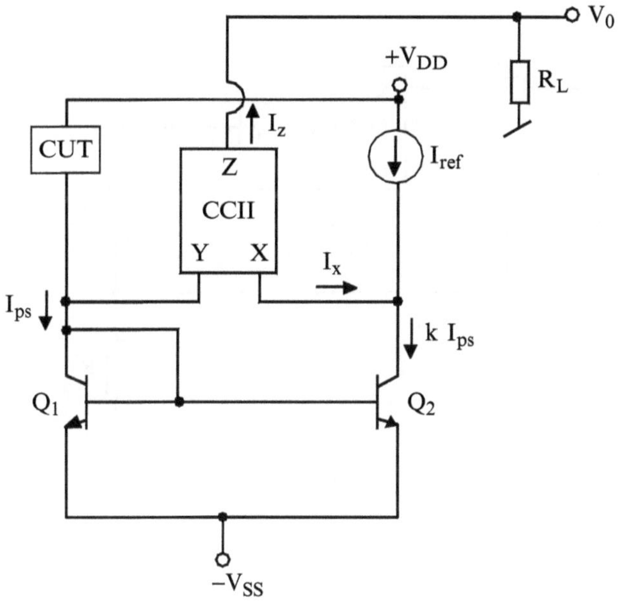

Fig. 8.23 BICS proposed by Hatzopoulos et al. [10]

8.9 Schmitt Trigger and Waveform Generators Using CCs

Because a translinear CCII+ has its input stage comprised of a four transistor mixed translinear cell and the output current at the Z-terminal is a replica of the current into X-terminal which is obtained by using a pair of current mirrors, it turns out that the output current is a *Sinh* function of the differential input $V_y - V_x$. As a consequence, the maximum charging current for the compensating capacitance connected at Z-terminal (which in the absence of an external capacitor, is simply the parasitic output capacitance looking into terminal-Z which is typical in the range of 4–5 pF) is unlimited and theoretically infinite. This results in a very high slew rate in case of CCII as compared to conventional op-amps. As an example for CCII+ implemented from AD844 the slew rate is as high as 2,000 V/μs as against a very modest figure of 0.5 V/μs for the popular μA741op-amp.

In view of the above, it is obvious that a square wave generator of CCII+ is expected to be much superior to that made from an op-amp; alternatively the non-sinusoidal wave form generator made from CCII+ can be expected to perform adequately for much higher frequency ranges than possible with op-amps. Motivated by this superiority of CCII+ over the conventional op-amp, several researchers have paid attention on devising novel realization of Schmitt triggers and square/triangular wave form generators using CCIIs. In this section, we include a discussion about some prominent ideas and circuits in this direction.

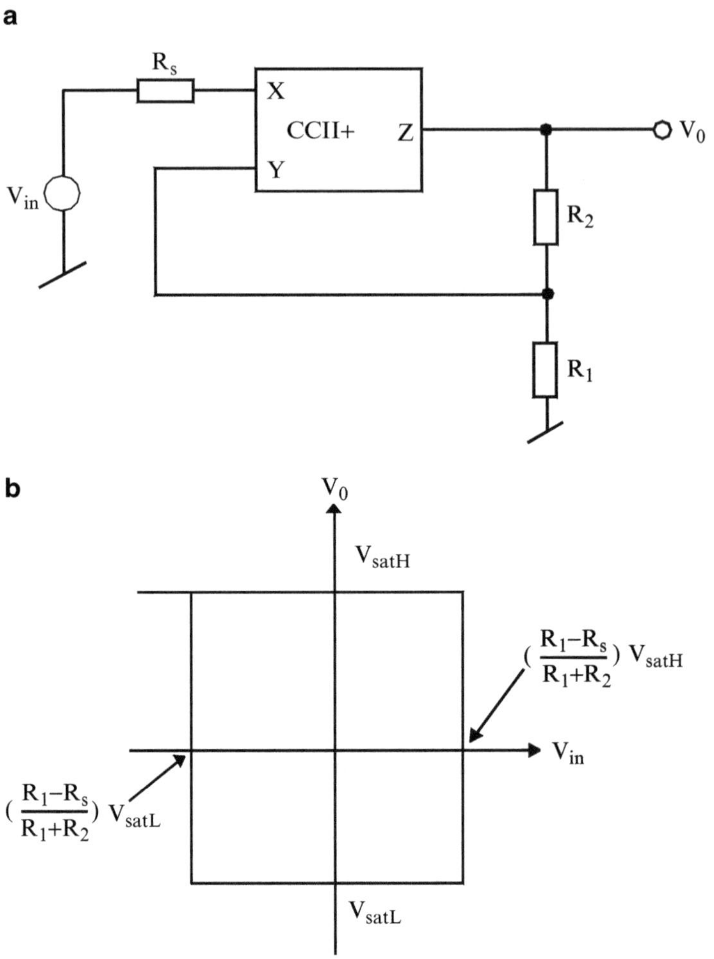

Fig. 8.24 (**a**) Schmitt trigger using CCII+ introduced by Cataldo et al. (**b**) its transfer characteristic (Adapted from [9] © 1995 Wiley)

The first ever Schmitt trigger using a single CCII+ along with two resistors was introduced by Cataldo et al. in [9] which is shown in Fig. 8.24a.

In this circuit, the re-generative feedback required to make a Schmitt trigger is created by the potential divider made from the two resistors. The operation of this circuit can be explained as follows. A decrease in the output voltage leads to a decrease in v_y; since $v_x = v_y$ therefore, v_x also deceases. Since $i_x = (V_{in} - V_x)/R_s$, this implies that as a consequence i_x will be increased. In turn i_z also increases by the same amount as i_x which results in the increase in the voltage $v_{z.}$. However, the maximum voltage V_0 is limited by the two power supply voltages hence, the circuit output will exhibit only two stable states, V_{satH} and V_{satL}, corresponding to which the two output saturation currents are found to be [9]:

$$I_{satH} = -\frac{V_{satH}}{R_1 + R_2} \tag{8.34}$$

$$I_{satL} = \frac{V_{satL}}{R_1 + R_2} \tag{8.35}$$

The two threshold voltages can be found by assuming first that v_z is set to the stable state V_{satH}.To change this stable state, the current i_x must satisfy the condition $i_x > i_z$, which implies:

$$\frac{v_{in} - v_y}{R_s} > -\frac{V_{satH}}{R_1 + R_2} \tag{8.36}$$

Hence, higher threshold voltage (V_{TH}) is given by:

$$V_{TH} = \frac{R_1 - R_s}{R_1 + R_2} V_{satH} \tag{8.37}$$

Using a similar procedure, it can be determined that the lower threshold voltage (V_{TL}) is given by:

$$V_{TL} = \frac{R_1 - R_s}{R_1 + R_2} V_{satL} \tag{8.38}$$

The transfer characteristics of this Schmitt trigger using the above derivation can be easily deduced to be as shown in Fig. 8.24b.

This Schmitt trigger leads to a square wave generator when a capacitor is connected from terminal-X to ground and the resistor R_s is eliminated and in its place the internal parasitic resistance of the CCII+ is accounted for. This square wave generator was proposed by Abuelma'atti and Al-Absi [17] and is shown in Fig. 8.25

It is straight forward to determine that the frequency of square wave generated from this circuit would be given by:

$$f = \frac{1}{2C_0 R_x l_n \left(2\frac{R_1}{R_x} - 1\right)} \tag{8.39}$$

A modified two CCII+ based Schmitt trigger/pulse squaring circuit was proposed by Srinivasulu [19] which is shown in Fig. 8.26.

The operation of this circuit can be explained as follows. Assume that the input V_{in} is positive rising triangular wave changing from $-V_{in}$ to $+V_{in}$. Also assume during this transaction the current in the terminal-X_1 changes from $-i_{x1}$ to $+i_{x1}$ whereas the Z-port current rises from $-i_{z1}$ to $+i_{z1}$. It is further assumed that at node 'b' the current changes from $+V_T/R_3$ to $-V_T/R_3$ and current at port-X_2 of CCII+ (2) is rising from $+i_{x2}$ to $-i_{x2}$ i.e. $+V_{x2}/R_2$ to $-V_{x2}/R_2$ and furthermore the Z-port

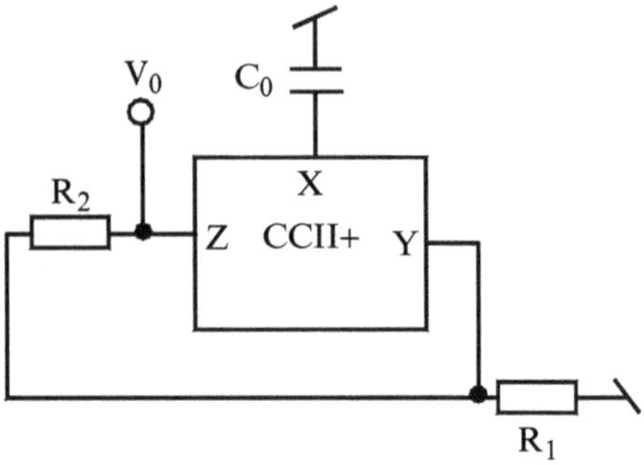

Fig. 8.25 CCII+ based square wave generator proposed by Abuelma'atti and Al-Absi [17]

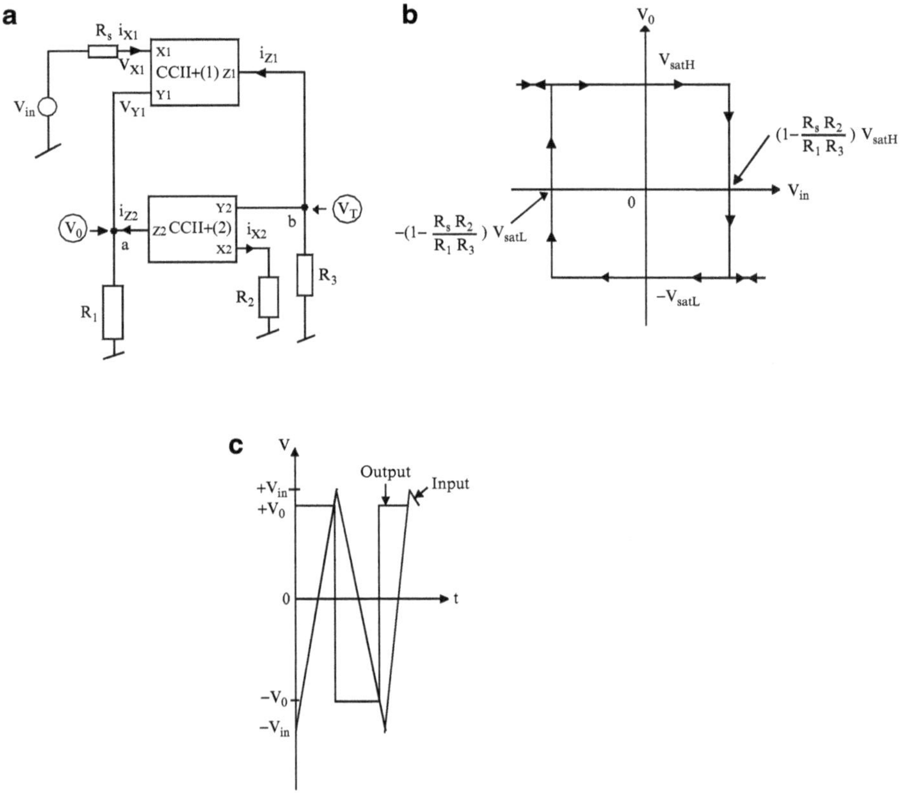

Fig. 8.26 Schmitt trigger/pulse squaring circuit proposed by Srinivasulu (Adapted from [19] © 2010 Wiley) (**a**) The circuit (**b**) its transfer characteristic (**c**) input/output waveforms of the trigger circuit (**d**) Sine wave form input and square wave output: x-axis 5 ms/div and y-axix 5 V/div. (**e**) Experimentally observed hysteresis characteristics

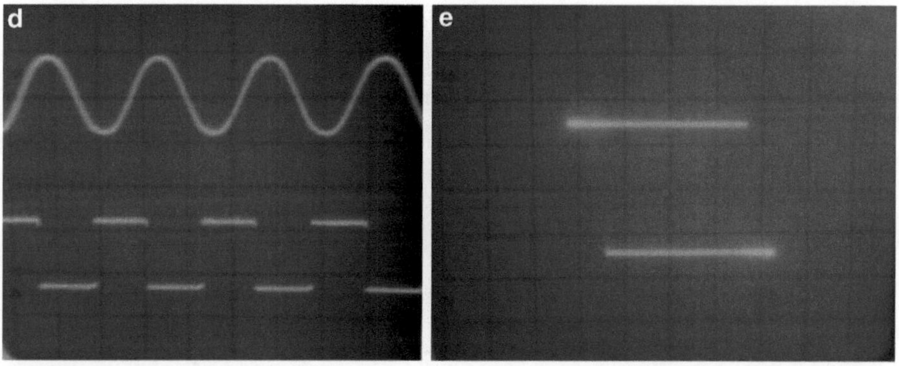

Fig. 8.26 (continued)

current of CCII+ (2) rising from $+i_{z2}$ to $-i_{z2}$. At this instant the output voltage at node 'a' is $+V_0$ (during the previous cycle) when the input current $i_{x1} = (V_{in-}V_{x1})/R_s \geq (R_2/R_1R_3)V_0$ at node 'a', the output current direction at port Z_1 changes to $+i_{z1}$ this is the upper trigger point. Now the current at node 'b' $-V_T/R_3$ and also $-i_{x2} = -V_{x2}/R_2$ and output current at port Z_2 is $-i_{z2}$. The output voltage at node 'a' is $-V_0$ which remains negative until the input triangular waveform reaches a value $-i_{x1} \geq (-V_{in-}V_{x1})/R_s$.

When input current $-i_{x1} \geq (-R_2/R_1R_3)V_0$ at node 'a', the direction of the port Z_1 current of CCII+ (1) changes to $-i_{z1}$ this is the lower trigger point. Now the node 'b' current is $+V_{T/}R_3$ and current at Port X_2 of CCII+ (2) is $+i_{x2} = +V_{x2}/R_2$ and the output current at port Z_2 for CCII+ (2) is $+i_{z2}$. The net output voltage at node 'a' is $+V_0$ which continues to remain positive until the input triangular waveform reaches a value of $+i_{x1}$ ($\geq$ $(V_{in-}V_{x1})/R_s$) at port X_1 of CCII+ (1). This cycle goes on repeating itself thereby generating a square waveform at the output as shown in Fig. 8.26c.

Some experimentally observed results for the circuit are shown in Fig. 8.26d, e for which AD844 were used to realize CCII+ biased with ± 6 V supply with component values taken as $R_1 = R_2 = 10$ KΩ, $R_s = 8$ KΩ, $R_2 = 7$ KΩ.

The circuit of Fig. 8.27a was proposed by Marcellis et al. [22] and consists of a voltage to current converter using CCII1 along with resistor R_1 and a hysteresis current comparator consisting of CCII along with resistor R_5 and R_6 and. The circuit operation can be explained as follows.

The CCII generates a square wave signal V_{out} converting the saturated current Iz_2 through R_5 and R_6 into saturation voltage $V_{out} = \pm V_{sat}$. A reduced version of this output through the potential divider consisting of R_5 and R_6 goes as input to Y_1 of CCII1. This provides a current I_{z1} part of which charges the capacitor C, in fact the circuit composed of C and R_3 acts as a differentiator which differentiates the square wave current I_{z1} and thereby generating an exponential voltage at node D (V_D) this signal is converted into a similar shaped current I_{x2} which compared with the saturation current I_{z2} thereby producing the square wave output V_{out} (Fig. 8.27b).

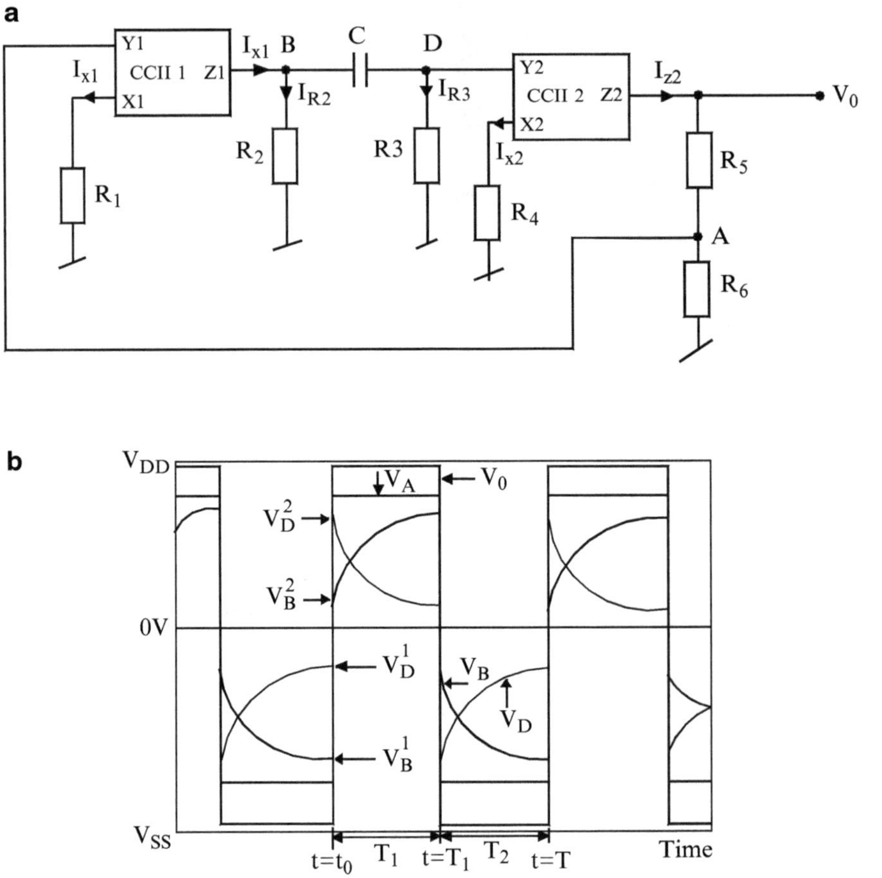

Fig. 8.27 CCII-based oscillator proposed by Marcellis et al. (**a**) The circuit (**b**) voltage signals behavior at circuit nodes (Adapted from [22] © 2011 Wiley)

Using a straight forward but lengthy analysis [22], it is found that the oscillation frequency of this circuit is given by:

$$f_0 = \frac{1}{2C(R_2 + R_3)\ln\left[\frac{2R_2R_3R_6 - R_1R_4(R_2+R_3)}{R_1R_4(R_2+R_3)}\right]} \tag{8.40}$$

The workability of this circuit has been verified in [22] using a CMOS CCII+ topology implemented in 0.35 μm technology and has also been verified by realizing CCII+ using AD844 biased with ±15 V DC power supplies. From the experimental results it has been found that a frequency range from about 390 kHz to 780 mHz is feasible exhibiting a good linearity over a frequency range of about five decades when the capacitor C was changed from 1 pF to 1 μF.

A much simpler circuit capable of generating a square wave as well as rectangular wave is obtainable using square wave generator of Fig. 8.25 followed by a non-inverting integrator made from a CCII+ along with a resistor and capacitor. This circuit was proposed in [23] by Almashary and Alhokail and is shown in Fig. 8.28.

The expression of the oscillation frequency of this circuit is given by the same expression as in Eq. (8.39) for $C_0 = C_1$.

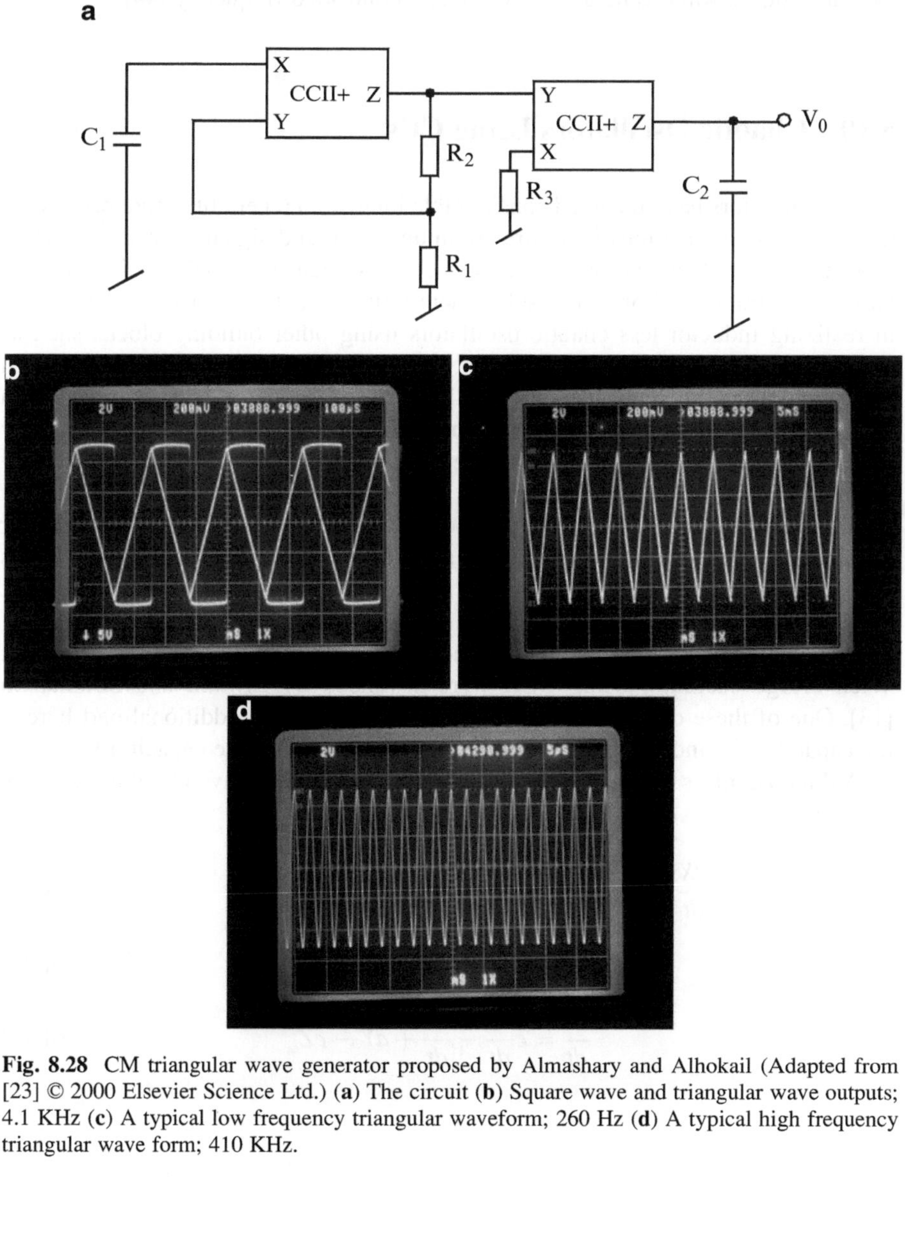

Fig. 8.28 CM triangular wave generator proposed by Almashary and Alhokail (Adapted from [23] © 2000 Elsevier Science Ltd.) (**a**) The circuit (**b**) Square wave and triangular wave outputs; 4.1 KHz (**c**) A typical low frequency triangular waveform; 260 Hz (**d**) A typical high frequency triangular wave form; 410 KHz.

Some sample experimental results from [23] based upon the implementation of the circuit using AD844 for CCII+, with passive component values as: $R_1 = 280\ \Omega$ and $C_1 = 1$ uF are shown in Fig. 8.28d, e. The frequency of the output waveform was found to be changeable through C_1 and/or R_1 while the amplitudes can be varied through C_2 and/or R_3.

Interested readers are referred to [28, 42–44] for some other contributions on this topic such as, design of low-power relaxation oscillators [28], the design of Schmitt triggers with controllable hysteresis [42], tunable CC-based relaxation oscillator [43] and square wave generator with voltage-controlled frequency [44].

8.10 Chaotic Oscillators Using CCs

Chaos generators have received considerable interest in recent literature because of their possible applications in secure communication and signal encryption. While most earlier implementations of chaotic circuits such as Chua's oscillator were based upon traditional op-amps, subsequently there has been considerable interest in realizing inductor-less chaotic oscillators using other building blocks such as CCIIs and CFOAs because of the several advantages provided by such implementations such as: canonic realizability of inductor simulation using a minimum possible number of external passive components without requiring any component matching, availability of the state variables such as inductive current through the Z terminals of the CCIIs and CFOAs, relatively high frequency applicability and ease of realizing the required simulated inductors and non-linear negative resistors etc. As a consequence, a variety of chaotic oscillators using CCIIs have been advanced in the literature. In the following we discuss some prominent CCII-based chaotic oscillators.

Two simple chaotic oscillators obtained by simple modification of the classical Wien bridge oscillator using CCII were advanced by Elwakil and Soliman in [13]. One of these circuits is shown here in Fig. 8.29. The additional part here is the capacitor C_3 and the JFET-based resistor in series with the capacitor C_1.

A dimensionless state-space description of this circuit is given by the following equations:

$$\frac{dX}{dt} = \frac{\alpha}{2}\begin{cases} (Z - X) & (Z - X) \leq 1 \\ 1 & (Z - X) > 1 \end{cases} \tag{8.41}$$

$$\frac{dY}{dt} = aY - bZ \tag{8.42}$$

$$\frac{dZ}{dt} = c\frac{dY}{dt} - \frac{dX}{dt} + dY - eZ \tag{8.43}$$

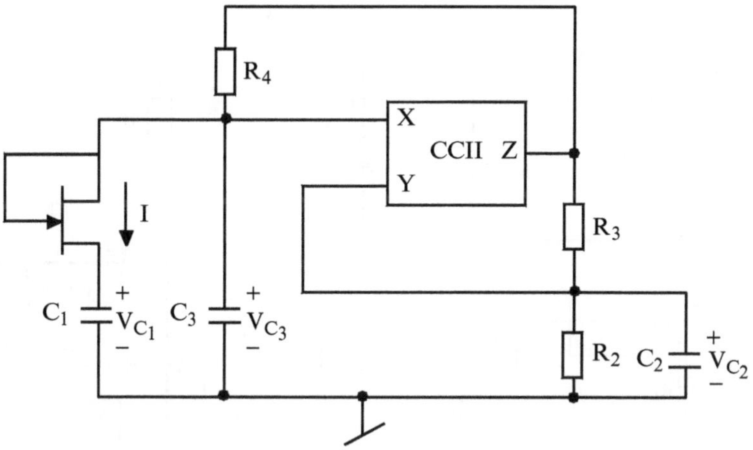

Fig. 8.29 A classical Wien Bridge oscillator converted into a chaotic oscillator as proposed by Elwakil and Soliman [13]

where the no-dimensional variables and other assumed relations are: $X = V_{C1}/V_T$, $Y = V_{C2}/V_T$ and $Z = V_{C3}/V_T$, $\alpha = R_2/R_J$, $t_n = t/R_2 C_2$ and $C_1 = C_3 = 2C$, $C_2 = C$ and the JFET is characterized by the following v-i relation:

$$I = \frac{1}{R_J}\begin{cases} V_{GD} & V_{GD} \geq V_T \\ V_T & V_{GD} < V_T \end{cases} \tag{8.44}$$

where R_J is the JFET small signal resistance at the operating point, V_{GD} is the JFET gate to drain voltage, V_T is the threshold voltage approximately equal to minus 0.66 V for the JFET J2N4338. The various coefficients appearing in the dimensionless state space equations are given by:

$$a = \frac{R_4}{R_4 + R_3}\left(\frac{R_2}{R_X} - \frac{R_2 + R_3}{R_4} - 1\right),\ b = \frac{R_4}{R_4 + R_3}\left(\frac{R_2}{R_X} - \frac{R_2}{R_4}\right), \tag{8.45}$$

$$c = \frac{R_3}{2R_4},\ d = \frac{R_2}{2R_4}\left(\frac{R_4}{R_X} + \frac{R_3}{R_2} + 1\right) \text{ and } e = \frac{1}{2}\left(\frac{R_2}{R_4} + \frac{R_2}{R_X}\right) \tag{8.46}$$

The experimental tests of this circuit using AD844 CCII+ biased with ±9 V supply with $C_1 = C_3 = 2$ nF, $C_2 = 1$ nF, $R_2 = 5$ KΩ potentiometer, $R_3 = 500$ Ω and R_4 as a 2 KΩ potentiometer demonstrated chaotic behavior.

Chua's circuit is one of the most investigated chaotic oscillator which consists of a third order linear part composed of two grounded capacitors, a grounded inductor and a resistor along with a nonlinear negative resistor popularly known as Chua's diode characterized by a three segment piecewise linear v-i characteristic having specified slopes of the three segments. A CCII+ based Chua's oscillator was presented by Sanchez-Lopez et al. in [20], which is shown here in Fig. 8.30.

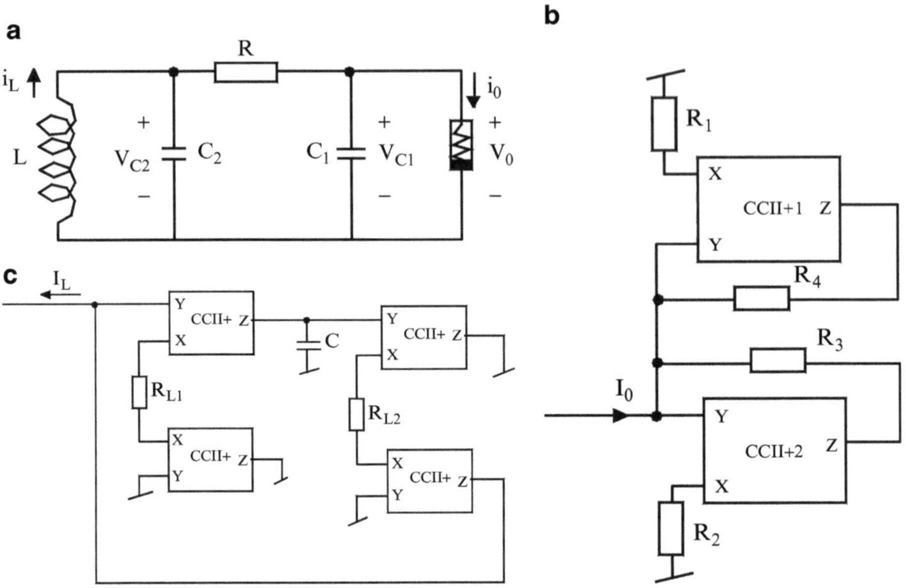

Fig. 8.30 Chua's circuits and its CCII+ based implementations as proposed by Sanchez-Lopez et al. (**a**) Chua's circuit (**b**) Chua's diode (**c**) Simulating inductance using CCII+ s [20]

In principle many different implementations of Chua's oscillator can be obtained depending upon how one realizes the Chua's diode and the simulated inductor. A Chua's diode implementation using two CCII+ and four resistors (analogous to the one devised by Senani-Gupta in [24]) has been shown in Fig. 8.30b. On the other hand a simulating inductance circuit using only CCII+ elements and requiring a bare minimum of only three passive components namely two resistors and a grounded capacitor is shown in Fig. 8.30c.

With all the CCII+ s realize from AD844 this circuit exhibits a double scroll attractor for the following component values and bias voltages. CCII+1 biased with +5 V and CCII+ 2 biased with 12 V; the values of resistors as $R_1 = 6.76$ KΩ, $R_2 = 3.91$ KΩ, $R_3 = 0.47$ KΩ, $R_4 = 10$ KΩ, $RL_1 = R_{L2} = 3.6$ KΩ, $C = 4.7$ nF, $C_1 = 4.7$ nF, $C_2 = 47$ nF and $R = 2{,}771$ Ω.

Elwakil and Kennedy in [14] demonstrated that three-phase sinusoidal oscillator based on current conveyors can also be modified to become chaotic oscillator using a circuit called diode-inductor composite. One such implementation is shown in Fig. 8.31. The basic circuit is essentially a cascade of three identical first order low pass sections such that the diode-inductor composite has been inserted in series with resistor R_4. In principle the diode-inductor composite could have been inserted in series with any of three resistors R_4, R_5 and R_6. It may be noted that CCII− can be realized from two CCII+ s.

The chaotic behavior of this circuit has been confirmed for the following set of component values: $C_1 = C_2 = C_3 = C = 50$ pF, $R_1 = R_2 = R_3 = R = 15$ KΩ, $R_4 = R_5 = R_6 = KR$, $K = 1/3$, $L = 100$ mH, $R_x = 1$ KΩ, $R_z = 5$ KΩ and the DC

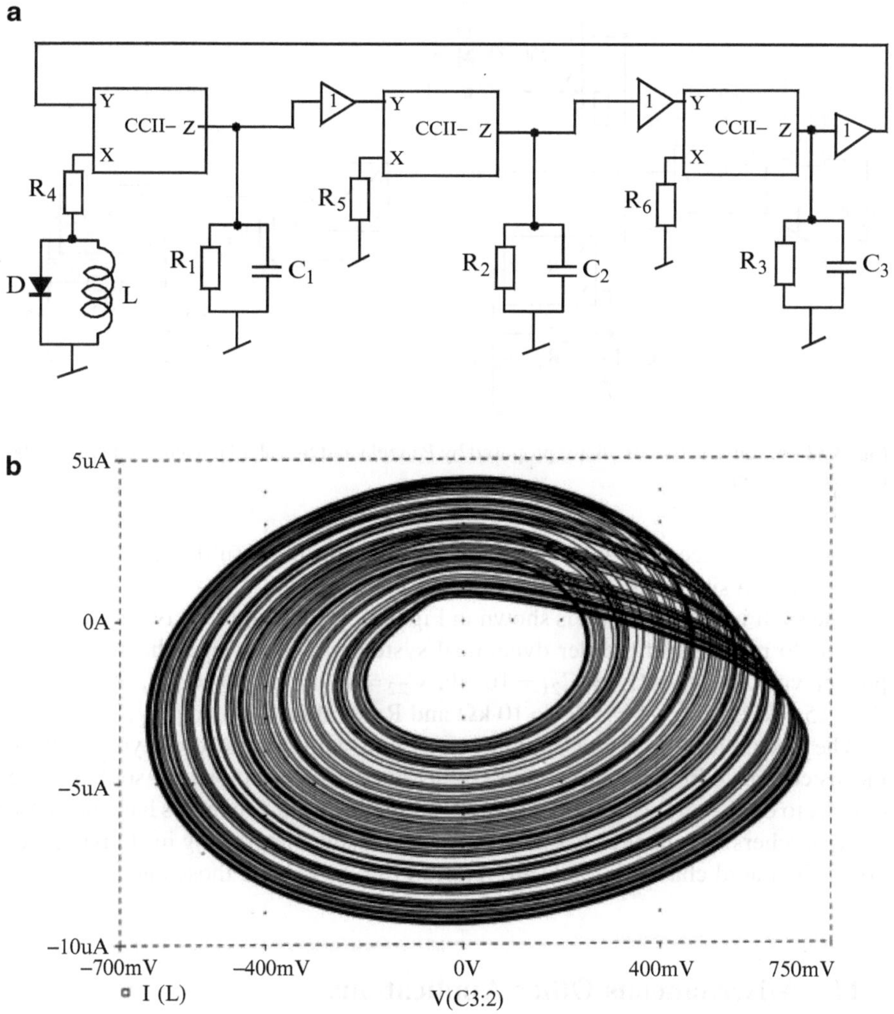

Fig. 8.31 A three phase oscillator converted into a chaotic oscillator using diode inductor composite proposed by Elwakil and Kennedy (Adapted from [14] © 1991 Elsevier Science Ltd.) (**a**) The circuit (**b**) PSPICE generated V_{c3}-I_L trajectory

biasing supply for CCII− taken as ± 9 V. The circuit (Fig. 8.31b) was tuned to obtain different trajectories by adjusting R_z to different values. The authors have claimed that using the same technique multiphase oscillators such as six phase oscillators can also be converted into chaotic oscillators.

Petrzela and Slezak in [18] demonstrated that several novel oscillators consisting of only five CCII+ s can be devised to generate conservative chaotic attractors. In fact they presented a class of autonomous deterministic dynamical systems with

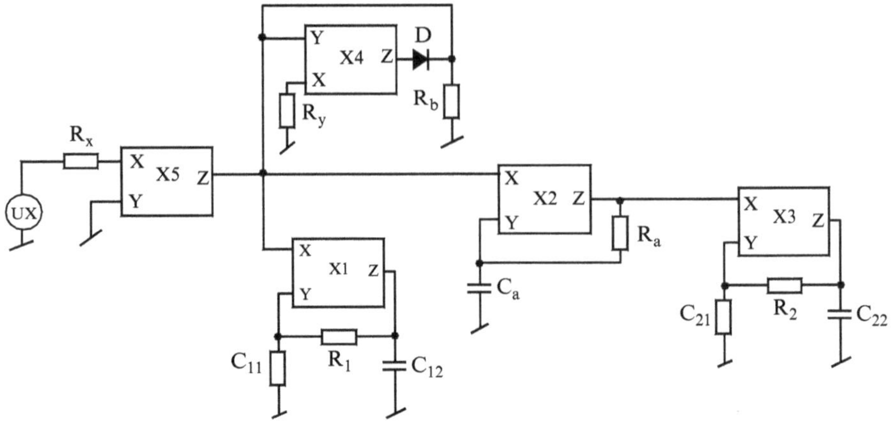

Fig. 8.32 A conservative oscillator proposed by Petrzela and Slezak (Adapted from [18] © 2008 Radioengineering)

two segment peace-wise linear element such that in the resulting circuit chaos is observed by a simple change of an external DC voltage.

One such implementation is shown in Fig. 8.32 which employs two lossy FDNR circuits to realize a fifth order dynamical system which become chaotic for component values: $C_{11} = C_{12} = C_{21} = 10$ nF, $C_{22} = 47$ nF, $C_a = 10$ nF, $R_1 = 81$ kΩ, $R_2 = 15.5$ kΩ, $R_a = 1.4$ kΩ, $R_b = 10$ kΩ and $R_y = 5$ kΩ.

The discussion of chaotic circuits realizable with current conveyors is by no means complete in this section whose objective was only to present some selected circuits to demonstrate a variety of ways in which current conveyors have been used by researchers to generate chaotic signals. The readers interesting in studying more about CC based chaotic circuits are referred to [25–44] and those cited therein.

8.11 Miscellaneous Other Applications

Other than the applications of CCs highlighted above, there are several other interesting applications which are briefly commented in the following. In [25] Abuelma'atti has demonstrated how CCs can be used to generate ASK/FSK/PSK/QAM waveforms useful in communications. Bruun and Haxthausen in [26] have employed CC to design of an EMG amplifier with shut down control. A new approach to designing active MMIC tuning element using CCII has been presented by Sinsky and Westgate in [27]. In [29], Sampiepro and Bertuccio have designed a zero power CC for DC stabilization and system rest of fast current pulse amplifiers. Vlassis and Siskos [30] have employed CCs to design an interfacing circuit for piezo-resistive pressure sensors with frequency output. A CC-based low voltage CMOS current source for space plaza probe has been designed by Rajput and Jamuar in [32]. Lastly, Acharya in [33] has designed a CC-based electrometer.

Concluding Remarks
Whereas, undoubtedly, the major attention of researchers working on CCs has been on their use in realizing linear circuits such as variable gain amplifiers, integrators, differentiators, filters, oscillators and simulated impedances, their applications in several non-linear circuits, relaxation oscillators and quite unusual areas have also been explored by a number of researchers.

In this chapter, some significant developments on the applications of CCs in realizing a variety of non-linear and other applications have been explored. It is however, felt that the work reported in this chapter indicates that there is enough scope for extending the application domain of the CCs beyond that included in this chapter.

References

1. Liu SI, Wu DS, Tsao HW, Wu J, Tsay JH (1993) Nonlinear circuit applications with current conveyors. IEE Proc G 140:1–6
2. Liu SI, Hwang YS, Tsay JH (1993) CCII-based fuzzy membership function and MAX/MIN circuits. Electron Lett 29:116–118
3. Toumazou C, Lidgey FJ, Chattong S (1994) High frequency current conveyor precision full-wave rectifier. Electron Lett 30:745–746
4. Hayatleh K, Porta S, Lidgey FJ (1994) Temperature independent current conveyor precision rectifier. Electron Lett 30:2091–2093
5. Stiurca D (1995) Truly temperature independent current conveyor precision rectifier. Electron Lett 31:1302–1303
6. Wilson B, Mannama V (1995) Current-mode rectifier with improved precision. Electron Lett 31:247–248
7. Khan AA, El-Ela MA, Al-Turaigi MA (1995) Current mode precision rectification. Int J Electron 79:853–859
8. Liu SI (1995) Square-rooting and vector summation circuits using current conveyors. IEE Proc Circ Devices Syst 142:223–226
9. Cataldo GD, Palumbo G, Pennisi S (1995) A schmitt trigger by means of a CCII+. Int J Circ Theor Appl 23:161–165
10. Hatzopoulos AA, Siskos S, Laopoulos TH (1997) Current conveyor based test structures for mixed signal circuits. IEE Proc Circ Devices Syst 144:213–217
11. Christophe P, Abouchi N, Grisel R, Chante JP (1998) A current conveyor-based high-frequency analog switch. IEEE Trans Circ Syst-I 45:298–300
12. Surakampontorn W, Anuntahirunrat K, Riewruja V (1998) Sinusoidal frequency doubler and full-wave rectifier using translinear current conveyor. Electron Lett 34:2077–2079
13. Elwakil AS, Soliman AM (1999) Current conveyor chaos generators. IEEE Trans Circ Syst-I 46:393–398
14. Elwakil AS, Kennedy MP (1999) Three-phase oscillator modified for chaos. Microelectron J 30:863–867
15. Monpapassorn A (2002) An analogue switch using a current conveyor. Int J Electron 89:651–656
16. Maundy B, Gift S (2005) Novel pseudo-exponential circuits. IEEE Trans Circ Syst-II 52:675–679
17. Abuelma'atti MT, Al-Absi MH (2005) A current conveyor-based relaxation oscillator as versatile electronic interface for capacitive and resistive sensors. Int J Electron 92:473–477

18. Petrzela J, Slezak J (2008) Conservative chaos generators with CCII+ based on mathematic model of nonlinear oscillator. Radioengineering 17:19–24
19. Srinivasulu A (2011) A novel current conveyor-based schmitt trigger and its application as a realization oscillator. Int J Circ Theor Appl 39:679–686
20. Sanchez-Lopez C, Munoz-Pacheco JM, Carbajal-Gomez VH, Trejo-Guerra R, Ramirez-Soto C, Echeverria-Solis S, and Tlelo-Cuautle E (2011) Design and applications of continuous-time chaos generators. In: Tlelo Cuautle E (ed) Chaotic systems book. Intech, Croatia 10:227–254
21. Monpapassorn A (2013) Low output impedance dual CCII full-wave rectifier. Int J Electron 100:648–654
22. Marcellis AD, Carlo CD, Ferri G, Stornelli V (2013) A CCII-based wide frequency range square waveform generator. Int J Circ Theor Appl 41:1–13
23. Almashary B, Alhokail H (2000) Current-mode triangular wave generator using CCIIs. Microelectron J 31:239–243
24. Senani R, Gupta SS (1998) Implementation of Chua's chaotic circuit using current feedback op-amps. Electron Lett 34:829–830
25. Abuelma'atti MT (2002) New ASK/FSK/PSK/QAM wave generator using a single current-controlled multiple output current conveyor. Int J Electron 89:35–43
26. Bruun E, Haxthausen EU (1991) Current conveyor based EMG amplifier with shutdown control. Electron Lett 27:2172–2174
27. Sinsky JH, Westgate CR (1996) A new approach to designing active MMIC tuning element using second-generation current conveyor. IEEE Microw Guided wave Lett 6:326–328
28. Popovic J, Pavasovic A, Vasiljevic D (1997) Low-power CMOS current-conveyor relaxation oscillators. IEEE Trans Ultra Ferro Freq Cont 44:895–901
29. Sampietro M, Bertuccio G (1998) Zero-power current conveyor for DC stabilization and system reset of fast current pulse amplifiers. Electron Lett 34:1801–1802
30. Vlassis S, Siskos S (2000) An interfacing circuit for piezo-resistive pressure sensors with frequency output. Int J Electron 87:119–127
31. Monpapassorn A (2000) A method for biasing a temperature independent current conveyor precision rectifier. Thammasat Int J Sc Tech 5:10–15
32. Rajput SS, Jamuar SS (2001) CCII based low voltage CMOS current sources for space plasma probes. Measure Sc Tech 12:N39–N43
33. Acharya YB (2005) Current conveyor based electrometer. Int J Electron Commun (AEU) 59:413–416
34. Maitreechit S, Monpapassorn A (2005) A full-wave rectifier using a current conveyor and current mirrors with improved efficiency. Thammasat Int J Sc Tech 10:21–27
35. Gift SJG (2005) New precision rectifier circuits with high accuracy and wide bandwidth. Int J Electron 92:601–617
36. Djukie SR (2008) Full-wave current conveyor precision rectifier. Serbian J Electr Eng 5:263–271
37. Kumngern M, Dejhan K (2008) Current conveyor-based versatile precision rectifier. WSEAS Trans Circ Syst 7:1070–1079
38. Dukic S (2009) Temperature independent current conveyor precision full-wave rectifier for low-level signal. Ser Elec Energ 22:117–123
39. Guerra RT, Cuautle ET, Hernandez C, Lopez CS (2009) Chaotic communication system using Chua's oscillators realized with CCII+ s. Int J Bifurcation Chaos 19:4217
40. Beg P, Khan IA, Maheshwari S (2012) Biphase amplifier based precision rectifiers using current conveyors. Int J Comput Appl 42:14–18
41. Petrovic PB (2013) A new precision peak detector/full-wave rectifier. J Sig Inform Process 4:72–81
42. Misurec J, Koton J (2012) Schmitt trigger with controllable hysteresis using current conveyors. Int J Adv Telecomm Electron Sig Syst 1:26–30
43. Srinivasulu A (2012) Current conveyor based relaxation oscillator with tunable grounded resistor/capacitor. Int J Design Anal Tools Integr Circ Syst 3:1–7
44. Kubanek D, Khateb F, Vrba K (2013) Square wave generator with voltage-controlled frequency based on universal current conveyor. Przeglad Elektrotechniczny R 89:191–194

Part III
Different Variants of Current Conveyors, Their Implementations and Applications

Chapter 9
Second Generation Controlled Current Conveyors (CCCII) and Their Applications

Abstract An interesting method of realizing electronically-controllable functions is through the use of the translinear current conveyor which is implemented from a *mixed translinear cell* (MTC). This implementation exhibits an input resistance looking into terminal X which is electronically-controllable and hence, it is referred to as the Second generation Controlled Current Conveyor (CCCII). The objective of this chapter is to present significant developments on the hardware implementation and applications of electronically-controllable analog circuits using CCCIIs as active elements.

9.1 Introduction

Operational-transconductance-amplifier-capacitor (OTA-C) circuits have been the popular choice for synthesizing electronically-controllable universal voltage mode and current-mode biquads, sinusoidal oscillators as well as relaxation oscillators capable of generating non-sinusoidal signals such as square wave, triangular wave etc.

With the advent of the second generation controlled current conveyors (CCCII) [1–132] which provide electronic control of the X-port input resistance R_x through an external bias current, CCCIIs have attracted lot of attention of researchers and circuit designers around the world as alternative building blocks to synthesize a variety of electronically-controllable signal processing and signal generation circuits. The electronic-controllability in a CCCII emanates from a specific circuit architecture known as a *mixed translinear cell* (MTC) which forms the input cell of a typical CCII architecture. This MTC, apart from the electronic-control, also empowers the CCCIIs with higher slew rate and larger input signal handling capability. The above mentioned features have been the primary reasons for stimulating a lot of research on the use of CCCIIs in the synthesis of a variety of linear as well as non-linear analog signal processing circuits, for instance, see [1, 2] and the references cited therein.

This chapter focuses on the synthesis of various electronically-controllable signal processing/signal generation circuits such as simulated impedances,

© Springer International Publishing Switzerland 2015
R. Senani et al., *Current Conveyors*, DOI 10.1007/978-3-319-08684-2_9

current-controlled grounded and floating resistors, summers, subtractors, instrumentation amplifiers, universal CM/VM/mixed-mode biquad filters, sinusoidal signal generators, precision rectifiers, frequency doubler, multipliers/dividers, squarers and square rooters etc. Significant developments taken place on the design of CCCIIs in bipolar/CMOS/BiCMOS technologies have also been elaborated.

9.2 Bipolar/CMOS/BiCMOS CCCIIs

In this section, we take a look into the various bipolar/CMOS and BiCMOS architectures of the CCCIIs evolved in literature. In order to simplify the circuit diagram representations which employ a number of bipolar/MOS current mirrors (CM) and current repeaters (CR), we would be using the symbolic notations of the CMs and CRs as given in Fig. 9.1.

CCCII is the second generation CC in which the finite non-zero input resistance at port-X (R_x) is taken into consideration. The basic translinear CCCII circuit proposed by Fabre et al. [1, 2] employs a mixed-translinear-cell (MTC) comprising of transistors Q_1-Q_2-Q_3-Q_4 and is shown in Fig. 9.2.

A straight forward analysis of the circuit of Fig. 9.2 shows that it is characterized by:

$$i_y = 0 \tag{9.1}$$

$$\frac{\nu_y - \nu_x}{V_T} = \text{Sin}\, h^{-1}\left(\frac{-i_x}{2I_B}\right) \cong -\frac{i_x}{2I_B}, \text{for}\, i_x << 2I_B \tag{9.2}$$

$$\text{or}\, \nu_x \approx \nu_y + i_x R_x\, \text{where}\, R_x = \frac{V_T}{2I_B}\, \text{and}\, i_z = i_x \tag{9.3}$$

From equation (9.3), it follows that R_x is *electronically-tunable* through the external DC bias current I_B. With $\pm V = \pm 2.5$ V, this circuit exhibits a voltage gain of 0.9984, a current gain of 1.022 with a 3-dB bandwidth of about 615 MHz [2]. The symbolic notation and simplified equivalent circuit of a CCCII are shown in Fig. 9.3.

In the following sections, we demonstrate how a CCCII+ or CCCII− or a combination of both can be used to realize a number of useful electronically-controllable linear as well as non-linear circuits. It may be mentioned that from the circuit of Fig. 9.2, a CCCII− can be obtained by breaking the link at the junction of Z-terminal and adding one more pair of current mirrors in a cross-coupled manner.

A CCCII can also be realized by CMOS architecture analogous to that of Fig. 9.2 as shown Fig. 9.4 which was introduced by Chaisricharoen et al. [3].

The basic translinear cell consisting of M_{n1}, M_{n2}, M_{p5}, M_{p6} in the above CMOS implementation is analogous to the bipolar MTC. This cell is redrawn as a core of

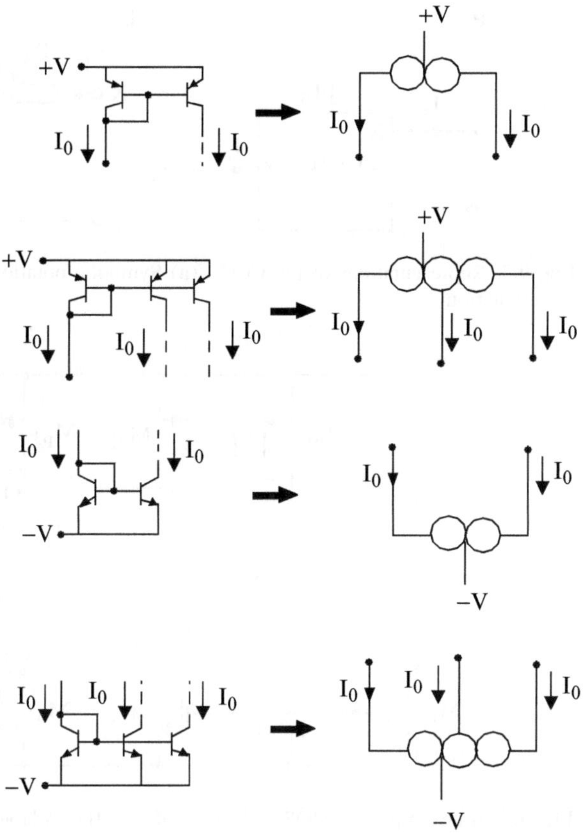

Fig. 9.1 Symbolic notations of the current mirrors and current repeaters

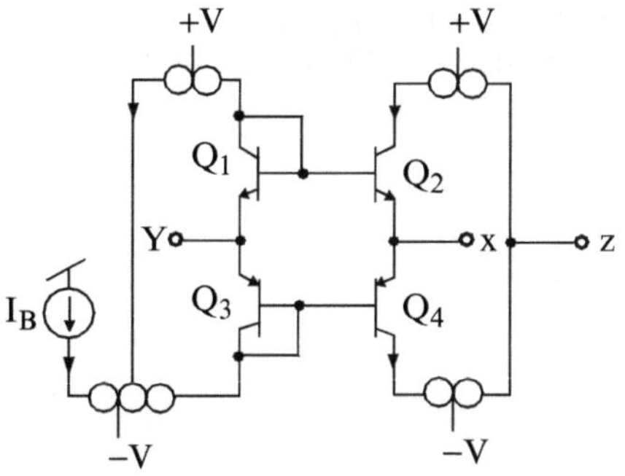

Fig. 9.2 Bipolar circuit architecture of the CCCII+ [1, 2]

translinear CMOS CCCII and is shown in Fig. 9.5. To get a linear relationship between V_X, V_Y and i_x, the following condition should be satisfied:

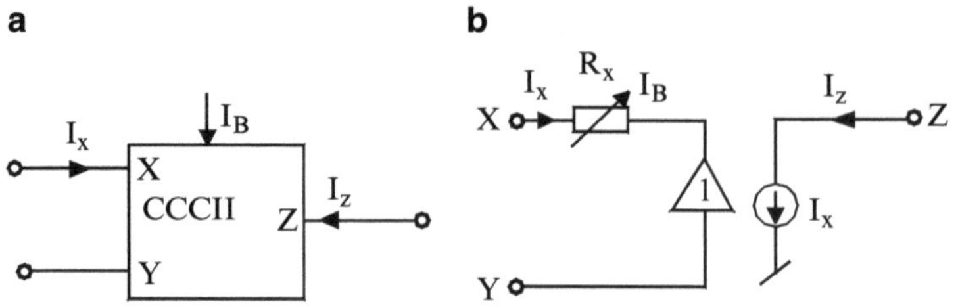

Fig. 9.3 Representations of the CCCII (**a**) Symbolic notation (**b**) Simplified equivalent circuit representation

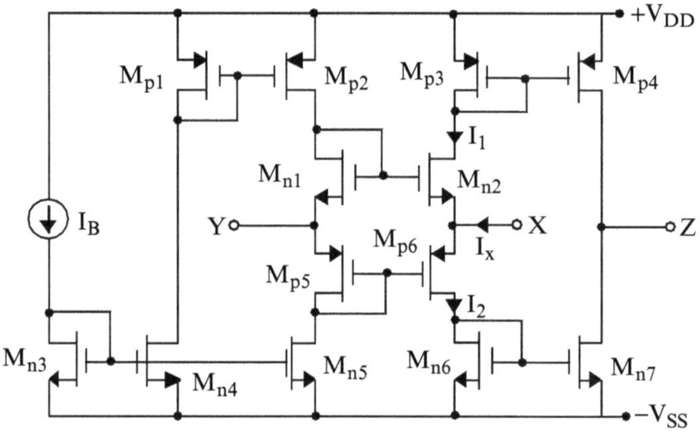

Fig. 9.4 An exemplary CMOS realization of CCCII+ (Adapted from [3] © 2010 Elsevier)

Fig. 9.5 Core of translinear CMOS CCCII (Adapted from [3] © 2010 Elsevier)

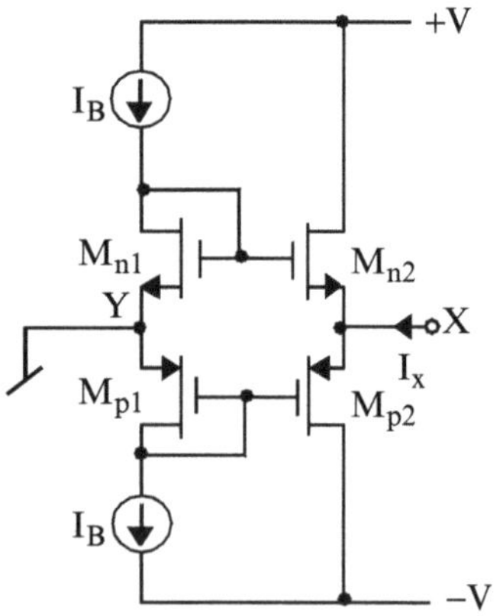

$$\frac{\mu_p W_p}{L_p} \cong \frac{\mu_n W_n}{L_n} \tag{9.4}$$

An analysis of the circuit of Fig. 9.5 gives the characterizing equations of the circuit as:

$$V_X \cong V_Y + \frac{i_X}{\sqrt{2I_B C_{Ox}}\left(\sqrt{\frac{\mu_p W_p}{L_p}} + \sqrt{\frac{\mu_n W_n}{L_n}}\right)} \tag{9.5}$$

$$R_x \cong \frac{1}{\sqrt{2I_B C_{Ox}}\left(\sqrt{\frac{\mu_p W_p}{L_p}} + \sqrt{\frac{\mu_n W_n}{L_n}}\right)} \tag{9.6}$$

where Rx is the transresistance at port-X.

Thus, in this case also, R_x is controllable by DC-bias current I_B. For further detailed analysis of this circuit, the reader is referred to [3]. Several other interesting CMOS-based CCCII configurations have also been advanced, for instance, see [4–7].

A 2GHz Bi-CMOS CCCII in standard 0.8 μm BiCMOS technology was proposed by Seguin and Fabre in [8] which can be equivalently shown as in the circuit of Fig. 9.6.

In this circuit, the signal processing portion is made of only NPN transistor Q_1-Q_6 and the DC biasing circuit is composed of a P-MOS repeater and N-MOS

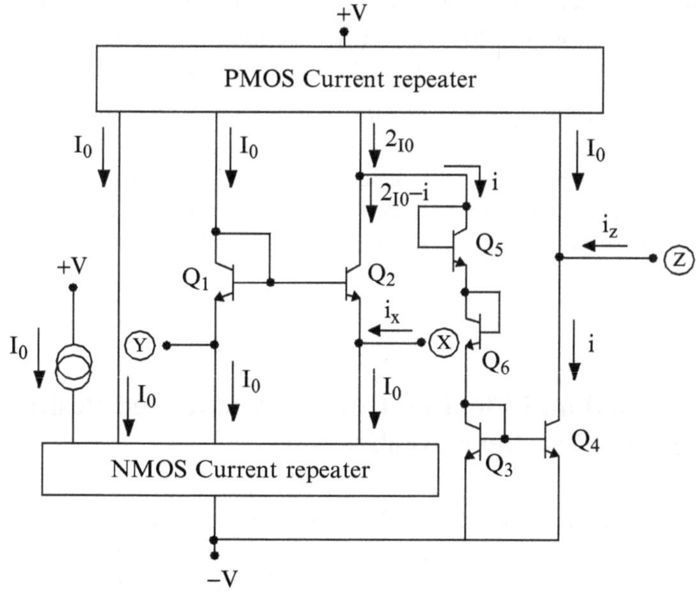

Fig. 9.6 Bi-CMOS CCCII architecture [8]

current repeater such that, as shown in the diagram, all the bias currents are equal to I_o except one which has been arranged to be $2I_0$. From the various currents marked on the diagram, it can be easily worked out that $I_y = 0$, $V_x = V_y + R_x i_x$ where $R_x = V_T/I_0$ and that $i_z = i_x$.

This circuit was implemented in 0.8 μm Bi-CMOS technology which incorporates vertical NPN and lateral PNP transistors with transition frequencies (f_T) of the order of 11 GHz and 500 MHz, respectively. For the N-MOS transistors W/L was taken as 40 μm/3 μm whereas for all the P-MOS transistors the aspect ratio was 20 μm/1.1 μm except for one it was required to supply a current of $2I_0$ in which case the aspect ratio was 40 μm/1.1 μm. It was found that with these values, the CCCII when biased by ± 2.2 V supply and $I_0 = 500$ μA, was characterized by $R_x = 80\ \Omega$ and 3 dB bandwidth of the current gain and voltage gain was more than 2 GHz.

The significant advantages accrued in circuit design employing CCCIIs can be summarized as follows:

(i) All the functions can be realized by using only CCCIIs and capacitors; no external resistors are needed. Due to the complete absence of the resistors and availability of electronic-controllability of the parameters of interest, CCCII-based circuits are highly suitable for IC implementation.
(ii) The parameters of interest of the realized functional circuits can be controlled by varying external DC bias currents.
(iii) CCCII-based circuits can be operated from low power supply voltages, typically ±2.5 V DC, in contrast to op-amp circuits which typically require ±6V to ± 15 V DC power supplies.
(iv) Bipolar CCCIIs based upon the four-transistor mixed translinear cell have much higher slew rate typically, several hundred V/μs as compared to op-amps (such as uA741) having a very modest 0.5 V/μs. Thus, CCCII-based circuits can be operated at relatively much higher frequencies.

Although a majority of the applications described in the following deal with bipolar CCCIIs, however, there are a number of applications which are based upon CMOS CCCIIs in which case the dependence of the various parameters on the external bias currents is different. It may be noted that R_x is inversely proportional to I_B in case of the bipolar CCCII whereas it is inversely proportional to the square root of I_B in case of CMOS CCCIIs.

9.3 Grounded and Floating Current-Controlled Positive/ Negative Resistance Realization

If terminal-Y is grounded and terminal-Z is also grounded as shown in Fig. 9.7, the circuit realizes a current-controlled resistance. Looking into terminal-X, the equivalent input impedance is then found to be $R_{in} = \frac{V_T}{2I_B}$.As an example, if V_T is taken as

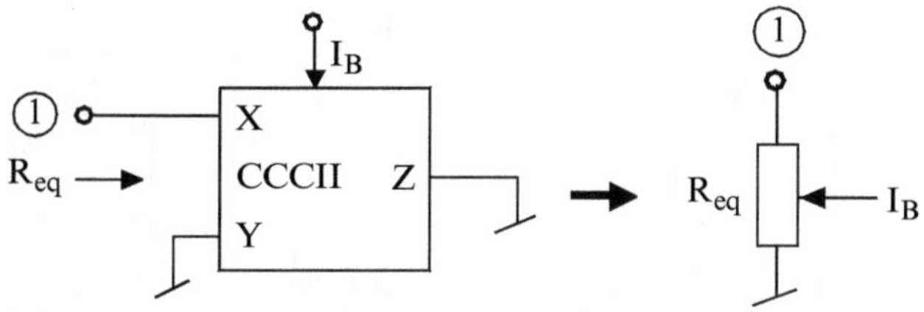

Fig. 9.7 Current-controlled grounded resistance simulation using CCCII+

25mV and I_B is changed from 0.1 μA to 1 mA, the equivalent resistance will change from 12.5 Ω to 125 KΩ.

Senani et al. in [9] presented a circuit for realizing a floating current controlled positive resistance (FCCPR) which simulates a floating resistance of value $R_{1-2} = V_T/2I_B$ between ports 1 and 2. A circuit analogous to the circuit of [9], as the floating counterpart of the one shown in Fig. 9.7 is shown in Fig. 9.8a which realizes a FCCPR. A slightly modified (new) form is shown in Fig. 9.8b which, however, realizes a FCCPR of value $V_T/4I_B$.

A grounded current-controlled negative resistance can be realized from the CCCII+ [2] by shorting terminals Y and Z, grounding terminal-X and then looking into terminal-Y as shown in Fig. 9.9. In this case, the realized resistance is given by $R_{in} = -\frac{V_T}{2I_B}$.

The above circuit can be converted into a two ports realizing a floating current controlled negative resistance (FCCNR) and FCCPR as shown in Fig. 9.10. While the circuit of Fig. 9.10a realizes FCCNR of value R1–2 = −VT/IB that of Fig. 9.10b, due to the cross coupling at the Z outputs, realizes the same value of FCCPR.

A new design for electronically-tunable grounded/floating resistors was proposed by Pawarangkoon and Kiranon in [10] based upon a multiple output CCCII characterized by the following equations: Iy = 0, Vx = Vy + RxIx, Iz1 = −Ix, Iz2 = KIx and Iz3 = −KIx where K was equal to ratio of two bias currents i.e K = I1/I2. This modified CCCII structure is shown in Fig. 9.11.

This modified CCCII can be configured either as a floating resistance or as a grounded resistor as shown in Fig. 9.12

In either case, the equivalent resistance realized is given by:

$$R_{eq} = \frac{V_T}{2I_0}\left(\frac{1}{1 - (I_1/I_2)}\right) \tag{9.7}$$

From the above expression, it is seen that by appropriate selection of the external DC bias currents I1 and I2, the realized Req can be made positive or negative and that after this setting, the value of the realized resistance can be continuously tuned by varying the DC bias current I0. The workability of this circuit has been confirmed by realizing the entire circuit with bipolar transistors NR100N and PR100N.

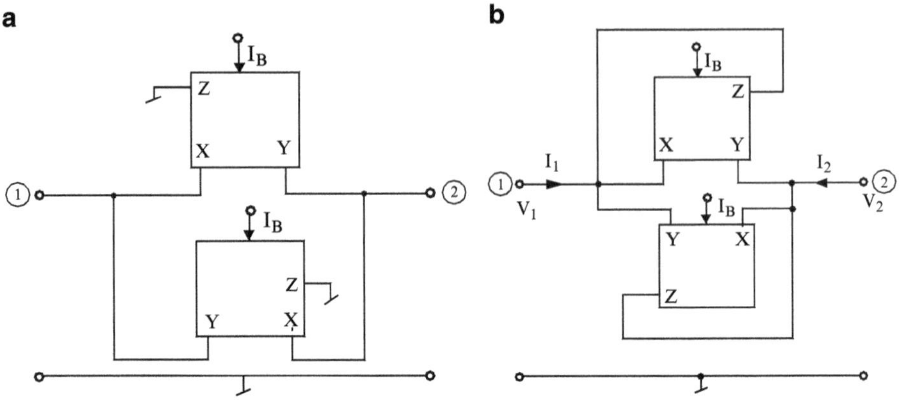

Fig. 9.8 Floating current controlled positive resistance (FCCPR) configurations

Fig. 9.9 A negative current-controllable grounded resistance

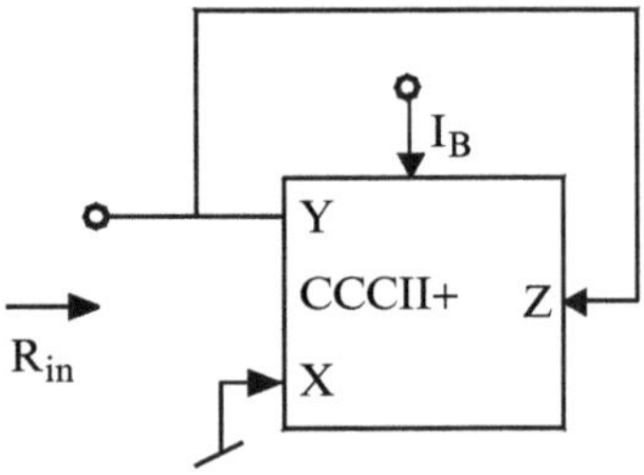

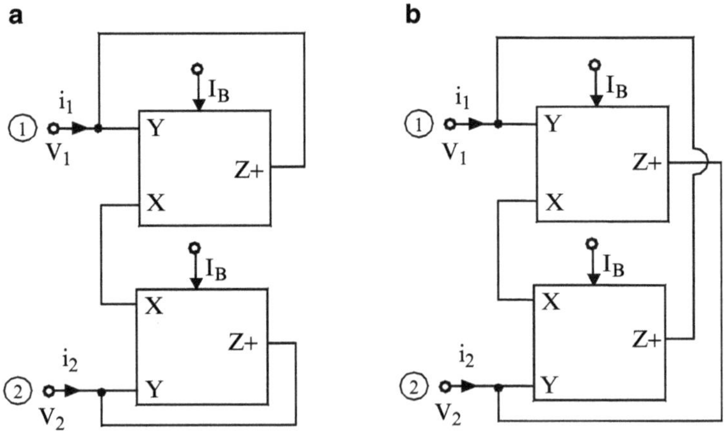

Fig. 9.10 FCCNR and FCCPR derived from the circuit of Fig 9.9

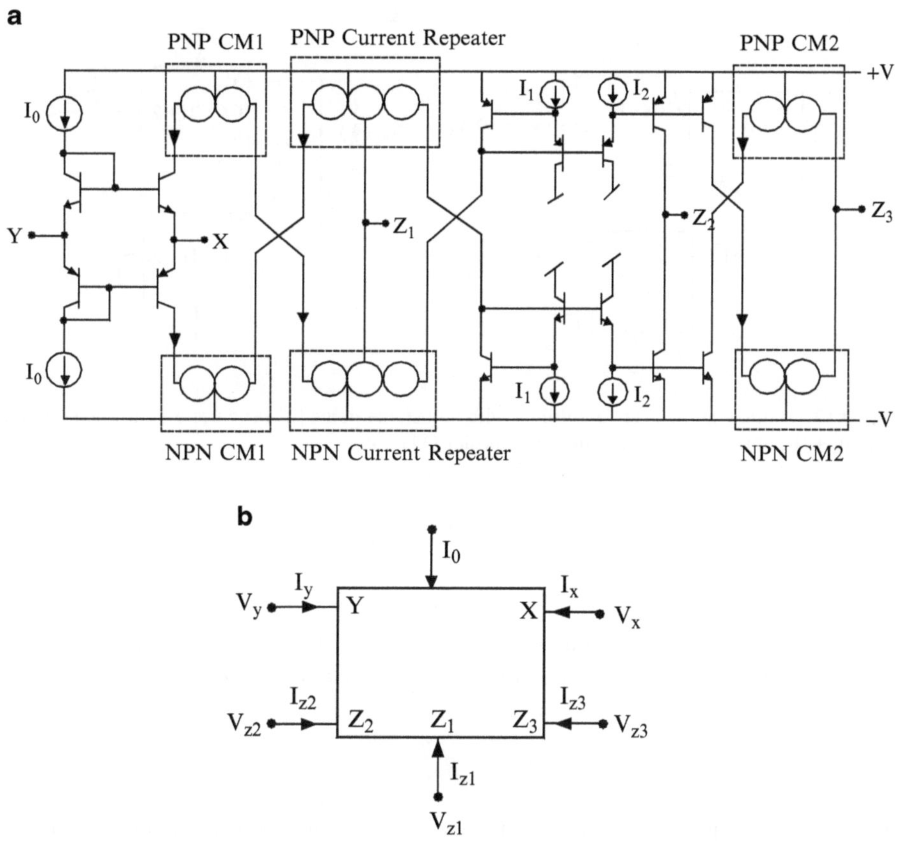

Fig. 9.11 (a) A multi-output CCCII (MO-CCCII) with controlled current gains (b) Symbolic notation of MO-CCCII [10]

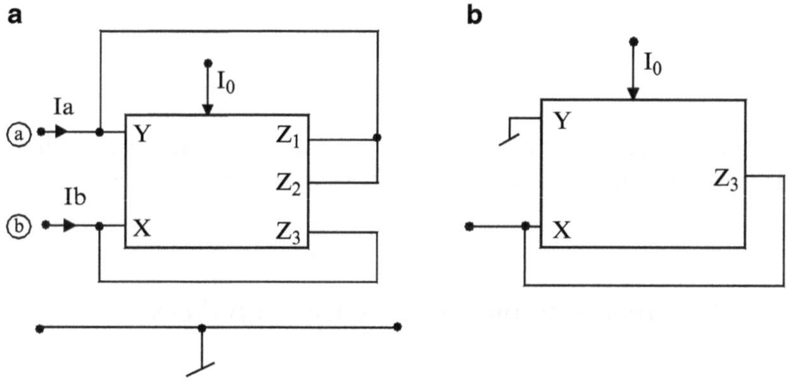

Fig. 9.12 Electronically-tunable resistors [10] (a) floating resistor (b) grounded resistor

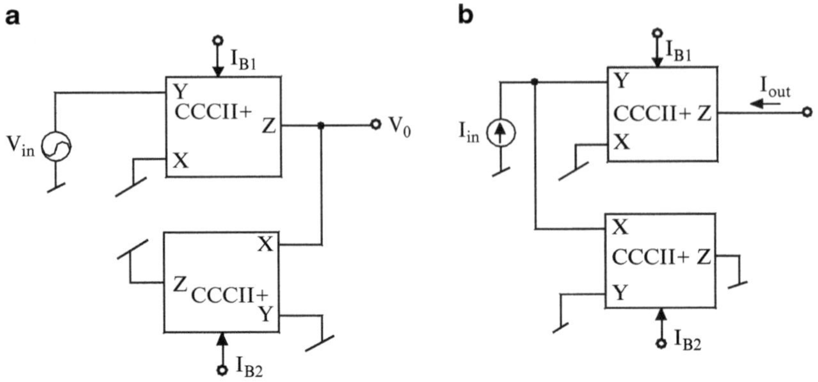

Fig. 9.13 Voltage and current amplifiers with adjustable gains [2] (**a**) current-controlled voltage amplifier (**b**) current-controlled current amplifier [11]

9.4 Current Controlled VM/CM Amplifiers

The minimum number of resistors needed to make a VCVS is two. It is apparent that with R_x of CCCII accounted for; only two CCCIIs should suffice for this purpose. One such circuit is shown in Fig. 9.13a which is characterized by:

$$\frac{V_{out}}{V_{in}} = \frac{R_{x2}}{R_{x1}} = \frac{I_{B1}}{I_{B2}} \tag{9.8}$$

Thus, the voltage gain is electronically-controllable by I_{B1} and/or I_{B2} and is temperature-insensitive since the parameter V_T is cancelled out!

A current amplifier with current gain equal to the ratio of the two external bias currents can also be realized by appropriate connections of only two CCCIIs as shown in Fig. 9.13b. In this case, the current gain can be expressed as:

$$\frac{I_{out}}{I_{in}} = -\frac{R_{x1}}{R_{x2}} = -\frac{I_{B1}}{I_{B2}} \tag{9.9}$$

It may be noted that this circuit also offers the desirable properties of low input impedance, high output impedance and temperature-insensitive gain.

9.5 Active-Only Summing/Difference Amplifiers

CCCIIs make it possible to realize sum and difference circuits using only active devices without requiring even a single resistor [12]. One such circuit is shown in Fig. 9.14. The output voltage of this circuit is given by:

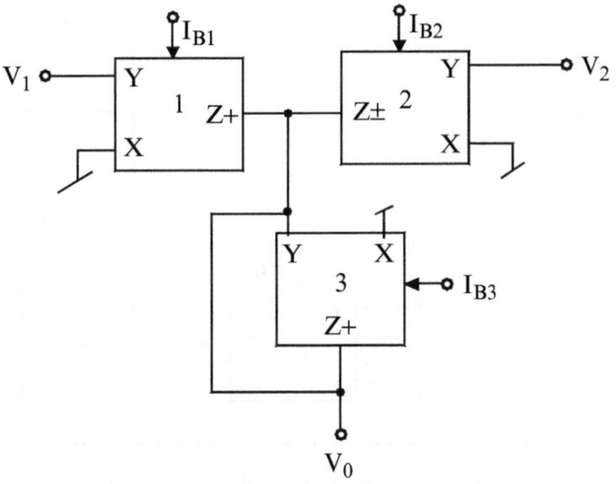

Fig. 9.14 Sum and difference amplifiers proposed by Maheshwari [12]

$$V_o = -R_{x3}\left(\frac{V_1}{R_{x1}} \pm \frac{V_2}{R_{x2}}\right) \tag{9.10}$$

From the above equation it is clear that with the choice for CCCII as *positive*, a summing amplifier is realized and with the choice for CCCII as *negative*, a difference amplifier is realized.

9.6 Instrumentation Amplifiers

A traditional instrumentation amplifier typically requires as many as three op-amps along with seven resistors out of which, at least four are required to be identical. This circuit although provides a single-resistance-controllable variable gain with infinite input impedance at both the input terminals and zero output impedance at the output terminal, but suffers from the conflict between the maximum gain and the available bandwidth. This *gain-bandwidth-conflict* can be avoided by using either a three-CCII-based circuit or a three current feedback op-amp (CFOA) based circuit. Both these circuits use bare minimum of (only two) resistors for varying the voltage gain. Using CCCIIs however, an instrumentation amplifier can be made with only three active building blocks *without requiring any external resistors*. In addition, the voltage gain can be controlled through the external DC bias currents. One such circuit proposed by Maheshwari [13] is shown in Fig. 9.15 which is characterized by:

$$V_0 = \frac{2R_{x3}}{(R_{x1} + R_{x2})}(V_1 - V_2) = \frac{2I_{B1}I_{B2}}{I_{B3}(I_{B1} + I_{B2})}(V_1 - V_2) \tag{9.11}$$

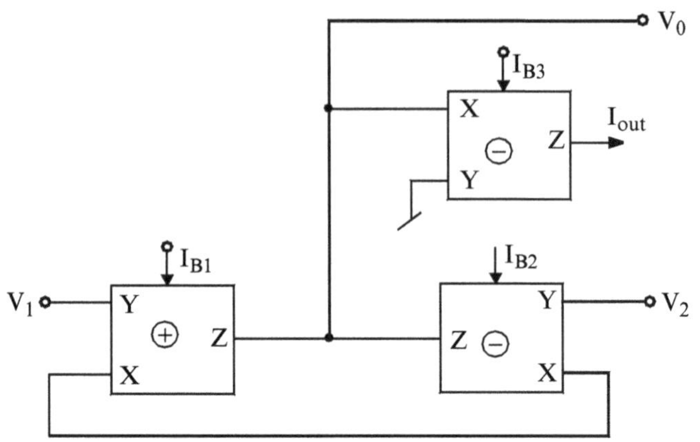

Fig. 9.15 Active-only instrumentation amplifier with electronically-controllable voltage gain [13]

It may be observed that the gain of this instrumentation amplifier is also temperature-insensitive since V_T is cancelled out.

An instrumentation amplifier realizable with only two CCCIIs was proposed by Ercan et al. in [14] which is shown in Fig. 9.16. A routine analysis of this circuit reveals that the differential mode gain for this circuit is given by:

$$A_{dm} = \frac{R_A}{2R_x}\left(\frac{1}{(1 + sC_bR_A)}\right) \tag{9.12}$$

where the equivalent resistance (R_A) of the active resistor realized by the two MOSFETs is given by:

$$R_A = \frac{1}{\mu_n C_{ox}(W/L)(V_C - 2V_{TH})} \tag{9.13}$$

and C_b is the output node capacitance

$$C_b = C_A + C_{Z1} \tag{9.14}$$

If the non-ideal current gain and voltage gain of the CCCIIs are taken as α and K respectively, the common-mode gain is given by:

$$A_{cm} = \frac{v_0}{v_{cm}} = \frac{\alpha(K_1 - K_2)R_A}{2R_x(1 + sC_bR_A)} \tag{9.15}$$

From the equation (9.15), it is clear that for nominal value of α and K_1, K_2, $A_{cm} \to 0$ and consequently, CMRR also $\to \infty$. Also, the bandwidth can be fixed by fixing R_A while the gain can be electronically-controlled through R_x.

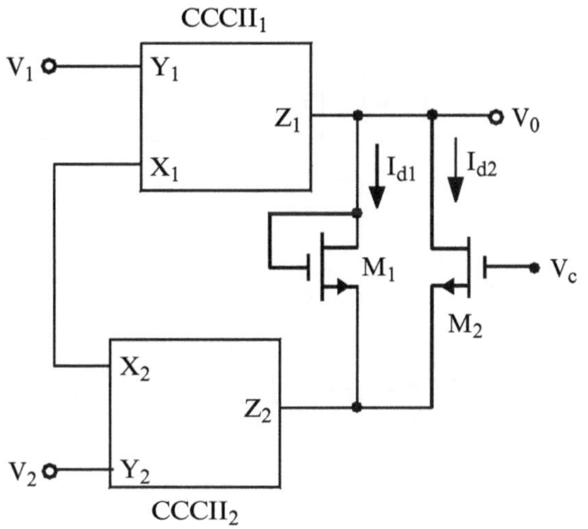

Fig. 9.16 Instrumentation amplifier proposed by Ercan et al. [14]

It may also be seen that while the voltage gain of the circuit is controllable by $R_A/2R_x$ and therefore, it can be controlled by the external DC bias current I_B of the two CCCIIs and/or by the DC control voltage V_c of the MOSFET M_2.

The SPICE simulation results of the circuit, using the parameters of 0.35 μm TSMC CMOS technology, with CCCII biased from ±3.3 V DC power supplies have exhibited a CMRR as high as 142 dB, a bandwidth of 70 MHz for a gain of 45 with total power dissipation as low as 519 μW. This instrumentation amplifier appears to be particularly useful as a part of many sensitive bio-medical applications that specifically require a high CMRR.

9.7 Electronically-Tunable Grounded/Floating Synthetic Impedances and Related Circuits

Since electronically-controlled resistances with both positive and negative values, in grounded as well as floating forms, are realizable by CCCIIs, it is natural to think about realizing various other kinds of impedances with the feature of electronic-tunability using CCCIIs. In this section, we include some prominent works done in this direction.

Kiranon and Pawarangkoon in [15] presented a four CCCII+ based floating inductance simulation circuit which employs no more than two resistors and a grounded capacitor to simulate a lossless floating inductance of value $L_{eq} = C\,R_1R_2$. Since both the resistors were connected between X-terminals of the two CCCII+ s, they can be easily replaced and in their place the intrinsic input resistances R_x of the CCCIIs could be considered. Replacing one resistor at a time Kiranon and

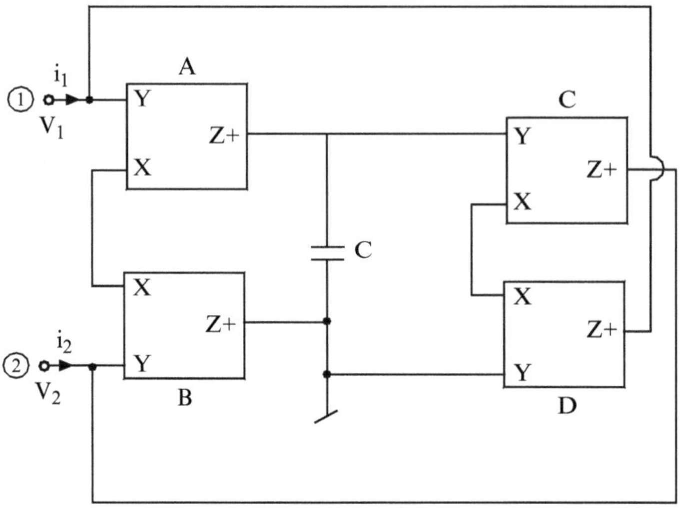

Fig. 9.17 Electronically-controlled lossless floating inductance derived from the circuit of Kiranon and Pawarangkoon [15]

Pawarangkoon [15] presented two modified FI simulation circuits realizing a current controlled floating inductance of values $L_a = \frac{CR_1V_T}{2I_0}, L_b = \frac{CR_2V_T}{2I_0}$ respectively. It may, however, be noted that there is no restriction on removing both the external resistors R_1 and R_2 and considering in their place the intrinsic parasitic resistances of the CCCIIs. When this is done this result in the circuit shown in Fig. 9.17

This version of the circuit simulates a floating inductance between ports 1 and 2 having value

$$L_{1-2} = \left(\frac{V_T^2}{4I_0^2}\right)C \tag{9.16}$$

If dual outputs CCCIIs are permitted, Yuce et al. [16] demonstrated that only three CCCIIs are sufficient to realize a lossless floating inductance along with a grounded capacitor for both positive inductor and negative inductor. Two such circuits are shown in Fig. 9.18a, b.

For the circuit Fig. 9.18a, a straight forward analysis shows that its Y-matrix is given by:

$$\begin{bmatrix} I_1 \\ I_2 \end{bmatrix} = \frac{1}{sCR_{x1}(R_{x2}+R_{x3})}\begin{bmatrix} 1 & -1 \\ -1 & 1 \end{bmatrix}\begin{bmatrix} V_1 \\ V_2 \end{bmatrix} \tag{9.17}$$

Therefore, the circuit simulates a positive floating inductance of value $L_{eq} = C\,R_{x1}$ $(R_{x2} + R_{x3})$.

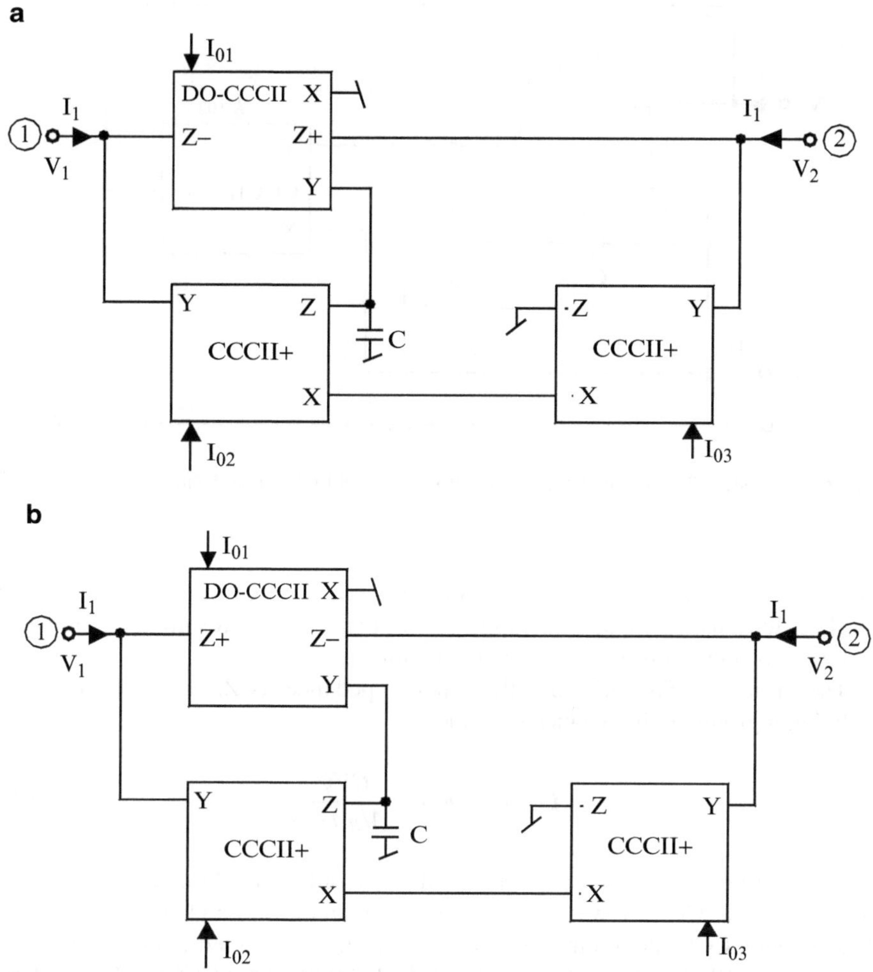

Fig. 9.18 Floating inductance simulators [16] (**a**) Positive floating inductor (**b**) Negative floating inductor

On the other hand, the Y-matrix for the circuit of Fig. 9.18b is of the same form as the above but with a negative sign thereby representing a negative floating inductance of value $L_{eq} = -C\, R_{x1}(R_{x2} + R_{x3})$.

Yet another circuit requiring only thee active elements to simulate a lossless floating inductance was presented by Khan and Zaidi [17] which employs two CCCIIs and one voltage buffer as shown in Fig 9.19.

By straight forward analysis, the Y- matrix of this circuit is found to be:

$$[Y] = \frac{1}{sCR_{x1}R_{x2}}\begin{bmatrix} 1 & -1 \\ -1 & 1 \end{bmatrix};\ \text{where}\, R_{xi} = \frac{V_T}{2I_{0i}},\ \text{i} = 1, 2 \tag{9.18}$$

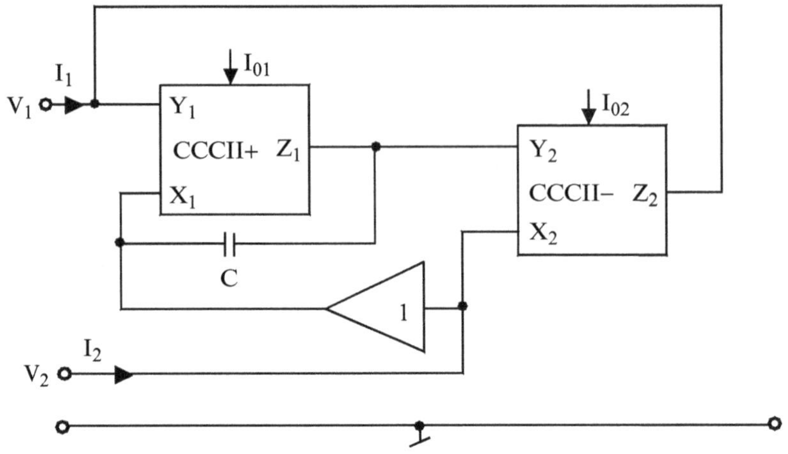

Fig. 9.19 Lossless floating inductance simulator proposed by Khan and Zaidi [17]

In the Sedra and Smith's two CCII-based gyrator, if CCCII± are used instead of CCII±, along with a capacitor C, as shown in Fig. 9.20, the resulting circuit would realize a current-controlled grounded inductor.

The circuit of Fig. 9.20 has the input impedance as $Z_{in} = sCR_{x1}R_{x2}$ thereby realizing a grounded inductance of value:

$$L = CR_{x1}R_{x2} = \frac{CV_T^2}{4I_{B1}I_{B2}} \tag{9.19}$$

That a floating inductance can be realized using only a single grounded capacitor along with four CCCII+ was shown by Kiranon and Pawarangkoon in [15]. Other dual output CCCII-based circuits were subsequently presented in [18] and [19]. A circuit from [18] which employs only two dual outputs (DO) CCCIIs (DO-CCCIIs) to realize lossless floating inductance is shown in Fig. 9.21.

The circuit is described by the following equations:

$$I_1 = -I_2 = sCR_{x1}R_{x2}(V_1 - V_2) \tag{9.20}$$

so that the inductance value realized between ports 1 and 2, is given by:

$$L = CR_{x1}R_{x2} = \frac{CV_T^2}{4I_{B1}I_{B2}} \tag{9.21}$$

It may be noted that equivalent L is controllable by any or both of the two external DC bias currents I_{B1} and I_{B2}.

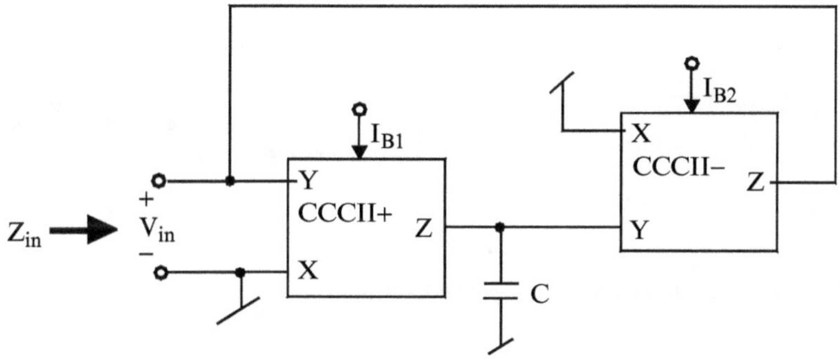

Fig. 9.20 Electronically-tunable grounded inductance

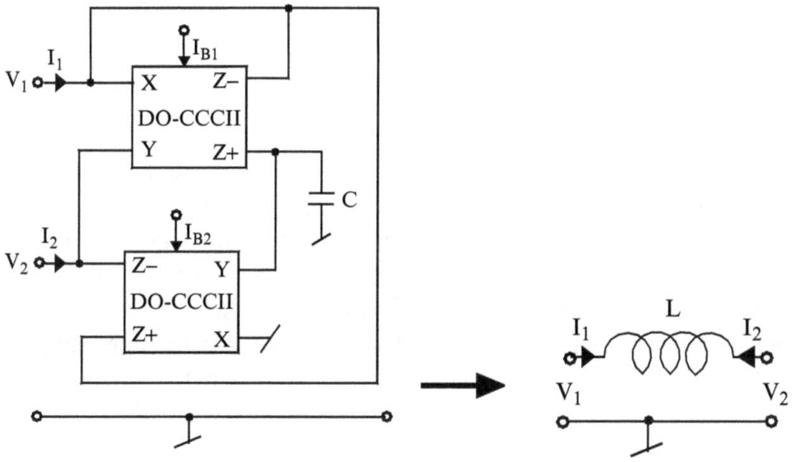

Fig. 9.21 Current-controlled floating inductance proposed by Sedef et al. [18]

A negative floating inductor can be realized from the above configuration (though not so recognized in [18]) by exchanging the Z+ and Z− terminals of second DO-CCCII as shown in Fig. 9.22 in which case the equivalent floating inductance is given by:

$$L = -C\,R_{x1}(R_{x2} + R_{x3}) \tag{9.22}$$

We now present two more circuits proposed by Abuelma'atti and Tasadduq [20], one of which simulates a floating FDNR and the other one a floating synthetic

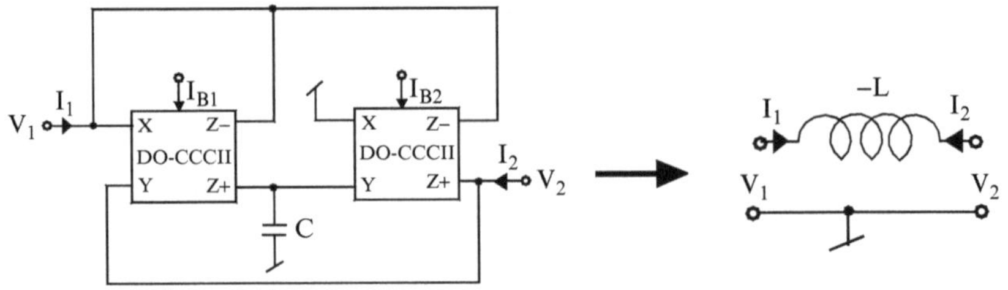

Fig. 9.22 Current-controlled floating negative inductance

capacitance. By a straight forward analysis, the circuit of Fig. 9.23a has the input impedance given by:

$$Z_{in} = R_{x1} + R_{x2} + \frac{Z_1 Z_2}{R_{x3}} \tag{9.23}$$

From this expression, it is seen that with $Z_1 = 1/sC_1$ and $Z_2 = R_2$, the circuit can realize a resistance in series with a grounded capacitor whose value is electronically-controllable by the external current I_{B3}. On the other hand, with Z_1 and Z_2 both taken as capacitors, the circuit would realize a resistance in series with a grounded FDNR whose value would be controllable by I_{B3}.

A floating capacitance multiplier employing grounded capacitor is shown in Fig. 9.23b. This circuit simulates a floating capacitance in series with a resistance with parameter values as $C_{eq} = C_1 \frac{R_{x4}}{R_{x3}}$ and $R_{eq} = (R_{x1} + R_{x2})$.

Synthetic transformer may be useful element for constructing stagger-tuned filters. Compared to passive transformers, an active transformer provides the advantages of integratability, easy adjustment of the coupling co-efficient and elimination of any magnetic interference (because they do not employ any inductive component). A circuit using CCCIIs, to simulate a synthetic transformer, derived from a circuit proposed by Yuce and Minaei [21], is shown in Figs. 9.24 and 9.27

This circuit is equivalent to an inductive-T network with two self-inductances L_1 and L_2 and mutual inductance M such that the two ports can be modeled by the following matrix equation:

$$\begin{bmatrix} V_1 \\ V_2 \end{bmatrix} = \begin{bmatrix} sL_p & sM_{12} \\ sM_{21} & sL_s \end{bmatrix} \begin{bmatrix} I_1 \\ I_2 \end{bmatrix} \tag{9.24}$$

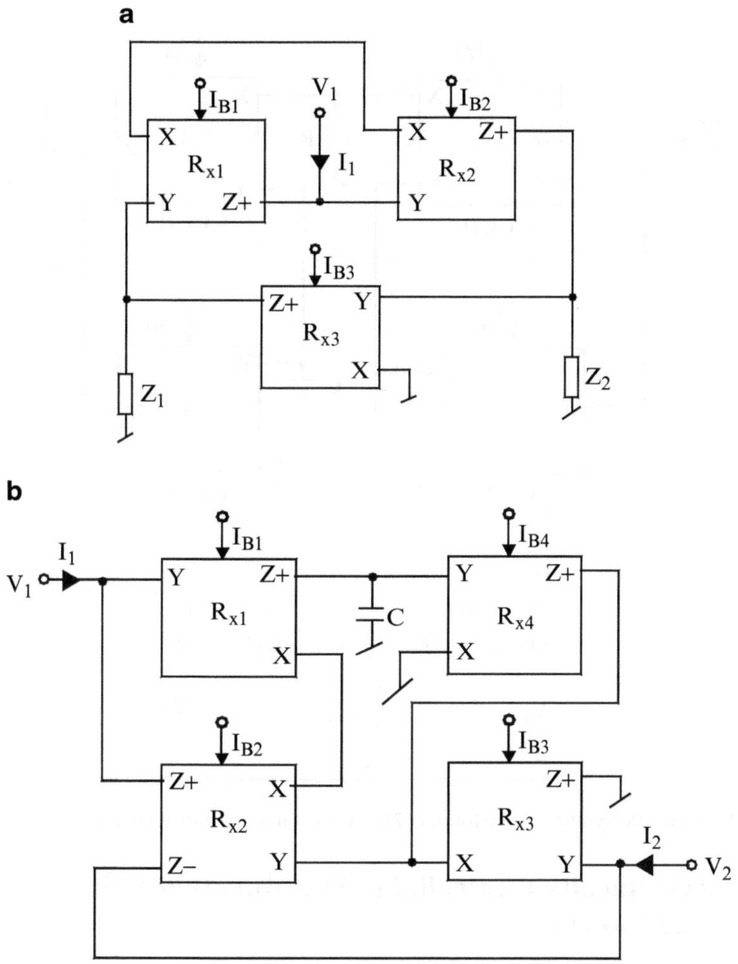

Fig. 9.23 Electronically-controllable Resistance, FDNR and capacitance simulators (Adapted from [20] ©1999 Elsevier) (**a**) Circuit for floating FDNR simulation (**b**) Circuit for floating capacitance multiplier

where $L_p = L_1 + M_{11}$ and $L_s = L_2 + M_{22}$. The various equivalent inductances are given by:

$$L_1 = R_{x1}R_{x2}C_1; L_2 = R_{x3}R_{x4}C_2; M_{11} = M_{12} = R_{x2}R_5C_1 \text{ and } M_{21} = M_{22} = R_{x4}R_5C_2 \tag{9.25}$$

It may be noted that the only a single resistor R_5 is used in the circuit which can be replaced by another CCCII+ configured as a current-controlled resistor.

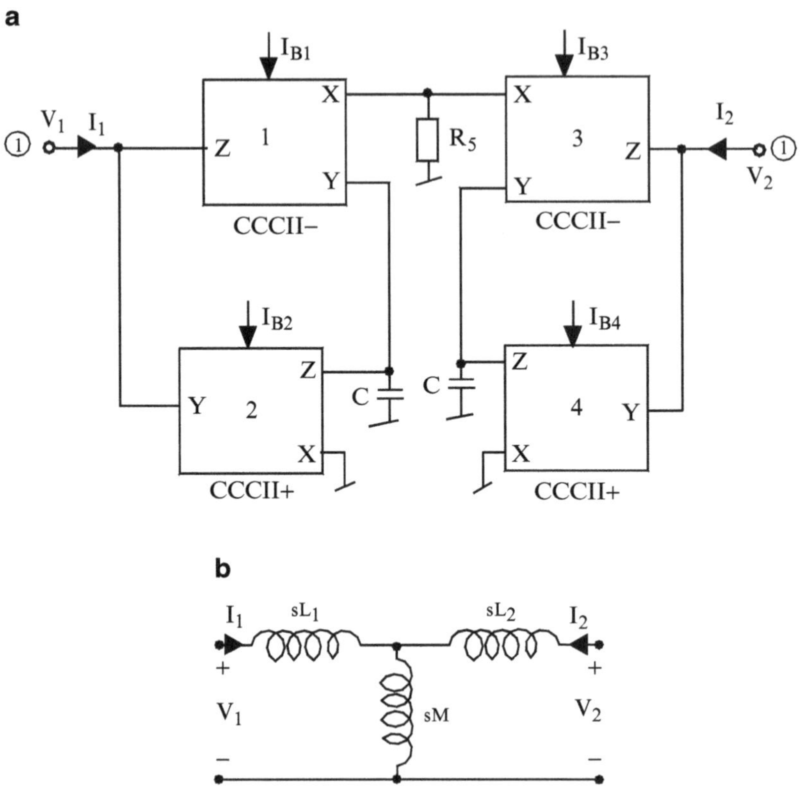

Fig. 9.24 A tunable synthetic transformer [21] (**a**) the circuit (**b**) its equivalent

9.8 Electronically-Controllable Multifunction Voltage Mode Biquad

Whereas a majority of works have dealt with CM biquads, a number of voltage-mode biquads using CCCIIs have also been advanced by various researchers. We now show a very compact circuit, realizable with only two CCCIIs along with a minimum number of capacitors, as proposed by Parveen et al. [22]. This circuit is shown in Fig. 9.25 and can realize all the five standard filter functions by taking different conditions/relationship between the various inputs applied to the circuit as shown in Fig. 9.25.

An analysis of this circuit reveals that the output voltage of this circuit is given by:

$$V_0 = \frac{s^2 V_4 + s\left(\frac{1}{R_{x2}C_2}V_3 - \frac{1}{R_{x2}C_2}V_2 + \frac{1}{C_2R_{x1}}V_1\right) + \frac{V_1}{R_{x1}R_{x2}C_1C_2}}{s^2 + s\frac{1}{C_2R_{x1}} + \frac{1}{R_{x1}R_{x2}C_1C_2}} \tag{9.26}$$

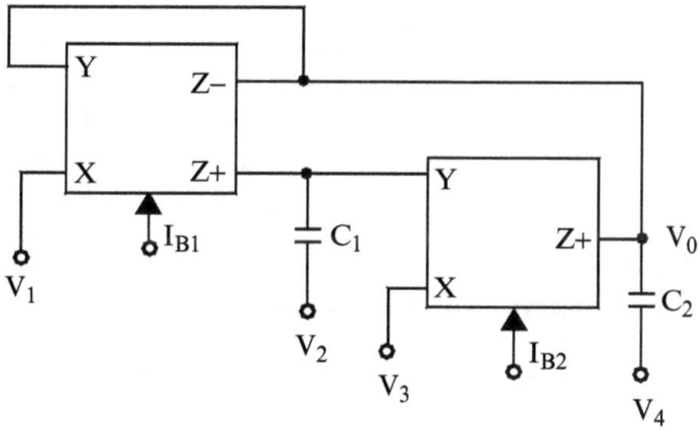

Fig. 9.25 Universal voltage mode biquad filter proposed by Parveen et al. [22]

The various filter responses can be determined through the following selection of the inputs:

Type of filter	Condition of realization
HPF	$V_4 = V_{in}$, $V_1 = V_2 = V_3 = 0$
Non-inverting HPF	$V_3 = V_{in}$, $V_1 = V_2 = V_4 = 0$
Inverting BPF	$V_2 = V_{in}$, $V_1 = V_3 = V_4 = 0$
LPF	$V_1 = V_2 = V_{in}$, $V_3 = V_4 = 0$, and $R_{x1} = R_{x2}$
Band stop filter (BSF)	$V_1 = V_2 = V_4 = V_{in}$, $V_3 = 0$, and $R_{x1} = R_{x2}$
APF	$V_1 = V_2 = V_4 ,= V_{in}$, $V_3 = 0$, and $R_{x1} = 2R_{x2}$

In case of HPF, and BPF responses, one does not require any matching constraints. On the other hand, the constraints in LPF, BSF and APF are simple to satisfy through design, particularly in monolithic implementations, where matching of the devices is not difficult to achieve. The pole frequency ω_o and the pole-Q_0 of the filter are given by:

$$\omega_o = \frac{1}{\sqrt{R_{x1}R_{x2}C_1C_2}},\ Q_0 = \sqrt{\frac{R_{x1}}{R_{x2}}.\frac{C_1}{C_2}} \tag{9.27}$$

Universal VM biquad of Fig. 9.25, although realizes all the five filtering functions, however, in case of LPF and BSF, it requires a matching condition Rx1 = Rx2 and in case of APF, it requires Rx1 = 2Rx2.

An alternative circuit which although employs same number of CCCIIs and only two equal-valued capacitors but does not require such conditions for realizing various functions was presented by Ranjan and Paul in [23] and is shown here in Fig. 9.26.

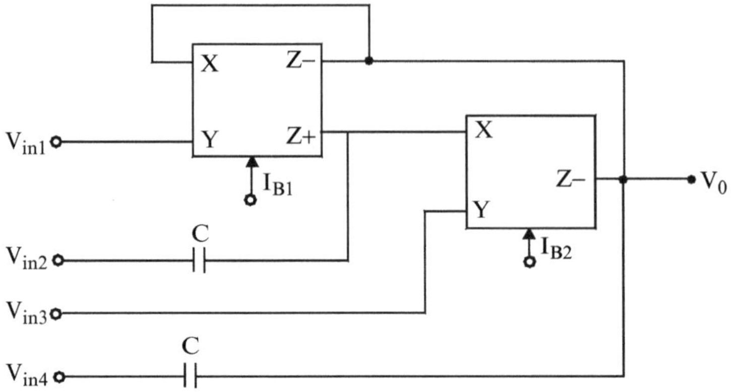

Fig. 9.26 A voltage mode universal biquad presented by Ranjan and Paul [23]

The output voltage (V_{out}) of this circuit is obtained as:

$$V_{out} = \frac{V_{in1} + sCR_{x1}V_{in2} - sCR_{x1}V_{m3} + \left(s^2C^2R_{x1}R_{x2} + sCR_{x1}\right)V_{in4}}{s^2C^2R_{x1}R_{x2} + sCR_{x1} + 1} \tag{9.28}$$

An inspection of the equation (9.28) reveals that a LPF is realizable by choosing Vin2 = Vin3 = Vin 4 = 0; a HPF is realized by choosing Vin1 = Vm2 = 0 and Vin3 = Vm4; a BPF is realizable by choosing Vin1 = Vin2 = 0; a band-reject is realizable with Vin2 = 0 and making Vin1 = Vin3 = Vin4 and finally, an APF is realizable by choosing Vin1 = Vin3 = Vin4 = Vin but Vin2 = −Vin.

Although it is claimed that this circuit removes most of the short comings of the other previously known CCCII-based topologies, an apparent difficulty is that in case of APF, it does require an additional inverting amplifier.

From the available VM universal biquads, it appears that any circuit which can realize all the five functions completely without any constraints, while using no more than two or three CCCIIs, has still not appeared in the literature and hence, constitutes an interesting problem for investigation.

9.9 Current-Mode Universal Biquad Filters

A large number of universal current-mode biquad filter circuits have been proposed by various authors using CCCII as active elements. In the following, we present some exemplary circuits possessing some interesting and useful properties.

The classical Kerwin, Huelsman and Newcomb (KHN) biquad has been a very popular and versatile biquad filter structure which has been translated into almost every new technology and using almost every new building block over the past two decades. Altuntas and Toker [24] have presented KHN-equivalent biquads employing CCCIIs for both VM and CM filter designs. In the following, we present a CCCII-based implementation of the CM KHN-equivalent biquad (see Fig. 9.27).

A routine analysis of this circuit reveals that the three transfer functions of interest for this circuit are given by:

$$\frac{I_{out}}{I_{i1}} = \frac{\frac{1}{C_1 C_2 R_{x1} R_{x2}}}{D(s)};\ \frac{I_{out}}{I_{i2}} = \frac{-\frac{s}{C_1 R_{x1}}}{D(s)} \text{and} \frac{I_{out}}{I_{i3}} = \frac{s^2}{D(s)} \tag{9.29}$$

Where

$$\mathrm{D(s)} = s^2 + s\frac{R_{x3}}{C_1 R_{x1} R_{x5}} + \frac{R_{x3}}{R_{x4} C_1 C_2 R_{x1} R_{x2}} \tag{9.30}$$

so that the parameters ω_0 and Q_0 are given by:

$$\omega_0 = \sqrt{\frac{R_{x3}}{R_{x4} C_1 C_2 R_{x1} R_{x2}}} \text{ and } Q_0 = R_{x5}\sqrt{\frac{C_1 R_{x1}}{C_2 R_{x2} R_{x3} R_{x4}}} \tag{9.31}$$

Thus, the circuit can realize three basic functions as above. In addition, a notch response is realizable for the case when $I_{i1} = I_{i3} = I_i$ along with $I_{i2} = 0$ as well as $R_{x3} = R_{x4}$. Lastly, an all pass response is realizable for $I_{i1} = I_{i2} = I_{i3} = I_i$ along with $R_{x3} = R_{x4}$.

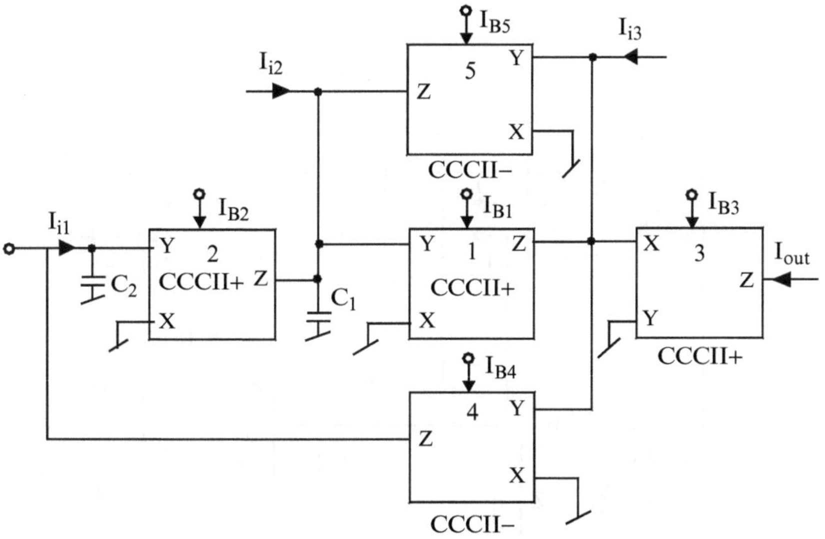

Fig. 9.27 A CM KHN-equivalent biquad using CCCIIs presented by Altuntas and Toker (Adapted from [24] © 2002, Urban and Fischer Verlag)

Due to being a multiple-input-single-output (MISO)-type configuration, the above mentioned KHN biquad suffers from the requirement of identical currents to be applied in case of notch and all pass filters which will call for an additional active element namely, a multiple output current follower (MOCF). Furthermore, another drawback of the circuit is that additional realization conditions are needed for designing both notch and all pass filters.

The above difficulties are easily over come in a design known as single-input-multiple-output (SIMO)-type configuration and a number of such circuits using CCCIIs have been evolved by various researchers. A circuit proposed by Wang et al. [25] is particularly appealing and is shown here in Fig. 9.28.

It is interesting to note that this circuit has only four CCCIIs (of course, three of them have multiple output terminals) and five CM outputs each one of which is available from a high output impedance terminal and each one realizing a specific filter function as shown in the following.

$$\frac{I_{01}}{I_{in}} = K\frac{1}{D(s)}; \frac{I_{02}}{I_{in}} = \frac{KsT_2}{D(s)}; \frac{I_{03}}{I_{in}} = K\frac{s^2T_1T_2}{D(s)}; \tag{9.32}$$

$$\frac{I_{04}}{I_{in}} = K\frac{s^2T_1T_2 + 1}{D(s)}; \frac{I_{05}}{I_{in}} = K\frac{s^2T_1T_2 - sT_2 + 1}{D(s)} \tag{9.33}$$

Where

$$\mathrm{D(s)} = s^2T_1T_2 + sT_2K + 1 \tag{9.34}$$

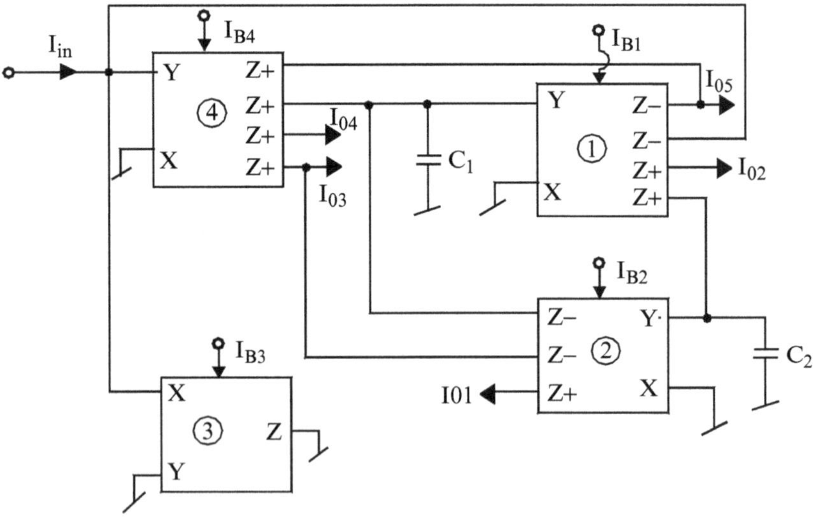

Fig. 9.28 SIMO type CM universal biquad using CCCII proposed by Wang et al. (Adapted from [25] © 2008, Birkhauser)

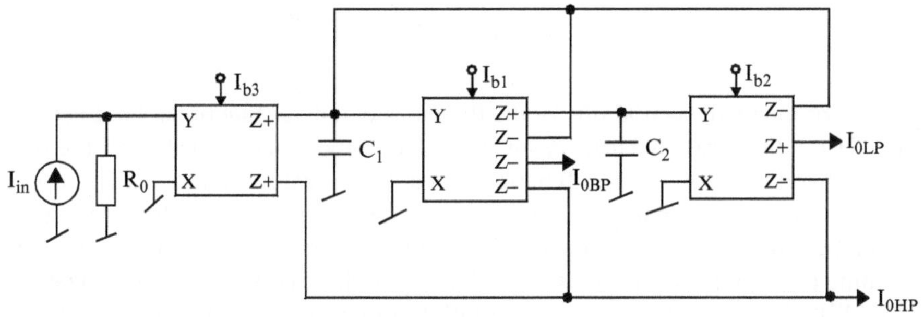

Fig. 9.29 Current-controlled CM universal biquad proposed by Senani et al. (Adapted from [26] © 2004 IEICE)

In the above expressions $T_1 = R_{x1}C_1$; $T_2 = R_{x2}C_2$ and $K = R_{x3}/R_{x4}$ Thus, an attractive feature of this circuit is that no component-matching or realization constraints have been encountered in case of notch and all pass filter realizations. Using bipolar transistors PR100N and NP100N to realize a standard translinear CCCII with multiple outputs, with DC bias voltages taken as ±3 V, it has been possible to design the circuit to realize various filters with f_0 of the order of 1 MHz and the tunability of f_0 and Q_0 has been possible by changing the various DC bias currents appropriately, typically in the range 10–300 μA.

A current-mode universal active filter with one input and multiple outputs was proposed by Senani et al. [26] which is shown in Fig. 9.29.

The three responses realized by this circuit are given by:

$$\frac{I_{OLP}}{I_{in}} = \frac{H_{OLP}\,\omega_0^2}{D(s)};\ \frac{I_{OBP}}{I_{in}} = -\frac{H_{OBP}\left(\frac{\omega_0}{Q_0}\right)s}{D(s)}\ and\ \frac{I_{OHP}}{I_{in}} = \frac{H_{OHP}s^2}{D(s)} \tag{9.35}$$

where

$$H_{OHP} = -H_{OBP} = H_{OLP} = \frac{R_0}{R_{x3}} \tag{9.36}$$

and

$$\mathrm{D(s)} = s^2 + \frac{1}{R_{x1}C_1}s + \frac{1}{R_{x1}R_{x2}C_1C_2} \tag{9.37}$$

Using $R_{xi} = \frac{V_T}{2I_{bi}}$; $i = 1-3$, the parameters of the realized filters are found to be:

$$\omega_0 = \frac{2}{V_T}\sqrt{\frac{I_{b1}I_{b2}}{C_1C_2}};\ \frac{\omega_0}{Q_0} = \frac{2I_{b1}}{V_TC_1};\ Q_0 = \sqrt{\frac{I_{b2}C_1}{I_{b1}C_2}};$$

$$H_{OHP} = -H_{OBP} = H_{OLP} = \frac{2R_0 I_{b3}}{V_T} \tag{9.38}$$

Because of the ready availability of the three basic transfer functions (i.e. LPF, BPF and HPF) with correct polarities as well as due to the equal values of all the three gains, a band stop filter (BSF) function is realizable just by joining LPF and HPF outputs. On the other hand, an APF is realizable by joining all the three output terminals. It may be noted that a novel feature of this circuit is that no component-matching or equality constraint is needed in any of these two additional realizations.

In [27], Chen and Chu presented a multiple-input-multiple-output (MIMO)-type universal filter which, in contrast to the KHN equivalent biquad described above, needs only three CCCIIs (rather than five, as employed in KHN equivalent circuit). This circuit is shown in Fig. 9.30.

Chen and Chu [27] have demonstrated two possible modes of operation of this circuit out of which, the following appears to be interesting because of the reason that in this case, all the five generic filters function can be realized from the same circuit without requiring any cancellation-constraints. The five functions realized by this circuit are given by:

$$\frac{I_{01}}{I_{in}} = \frac{-\frac{s}{C_1 R_{x2}}}{D(s)};\ \frac{I_{02}}{I_{in}} = \frac{-\frac{1}{C_1 C_2 R_{x2} R_{x3}}}{D(s)};\ \frac{I_{03}}{I_{in}} = \frac{s^2 + \frac{1}{C_1 C_2 R_{x2} R_{x3}}}{D(s)} \tag{9.39}$$

$$\frac{I_{HP}}{I_{in}} = \frac{I_{02}}{I_{in}} + \frac{I_{03}}{I_{in}} = \frac{s^2}{D(s)} \tag{9.40}$$

$$\frac{I_{AP}}{I_{in}} = \frac{I_{01}}{I_{in}} + \frac{I_{03}}{I_{in}} = \frac{s^2 - \frac{1}{C_1 R_{x2}} s + \frac{1}{C_1 C_2 R_{x2} R_{x3}}}{D(s)} \tag{9.41}$$

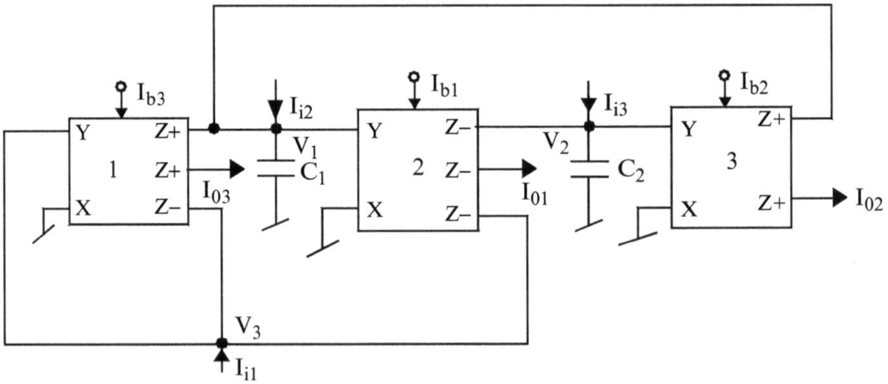

Fig. 9.30 MIMO type CM universal filter proposed by Chen and Chu (Adapted from [27] © 2009 IEICE)

where

$$D(s) = s^2 + \frac{s}{C_1 R_{x2}} + \frac{1}{C_1 C_2 R_{x2} R_{x3}} \tag{9.42}$$

SPICE simulations based upon a CMOS DO-CCCII (a multiple output version of the circuit of Fig. 9.4) using TSMC 0.35 μm CMOS process have shown that f_0 in the order of 1.27 MHz is conveniently realizable. It has also been demonstrated that compared to other realizations, this circuit simultaneously offers the advantageous features of employment of both grounded capacitors, high output impedance good for cascadability, no need to employ inversion of input current or requirement of double of the input signals to realize all five generic filters and no component-matching constraints.

Another three CCCIIs based circuit which also employs both grounded capacitors without requiring any passive component matching and still realizing all the five generic filter function is the SIMO-type universal biquad of Maheshwari and Khan [28] which is shown in Fig. 9.31.

Note that this circuit has a resistive (although variable due to the DC bias current I_{03} being variable) input impedance and provides all the five outputs from high output impedance nodes.

The five transfer functions realized by the circuit are given by:

$$\frac{I_{LP}}{I_{in}} = \frac{1/R_{x1}R_{x2}C_1C_2}{D(s)}, \frac{I_{BP}}{I_{in}} = \frac{s/R_{x1}C_2}{D(s)}, \frac{I_{BR}}{I_{in}} = -\frac{(s^2 + 1/R_{x1}R_{x2}C_1C_2)}{D(s)},$$
$$\frac{I_{HP}}{I_{in}} = \frac{-s^2}{D(s)}, \frac{I_{AP}}{I_{in}} = -\frac{(s^2 - s/R_{x1}C_2 + 1/R_{x1}R_{x2}C_1C_2)}{D(s)}, \tag{9.43}$$

Where

$$D(s) = s^2 + \frac{s}{R_{x1} C_2} + \frac{1}{R_{x1} R_{x2} C_1 C_2} \tag{9.44}$$

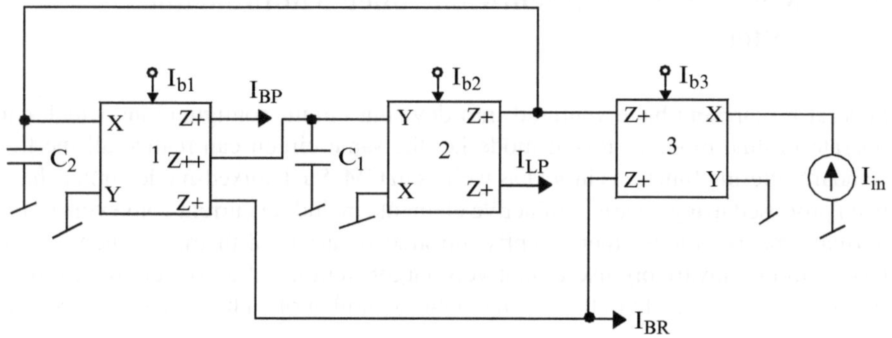

Fig. 9.31 A SIMO type universal CM biquads proposed by Maheshwari and Khan [28]

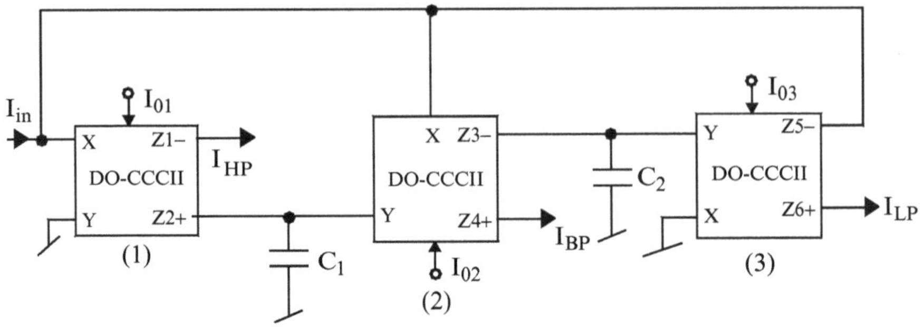

Fig. 9.32 SIMO type universal biquad filter proposed by Yuce (Adapted from [29] © 2009 Elsevier Ltd.)

Yet another SIMO-type universal CM biquad realizable with only three CCCIIs and two grounded capacitors was reported by Yuce [29] which requires all CCCIIs of the same kind; all three are DO-CCCIIs. This circuit simultaneously provides the advantages of electronic-tunability, use of a minimum number of components, lack of component-matching constraints, low input and high out impedances and requiring the least possible chip area as compared to the various other three-CCCII-based alternatives due to employment of three identical DO-CCCIIs (Fig. 9.32).

The ω_0 and Q_0 for all the five responses are given by:

$$\omega_0 = \sqrt{\frac{1}{C_1 C_2 R_{x2} R_{x3}}}; Q_0 = \sqrt{\frac{C_1 R_{x2}}{C_2 R_{x3}}} \tag{9.45}$$

from where it can be seen that both the parameters can be adjusted arbitrarily; by simultaneously changing R_{x2} and R_{x3} while keeping R_{x2}/R_{x3} constant and vise-versa. Also, the circuit realizes a non-inverting low pass and inverting band pass, a non-inverting high pass, a notch filter by adding low pass and high pass output currents and all pass by adding low pass, band pass and high pass current outputs.

9.10 Mixed-Mode Current-Controlled Multifunction Filters

Several researchers have ventured into devising circuit configurations which can operate in dual mode or mixed mode i.e. the same circuit can realize all the five standard filtering functions in VM as well as in CM. Such mixed-mode circuits have been proposed using a variety of active elements including current conveyors of the normal type as well as other recently introduced variety of them. All such mixed-mode circuits invariably use a relatively larger number of active building blocks (typically seven or eight of them). Of course, a number of authors have succeeded in reducing the number of building blocks considerably by incorporating a circuit technique in which one of the inputs is added to the output node directly.

The summation of one of the input current signal (say I_3) on the output node itself raises an important question and implies the (hidden) requirement of one more active element to make the circuit 'practical' which is generally not been recognized by the concerned authors who have reported earlier several biquad filter circuits employing this kind of summing of an input at the output node.

Consider now the following:

(i) Looking into the node from where the output I_{out} is being taken, the output impedance should be very high (ideally infinite); since input I_3 is also applied at the same node, the input impedance looking into the same node (as seen by I_3) should be low (ideally zero). Obviously the same node cannot exhibit these two absolutely contradictory requirements of impedance levels.

(ii) The above mentioned anomalous situation can be corrected if input I_3 is added to the output node through a current follower (CF) exhibiting low input impedance and high output impedance. By doing this, I_3 will see a low input impedance (that of this additional CF)) and at the output terminal now, one has the effective output impedance equal to the parallel combination of the output impedance of the active building block employed and the output impedance of this additional CF which can be still considered to be high if the employed building block and CF both are designed to have very high output impedances.

(iii) The addition of the 'input current' from a current source and the 'output current' of the building block considered, as employed in a number of circuits earlier for instance see [80, 93, 99, 101, 107, 116, 120, 123, 131], might work in SPICE simulations but the circuits of the quoted type has the flaw and anomalous situation pointed out in (i) above. In practice, such an addition should require one more active element (a CF) thereby increasing the total count of active building blocks employed by 'one'.

In the opinion of the present authors, the type 'addition of an input current at the output node', in the circuits quoted above and reported earlier 'misleadingly' conceals this fact.

Such mixed-mode circuits have been developed around CCCIIs also and several varieties have so far been developed for instance, see [30, 102, 117, 132]. In this section, we include one such mixed-mode universal biquad configuration which uses no more than four CCCIIs and can realize various filtering functions in VM as well as in CM (Fig. 9.33).

From straight forward analysis of this circuit it is found that the various current and voltage outputs of this circuit are given by the following equations:

$$V_{out1} = \frac{N_1(s)}{D(s)}, V_{out2} = -\frac{N_2(s)}{D(s)}, V_{out3} = \frac{V_{in1} - V_{out2}}{sC_2R_{x2}} + \frac{1}{sC_2}I_{in2} \tag{9.46}$$

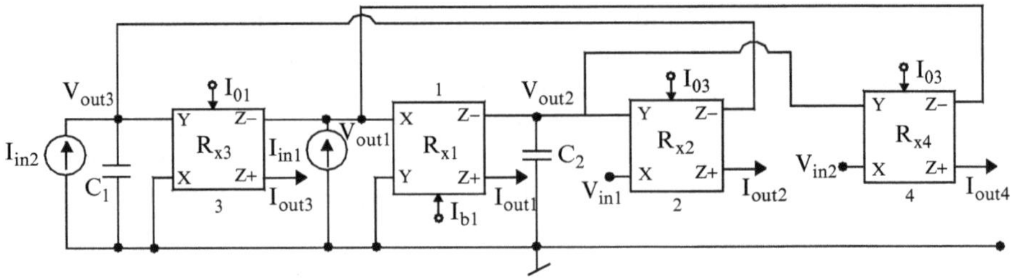

Fig. 9.33 A mixed mode current controlled CCCII based filter proposed by Abuelma'atti [30]

$$I_{out1}=\frac{1}{R_{x1}}\frac{N_1(s)}{D(s)},I_{out2}=\frac{1}{R_{x2}}\frac{N_3(s)}{D(s)},I_{out3}=\frac{V_{out3}}{R_{x3}}I_{out4}=\frac{1}{R_{x4}}\frac{N_4(s)}{D(s)} \tag{9.47}$$

where

$$N_1(s)=s^2C_1C_2R_{x1}R_{x2}R_{x3}I_{in1}+sC_1R_{x1}R_{x2}I_{in2}-sC_1R_{x1}V_{in1}$$
$$-s^2C_1C_2\frac{R_{x1}R_{x2}R_{x3}}{R_{x4}}V_{in2} \tag{9.48}$$

$$N_2(s)=sC_2R_{x2}R_{x3}I_{in1}+R_{x2}I_{in2}-V_{in1}-sC_2\frac{R_{x2}R_{x3}}{R_{x4}}V_{in2} \tag{9.49}$$

$$N_3(s)=sC_2R_{x2}R_{x3}I_{in1}+R_{x2}I_{in2}+\left(sC_2\frac{R_{x2}R_{x3}}{R_{x4}}+s^2C_1C_2R_{x2}R_{x3}\right)V_{in1}$$
$$-sC_2\frac{R_{x2}R_{x3}}{R_{x4}}V_{in2} \tag{9.50}$$

$$N_4(s)=sC_2R_{x2}R_{x3}I_{in1}+R_{x2}I_{in2}-V_{in1}+\left(1+s^2C_1C_2R_{x2}R_{x3}\right)V_{in2} \tag{9.51}$$

and

$$D(s)=1+sC_2\frac{R_{x2}R_{x3}}{R_{x4}}+s^2C_1C_2R_{x2}R_{x3} \tag{9.52}$$

The authors have described five special cases of this arrangement to explore the realization of many different kinds of filter functions. However, for the sake of brevity we elaborate only one case which appears to be more interesting than the remaining ones. Taking $I_{in1}=0=I_{in2}$ and Vin1 = 0, the circuit can realize (i) inverted HPFs from V_{out1} and I_{out1} (ii) non-inverted BPF from V_{out2} and inverted BPF from I_{out2} (iii) non-inverted notch filter (NF) from I_{out4} and (iv) inverted LPF from V_{out3} and I_{out3}.

An important point to be noted is that the circuit has the feasibility of controlling ω_0 by adjusting the biasing current I_{b2} and/or I_{b3} without disturbing the parameter ω_0/Q_0, while, on the other hand the parameter ω_0/Q_0 can be controlled by adjusting the bias current I_{b4} without disturbing the parameter Ω_0.

An important deficiency of this circuit is that it does not provide an all pass filter response either in voltage mode or current mode.

In fact, from the inspection of available mixed-mode biquad circuits using CCCIIs it appears that a completely universal circuit capable of realizing all the five basic filtering functions in both VM and CM without any constraints with electronic controllability of the filter parameters while employing a reasonably small number of CCCIIs has still not been discovered so far. This, therefore, constitutes an interesting problem which is open to investigation.

9.11 Tunable Ladder Filters Using Multiple-output CCCIIs

Whereas a majority of publications dealing with CCCII-based filters have dealt with universal VM filters or CM filters, comparatively less attention appears to have been given on the design of higher order filters using CCCIIs and only earlier works [31–36] can be cited in this context.

In this section, we elaborate only two specific methodologies of designing CCCIIs based higher order filters; namely the FDNR based approach of Ozoguz and Toker [31] and the signal flow graph based method of Jiraseree-amornkun and Surakampontorn [32].

The method of Ozoguz and Toker [31] is based upon a new two CCCII-based realization of a series FDNR-R equivalent circuit and a known realization of current controlled floating resistance of Fig. 9.12a. The method is explained here for the design of a ladder type electric filter for which the LC prototype shown in Fig. 9.34a.

Applying Bruton's transformation, the LC ladder filter is transformed into RCD ladder of Fig. 9.34b, the conversion of this circuit requires three floating resistances to be simulated by CCCIIs and simulation of two shunt RD branches by CCCIIs based circuit. Ozoguz and Toker [31] in their paper have shown that a floating resistance can be simulated by only a single CCCII as can be seen in the circuit of Fig. 9.34c whereas two CCCIIs and two grounded capacitor are required for the FDNR-R branch as per their preposition. The resulting circuit is shown in Fig. 9.34c. It maybe seen that it appears to be quite economical design in the sense that the total number of active building blocks namely CCCIIs is exactly equal to the number of reactive elements in the original LC prototype. Furthermore, another advantageous feature of this circuit is that it employs all grounded capacitors as preferred for integrated circuit implementation. Lastly, because of the electronic controllability of R_x of the various CCCIIs, the circuit has excellent tuning capability.

Jiraseree-amornkun and Surakampontorn [32] presented a signal flow graph based methodology to realize continuous time current mode tunable ladder filters of any order using multi output CCCIIs (MCCCIIs) and grounded capacitors.

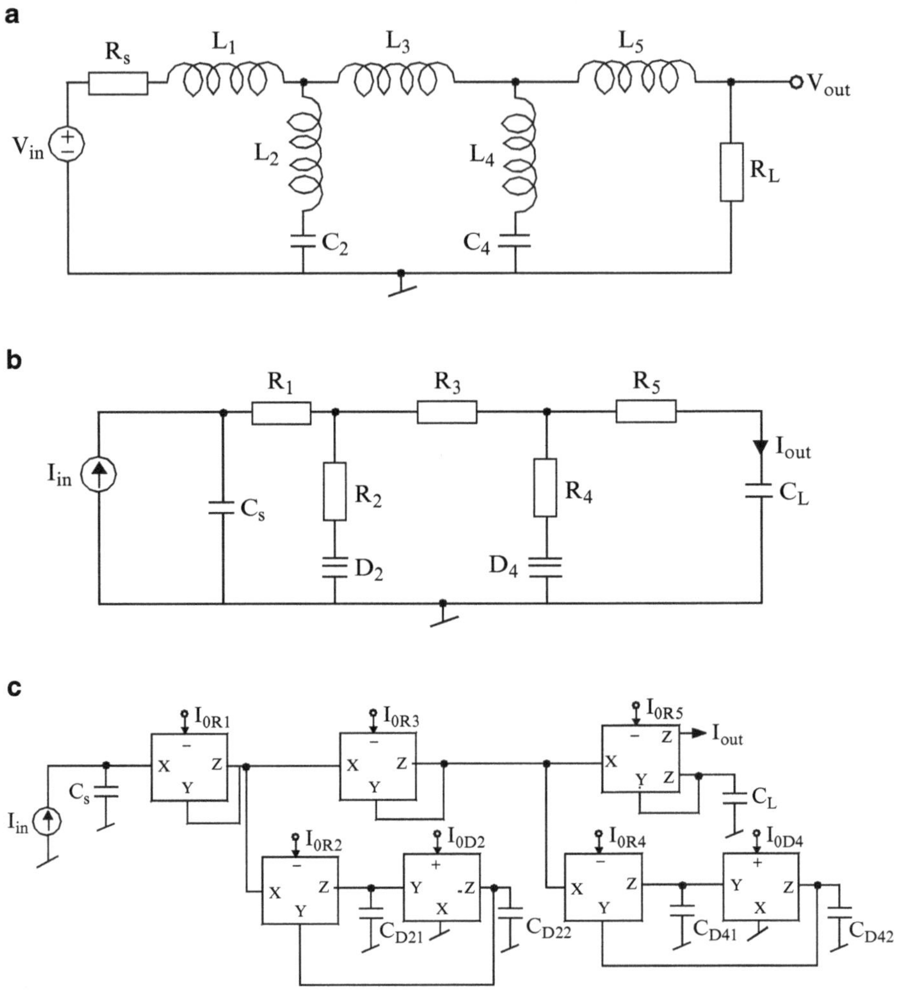

Fig. 9.34 CM ladder type fifth order electric filter using CCCIIs [31] (**a**) LC prototype (**b**) FDNR version (**c**) Complete CCCIIs based circuit implementation

The technique presented by them is based upon simulating the signal flow graph (SFG) of each branch current from passive filter prototype. In a passive ladder the various elements can be broadly classified into four distinct categories (i) single passive element which could be a shunt capacitance or a floating inductance which are characterized by similar equations or a shunt inductor and a series capacitor which also have similar characterizations (ii) single shunt element with a source or load resistor or a single series element with source or load resistor (iii) double passive elements such as shunt parallel LC, shunt series LC, series LC in series arm and parallel LC in series arm (iv) double passive elements with source of load resistor which can have four possible forms, a shunt resistor in parallel with in shunt

LC, a shunt resistor along with a series LC, a shunt resistor with a series LC in parallel with this and a shunt resistor along with a parallel LC in series arm.

For each of elements described in above four categories MCCCIIs based realizations have been developed by simulating the node equations of the passive block by MCCII based circuit in conjunction with grounded capacitor(s). For example the MCCCII based realizations of the four single passive element blocks and realizations of single passive element along with source or load resistor have been shown in Fig. 9.35.

For the sake of brevity MCCCII based realizations of the remaining two classes are not reproduced here, the interested reader may refer to [32] for their details.

As an illustrative example, we present the essence of the methodology by considering the example of the third order butter worth low pass LC ladder filter and its current mode implementation. The prototype is shown in Fig. 9.36 from where it can be seen that this circuit can be considered to be a cascade of three sub-sections.

Replacing each subsection by its corresponding MCCCII based realization and then connecting them appropriately leads to the MCCCIIs based implementation which is shown in Fig. 9.37.

Using this technique it has been shown that filter design with cut off frequency of 5 MHz has been found to be easily tunable (by changing the bias current of MCCCIIs) from 2.5 to 10 MHz with frequency accuracy within about 6 % and with characteristic quite close to that of the passive prototype except beyond 100 MHz where the frequency response begins to deviate due to the frequency limitations of the MCCCIIs.

Jiraseree-amornkun and Surakampontorn [32] have also successfully applied this approach to realize current mode third order elliptic low pass ladder filter employing five MCCCIIs with four grounded capacitors and have shown that because this technique derives each elements with one by one replacement circuit, it is possible to reduce the total number of MCCCIIs from five to three at the cost of employing one floating capacitor in the design.

In general this method is not only quite simple but covers all types of passive LC ladder prototypes. For designing an nth-order all-pole filter only n MCCCIIs are needed whereas (2n-1) and (2n-2) MCCCIIs are required to design odd and even nth-order elliptic filters respectively, when implementing only grounded capacitors are required.

9.12 Current-Controlled Sinusoidal Oscillators

There have been a number of proposals for realizing CC based sinusoidal oscillators, however, it is known that to realize such an oscillator, employing both grounded capacitors, at least two CCs along with two capacitors and three resistors are needed. Exploiting the current-controllable R_x of the CCCIIs, it is interesting to note that an electronically controllable sinusoidal oscillator can be realized with only two CCCIIs along with both grounded capacitors as desirable for integrated circuit implementation. The first such circuit was proposed by Horng [37] and is

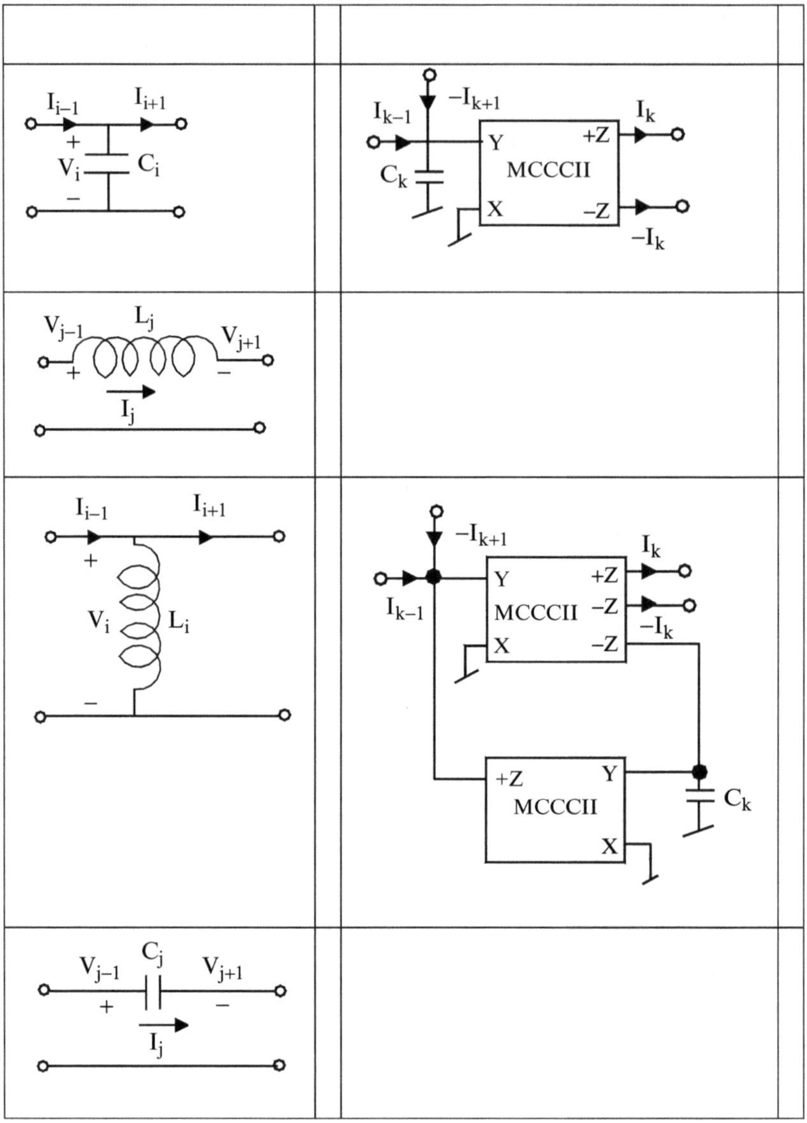

Fig. 9.35 Realization of various elements or combinations thereof (Adapted from [32] © 2007 Elsevier GmbH)

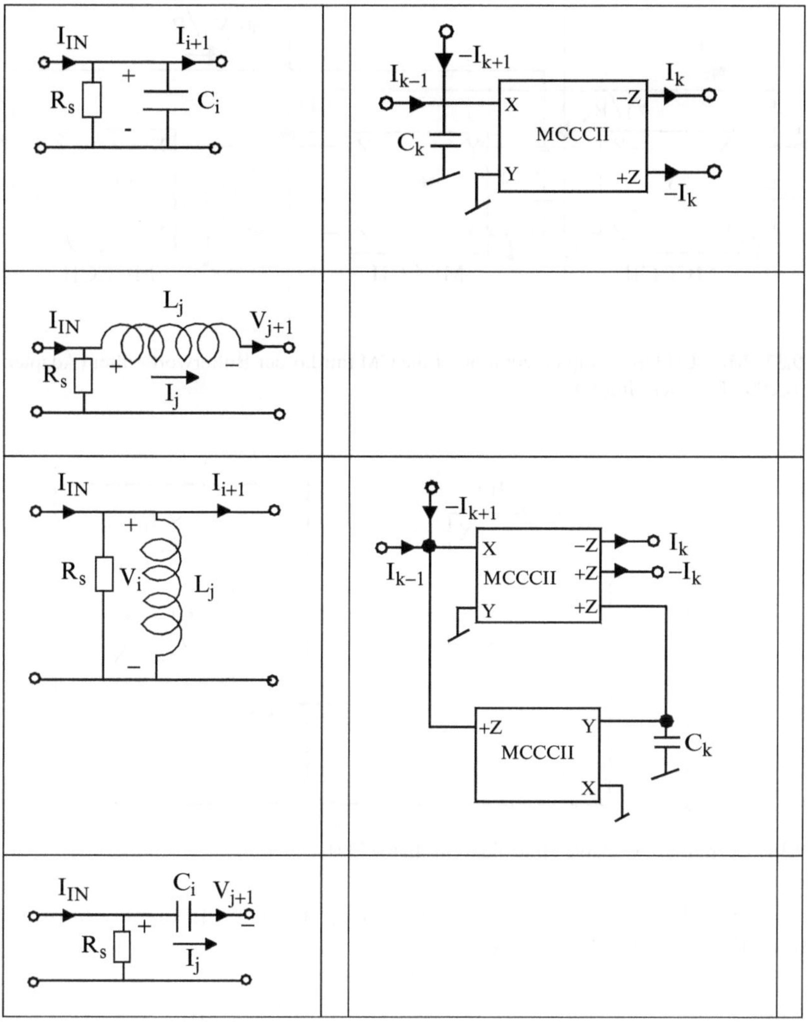

Fig. 9.35 (continued)

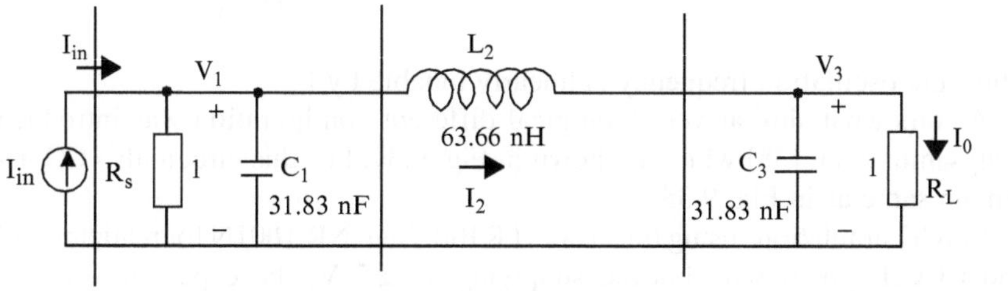

Fig. 9.36 Prototype CM third order Butterworth filter (Adapted from [32] © 2007 Elsevier GmbH)

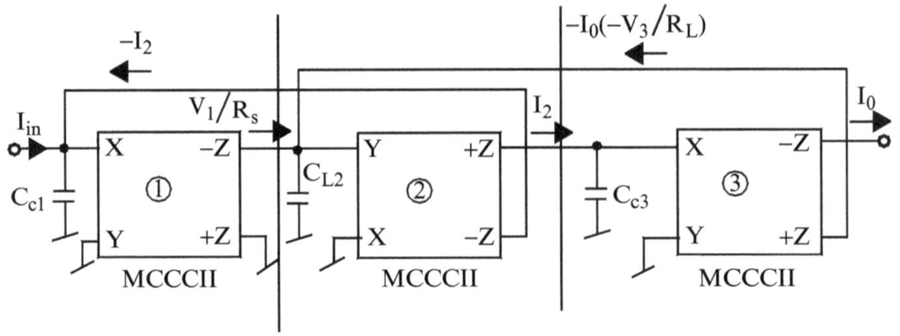

Fig. 9.37 MCCCII based implementation of the CM third order Butterworth filter (Adapted from [32] © 2007 Elsevier GmbH)

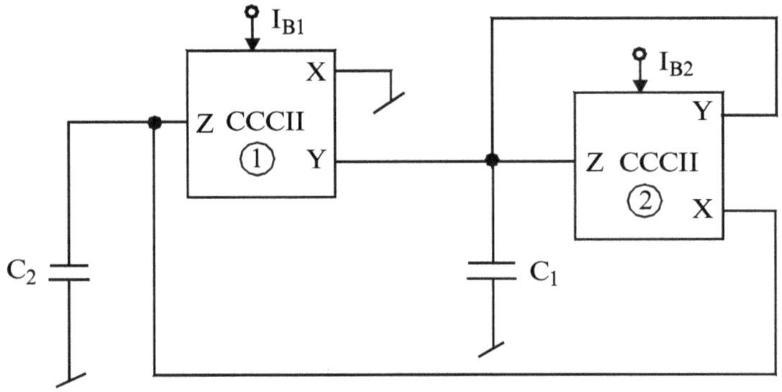

Fig 9.38 Current-controllable sinusoidal oscillator [37]

shown in Fig. 9.38. The condition of oscillation (CO) and frequency of oscillation (FO) for this circuit are given by:

$$\text{CO}:\ \text{C}_1 = C_2\,\text{and}\,\text{FO}: f_0 = \frac{1}{\pi V_T}\sqrt{\frac{I_{B1}I_{B2}}{C_1C_2}} \tag{9.53}$$

$$\text{with}\ \text{I}_{\text{B1}} = \text{I}_{\text{B2}} = \text{I}_{\text{B}},\ \text{one obtains}\ f_0 = \frac{I_B}{\pi V_T}\sqrt{\frac{1}{C_1C_2}} \tag{9.54}$$

Thus, the oscillation frequency is linearly tunable by I_B.

A somewhat similar yet topological different configuration was introduced by Fongsamut et al. [38] which is shown in Fig. 9.39. For this circuit also CO and FO remain same as in Fig. 9.38.

SPICE simulations using transistors PR100N and NR 1001N for realizing CCCII+ and CCCII− both with biasing supplying as ±5 V, the capacitor values taken $C_1 = 11$ nF, $C_2 = 10$ nF and the external DC bias current $I_0 = I_{01} = I_{02}$ was varied

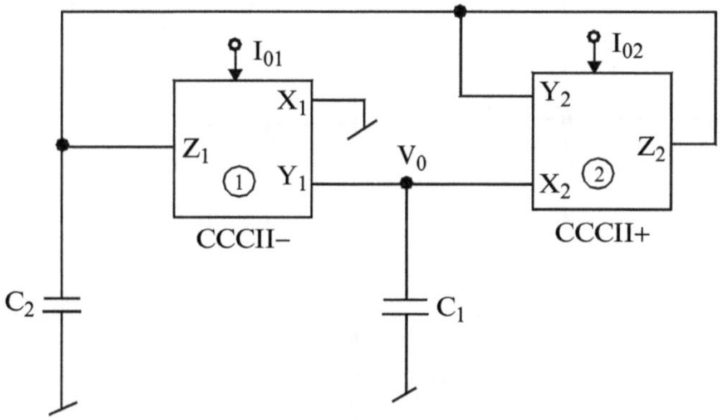

Fig 9.39 An alternative form of current-controllable sinusoidal oscillator [38]

from 0.1 to 100 μA have shown excellent correspondence between the theoretical frequency and that obtained by SPICE simulations from about 100 Hz to 10 KHz.

While the circuits described in Figs. 9.38 and 9.39 employ only two CCCIIs besides a canonic a number of grounded capacitors the circuits have control over the CO through variable capacitors which is not very convenient option. A better situation would be to have control of the condition of oscillation through an external bias current or the gain of an active element which may be extremely programmable; in the following we present two such circuits.

Kiranon et al. in [39] presented a current controlled oscillator based on two translinear CCCIIs and a current mirror having current gain α as shown in Fig. 9.40. By straight forward analysis the CO is found to be $\alpha = 1 + (R_{x2}/R_{x1})$ whereas the FO is found to be:

$$\text{FO} : f_0 = \frac{I_0}{\pi V_T \sqrt{C_1 C_2}} \tag{9.55}$$

It is thus seen that f_0 is linearly controllable by external bias current I_0 where $I_{01} = I_{02} = I_0$.

SPICE simulations based on bipolar transistor PR200N and NR 200N with the circuit biased from ±2.5 V supply have demonstrated excellent linear tunability of FO over a range of three to four decades when I_0 was varied from 0.1 μA to 1 mA.

A drawback of above mentioned circuit is that the CO is fixed by setting the required value of the current gain α which in term is controlled by the emitter area ratio of the current amplifier hence no control is possible to be exercised on the amplitude of oscillation.

An alternative circuit which overcomes this difficulty was presented by Abuelma'atti and Tasadduq in [40] which also uses three active elements, all of which are CCCIIs along with two grounded capacitors. This circuit is shown in Fig. 9.41 and is governed by the following equations:

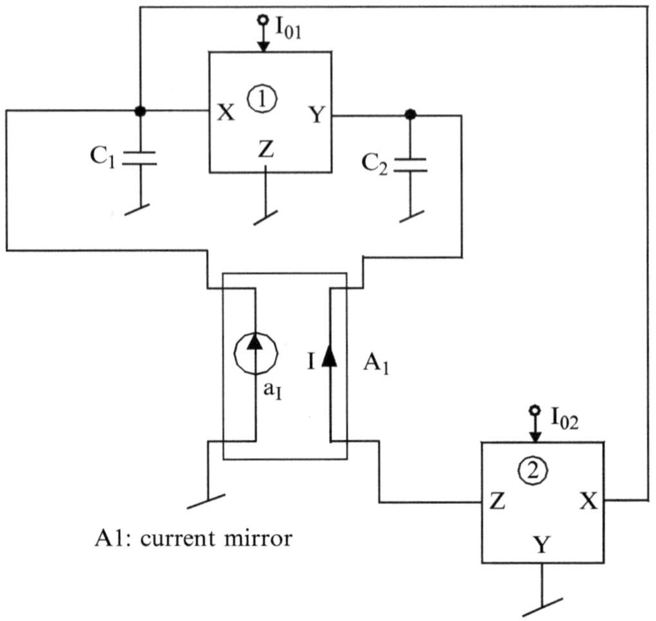

Fig. 9.40 Current controlled oscillator proposed by Kiranon et al. [39]

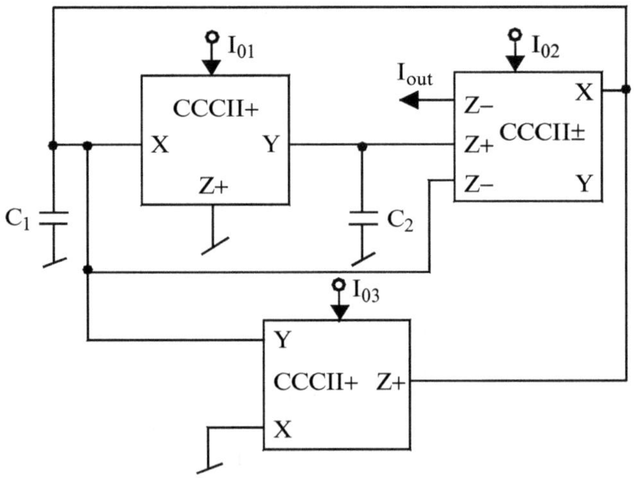

Fig. 9.41 Current control oscillator using only CCCIIs proposed by Abuelma'atti and Tasadduq [40]

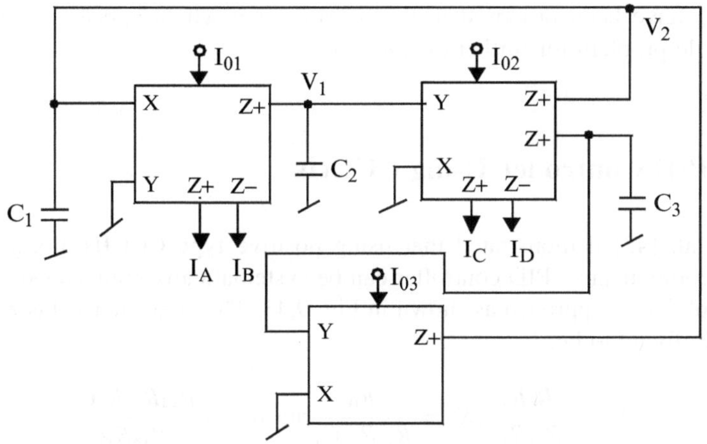

Fig. 9.42 A third order quadrature oscillator using CCCIIs proposed by Maheshwari [41]

$$\mathrm{CO}: I_{03} = I_{01} \text{ and } \mathrm{FO}: f_0 = \frac{1}{\pi V_T}\sqrt{\frac{I_{01} I_{02}}{C_1 C_2}} \tag{9.56}$$

It is, therefore, seen that CO can be controlled by I_{03} whereas f_0 is controllable through I_{02}. SPICE simulation results have confirmed very good correspondence between frequency and controlling current I_{02}.

Quadrature oscillators have attracted lot of attention in analog circuits literature because of their wide spread use in communication and instrumentation systems. Quadrature oscillators have also been realized by using CCs by a number of researchers. In the following we present an exemplary third order quadrature oscillator which is capable of providing two voltage outputs which are in quadrature as well as four current outputs using CCCIIs. Two of which are also in quadrature the circuit enjoys non-interactive frequency control while employing three grounded capacitors. This circuit is shown in Fig. 9.42.

With identical capacitors the CO and FO for this circuit are given by:

$$\mathrm{CO}: \mathrm{R_{x1}} = \mathrm{R_{x3}} \,\mathrm{and FO}: f_0 = \frac{1}{2\pi C}\sqrt{\frac{1}{R_{x2} R_{x3}}} \tag{9.57}$$

Using SPICE simulation based upon a CMOS CCCIIs (see Fig. 3 of [41]) biased from ±2.5 V using 0.5 μm CMOS process parameters. It has been confirmed that the circuit can function quite well at oscillation frequency of the order of 4 MHz with THD of <2 % for current outputs and <1% for voltage outputs.

There appears to be enough scope in devising new topologies for evolving fully uncoupled oscillators using CCCIIs which provide independent control of CO through separate external bias currents and may provide linear control of FO through simultaneous variation of two external bias currents. Furthermore, any CMOS CCCIIs based circuit capable of providing linear control of FO does not appear to have been proposed in the literature yet which appears to be another worthwhile problem for further exploration.

9.13 PID Controller Using CCCIIs

Erdal et al. [42] demonstrated that using positive type CCCIIs along with two passive components a PID controller can be systematically realized using a signal flow graph based approach as shown in Fig. 9.43. The controller gains K_P, K_I, K_D can easily found to be:

$$K_P = \frac{R_0 R_1}{R_{x1} R_{x6}}, K_I = \frac{R_0}{R_{x2} R_{x7} C_1} \text{and} K_D = \frac{R_{x4} R_{x5} R_o C_2}{R_{x3} R_{x8}} \tag{9.58}$$

A notable feature of this realization is that all the three controller gains can assigned the required values by adjusting the control current I_{bi} of selected CCCII. Since all of them depend upon the X-terminal input resistances of the CCCIIs. This PID circuit has been shown to be very useful for a stable control, very wide frequency band which is almost 1MHz by appropriate selections of parameters values as compared to PID controllers realized from conventional op-amps.

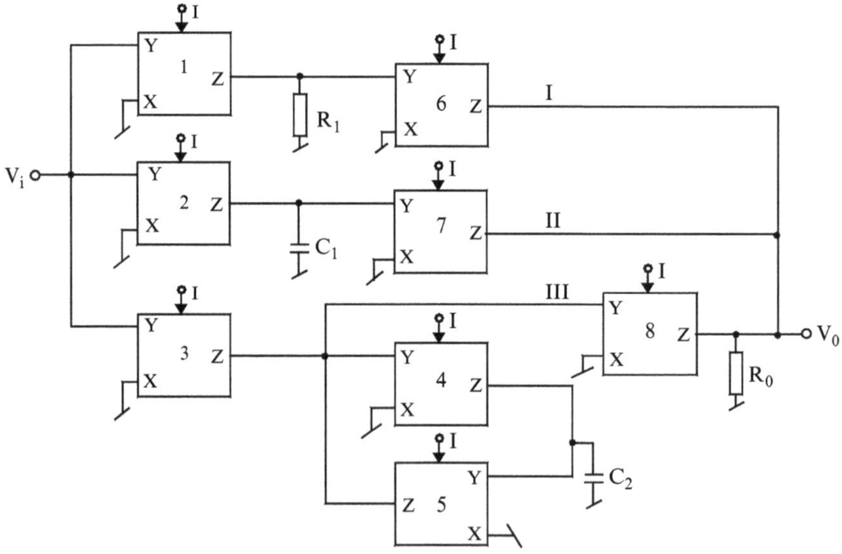

Fig. 9.43 A completely adjustable PID controller [42]

9.14 CCCII-Based Precision Rectifiers

Full wave precision rectifiers have been proposed in literature using a variety of building blocks such as op-amps, current conveyors and current feedback op-amps. A circuit to realize a precision rectifier using CCCIIs was presented in [43] which employ a pair of complimentary MOS transistors and two resistors in addition. This circuit is shown in Fig. 9.44.

The lower CCCII provides a current output proportional to the input voltage. This creates a controlled voltage at the gates of the two MOSFETs in conjunction with the resistor R which is so selected that both the transistors remain in saturation even for small input signals. Depending upon whether V_{in} is positive or negative, the gate voltage V_G would be either +V or –V (this requires that the ratio R/R_{x1} is large which can be achieved by proper choice of I_{B1} and R). Here $\pm$V denotes the DC biasing supply voltage of the CCCIIs. An analysis of this circuit shows:

For

$$V_{in} > 0, V_G = +V, \ V_o = \frac{2I_{B2}R_L}{V_T} V_{in} \tag{9.59}$$

and for

$$V_{in} < 0: \ V_G = -V, \ V_o = -\frac{2I_{B2}R_L}{V_T} V_{in} \tag{9.60}$$

An important merit of this circuit is that the output amplitude of this rectifier is electronically controllable by external bias current I_{B2}.

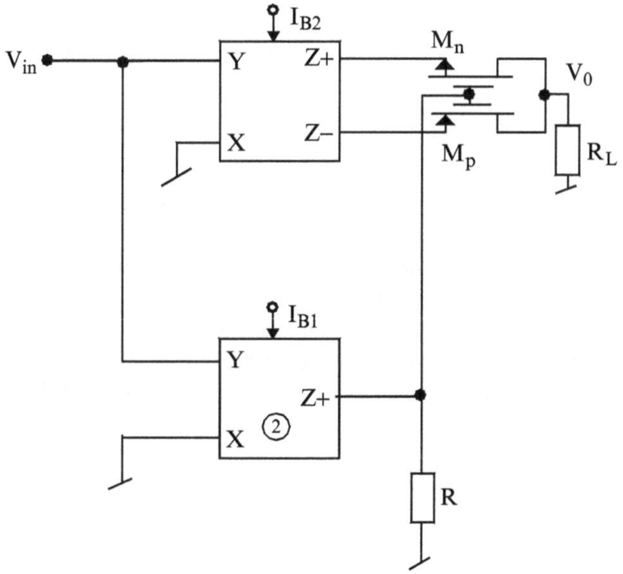

Fig. 9.44 CCCII based precision rectifier proposed by Maheshwari and Khan [43]

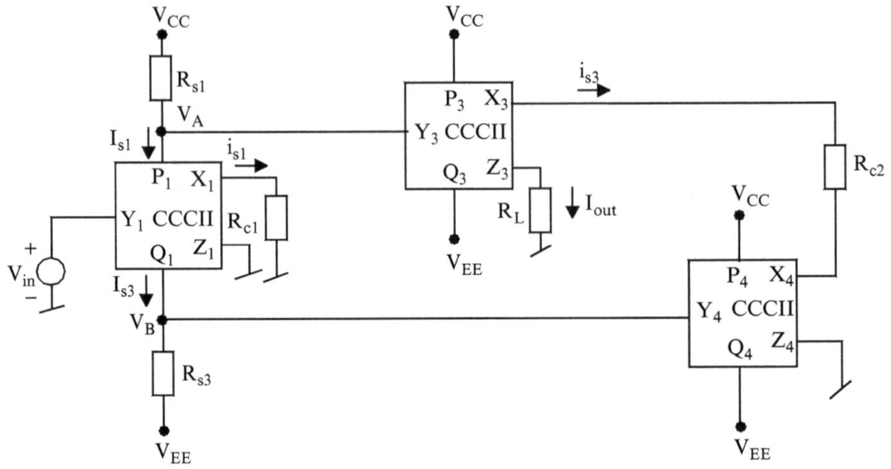

Fig. 9.45 Frequency doubler proposed in [44]

Frequency doubler is widely used in analog signal processing, communication and instrumentation systems. Frequency doubler is often made from analog multipliers. However, Anuntahirunrat et al. [44] have demonstrated that by imposing appropriate design conditions, the CCCIIs based circuit shown in Fig. 9.45 can accomplish both these functions from the same structure.

It has been found [45] that an approximate expression for the output current of this circuit is given by:

$$i_{out} \cong k_5 - (I_{DC} + I_{2\omega} \cos 2\omega t) \tag{9.61}$$

$$k_5 = k_4(V_{CC} - V_{EE}) - 5I_1 k_4 R_S;\ \ k_4 = \frac{1}{R_{C2} + 2R_x} \tag{9.62}$$

$$I_{DC} \cong 4I_1 k_4 R_S \left(1 + \frac{1}{4} k_3\right);\ I_{2\omega} \cong -I_1 R_s K_3 K_4;\ K_3 = \frac{I_m^2}{(2I_1)^2} \tag{9.63}$$

From the above equations, it is clear that by appropriately selecting component values and parameters, the circuit can generate a cosine wave of frequency two ω from an input current $i_{x1} = I_m \sin \omega t$ derived from the input V_m.

Siripruchyanun and Jaikla demonstrated in [45] that three CCCIIs are enough to realize a current mode precision full wave rectifier with a couple of advantageous features. Their preposition is shown in Fig. 9.46.

The analysis of this circuit can be carried out by noting the following:

$$V_{y2} = V_{y3} = V_{x1} = R_{x1} i_{x1} = \frac{I_{con} V_T}{4I_{B1}};\ i_{z2} = v_{y2} \frac{2I_{B2}}{V_T};\ i_{z3} = v_{y3} \frac{2I_{B3}}{V_T} \tag{9.64}$$

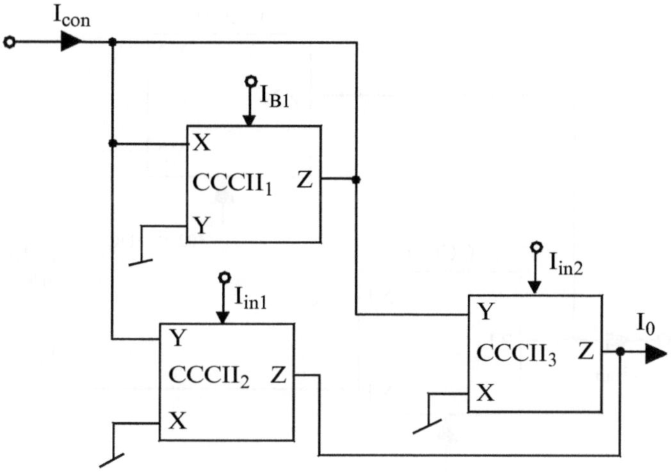

Fig. 9.46 Current mode full wave rectifier [45]

From the above equation we can derive the following equations for the resultant output current of the circuit.

$$i_{z2} = \begin{cases} \dfrac{I_{con}}{2I_{B1}} I_{in} & if \quad I_{in} > 0 \\ 0 & if \quad I_{in} < 0 \end{cases}; \; i_{z3} = \begin{cases} 0 & if \quad I_{in} > 0 \\ \dfrac{I_{con}}{2I_{B1}} I_{in} & if \quad I_{in} < 0 \end{cases} \tag{9.65}$$

$$I_o = i_{z2} + i_{z3} = \frac{I_{con}}{2I_{B1}} |\mathrm{I_{in}}| \tag{9.66}$$

From equation (9.66) it is, therefore, confirmed that the circuit can, indeed, perform the function of a precision full wave rectifier in current mode whose output magnitude can be controlled by control current I_{con} and which is temperature independent.

9.15 Current-Mode Multiplier/Divider Using CCCIIs

A current mode multiplier employing only two CCCIIs was proposed by Abuelma'atti and Al-Qahtani [46] which is shown here in Fig. 9.47.

Applying straight forward analysis, the output current of this circuit is given by:

$$i_{out}(t) = \frac{i_1(t) i_2(t)}{I_{DC}} \tag{9.67}$$

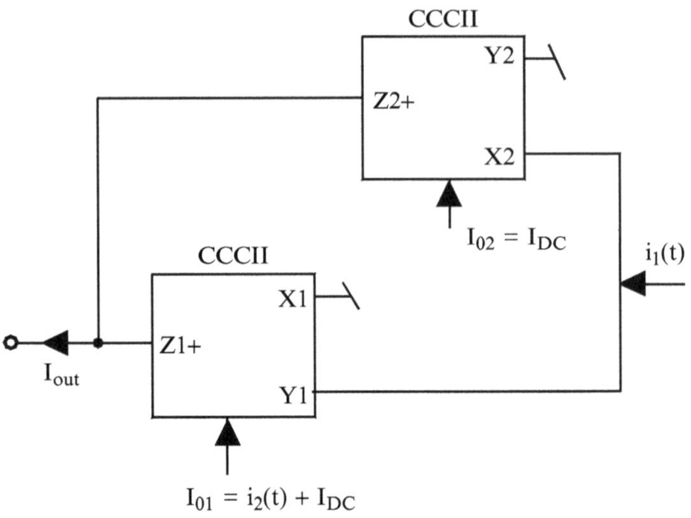

Fig. 9.47 A CM multiplier proposed by Abuelma'atti and Al-Qahtani [46]

Thus, from the above equation it is seen that the circuit indeed performs the function of four quadrant multiplication if $i_1(t)$ and $i_2(t)$ are the input signals while for input signals $i_1(t)$ (or $i_2(t)$) and I_{DC}, the circuit is a divider.

Yuce in [47] demonstrated that the same function can be performed even from a single CCCII-based circuit as shown in Fig. 9.48.

The analysis of this circuit shows that its operation is characterized by the following equation.

$$i_{out} = \left(\frac{2R}{V_T}\right) i_1(t)[i_2(t) + I_{DC}] + i_3(t) \qquad (9.68)$$

If $i_3(t)$ is selected such that $i_3(t) = -\left(\frac{2R}{V_T}\right) I_{DC} i_1(t)$, then the output is given by:

$$i_{out} = \left(\frac{2R}{V_T}\right) i_1(t) i_2(t) \qquad (9.69)$$

From the circuits for carrying out the multiplier/divider operations, it is clear that there is enough scope of devising newer circuits in this area there appears to be enough scope for evolving novel circuits without requiring elaborate realization conditions and/or restrictive conditions.

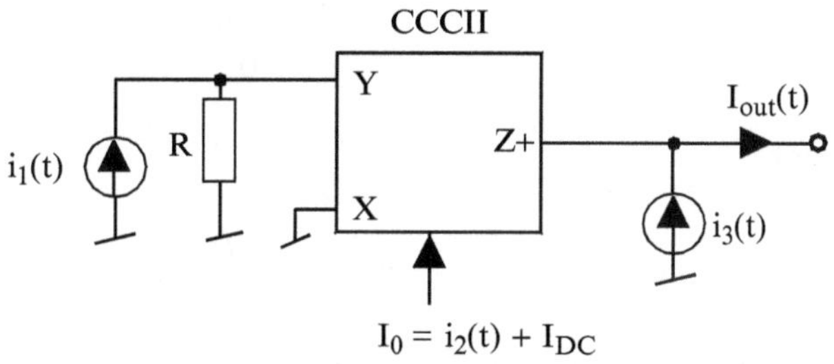

Fig. 9.48 A CM multiplier using only one CCCII proposed by Yuce [47]

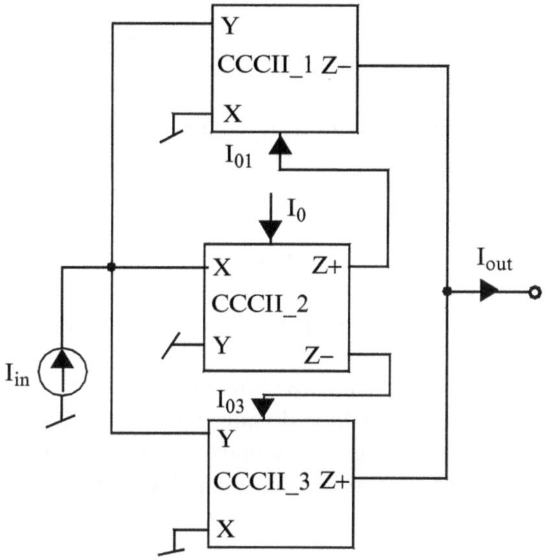

Fig. 9.49 A CM squarer using only CCCIIs (Adapted from [48] © 2009 Elsevier GmbH)

9.16 Squaring/Square Rooting Circuits

A current mode squarer using three CCCIIs was presented by Petchakit et al. [48] and is shown in Fig. 9.49. The analysis of this circuit can be carried out as follows:

$$I_{out} = i_{z1} + i_{z3} = i_{x1} + i_{x3} = \frac{V_{y1}}{R_{x1}} + \frac{V_{y3}}{R_{x3}} = \frac{V_{x2}}{R_{x1}} + \frac{V_{x2}}{R_{x3}}$$
$$= I_{in}R_{x2}\left(\frac{1}{R_{x1}} + \frac{1}{R_{x3}}\right) \tag{9.70}$$

$$i_{Z1} = \begin{cases} I_{in}^2/I_0 & I_{in} > 0 \\ 0 & I_{in} \leq 0 \end{cases} \tag{9.71}$$

and

$$i_{Z3} = \begin{cases} I_{in}^2/I_0 & I_{in} < 0 \\ 0 & I_{in} \geq 0 \end{cases} \tag{9.72}$$

The CCCII-I and the CCCII-3 alternate for positive and negative input cycle. The final output is then obtained as:

$$I_{out} = \frac{I_{in}^2}{I_0} \tag{9.73}$$

From the above it is clear that the circuit is a true four quadrant squaring circuit with no constraint in the polarity of the input signal.

Square rooting circuits using op-amps and MOSFETs suffer from the drawbacks of high frequency limitation finite gain bandwidth product of the op-amps and are generally not suitable for IC implementation. A number of square rooting circuits using CCs have also been advanced a discussion about which has been included in Chap. 8 of this monograph. These circuits however use MOSFETs in non-saturation regions.

In the following we show that using CCCIIs interesting square rooting circuits can be evolved which do not require any MOSFETs and can perform the operation of square rooting for a voltage input signal or current input signal.

A square rooting circuit capable of providing a voltage output proportional to square root of input voltage using only two CCCIIs was proposed by Dejhan and Netbut in [49] and Netbut et al. in [50] and is shown in Fig. 9.50. In this circuit, the variable resistor R_{eq} is realized by the mixed translinear cell with terminal-Y grounded and input impedance seen into terminal-X as shown in Fig. 9.51.

The variable resistance R_{eq} employed in the above circuit is realized by the circuit shown in the Fig. 9.51.

Using the usual characterizations of the CCCII and a straight forward analysis taking $I_{01} = I_{02} = I_0$ and $R_x = V_T/2I_0$ shows that the relation of V_{out} is given by (with $Rc_1 = R_{c2}$):

$$V_{out} = \sqrt{V_T/2}\sqrt{V_{in}} \tag{9.74}$$

Therefore, from equation (9.74), it is clear that the circuit functions as a square rooter.

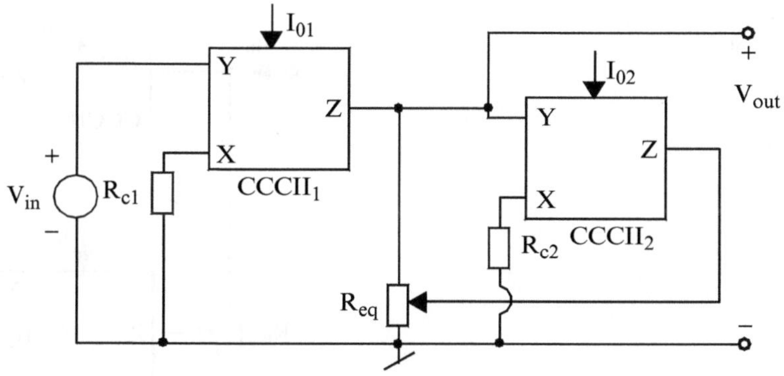

Fig. 9.50 A square rooting circuit proposed by Dejhan and Netbut [49], Netbut et al. [50]

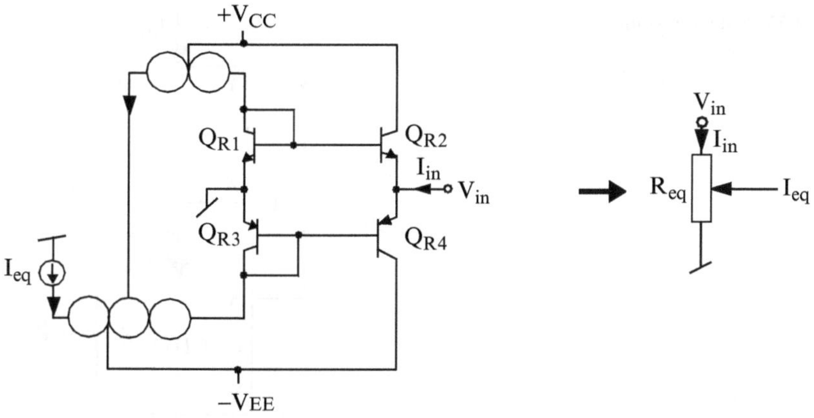

Fig. 9.51 A current-controlled resistance [49, 50]

A square rooter in CM was also proposed by Dejhan and Netbut in [49] and in [50] by Netbut et al.. This circuit is shown in Fig. 9.52. The expression for the output current of this circuit is given by:

$$I_{out} = \sqrt{I_{01}}\sqrt{I_{in}} \tag{9.75}$$

Thus, the circuit of Fig. 9.52 yields the output current which is square-root of the input current signal.

It is interesting to note that while the performance of the VM circuits is temperature dependent that of the CM circuits is temperature independent. Furthermore it can be seen that out of the two circuits the CM square rooter can be realized exclusively by bipolar transistors and hence more suitable for IC implementation. The workability of both the structures has however being established by SPICE simulation results using BJTs PR100N and NR 100N. It has been shown that the

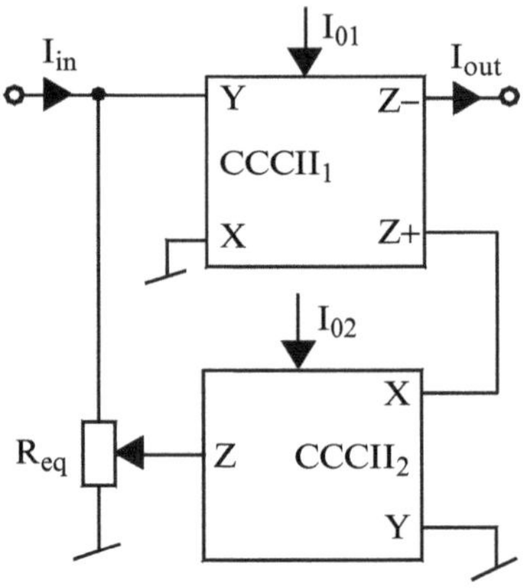

Fig. 9.52 A CM square rooter [49, 50]

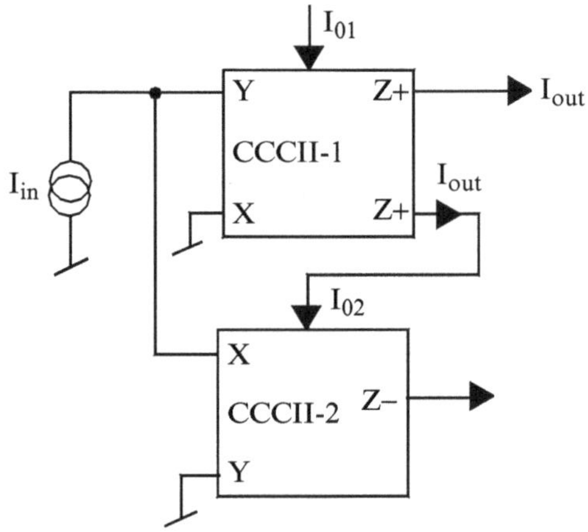

Fig. 9.53 Square rooting circuit proposed by Petchakit et al. (Adapted from [48] ©2009 Elsevier GmbH)

VM circuit operated from a power supply of ±2.5 V has an input range from 0 to 1.2 V, 3 dB bandwidth of 60 MHz and power consumption of 50.4 mW. On the other hand the CM square rooter operating from ±2.5 V exhibited an input range of 0–2 mA, 3 dB bandwidth of 12 MHz and power consumption of only 5.05 mW.

Another square rooting circuit was presented by Petchakit et al. [48] is shown in Fig. 9.53.

The analysis of this circuit shows that:

$$I_{out} = \frac{R_{x2}}{R_{x1}}.I_{in}, \tag{9.76}$$

It follows that

$$I_{out}^2 = I_{01} I_{in} \,\text{or}\, I_{out} = \sqrt{I_{01}}\sqrt{I_{in}} \tag{9.77}$$

9.17 ASK/FSK/PSK/QAM Wave Generator

Amplitude- shift/frequency-shift/phase- shift keying/quadrature amplitude moderator find numerous applications in Digital Communication systems whereas an ASK generator can be realized from a sine-wave oscillator along with a programmable amplifier, a FSK generator requires a number of switches along with a number of standard sine-wave oscillators. Alternatively an FSK generator can also be realized from single sine-wave generator along with binary weighted switched capacitors or resistors. Lastly a PSK generator requires a sine-wave oscillator, phase shifters and switches. In [51] Abuelma'atti demonstrated that ASK, FSK, PSK, QPSK and QAM waveforms can be generated using only a single multiple output current controlled current conveyor along with a few RC components and a number of switches. The interested readers are referred to [51] for further details.

9.18 Advances in the Realization of Bipolar/CMOS/Bi-CMOS CCCIIs

In this section, we bring out a number of recent advances made on the implementation of CCCIIs in bipolar/CMOS/Bi-CMOS technologies.

Barthelemy and Fabre [52] proposed a second generation current controlled conveyor with negative intrinsic resistance whose value is given by $R_x = -\frac{V_T}{2I_B}$ for which the circuit implementation Fig. 9.54 was proposed. It was shown that with the circuit realized from ALA200 transistor arrays from ATT, with = ±1.5 V DC power supplies the circuit exhibits bandwidth higher than 40 MHz and a total power consumption only 1 mW for a bias current of 50 μA. It was shown that this building block was particularly useful in realizing a floating negative controlled resistance.

In [53] Minaei et al. developed two new realizations of CCCII one in CMOS and the other in Bipolar technology, which were claimed to provide good linearity, very high input impedance at port-Y, high output impedance at port- Z and highly accurate current gain.

Chunhua et al. in [54] studied a new CCCII architecture in which all the current mirrors in the CCCII architecture were replaced by Wilson current mirrors as shown in the schematic for Fig. 9.55. In this circuit, transistors Q_1–Q_4 constitute the basic mixed translinear cell whereas all the current mirrors and current repeaters have been realized by Wilson mirror based implementations as shown in the diagram.

The test results of this circuit using transistors TEC9014C and TEC9015C with I_B of 13 μA shows that the circuit exhibits a good correspondence between input current i_x and output current i_{z+} and i_{z-} has higher input impedance than the conventional design and offers a bandwidth of about 50 MHz.

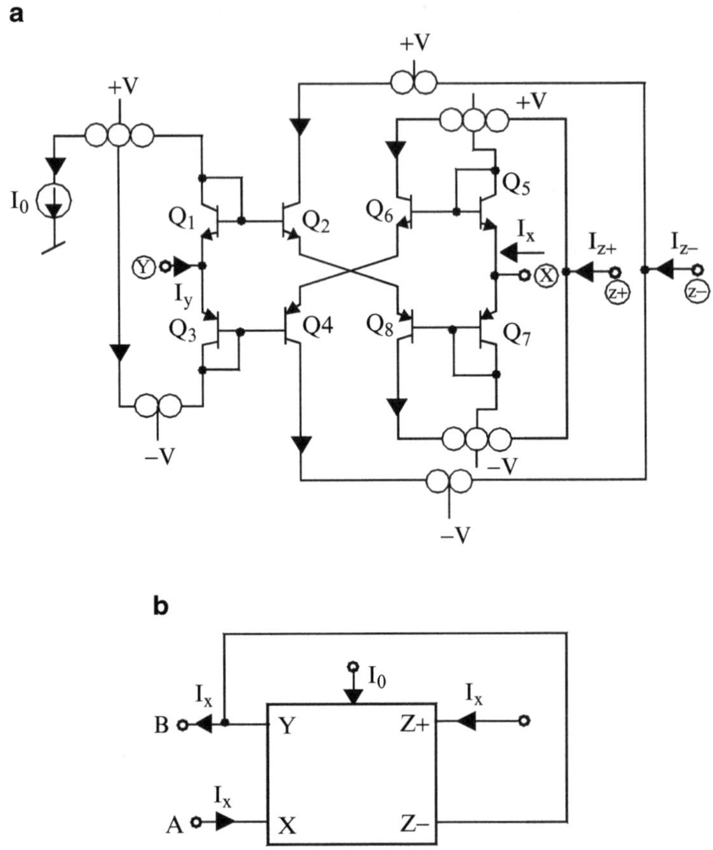

Fig. 9.54 (**a**) Implementation of the CCCII± with negative intrinsic resistance (**b**) circuit connections for realizing a floating negative controlled resistance [52]

Whereas most common CCCIIs are built around the four transistors mixed translinear cell, a number of authors also attempted alternative architectures based upon differential pair with biasing circuitry and mechanism of duplicating the X- port current through CMOS transistors. Two such architectures proposed by Erkan and Alci for realizing a CCCII with positive R_x and CCCII with negative R_x [55] are shown in Figs. 9.56 and 9.57 respectively.

In the first circuit, by a straight forward analysis, it can be easily verified that relationship between current I_x and the differential input voltage (V_x-V_y) is given by:

$$I_x = I_B \tanh\left(\frac{V_x - V_y}{2V_T}\right) \tag{9.78}$$

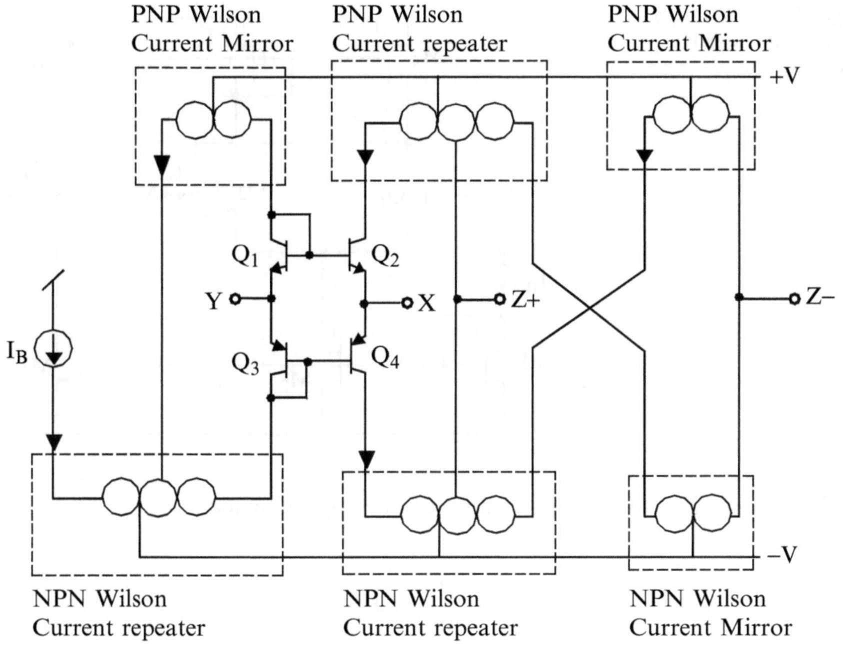

Fig. 9.55 A CCCII circuit based upon Wilson current mirrors [54]

Fig. 9.56 A BiCMOS CCCII with positive Rx [55]

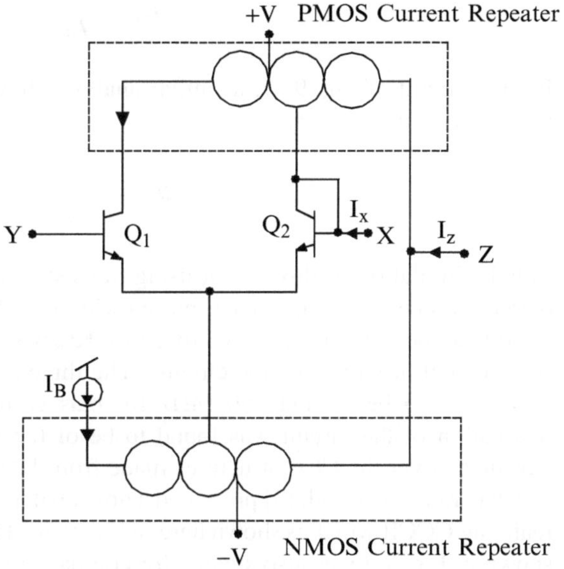

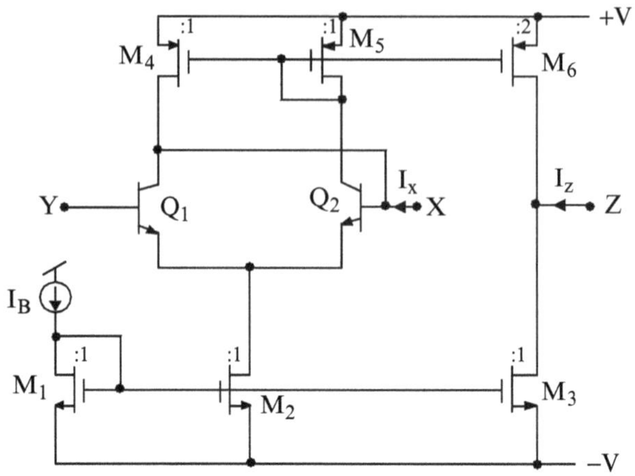

Fig. 9.57 A BiCMOS CCCII with negative Rx [55]

If (V_x-V_y) <<2V_T the above equation can be approximated as:

$$I_x = I_B \frac{V_x - V_y}{2V_T} \tag{9.79}$$

which implies that with Y-terminal grounded the input resistance looking into the terminal-X is given by:

$$R_x = \frac{2V_T}{I_B} \tag{9.80}$$

For the circuit of Fig. 9.57, a similar analysis shows that in this case the intrinsic resistance R_x is given by:

$$R_x = -\frac{2V_T}{I_0} \tag{9.81}$$

SPICE simulation of this circuit using transistors of the type AMS S35 BiCMOS 0.35 μm process with the CCCII biased with ±1.5 V, the parasitic resistance of the X-port can be tuned from ±1.03 to ±51.6 KΩ when the biasing current I_B changes from 1 to 50 μA for both the circuits. The three dB bandwidth of these structures was found to be around 550 MHz for bias current of $I_B = 10$ μA. The power dissipation of the circuit was found to be of the order of 75 μW which is clear advantage over the CCC structures made from BJTs exclusively.

Chaisricharoen et al. [3] presented a novel balanced differential pair structure for realizing CCCII which is shown here in Fig. 9.58. The classical translinear structure shown in Fig. 9.2 was also studied for comparison.

From detailed mathematical and simulation study it has been found that for a DC bias current 50 μA. this architecture exhibits a current gain of 0.991, a voltage gain of 0.9915 with 3 dB bandwidth of both in excess of 450 MHz and the X-port input

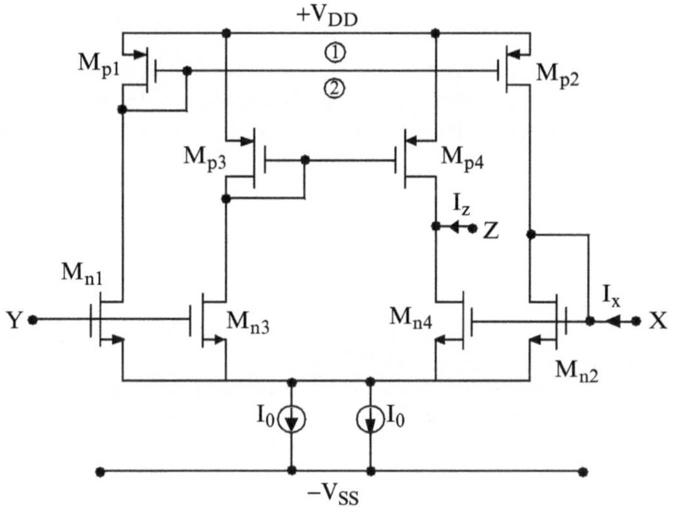

Fig. 9.58 A CMOS CCCII using balanced differential structure [3]

resistance of 2.75 KΩ. An approximate expression for the input resistance R_x as a function of the DC bias current I_0 is given by:

$$R_x \cong \frac{1}{\sqrt{I_0 \mu_n C_{ox} \frac{W_D}{L_D}}} \tag{9.82}$$

where W_D/L_D is the aspect ratio of the two identical and MOS differential pair transistors. This study reveals that the balanced differential pair CCCII is a good element when almost unity and wideband current and voltage transfer gains are of interest.

More recently, CMOS current controlled fully balanced current conveyor architecture was proposed by Chunhua et al. [4] whereas a CMOS-based circuit of DO-CCCII using 65 nm CMOS technology was introduced by Abbas et al. [5].

Concluding Remarks

Although traditionally operational transconductance amplifiers (OTAs) or CMOS transconductors with controllable transconductance values have been the main elements for designing electronically-controllable analog circuits, during the past 15 years, the so-called Current-Controlled Current Conveyors (CCCIIs) have also attracted considerable attention in devising various electronically-controllable linear as well as nonlinear functions. This chapter has presented some of the prominent works done in this area. From a comprehensive survey of references, it has been revealed that the work on the

(continued)

(continued)
various applications of these building blocks is far from being complete and it is clear that numerous newer possibilities are yet to be unfolded and are waiting to be discovered. This, however, warrants further investigations and research on this exciting area.

References

1. Fabre A, Saaid O, Wiest F, Boucheron C (1995) Current controlled bandpass filter based on translinear conveyor. Electron Lett 31:1727–1728
2. Fabre A, Saaid O, Wiest F, Boucheron C (1996) High frequency applications based on a new current controlled conveyor. IEEE Trans Circ Syst-I 43:82–91
3. Chaisricharoen R, Chipipop B, Sirinaovakul B (2010) CMOS CCCII: structures, characteristics, and considerations. Int J Electron Commun (AEU) 64:540–557
4. Chunhua W, Qiujing Z, Haiguang L (2009) CMOS current controlled fully balanced current conveyor. J Semiconductors 30:075009-1–075009-6
5. Abbas Z, Scotti G, Olivieri M (2011) Current controlled current conveyor (CCCII) and application using 65 nm CMOS technology. World Acad Sci Eng Tech 55:935–939
6. Zhang Q, Wang C, Sun J, Du S (2011) A new type of current conveyor and its application in fully balanced differential current mode elliptic filter design. J Electr Eng 62:126–133
7. Kapur G, Mittal S, Markan CM, Pyara VP (2012) Design of analog field programmable CMOS current conveyor. Sci J Circ Syst Sig Process 1:9–21
8. Seguin F, Godara B, Alicalapa F, Fabre A (2004) 2.2 GHz all n-p-n second-generation controlled conveyor in pseudo class AB using 0.8-μm BiCMOS technology. IEEE Trans Circ Syst–II 51:369–373
9. Senani R, Singh AK, Singh VK (2004) A new floating current-controlled positive resistance using mixed translinear cells. IEEE Trans Circ Syst-II 51:374–377
10. Pawarangkoon P, Kiranon W (2004) Electronically-tunable floating resistor. Int J Electron 91:665–673
11. Saaid O, Fabre A (1996) Class AB current-controlled resistor for high performance current-mode applications. Electron Lett 32:4–5
12. Maheshwari S (2005) Additional summing/difference amplifiers using CCCIIs. J Active Passive Electron Devices 1:159–162
13. Maheshwari S (2002) High CMRR wide bandwidth instrumentation amplifier using current controlled conveyors. Int J Electron 89:889–896
14. Ercan H, Tekin SA, Alci M (2012) Voltage and current controlled high CMRR instrumentation amplifier using CMOS current conveyors. Turk J Electron Eng Comput Sci 20:547–556
15. Kiranon W, Pawarangkoon P (1997) Floating inductance simulation based on current conveyors. Electron Lett 33:1748–1749; Comment Abuelma'atti MT Floating inductance simulation based on current conveyors. Electron Lett 34: 1037–1038
16. Yuce E, Minaei S, Cicekoglu O (2006) Resistorless floating immittance function simulators employing current controlled conveyors and a grounded capacitor. Electr Eng 88:519–525
17. Khan IA, Zaidi MH (2003) A novel ideal floating inductor using translinear conveyors. Active Passive Electron Comp 26:87–89
18. Sedef H, Sagbas M, Acar C (2008) Current-controllable fully-integrated inductor simulator using CCCIIs. Int J Electron 95:425–429
19. Minaei S, Yuce E (2008) Realization of tunable active floating inductance simulators. Int J Electron 95:27–37

20. Abuelma'atti MT, Tasadduq NA (1999) Electronically tunable capacitance multiplier and frequency-dependent negative-resistor simulator using the current-controlled current conveyor. Microelectron J 30:869–873
21. Yuce E, Minaei S (2008) Electronically tunable simulated transformer and its application to stagger-tuned filter. IEEE Trans Instrum Meas 57:2083–2088
22. Parveen T, Ahmed MT, Khan IA (2009) A canonical voltage-mode universal CCCII-C filter. J Active Passive Electron Devices 4:7–12
23. Ranjan A, Paul SK (2011) Voltage mode universal biquad using CCCII. Active Passive Electron Comp 2011:1–5, ID 439052
24. Altuntas E, Toker A (2002) Realization of voltage and current mode KHN biquads using CCCIIs. Int J Electron Commun(AEU) 56:45–49
25. Wang C, Liu H, Zhao Y (2008) A new current mode current controlled universal filter based on CCCII(±). Circ Syst Sig Process 27:673–682
26. Senani R, Singh VK, Singh AK, Bhaskar DR (2004) Novel electronically controllable current-mode universal biquad filter. IEICE Electron Express 1:410–415
27. Chen HP, Chu PL (2009) Versatile universal electronically tunable current-mode filter using CCCIIs. IEICE Electron Express 6:122–128
28. Maheshwari S, Khan IA (2004) Novel cascadable current-mode translinear-C universal filter. Active Passive Electron Comp 27:215–218
29. Yuce E (2009) Current-mode electronically tunable biquadratic filters consisting of only CCCIIs and grounded capacitors. Microelectron J 40:1719–1725
30. Abuelma'atti MT (2003) A novel mixed mode current controlled current conveyor based filter. Active Passive Electron Comp 26:185–191
31. Ozoguz S, Toker A (2002) Tunable ladder-type realization of current mode elliptic filters. Int J Electron Commun 56:193–199
32. Jiraseree-amornkun A, Surakampontorn W (2008) Efficient implementation of tunable ladder filters using multi-output current controlled conveyors. Int J Electron Commun 62:11–23
33. Yuce E, Minaei S (2008) On the realization of high-order current mode filter employing current controlled conveyors. Comput Electr Eng 34:165–172
34. Koksal M, Sagbas M (2008) A versatile signal flow graph realization of a general current transfer function. Int J Electron Commun 62:33–40
35. Ranjan A, Paul SK (2011) Nth order voltage mode active-C filter employing current controlled current conveyors. Circ Syst 2:85–90
36. Liu SI, Yang CY (1996) Higher-order immittance function synthesis using CCIIIs. Electron Lett 32:2295–2296
37. Horng JW (2001) A sinusoidal oscillator using current-controlled current conveyors. Int J Electron 88:659–664
38. Fongsamut C, Anuntahirunrat K, Kumwachara K, Surakampontorn W (2006) Current conveyor-based single element controlled and current controlled sinusoidal oscillators. Int J Electron 93:467–478
39. Kiranon W, Kesorn J, Wardkein P (1996) Current controlled oscillator based on translinear conveyors. Electron Lett 32:1330–1331
40. Abuelma'atti MT, Tasadduq NA (1998) A novel current controlled oscillator using translinear current conveyor. Frequenz 52:123–124
41. Maheshwari S (2010) Current mode third order quadrature oscillator. IET Circ Devices Syst 4:188–195
42. Erdal C, Kuntman H, Kafali S (2004) A current controlled conveyor based proportional-integral-derivative (PID) controller. J Electr Electron Engg 4:1243–1248
43. Maheshwari S, Khan IA (2005) An integrable precision rectifier with current controlled output. J Active Passive Electron Devices 1:177–182
44. Anuntahirunrat K, Tangsrirat W, Riewruja V, Surakampontorn W (2004) Sinusoidal frequency doubler and full-wave rectifier based on translinear current-controlled current conveyors. Int J Electron 91:227–239

45. Siripruchyanun M, Jaikla W (2006) A novel precision current mode full wave rectifier and class B push-pull current amplifier using BiCMOS current controlled current conveyors. J KMITNB 16:1–7
46. Abuelma'atti MA, Al-qahtani MA (1998) A current-mode current-controlled current-conveyor-based analogue multiplier/divider. Int J Electron 85:71–77
47. Yuce E (2008) Design of a simple current-mode multiplier topology using a single CCCII+. IEEE Trans Instrum Meas 57:631–637
48. Petchakit W, Lorsawatsiri A, Kiranon W, Wongtaychathum C, Wardkein P (2010) Current mode squaring, square-rooting and vector summation circuits. Int J Electron Commun (AEU) 64:443–449
49. Dejhan K, Netbut C (2007) New simple square-rooting circuits based on translinear current conveyor. Int J Electron 94:707–723
50. Netbut C, Kumngern M, Prommee P, Dejhan K (2007) New simple square-rooting circuits based on translinear current conveyors. ECTI Trans Electr Eng Electro Commun 5:10–17
51. Abuelma'atti MT (2002) New ASK/FSK/PSK/QAM wave generator using a single current controlled multiple output current conveyor. Int J Electron 89:35–43
52. Barthelemy H, Fabre A (2002) A second generation current-controlled conveyor with negative intrinsic resistance. IEEE Trans CAS-I 49:63–65
53. Minaei S, Kaymak D, Kuntman H (2002) New high performance realization for current-controlled conveyor (CCCII). J Electric Electron 2:547–553
54. Chunhua W, Qiujing Z, Wei Y (2007) A second current controlled current conveyor realization using Wilson current mirror. Int J Electron 94:699–706
55. Ercan H, Alci M (2013) A new design for BiCMOS controlled current conveyor. Elektronika IR Elektrotechnika 19:56–60
56. Fabre A, Saaid O, Wiest F, Boucheron C (1997) Low power current-mode second-order band pass IF filter. IEEE Trans Circ Syst-II 44:436–446
57. Abuelma'atti MT, Al-Zaher HA (1997) Nonlinear performance of the mixed translinear loop. Int J Electron 83:467–471
58. Kiranon W, Kesorn J, Sangpisit W, Kamprasert N (1997) Electronically tunable multifunctional translinear-C filter and oscillator. Electron Lett 33:573–574
59. Abuelma'atti MT, Abed SM (1997) A new current controlled multiple output current conveyor. Proc Natl Sci Counc ROC (A) 21:616–622
60. Abuelma'atti MT, Tasadduq NA (1998) Universal current-controlled current mode filters using the current controlled conveyor. Proc Natl Sci Counc ROC (A) 22:358–361
61. Abuelma'atti MT, Tasadduq NA (1998) Universal current-controlled current mode filters using the multiple-output translinear current conveyor. Frequenz 52:252–254
62. Abuelma'atti MT, Al Qahtani M (1998) Universal current-controlled current mode filters with three inputs and one output using the current controlled conveyor. Active Passive Electron Comp 21:33–41
63. Abuelma'atti MT, Tasadduq NA (1998) A novel single-input multiple-output current mode current controlled universal filter. Microelectron J 29:901–905
64. Abuelma'atti MT (1998) Comment: floating inductance simulation based on current conveyors. Electron Lett 34:1037–1038
65. Abuelma'atti MT, Al-Qahtani MA (1999) On the realization of the current controlled current-mode amplifier using the current controlled conveyor. Int J Electron 86:1333–1340
66. Acar C, Ozoguz S (1999) A new versatile building block: current differencing buffered amplifier suitable for analog signal processing filters. Microelectron J 30:157–160
67. Khan IA, Zaid MH (2000) Multifunctional translinear-C current mode filter. Int J Electron 87:1047–1051
68. Barthelemy H, Fabre A (2000) A new floating controlled resistance operating in class AB. IEEE Trans Circ Syst-I 47:67–72
69. Seguin F, Fabre A (2001) 2 GHz controlled current conveyor in standard 0.8 μm BiCMOS technology. Electron Lett 37:329–330

70. Minaei S, Cicekoglu O, Kuntman H, Turkoz S (2001) High output impedance current-mode low pass, band pass and high pass filters using current controlled conveyors. Int J Electron 88:915–922
71. Minaei S, Turkoz S (2001) Erratum 'New current mode current controlled universal filter with single input and three outputs'. Int J Electron 88:955
72. Minaei S, Cicekoglu O, Kuntman H, Turkoz S (2002) Electronically tunable, active only floating inductance simulation. Int J Electron 89:905–912
73. Maheshwari S (2003) Active-only current controlled summing/difference amplifiers using CCCIIs. Active Passive Electron Comp 26:231–234
74. Maheshwari S, Khan IA (2003) Simple first-order translinear-C current mode all pass sections. Int J Electron 90:79–85
75. Shah NA, Iqbal SZ (2003) Current mode active only universal filter. Int J Electron 90:407–411
76. Maheshwari S (2003) Electronically-tunable quadrature oscillator using translinear conveyors and grounded capacitors. Active Passive Electron Comp 26:193–196
77. Maheshwari S (2004) New voltage and current mode APS using current controlled conveyor. Int J Electron 91:735–743
78. Minaei S, Turkoz S (2004) Current mode electronically tunable universal filter using only plus-type current controlled conveyors and grounded capacitors. ETRI J 26:292–296
79. Sagbas M, Fidanboylu K (2004) Electronically-tunable current-mode second-order universal filter using minimum elements. Electron Lett 40:2–3
80. Chang CM, Al-Hashimi BM, Ross JN (2004) Unified active filter biquad structures. IEE Proc Circ Devices Syst 151:273–277
81. Shah NA, Rather MF, Iqbal SZ (2005) Electronically-tunable high output impedance current mode universal filter. J Active Passive Electron Devices 1:163–169
82. Pandey N, Paul SK, Bhattacharyya A, Jain SB (2005) A novel current controlled current mode universal filter: SITO approach. IEICE Electron Express 2:451–457
83. Yuce E, Minaei S, Metin B (2005) Comment Electronically tunable current mode second order universal filters using minimum elements. Electron Lett 41:453
84. Senani R, Singh VK, Singh AK, Bhaskar DR (2005) Tunable current-mode universal biquads employing only three MOCCs and all grounded passive elements: additional new realizations. Frequenz 59:220–224
85. Maheshwari S, Khan IA (2005) High performance versatile translinear-C universal filter. J Active Passive Electron Devices 1:41–51
86. Zouaoui-Abouda H, Fabre A (2006) New high-value floating controlled resistor in CMOS technology. IEEE Trans Instrum Meas 55:1017–1020
87. Tangsrirat W, Surakampontorn W (2006) Electronically-tunable current-mode universal filter employing only plus-type current-controlled conveyors and grounded capacitors. Circ Syst Sig Process 25:701–713
88. Shah NA, Rather MF, Iqbal SZ (2006) SIFO electronically tunable current-mode cascadable active only universal filter. J Active Passive Electron Devices 8:327–334
89. Shah NA, Rather MF, Iqbal SZ (2006) SITO electronically tunable high output impedance current mode universal filter. Analog Integr Circ Sig Process 47:335–338
90. Yuce E, Minaei S, Cicekoglu O (2006) Universal current-mode active-C filter employing minimum number of passive elements. Analog Integr Circ Sig Process 46:169–171
91. Metin B, Minaei S, Cicekoglu O (2007) Enhanced dynamic range analog filter topologies with a notch/all pass circuit example. Analog Integr Circ Sig Process 51:181–189
92. Siripruchyanun M, Jaikla W (2007) Three input single-output electronically controllable dual-mode universal biquad filter using DO-CCCIIs. Active Passive Electron Comp Article ID 36849
93. Chang CM, Huang TH, Tu SH, Hou CL, Horng JW (2007) Universal active current filter using single second-generation current controlled conveyor. Int J Circ Syst Sig Process 1:194–198

94. Minaei S, Yuce E (2007) Current-mode active–C filter employing reduced number of CCCII+ s. J Circ Syst Comput 16:507–515
95. Maheshwari S (2007) A new current-mode current-controlled all-pass section. J Circ Syst Comput 16:181–189
96. Yuce E (2007) Comment on SITO electronically tunable high output impedance current-mode universal filter. Analog Integr Circ Sig Process 50:271–272
97. Keskin AU, Cam U (2007) Insensitive high-output impedance minimum configuration SITO type current mode biquad using dual-output current conveyors and grounded passive. Comp Int J Electron Commun 61:341–344
98. Yuce E, Tokat S, Minaei S, Cicekoglu O (2007) Stability problems in universal current mode filters. Int J Electron Commun (AEU) 61:580–588
99. Tangsrirat W (2007) Current-tunable current mode multifunction filter based on dual output current controlled conveyors. Int J Electron Commun (AEU) 61:528–533
100. Tangsrirat W, Surakampontorn W (2007) High output impedance current mode universal filter employing dual-output current controlled conveyors and grounded capacitors. Int J Electron Commun(AEU) 61:127–131
101. Tsukutani T, Sumi Y, Yabuki N (2007) Versatile current-mode biquadratic circuit using only plus type CCCIIs and grounded capacitors. Int J Electron 94:1147–1156
102. Pandey N, Paul SK, Jain SB, Bhattacharyya A (2007) Generalized mixed mode universal filter realization using current controlled conveyors. HAIT J Sci Eng: 1–17
103. Maheshwari S, Khan IA (2007) Novel voltage/current mode translinear-C quadrature oscillator. J Active Passive Electron Devices 2:235–239
104. Tangsrirat W (2007) Current-tunable current-mode multifunction filter based on dual-output current- controlled conveyors. Int J of Electron Commun (AEU) 61:528–523
105. Pandey N, Paul SK, Jain SB (2008) Voltage mode universal filter using two plus type CCIIs. J Active Passive Electron Devices 3:165–173
106. Pandey N, Paul SK, Jain SB (2008) New high-input impedance voltage-mode universal biquad: multi input multi output. J Active Passive Electron Devices 3:93–100
107. Yuce E, Kircay A, Tokat S (2008) Universal resistorless current mode filters employing CCCIIs. Int J Circ Theor Appl 36:739–755
108. Oztayfun S, Kilinc S, Celebi A, Cam U (2008) A new electronically tunable phase shifter employing current controlled current conveyors. Int J Electron Commun 62:228–231
109. Yuce E, Minaei S (2008) Universal current-mode filters and parasitic impedance effects on the filter performances. Int J Circ Theor Appl 36:161–171
110. Pandey N, Paul SK (2008) A novel electronically tunable sinusoidal oscillator based on CCCII (−IR). J Active Passive Electron Devices 3:135–141
111. Biolek D, Senani R, Biolkova V, Kolka Z (2008) Active elements for analog signal processing: classification, review and new proposals. Radioeng J 17:15–32
112. Siripruchyanun M, Chanapromma C, Silapan P, Jaikla W (2008) BiCMOS current-controlled current feedback amplifier (CC-CFA) and its applications. WSEAS Trans Electron 5:203–219
113. Siripruchyanun M (2008) A current-mode analog multiplier/divider based on CCCDTA. Int J Electron Commun (AEU) 62:223–227
114. Soliman AM (2009) Comment on: realization of voltage and current mode KHN biquads using CCCII. Int J Electron Commun 63:877
115. Siripruchyanun M, Jaikla W (2009) Cascadable current-mode biquad filter and quadrature oscillator using DO-CCCIIs and OTA. Circ Syst Sig Process 28:99–110
116. Tangsrirat W, Surakampontorn W (2009) Low component current mode universal filter using current controlled conveyors and grounded capacitors. J Active Passive Electron Devices 8:259–264
117. Shah NA, Rather MF (2009) Electronically tunable current mode/mixed-mode/voltage mode multifunction active only biquads. J Active Passive Electron Devices 8:223–235
118. Lahiri A (2010) Current controlled inductor simulation using dual-output current controlled conveyors. J Active Passive Electron Devices 5:191–195

119. Pandey N, Paul SK (2010) Digitally switched current mode universal filter. J Active Passive Electron Devices 5:197–208
120. Kummgern M, Jongchanachavawat W, Dejhan K (2010) New electronically tunable current mode universal biquad filter using translinear current conveyors. Int J Electron 97:511–523
121. Sotner R, Slezak J, Dostal T, Petrzela J (2010) Universal tunable current mode biquad employing distributed feedback structure with MO-CCCII. J Electr Eng 61:52–56
122. Sagbas M, Ayten UE, Sedef H, Koksal M (2011) Reply to comment on "Electronically tunable floating inductance simulator". Int J Electron Commun(AEU) 66:86–88
123. Tangsrirat W, Channumsin O (2011) Minimum-component current-mode universal filter. Ind J Pure Appl Physics 49:137–141
124. Wang C, Xu J, Keskin AU, Du S, Zhang Q (2011) A new current mode current controlled SIMO type universal filter. Int J Electron Commun 65:231–234
125. Lahiri A (2011) Deriving (MO) (I) CCCII based second order sinusoidal oscillators with non-interactive tuning laws using state variable method. Radioengineering 20:349–353
126. Saied AB, Salem SB, Masmoudi DS (2011) A new CMOS current controlled quadrature oscillator based on a MCCII. Circ Syst 2:269–273
127. Yasin MY, Gopal B (2011) High frequency oscillator design using a single 45 nm CMOS current controlled current conveyor (CCCII+) with minimum passive components. Circ Syst 2:53–59
128. Fakhfakh M, Pierzchala M, Rodanski B (2012) On the design of active inductors with current-controlled voltage sources. Analog Integr Circ Sig Process 73:89–98
129. Metin B, Pal K, Cicekoglu O (2012) A new approach for high-input impedance in voltage mode filters using first-generation current conveyor in place of second-generation current conveyor. Int J Electron 99:131–139
130. Yildiz HA, Toker A, Ozoguz S (2013) Biquadratic filter applications using a fully differential active-only integrator. Radioengineering 22:3–13
131. Srisakultiew S, Lawanwisut S, Siripruchyanun M (2013) A current mode electronically controllable multifunction biquadratic filter using CCCIIs. Int J Adv Telecommun Electron Sig Syst 2:45–50
132. Pandey N, Paul SK (2013) Mixed mode universal filter. J Circ Syst Comput 22:1250064-1–1250064-10

Chapter 10
Varieties of Current Conveyors

Abstract This chapter describes numerous varieties of CCs evolved till date. In each case, the characterization of the building block is given, its salient features have been outlined and wherever necessary and relevant, the most significant applications/advantageous features have also been highlighted.

10.1 Introduction

The initial focus of researchers was dominantly on investigating the applications and hardware implementation of only the basic types of current conveyors introduced by Sedra and Smith namely CCI and CCII, as evidenced by the numerous publications during 1968 till about 1990 which are included in several chapters of this monograph and have been well documented in several tutorial/ review papers on Current Conveyors. With the advent of a number of possible hardware implementations of CCI/CCII, it was soon understood that creating multiple outputs having similar or complementary polarity was quite easy thereby rendering further flexibility to CCI± as well as CCII±. Subsequently, many researchers started contemplating different varieties of current conveyors with modified terminal characteristics between X, Y and Z ports and/or increasing the number of X, Y and Z terminals or some other generalizations, thereby leading to a variety of new types of current/voltage conveyors, see [1–73].

In this chapter, we enumerate many different varieties of current conveyors introduced till date, discuss their models/implementations and also highlight the areas of applications or characteristic features due to which the new variants offer significant advantages over the basic types of Current Conveyors.

10.2 Different Variants of the Current Conveyors

In the following, we present the new variants of the current conveyors evolved so far.

© Springer International Publishing Switzerland 2015
R. Senani et al., *Current Conveyors*, DOI 10.1007/978-3-319-08684-2_10

10.2.1 Current Voltage Conveyor

In 1981, Filanovsky and Stromsmoe [1] introduced the *Current Voltage Conveyor* (CVC) as an extension to the CCI by adding another auxiliary voltage-controlled-voltage-source (VCVS) as the fourth port, thereby leading to the following characterization:

$$\begin{bmatrix} v_w \\ v_x \\ i_y \\ i_z \end{bmatrix} = \begin{bmatrix} 0 & 0 & \pm 1 & 0 \\ 0 & 0 & \pm 1 & 0 \\ 0 & \pm 1 & 0 & 0 \\ 0 & \pm 1 & 0 & 0 \end{bmatrix} \begin{bmatrix} i_w \\ i_x \\ v_y \\ v_z \end{bmatrix} \tag{10.1}$$

Two versions of the CVC called CVC1 and CVC2 were postulated which were shown to be useful elements for the realization of controlled sources, negative impedance converters and other active elements. It was shown that the CVC could be realized by the OA-OTA based synthesis as done previously for CC realization by Senani [2]. The Filanovsky and Stromsmoe implementation of the CVC from [1] is shown in Fig. 10.1.

This circuit can realize a CVCI subject to the conditions $GR_2 = 1$ and $GR_1 = 1$ with R_4 open circuited. It can be readily seen that an appropriate change of polarities of two OTAs will result in the realization of various controlled sources.

It was also demonstrated by Filanovsky and Stromsmoe [1] that the CVC can realize a variety of CCs as special cases, namely: CVC-I by including ports 1, 2, 3, 4; CCI by including ports 1, 2, 3; CCII by including ports 2, 3; VC-I by including ports 1, 3, 4; VC-II by including ports 1, 4 and CVC-II by including ports 2, 4.

Subsequently, Dostal and Pospisil [3] systematically demonstrated that a variety of new voltage conveyors are basically special cases of a family of three port immittance converters.

10.2.2 Generalized Current Conveyor

The generalized current conveyor (GCC) was defined by Senani [4] in 1984 as a grounded three-port active building block characterized by the hybrid matrix:

$$\begin{bmatrix} i_y \\ v_x \\ i_z \end{bmatrix} = \begin{bmatrix} 0 & h_{12} & 0 \\ h_{21} & 0 & 0 \\ 0 & h_{32} & 0 \end{bmatrix} \begin{bmatrix} v_y \\ i_x \\ v_z \end{bmatrix} \tag{10.2}$$

where the hybrid parameters involved may be either positive or negative. Two possible implementations of this building block are shown in Fig. 10.2.

It was shown that a GCC characterized by $h_{12} < 0$, $h_{32} < 0$ and $h_{21} = 1$ in conjunction with an active-RC network derived from a given grounded immittance

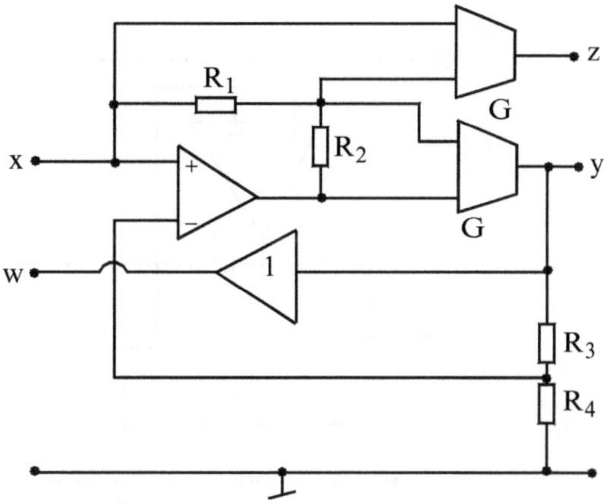

Fig. 10.1 Filanovsky and Stromsmoe implementation of the CVC [1]

simulation network (i.e. a one-port having $Y_{11} = y(s)$) can be used to realize the same type of immittance function in floating form (i.e., a two-port characterized by $y_{11} = -y_{12} = -y_{21} = y_{22} = k\ y(s)$, k being a positive scaling factor) with several advantages over previously known methods of floating immittance synthesis, as per the schematic shown in Fig. 10.3, where the network N_1 represents a unilateral floating impedance network characterized by $y_{11} = -y_{12} = y(s)$; $y_{21} = y_{22} = 0$.

10.2.3 Operational Floating Conveyor

The operational floating Conveyor (OFC) was introduced by Toumazou et al. [5] in 1991. It is a four-port active building block characterized by the following hybrid matrix:

$$\begin{bmatrix} v_x \\ i_y \\ v_w \\ i_z \end{bmatrix} = \begin{bmatrix} 0 & 1 & 0 & 0 \\ 0 & 0 & 0 & 0 \\ Z_t & 0 & 0 & 0 \\ 0 & 0 & -1 & 0 \end{bmatrix} \begin{bmatrix} i_x \\ v_y \\ i_w \\ v_z \end{bmatrix} \tag{10.3}$$

The output voltage at port-w is obtained by multiplying the current at the x-port by the trans-impedance Z_t, the voltage at port-x follows the voltage at port-y which has infinite input impedance so that the current $i_y = 0$ and finally, the current flowing into port-w is conveyed to port-z. An exemplary translinear implementation of the OFC is shown in Fig. 10.4. In this circuit diagram, as well as elsewhere subsequently, the various current mirrors and current repeaters are shown by their standard symbolic notations consisting of two or more circles touching each other with connection to positive/negative power supply rail(s), with input terminal shown with an arrow mark and the terminal(s) without arrow mark being the output terminal(s).

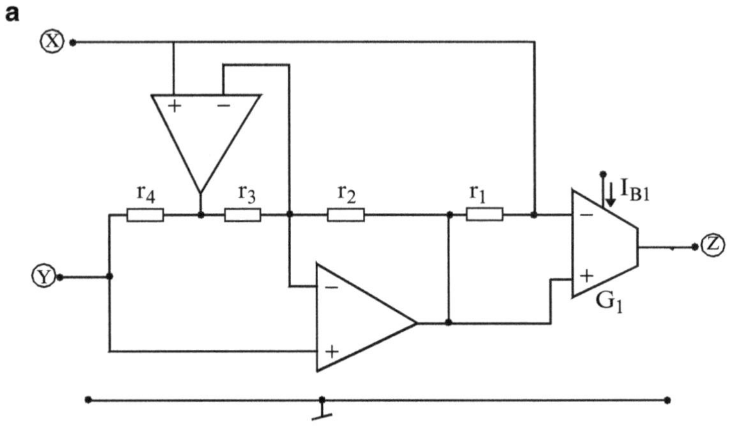

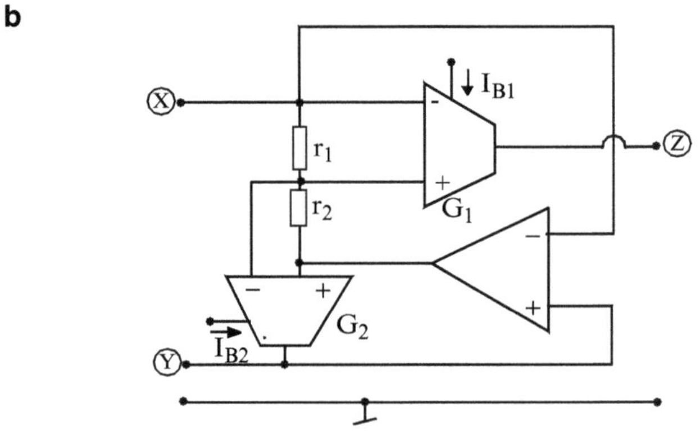

Fig. 10.2 Two possible circuit implementations of the GCC proposed by Senani in [4] (**a**) $h_{12} = -r_1r_3/r_2r_4$, $h_{21} = 1$, $h_{32} = -Gr_1$ (**b**) $h_{12} = -G_2r_2$, $h_{21} = 1$, $h_{32} = -G_1r_1$

Fig. 10.3 Scheme of using the GCC as an immittance-floatator [4]

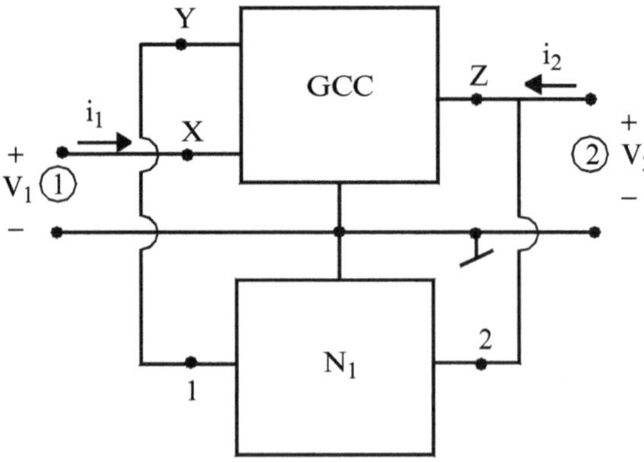

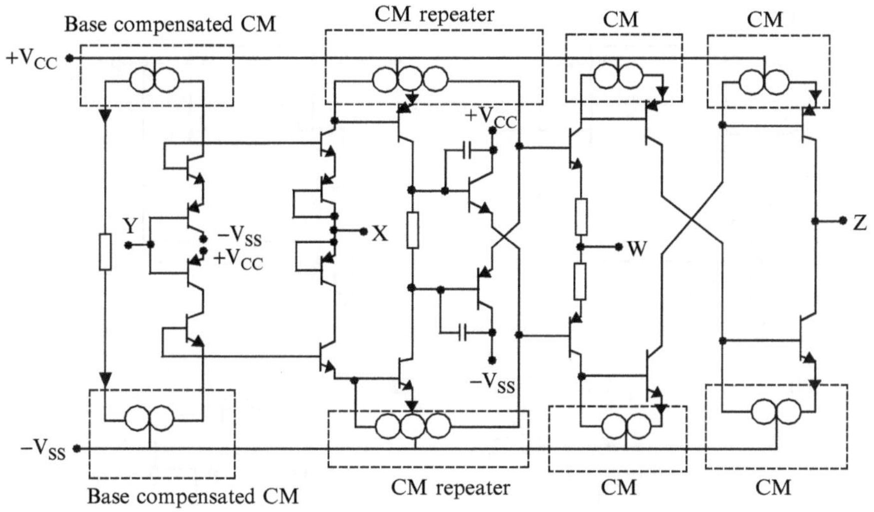

Fig. 10.4 Equivalent circuit of an exemplary CMOS implementation of the OFC proposed by Toumazou et al. [5]

The OFC is particularly useful in the realization of controlled sources with a minimum possible number of external passive elements. For this purpose, it is used in closed loop with current feedback from node w to x achieving a current feedback resulting in the gain-bandwidth decoupling while the translinear cell between ports y and x ensures a high slew rate due to the Sinh characteristics between the differential input (v_y-v_x) and the current i_x. Thus, the various controlled sources can be realized with the OFC using a bare minimum number of passive components, achieving the ideal input and output impedances and with each configuration exhibiting the pleasing property of gain bandwidth independence. The various realizations of the controlled sources using a single OFC, possessing these properties, are shown in Fig. 10.5.

10.2.4 Third Generation Current Conveyor

The third generation current conveyor (CCIII) was introduced by Fabre in 1995 [6] as a new helpful active element characterized by the hybrid matrix:

$$\begin{bmatrix} i_y \\ v_x \\ i_z \end{bmatrix} = \begin{bmatrix} 0 & -1 & 0 \\ 1 & 0 & 0 \\ 0 & \pm 1 & 0 \end{bmatrix} \begin{bmatrix} v_y \\ i_x \\ v_z \end{bmatrix} \tag{10.4}$$

The CCIII was conceived to be useful element to be used as the input cell of probes

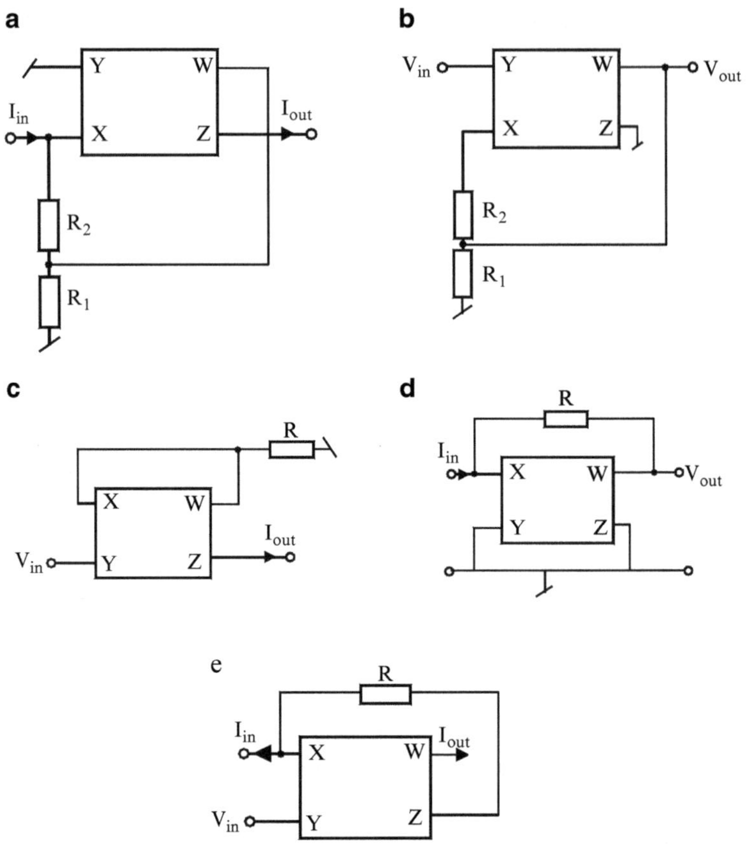

Fig. 10.5 Realization of all the basic amplifiers using OFC [5] (**a**) Current amplifier $I_{out}/I_{in} = [1+(R_2/R_1)]$ (**b**) Voltage amplifier $V_{out}/V_{in} = [1+(R_1/R_2)]$ (**c**) Transconductance amplifier $I_{out}/V_{in} = 1/R$ (**d**) Trans impedance amplifier $V_{out}/I_{in} = R$ (e) Current conveyor $i_y = 0$, $V_x = V_{in}$, $I_{out} = I_{in}$

and current measuring devices. It can be used to sense the output current flowing to ground or to create a floating current source as shown in Fig. 10.6.

A CCIII can be realized from two CCII+ s as per the schematic of Fig. 10.7.

10.2.5 *Differential-Difference Current Conveyor*

Chiu et al. [7] in 1996 presented the concept of the so-called differential difference current conveyor which is inspired by the earlier concept of differential difference amplifiers [8]. A DDCC has three Y-terminals leading to $V_x = (V_{y1} - V_{y2} + V_{y3})$ with all Y-terminals having ideally infinite input impedance leading to $i_{y1} = 0$, $i_{y2} = 0$,

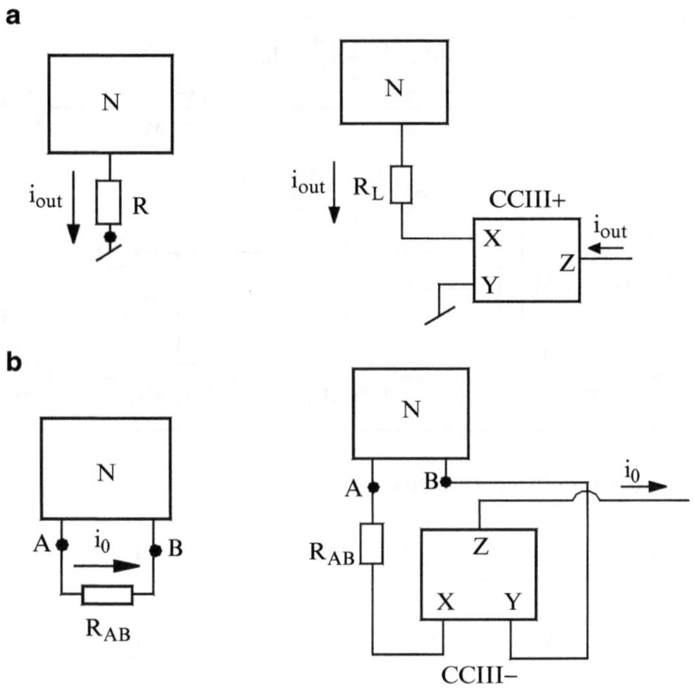

Fig. 10.6 General configuration for a current processing circuit and output currents [6] (**a**) Output current flowing to ground (**b**) Floating output current

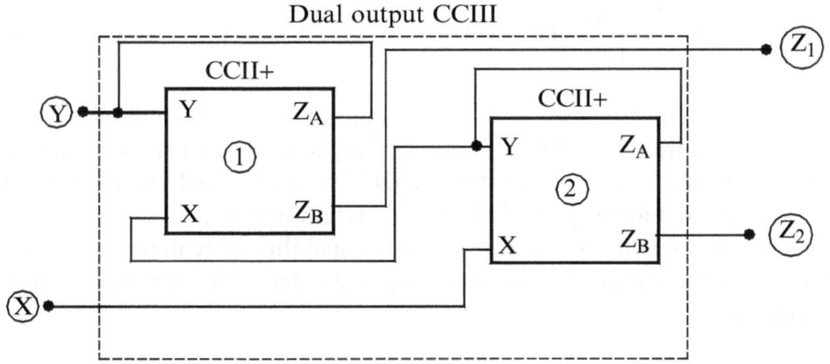

Fig. 10.7 Implementation of a dual output CCIII [6]

$i_{y3} = 0$ and as usual $i_z = \pm i_x$. An exemplary CMOS implementation of the DDCC is shown in Fig. 10.8. Other implementations can be devised from existing CMOS CCII structures such as [9]. In the circuit of Fig. 10.8, the two differential stages composed of M_1-M_2 and M_3-M_4 and a current mirror M_5-M_6 which converts the

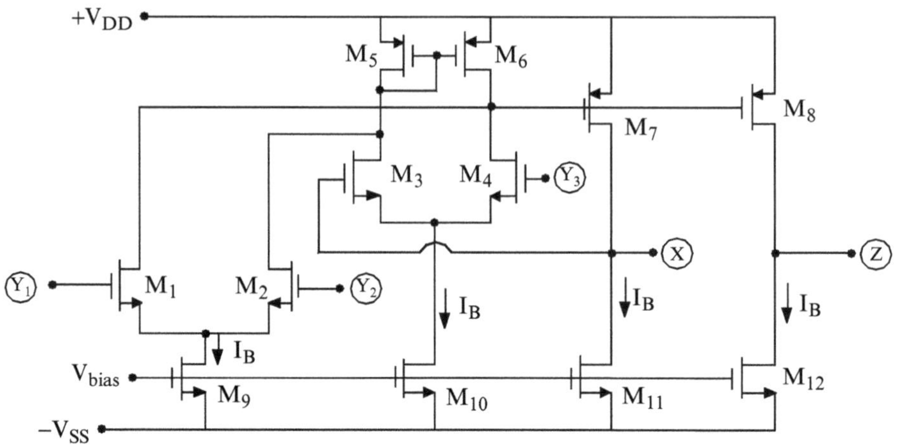

Fig. 10.8 An exemplary CMOS DDCCII+

differential current two a single ended output current in the MOSFET M_7. The output voltage of this amplifier is given by:

$$V_X = A_0[(V_{Y1} - V_{Y2}) - (V_{G3} - V_{Y3})] \quad (10.5)$$

where A_0 is the open loop gain of the amplifier and V_{G3} is gate voltage of MOSFET M_3.

The relation between the four voltages can be expressed as:

$$V_X = \frac{A_0}{A_0 + 1}(V_{Y1} - V_{Y2} + V_{Y3}) \cong V_{Y1} - V_{Y2} + V_{Y3} \quad (10.6)$$

Lastly, it can be readily seen that both i_x and i_z flow simultaneously and equally towards or away from the DDCC thereby leading to $i_z = +i_x$ and the circuit realizing a DDCC+. It may be mentioned that a DDCC− can be easily realized by appropriate addition of another pair of MOSFET current mirrors.

An attractive feature of the DDCC is its capability of realizing summation or subtraction of the signals without requiring any external resistor as reflected from the equation:

$$V_X = (V_{Y1} - V_{Y2} + V_{Y3}) \quad (10.7)$$

That DDCCIIs are particularly attractive for evolving single resistance controlled sinusoidal oscillators employing all grounded passive elements, has been demonstrated in [10]. On the other hand Current-controllable versions of DDCC, known as CCDDCC implementable in CMOS as demonstrated in [11], are useful in devising grounded-capacitor based floating capacitance multiplier and Current-mode universal filter [11].

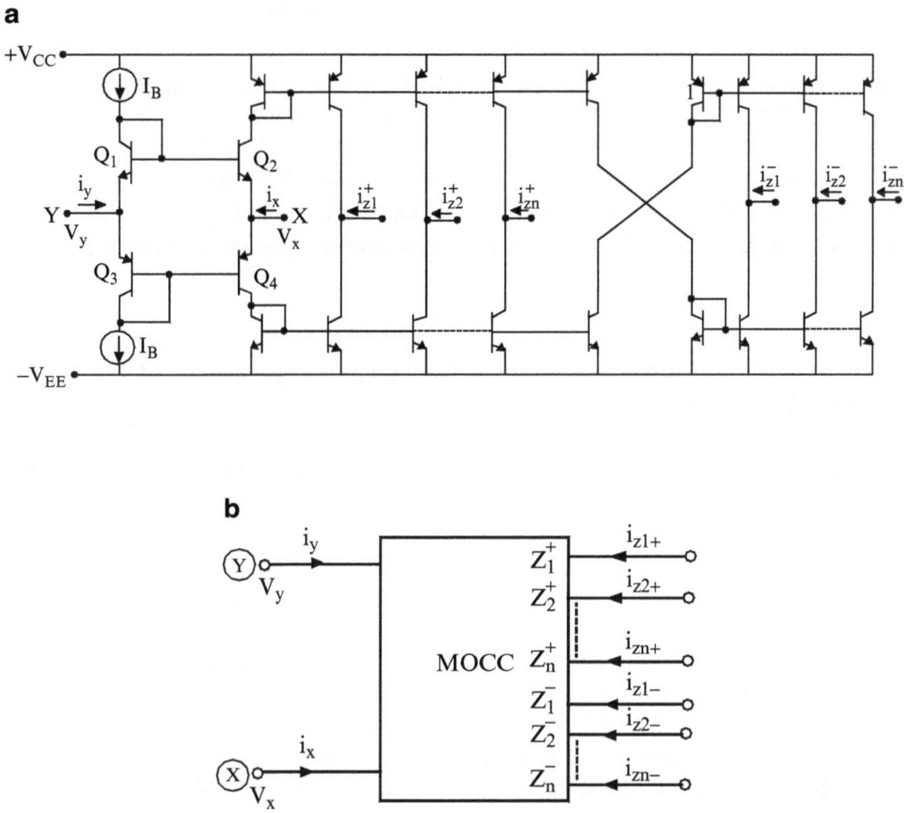

Fig. 10.9 The MOCC (**a**) An examplary bipolar implementation (**b**) Symbolic notation

10.2.6 Multiple-Output Current Conveyor

Having a multiple output CCII (MOCC) with multitude of identical as well as complementary outputs is particularly useful in current mode signal processing/ signal generation; see [12–16].

A MOCC can be realized from a bipolar or CMOS implementation of a CCII by adding additional transistors and current mirrors. An exemplary bipolar MOCC is shown in Fig. 10.9a, where the additional part creating multiple copies can be easily identified. Its symbolic notation is shown in Fig. 10.9b from where a MOCC is characterized by the following terminal equations: $i_y = 0$, $v_x = v_y$, $i_{z1+} = i_{z2+} = i_{z3+} = \ldots. = i_{zn+} = = +i_x$ and $i_{z1-} = i_{z2-} = i_{z3-} = \ldots. = i_{zn-} = = -i_x$.

MOCCs are particularly useful for realizing current mode universal biquad filters and also in higher order ladder filter design based upon the simulation of its node equations, for instance, see [13, 14].

10.2.7 Differential-Voltage Current Conveyor

When Elwan and Soliman introduced a CMOS differential voltage current conveyor (DVCC) in 1997 [17] it went almost unnoticed that the basic idea of the DVCC was introduced by Pal [18] as early as in 1989 itself and had demonstrated its advantage in realizing lossless floating inductors. The DVCC is a four port building block with there being two Y-terminals with flexibility that a differential Y-input can be applied. A DVCC is, thus, characterized by the terminal equations:

$$i_{y1} = i_{y2} = 0, \; v_x = \left(v_{y1} - v_{y2}\right) \text{and } i_z = \pm i_x \tag{10.8}$$

In retrospection, a DVCC can also be looked upon as a special case of DDCC with the third Y-input grounded.

In the available literature, a number of CMOS realizations of the DVCC have been proposed, for example see, [17, 19–23]. As demonstrated by Elwan and Soliman in [17], the DVCCs are particularly attractive in realizing grounded to floating positive impedance converter, grounded to floating negative impedance converter, floating generalized negative impedance converter and floating generalized positive impedance inverter/gyrator and MOS transconductor, to name a few (see Fig. 10.10).

In addition to the above quoted applications, DVCCs have been found to be equally efficient in realizing instrumentation amplifiers, floating inductors and FDNRs, universal VM/CM filters and relaxation oscillators and waveform generators [17–25]. It is interesting to note that the adjoint of a DVCC is an element called balance output CCII (BOCCII) [24] characterized by $i_y = 0$, $v_x = v_y$, $i_{z+} = +i_x$ and $i_{z-} = -i_x$.

10.2.8 Inverting CCIIs

In 1999, Awad and Soliman [26] introduced the *inverting* second generation current conveyors as the *missing building blocks* characterized by $v_x = -v_y$, $i_y = 0$ and $i_z = \pm i_x$ thereby leading to two varieties of inverting current conveyors namely ICCII+ and ICCII−.

It is interesting to point out that it went completely unnoticed by Awad and Soliman, as well as most of the other researchers who have been using ICCII+ as a building block that this variant of the CCII was already conceived much earlier by Chong and Smith (see reference [5] of Chap. 6) wherein they called this version of CC as the new type of CC, termed CCIIB which was employed by them in realizing one of the special cases of their single-CC biquads. Nevertheless, Awad and Soliman were the first to demonstrate that the new pathological elements *current-mirror* and *voltage-mirror* introduced by them together with the *nullator* and the *norator* were sufficient to model CCII+, CCII−, ICCII+ and ICCII− without

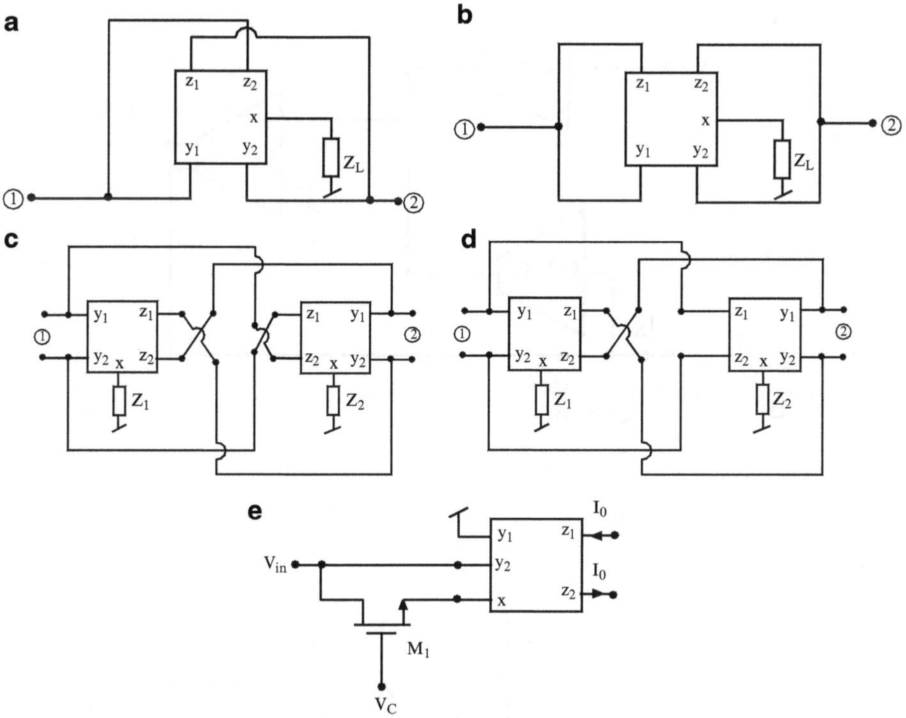

Fig. 10.10 Some applications, as demonstrated by Elwan and Soliman in [17], where DVCCIIs are particularly attractive (**a**) grounded to floating positive impedance converter (**b**) grounded to floating negative impedance converter (**c**) floating generalized negative impedance converter (**d**) floating generalized positive impedance inverter/gyrator (e) MOS transconductor

requiring any resistors. They were also the first to present CMOS implementations of the ICCII elements.

The pathological models of ICCII in terms of the pathological elements, nullator, norator, current mirror and voltage mirrors together with their relationship with CCII has been dealt in [27] while a number of CMOS implementations of the ICCII have been presented in [28, 29]. Novel generations of ICCs have been discussed in [30].

In [26] Awad and Soliman presented two different CMOS realization of the ICCII based upon the employment of two single input single output CMOS transconductors and one single input dual output CMOS transconductor as per the schematic of Fig. 10.11a which is shown in Fig. 10.11b

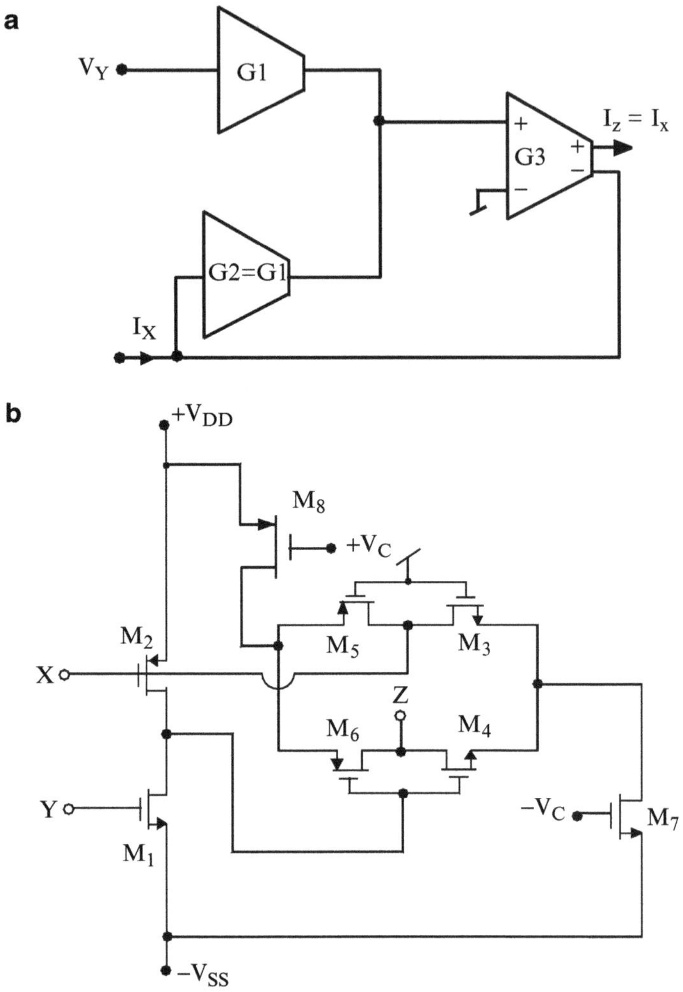

Fig. 10.11 Inverting CCII realizations [26] (**a**) implementation based upon transconductor realizations (**b**) A CMOS realization of the ICCII

10.2.9 Inverting Third Generation Current Conveyors

Extending the concept of CCIII further, inverting CCIII of both polarities, namely, ICCIII+ and ICCIII− were introduced by Sobhy and Soliman in [31]. The ICCIII elements are characterized by $I_y = -I_x$, $V_x = -V_y$, $I_{Z+} = -I_x$ and $I_{Z-} = -I_x$. It has been shown in [31] that ICCIII± are particularly attractive in realizing V to I converters, integrators, filters and oscillators- all providing electronic control of various parameters.

10.2.10 Differential-Current Voltage Conveyor

In 1999, Salama and Soliman [32] introduced a new variant of CC which they chose to call a *Differential current voltage conveyor* (DCVC) characterized by the following equations: $V_{x1} = 0$, $V_{x2} = 0$, $I_z = I_{x1} - I_{x2}$ and $V_o = V_z$. They also proposed a CMOS implementation of this block and showed that this is particularly useful for realizing electronically-controllable MOS-C quadrature oscillator suitable for designing digitally-controlled oscillators.

It is interesting to point out that in the same year 1999, Acar and Ozoguz [33] introduced a new versatile building block suitable for analog signal processing which the called *Current differencing buffered amplifier* (CDBA) which is characterized by exactly the same equations as specified above for the DCVC. Thus, it is clear that DCVC and CDBA are one and the same thing-a fact which does not appear to be so well recognized in the technical literature. The symbolic notation of the DCVC consists of a differential current conveyor (DCC) followed by a voltage buffer as shown in Fig. 10.12a. An exemplary CMOS realization of the DCVC is shown in Fig. 10.12b.

For CMOS implementation and some exemplary application circuits of DCVC, the readers are referred to [32, 34, 35] and the references cited therein. Also, yet another variant of DCVC called Current-controlled DCVC has been dealt with in [36].

10.2.11 Fully-Differential CCII

El-Adawy et al. in the year 2000 presented [37] another novel modification of the current conveyor termed *Fully differential second generation current conveyor* (FDCCII) which is eight terminal analog building block described by the matrix equation:

$$\begin{bmatrix} V_{X+} \\ V_{X-} \\ I_{Z+} \\ I_{Z-} \end{bmatrix} = \begin{bmatrix} 0 & 0 & 1 & -1 & 1 & 0 \\ 0 & 0 & -1 & 1 & 0 & 1 \\ 1 & 0 & 0 & 0 & 0 & 0 \\ 0 & 1 & 0 & 0 & 0 & 0 \end{bmatrix} \begin{bmatrix} I_{X+} \\ I_{X-} \\ V_{Y1} \\ V_{Y2} \\ V_{Y3} \\ V_{Y4} \end{bmatrix} \tag{10.9}$$

The symbolic notation of the FDCCII has been shown in Fig. 10.13a. The FDCCII has been since then used by a number of researchers, for instance, see [37–43] in various applications and a number of CMOS implementations have also been evolved. An equivalent of the exemplary circuit proposed by Kacar et al. [38] has

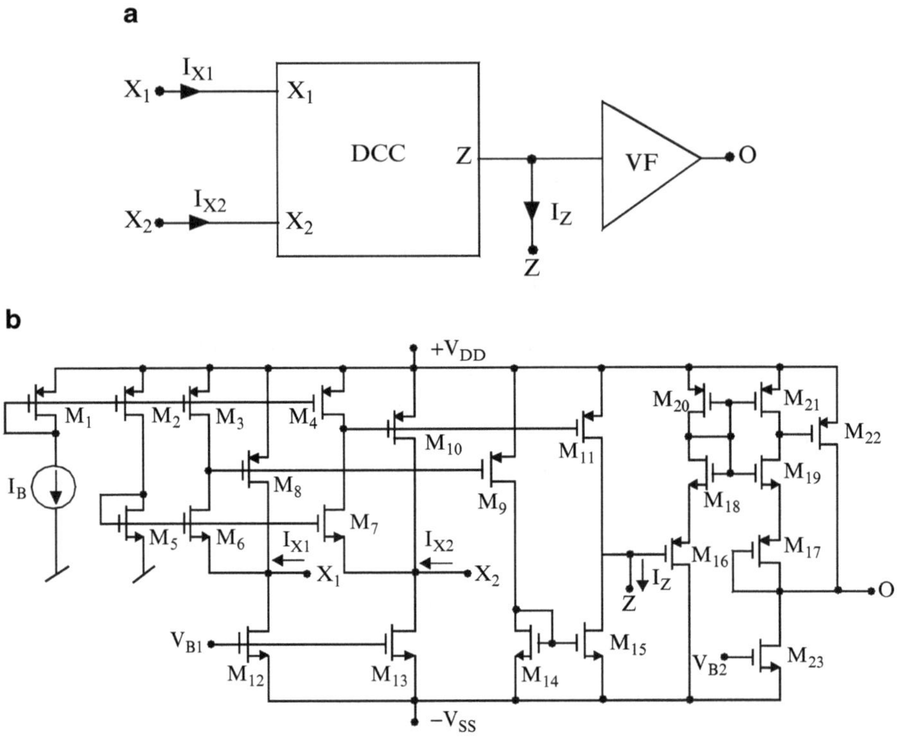

Fig. 10.12 DCVC implementation proposed by Salama et al. [34] (**a**) symbolic notation (**b**) An exemplary CMOS implementation (Adapted from [34] © Springer 2001)

been shown here in Fig. 10.13b where the blocks employed have been detailed in Fig. 10.13c, d.

It has been demonstrated by a number of researchers that FDCCII is practically attractive building block useful in mixed mode applications where fully differential signal processing is required. As novel applications of the FDCCII they presented its use in realizing a differential input balanced output transconductor, a four quadrant multiplier, a number of mixed mode universal second order filters and a fully differential filter; see [37–43].

10.2.12 General Three-Port Conveyors

In 1998, Soliman [44] proposed generalized Voltage Conveyor (GVC) and Generalized Current Conveyor (GCC) and showed that both of these elements as well as their special cases can be realized with only two normal types of Current Conveyors and at most two resistors (for the general case).

A GVC is a three-port element characterized by the hybrid matrix:

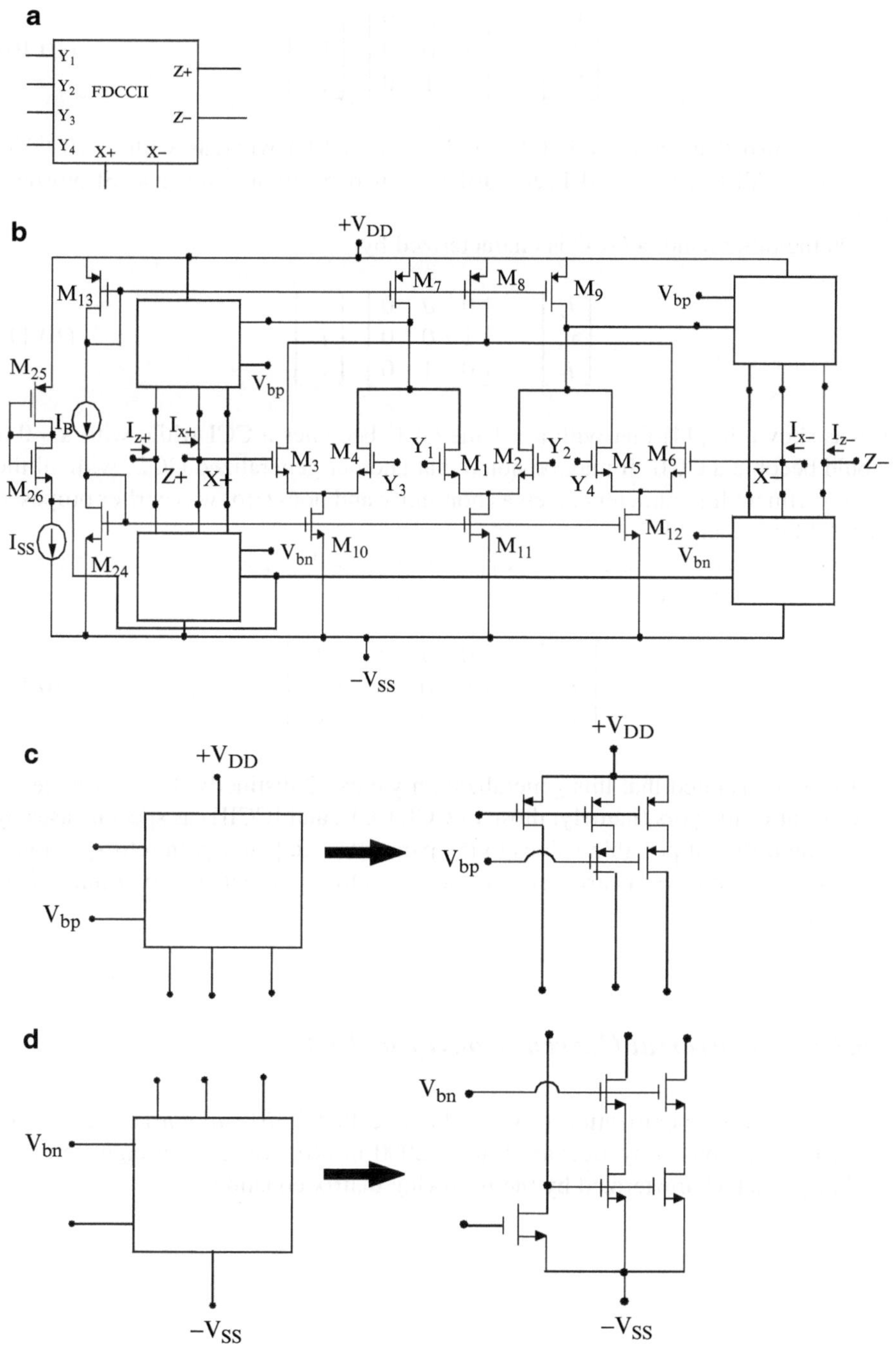

Fig. 10.13 Symbol of FDCCII and an exemplary CMOS implementation (**a**) The symbolic notation of FDCCII (**b**) an exemplary CMOS implementation [38]; (**c**), (**d**) The details of the various blocks used in the circuit of Fig. 10.13b

$$\begin{bmatrix} V_1 \\ I_2 \\ V_3 \end{bmatrix} = \begin{bmatrix} 0 & a & 0 \\ 1 & 0 & 0 \\ 0 & 1 & 0 \end{bmatrix} \begin{bmatrix} I_1 \\ V_2 \\ I_3 \end{bmatrix} \tag{10.10}$$

It was shown that with a = 1, GVC reduces to a VCI whereas with a = 0 GVC becomes a VCII which had been earlier defined by Filanovsky and Stromsmoe in [1].

On the other hand, a GCC is characterized by

$$\begin{bmatrix} i_y \\ v_x \\ i_z \end{bmatrix} = \begin{bmatrix} 0 & a & 0 \\ 1 & 0 & 0 \\ 0 & 1 & 0 \end{bmatrix} \begin{bmatrix} v_y \\ i_x \\ v_z \end{bmatrix} \tag{10.11}$$

It was shown in [44] that with a = 1 the GCC becomes a CCI while with a = 0 it would become a CCII. It may be noted that another generalized GCC, with all the three pertinent h-parameters taken as non-unity and non-zero, was earlier proposed in 1984 by Senani [4].

Another three port general conveyor was proposed by Biolek et al. in 2000 [45], which were characterized by the hybrid matrix:

$$\begin{bmatrix} V_x \\ I_y \\ I_z \end{bmatrix} = \begin{bmatrix} 0 & \alpha & 0 \\ \beta & 0 & 0 \\ \gamma & 0 & 0 \end{bmatrix} \begin{bmatrix} I_x \\ V_y \\ V_z \end{bmatrix} \tag{10.12}$$

It was demonstrated that this generalization yields 12 distinctly different varieties of Current Conveyors (namely, those of CCI, CCII, and ICCIII) as special cases by assigning different possible values to the parameters α, β and γ. In retrospection it may be seen that it was nearly same as the generalized current conveyor introduced by Senani in [4].

10.2.13 Universal Current Conveyor (UCC)

As an ultimate generalization of CCs, the so called *Universal Current Conveyor* (UCC) was proposed by Becvar et al. in 2000 in [46]. UCC is an eight terminal building block characterized by the following matrix equation.

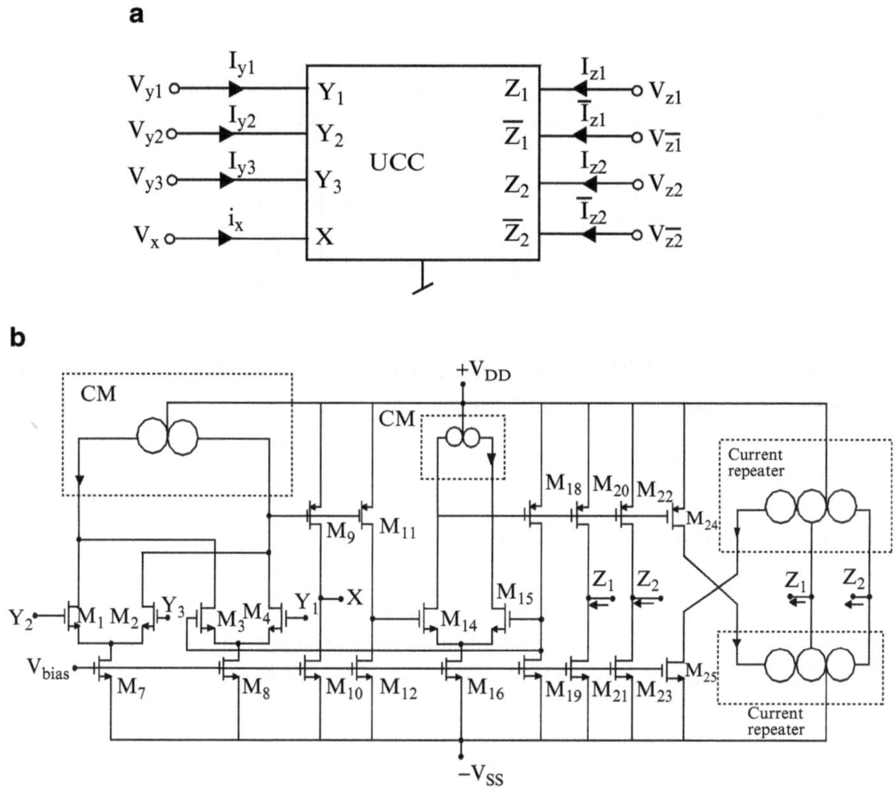

Fig. 10.14 Universal current conveyor symbol and implementation (**a**) Symbolic notation (**b**) A CMOS implementation by Becvar et al. in [46]

$$\begin{bmatrix} i_{y1} \\ i_{y2} \\ i_{y3} \\ v_x \\ i_{z1} \\ i_{z2} \\ i_{\overline{z1}} \\ i_{\overline{z2}} \end{bmatrix} = \begin{bmatrix} 0 & 0 & 0 & 0 & 0 & 0 & 0 & 0 \\ 0 & 0 & 0 & 0 & 0 & 0 & 0 & 0 \\ 0 & 0 & 0 & 0 & 0 & 0 & 0 & 0 \\ 1 & -1 & 1 & 0 & 0 & 0 & 0 & 0 \\ 0 & 0 & 0 & 1 & 0 & 0 & 0 & 0 \\ 0 & 0 & 0 & 1 & 0 & 0 & 0 & 0 \\ 0 & 0 & 0 & -1 & 0 & 0 & 0 & 0 \\ 0 & 0 & 0 & -1 & 0 & 0 & 0 & 0 \end{bmatrix} \begin{bmatrix} v_{y1} \\ v_{y2} \\ v_{y3} \\ i_x \\ v_{z1} \\ v_{z2} \\ v_{\overline{z1}} \\ v_{\overline{z2}} \end{bmatrix} \quad (10.13)$$

An exemplary implementation of the UCC is shown here in Fig. 10.14.

For the various applications of the UCCs, their inter-relationship with other forms of CCs and derivation of numerous known forms of CCs as special cases of the UCC, the reader is referred to [44–49].

10.2.14 Modified Inverting CCII

A modified third generation Current Conveyor (MCCIII) was advanced by Kuntman et al. in 2002 [50], with the proposed characterization as

$$\begin{bmatrix} I_y \\ V_x \\ I_{z1} \\ I_{z2} \end{bmatrix} = \begin{bmatrix} 0 & -1 & 0 & 0 \\ 1 & 0 & 0 & 0 \\ 0 & -2 & 0 & 0 \\ 0 & 1 & 0 & 0 \end{bmatrix} \begin{bmatrix} V_y \\ I_x \\ V_{z1} \\ V_{z2} \end{bmatrix} \tag{10.14}$$

It was shown that this modified version results in some flexibility in biquad filter design using a single-CCIII− based parallel RL simulator and helps in reducing the total number of passive component count.

10.2.15 Dual-X Current Conveyor

In 2002, Zeki and Toker proposed in [51] a *Dual-X current conveyor* (DXCCII) as a new building block suitable for electronically tunable continues time filter. A DXCCII is characterized by a following terminal equation.

$$\begin{bmatrix} I_Y \\ V_{XP} \\ V_{Xn} \\ I_{Zp} \\ I_{Zn} \end{bmatrix} = \begin{bmatrix} 0 & 0 & 0 \\ 1 & 0 & 0 \\ -1 & 0 & 0 \\ 0 & 1 & 0 \\ 0 & 0 & 1 \end{bmatrix} \begin{bmatrix} V_Y \\ I_{Xp} \\ I_{Xn} \end{bmatrix} \tag{10.15}$$

It was shown that since $V_{XP} = -V_{XN}$, if MOSFET is connecting between the two X terminals with a control voltage V_c applied at the gate terminal the square nonlinearities of the MOSFET drain current are cancelled out and it is, therefore, given by

$$I_D = \beta(V_C - V_{Tn})\left(V_{Xp} - V_{Xn}\right) \text{ where } \beta = \left(\frac{W}{L}\right)\mu_n C_{ox} \tag{10.16}$$

due to which the MOSFET behaves almost as a linear resistor as long as the MOSFET operates a triode region. The equivalent resistance value is given by:

$$R_{eq} = \frac{1}{\beta(V_c - V_{th})} \tag{10.17}$$

A CMOS realization of the DXCCII was also suggested by Zeki and Toker in [51]. Other interesting variants and applications of the DXCCII can be found in [52–55].

10.2.16 Fully-Balanced CCII

The notion of a fully- balanced Second generation Current Conveyor (FBCCII) was introduced by Alzaher et al. in 2003 in [56].

They demonstrated that such a building block can be constructed from a fully-balanced differential difference amplifier (FBDDA) through power supply current sensing technique. An FBCCII can be described by the same equations as a CCII+, however, all signals are differential i. e. $V_x = V_{xp} - V_{xn}$, $V_y = V_{yp} - V_{yn}$, $I_x = I_{xp} - I_{xn}$ and $I_z = I_{zp} - I_{zn}$.

It was emphasized that an FBCCII can be used to implement the fully differential version of any single ended architecture using conventional CCII. The operation and applications of a CMOS chip demonstrated the utility of the concept (Fig. 10.15).

10.2.17 Extended Current Conveyors

The extended Current Conveyors were introduced by Sato et al. in [55] as relaxed versions of the conventional CCs with current and voltage offsets allowed apart from the normal Current Controlled Current Sources and Voltage controlled voltage sources incorporated in the internal architecture of Current Conveyors. It was shown that this results in much simpler transistor level implementations of the extended CCs as compared to normal CCs. In most general form, the extended CC defined by them has the following characterization:

$$\begin{bmatrix} V_x \\ I_y \\ I_z \end{bmatrix} = \begin{bmatrix} 0 & 1 & 0 \\ \varphi & 0 & 0 \\ \pm\beta & 0 & 0 \end{bmatrix} \begin{bmatrix} I_x \\ V_y \\ V_z \end{bmatrix} + \begin{bmatrix} V_{offsetx} \\ I_{offsety} \\ I_{offsetz} \end{bmatrix} \tag{10.18}$$

Two specific examples would suffice to demonstrate as to how the number of MOSFETs is reduced in making the extended CCs with relaxed definitions. Consider the two exemplary MOSFET-based extended CCs from [55], which are reproduced here in Fig. 10.16.

The first circuit of Fig. 10.16a is perhaps the simplest CC which is realizable by only four MOSFETs and is characterized by:

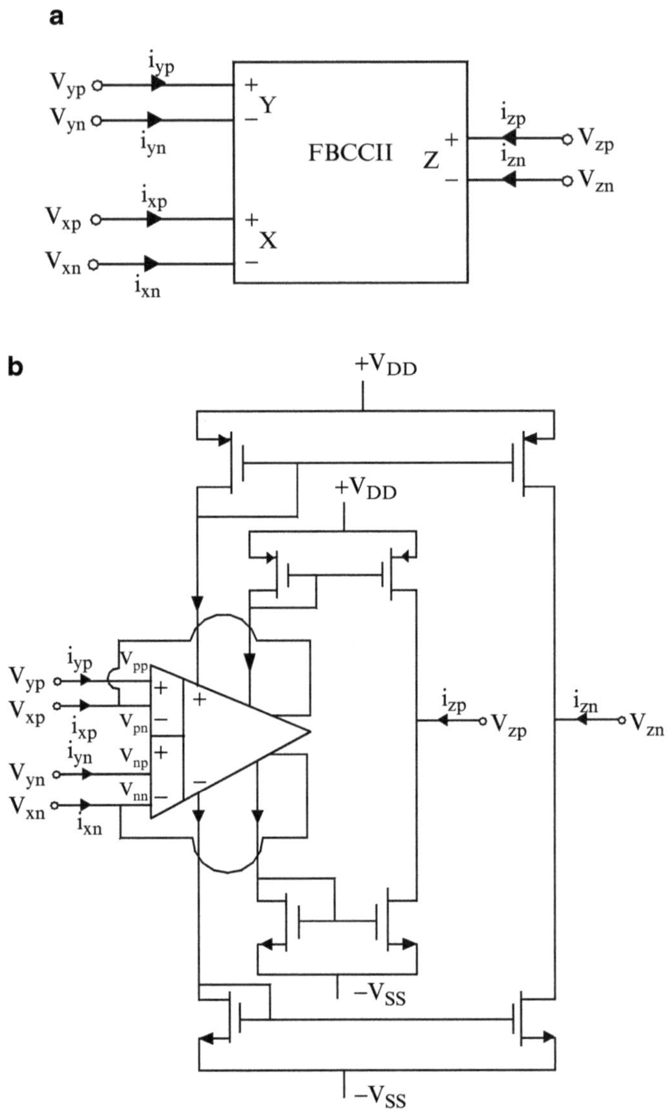

Fig. 10.15 Fully-balanced current conveyor (**a**) Symbolic notation (**b**) An exemplary implementation based upon the one presented by Alzaher et al. [56]

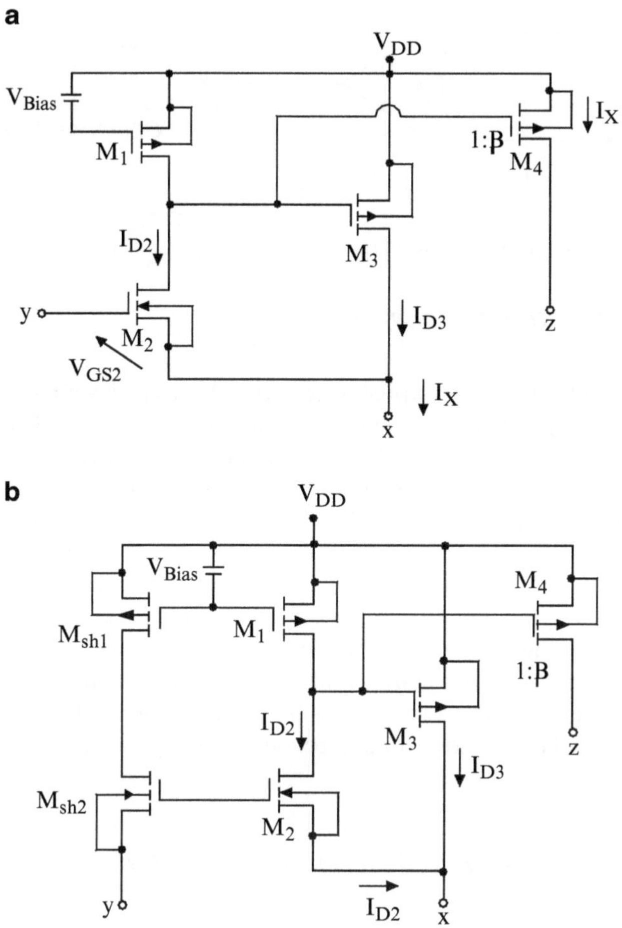

Fig. 10.16 Simple implementations of the extended current conveyors [55]

$$\begin{bmatrix} V_x \\ I_y \\ I_z \end{bmatrix} = \begin{bmatrix} 0 & 1 & 0 \\ 0 & 0 & 0 \\ \beta & 0 & 0 \end{bmatrix} \begin{bmatrix} I_x \\ V_y \\ V_z \end{bmatrix} + \begin{bmatrix} -V_{GS2} \\ 0 \\ -\beta I_{D2} \end{bmatrix} \tag{10.19}$$

while the second architecture of Fig. 10.16b uses no more than six MOSFETs (still a smaller number as compared to those required for implementing a normal CC) and has a characterization:

$$\begin{bmatrix} V_x \\ I_y \\ I_z \end{bmatrix} = \begin{bmatrix} 0 & 1 & 0 \\ 0 & 0 & 0 \\ \beta & 0 & 0 \end{bmatrix} \begin{bmatrix} I_x \\ V_y \\ V_z \end{bmatrix} + \begin{bmatrix} 0 \\ +I_{D2} \\ -I_{D2} \end{bmatrix} \tag{10.20}$$

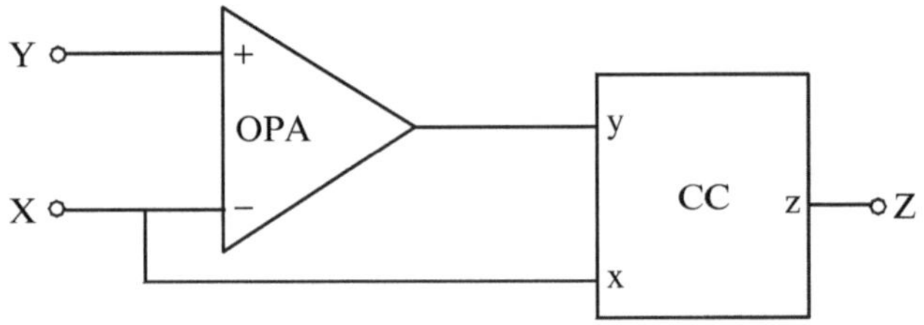

Fig. 10.17 The operational conveyor [57]

10.2.18 Operational Conveyor

Gift proposed in [57] the *Operational conveyor* (OC) as a new circuit configuration comprised of a hybrid arrangement of a current conveyor and an operational amplifier as shown in Fig. 10.17. The operational conveyor shows significant improvement over a current conveyor. In particular, the effect of the parasitic input resistance of the X port, R_x is significantly reduced by a factor A_0 (the maximum gain of the op-amp employed, at zero frequency) which results in the near elimination of the effect of the R_x from the system transfer function. The employment of high GBP of the op-amp enables the system dynamics to be determined by the current conveyor and not by the associated op-amp.

The utility of the OC was demonstrated by designing a current amplifier using LM351 op-amp and AD844 based CCII+ and it was observed that a highly accurate realization with a large bandwidth independent of the gain was obtainable.

10.2.19 Multiple-Input Differential CC (MIDCC)

Horng et al. [58] presented a new variety of CC called multi input differential current conveyor (MIDCC). The characterizing equation of the MIDCC can be written as:

$$\begin{bmatrix} i_{y+} \\ \cdots \\ i_{yn+} \\ i_{y1-} \\ \cdots \\ i_{ym-} \\ v_x \\ i_z \end{bmatrix} = \begin{bmatrix} 0 & \cdots & 0 & 0 & \cdots & 0 & 0 & 0 \\ \cdots & \cdots & \cdots & \cdots & \cdots & \cdots & \cdots & \cdots \\ 0 & \cdots & 0 & 0 & \cdots & 0 & 0 & 0 \\ 0 & \cdots & 0 & 0 & \cdots & 0 & 0 & 0 \\ \cdots & \cdots & \cdots & \cdots & \cdots & \cdots & \cdots & \cdots \\ 0 & \cdots & 0 & 0 & \cdots & 0 & 0 & 0 \\ 1 & \cdots & 1 & -1 & \cdots & -1 & 0 & 0 \\ 0 & \cdots & 0 & 0 & \cdots & 0 & \pm 1 & 0 \end{bmatrix} \begin{bmatrix} v_{y1+} \\ \cdots \\ v_{yn+} \\ v_{y1-} \\ \cdots \\ v_{ym-} \\ i_x \\ v_z \end{bmatrix} \quad (10.21)$$

It was shown that the new building blocks are, in fact, an adjoint of the so-called multiple output second generation current conveyors (MOCCII). The relation between MOCCII and MIDCC is illustrated in Fig. 10.18.

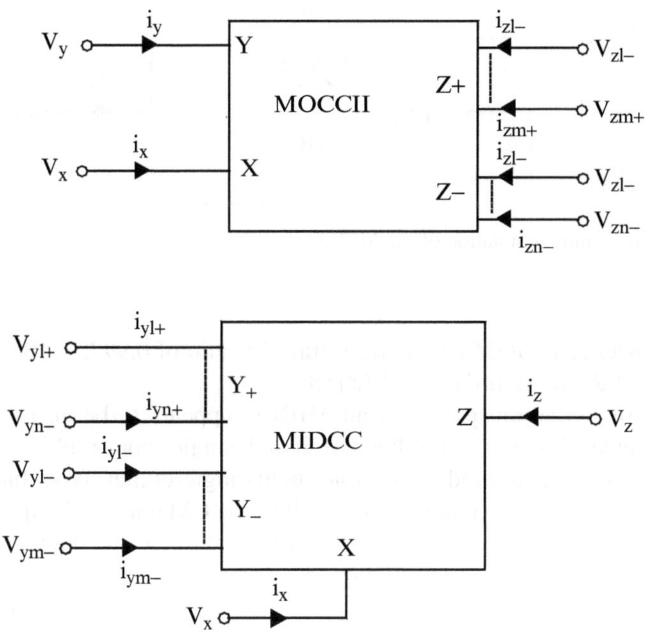

Fig. 10.18 The relationship between MIDCC and MOCC

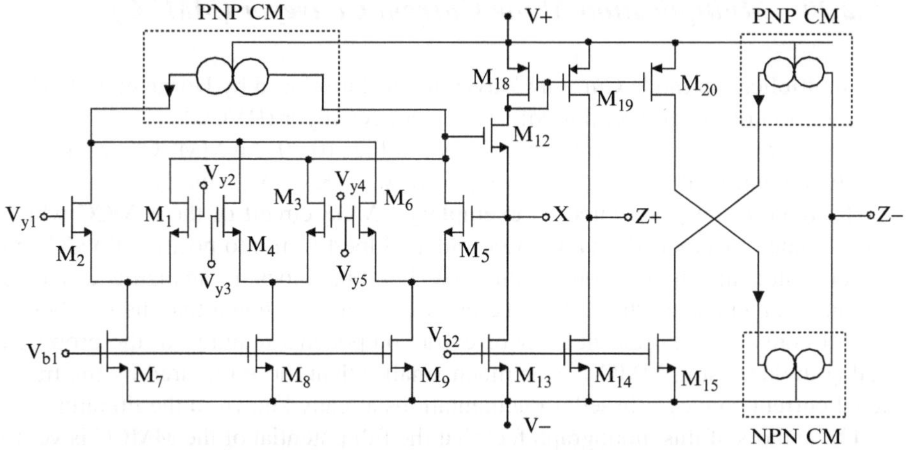

Fig. 10.19 An exemplary CMOS implementation of the MIDCC based upon the one proposed by Horng et al. [58]

A typical CMOS implementation of MIDCC employing N number of differential pairs, current mirrors and current repeater is shown in Fig. 10.19 and appears to be self-explanatory. When implemented in TSMC 0.35 μm technology, it exhibited

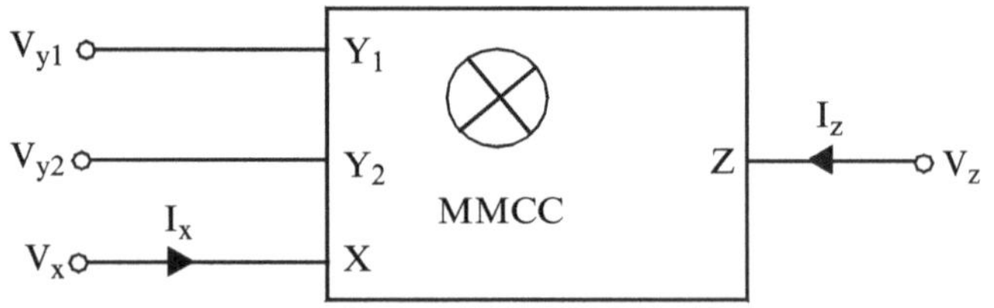

Fig. 10.20 The symbolic notation of the MMCC

voltage transfer ratio of 0.9228, current transfer ratio of 0.997, 3-dB bandwidth of 391 MHz and R_x of the order of 43 Ω [58].

Because of the adjoin property, an MIDCC appears to be an useful building block for converting a given MOCCII based single-input-multiple-output type biquad into a voltage mode, multiple-input-single-output type biquads using MIDCCs. In view of a large number of SIMO-type CM universal biquads available in the literature, it is obvious that this development opens the door for deriving as many MISO-type VM universal biquads using the adjoint relations between MOCCII and MIDCC. This, however, does not appear to have been exploited in the earlier literature yet and therefore, leaves a lot of work still to be carried out.

10.2.20 Multiplication-Mode Current Conveyor (MMCC)

The multiplication mode Current Conveyor was introduced by Horng et al. [59] as an extension of the differential voltage current conveyor (DVCC).

It block diagram representation is shown in Fig. 10.20. An MMCC is characterized by the terminal equations: $I_{Y1} = I_{Y2} = 0$, $V_X = KV_{Y1}V_{Y2}$ and $I_Z = I_X$.

Horng et al. [59] presented an exemplary CMOS circuit of the MMCC which was obtained by embedding a CMOS folded Gilbert Cell into positive DVCC and demonstrated the application of the new building block in realizing a square summer and an analog divider. In the latter case, it was shown that the number of current conveyors and the number of extra components required in the proposed analog divider using MMCCs was much smaller than those required by the traditional current conveyor based implementations already known in the literature.

The authors of this monograph feel that the full potential of the MMCC is yet to be realized. Although not discovered yet, however, it appears that MMCCs could yield novel solutions for realizing voltage-controlled oscillators, voltage-tunable filters and a host of other functional circuits which could be realized up till now only by a combination of op-amps and analog multipliers. Thus, there appears enough scope in devising new applications of the MMCCs.

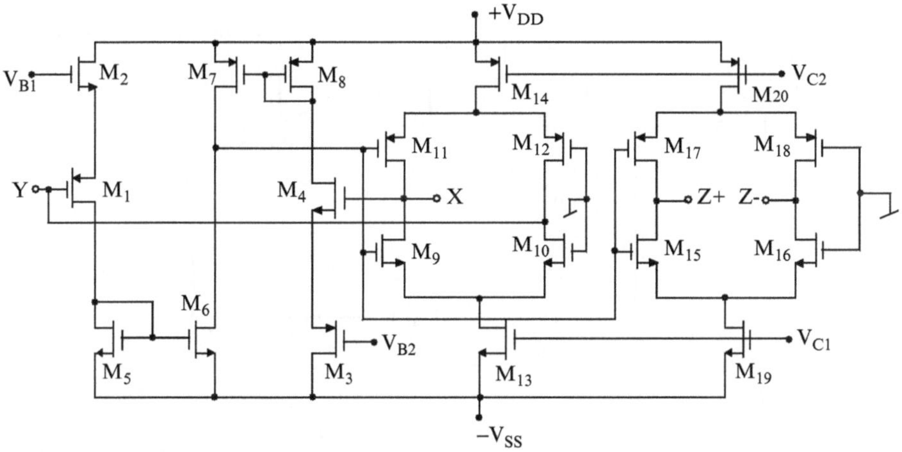

Fig. 10.21 An exemplary realization of the balanced output third generation current conveyor proposed by Sobhy and Soliman (Adapted from [31] © 2009 Springer)

10.2.21 Balanced-Output Third Generation Inverting CC

Balanced output third generation inverting current conveyor was introduced by Sobhy and Soliman in [31]. It combines the characteristic of an inverting current conveyor and the third generation current conveyor and is characterized by the following matrix equation.

$$\begin{pmatrix} I_Y \\ V_X \\ I_{Z+} \\ I_{Z-} \end{pmatrix} = \begin{pmatrix} 0 & -1 & 0 & 0 \\ -1 & 0 & 0 & 0 \\ 0 & 1 & 0 & 0 \\ 0 & -1 & 0 & 0 \end{pmatrix} \begin{pmatrix} V_Y \\ I_X \\ V_{Z+} \\ V_{Z-} \end{pmatrix} \tag{10.22}$$

A typical CMOS realization devised by the author of [31] is shown in Fig. 10.21. The applications, where balanced output third generation inverting current conveyor can be employed usefully, have been demonstrated in [31] and include the realization of voltage-to-current converter, MOSFET-C lossless integrators and biquad filters and MOSFET-C oscillators.

10.2.22 Voltage and Current Gain Second Generation Current Conveyor (VCG-CCII)

The voltage and current gain CCII (VCG-CCII) was introduced by Marcellis et al. in [60]. Its internal architecture can be understood by the schematic shown in Fig. 10.22. It is characterized by the following equations: $I_y = 0$ $V_x = h\,V_y$, $I_z = \pm k\,I_x$ where h and k are the voltage gain and current gain respectively.

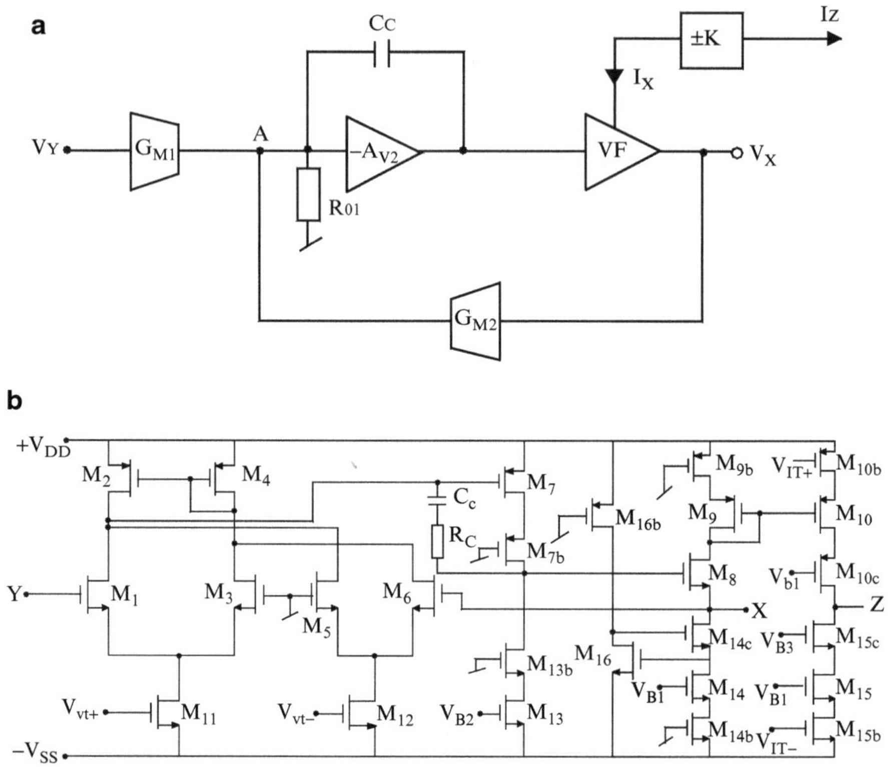

Fig. 10.22 The architecture and implementation of the VCG-CCII proposed by Marcellis et al. (**a**) Block diagram (**b**) An exemplary CMOS implementation (Adapted from [60] © 2009 Springer)

The three ports of the inherent CC are created as follows: A voltage follower between ports Y and X is created by two transconductance stages, an internal frequency compensation circuit and a voltage follower. Since the input front end is a transconductor, its very high input impedance would imply $I_y = 0$. The output current of the voltage follower, which constitutes $I_{x, \text{ is}}$ sensed through the technique of power-supply-current-sensing technique applied on this voltage follower and accordingly, the port Z is created. The two transconductors along with the compensating circuitry and VF ensure that the voltage gain between ports Y and X is electronically-controllable by the ratio of the transconductances, since $V_x/V_y = G_{m1}/G_{m2}$. The input impedance looking into the port Y is ideally infinite leading to $I_y = 0$.

For some other, digitally-controlled Current Conveyor versions, the reader is referred to [61, 62].

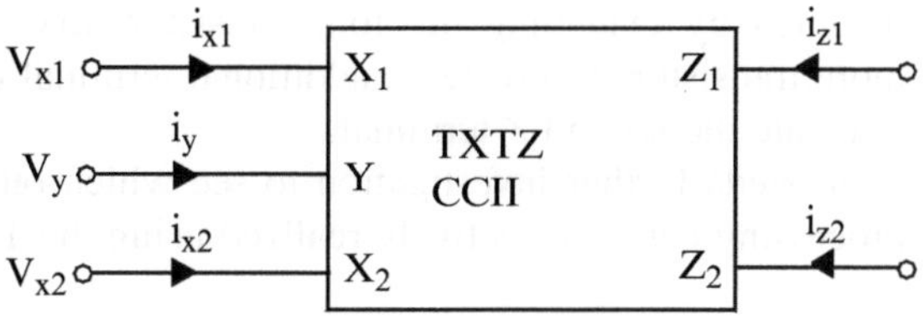

Fig. 10.23 The block diagram of the TXTZ CCII

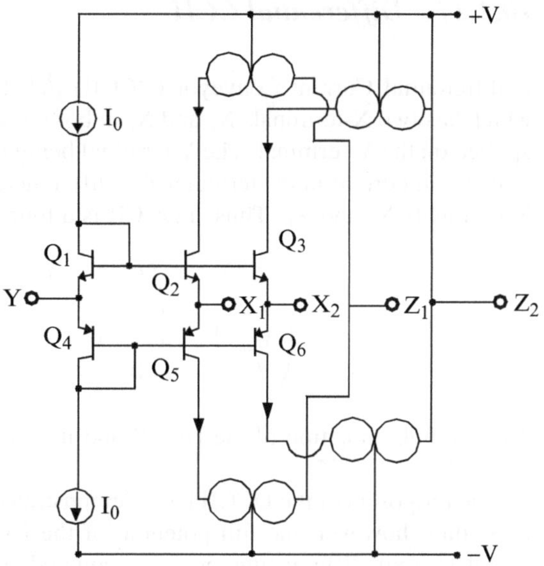

Fig. 10.24 Simplified representation of a CMOS implementation of the TXTZ-CCII as devised in [63]

10.2.23 TXTZ CCII

The two-X and two-Z second generation current conveyor (TXTZCCII), as symbolically shown in Fig. 10.23, was introduced by Fongsamut and Surakampontorn in [63] and is characterized by the following matrix:

$$\begin{bmatrix} i_y \\ v_{x1} \\ v_{x2} \\ i_{z1} \\ i_{z2} \end{bmatrix} = \begin{bmatrix} 0 & 0 & 0 & 0 & 0 \\ 1 & 0 & 0 & 0 & 0 \\ 1 & 0 & 0 & 0 & 0 \\ 0 & 1 & 0 & 0 & 0 \\ 0 & 0 & 1 & 0 & 0 \end{bmatrix} \begin{bmatrix} v_y \\ i_{x1} \\ i_{x2} \\ v_{z1} \\ v_{z2} \end{bmatrix} \quad (10.23)$$

Through two examples, the authors of [63] have demonstrated that a TXTZ-CCII is particularly useful in realizing single-element-controlled sinusoidal oscillator using a bare minimum of only four passive components (i.e. only two resistors and only two capacitors). A typical bipolar implementation is shown in Fig. 10.24 which is obtained by extending the four-transistor mixed-translinear-cell by adding two more transistors to create an additional terminal and using simple current mirrors to create the second Z terminal.

It needs further investigation to see which other applications in analog signal processing can be effectively realized using the TXTZ-CCII building blocks.

10.2.24 Differential CCII

A differential Current Conveyor (DCCII) [64, 65] is another special type of CC which has two X-terminals X_p and X_n at both of which the same voltage appears as applied on the Y-terminal. The Y terminal being a voltage input terminal has $I_y = 0$. The output current in the terminal Z is difference of the two currents applied to the X-terminals X_p and X_n. Thus, a DCCII is a four-port characterized by:

$$\begin{pmatrix} I_Y \\ V_{xp} \\ V_{xn} \\ I_Z \end{pmatrix} = \begin{pmatrix} 0 & 0 & 0 & 0 \\ 1 & 0 & 0 & 0 \\ 1 & 0 & 0 & 0 \\ 0 & 1 & -1 & 0 \end{pmatrix} \begin{pmatrix} V_Y \\ I_{Xp} \\ I_{xn} \\ V_Z \end{pmatrix} \tag{10.24}$$

The symbolic notation of the DCCII and its exemplary CMOS implementation is shown in Fig. 10.25.

The proposers of the DCCII have demonstrated its use in realizing a fist order all pass filter, however, the full potential of the DCCII as made available due to its useful current differencing property coupled with its relatively simple internal circuit architecture, does not appear to have been fully exploited yet.

10.2.25 Universal Voltage Conveyor

The concept of an n-port voltage conveyor (VC) was introduced by Cajka and Vrba in 2004 [47, 48]. In an n-port VC, the driving quantity is the independent voltage V_x applied to the input port X. This independent voltage V_x is *conveyed* to the output ports Z_1, Z_2.........Z_n with the $\pm$ polarities. It has a number of auxiliary ports labeled as Y_1, Y_2 ......Y_n, on which one can apply **n** number of independent node current inputs called I_{y1}, I_{y2}...I_{yn} (Fig. 10.26).

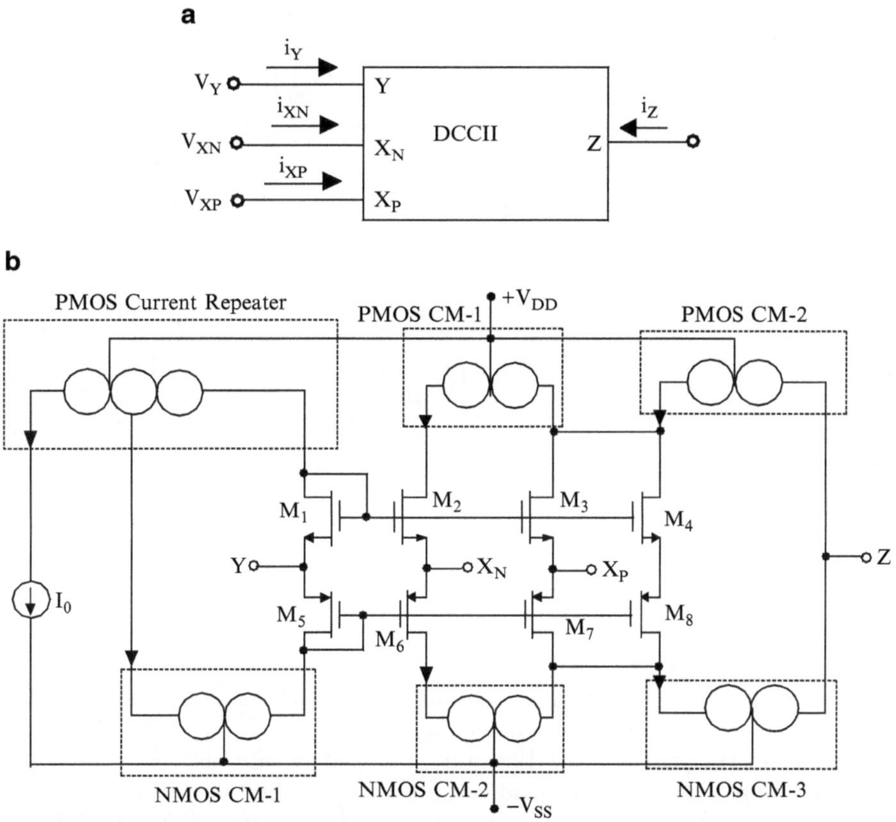

Fig. 10.25 The CMOS implementation of the DCCII proposed by Metin et al. [65] (**a**) The block diagram (**b**) simplified representation with all MOS current mirrors and repeaters represented symbolically

10.2.26 Floating Current Conveyors

Floating Current Conveyors (FCC) were introduced by Soliman and Saad in [66]. An FCC is a four-port element characterized by $V_y = V_x$, $I_y = 0$, $I_{Z+} = +I_x$ and $I_{Z\_} = -2\ I_x$. The floating CCII, thus, includes both CCII+ and CCII− as special cases. Similarly, floating ICCIIs have also been defined in [66]. These elements are easily realizable in CMOS and are useful in some situations, such as in realizing two output VCVS. Investigation of other applications is open to investigation.

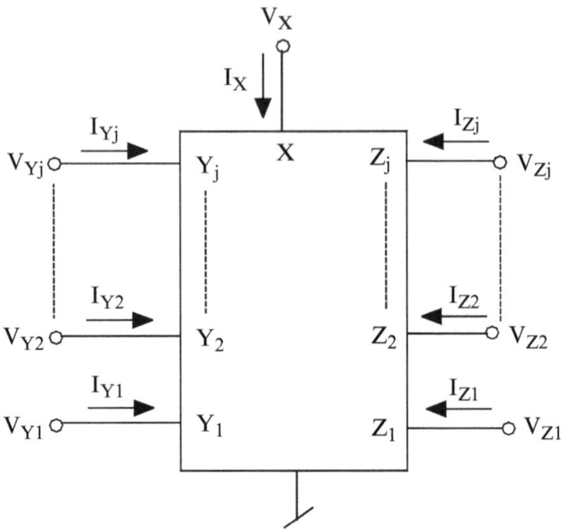

Fig. 10.26 The block diagram representation of a universal voltage conveyor

Concluding Remarks

In this chapter, we have described twenty six distinctly different varieties of current conveyors or extension thereof, which have been proposed by a number of researchers since 1981 till date. Undoubtedly, the basic CCII± are not only the most simple and versatile of the entire CC family; they are realizable practically by using commercially available off-the-shelf ICs (e.g. AD844, PA630, OPA2662, OPA660 and OPA860) and hence, are of more practical significance. Nevertheless, among the numerous variants described here, several of them possess quite interesting characterizing equations, which also appear to be useful from the view point of many specific applications, as the proposers of such blocks have highlighted.

All the variants of CCs described here have the potential of being fully integrated in bipolar or CMOS technology and are, therefore, of obvious interest from the view point of VLSI implementation.

Moreover, if there could be a really universal conveyor from where all the twenty six variants described here could be realized as special cases, it would have been perhaps the most versatile building blocks among all varieties described here in. Although in a number of cases, the proposers of the new types of CCs have often demonstrated that their proposal can realize several earlier types of current conveyors as special cases, a really universal conveyor from which all the above described over two dozen types of conveyors may be derivable as special cases has not been proposed so far. Obviously, if such

(continued)

(continued)

an universal current conveyor be produced as an IC it would be of immense use since the entire spectrum of application circuits evolved through the use of the above mentioned types of CCs, as well as those accumulated over the years using the basic CCs, can be then practically realizable using only one kind of off-the-shelf ICs. In the opinion of the authors of this monograph, this question does not appear to have been settled completely and conclusively yet and therefore, it appears that the basic question which still remains unresolved is; '*the* ***most universal conveyor****; what shall it be*?'

References

1. Filanovsky IM, Stromsmoe KA (1981) Current–voltage conveyor. Electron Lett 17:129–130
2. Senani R (1980) Novel circuit implementation of current conveyors using an O.A. and an O.T.A. Electron Lett 16:2–3
3. Dostal T, Pospisil J (1982) Current and voltage conveyors- a family of three-port immittance converters. ISCAS, Rome, pp 419–422
4. Senani R (1984) Novel application of generalized current conveyor. Electron Lett 20:169–170 (Errata: (1984) 20: 356)
5. Toumazou C, Payne A, Lidgey FJ (1991) Operational floating conveyor. Electron Lett 27:651–652
6. Fabre A (1995) Third-generation current conveyor: a new helpful active element. Electron Lett 31:338–339
7. Chiu W, Liu SI, Tsao HW, Chen JJ (1996) CMOS differential difference current conveyors and their applications. IEE Proc Circ Devices Syst 143:91–96
8. Sackinger E, Guggenbuhl W (1987) A versatile building block: the CMOS differential difference amplifier. IEEE J Solid State Circ 22:287–294
9. Surakampontorn W, Riewruja V, Kumwachara K, Dejhan K (1991) Accurate CMOS based circuit conveyors. IEEE Trans Instrum Meas 40:699–702
10. Gupta SS, Senani R (2001) CMOS differential difference current conveyors and their applications. IEE Proc Circ Devices Syst 148:335–336
11. Prommee P, Somdunyakanok M (2011) CMOS-based current-controlled DDCC and its applications to capacitance multiplier and universal filter. Int J Electron Commun(AEU) 65:1–8
12. Horng JW, Hou CL, Chang CM, Chung WY, Wei HY (2005) Voltage-mode universal biquadratic filters with one input and five outputs using MOCCIIs. Comput Electr Eng 31:190–202
13. Wu J, El Masry E (1996) Current-mode ladder filters using multiple output current conveyors. IEE Proc Circ Devices Syst 143:218–222
14. Hwang YS, Hung PT, Chen W, Liu SI (2002) Systematic generation of current-mode linear transformation filters based on multiple output CCIIs. Analog Integr Circ Sig Process 32:123–134
15. Abdalla KK (2013) Universal current-mode biquad employing dual output current conveyors and MO-CCCA with grounded passive elements. Circ Syst 4:83–88
16. Maheshwari S (2009) Current-mode third-order quadrature oscillator. IET Circ Devices Syst 4 (3):188–195
17. Elwan HO, Soliman AM (1997) Novel CMOS differential voltage current conveyor and its applications. IEE Proc Circ Devices Syst 144:195–200
18. Pal K (1989) Modified current conveyors and their applications. Microelectron J 20:37–40

19. Hassan TM, Mahmoud SA (2010) New CMOS DVCC realization and applications to instrumentation amplifier and active-RC filters. Int J Electron Commun (AEU) 64:47–55
20. Minaei S, Yuce E (2012) A simple schmitt trigger circuit with grounded passive elements and its application to square/triangular wave generator. Circ Syst Sig Process 31:877–888
21. Ibrahim MA, Minaei S, Kuntman H (2005) A 22.5 MHz current mode KHN-biquad using differential voltage current conveyor and grounded passive elements. Int J Electron Commun (AEU) 59:311–318
22. Horng JW, Hou CL, Chang CM, Chung WY (2006) Voltage-mode universal biquadratic filters with one input and five outputs. Analog Integr Circ Sig Process 47:73–83
23. Laoudias C, Psychalinos C (2014) Differential voltage current controlled current conveyor with low-voltage operation capability. Int J Electron 101:939–949
24. Soliman AM (2009) On the DVCC and the BOCCII as adjoint elements. J Circ Syst Comput 18:1017–1032
25. Pal K, Nigam MJ (2008) Novel active impedances using current conveyors. J Active Passiv Electron Devices 3:29–34
26. Awad IA, Soliman AM (1999) Inverting second generation current conveyors: the missing building blocks, CMOS realizations and applications. Int J Electron 86:413–432
27. Soliman AM, Saad RA (2010) Generation of second generation current conveyor (CCII) family from inverting second generation current conveyor (ICCII). Int J Electron 97:405–414
28. Soliman AM (2008) The inverting second generation current conveyors as universal building blocks. Int J Electron Commun (AEU) 62:114–121
29. Sobhy EA, Soliman AM (2007) Novel CMOS realizations of the inverting second-generation current conveyor and applications. Analog Integr Circ Sig process 52:57–64
30. Beevao D, Vrba K (2000) Novel generations of inverting current conveyor using universal current conveyor. Electron J Eng Tech 3:4
31. Sobhy EA, Soliman AM (2009) Novel CMOS realization of balanced-output third generation inverting current conveyor with applications. Circ Syst Sig Process 28:1037–1051
32. Salama K, Soliman A (1999) Novel MOS-C quadrature oscillator using the differential current voltage conveyor. In: 42nd IEEE midwest symposium on circuits and systems, Las Cruces, NM, vol 1, pp 279–282
33. Acar C, Ozoguz O (1999) A new versatile building block: current differencing buffered amplifier suitable for analog signal-processing filters. Microelectron J 30:157–160
34. Salama KN, Elwan HO, Soliman AM (2001) Parasitic-capacitance-insensitive voltage-mode MOSFET-C filters using differential current voltage conveyor. Circ Syst Sig Process 20:11–26
35. Soliman AM, Madian AH (2009) MOS-C KHN filter using voltage OP AMP CFOA, OTRA and DCVC. J Circ Syst Comp 18:733–769
36. Piyatat T, Tangsrirat W, Surakampontorn W (2005) Current-controlled differential current voltage conveyor and its applications, ECTI-Conference, Pathumthani, Thailand, pp 661–646
37. El-Adawy AA, Soliman AM, Elwan HO (2000) A novel fully differential current conveyor and applications for analog VLSI. IEEE Trans Circ Syst-II 47:306–313
38. Kacar F, Metin B, Kuntman H, Cicekoglu O (2010) A new high-performance CMOS fully differential second-generation current conveyor with application example of biquad filter realisation. Int j Electron 97:499–510
39. Alzaher HA, Elwan HO, Ismail M (2000) CMOS fully differential second-generation current conveyor. Electron Lett 36:1095–1096
40. Sobhy E, Soliman AM (2010) Realizations of fully differential voltage second generation current conveyor with an application. Int J Circ Theor Appl 38:441–452
41. Chang CM, Soliman AM, Swamy MNS (2007) Analytical synthesis of low-sensitivity high-order voltage-mode DDCC and FDCCII-grounded R and C All-pass filter structures. IEEE Trans Circ Syst-I 54:430–1443
42. Mahmoud SA (2006) New fully-differential CMOS second-generation current conveyor. ETRI J 28:495–501

43. Alzaher HA (2004) CMOS highly liner fully differential current conveyor. Electron Lett 40:214–216
44. Soliman AM (1998) Generalized voltage and current conveyors: practical realizations using CCII. IEICE Trans Fundam 5E–81:973–975
45. Biolek D, Vrba K, Cajka J, Dostal T (2000) General three-port current conveyor: a useful tool for network design. J Electr Eng 51:36–39
46. Becvar D, Vrba K, Zeman V, Musil V (2000) Novel universal active block: a universal current conveyor. ISCAS 3:471–474
47. Cajka J, Dostal T, Vrba K (2004) General view on current conveyors. Int J Circ Theor Appl 32:133–138
48. Cajka J, Vrba K (2004) The voltage conveyor may have in fact found its way into circuit theory. Int J Electron Commun (AEU) 58:244–248
49. Sponar R, Vrba K, Kubanek D (2005) Universal conveyor-novel active device suitable for analog signal processing. In: Proceedings of the 9th international conference on circuit, WSEAS, Wisconsin
50. Kuntman H, Cicekoglu O, Ozoguz S (2002) A modified third generation current conveyor, its characterization and applications. Frequenz 56:47–54
51. Zeki A, Toker A (2002) The dual-X current conveyor (DXCCII): a new active device for tunable continuous-time filters. Int J Electron 89:913–923
52. Beg P, Maheshwari S (2014) Generalized filter topology using grounded components and single novel active element. Circ Syst Sig Process. doi:10.1007/s00034-014-9807-4
53. Kacar F, Metin B, Kuntman H (2010) A new CMOS dual-X second generation current conveyor (DXCCII) with an FDNR circuit application. Int J Electron Commun (AEU) 64:774–778
54. Maheshwari S, Ansari MS (2012) Catalog of realizations for DXCCII using commercially available ICs and applications. Radioengineering 21:281–289
55. Sato T, Wada K, Takagi S, Fujii N (2002) Extension of current conveyor concept and its applications. IEICE Trans Fundam E85-A:414–421
56. Alzaher HA, Elwan H, Ismail M (2003) A CMOS fully balanced second-generation current conveyor. IEEE Trans Circ Syst-II 50:278–287
57. Gift SJG (2005) The operational conveyor and its application in an accurate current amplifier with gain-independent bandwidth. Int J Electron 92(1):33–47
58. Horng JW, Hou CL, Chang CM (2008) Multi-input differential current conveyor, CMOS realization and application. IET Circ Devices Syst 2:469–475
59. Hwang YS, Liu WH, Tu SH, Chen JJ (2009) New building block: multiplication-mode current conveyor. IET Circ Devices Syst 3:41–48
60. De Marcellis A, Ferri G, Guerrini NC, Scotti G, Stornelli V, Trifiletti A (2009) The VCG-CCII: a novel building block and its application to capacitance multiplication. Analog Integr Circ Sig Process 58:55–59
61. Khan IA, Nahhas AM (2012) Reconfigurable voltage mode phase shifter using low voltage digitally controlled CMOS CCII. Electr Electron Eng 2:226–229
62. Alzaher H, Tasadduq N, Al-Ees O, Al-Ammari F (2013) A complementary metal-oxide semiconductor digitally programmable current conveyor. Int J Circ Theor Appl 41:69–81
63. Fongsamut C, Surakampontorn W (2010) Minimal realization for single-element-controlled sinusoidal oscillators using single current conveyor. In: International symposium on communication and information technology, Tokyo, pp 196–199
64. Chavoshisani R, Hashemipour O (2011) Differential current conveyor based current comparator. Int J Electron Commun(AEU) 65:949–953
65. Metin B, Herencsar N, Vrba K (2012) A CMOS DCCII with a grounded capacitor based cascadable all-pass filter application. Radioengineering 21:718–724
66. Soliman AM, Saad RA (2009) On the introduction of new floating current conveyors. J Circ Syst Comput 18:1005–1016

67. Herencsar N, Koton J, Vrba K (2009) A new electronically tunable voltage-mode active- C phase shifter using UVC and OTA. IEICE Electron Express 6:1212–1218
68. Zhang Q, Wang C, Sun J, Du S (2011) A new type of current conveyor and its application in fully balanced differential current-mode elliptic filter design. J Electr Eng 62:126–133
69. Chunhua W, Yang L, Qiujing Z, Yu F (2010) Systematic design of fully balanced differential current-mode multiple-loop feedback filters using CFBCII. Radioengineering 19:185–193
70. Kacar F, Yesil A, Minaei S, Kuntman H (2014) Positive/negative lossy/lossless grounded inductance simulators employing single VDCC and only two passive elements. Int J Electron Commun 68:73–78
71. Prasad d, Ahmad J (2014) New electronically- controllable lossless synthetic floating inductance circuit using single VDCC. Circ Syst 5:13–17
72. Jerabek J, Sotner R, Vrba K (2014) TISO adjustable filter with controllable-gain voltage differencing current conveyor. J Electr Eng 65:137–143
73. Kumngern M, Khateb F, Phasukkit P, Tungjitkusolmun S, Junnapiya S (2014) ECCCII- based current-mode universal filter with orthogonal control of ω_o and Q_o. Radioeng J 23:687–696

Chapter 11
Other Building Blocks Having MTC or CC at Front-end and Their Applications

Abstract There have been a number of disclosures of new building blocks which have either the *mixed-translinear-cell* (MTC) or the CC at the front-end and more often than not have more than three terminals thereby providing additional flexibility and versatility in devising various functional circuits. This chapter presents a number of such building blocks. In each case, typical hardware implementation of such building blocks, along with their most significant applications, have also been highlighted.

11.1 Introduction

In this chapter, we discuss those additional building blocks[1] whose front-end consists of one or more MTC-type structures or CCs, thereby providing additional flexibility and versatility for analog circuit designs.

It may be recalled that, as explained Chap. 9 of this monograph, the electronic-tunability of R_x is manifested in a number of other analog building blocks of relatively more recent origin such as current-controlled current feedback amplifiers (CC-CFA), current-controlled current differencing buffered amplifiers (CC-CDBAs), current-controlled current differencing transconductance amplifiers (CC-CDTA), current-controlled current conveyors transconductance amplifiers (CCCCTA) and CCCIIs with negative intrinsic resistance etc. which also employ the MTC as a basic front-end unit in their internal circuit architecture.

All the considered building blocks are implementable in both bipolar and CMOS IC technologies and hence, the circuits employing them are attractive from the point of view of IC implementation.

[1] An exhaustive review of various existing analog circuit building blocks as well as several new possible building blocks was presented in: Biolek et al. [1].

© Springer International Publishing Switzerland 2015
R. Senani et al., *Current Conveyors*, DOI 10.1007/978-3-319-08684-2_11

11.2 CC-CFA

Since the internal architecture of the CFA[2] contains a CCII+ followed by a unity gain voltage buffer, it is, therefore, obvious that a CCCII+ followed by a voltage buffer would become a CC-CFA. This concept has been studied by Siripruchyanun et al. in [3] wherein using bipolar and MOS current mirrors several topologies of BiCMOS CCCIIs have been studied, ultimately resulting in Bi-CMOS CC-CFA. In addition, a number of interesting applications have been proposed some of which are shown here in Fig. 11.1.

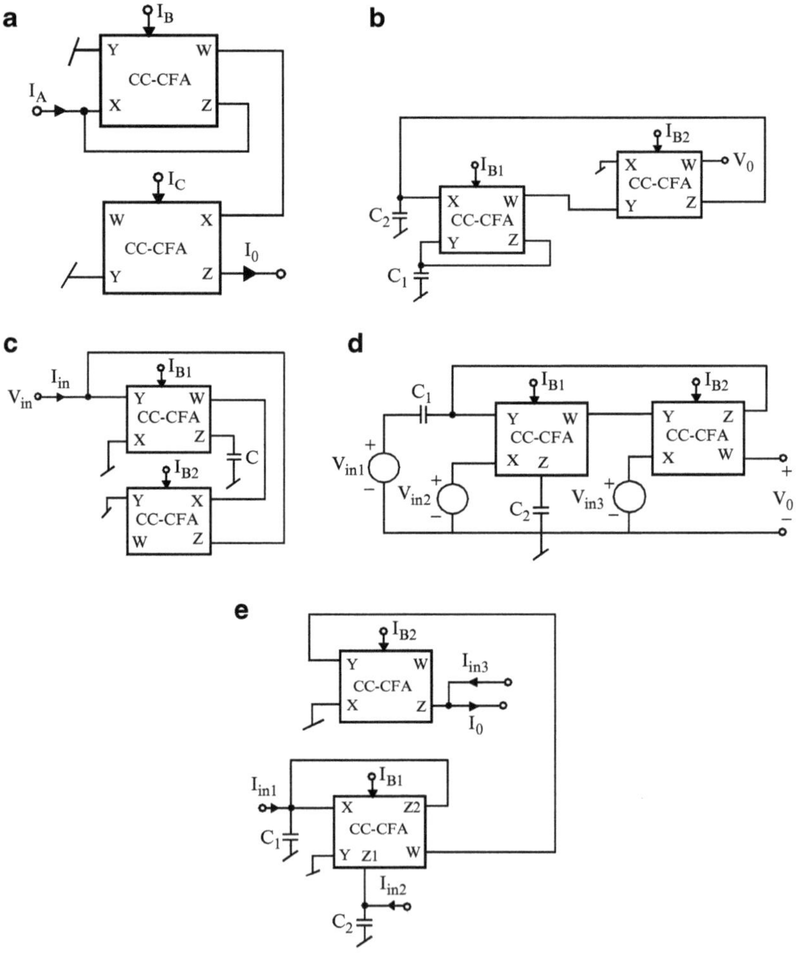

Fig. 11.1 The various applications of the CC-CFAs [3]

[2] For an interesting article demonstrating how the basic CCII+ (AD844) can be used to realize a variety of CCs and other related building blocks, see [2].

For the circuit of Fig. 11.1a, the I_0 is given by:

$$I_o = -\frac{I_A I_C}{2I_B} \tag{11.1}$$

Hence, the circuit would function as a current-mode multiplier/divider.

For the sinusoidal oscillator circuit of Fig. 11.1b, the CO and FO are given by:

$$C_1 = C_2; \qquad \omega_0 = \sqrt{\frac{1}{C_1 C_2 R_{x1} R_{x2}}} \tag{11.2}$$

Hence, an electronically controllable oscillator is realized. If the DC bias currents of the two CC-CFAs are taken equal i.e. $I_{B1} = I_{B2} = I_B$, it can be easily verified that ω_0 will become directly proportional to I_B and the circuit, thus, becomes a *linear* current-controlled oscillator.

In case of the grounded inductance circuit of Fig. 11.1c, the value of the simulated inductance is given by:

$$L = CR_{x1}R_{X2} = \frac{CV_T^2}{4I_{B1}I_{B2}} \tag{11.3}$$

For the circuit of Fig. 11.1d, the output voltage is given by:

$$V_o = \frac{V_{in1}s^2C_1C_2R_{x1}R_{x2} + V_{in2}(1 + sC_2R_{x2}) - V_{in3}sC_2R_{x1}}{s^2C_1C_2R_{x1}R_{x2} + sC_2R_{x2} + 1} \tag{11.4}$$

while in case of the circuit of Fig. 11.1e, the output current is given by:

$$I_o = \frac{I_{in1}sC_2R_{x2} + I_{in2} - I_{in3}(s^2C_1C_2R_{x1}R_{x2} + sC_2R_{x2} + 1)}{s^2C_1C_2R_{x1}R_{x2} + sC_2R_{x2} + 1} \tag{11.5}$$

From the Eqs. (11.4) and (11.5) it is easy to figure out that by choosing inputs appropriately all the five standard filter functions can be realized in both multi-input-single-output-type VM and CM filters of Fig. 11.1d and e respectively.

11.3 Four-Terminal-Floating-Nullor (FTFN)

The four-terminal-floating-nullor (FTFN) is a conceptual generalization of the nullor with all the four terminals being floating with the ground terminal being taken as external, as shown in Fig. 11.2. In concept, it is similar to an *operational floating amplifier* introduced by Huijsing [4, 5]. In the literature several

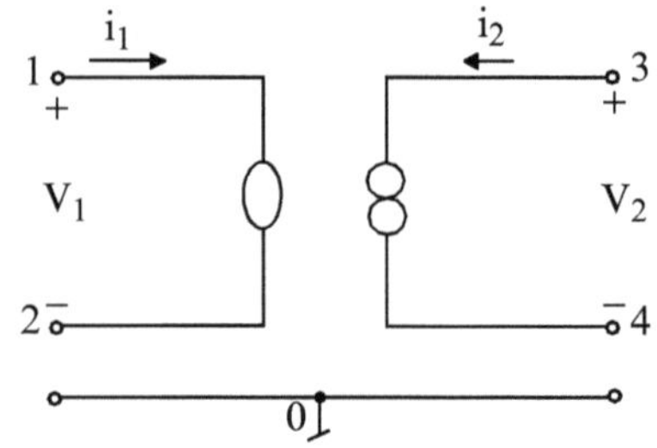

$$\begin{bmatrix} v_1 \\ i_1 \end{bmatrix} = \begin{bmatrix} 0 & 0 \\ 0 & 0 \end{bmatrix} \begin{bmatrix} v_2 \\ -i_2 \end{bmatrix}$$

Fig. 11.2 The symbolic notation of the FTFN and its transmission matrix

implementation of the FTFN have been introduced by Nordholt [6] Stevenson [7] and Senani [8] using commercially available Op-amps [7], using BJTs [4, 5] or using combination of an op-amp, an OTA and a few resistors [8]. However, a particular implementation of the FTFN which can be obtained by a composite connection of two CCIIs, first suggested by Senani in [8], has been particularly popular and has been employed by numerous researchers in verifying FTFN-based circuits. This implementation is based upon the identification that a CCII− is a three terminal nullor hence using the nullor identity shown in Fig. 11.2; two CCII− can be appropriately connected to constitute an FTFN. Incidentally, if CCII− are replaced by CCII+, the modified circuit still implements an FTFN and is now realizable with two AD844 ICs (see Fig. 11.2).

FTFN has been recognized in the literature as an universal building block and has been extensively used in the realization of universal filter, SRCOs, synthetic impedances, inverse filters and numerous others applications. However, the most significant application of the FTFNs has been in the areas of floating impedance simulation and generating equivalents of op-amp-RC sinusoidal oscillator configurations [9–20] (Fig. 11.3).

$$\begin{bmatrix} v_1 \\ i_1 \end{bmatrix} = \begin{bmatrix} 0 & 0 \\ 0 & 0 \end{bmatrix} \begin{bmatrix} v_2 \\ -i_2 \end{bmatrix}$$

An exhaustive bibliography on FTFNs and their applications in analog circuit analysis, synthesis and design has been presented in [21] whereas a concise treatment of major developments taken place in this area has been presented in Chap. 2 of the book [22].

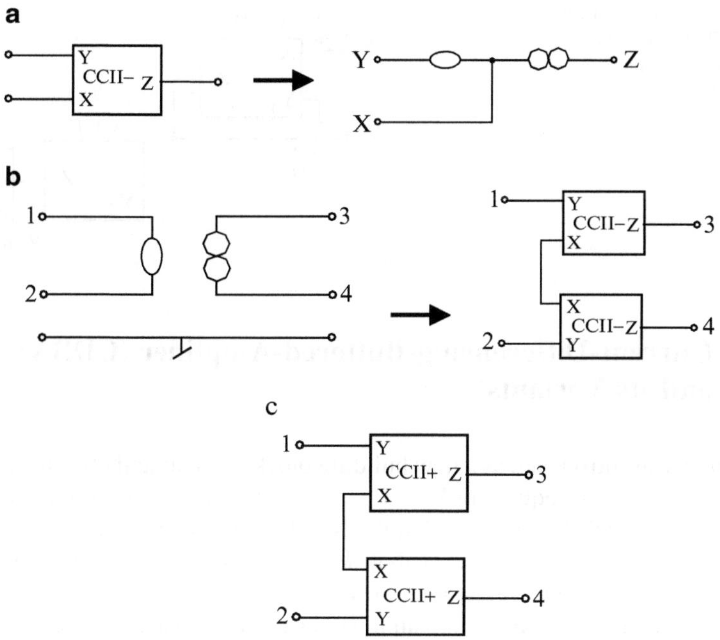

Fig. 11.3 Practical realization of FTFN using current conveyor (**a**) nullor equivalent of the CCII− (**b**) FTFN implementation using two CCII− (**c**) FTFN implementation using two CCII+ (realizable using two AD844s)

11.4 Operational Trans-Resistance Amplifier (OTRA)

An operational transresistance amplifier has two input terminals with input impedance being ideally zero thereby providing a *virtual ground* at both the inputs. The difference of the input currents applied at the two input terminals flows out of a node where the transresistance gain a set up by a resistance R_m thereby leading to the equation $V_O = R_m(I_1\text{-}I_2)$. When used in negative feedback, the infinite transresistance gain R_m forces the two inputs current to be equal-a property which can be considered to be the dual of that available in a voltage- mode op-amp. The internal circuit architecture of an OTRA employs two input cells which are essentially mixed-translinear-cells as employed in the front end of a CCII. Since the OTRA has virtual ground at its both inputs and has low impedance voltage output, most of the effect of the parasitic capacitances becomes ineffective. Due to these advantageous features, the OTRA has been employed extensively in realizing immittance simulator topologies, Schmitt trigger, monostable/bi-stable multivibrators, sinusoidal oscillators and SRCOs, for instance, see references [23–26] and those cited therein. A CC based implementation of the OTRA is shown in Fig. 11.4.

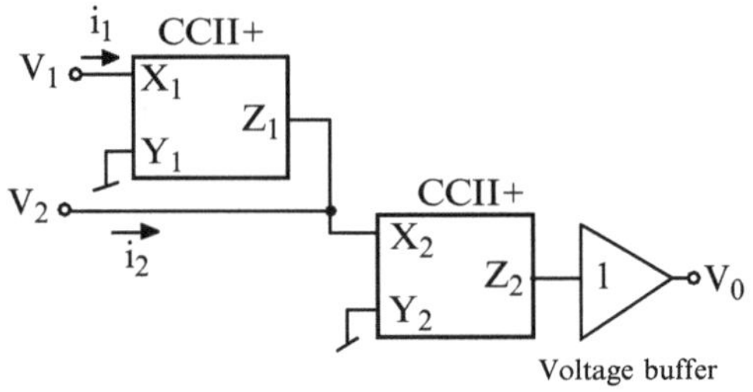

Fig. 11.4 Implementation of the OTRA using two CCIIs and a voltage buffer [23]

11.5 Current-Differencing-Buffered-Amplifier (CDBA) and Its Variants

The CDBA was introduced as a new building block by Acar and Ozoguz in [27] and is characterized by the equations $V_p = V_n = 0$, $i_z = i_p - i_n$, $V_w = V_z$. From the terminal equation of the CDBA, it is clear that its CCII-based implementation can be nearly same as that of the OTRA except that in the case of OTRA, the current output terminal is left open to facilitate $R_m \rightarrow 0$ whereas in case of the CDBA, this current output terminal is to be loaded into an impedance Z_L so that a voltage V_Z is created across it whose buffered version is then made available at the W terminal of the CDBA.

If a current-controlled CDBA (CC-CDBA) is needed, the same can be realized by the current sensing front-ends to be composed of the same mixed translinear cells as employed in CCCII. This applies to both bipolar and CMOS implementations of the CC-CDBAs. Consequently, a CC-CDBA can be implemented from CCCIIs as shown in Fig. 11.5.

An exemplary CDBA circuit exhibiting current-controllable input resistances at its both p and n input terminals is shown here in Fig. 11.6. A CC-CDBA is characterized by $V_p = R_p I_p$, $V_n = R_n I_n$, $I_z = I_p - I_n$, and $V_w = V_z$.

An interesting application of the CC-CDBA is in the realization of a very compact current-mode analog multiplier/divider which is shown in Fig. 11.7.

A routine analysis of the circuit reveals that the output current is given by

$$I_0 = \frac{I_A I_C}{2 I_B} \tag{11.6}$$

From the above, it is seen that one can engage the circuit either as a multiplier by taking I_A and I_C as the inputs or as a divider by taking I_A and I_B as inputs. However, it should be noted that in both the cases, only two-quadrant operation is possible.

Two other novel applications of CC-CDBA are in the realization of floating negative grounded and floating inductors. Two exemplary circuits to accomplish this are shown in Fig. 11.8 and have the novelty of employing only two active building blocks and a single grounded capacitor while providing electronically-controlled inductance value.

Fig. 11.5 The implementation of the CDBA using CCIIs and a voltage buffer

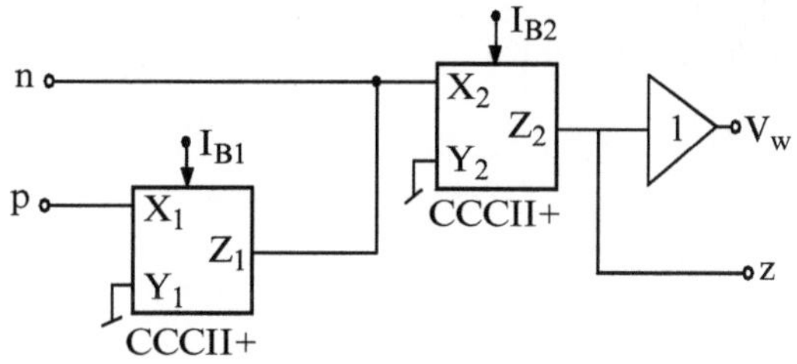

a

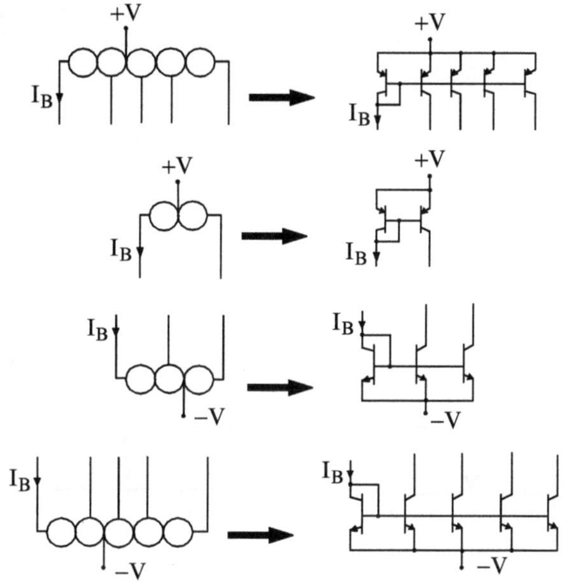

b

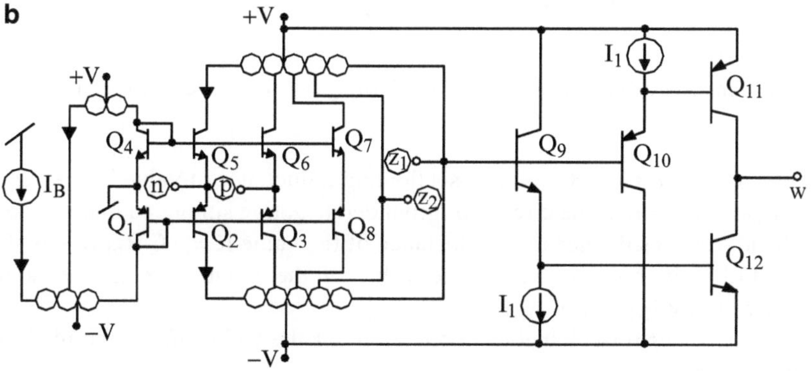

Fig. 11.6 Internal construction of a DO-CC-CDBA [28]

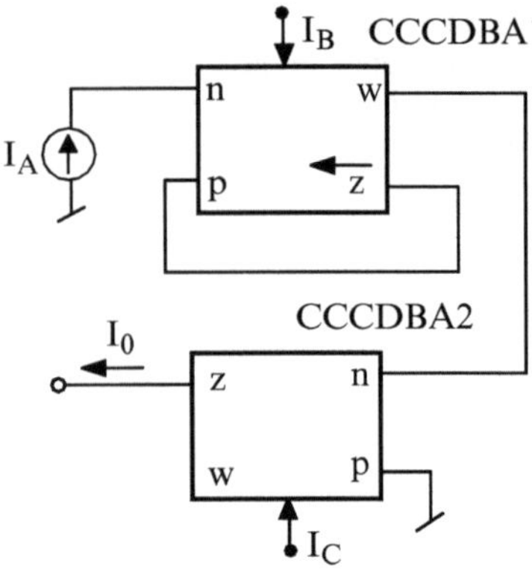

Fig. 11.7 Current mode multiplier/divider [29]

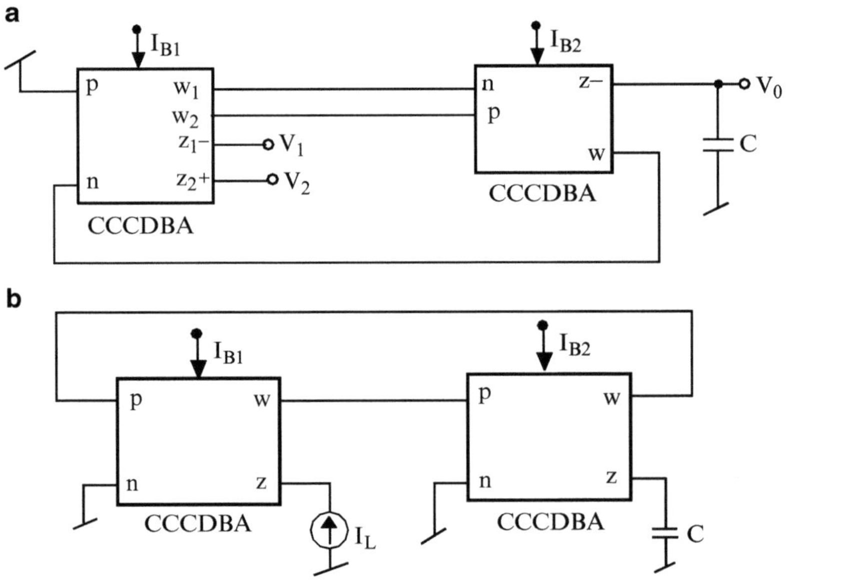

Fig. 11.8 Simulated negative inductors [30] (**a**) Floating negative inductor (**b**) grounded negative inductor

The circuit of Fig. 11.8a simulates a floating inductance of value $L = -C\,R_{n1}\,R_2$ where $R_{n2} = R_{p2} = R_2$. The circuit for grounded negative simulated inductor, on the other hand, realizes the negative inductance of the value $L = -C\,R_{p1}\,R_{p2}$ as shown in Fig. 11.8b. In both the cases, the value of the inductance can be adjusted electronically by I_{B1} and/or I_{B2}.

For some other miscellaneous applications of the CC-CDBAs, the readers are referred to [28–40].

11.6 Current Controlled Current-differencing Transconductance Amplifier (CC-CDTA)

The current differencing transconductance amplifier (CDTA) was introduced by Biolek [41] as a new versatile building block for current-mode analog signal processing. It is a five-port building block having two inputs and three outputs with the following characterizations. The two inputs labeled as p and n are low impedance terminals having virtual ground at their inputs. The difference of the two input currents I_p and I_n is made available from the high output impedance Z-terminal where an external impedance may be connected as a load thereby creating a voltage V_z which is given as input to an internal OTA which creates two current outputs called I_{x+} and I_{x-} which are respectively given by $I_{x+} = g_m V_z$ and $I_{x-} = g_m V_z$. This is shown in Fig. 11.9.

A CC-CDTA is an extension of the concept of CDTA and is characterized by $V_p = R_p I_p$, $V_n = R_n I_n$, $I_z = I_p$-I_n, $I_x = \pm g_m V_z$ where $R_p = R_n = V_T/2I_B$ and $g_m = I_C/2V_T$ which is shown in Fig. 11.10a.

A novel application of this implementation is the current mode multiplier/divider shown in Fig. 11.10b which has the output current given by

$$I_O = \frac{I_A I_C}{8 I_B} \tag{11.7}$$

From the above it is seen that I_O is result of either multiplication of I_A and I_C, or dividing I_A or I_C by I_B. However, a limitation of this circuit is that it is only two quadrant multiplier/divider. On the other hand, a merit of this circuit is that the functions performed by this circuit are temperature insensitive since the terms involving V_T have been cancelled out and do not appear in the output equation.

For a number of interesting applications of the CC-CDTAs see [41–73].

11.7 Current Controlled Current Conveyor Transconductance Amplifier (CCCC-TA)

This building block was proposed by Siripruchyanun and Jaikla [74] and is characterized by the following terminal equations: $I_y = 0$, $V_x = V_y + I_x R_x$, where $R_x = V_T/2I_{B1}$, $I_Z = I_x$ and $I_0 = \pm g_m V_z$ where $g_m = I_{B2}/2V_T$.

An exemplary implementation of this building block is shown in Fig. 11.11 from where it can be seen that the input front end of this building block also happens to be the MTC which results in the current controllability of its R_x.

An interesting application of this building block was demonstrated by Siripruchyanun and Jaikla [74] in realizing a universal current-mode biquad. The novelty of this circuit, shown here in Fig. 11.12, lies in using only a single active building block for realizing all the five standard filter responses with both grounded

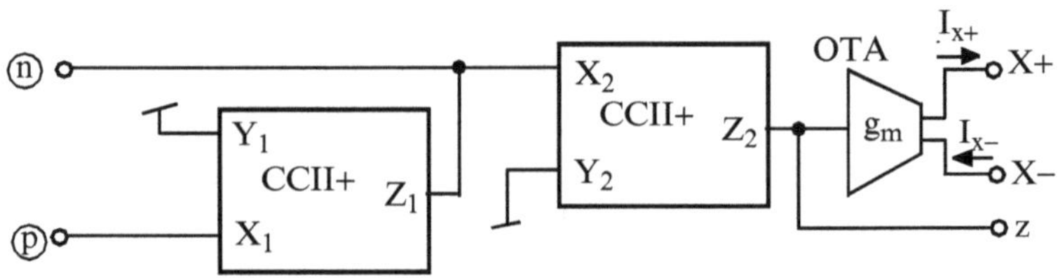

Fig. 11.9 CDBA implementation using CCIIs and OTA [41]

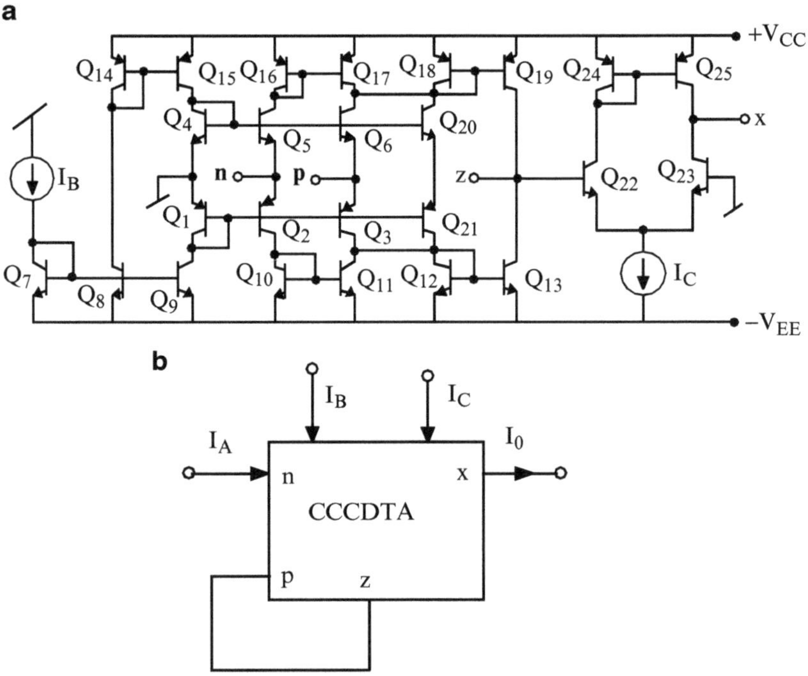

Fig. 11.10 Current controlled current differencing transconductance amplifier (**a**) CMOS CC-CDTA architecture (**b**) A CC-CDTA configured as a multiplier/divider (Adapted from [42] © 2008 Elsevier)

capacitors thereby making it highly suitable for integrated circuit implementation. This circuit can, thus, be seen to be a minimum-component biquad exhibiting these properties.

The output current of this circuit is given by

$$I_o = \frac{I_{in1}sC_2 + g_m I_{in2} - I_{in3}(s^2C_1C_2R_x + sC_2 + g_m)}{s^2C_1C_2R_x + sC_2 + g_m} \tag{11.8}$$

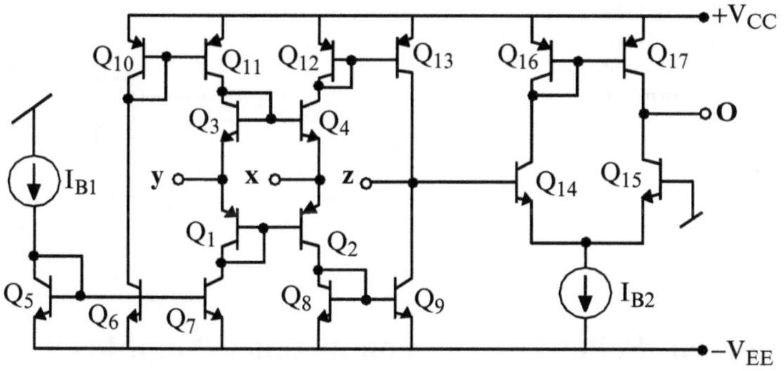

Fig. 11.11 Current controlled current conveyor transconductance amplifier (Adapted from [74] © 2007 Springer)

Fig. 11.12 Novel CM universal biquad using a single CCCC-TA (Adapted from [74] © 2007 Springer)

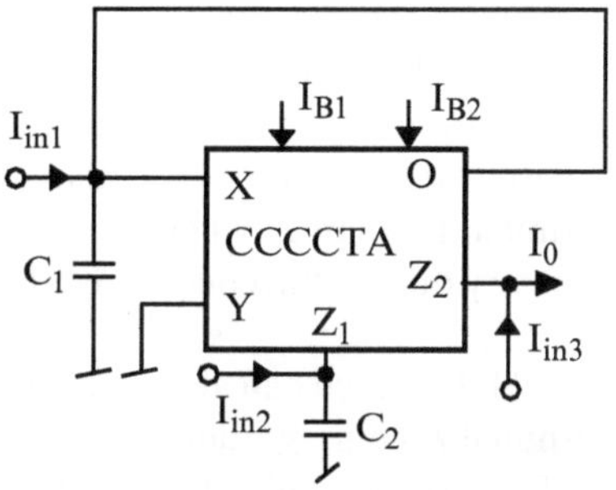

from where one can easily work out that, subject to different input conditions / selections, it would be possible to realize all the five standard functions with tuning of ω_o.

In Fig. 11.13, another novel application of the CCCC-TA in realizing an electronically-controllable grounded inductor is shown, which employs only one CCCC-TA along with a grounded capacitor as advantageous for IC implementation. The realized inductance value is given by

$$L_{eq} = \frac{CV_T^2}{I_{B1}I_{B2}} \tag{11.9}$$

For numerous other applications of CCCCTA the reader is referred to references [74–86].

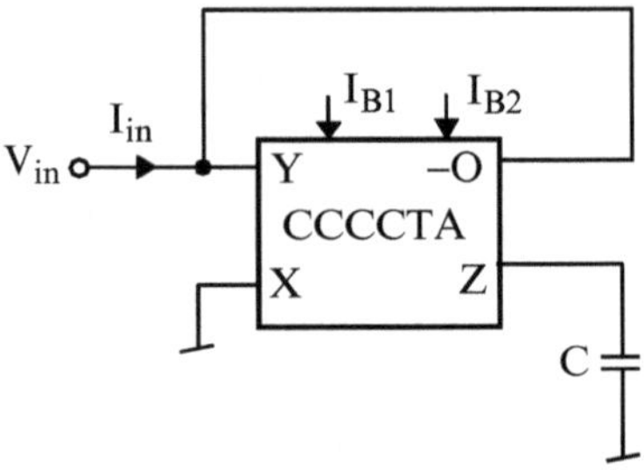

Fig. 11.13 Novel grounded inductance simulator using a single CCC-TA (Adapted from [74] © 2007 Springer)

11.8 Current Follower Transconductance Amplifier (CFTA)

A current follower transconductance amplifier (CFTAA) can be considered to be a reduced version of the CDTA as instead of *current-differencing unit* as the front end, it has a *current follower* as the front end which can be easily implemented from a CCII+ by grounding its Y terminal. The block diagram of the CFTA± can be drawn as shown in Fig. 11.14. Thus, it is a four terminal building block having one input terminal **f** and three outputs terminals **Z**, $\mathbf{x}^{+}$ and $\mathbf{x}^{-}$. The output current entering into terminal Z is the same as the one entering into the low impedance terminal f; whereas when terminal Z is terminated into an impedance, two complimentary output currents are available as i_x+ and i_{x-} which are respectively given by $i_{x+} = g_m V_z$ and $i_{x-} = g_m V_z$. A CC-based implementation of the CFTA is shown in Fig. 11.14 which can also be considered to be a special case of the CDTA with input terminal ***n*** being grounded.

For various applications of the CFTAs see [87–94]; for their other variants such as differential voltage controlled current conveyor transconductance amplifier (DVCCCTA), differential current controlled conveyor transconductance amplifier (DCCCTA), Z-copy Current Controlled Current inverting transconductance amplifiers (ZC-CCCITA) and Current controlled backward transconductance amplifier (CC-CBTA), the readers are referred to [95–106].

11.9 Current Through Transconductance Amplifier (CTTA)

A multiple output current controlled current through transconductance amplifier (MO-CC-CTTA) is a five terminal building block which is symbolically shown in Fig. 11.15a and is characterized by following matrix equation:

$$\begin{bmatrix} I_n \\ V_p \\ I_z \\ I_{x+} \\ I_{x-} \end{bmatrix} = \begin{bmatrix} -1 & 0 & 0 & 0 & 0 \\ R_i & 1 & 0 & 0 & 0 \\ 1 & 0 & 0 & 0 & 0 \\ 0 & 0 & 0 & g_{m1} & 0 \\ 0 & 0 & 0 & 0 & -g_{m2} \end{bmatrix} \begin{bmatrix} I_p \\ V_n \\ V_x \\ V_z \\ V_z \end{bmatrix} \tag{11.10}$$

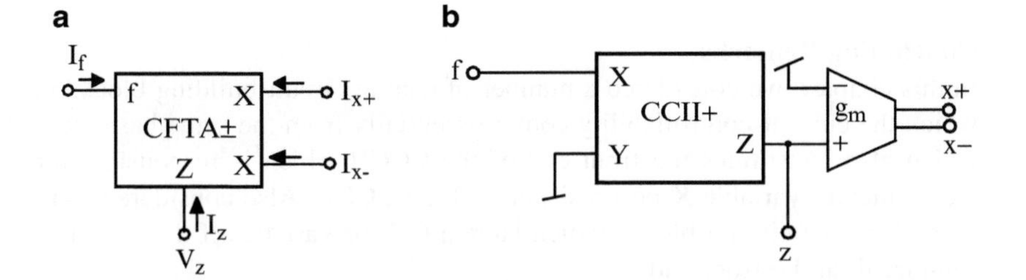

Fig. 11.14 CFTA realization (**a**) block diagram implementation (**b**) Implementation of the CFTA using CCII and OTA [87]

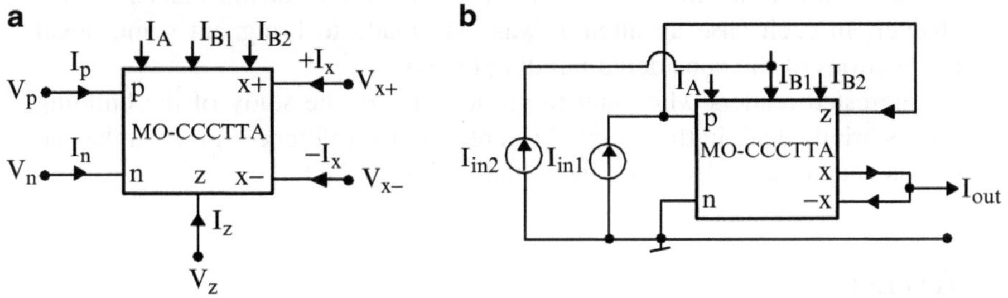

Fig. 11.15 The CTTA and its application (**a**) Block diagram (**b**) Current-mode multiplier/divider [107]

$$\text{where} R_i = \frac{V_T}{2I_A}; \; g_{m1} = \frac{I_{B1}}{2V_T}, g_{m2} = \frac{I_{B2}}{2V_T} \tag{11.11}$$

An interesting application of this block was proposed by Pisutthipong and Siripruchyanun [107] in realizing a current mode multiplier/divider as shown in Fig. 11.15b. A novel feature is this circuit is that does not require even a single external passive component and both the function are realized in the same circuit.

By a straight forward analysis of this circuit, taking cognizance of its characterizing equations, the current output of the circuit in terms of the two input currents is found to be

$$I_{out} = \frac{I_{in1} I_{in2}}{8I_A}. \tag{11.12}$$

From the above equation, it is clear the circuit functions as a four-quadrant multiplier because both the input currents can be either positive or negative. A novel feature of this circuit is that its performance is *temperature-insensitive*. Furthermore, if either I_{in1} or I_{in2} and I_A are the input currents, the proposed circuit can also function as a current-mode divider although in this case, since I_a cannot have negative value, the realized analog divider would be only a two-quadrant divider.

Concluding Remarks

In this chapter, we considered a number of analog circuit building blocks in which the current-controllability comes essentially from the same basic core which also constitutes the front end of the CCCII which is responsible for electronically-variable X-port resistance of the CCCII. Also considered were several other building blocks which have a CC (or varieties if the CC) as a sub-circuit at the front end.

With the inclusion of the building blocks of this chapter, together with the varieties of CCs introduced in Chap. 10, we believe that the set of *elements related to current conveyors* is considerably enlarged.

To highlight the importance of the building blocks introduced in this chapter, in each case an attempt was also made to bring out some novel applications of the considered building blocks.

Interested readers who want to go deeper into the study of the building blocks briefly dealt in this chapter, can refer to the references given at the end of this chapter and also those including in Appendix 1.

References

1. Biolek D, Senani R, Biolkova V, Kolka Z (2008) Active elements for analog signal processing: classification, review and new proposals. Radioengineering 17:15–32
2. Erkan Y, Minaei S, (2007) Realization of various active devices using commercially available AD844s and external resistors. Electron World 113:46–49
3. Siripruchyanun C, Silapan J (2008) BiCMOS current-controlled current feedback amplifier (CC-CFA) and its applications. WSEAS Trans Electron 6:203–219
4. Huijsing JH, Korte D (1977) Monolithic nullor-a universal active network element. IEEE J Solid State Circ 21:59–64
5. Huijsing JH (1990) Operational floating amplifier. IEE Proc 137:131–136
6. Nordholt EH (1982) Extending op-amp capabilities by using a current- source power supply. IEEE Trans Circ Syst 29:411–414
7. Senani R (1987) A novel application of four terminal floating nullor. Proc IEEE 75:1544–1546
8. Stevenson JK (1984) Two-way circuits with inverse transmission properties. Electron Lett 20:965–967
9. Hou CL, Yean R, Chang CK (1996) Single-element controlled oscillators using single FTFN. Electron Lett 32:2032–2033
10. Liu SI (1997) Single-resistance-controlled sinusoidal oscillator using two FTFNs. Electron Lett 33:1185–1186
11. Wang HY, Lee CT (1998) Realization of R-L and C-D immittances using single FTFN. Electron Lett 34:502–503
12. Abuelma'atti MT, Al-Zaher HA (1999) Current mode quadrature sinusoidal oscillators using two FTFNs. Frequenz 53:27–30
13. Abuelma'atti MT, Al-Zaher HA (1999) Current-mode sinusoidal oscillators using single FTFN. IEEE Trans Circ Syst-II 46:69–74
14. Cam U, Cicekoglu O, Kuntman H (2000) Universal series and parallel immittances simulators using four terminal floating nullors. Analog Integr Circ Sig Process 25:59–66

15. Cam U, Toker A, Cicekoglu O, Kuntman H (2000) Current-mode high output impedance sinusoidal oscillator configuration employing single FTFN. Analog Integ Circ Sig Process 24:231–238
16. Cam U, Cicekoglu O, Kuntman H (2001) Novel lossless floating immittance simulator employing only two FTFNs. Analog Integ Circ Sig Process 29:233–235
17. Lee CT, Wang H-Y (2001) Minimum realization for FTFN-based SRCO. Electron Lett 37:1207–1208
18. Shah NA, Malik MA (2005) High impedance voltage-and current-mode multifunction filters. Int J Electron Commun (AEU) 59:262–266
19. Malik MA (2010) Current/voltage-mode universal filter using FTFNs. J Franklin Inst 347:523–532
20. Fakhfakh M, Pierzchala M, Rodanski B (2012) On the design of active inductors with current-controlled voltage sources. Analog Integ Circ Sig Process 73:89–98
21. Kumar P, Senani R (2002) Bibliography on nullors and their applications in circuit analysis, synthesis and design. Analog Integr Circ Sig Process 33:65–76
22. Senani R, Singh AK, Kumar P, Sharma RK (2013) Nullors, their bipolar and CMOS implementations and applications in Analog circuit synthesis and design. Integr Circ Analog Sig Process: 31–59
23. Lo YK, Chien HC, Chiu HJ (2008) Switch-controllable OTRA-based bistable multivibrators. IET Circ Devices Syst 2:373–382
24. Chen JJ, Tsao HW, Liu SI, Chiu W (1995) Parasitic-capacitance-insensitive current-mode filters using operational transresistance amplifiers. IEE Proc Circ Devices Syst 142:186–192
25. Cam U, Kacar F, Cicekoglu O, Kuntman H, Kuntman A (2004) Novel Two OTRA-based grounded immittance simulator topologies. Analog Integ Circ Sig Process 39:169–175
26. Lo YK, Chien HC, Chiu HJ (2010) Current-input OTRA Schmitt trigger with dual hysteresis modes. Int J Circ Theor Appl 38:739–746
27. Acar C, Ozoguz S (1999) A new versatile building block: current differencing buffered amplifier suitable for analog signal processing filters. Microelectron J 30:157–160
28. Jaikla W, Siripruchyanun M (2006) Novel current-mode four –quadrant multiplier/divider based on current controlled CDBAs (CCCDBAs). In: ECTI-CON 2006 international conference, Ubon Ratchathani, Thailand, pp 509–512
29. Silapan P, Jaikla W, Siripruchyanun M (2007) High-performance Bi-CMOS current controlled CDBA and applications. (ISCIT-2007):40–43
30. Jaikla W, Siripruchyanun M (2006) Current controlled CDBAs based-novel floating and grounded negative inductance simulators. In: The 21st international technical conference on circuits and systems, computer and communication, Chiangmai, Thailand, pp III701–704
31. Lahiri A, Chowdhury A (2011) Four quadrant analog multiplier using dual-current-controlled current differencing buffered amplifier. J Circ Syst Comput 20:223–231
32. Koksal M, Sagbas M (2007) General synthesis procedure for nth-order current transfer function using CDBA. Frequenz 61:94–101
33. Tangsrirat W, Surakampontorn W (2009) Electronically tunable quadrature oscillator using current-controlled current differencing buffered amplifiers. J Active Passiv Electron Devices 4:163–174
34. Tangsrirat W, Prasertsom D, Surakampontorn W (2008) Low-voltage digitally controlled current differencing buffered amplifier and its application. Int J Electron Commun (AEU) 63:249–258
35. Biolek D, Bajer J, Biolkova V, Kolka Z, Kubicek M (2011) Z copy-controlled gain-current differencing buffered amplifier and its applications. Int J Circ Theor Appl 39:257–274
36. Metin B, Pal K, Cicekoglu O (2011) CMOS-controlled inverting CDBA with a new all-pass filter application. Int J Circ Theor Appl 39:417–425
37. Khateb F, Jaikla W, Kubanek D, Khatib N (2013) Electronically tunable voltage-mode quadrature oscillator based on high performance CCCDBA. Analog Integ Circ Sig Process 74:499–505

38. Biolek D, Lahiri A, Jaikla W, Siripruchyanun BB (2013) Realizations of electronically tunable voltage-mode/current-mode quadrature Sinusoidal oscillator using ZC-CG-CDBA. Microelectron J 42:1116–1123
39. Bashir SA, Shah NA (2012) Voltage-mode universal filter using current differencing copy current controlled current differencing buffered amplifier. Int J Electron Electr Eng 2:229–234
40. Dabbous SBA, Alsibai Z (2014) Ultra low voltage low power bulk driven Z copy current controlled current differencing buffered amplifier. Int J Electron Elect Eng 2:229–234
41. Biolek D (2003) CDTA-building block for current-mode analog signal processing. In: Proceedings of ECCTD 03, Krakow, Poland, pp 397–400
42. Siripruchyanun M, Jaikla W (2008) A current-mode analog multiplier/divider based on CCCDTA. Int J Electron Commun 62:223–227
43. Sakul C, Jaikla W, Dejhan K (2011) New resistorless current-mode quadrature oscillators using 2 CCCDTAs and grounded capacitors. Radioengineering 20:594–599
44. Li Y (2012) A new single MCCCDTA based Wien-bridge oscillator with AGC. Int J Electron Commun(AEU) 66:153–156
45. Silapan P, Siripruchyanun M (2011) Fully and electronically controllable current-mode Schmitt trigger employing only single MO-CCCDTA and their applications. Analog Integr Circ Sig Process 68:111–128
46. Jaikla W, Lahiri A (2012) Resistor-less current-mode four-phase quadrature oscillator using CCCDTAs and grounded capacitors. Int J Electron Commun (AEU) 66:214–218
47. Jaikla W, Siripruchyanun M (2009) High output impedance current-mode universal biquadratic filter using CCCDTAs. J KMUTNB 19:136–143
48. Siripruchyanun M, Jaikla W (2009) Current-controlled current differencing transconductance amplifier and applications in continuous-time signal processing circuits. Analog Integr Circ Sig Process 61:247–257
49. Siripruchyanun M, Jaikla W (2008) Electronically controllable current-mode universal biquad filter using single DO-CCCDTA. Circ Syst Sig Process 27:113–122
50. Siripruchyanun M, Jaikla W (2008) CMOS current-controlled current differencing transconductance amplifier and applications of analog signal processing. Int J Electron Commun 62:277–287
51. Silapan P, Srisakul T, Jaikla W, Siripruchyanun M (2007) CCCDTA-based filtillator. EECON-30:909–912
52. Siripruchyanun M, Jaikla W (2007) Implementation of CMOS current controlled current differencing transconductance amplifier and its applications. IEEE CAS Society Thailand: 3–7
53. Siripruchyanun M, Jaikla W (2007) Realization of current controlled current differencing transconductance amplifier (CCCDTA) and its applications. ECTI Trans Electr Eng Electron Commun 5:41–50
54. Jaikla W, Siripruchyanun M (2006) Current controlled current differencing transconductance amplifier (CCCDTA): a new building block and its applications. In: The 2006 ECTI international conference, Ubon Ratchathani, Thailand, pp 348–351
55. Jaikla W, Siripruchyanun M (2006) A versatile quadrature oscillator and universal biquad filter using dual-output current controlled current differencing transconductance amplifier. ISCIT 2006:1072–1075
56. Jaikla W, Siripruchyanun M (2007) 3 Input/2 Output current-mode universal biquad filter using single DO-CCCDTA. ECTI-CON: 69–72
57. Jaikla W, Prommee P (2011) Electronically tunable current-mode multiphase sinusoidal oscillator employing CCCDTA-based all pass filters with only grounded passive elements. Radioengineering 20:594–599
58. Xu J, Wang C, Jin J, Xia Z (2014) Low-voltage high-linearity wideband current differencing transconductance amplifier and its application on current-mode active filter. Radioengineering 23:512–522

59. Pandey N, Pandey R (2013) Current mode full-wave rectifier based on a single MZC-CDTA. Active Passive Electron Comp 2013:1–5
60. Khateb F, Vavra J, Biolek D (2010) A novel current-mode full-wave rectifier based on one CDTA and two diodes. Radioengineering 19:437–445
61. Li Y (2012) A novel current-mode multiphase sinusoidal oscillator using MO-CDTAs. Int J Electron 99:477–489
62. Jin J, Wang CH (2014) Current-mode universal filter and quadrature oscillator using CDTAs. Turk J Electr Eng Comp Sci 22:276–286
63. Jin J (2013) Current-mode resistorless SIMO universal filter and four-phase quadrature oscillator. Int J Electr Electron Sci Eng 7:96–101
64. Jin J, Wang C (2012) Single CDTA-based current-mode quadrature oscillator. Int J Electron Commun 66:933–936
65. Jin J, Wang C (2014) CDTA-based electronically tunable current-mode quadrature oscillator. Int J Electron 101:1086–1095
66. Tangsrirat W, Pukkalanun T, Mongkolwai P, Surakampontorn W (2011) Simple current-mode analog multiplier, divider, square-rooter and squarer based on CDTAs. Int J Electron Commun 65:198–203
67. Tangsrirat W, Tanjaroen W, Pukkalanun T (2009) Current-mode multiphase sinusoidal oscillator using CDTA-based all pass sections. Int J Electron 63:616–622
68. Prasad D, Bhaskar DR, Singh AK (2008) Realization of single resistance controlled sinusoidal oscillator: a new application of CDTA. WSEAS Trans Electron 5:257–259
69. Biolek D, Hancioglu E, Keskin AU (2008) High performance current differencing transconductance amplifier and its application in precision current mode rectification. Int Electron Commun 62:92–96
70. Lahiri A, Chowdhary A (2009) A novel first-order current mode all pass filter using CDTA. Radioengineering 18:300–305
71. Tuntrakool A, Supavarsuwat P, Jaikla W (2012) Current-tunable current-mode first order all pass section with gain controllability using CCCDTAs. Recent Adv Commun Circ Technology Innovation, Paris, France: 208–212
72. Chien HC, Wang JM (2013) Dual mode resistorless sinusoidal oscillator using single CCCDTA. Microelectron J 44:216–224
73. Jaikla W, Khateb F, Siripongdee S, Supavarasuwat P (2013) Electronically tunable current mode biquad filter employing CCCDTA and grounded capacitors with low input and high output impedance. Int J Electron J Commun 67:1005–1009
74. Siripruchyanun M, Jaikla W (2008) Current controlled current conveyor transconductance amplifier (CCCCTA): a building block for analog signal processing. Electr Eng 90:443–453
75. Siripruchyanun M, Silapan P, Jaikla W (2009) Realization of CMOS current controlled current conveyor transconductance amplifier (CCCCTA) and its applications. J Active Passiv Electron Devices 4:35–53
76. Siripruchyanun M, Phattanasak M, Silapan P, Jaikla W (2007) Design of CMOS current controlled current conveyor transconductance amplifier (CCCCTA) and its application. J KMITNB 17:2
77. Silapan P, Tanaphatsiri C, Siripruchyanun M (2008) Current controlled CCTA based-novel grounded capacitance multiplier with temperature compensation. In: Proceedings of IEEE Asia-Pacific conference on circuits and systems, Uttaradit Rajabhat University, Uttaradit, pp 1490–1493
78. Maheshwari S, Singh SV, Chauhan DS (2011) Electronically tunable low-voltage mixed-mode universal biquad filter. IET Circ Devices Syst 5:149–158
79. Mangkalakeeree S, Duangmalai D, Siripruchyanun M (2009) Current-mode KHN filter using single CC-CCTA. In: The 7th PSU engineering conference, Songkla, Thailand, pp 306–309
80. Jaikla W, Siripruchyanun M (2008) Voltage-mode universal filter based on single CCCCTA. In: The 6th PSU engineering conference, Songkla, Thailand, 242–245

81. Jaikla W, Siripongdee S, Suwanjan P (2012) MISO current-mode biquad filter with independent control of pole frequency and quality factor. Radioengineering 21:886–891
82. Jaikla W, Noppakarn A, Lawanwisut S (2012) New gain controllable resistor-less current-mode first order-all pass filter and its application. Radioengineering 21:312–316
83. Singh SV, Maheshwari S (2012) Current-processing current-controlled universal biquad filter. Radioengineering 21:317–323
84. Lawanwisut S, Siripruchyanun M, Jaikla W (2013) An active only temperature-insensitive current-mode biquad filter based on differentiator structures employing CCCCTAs. Circ Syst 4:280–286
85. Tomar RS, Singh SV, Chauhan DS (2013) Current processing current tunable universal biquad filter employing two CCTAs and two grounded capacitors. Circ Syst 4:443–450
86. Sotner R, Jerabek J, Prokop R, Vrba K (2011) Current gain controlled CCTA and its application in quadrature oscillator and direct frequency modulator. Radioengineering 20:317–326
87. Herencsar N, Koton J, Vrba K (2010) Realization of current-mode KHN equivalent biquad using current follower transconductance amplifiers (CFTAs). IEICE Trans Fundam 93-A:1816–1819
88. Herencsar N, Koton J, Vrba K, Misurec J (2009) A novel current-mode SIMO type universal filter using CFTAs. Contemp Eng Sci 2:59–66
89. Herencsar N, Koton J, Vrba K, Lahiri A, Cicekoglu O (2010) Current-controlled CFTA based current-mode SITO universal filter and quadrature oscillator. In: International conference on applied electron, University of West Bohemia, Pilsen
90. Tangsrirat W (2014) SFG synthesis of general high-order all-pass and all-pole current transfer functions using CFTAs. Sci World J 2014:Article ID 271926,8p
91. Herencsar N, Vrba K, Koton J, Lahiri A (2010) Realization of single resistance controlled quadrature oscillators using a generalized current follower transconductance amplifier and a unity gain voltage follower. Int J Electron 97:897–906
92. Li Y (2012) A series of new circuits based in CFTAs. Int J Electron Commun 66:587–592
93. Nie X, Pan Z (2013) Multiple-input single output and high output impedance current mode biquadratic filter employing five modified CFTAs and two grounded capacitors. Microelectron J 44:802–806
94. Li Y (2014) Electronically tunable current biquadratic filter and four phase quadrature oscillator. Microelectron J 45:330–335
95. Pandey N, Kumar P, Choudhary J (2013) Current controlled differential difference current conveyor transconductance amplifier and its application as wave active filter. Hindawi Publishing Corporation, NY, USA
96. Channumsin O, Pukkalanun T, Tangsrirat W (2012) Voltage-mode universal filter with one input and five outputs using DDCCTAs and all-grounded passive components. Microelectron J 43:555–561
97. Herencsar N, Lahiri A, Koton J, Vrba K, Metin B (2012) Realization of resistorless lossless positive and negative grounded inductor simulators using single ZC-CCCITA. Radioengineering 21:264–272
98. Jaikla W, Siripruchyanun M, Lahiri A (2011) Resistorless dual-mode quadrature sinusoidal oscillator using a single active building block. Microelectron J 42:135–140
99. Channumsin O, Tangsrirat W (2013) Single-input four-output voltage-mode universal filter using single DDCCTA. Microelectron J 44:1084–1091
100. Pandey N, Arora S, Takkar R, Pandey R (2012) DVCCCTA-based implementation of mutually coupled circuit. ISRN Electronics Article ID 303191, 6p
101. Pandey N, Kumar P (2011) Realization of resistorless wave active filter using differential voltage current controlled conveyor transconductance amplifier. Radioengineering 20:911–916

102. Chien HC, Chen YC (2014) CMOS realization of single resistance controlled and variable frequency dual mode sinusoidal oscillators employing a single DVCCTA with all grounded passive components. Microelectron J 45:226–238
103. Tangsrirat W, Channumsin O, Pukkalanun T (2013) Resistorless realization of electronically tunable voltage-mode SIFO type universal filter. Microelectron J 44:210–215
104. Laoudias C, and Psychalinos C (2013) Comparative study of resistor less filters using ifferential voltage current controlled current feedback operational amplifiers and differential voltage current controlled current conveyors. ISRN Electron 2013, Article ID 910170, 7p
105. Ayten UE, Sagbas M, Sedef H (2012) Electronically tunable sinusoidal oscillator circuit with current and voltage output. Int J Electron 99:1133–1144
106. Ayten UE, Sagbas M, Sedef H (2010) Current mode leapfrog ladder filters using a new active building block. Int J Electron Commun 64:503–511
107. Pisutthipong N, Siripruchyanun M (2009) A novel simple current-mode multiplier/divider employing only single multiple-output current controlled CTTA. In: Proceedings of TENCON IEEE region 10th annual international conference, University of Technology, Khonkaen

Part IV
Second Generation Applications: Realization of Various Linear/Nonlinear Functions Using Other Types of Current Conveyors

Chapter 12
Analog Filter Design Revisited: Circuit Configurations Using Newer Varieties of CCs

Abstract This chapter presents some selected universal biquad filter realizations in Voltage-mode, Current- mode and mixed–mode, using a variety of new types of CCs reported in the technical literature during the past two decades or more. In each category, we have chosen only a few representative configurations which, in our opinion, exhibit features superior to the other available alternatives.

12.1 Introduction

In Chap. 6 of this monograph we have presented significant contributions made by various researchers in devising universal biquad filter configurations in voltage-mode (VM) and current-mode (CM) using the normal kind of CCs. The circuits described therein can, therefore, be practically implemented with the commercially available IC CCs, for example, by using AD844 (one for CCII+ and two for CCII−), op-amps uA741/LF356 and transistor arrays CA9086, PA630 or OPA660/860. Thus, all the circuits described in Chap. 6 are physically realizable using of-the-shelf IC components.

In the available literature, there exist a large number of other varieties of current conveyors proposed by various researchers from time to time, which have already been dealt with in Chap. 10 of this monograph. Although such variants of CCs are not commercially available as of now, however, if one tries to build them with the available ICs, it can be done *albeit* at the expense of requiring several such IC components and passive elements. Since the proposers of such building blocks and other researchers have also devised a number of CMOS-based hardware implementations of these, it is obvious that the circuits using such variants of CCs are implicitly intended and are most suitable for fully-integrated circuit designs rather than discrete application. Since the circuits using the different varieties of CCs do offer a number of significant advantages, such as, a reduced number of active elements employed, greater functionality and improved characteristic features, apart from their suitability for VLSI implementation using bipolar/CMOS/BiCMOS technologies, they are naturally of considerable interest and therefore, worth considering for studies and investigations.

© Springer International Publishing Switzerland 2015
R. Senani et al., *Current Conveyors*, DOI 10.1007/978-3-319-08684-2_12

In this chapter, therefore, we revisit the problem of realizing universal biquad filters but the main focus is now on configurations using numerous variants of CCs such as, DVCC [1–30], DDCC [31–69], FDCCII [70–87], ICCII [88–100] DCVC [101, 102], CCIII [103–110], DXCCII [111–116], CFCCII [117–121], OFCC [122, 123], DOCCs/MOCCs [125–195] and the so-called universal current conveyor (UCC) [124, 196–202].

12.2 Filter Design Using Different Varieties of CCs

In the various subsections of this chapter, we undertake a discussion of some prominent configurations evolved during the past several years to realize universal biquad filters in VM, CM, dual-mode or mixed-mode from among those published in [1–212].

12.2.1 Filter Design Using DVCCs

Since a DVCC has differential input capability with infinite input impedance looking into its two y-terminals, it becomes possible to realize a number of functions with not only the least possible number of external passive components but also with reduced requirements of the number of active components. Before enumerating a number of interesting topologies for universal biquad designs, we present here one such example.

The example is that of an instrumentation amplifier which, when designed with normal CCs requires two resistors only one of which is grounded but needs as many as three CCs; by contrast, the same function can be performed with only one DVCC and with a minimum of only two resistors both of which are grounded as shown in Fig. 12.1

An ideal DVCC+ is characterized by the following hybrid matrix:

$$\begin{bmatrix} I_{Y1} \\ I_{Y2} \\ V_X \\ I_Z \end{bmatrix} = \begin{bmatrix} 0 & 0 & 0 & 0 \\ 0 & 0 & 0 & 0 \\ 1 & -1 & 0 & 0 \\ 0 & 0 & 1 & 0 \end{bmatrix} \begin{bmatrix} V_{Y1} \\ V_{Y2} \\ I_X \\ V_Z \end{bmatrix} \tag{12.1}$$

Assuming the characterization of equation (12.1), the expression for the output voltage in terms of differential input voltage for the circuit of Fig. 12.1a is given by:

$$V_{out} = \frac{R_2}{R_1} V_{id} \tag{12.2}$$

It is obvious that the circuit can realize both inverting and non-inverting amplifiers by grounding V_{Y1} or V_{Y2}. Furthermore, the circuit is expected to exhibit constant 3 dB-bandwidth for different values of the gains. That it is indeed so, has been demonstrated by Hassan and Mahmoud in [1] by implementing the above circuit

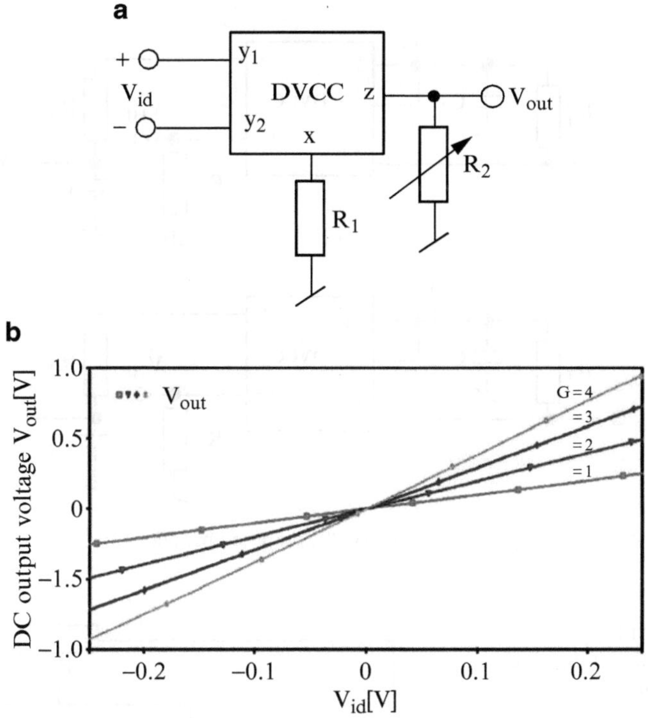

Fig. 12.1 Instrumentation amplifier presented by Hassan and Mahmoud (**a**) Circuit configuration (**b**) Transfer characteristics (Adapted from [1] © 2008 Elsevier GmbH)

with CMOS DVCC implementable in 0.25 μm CMOS technology with TMSC process parameters with DVCC biased from a DC power dual power supply of ±1.5 V. Figure 12.1b shows the transfer characteristics of this instrumentation amplifier when gain is varied from 1 to 4. SPICE simulations show that the 3-dB bandwidth of this circuit, even when the gain is varied, remains essentially constant around 8 MHz.

We now present some selected DVCC-based filter configurations employing one to four DVCCs and realizing filters in VM or CM exhibiting different characteristic features.

Single-DVCC Yuce biquads Figure 12.2 shows two VM multifunction filter configurations which simultaneously realize second-order LPF, HPF and BPF from the same circuit topology employing single dual output DVCC.

A straightforward circuit analysis of Fig. 12.2a and b yields the following voltage transfer functions:

$$H_{LP} = \frac{1 + sC_2(R_2 - R_1)}{D(s)}; H_{BP} = \mp\frac{sC_1R_1}{D(s)}; H_{HP} = \frac{s^2C_1C_2R_1R_3}{D(s)} \tag{12.3}$$

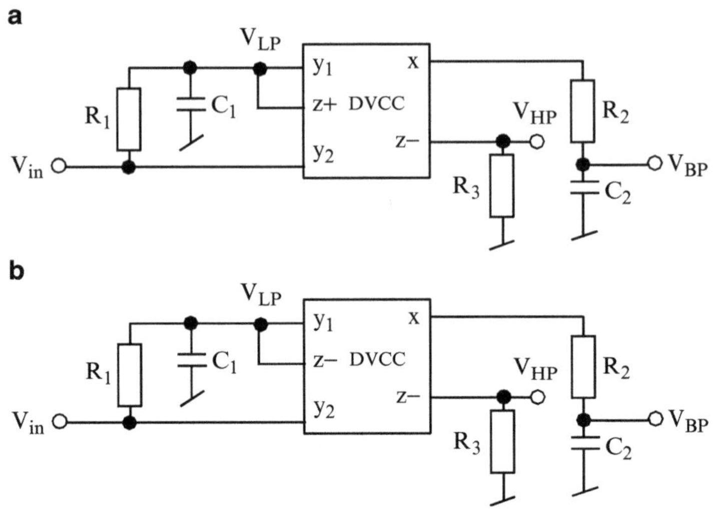

Fig. 12.2 VM biquads employing single DVCC, proposed by Yuce [2]

Fig. 12.3 VM biquadratic filter proposed by Tangsrirat and Channumsin [4]

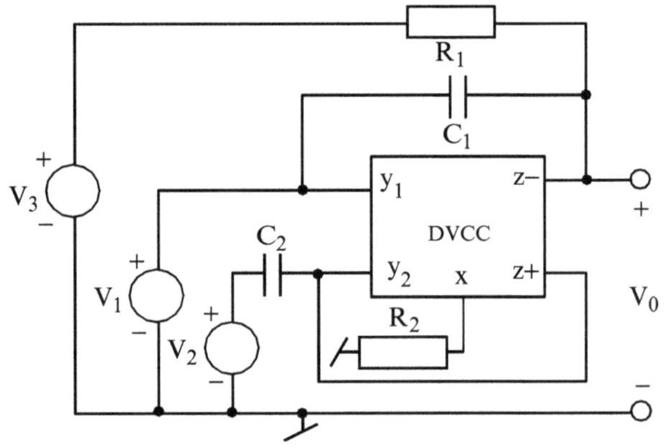

$$\text{where } D(s) = s^2 C_1 C_2 R_1 R_2 + s\{C_1 R_1 + C_2(R_2 - R_1)\} + 1 \tag{12.4}$$

The minus and plus signs of the BPF responses correspond to Fig. 12.2a, b respectively. The ideal filter responses in both the structures can easily be obtained by selecting $R_2 = R_1$.

Tangsrirat-Channumsin biquad A VM multifunctional biquadratic filter proposed by Tangsrirat and Channumsin [4] using a DVCC along with a minimum number of passive elements is shown in Fig. 12.3.

Using ideal DVCC, a routine circuit analysis gives the following output voltage in terms of the various input voltages:

$$V_0 = \frac{\left\{s^2 + \frac{s}{R_2}\left(\frac{1}{C_2} - \frac{1}{C_1}\right)\right\}V_1 + \frac{s}{C_1R_2}V_2 + \left(\frac{s}{C_1R_1} + \frac{1}{C_1C_2R_1R_2}\right)V_3}{D(s)} \tag{12.5}$$

where
$$\mathrm{D(s)} = s^2 + s\left(\frac{1}{R_1C_1} + \frac{1}{R_2C_2}\right) + \frac{1}{C_1C_2R_1R_2} \tag{12.6}$$

From equations (12.5) and (12.6), the following filter responses can be obtained: LPF: if $V_1 = 0$ (grounded), $-V_2 = V_3 = V_{in}$ (input voltage) and $R_1 = R_2$; BPF: $V_1 = 0 = V_3$, $V_2 = V_{in}$ and $(R_2/R_1 + C_1/C_2) = 1$; HPF: $V_2 = 0 = V_3$, $V_1 = V_{in}$, $R_1 = R_2$ and $C_1 = C_2$; Notch: $V_1 = -V_2 = V_3 = V_{in}$, $R_1 = R_2$ and $C_1 = C_2$ and APF: $V_1 = -V_2 = V_3 = V_{in}$ and $(R_2/R_1 + C_1/C_2) = 1$

An examination of the above filter realizations reveals that the circuit suffers from the realization conditions/constraints as well as non-availability of high input impedance, although it has the advantage of employing only one active element.

Horng-Jhao biquad Horng and Jhao [5] have presented a VM MIMO-type universal biquad using a DVCC, two resistors and two capacitors as shown in Fig. 12.4.

The circuit can realize all the standard second-order filter responses without altering the circuit topology. Assuming ideal DVCC, the three voltage outputs can be obtained in terms of four inputs as follows:

$$V_{01} = \frac{1}{D(s)}\left[\left\{\left(C_1C_2s^2 + sC_2\left(\frac{1}{R_1} - \frac{1}{R_2}\right)\right\}V_{in3} + sC_1\left(\frac{1}{R_2}\right)V_{in2} + \left(\frac{1}{R_1R_2}\right)V_{in1} - \left(sC_1 + \frac{1}{R_1}\right)\frac{1}{R_2}V_{in4}\right] \tag{12.7}$$

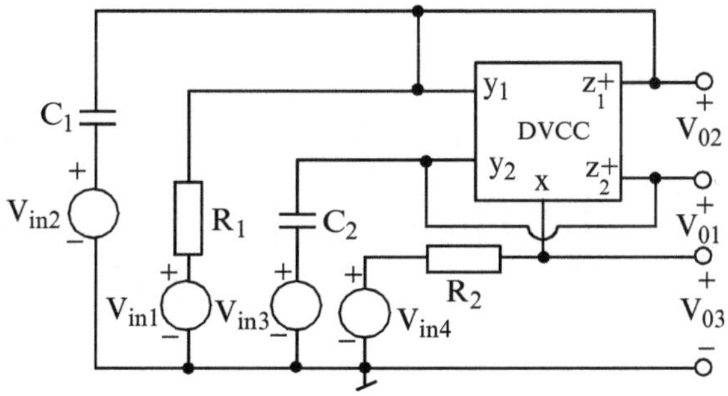

Fig. 12.4 VM biquad using single DVCC [5]

$$V_{02} = \frac{1}{D(s)}\left[\left\{\left(C_1C_2s^2 + sC_1\left(\frac{1}{R_2}\right)\right\}V_{in2} - sC_2\left(\frac{1}{R_2}\right)(V_{in3} + V_{in4}) + \left(sC_2 + \frac{1}{R_2}\right)\frac{1}{R_1}V_{in1}\right] \quad (12.8)$$

$$V_{03} = \frac{1}{D(s)}\left[-\left\{\left(C_1C_2s^2 + sC_2\left(\frac{1}{R_1}\right)\right\}V_{in3} + s^2C_1C_2V_{in2} + sC_2\left(\frac{1}{R_1}\right)V_{in1} + \left(sC_1 + \frac{1}{R_1} - sC_2\right)\frac{1}{R_2}V_{in4}\right] \quad (12.9)$$

$$\text{where}\quad D(s) = s^2C_1C_2 + s\left\{\left(\frac{1}{R_2}\right)(C_1 - C_2) + C_2\frac{1}{R_1}\right\} + \frac{1}{R_1R_2} \quad (12.10)$$

From equations (12.7) to (12.10), the authors have described as many as thirteen cases however, it is obvious that because the circuit is minimal (i.e., employment of only two resistors and two capacitors) the requirement of conditions/constraints cannot be eliminated to realize the various filter responses. In view of this, therefore, we present in the following five exemplary special cases that lead to realization of five standard filter functions.

LPF, HPF and Notch can be realized at V_{01}, V_{02} and V_{03} respectively for $V_{in1} = V_{in3} = 0$ (grounded), $V_{in2} = V_{in4} =$ input voltage and $C_1 = C_2$. If $V_{in1} = V_{in2} = 0$ (grounded), $V_{in3} = V_{in4} =$ input voltage, a BPF is realized at V_{02}, and if $V_{in2} = V_{in4} = 0$ (grounded), $V_{in1} = V_{in3} =$ input voltage and $2C_2/R_2 = 2C_2/R_1 + C_1/R_2$, an APF is realized at V_{01} and a HPF is realized at V_{03}.

Chen-Shen biquad A versatile MIMO-type VM biquadratic filter circuit was proposed by Chen and Shen [6] employing two DVCCs with six passive elements and is shown in Fig. 12.5. The circuit has the flexibility to be used as either a three-input two-output universal filter or single-input five output filter without altering the circuit topology. The filter configuration also offers orthogonal control of ω_0 and Q_0 without any conditions/constraints.

Assuming ideal DVCCs, a routine circuit analysis of Fig. 12.5 yields the following transfer functions:

$$V_{01} = \frac{\left[sC_2\left(\frac{1}{R_1R_3}\right)(V_{i1} + V_{i2}) + \left(sC_2 + \frac{1}{R_3}\right)\left(\frac{1}{R_1R_2}\right)V_{i3}\right]}{D(s)} \quad (12.11)$$

$$V_{02} = \frac{\left[\left(\frac{1}{R_1R_2R_3}\right)(V_{i1} + V_{i2}) - sC_1\left(\frac{1}{R_3R_2}\right)V_{i3}\right]}{D(s)} \quad (12.12)$$

$$V_{03} = \frac{\left[\left\{s^2C_1C_2\left(\frac{1}{R_3}\right) + \frac{1}{R_1R_2R_3}\right\}(V_{i1}) - sC_2\left(\frac{1}{R_1R_2}\right)V_{i2} + s^2C_1C_2\left(\frac{1}{R_2}\right)V_{i3}\right]}{D(s)} \quad (12.13)$$

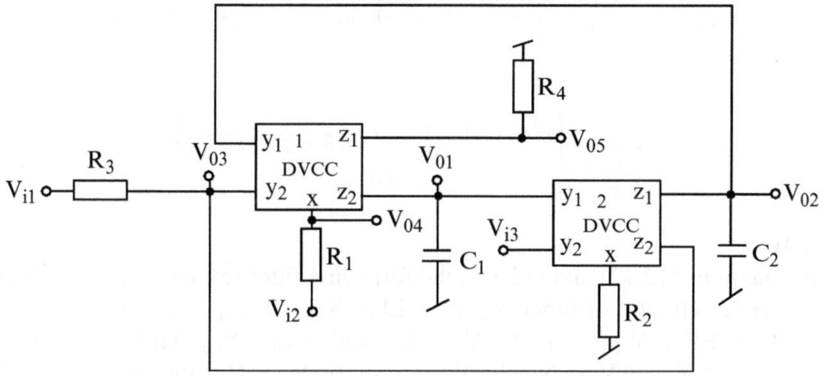

Fig. 12.5 VM biquad filter using two DVCCs proposed by Chen and Shen [6]

$$V_{04} = \frac{\left[-s^2C_1C_2\left(\frac{1}{R_3}\right)V_{i1} + \left\{sC_2\left(\frac{1}{R_1R_2}\right) + \frac{1}{R_1R_2R_3}\right\}V_{i2} - \left\{s^2C_1C_2\left(\frac{1}{R_2}\right) + sC_1\left(\frac{1}{R_2R_3}\right)V_{i3}\right\}\right]}{D(s)} \tag{12.14}$$

$$V_{05} = \left(\frac{R_4}{R_1}\right)\frac{\left[-s^2C_1C_2\left(\frac{1}{R_3}\right)(V_{i1} + V_{i2}) - \left\{s^2C_1C_2\left(\frac{1}{R_2}\right) + sC_1\left(\frac{1}{R_2R_3}\right)V_{i3}\right\}\right]}{D(s)} \tag{12.15}$$

$$\text{where } \mathrm{D(s)} = s^2C_1C_2\left(\frac{1}{R_3}\right) + sC_2\left(\frac{1}{R_1R_2}\right) + \frac{1}{R_1R_2R_3} \tag{12.16}$$

Depending upon the appropriate choice(s) of the input signal(s), the various filter responses can be obtained through the following two cases:

Case I

(a) If we choose $V_{i2} = V_{i3} = 0$ and $V_{i1} = V_{in}$ = input signal, then the following filter functions can be obtained:

$$\frac{V_{01}}{V_{in}} = \frac{sC_2\left(\frac{1}{R_1R_3}\right)}{D(s)}; \frac{V_{02}}{V_{in}} = \frac{\left(\frac{1}{R_1R_2R_3}\right)}{D(s)}; \frac{V_{03}}{V_{in}} = \frac{\left\{s^2C_1C_2\left(\frac{1}{R_3}\right) + \frac{1}{R_1R_2R_3}\right\}}{D(s)}; \frac{V_{04}}{V_{in}} = \frac{-s^2C_1C_2\left(\frac{1}{R_3}\right)}{D(s)} \text{ and } \frac{V_{05}}{V_{in}} = \left(\frac{R_4}{R_1}\right)\frac{-s^2C_1C_2\left(\frac{1}{R_3}\right)}{D(s)} \tag{12.17}$$

Thus, from the above equations, it is clear that a BPF, LPF, and notch can be realized at V_{01}, V_{02}, and V_{03} respectively while two HPF functions can be obtained at V_{04} and V_{05} respectively.

(b) if we select $V_{i1} = V_{i2} = Vin$ and $V_{i3} = 0$, an APF can be realized at V_{03} i.e.

$$\frac{V_{03}}{V_{in}} = \frac{\left\{s^2 C_1 C_2\left(\frac{1}{R_3}\right) - sC_2\left(\frac{1}{R_1 R_2}\right) + \frac{1}{R_1 R_2 R_3}\right\}}{D(s)} \tag{12.18}$$

Case II

From equations (12.12) and (12.13), the following filter responses can be obtained by proper selection(s) of input signal(s): LPF: $V_{i3} = 0$, $V_{i1} = V_{in}$, or $V_{i2} = V_{in}$ and $V_{out} = V_{02}$; BPF: $V_{i1} = V_{i3} = 0$, $V_{i2} = V_{in}$ and $V_{out} = V_{03}$; HPF: $V_{i1} = V_{i2} = 0$, $V_{i3} = V_{in}$ and $V_{out} = V_{03}$; Notch: $V_{i2} = V_{i3} = 0$, $V_{i1} = V_{in}$ and $V_{out} = V_{03}$; APF: $V_{i3} = 0$, $V_{i1} = V_{i2} = V_{in}$, and $V_{out} = V_{03}$

The angular frequency (ω_0) and quality factor (Q_0) of all the filter responses is given by:

$$\omega_0 = \sqrt{\frac{1}{C_1 C_2 R_1 R_2}} \text{ and} \tag{12.19}$$

$$Q_0 = \left(\frac{1}{R_3}\right)\sqrt{\frac{R_1 R_2 C_1}{C_2}} \tag{12.20}$$

Thus, ω_0 and Q_0 can be controlled by R_1 and /or R_2 and R_3, in that order.

Tsukutani-Sumi-Yabuki biquad Figure 12.6 shows a CM biquad filter structure employing two plus type DO-DVCCs and grounded passive elements. The configuration is capable of realizing LPF, BPF, HPF, notch and APF fitter responses by proper selection of input and output terminals. The filter parameters ω_0 and Q_0 are orthogonally controllable. Since all passive elements are grounded, the configuration is attractive for IC implementation in both bipolar and CMOS technologies.

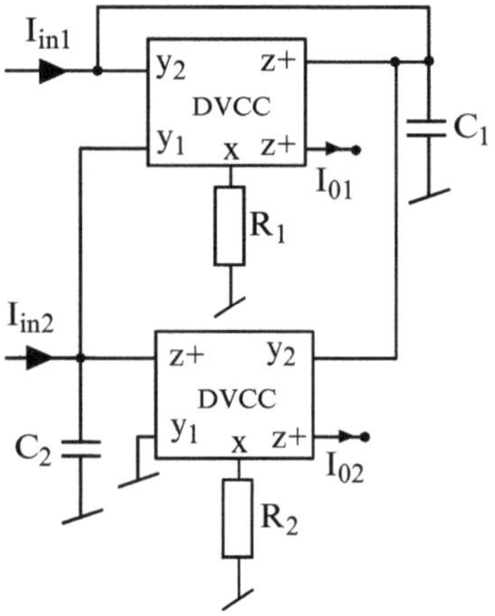

Fig. 12.6 CM biquadratic filter proposed by Tsukutani et al. [7]

Assuming ideal DO-DVCCs, a routine circuit analysis yields:

$$I_{01} = \frac{-(sC_2R_2 + 1)I_{in1} + sC_1R_2I_{in2}}{s^2C_1C_2R_1R_2 + sC_2R_2 +} 1 \tag{12.21}$$

$$I_{02} = -\frac{(sC_2R_1)I_{in1} + I_{in2}}{s^2C_1C_2R_1R_2 + sC_2R_2 +} 1 \tag{12.22}$$

From the above equations, the following CM filter functions can be obtained:

If $I_{in1} = 0$, $I_{in2} = I_{in}$;

$$\text{If } I_{m1} = 0,\ I_{m2} = Im;\ \frac{I_{02}}{I_{in}}\Big|_{LPF} = -\frac{1}{s^2C_1C_2R_1R_2 + sC_2R_2 + 1} \tag{12.23}$$

$$\text{For } I_{in2} = 0,\ I_{in1} = I_{in},\ \text{and } R_1 = R_2;\ \frac{I_{02}}{I_{in}}\Big|_{BPF}$$

$$= -\frac{sC_2R_1}{s^2C_1C_2R_1R_2 + sC_2R_2 + 1} \tag{12.24}$$

$$\text{If } I_{03} = (I_{in} + I_{01}),\ I_{in1} = I_{in},\ \text{and } I_{in2} = 0\ ;\frac{I_{03}}{I_{in}}\Big|_{HPF}$$

$$= \frac{s^2C_1C_2R_1R_2}{s^2C_1C_2R_1R_2 + sC_2R_2 + 1} \tag{12.25}$$

If $I_{04} = (I_{in} + I_{02})$, $I_{in1} = I_{in}$, $I_{in2} = 0$ and $R_1 = R_2$;

$$\frac{I_{04}}{I_{in}}\Big|_{Notch} = \frac{s^2C_1C_2R_1R_2 + 1}{s^2C_1C_2R_1R_2 + sC_2R_2 + 1} \tag{12.26}$$

and if $I_{05} = (I_{04} + I_{02})$, $I_{in1} = I_{in}$, $I_{in2} = 0$ and $R_1 = R_2$;

$$\frac{I_{05}}{I_{in}}\Big|_{APF} = \frac{s^2C_1C_2R_1R_2 - sC_2R_2 + 1}{s^2C_1C_2R_1R_2 + sC_2R_2 + 1} \tag{12.27}$$

For $R_1 = R_2 = R$, the filter parameters ω_0 and Q_0 are given by:

$$\omega_0 = \frac{1}{R}\sqrt{\frac{1}{C_1C_2}} \text{ and } Q_0 = \sqrt{\frac{C_2}{C_1}} \tag{12.28}$$

Hence, parameters ω_0 and Q_0 are orthogonally controllable.

The validity of the proposed configuration has been confirmed through PSPICE simulations using model parameters from 0.5 μm CMOS technology obtained through MOSIS.

Horng generalized first order filter The circuit shown in Fig. 12.7 is a DVCCs-based high input impedance VM first-order LPF, HPF and APF configuration using

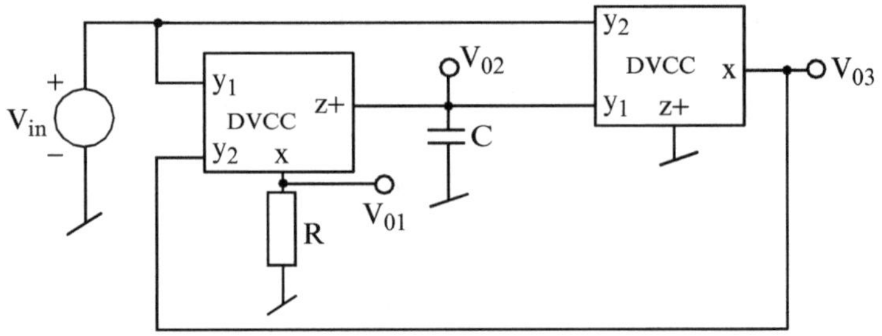

Fig. 12.7 DVCCs-based first-order filters proposed by Horng [8]

two grounded passive elements and without any component matching condition. Assuming ideal DVCCs, the three VM filter transfer functions obtained as:

$$\text{HPF}:\frac{V_{01}}{V_{in}}=\frac{2sCR}{sCR+1}\text{LPF}:\frac{V_{02}}{V_{in}}=\frac{2}{sCR+1}\text{and BPF}:\frac{V_{03}}{V_{in}}=\frac{-sCR+1}{sCR+1}; \tag{12.29}$$

Khateb- Khatib-Koton biquad A SIMO-type VM multifunction filter configuration based upon the idea of Akerberg-Mossberg [203] structure has been presented by Khateb et al. [11] which employs three DVCCs and six grounded passive elements.

A circuit analysis of Fig. 12.8 reveals the following transfer functions:

$$\frac{V_{LP}}{V_{IN}}=\frac{\left(\frac{R_4}{C_1C_2R_1R_2R_3}\right)}{D(s)};\frac{V_{BP}}{V_{IN}}=\frac{s\left(\frac{R_4}{C_2R_1R_2}\right)}{D(s)};\frac{V_{HP}}{V_{IN}}=\frac{s^2\left(\frac{R_4}{R_1}\right)}{D(s)} \tag{12.30}$$

$$\frac{V_{BR}}{V_{IN}}=\frac{\left(s^2+\frac{1}{C_1C_2R_3R_2}\right)}{D(s)} \tag{12.31}$$

where

$$\text{D(s)}=s^2+\frac{R_4}{C_2R_1R_2}+\frac{1}{C_1C_2R_3R_2} \tag{12.32}$$

The filter parameters namely Q_0 and ω_0 can be obtained as:

$$Q_0=\frac{R_1}{R_4}\sqrt{\frac{C_2R_2}{C_1R_3}}\text{ and }\omega_0=\sqrt{\frac{1}{C_1C_2R_2R_3}} \tag{12.33}$$

Thus, for $R_2=R_3$, Q_0 can be controlled by R_1 and/or R_4 without disturbing ω_0, and ω_0 can also be tuned independently by simultaneously varying R_2 and R_3 ($R_2=R_3$). To confirm the workability of the proposed structure, simulation results using 0.18 μm CMOS technology have been provided [11].

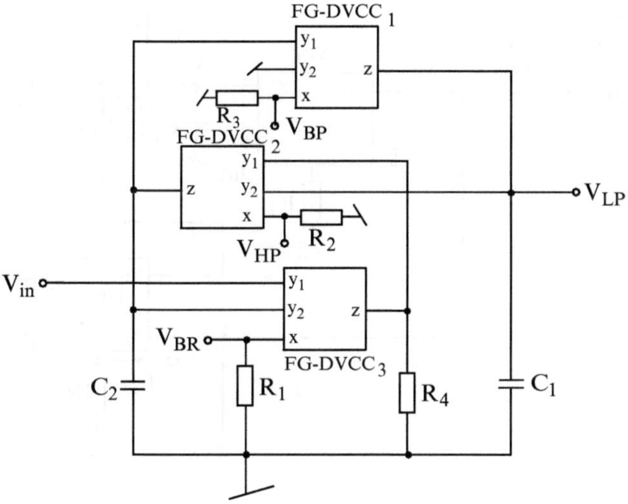

Fig. 12.8 VM multifunction filter configuration proposed by Khateb et al. (Adapted from [11] © 2011 Elsevier Ltd.)

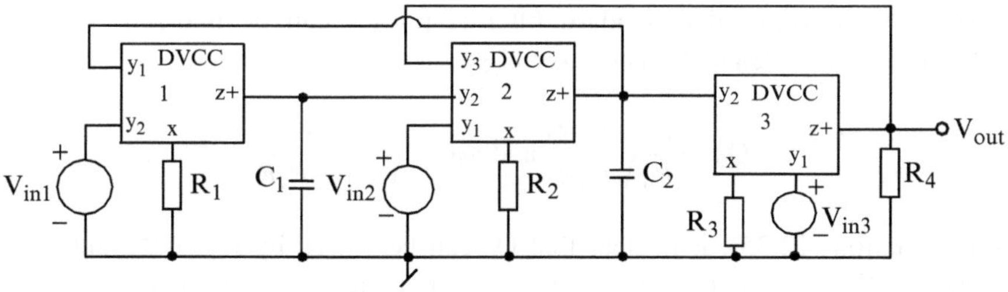

Fig. 12.9 VM universal filter proposed by Horng et al. [12]

Horng-Hsu-Tseng biquad

In Fig. 12.9 Horng et al. [12] have presented a VM MISO-type high input impedance universal biquad filter circuit which employs three DVCCs and six passive elements. The parasitic resistance of the X-terminal of the DVCC can be incorporated directly in the main resistances of each DVCC.

Assuming ideal DVCCs, the expression for output voltage in terms of input voltages can be given by:

$$V_{out} = \left(\frac{R_4}{R_3}\right) \frac{\left(s^2 + \frac{1}{C_1C_2R_1R_2}\right)V_{in3} - s\left(\frac{1}{C_2R_2}\right)V_{in2} - \left(\frac{1}{C_1C_2R_1R_2}\right)V_{in1}}{s^2 + s\left(\frac{R_4}{C_2R_2R_3}\right) + \left(\frac{1}{C_1C_2R_1R_2}\right)} \tag{12.34}$$

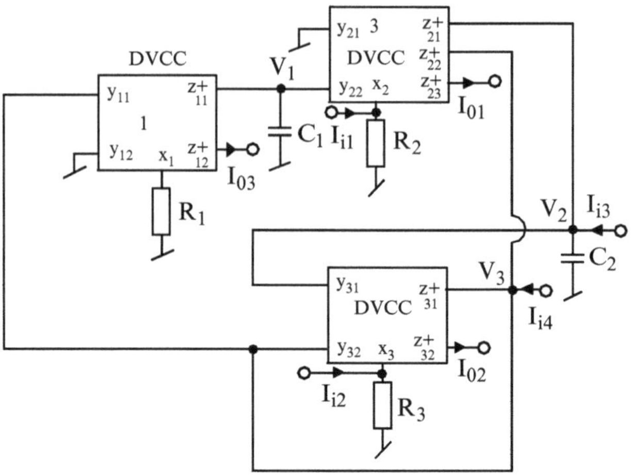

Fig. 12.10 DVCC-based tunable CM universal filter proposed by Chen (Adapted from [13] © 2011 Elsevier GmbH)

From the above equation (12.34), the various filter responses can be obtained as:

LPF: $V_{in2} = V_{in3} = 0$, and $V_{in1} = V_{in}$; BPF: $V_{in1} = V_{in3} = 0$, and $V_{in2} = V_{in}$; Notch: $V_{in1} = V_{in2} = 0$, and $V_{in3} = V_{in}$; HPF: $V_{in2} = 0$ and $V_{in1} = V_{in3} = V_{in}$; and APF: $V_{in1} = 0$ and $V_{in3} = V_{in2} = V_{in}$, and $R_3 = R_4$.

Thus, all second-order standard filter responses can be realized. The filter parameters namely Q_0 and ω_0 are given by:

$$Q_0 = \frac{R_3}{R_4}\sqrt{\frac{C_2R_2}{C_1R_1}} \text{ and } \omega_0 = \sqrt{\frac{1}{C_1C_2R_1R_2}} \tag{12.35}$$

From equation (12.35), it is clear that Q_0 can be varied by R_3 or R_4 and ω_0 can independently be tuned by R_1 or R_2 provided they are varied simultaneously thereby keeping Q_0 unchanged.

Chen biquad Figure 12.10 shows a CM MIMO-type versatile tunable DVCC-based universal filter proposed by Chen [13]. The filter circuit employs three plus-type DVCCs and five grounded passive elements. The structure can be used as either a single-input and three-output or four-input and two-output circuit. As a single-input, three-output, the configuration can realize BPF, HPF and notch responses simultaneously while as a four-input, two-output circuit, the circuit can yield both inverting and non-inverting LPF, BPF, HPF, notch and APF filter functions simultaneously without any component matching condition. The proposed structure offers orthogonal controllability of ω_0 and Q_0 through grounded passive elements.

By straight forward circuit analysis of Fig. 12.10, the following CM transfer functions can be realized:

$$I_{01} = \frac{-s^2C_1C_2R_1R_2I_{i1} + sC_2R_3I_{i2} - I_{i3} - sC_2R_3I_{i4}]}{s^2C_1C_2R_1R_2 + sC_2R_3 + 1} \tag{12.36}$$

$$I_{02} = \frac{s^2C_1C_2R_1R_2I_{i1} - sC_2R_3I_{i2} + I_{i3} - (s^2C_1C_2R_1R_2 + 1)I_{i4}}{s^2C_1C_2R_1R_2 + sC_2R_3 + 1} \tag{12.37}$$

$$I_{03} = \frac{-(s^2C_1C_2R_3R_2 + sC_1R_2)I_{i1} - s^2C_1C_2R_2R_3I_{i2} + sC_1R_2I_{i3} + s^2C_1C_2R_2R_3I_{i4}}{s^2C_1C_2R_1R_2 + sC_2R_3 + 1} \tag{12.38}$$

Now, by appropriate selection of the various input signals, the following two cases can be considered to yield various filter functions:

Case I

If $I_{i1} = I_{i2} = I_{i3} = 0$ (open circuited) and $I_{i4} = I_{in}$, the following three CM transfer functions can be derived as:

$$\frac{I_{01}}{I_{in}} = \frac{-sC_2R_3}{D(s)}; \quad \frac{I_{02}}{I_{in}} = \frac{-(s^2C_1C_2R_1R_2 + 1)}{D(s)} \tag{12.39}$$

and

$$\frac{I_{03}}{I_{in}} = \frac{s^2C_1C_2R_2R_1}{D(s)} \tag{12.40}$$

where

$$D(s) = s^2C_1C_2R_1R_2 + sC_2R_3 + 1 \tag{12.41}$$

Thus, from above equations, it is clear that an inverting BPF can be obtained at I_{01}, an inverting notch can be realized at I_{02} and a non-inverting HPF can be obtained at I_{03}. Hence, the circuit is a SIMO-type filter in this mode.

Case-II

If $I_{i1} = I_{i2} = I_{i4} = 0$ (open circuited) and $I_{i3} = I_{in}$, the inverting and non-inverting LPF responses can be realized at I_{01} and I_{02} simultaneously; if $I_{i1} = I_{i3} = I_{i4} = 0$ (open circuited) and $I_{i2} = I_{in}$, the non-inverting and inverting BPF responses can be obtained at I_{01} and I_{02} simultaneously; if $I_{i3} = I_{i2} = I_{i4} = 0$ (open circuited) and $I_{i1} = I_{in}$, the inverting and non-inverting HPF responses can be realized at I_{01} and I_{02} simultaneously; if $I_{i2} = I_{i4} = 0$ (open circuited) and $I_{i1} = I_{i3} = I_{in}$, the inverting and non-inverting notch responses can be realized at I_{01} and I_{02} simultaneously and if $I_{i1} = I_{i2} = I_{i3} = I_{in}$ and $I_{i4} = 0$ (open circuited), the inverting and non-inverting APF responses can be realized at I_{01} and I_{02} simultaneously.

The filter parameters ω_0 and Q_0 of the proposed filter circuit can be obtained as:

$$\omega_0 = \sqrt{\frac{1}{C_1C_2R_1R_2}} \text{ and } Q_0 = \left(\frac{1}{R_3}\right)\sqrt{\frac{R_1R_2C_1}{C_2}} \tag{12.42}$$

From the equation (12.42), it is clear that Q_0 can be controlled without effecting ω_0, but the vice versa is not feasible.

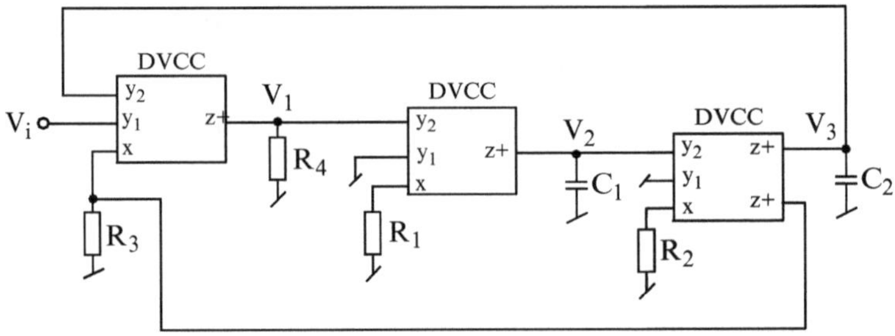

Fig. 12.11 DVCC-based VM KHN-equivalent using grounded passive elements proposed by Soliman (Adapted from [14] © 2008 Wiley)

Soliman's KHN-equivalent biquad using DVCC In [14] Soliman derived a VM DVCC-based KHN-biquad employing grounded passive components which is shown here in Fig. 12.11. A straight forward circuit analysis of this circuit yields the following VM transfer functions (for $R_4 = R_3$):

$$\frac{V_1}{V_i} = \frac{s^2 C_1 C_2 R_1 R_2}{D(s)}; \frac{V_2}{V_i} = -\frac{sC_2 R_2}{D(s)} \tag{12.43}$$

and

$$\frac{V_3}{V_i} = \frac{1}{D(s)} \tag{12.44}$$

where

$$D(s) = s^2 C_1 C_2 R_1 R_2 + sC_2 R_3 + 1 \tag{12.45}$$

The parameters ω_0 and Q_0 are obtained as:

$$\omega_0 = \sqrt{\frac{1}{C_1 C_2 R_1 R_2}} \text{ and } Q_0 = \left(\frac{1}{R_3}\right)\sqrt{\frac{R_1 R_2 C_1}{C_2}} \tag{12.46}$$

Thus, Q_0 can be independently tuned by the two equal resistors R_3 having fixed ω_0 by R_1 and R_2

Ibrahim-Minaei-Kuntman biquad A 22.5 MHz CM DVCC-based KHN-equivalent filter structure was proposed by Ibrahim et al. [15] using all grounded passive elements which is shown in Fig. 12.12.

The filter circuit offers BPF, HPF and LPF filter responses simultaneously. By proper connection of output currents notch and APF responses can also be obtained without any additional active element(s). All the outputs are available from the high

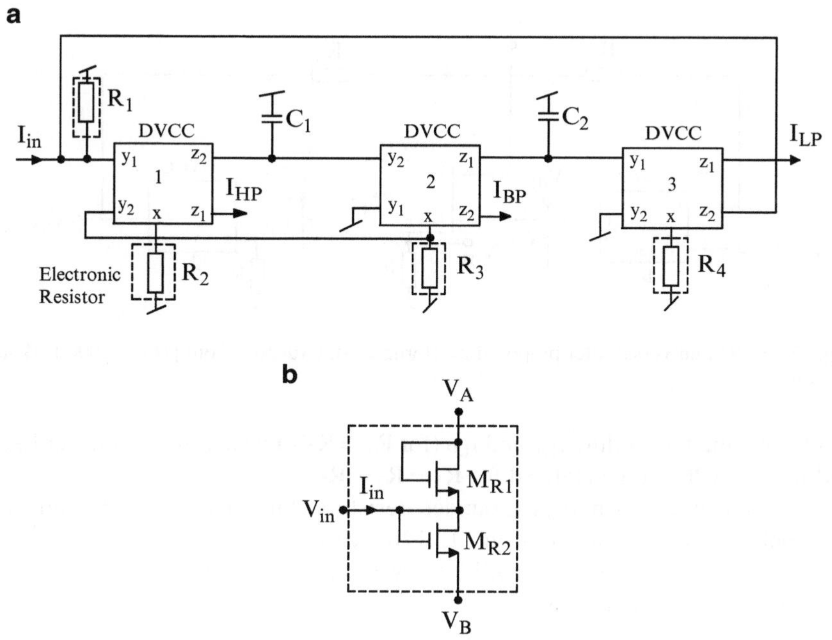

Fig. 12.12 (**a**) DVCC-based CM KHN-equivalent (**b**) realization of electronic resistor proposed by Ibrahim et al. (Adapted from [15] © 2004 Elsevier GmbH)

output impedance nodes which is useful for easy cascading. A routine circuit analysis of Fig. 12.12a gives the following CM transfer functions:

$$\frac{I_{HP}}{I_{in}} = \frac{s^2\left(\frac{R_1}{R_2}\right)}{D(s)};\ \frac{I_{BP}}{I_{in}} = \frac{s\left(\frac{R_1}{C_1 R_2 R_3}\right)}{D(s)} \tag{12.47}$$

$$\frac{I_{LP}}{I_{in}} = \frac{\left(\frac{R_1}{R_2 R_3 R_4 C_1 C_2}\right)}{D(s)} \tag{12.48}$$

where

$$D(s) = s^2 + s\left(\frac{1}{C_1 R_2}\right) + \frac{R_1}{C_1 C_2 R_2 R_3 R_4} \tag{12.49}$$

The filter parameters ω_0 and Q_0 are determined as:

$$\omega_0 = \sqrt{\frac{R_1}{C_1 C_2 R_3 R_2 R_4}} \text{ and } Q_0 = \sqrt{\frac{R_1 R_2 C_1}{R_3 R_4 C_2}} \tag{12.50}$$

Thus, the angular frequency (ω_0) can be controlled without disturbing quality factor (Q_0) by adjusting R_2 and R_4 simultaneously keeping (R_2/R_4) constant. A notch filter

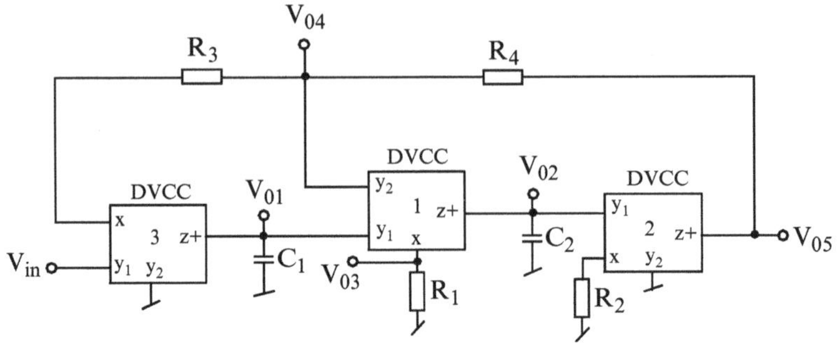

Fig. 12.13 VM universal filter proposed by Horng et al. (Adapted from [16] © 2006 Birkhauser Boston)

can be obtained by adding I_{HP} and I_{LP} (for $R_1 = R_2$). Further, if we connect I_{HP}, I_{LP} and I_{BP}, an APF can be realized for $R_1 = R_2 = R_3$.

Since the circuit employs grounded resistors, this enables simulation of the electronic resistor as shown in Fig. 12.12b. Assuming saturation region operation of diode connected matched MOSFETs, MR_1 and MR_2 in Fig. 12.12b, its equivalent resistance is obtained as:

$$R_{eq} = \frac{V_{in}}{I_{in}} = \frac{L}{2\mu_n C_{ox} W(V_A - V_T)}; \text{ for } V_B = -V_A \tag{12.51}$$

where symbols have their usual meanings.

Horng- Hou-Chang-Chou-Lin biquad A high input impedance voltage-mode universal biquadratic filter using DVCCs as shown in Fig. 12.13 was proposed by Horng et al. [16] which can realize all five standard second-order filters with grounded capacitors.

By a routine circuit analysis, the various filter responses (assuming ideal DVCCs) are given by:

$$\frac{V_{01}}{V_{in}} = \frac{\left(\frac{1}{C_1 C_2 R_1 R2}\right)}{D(s)}; \frac{V_{02}}{V_{in}} = \frac{-s\left(\frac{1}{C_2 R_1}\right)}{D(s)}; \frac{V_{03}}{V_{in}} = \frac{-s^2}{D(s)}; \frac{V_{04}}{V_{in}}$$
$$= \frac{s^2 + \left(\frac{1}{C_1 C_2 R_1 R_2}\right)}{D(s)} \text{ and } \frac{V_{05}}{V_{in}}$$
$$= \frac{s^2 - s\left(\frac{R_3}{C_2 R_1 R_2}\right) + \left(\frac{1}{C_1 C_2 R_1 R_2}\right)}{D(s)} \text{ for } R_3 = R_4 \tag{12.52}$$

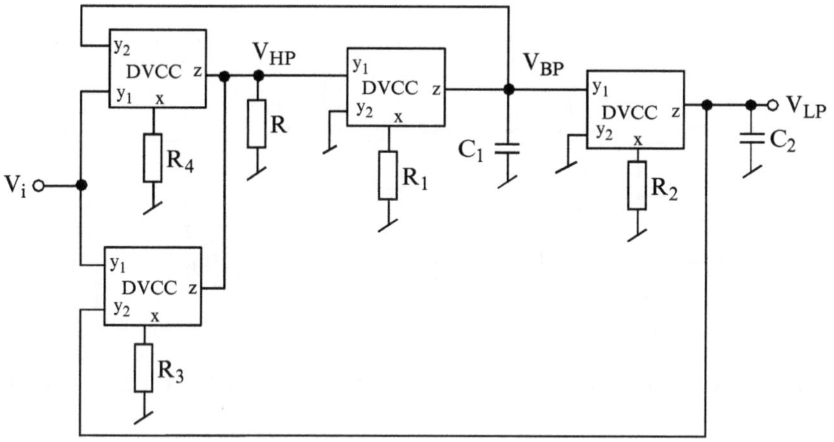

Fig. 12.14 SIMO-type VM biquad filter proposed by Hassan and Mahmoud (Adapted from [1] © 2008 Elsevier GmbH)

where

$$D(s) = s^2 + \frac{R_3}{C_2R_1R_2} + \frac{1}{C_1C_2R_1R_2} \tag{12.53}$$

Thus, LPF, BPF, HPF, notch and APF can be realized at V_{01}, V_{02}, V_{03}, V_{04}, and V_{05} respectively. The expressions for ω_0 and Q_0 are given by:

$$\omega_0 = \sqrt{\frac{1}{C_1C_2R_1R_2}} \text{ and } Q_0 = \left(\frac{1}{R_3}\right)\sqrt{\frac{R_1R_2C_2}{C_1}} \tag{12.54}$$

Hence, ω_0 and Q_0 can be adjusted orthogonally. The configuration does not need any component-matching conditions except for APF ($R_3 = R_4$). To demonstrate the workability of the proposed biquad filter PSPICE simulations were carried out using CMOS realization of DVCC from [204] employing 0.35 μm CMOS process parameters from TSMC.

Hassan-Mahmoud biquad Another high input impedance SIMO-type VM biquad filter employing four DVCCs, two grounded capacitors and five grounded resistors was proposed by Hassan and Mahmoud [1] which is shown here in Fig. 12.14.

Assuming ideal DVCCs, the following VM transfer functions are determined using a routine circuit analysis:

$$\frac{V_{HP}}{V_i} = \frac{\left\{s^2R(R_3 + R_4)/R_3R_4\right\}}{s^2 + s\left(\frac{R}{C_1R_1R_4}\right) + \left(\frac{R}{C_1C_2R_1R_2R_3}\right)} \tag{12.55}$$

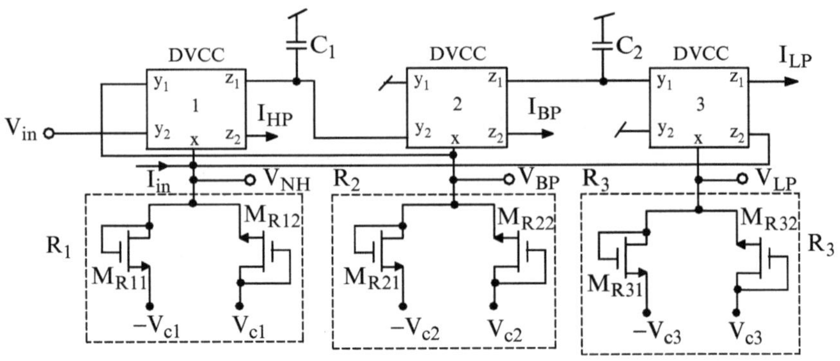

Fig. 12.15 Mixed-mode KHN- biquad proposed by Minaei and Ibrahim (Adapted from [17] © 2008 Wiley)

$$\frac{V_{BP}}{V_i} = \frac{\{sR(R_3 + R_4)/R_1C_1R_3R_4\}}{s^2 + s\left(\frac{R}{C_1R_1R_4}\right) + \left(\frac{R}{C_1C_2R_1R_2R_3}\right)} \tag{12.56}$$

$$\frac{V_{LP}}{V_i} = \frac{\{R(R_3 + R_4)/R_1R_2C_1C_2R_3R_4\}}{s^2 + s\left(\frac{R}{C_1R_1R_4}\right) + \left(\frac{R}{C_1C_2R_1R_2R_3}\right)} \tag{12.57}$$

From the above equations, the parameters ω_0 and Q_0 can be obtained as:

$$\omega_0 = \sqrt{\frac{R}{C_1C_2R_1R_2R_3}} \text{ and } Q_0 = R_4\sqrt{\frac{R_1C_1}{RC_2R_2R_3}} \tag{12.58}$$

Hence, Q_0 can be controlled independently by R_4.

Minaei-Ibrahim biquad In 2009, Minaei and Ibrahim proposed a mixed-mode KHN-biquad using three DVCCs and five grounded passive elements suitable for direct cascading [17]. Their circuit is shown here in Fig. 12.15.

The filter circuit is capable of providing BPF, HPF and LPF responses simultaneously in CM operation and notch, BPF and LPF simultaneously in VM operation. The notch and APF responses are also realizable by summing appropriate output currents. Higher-order filters can be realized because of the low input and high output impedance of the circuit for current signals and the high input and low output impedance for the voltage signals. Assuming ideal DVCCs, an analysis of the circuit of Fig. 12.15 gives the following transfer functions:

(i) CM operation ($V_{in} = 0$ (grounded)):

$$\frac{I_{HP}}{I_{in}} = \frac{s^2}{D(s)}; \frac{I_{BP}}{I_{in}} = -\frac{s\left(\frac{1}{C_1R_2}\right)}{D(s)} \text{ and } \frac{I_{LP}}{I_{in}} = \frac{\left(\frac{1}{C_1C2R_2R_3}\right)}{D(s)} \tag{12.59}$$

A notch (I_{NH}) response can be obtained by $I_{NH} = (I_{LP} + I_{HP})$ and an APF (I_{AP}) response can be realized by $I_{AP} = (I_{LP} + I_{BP} + I_{HP})$ for $R_1 = R_2$.

(ii) VM operation ($I_{in} = 0$ (open circuited)):

$$\frac{V_{NH}}{V_{in}} = -\frac{s^2 + \left(\frac{1}{C_1 C2 R_2 R_3}\right)}{D(s)}; \frac{V_{BP}}{V_{in}} = \frac{s\left(\frac{1}{C_1 R1}\right)}{D(s)}; \frac{V_{LP}}{V_{in}} = \frac{\left(\frac{1}{C_1 C2 R_1 R_2}\right)}{D(s)} \tag{12.60}$$

(iii) Transadmittance (TAM) mode operation ($I_{in} = 0$ (open circuited)):

$$\frac{I_{HP}}{V_{in}} = \frac{s^2\left(\frac{1}{R_1}\right)}{D(s)}; \frac{I_{BP}}{V_{in}} = -\frac{s\left(\frac{1}{C_1 R_1 R_2}\right)}{D(s)}; \frac{I_{LP}}{V_{in}} = \frac{\left(\frac{1}{C_1 C_2 R_1 R_2 R_3}\right)}{D(s)} \tag{12.61}$$

(iv) Transimpedance (TIM) mode operation ($V_{in} = 0$):

$$\frac{V_{BP}}{I_{in}} = \frac{s\left(\frac{1}{C_1}\right)}{D(s)} : \frac{V_{LP}}{I_{in}} = \frac{\left(\frac{1}{C_1 C2 R_2}\right)}{D(s)} \tag{12.62}$$

where

$$\mathrm{D(s)} = s^2 + s\left(\frac{1}{C_1 R_1}\right) + \frac{1}{C_1 C_2 R_2 R_3} \tag{12.63}$$

The filter parameters namely, ω_0, Q_0 and bandwidth ($\frac{\omega_0}{Q_0}$) can be obtained as:

$$\omega_0 = \sqrt{\frac{1}{C_1 C_2 R_2 R_3}};\ Q_0 = R_1\sqrt{\frac{C_1}{R_2 R_3 C_2}} \text{and} \frac{\omega_0}{Q_0} = \frac{1}{C_1 R_1} \tag{12.64}$$

From the above equation, it is seen that Q_0 and ($\frac{\omega_0}{Q_0}$) can be controlled by R_1 without disturbing ω_0.

In Fig. 12.15, each resistor R_i (i = 1, 2, 3) is simulated by two diode-connected MOSFETs (M_{Ri1} and M_{Ri2}) which can be controlled by their corresponding control voltages V_{Ci} and $-V_{Ci}$. The value of the simulated resistance R_i (assuming identical devices with saturation region operation) can be given by:

$$R_i = \frac{L}{2\mu C_{ox} W(V_{Ci} - V_T)} \tag{12.65}$$

where symbols have their usual meanings.

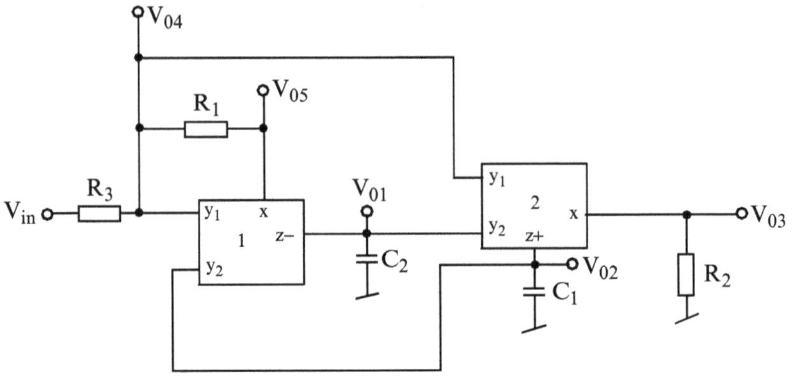

Fig. 12.16 VM SIMO-type universal biquad filter proposed by Horng (Adapted from [18] © 2010 Springer Science + Buisiness Media)

Horng's universal SIMO VM biquad In 2010, Horng proposed a universal VM SIMO-type biquad employing only two DVCCs, two grounded capacitors and three resistors [18] which is shown in Fig. 12.16.

In this circuit, the filter parameters ω_0 and Q_0 are orthogonally controllable. The various transfer functions realized by this circuit (Fig. 12.16), assuming ideal DVCCs are given by:

$$\frac{V_{01}}{V_{in}}=\frac{\left(\frac{1}{C_1C_2R_1R_2}\right)}{D(s)};\frac{V_{02}}{V_{in}}=\frac{\left(\frac{s}{C_1R_2}\right)}{D(s)};\frac{V_{03}}{V_{in}}=\frac{s^2}{D(s)};$$
$$\frac{V_{04}}{V_{in}}=\frac{\left(s^2+\frac{1}{C_1C_2R_1R_2}\right)}{D(s)};\frac{V_{05}}{V_{in}}=\frac{\left(s^2-\frac{s}{C_1R_2}+\frac{1}{C_1C_2R_1R_2}\right)}{D(s)} \tag{12.66}$$

where

$$\mathrm{D(s)}=s^2+s\left(\frac{R_3}{C_1R_1R_2}\right)+\left(\frac{1}{C_1C_2R_1R_2}\right) \tag{12.67}$$

From the above equations it may be seen that a LPF, BPF, HPF, notch can be realized at V_{01}, V_{02}, V_{03}, V_{04} respectively simultaneously and an APF at V_{05} for $R_1 = R_3$ can also be obtained. The filter parameters namely, ω_0 and Q_0 can be given as:

$$\omega_0=\sqrt{\frac{1}{C_1C_2R_2R_1}}\,\mathrm{and}\,Q_0=\frac{1}{R_3}\sqrt{\frac{C_1R_1R_2}{C_2}} \tag{12.68}$$

Thus, Q_0 can be controlled independently by R_3.

12.2.2 Filter Design Using DDCC

A DDCC can be considered to be an extended version of DVCC with one additional input terminal thereby making $V_x = (V_{y1} - V_{y2} + V_{y3})$. This provides the feasibility of adding and subtracting signals without requiring any external resistor(s) or additional hardware and thereby offers additional degrees of freedoms in a number of applications. A DDCC is characterized by the following hybrid matrix:

$$\begin{bmatrix} I_{Y1} \\ I_{Y2} \\ I_{Y3} \\ V_X \\ I_Z \end{bmatrix} = \begin{bmatrix} 0 & 0 & 0 & 0 & 0 \\ 0 & 0 & 0 & 0 & 0 \\ 0 & 0 & 0 & 0 & 0 \\ 1 & -1 & 1 & 0 & 0 \\ 0 & 0 & 0 & \pm 1 & 0 \end{bmatrix} \begin{bmatrix} V_{Y1} \\ V_{Y2} \\ V_{Y3} \\ I_X \\ V_Z \end{bmatrix} \tag{12.69}$$

where plus and minus signs represent DDCC+ and DDCC− respectively.

Ibrahim-Kuntman-Cicekoglu all pass filter As an example, we present a canonic implementation of first order APF in Fig. 12.17b which has been possible because of capability of the DDCC and is a special case of the general schematic of Fig. 12.17a.

Using ideal DDCC+ as shown in Fig. 12.17, a routine circuit analysis yields the following voltage transfer function:

$$\frac{V_o}{V_i} = \frac{sCR - 1}{sCR + 1} \tag{12.70}$$

and the phase is given by

$$\Phi(\omega) = 180 - 2\tan^{-1}(\omega CR) \tag{12.71}$$

An alternative APF by Horng-Hou-Chang-Lin-Shiu-Chiu Another first-order APF using same number of active and passive elements as in Fig. 12.17 was proposed by Horng et al. [37]. However the circuit suffers from a lack of high input impedance. A routine circuit analysis of Fig. 12.18 yields the following voltage transfer function:

$$\frac{V_{out}}{V_i} = \frac{1 - sCR}{1 + sCR} \tag{12.72}$$

From equation (12.73), the phase of this first-order APF can be determined as:

$$\Phi(\omega) = -2\tan^{-1}(\omega CR) \tag{12.73}$$

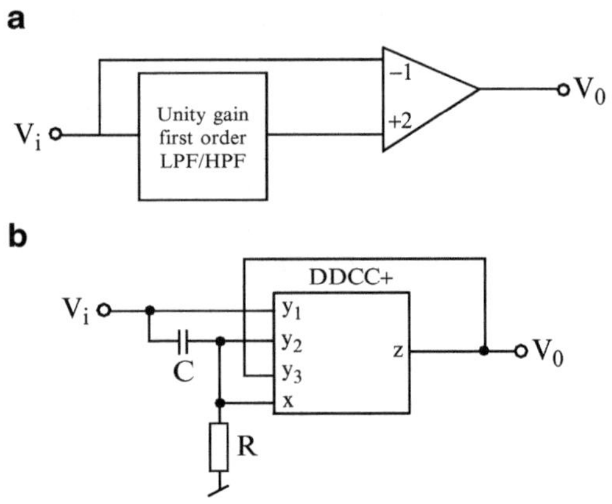

Fig. 12.17 (**a**) A general scheme of realizing an all-pass filter (**b**) A canonical VM APF proposed by Ibrahim et al. (Adapted from [32] © 2003 Birkhauser Boston)

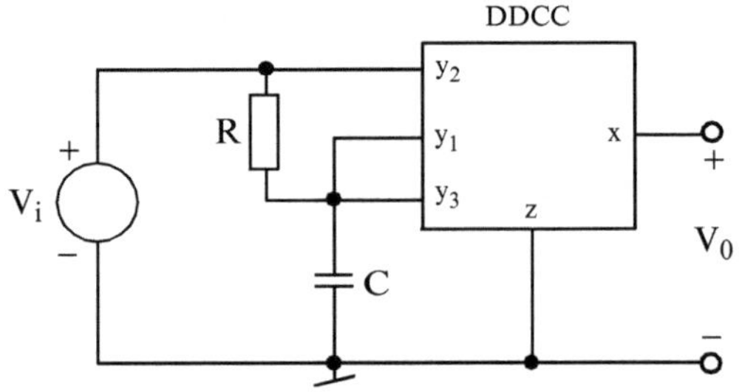

Fig. 12.18 APF proposed by Horng et al. [37]

Chen's SIMO-type VM biquad Figure 12.19 shows a SIMO-type VM universal filter employing two DDCCs and five passive elements proposed by Chen.

Assuming ideal DDCCs, a routine circuit analysis of Fig. 12.19 gives the following VM transfer functions:

$$\frac{V_{01}}{V_{in}} = \frac{\left(-\frac{1}{C_1C_2R_1R_2}\right)}{D(s)};\ \frac{V_{02}}{V_{in}} = \frac{s\left(\frac{1}{C_2R_2}\right)}{D(s)};\ \frac{V_{03}}{V_{in}} = \frac{s^2 + \left(\frac{1}{C_1C_2R_1R_2}\right)}{D(s)};$$
$$\frac{V_{04}}{V_{in}} = \frac{s^2}{D(s)} \quad \text{and} \quad \frac{V_{05}}{V_{in}} = \frac{\left\{s^2 - s\left(\frac{1}{C_2R_2}\right) + \frac{1}{R_1R_2C_1C_2}\right\}}{D(s)} \tag{12.74}$$

where

$$D(s) = s^2 + s\left(\frac{R_3}{C_2R_1R_2}\right) + \frac{1}{C_1C_2R_1R_2} \tag{12.75}$$

From the above expressions, it can be seen that an APF can be realized with $R_3 = R_1$. The component matching conditions can be eliminated if capacitor C_1

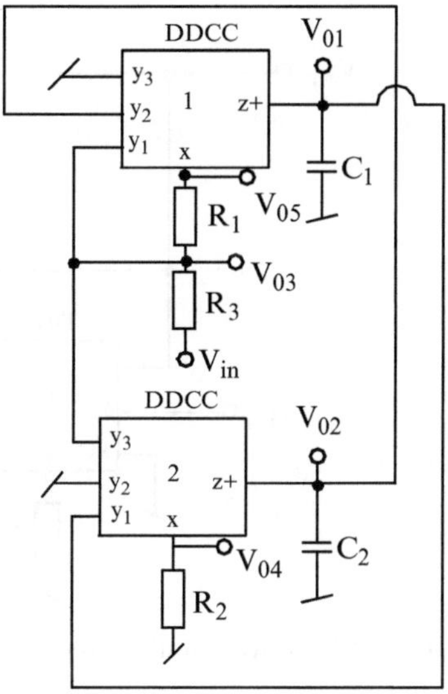

Fig. 12.19 VM universal biquad proposed by Chen (Adapted from [40] © 2007 Springer Science + Business Media)

can be made floating and inserting V_{in} into the floating terminals of the capacitor C_1 and output is being taken at V_{03}. The ω_0 and Q_0 for all the cases can be given by:

$$\omega_0 = \sqrt{\frac{1}{C_1 C_2 R_1 R_2}} \quad \text{and} \quad Q_0 = \frac{1}{R_3}\sqrt{\frac{C_2 R_2 R_1}{C_1}} \tag{12.76}$$

Thus, ω_0 and Q_0 can be controlled orthogonally.

Chiu-Horng biquad In 2007, Chiu and Horng [41] proposed a high input and low output impedance VM MISO-type universal biquadratic filter structure using three DDCCs, two grounded capacitors and two grounded resistors which is shown here in Fig. 12.20.

The circuit of Fig. 12.20 can realize all the five second-order standard fitter responses without any component matching condition. Assuming ideal DDCCs, the output voltage in terms of input voltages can be expressed as:

$$V_{out} = \frac{s^2 V_{in3} - s\left(\frac{V_{in2}}{C_2 R_2}\right) + \frac{V_{in1}}{C_1 C_2 R_1 R_2}}{s^2 + s\left(\frac{1}{C_2 R_2}\right) + \frac{1}{C_1 C_2 R_1 R_2}} \tag{12.77}$$

The various filter responses can now be realized from the above equation as follows:

HPF: if $V_{in1} = V_{in2} = 0$ (grounded) and $V_{in3} = V_{in}$; BPF: if $V_{in1} = V_{in3} = 0$ (grounded) and $V_{in2} = V_{in}$; LPF: if $V_{in3} = V_{in2} = 0$ (grounded) and $V_{in1} = V_{in}$; Notch: if $V_{in2} = 0$ (grounded) and $V_{in3} = V_{in} = V_{in1}$and APF: if

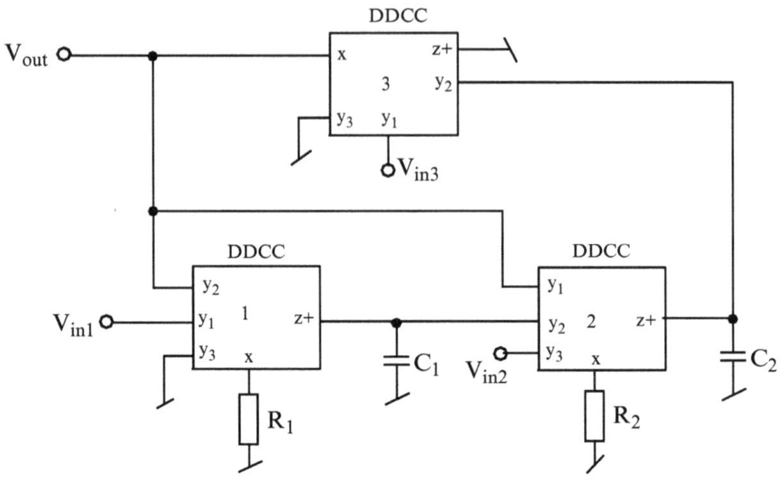

Fig. 12.20 Universal filter proposed by Chiu and Horng [41]

$V_{in1} = V_{in2} = V_{in3} = V_{in}$. Hence, the circuit can realize all the filter responses. The filter parameters ω_0 and Q_0 for all the cases can be given by:

$$\omega_0 = \sqrt{\frac{1}{C_1 C_2 R_1 R_2}} \text{ and } Q_0 = \sqrt{\frac{C_2 R_2}{C_1 R_1}} \tag{12.78}$$

An extended version of this circuit with the provision of multiple outputs thereby introducing additional degrees of freedom while employing same number of active and passive elements as in Fig. 12.20 has been subsequently described by Horng in [46].

Ibrahim-Kuntman biquad A dual output DDCC (DO-DDCC) based filter circuit for realizing differential VM KHN biquad with high common mode rejection ratio (CMRR) and high input impedance was presented by Ibrahim and Kuntman [34]. The configuration (shown in Fig. 12.21) uses three DO-DDCCs, two grounded capacitors and five grounded resistors. The various filter functions in differential mode (A_{dm}) are being given by:

$$A_{dm}HP = \frac{\left(\frac{R_1+R_4}{R_1}\right)s^2}{D(s)};\ A_{dm}BP = \frac{\left(\frac{2}{R_1C_1}\right)s}{D(s)};\ A_{dm}LP = \frac{\left(\frac{2}{R_1C_1R_2C_2}\right)}{D(s)} \tag{12.79}$$

where

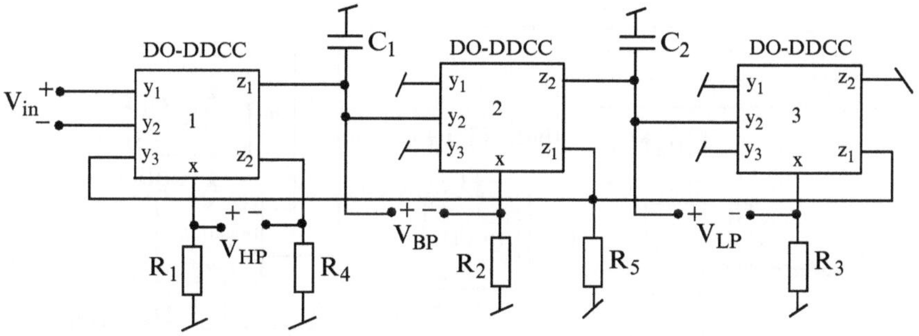

Fig. 12.21 DO-DDCC based differential VM KHN biquad proposed by Ibrahim and Kuntman (Adapted from [34] © 2004 Elsevier)

$$D(s) = s^2 + s\left(\frac{R_5}{C_1R_1R_2}\right) + \frac{R_5}{C_1C_2R_1R_2R_3} \tag{12.80}$$

The filter parameters ω_0 and Q_0 are orthogonally tunable and are given by:

$$\omega_0 = \sqrt{\frac{R_5}{C_1C_2R_1R_2R_3}} \text{ and } Q_0 = \sqrt{\frac{C_1R_1R_2}{R_5R_3C_2}} \tag{12.81}$$

Lee's mixed-mode biquad A mixed-mode fully cascadable universal biquad filter configuration was proposed by Lee [51] in 2011 employing three DDCCs, two grounded capacitors and three grounded resistors. The presented structure is capable of providing inverting, non-inverting and differential types all standard filter responses from the same topology and is suitable for cascading in all four possible modes.

Assuming ideal DDCCs, a routine analysis of the circuit of Fig. 12.22, yields:

$$I_{out} = \frac{N_1(s)/R_2 + N_3(s)}{D(s)} \tag{12.82}$$

and

$$V_{out} = \frac{-N_1(s) + N_2(s)R_2}{D(s)} \tag{12.83}$$

where

$$N_1(s) = s^2C_1C_2(V_{in1} - V_{in2}) - sC_1(V_{in3} - V_{in4})/R_1 + (V_{in5} - V_{in6})/R_1R_3, \tag{12.84}$$

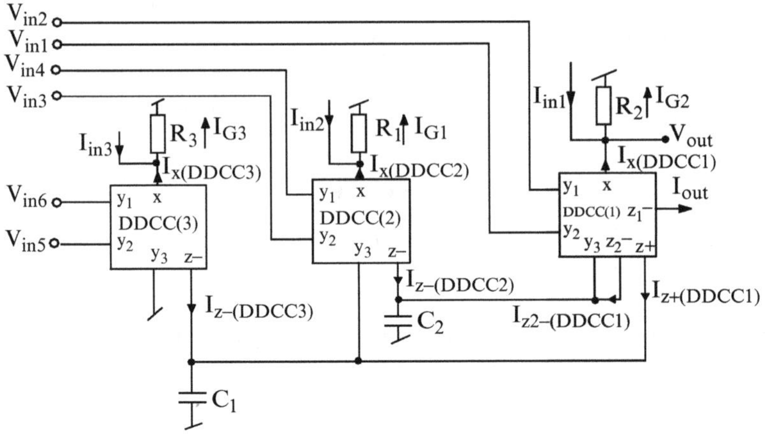

Fig. 12.22 Mixed-mode VM universal biquad proposed by Lee [51]

$$N_2(s) = sC_1(I_{in1} + I_{in2})/R_2 + (I_{in1} - I_{in3})/R_1R_2 \tag{12.85}$$

$$N_3(s) = s^2C_1C_2I_{in1} - s(C_1I_{in2})/R_2 + I_{in3}/R_1R_2 \tag{12.86}$$

and

$$\mathrm{D(s)} = s^2C_1C_2 + sC_1/R_2 + 1/R_1R_2 \tag{12.87}$$

In [51], it has been shown that the circuit can realize a large variety of filter responses, in CM, VM transadmittance mode and transimpedance mode. For the details of all these special cases, the reader is referred to [51].

Tangsrirat-Channumsin-Pukkalanun biquad In 2013 Tangsrirat et al. [65] proposed a high input impedance resistorless realization of electronically tunable VM single input five outputs (SIFO) type universal filter configuration using three DDCC transconductance amplifiers (DDCCTAs) and two grounded capacitors.

The circuit offers the realization of all the standard filter functions simultaneously without any component matching condition. An ideal DDCCTA can be characterized by the following hybrid matrix:

$$\begin{bmatrix} i_{y1} \\ i_{y2} \\ i_{y3} \\ v_x \\ i_z \\ i_O \end{bmatrix} = \begin{bmatrix} 0 & 0 & 0 & 0 & 0 & 0 \\ 0 & 0 & 0 & 0 & 0 & 0 \\ 0 & 0 & 0 & 0 & 0 & 0 \\ 1 & -1 & 1 & 0 & 0 & 0 \\ 0 & 0 & 0 & 1 & 0 & 0 \\ 0 & 0 & 0 & 0 & g_m & 0 \end{bmatrix} \begin{bmatrix} v_{y1} \\ v_{y2} \\ v_{y3} \\ i_x \\ v_z \\ v_O \end{bmatrix} \tag{12.88}$$

where g_m is the transconductance of the DDCCTA. A routine circuit analysis of Fig. 12.23 yields the following VM transfer functions:

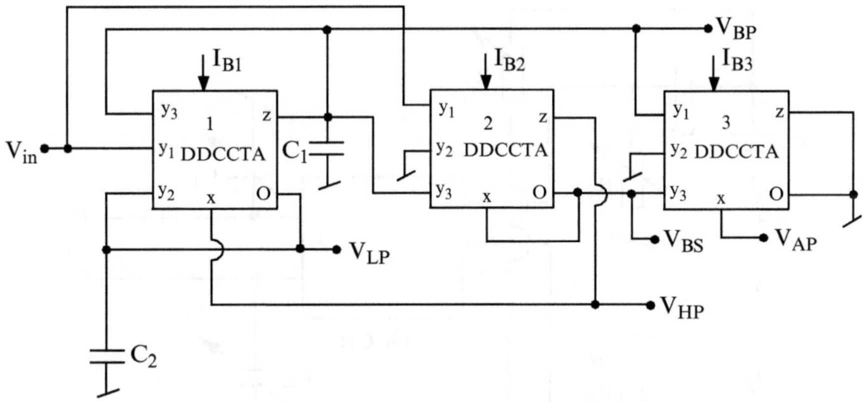

Fig. 12.23 DDCCTA-based resistorless universal filter proposed by Tangsrirat et al. (Adapted from [65] © 2013 Elsevier)

$$\frac{V_{LP}}{V_{in}} = \frac{(g_{m1}g_{m2}/C_1C_2)}{D(s)}; \frac{V_{HP}}{V_{in}} = \frac{s^2}{D(s)}; \frac{V_{BP}}{V_{in}} = \frac{-s(g_{m2}/C_1)}{D(s)} \tag{12.89}$$

$$\frac{V_{AP}}{V_{in}} = \frac{s^2 - s(g_{m2}/C_1) + (g_{m1}g_{m2}/C_1C_2)}{D(s)} \tag{12.90}$$

and

$$\frac{V_{Notch}}{V_{in}} = \frac{s^2 + (g_{m1}g_{m2}/C_1C_2)}{D(s)} \tag{12.91}$$

where

$$\mathrm{D(s)} = s^2 + s(g_{m2}/C_1) + (g_{m1}g_{m2}/C_1C_2) \tag{12.92}$$

The values of filter parameters ω_0 and Q_0 can be given by:

$$\omega_0 = \sqrt{\frac{g_{m1}g_{m2}}{C_1C_2}} \text{ and } Q_0 = \sqrt{\frac{C_1g_{m1}}{g_{m2}C_2}} \tag{12.93}$$

Thus, ω_0 can be controlled electronically by varying g_{m1} and g_{m2} simultaneously while maintaining the ratio of g_{m1}/g_{m2} constant.

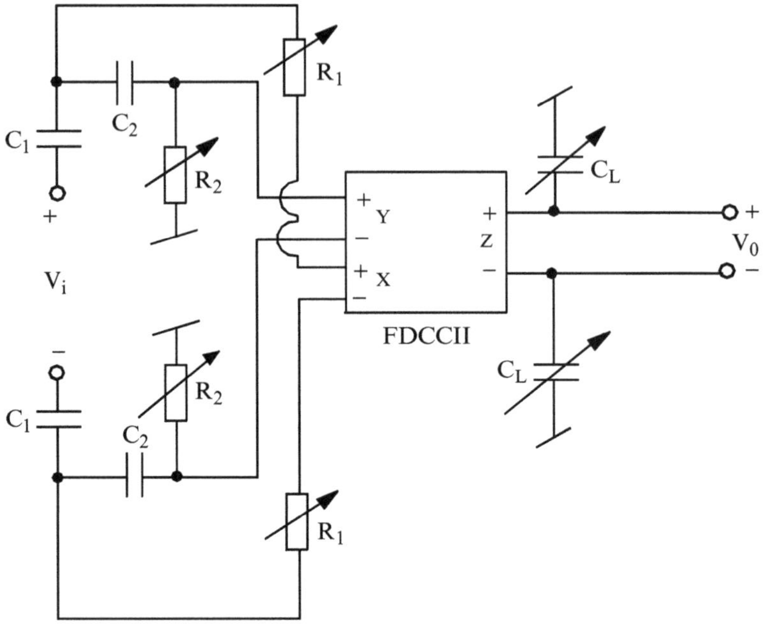

Fig. 12.24 AC-coupled LPF proposed by Alzaher et al. [70]

12.2.3 Filter Design Using FDCCII

The fully differential second generation CC (FDCCII) was advanced as a versatile building block suitable for designing fully differential filters. However, with the availability of extended versions of the FDCCII having as many as eight terminals one may expect that it may be possible to use FDCCII to perform various analog signal processing/ signal generation functions possibly with a bare minimum passive components. That it is indeed so has been amply demonstrated by a number of researchers see [70–87]. In the following we present some selected VM and CM universal biquads using FDCCII which re-affirm the above assertion.

Alzaher-Elwan-Ismail fully-differential biquad Alzaher et al. [70] demonstrated that a fully differential version of a classical Sallen-Key filter using FDCCII exhibits low noise, high linearity and low power consumption, thereby making it suitable for implementing the baseband filter for wide-band code division multiple access (WCDMA) direct conversion integrated wireless receivers. This topology is shown in Fig. 12.24 and is characterized by the following transfer function:

$$\frac{V_0}{V_i} = \frac{\frac{1}{C_L C_2 R_1 R_2}}{s^2 + s\left(\frac{1}{C_1 R_2} + \frac{1}{C_2 R_2}\right) + \frac{1}{C_1 C_2 R_1 R_2}} \tag{12.94}$$

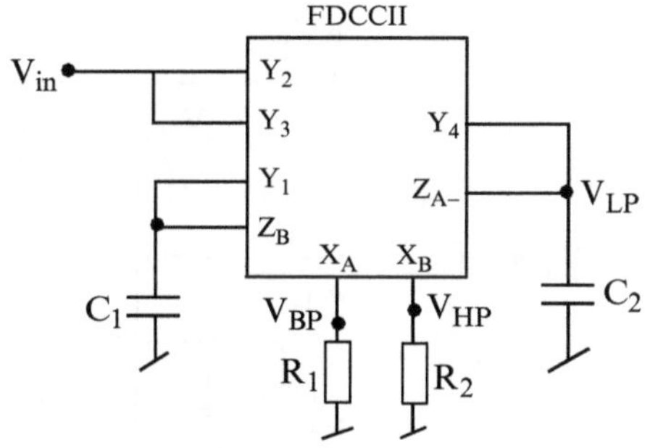

Fig. 12.25 VM universal filter using single FDCCII presented by Kacar and Yesil (Adapted from [79] © 2010 Springer Science + Business Media)

Experimental and simulation results obtained from a fabricated chip demonstrated that a WCDMA filter having a programmable bandwidth around 2.1 MHz and a variable gain of 50 dB is attainable.

Kacar-Yesil biquad Figure 12.25 shows a high input impedance SIMO-type VM universal biquadratic filter configuration using single FDCCII along with four passive elements (two grounded capacitors and two grounded resistors) proposed by Kacar and Yesil in [79].

The configuration can realize three filter responses simultaneously. The ideal port characteristics of FDCCII are given by:

$$\begin{bmatrix} V_{XA} \\ V_{XB} \\ I_{ZA+} \\ I_{ZA-} \\ I_{ZB+} \\ I_{ZB-} \end{bmatrix} = \begin{bmatrix} 1 & -1 & 1 & 0 & 0 & 0 \\ -1 & 1 & 0 & 1 & 0 & 0 \\ 0 & 0 & 0 & 0 & 1 & 0 \\ 0 & 0 & 0 & 0 & -1 & 0 \\ 0 & 0 & 0 & 0 & 0 & 1 \\ 0 & 0 & 0 & 0 & 0 & -1 \end{bmatrix} \begin{bmatrix} V_{Y1} \\ V_{Y2} \\ V_{Y3} \\ V_{Y4} \\ I_{XA} \\ I_{XB} \end{bmatrix} \tag{12.95}$$

An analysis of the circuit of Fig. 12.25 (assuming ideal FDCCII as characterized by the above equation (12.95) yields the following VM transfer functions:

$$\frac{V_{LP}}{V_{in}} = -\frac{\left(\frac{1}{C_1C_2R_1R2}\right)}{D(s)};\frac{V_{BP}}{V_{Iin}} = \frac{s\left(\frac{1}{C_1R_2}\right)}{D(s)}\text{ and}\frac{V_{HP}}{V_{in}} = \frac{s^2}{D(s)} \tag{12.96}$$

where

$$\mathrm{D(s)} = s^2 + s\left(\frac{1}{C_1R_2}\right) + \frac{1}{C_1C_2R_1R_2} \tag{12.97}$$

Thus, the circuit has the capability to realize three basic filtering functions.

Lee-Chang mixed-mode biquad Lee and Chang [75] proposed a mixed-mode biquad filter employing single FDCCII, two grounded capacitors and three resistors as shown here in Fig. 12.26.

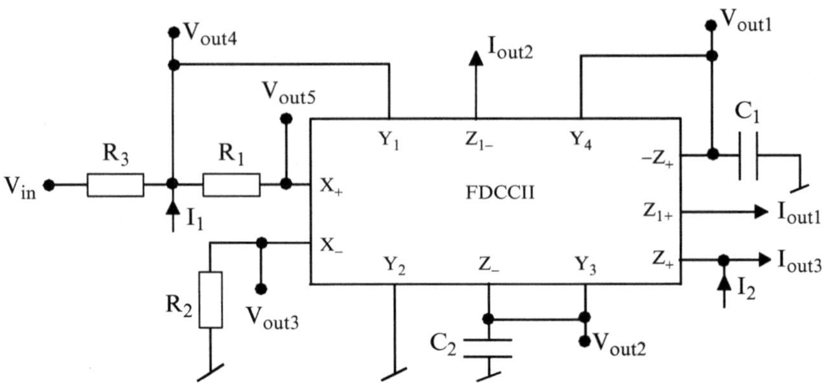

Fig. 12.26 Mixed-mode filter configuration proposed by Lee and Chang (Adapted from [75] © 2008 Elsevier GmbH)

Assuming the type of FDCCII employed to be characterized by the following matrix:

$$\begin{bmatrix} V_{X+} \\ V_{X-} \\ I_{-Z+} \\ I_{Z+} \\ I_{Z1+} \\ I_{Z-} \\ I_{Z1-} \end{bmatrix} = \begin{bmatrix} 0 & 0 & 1 & -1 & 1 & 0 \\ 0 & 0 & -1 & 1 & 0 & 1 \\ -1 & 0 & 0 & 0 & 0 & 0 \\ 1 & 0 & 0 & 0 & 0 & 0 \\ 1 & 0 & 0 & 0 & 0 & 0 \\ 0 & 1 & 0 & 0 & 0 & 0 \\ 0 & 1 & 0 & 0 & 0 & 0 \end{bmatrix} \begin{bmatrix} I_{X+} \\ I_{X-} \\ V_{Y1} \\ V_{Y2} \\ V_{Y3} \\ V_{Y4} \end{bmatrix} \tag{12.98}$$

a routine circuit analysis for the three different cases yields:

Case-I if $I_1 = I_2 = 0$ (open circuited) and input signal is V_{in}, this circuit is capable of providing simultaneously five generic responses in VM, BPF and HPF in transconductance mode. The various transfer functions are given by:

$$\frac{V_{out1}}{V_{in}} = \frac{\frac{1}{C_1C_2R_1R_2}}{D(s)};\ \frac{V_{out2}}{V_{in}} = \frac{-s\left(\frac{1}{C_2R_2}\right)}{D(s)};\ \frac{V_{out3}}{V_{in}} = \frac{-s^2}{D(s)} \tag{12.99}$$

$$\frac{V_{out4}}{V_{in}} = \frac{s^2 + \left(\frac{1}{C_1C_2R_1R_2}\right)}{D(s)} \tag{12.100}$$

and

$$\frac{V_{out5}}{V_{in}} = \frac{s^2 - s\left(\frac{1}{C_2R_2}\right) + \frac{1}{C_1C_2R_1R_2}}{D(s)}; for R_1 = R_3 \tag{12.101}$$

$$\frac{I_{out1}}{V_{in}} = \frac{-s\left(\frac{1}{C_2R_1R_2}\right)}{D(s)} \text{ and } \frac{I_{out2}}{V_{in}} = \frac{-s^2\left(\frac{1}{R_2}\right)}{D(s)} \tag{12.102}$$

where

$$\mathrm{D(s)} = s^2 + s\left(\frac{R_3}{C_2R_1R_2}\right) + \frac{1}{C_1C_2R_1R_2} \tag{12.103}$$

Case-II if Vin = 0 (grounded), $I_2 = 0$ (open circuited) and $I_1 = I_{in}$, the various transfer functions in transresistance mode and CM are given by:

$$\frac{V_{out1}}{I_{in}} = \frac{\left(\frac{R_3}{C_1C_2R_1R_2}\right)}{D(s)}; \frac{V_{out2}}{I_{in}} = \frac{-s\left(\frac{R_3}{C_2R_2}\right)}{D(s)}; \frac{V_{out3}}{I_{in}} = \frac{-s^2R_3}{D(s)} \tag{12.104}$$

$$\frac{V_{out4}}{I_{in}} = \frac{s^2R_3 + \left(\frac{R_3}{C_1C_2R_1R_2}\right)}{D(s)} \tag{12.105}$$

$$\frac{V_{out5}}{I_{in}} = \frac{s^2R_3 - s\left(\frac{R_3}{C_2R_2}\right) + \frac{R_3}{C_1C_2R_1R_2}}{D(s)}; for R_1 = R_3 \tag{12.106}$$

$$\frac{I_{out1}}{I_{in}} = \frac{-s\left(\frac{R_3}{C_2R_1R_2}\right)}{D(s)} \text{ and } \frac{I_{out2}}{I_{in}} = \frac{-s^2\left(\frac{R_3}{R_2}\right)}{D(s)} \tag{12.107}$$

where

$$\mathrm{D(s)} = s^2 + s\left(\frac{R_3}{C_2R_1R_2}\right) + \frac{1}{C_1C_2R_1R_2} \tag{12.108}$$

Case-III if $V_{in} = 0$ (grounded), $I_2 = I_1 = I_{in}$ then in addition to transfer functions realized in case-II, the configuration gives the following transfer functions in CM:

$\frac{I_{out3}}{I_{in}} = \frac{s^2 + \left(\frac{1}{C_1C_2R_1R_2}\right)}{D(s)}$ and an APF can be obtained by connecting I_{out1} and I_{out3}, i.e.,

$$\frac{I_{out3} + I_{out1}}{I_{in}} = \frac{s^2 - s\left(\frac{R_3}{C_2R_1R_2}\right) + \left(\frac{1}{C_1C_2R_1R_2}\right)}{D(s)} \tag{12.109a}$$

and a LPF can be realized by adding I_{out2} and I_{out3}, i.e.

$$\frac{I_{out3} + I_{out2}}{I_{in}} = \frac{\frac{1}{C_1C_2R_1R_2}}{D(s)} \tag{12.109b}$$

where D(s) remains same as above.

The ω_0 and Q_0 for all the cases can be given by

$$\omega_0 = \sqrt{\frac{1}{C_1C_2R_1R_2}} \text{ and } Q_0 = \frac{1}{R_3}\sqrt{\frac{C_2R_2R_1}{C_1}} \tag{12.109c}$$

Thus, ω_0 and Q_0 can be controlled orthogonally.

12.2.4 Filter Design Using ICCII

The so-called inverting CCII (ICCII) has the same characteristic as a CCII but has $v_x = -v_y$. This opens new possibilities for constructing universal CM and VM biquads as well as mixed-mode biquads using ICCII. In this section, we include some selected universal biquads configurations using ICCIIs. The departure $v_x = -v_y$ from the characteristic of normal CCII+ is particularly useful in non-linearity cancellation of the $v_D - i_D$ characteristics of the MOSFET if its drain and source are connected between Yand X terminals respectively of the ICCII. This immediately hints to the possibility of constituting a linear, external voltage-controlled MOSFET-C integrator with a capacitor connected from the Z-terminal to ground. Thus, interesting MOSFET-C filters could be easily conceived using ICII and MOSFETs.

ICCII is a five terminal active element and ideally characterized by the following matrix equations:

$$\begin{bmatrix} I_Y \\ V_X \\ I_Z \end{bmatrix} = \begin{bmatrix} 0 & 0 & 0 \\ -1 & 0 & 0 \\ 0 & \pm 1 & 0 \end{bmatrix} \begin{bmatrix} V_Y \\ I_X \\ V_Z \end{bmatrix} \tag{12.110}$$

where plus sign indicates ICCII+ and minus sign corresponds to ICCII− respectively.

We now present a number of interesting biquad filter configurations using ICCII elements.

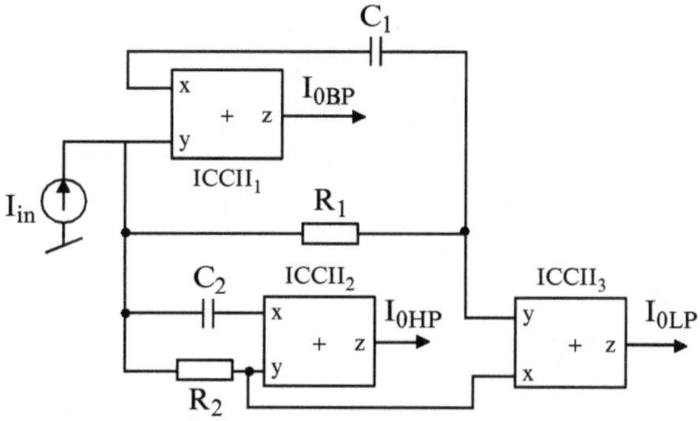

Fig. 12.27 ICCII-based CM universal filter proposed by Ozoguz et al. [88]

Ozoguz-Toker-Cicekoglu biquad Figure 12.27 shows a SIMO-type CM universal biquad proposed by Ozoguz et al. [88] employing three plus type ICCIIs, two capacitors and two resistors which is based on the implementation of first-order all pass sections.

A routine analysis of this configuration shows that it realizes the following three CM transfer functions:

$$\frac{I_{0LP}}{I_{in}} = -\frac{1}{D(s)};\ \frac{I_{0BP}}{I_{in}} = -\frac{sC_1R_2}{D(s)} \text{ and } \frac{I_{0HP}}{I_{in}} = -\frac{s^2C_1C_2R_1R_2}{D(s)} \tag{12.111}$$

$$\text{where } \mathrm{D(s)} = s^2C_1C_2R_1R_2 + sC_1R_2 + 1 \tag{12.112}$$

Thus, the configuration is capable of realizing LPF, BPF and HPF simultaneously. Further, by adding the current outputs I_{0LP} and I_{0BP}, a notch current output can be obtained and if $ICCII_1$ is replaced by ICCII− and all three output currents are added to create another output current, an APF can be realized.

Minaei-Yuce-Cicekoglu biquad The circuit in Fig. 12.28 is a single input six outputs VM biquad filter using one dual-output ICCII (DO-ICCII), one triple output ICCII (TO-ICCII), two grounded capacitors and four resistors. This filter structure proposed by Minaei et al. [89] can provide simultaneously both inverting and non-inverting LPF, BPF and HPF filter functions without any component matching condition, along with orthogonal control of ω_0 and Q_0. Assuming ideal ICCIIs, an analysis of the circuit of Fig. 12.28 yields the following VM transfer functions:

$$\text{LPF}: \frac{V_1}{V_{in}} = \frac{\frac{1}{C_1C_2R_1R_2}}{D(s)};\ \frac{V_2}{V_{in}} = -\frac{\frac{1}{C_1C_2R_1R_2}}{D(s)}; \tag{12.113}$$

$$\text{BPF}: \frac{V_3}{V_{in}} = \frac{\frac{s}{C_2R_1}}{D(s)};\ \frac{V_4}{V_{in}} = -\frac{\frac{s}{C_2R_1}}{D(s)}; \tag{12.114}$$

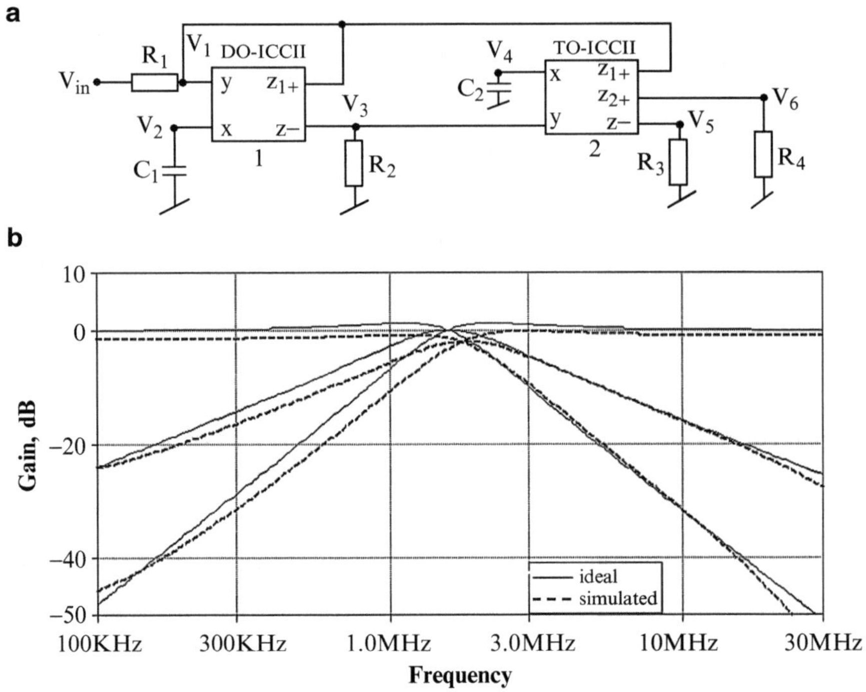

Fig. 12.28 ICCII-based VM biquadratic filter proposed by Minaei et al. (Adapted from [89] ©2006 Birkhauser Boston) (**a**) Circuit configuration (**b**) SPICE generated frequency responses of various filters

$$\text{HPF}: \frac{V_5}{V_{in}} = \left(\frac{R_3}{R_1}\right)\frac{s^2}{D(s)}; \frac{V_5}{V_{in}} = -\left(\frac{R_4}{R_1}\right)\frac{s^2}{D(s)}; \tag{12.115}$$

where

$$\text{D(s)} = s^2 + s\left(\frac{1}{C_2R_2}\right) + \left(\frac{1}{C_1C_2R_1R_2}\right) \tag{12.116}$$

The filter parameters ω_0 and Q_0 are given by:

$$\omega_0 = \sqrt{\frac{1}{C_1C_2R_1R_2}} \text{ and } Q_0 = \sqrt{\frac{C_2R_2}{C_1R_1}} \tag{12.117}$$

Thus ω_0 and Q_0 can be controlled orthogonally by varying R_1, R_2 or C_1, C_2 simultaneously.

The circuit was simulated in SPICE using 0.35 μm TSMC CMOS technology process parameters using the CMOS ICCII of Fig. 2 of [89], biased with ± 5 V Dc power supplies with passive components values chosen to realize various filters for

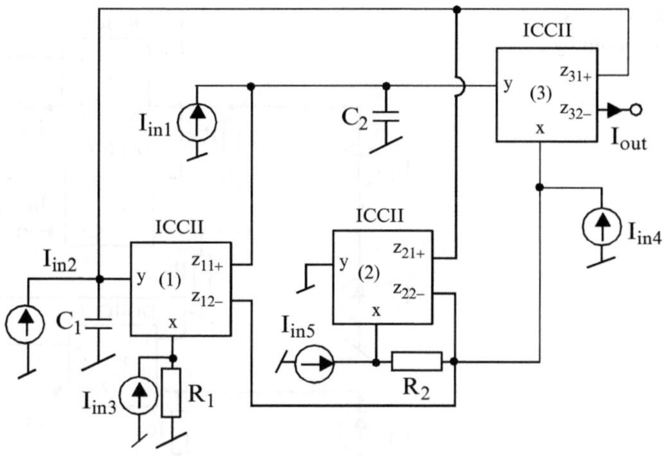

Fig. 12.29 CM universal filter proposed by Horng [97]

f_o of 1.59 MHz. The SPICE generated frequency responses of LP, BP and HP are shown in Fig. 12.28b which confirm the validity of the circuit of Fig. 12.28a.

Horng's MISO-type CM biquad In 2011 Horng [97] proposed a MISO-type CM universal biquad structure as shown in Fig. 12.29 employing three ICCIIs, two grounded capacitors and two resistors.

The configuration can yield all five second-order filter responses without any critical matching of elements. Assuming ideal ICCIIs, a routine circuit analysis gives the output current in terms of input currents as:

$$I_{out} = \frac{s^2(I_{in3} + I_{in4} + I_{in5}) + s\left(\frac{1}{R_1C_1}\right)(I_{in2} - I_{in5}) + \left(\frac{1}{C_1C_2R_1R_2}\right)(I_{in1} + I_{in4} + I_{in5})}{s^2 + s\left(\frac{1}{R_1C_1}\right) + \left(\frac{1}{C_1C_2R_1R_2}\right)} \tag{12.118}$$

From equation (12.118), it is seen that the various filter responses can be obtained as follows from this circuit:

HPF: if $I_{in1} = I_{in2} = I_{in4} = I_{in5} = 0$ (open circuited) and $I_{in} = I_{in3}$;
BPF: if $I_{in1} = I_{in3} = I_{in4} = I_{in5} = 0$ (open circuited) and $I_{in} = I_{in2}$;
LPF: if $I_{in3} = I_{in2} = I_{in4} = I_{in5} = 0$ (open circuited) and $I_{in} = I_{in1}$;
Notch: if $I_{in1} = I_{in2} = I_{in3} = I_{in5} = 0$ (open circuited) and $I_{in} = I_{in4}$; and
APF: if $I_{in1} = I_{in2} = I_{in4} = I_{in3} = 0$ (open circuited) and $I_{in} = I_{in5}$.

Chen's MIMO-type CM biquad Another MIMO-type CM filter configuration was proposed by Chen [98] using three DO-ICCIIs and five grounded passive elements. The circuit is shown in Fig. 12.30.

By straightforward circuit analysis of Fig. 12.30 gives the following CM transfer functions:

Fig. 12.30 MIMO-type CM biquad proposed by Chen [98]

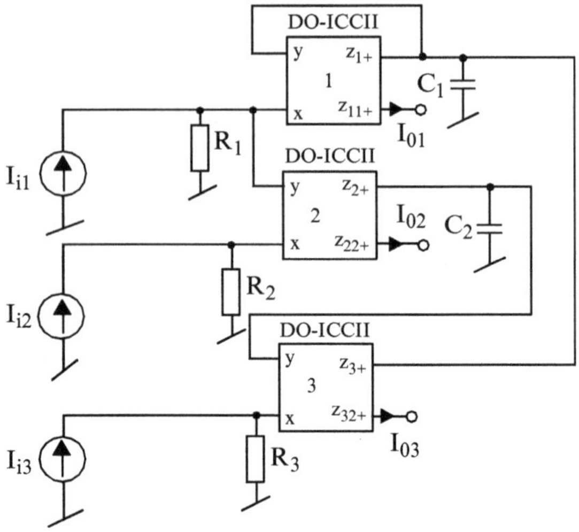

$$I_{01}=\left[\frac{C_1C_2R_1R_2R_3}{D(s)}-\left\{s^2+\frac{1}{C_1C_2R_3R_2}\right\}I_{i1}-\left(\frac{1}{C_1C_2R_1R_3}\right)I_{i2}+\left(\frac{s}{C_1R_1}\right)I_{i3}\right] \tag{12.119}$$

$$I_{02}=\frac{C_1C_2R_1R_2R_3}{D(s)}\left[\left(\frac{-s}{C_1R_2}\right)I_{i1}-\left\{s^2+\frac{s}{C_1R_1}\right\}I_{i2}-\left(\frac{s}{C_1R_2}\right)I_{i3}\right] \tag{12.120}$$

$$I_{03}=\frac{C_1C_2R_1R_2R_3}{D(s)}\left[\frac{1}{C_1C_2R_3R_2}I_{i1}+\left(\frac{s}{C_2R_3}+\frac{1}{C_1C_2R_1R_3}\right)I_{i2}-\left(s^2+\frac{s}{C_1R_1}\right)I_{i3}\right] \tag{12.121}$$

where

$$\mathrm{D(s)}=s^2C_1C_2R_1R_2R_3+sC_2R_2R_3+R_1 \tag{12.122}$$

Case-I

If $I_{i2}=I_{i3}=0$ (open circuited) and $I_{i1}=I_{in}$, the following three CM transfer functions namely notch, BPF and LPF can be realized simultaneously:

$$\frac{I_{01}}{I_{in}}=\frac{-(s^2C_1C2R_3R_2R_1+R_1)}{D(s)};\frac{I_{02}}{I_{in}}=\frac{-(sC_2R_1R_3)}{D(s)};\ \frac{I_{03}}{I_{in}}=\frac{R_1}{D(s)} \tag{12.123}$$

By adding I_{01} and I_{03} an inverting HPF can be obtained as follows

$$\frac{I_{HP}}{I_{in}}=\frac{-s^2C_1C2R_3R_2R_1}{D(s)} \tag{12.124}$$

Further, if $I_{i2}=0$ and $I_{i1}=I_{i3}=I_{in}$, an inverting APF is easily obtainable from I_{01} as given by the following CM transfer function:

$$\frac{I_{AP}}{I_{in}}=\frac{I_{01}}{I_{in}}=\frac{-\{s^2C_1C2R_3R_2R_1-sC_2R_2R_3+R_1\}}{D(s)} \tag{12.125}$$

Case-II

BPF: if $I_{i1}=I_{i2}=0$ (open circuited), $I_{in}=I_{i3}$ and $I_{out}=I_{01}$; LPF: if $I_{i1}=I_{i3}=0$ (open circuited), $I_{in}=I_{i2}$ and $I_{out}=I_{01}$; Notch: if $I_{i2}=I_{i3}=0$ (open circuited), $I_{in}=I_{i1}$ and $I_{out}=I_{01}$; LPF: if $I_{i2}=I_{i3}=0$ (open circuited), $I_{in}=I_{i1}$ and $I_{out}=I_{03}$; APF: if $I_{i2}=0$ (open circuited), $I_{in}=I_{i1}=I_{i3}$ and $I_{out}=I_{01}$; HPF: if $I_{i2}=I_{i3}=0$ (open circuited), $I_{in}=I_{i1}$ and $I_{out}=(I_{03}+I_{01})$.

The filter parameters ω_0 and Q_0 are given by:

$$\omega_0=\sqrt{\frac{1}{C_1C_2R_3R_2}}\,\text{and}\,Q_0=R_1\sqrt{\frac{C_1}{C_2R_2R_3}} \tag{12.126}$$

Thus, from the equation (12.126) it is clear that ω_0 and Q_0 can be controlled orthogonally.

It is, thus, seen that the circuit has the flexibility to be used as either a single-input three-outputs or three-input two-outputs. Single-input three-output operation yields LPF, BPF and notch responses simultaneously while HPF can also be realized by adding appropriate output currents. Three-input two-output operation offers the realization of all five generic filter functions by proper selection of input currents. The filter parameters ω_0 and Q_0 can be controlled orthogonally.

From the SPICE simulations, the proposed analog filter was found to be operating satisfactorily while operating at 3.183 MHz and with power dissipations of the order of 0.433 mW and the chip area of the filter was 0.022 mm^2. The functionality of the design was also verified [98] by the post-layout simulations.

Ibrahim's SIMO-type CM biquad An electronically-tunable SIMO-type CM biquad filter configuration suitable for easy cascading employing two DO-ICCIIs, one triple output ICCII (TO-ICCII), two grounded capacitors and two MOSFETs as resistors was proposed by Ibrahim in [99]. The presented circuit configuration can realize LPF, BPF and HPF responses simultaneously from the high output impedance nodes without any component matching condition. The other two responses namely, notch and APF can also be obtained by connecting appropriate output currents. The filter parameters ω_0 and Q_0 can be controlled electronically through the bias voltages of the MOSFETs. Assuming triode region operation of the MOSFETs, their resistance values is given by:

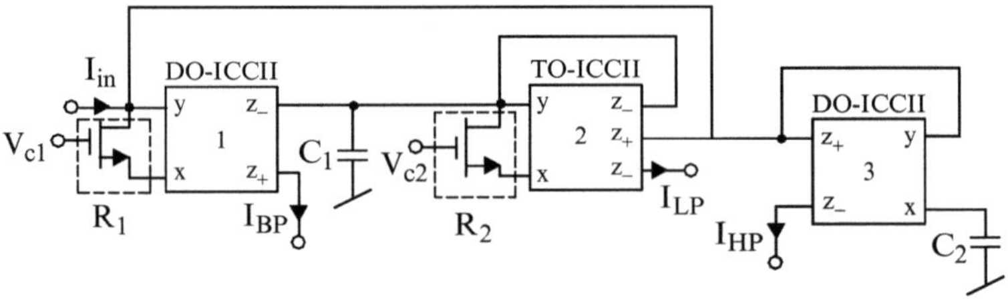

Fig. 12.31 SIMO-type CM biquad filter proposed by Ibrahim [99]

$$R = \frac{1}{K_n(V_C - V_{th})} \tag{12.127}$$

where $K_n = \mu_n C_{ox}\left(\frac{W}{L}\right)$, V_{th} = threshold voltage, V_C = control voltage, μ_n = surface mobility, C_{ox} = gate capacitance per unit area, W = channel width and L = channel length.

The various CM transfer functions, assuming ideal ICCIIs for the circuit of Fig. 12.31 are found to be:

$$\frac{I_{BP}}{I_{in}} = -\frac{\left(\frac{2}{C_2R_1}\right)s}{D(s)}; \frac{I_{LP}}{I_{in}} = \frac{\left(\frac{4}{C_1R_2C_2R_1}\right)}{D(s)}; \frac{I_{HP}}{I_{in}} = \frac{s^2}{D(s)} \tag{12.128}$$

where

$$\mathrm{D(s)} = s^2 + s\left(\frac{2}{C_2R_1}\right) + \frac{4}{C_1C_2R_1R_2} \tag{12.129}$$

The filter parameters ω_0 and Q_0 can be obtained as:

$$\omega_0 = \sqrt{\frac{4}{C_1C_2R_1R_2}}, \ \mathrm{Q_0} = \sqrt{\frac{R_1C_2}{C_1R_2}}, \tag{12.130}$$

Thus, ω_0 and Q_0 can be controlled electronically by varying the bias voltages of MOSFETs (i.e., changing R_1 and R_2). The notch and APF can also be realized by appropriate connection of output currents as given by the following expressions:

Notch:

$$\text{if}\, I_{out} = (I_{LP} + I_{HP}) = I_{BS};\ \text{APF} : \text{if}\ I_{out} = (I_{LP} + I_{BP} + I_{HP}) = I_{AP} \tag{12.131}$$

12.2.5 Filter Design Using DCVC or CDBA

The differential current voltage conveyor (DCVC) or current differencing buffered amplifier (CDBA) is an extended CC wherein the current to the Z-port is equal to

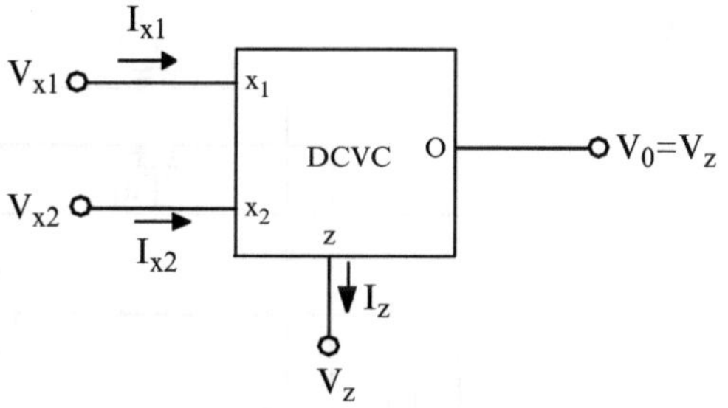

Fig. 12.32 Symbolic notation of the DCVC or CDBA

the difference of two currents at its two low input-impedance terminals. Its symbolic notation is shown in Fig. 12.32. A DCVC is a four terminal analog building block which is characterized by the following set of equations:

$$V_{X1} = V_{X2} = 0,\ I_Z = (I_{X1} - I_{X2}) \text{ and } V_O = V_Z \tag{12.132}$$

In view of the departure from the normal variety of CCs, this results in new possibilities of realizing universal biquads filters. It must be however, recognized that a DCVC or CDBA is easily realizable from three normal types of $CCII_s$+ elements or AD844s. On the other hand, a number of CMOS-compatible implementations of the CDBAs have been advanced by a number of researchers.

Although the literature is flooded with a variety of biquads using CDBAs, in the following, we present some exemplary biquads using these building blocks.

Salama-Elwan-Soliman biquad In 2001, Salama et al. [101] have proposed a universal biquad filter configuration employing two DCVCs, two grounded capacitors, one floating capacitor and ten MOSFETs (see Fig. 12.33). The filter structure is suitable for higher-order filter realization by cascading second-order sections. Also most of the parasitic capacitances can be eliminated because both the input terminals are virtually grounded and parasitic capacitances at Z-terminals can be absorbed in C_1 and C_2.

Assuming ideal DCVCs and operation of all MOSFETs in triode region, a routine circuit analysis of Fig. 12.33 reveals the following VM transfer function:

$$\frac{V_o}{V_i} = \frac{s^2\left(\frac{C_3}{C_2}\right) + s\left(\frac{1}{C_2R_4}\right) + \frac{1}{C_1C_2R_3R_2}}{s^2 + s\left(\frac{1}{C_2R_5}\right) + \frac{1}{C_1C_2R_1R_2}} \tag{12.133}$$

where

a

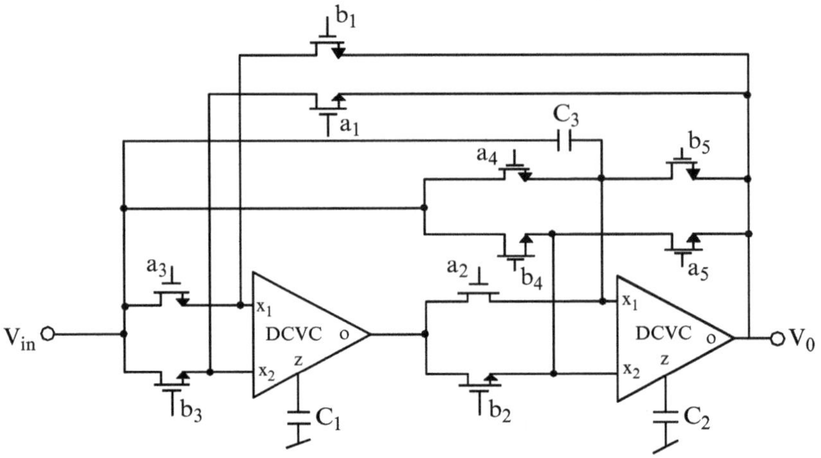

b

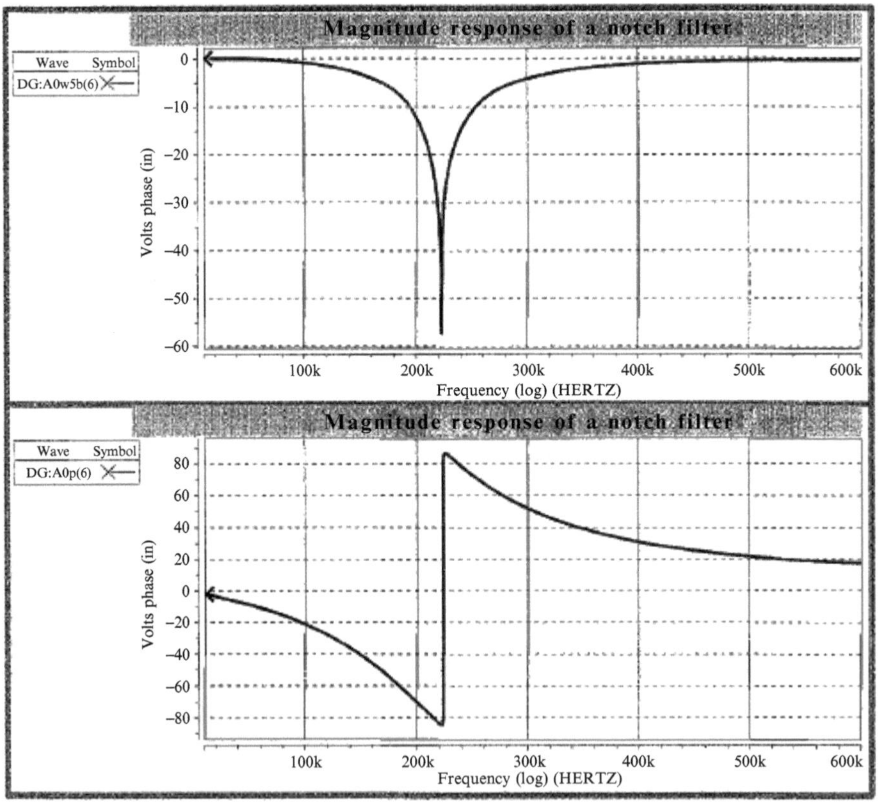

Fig. 12.33 Universal biquad filter using DCVC proposed by Salama et al. (Adapted from [101] © 2001Springer)

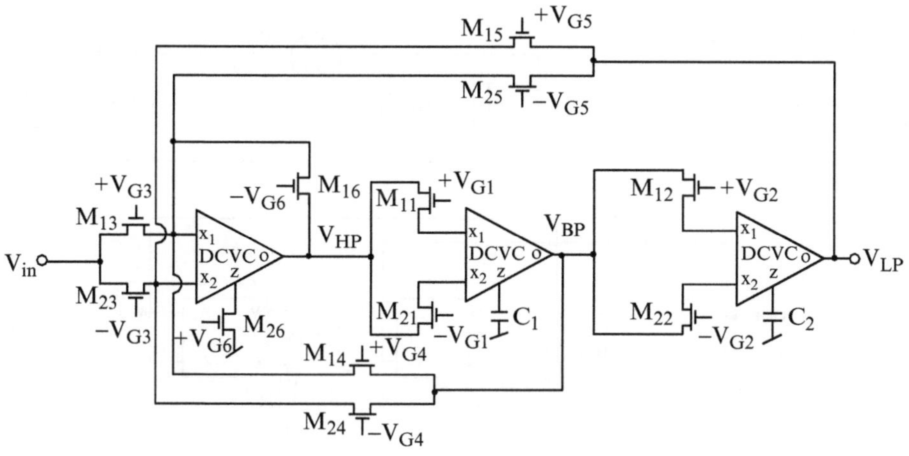

Fig. 12.34 KHN-equivalent employing DCVCs and MOS-C [102]

$$R_1 = \frac{1}{K(V_{a1} - V_{b1})}, R_2 = \frac{1}{K(V_{a2} - V_{b2})}, R_3 = \frac{1}{K(V_{a3} - V_{b3})},$$
$$R_4 = \frac{1}{K(V_{a4} - V_{b4})}, R_5 = \frac{1}{K(V_{a5} - V_{b5})} \text{ and } K = \mu_n C_{ox}\left(\frac{W}{L}\right) \tag{12.134}$$

The filter parameters ω_0 and Q_0 can be derived as:

$$\omega_0 = \sqrt{\frac{1}{C_1 C_2 R_1 R_2}}, Q_0 = R_5 \sqrt{\frac{C_2}{C_1 R_1 R_2}} \tag{12.135}$$

Thus, Q_0 can be tuned independently by varying R_5 without disturbing ω_0.

From equation (12.133), the various filter responses can be obtained as follows: HPF: $R_3 = R_4 = \infty$ (i.e., $V_{a3} = V_{b3}$ and $V_{a4} = V_{b4}$); BPF: $R_3 = \infty$ (i.e., $V_{a3} = V_{b3}$ and $C_3 = 0$); LPF: $R_4 = \infty$ (i.e., $V_{a3} = V_{b3}$ and $C_3 = 0$); Notch: $R_3 = R_1$, $R_4 = \infty$ and $C_3 = C_2$) and APF: $R_3 = R_1$, $R_4 = -R_5$ and $C_3 = C_2$).

A sample result of the SPICE simulation of the circuit is shown in Fig. 12.33b which shows the magnitude and phase response of the notch filter designed from the circuit.

Soliman-Madian MOS-C biquad Another application of DCVC in the realization of equivalent form of KHN- biquad is shown in Fig. 12.34.

Starting from the conventional op-amp-based biquad the derivation of this MOS-C version using DCVC and using dual MOSFET resistors is straight forward. The circuit realizes a LPF, BPF and HPF functions at the three output terminals of the DCVCs employed with the parameters of the realized filters being electronically-controllable through the various external gate control voltages. More detailed information about the circuit can be found in [102].

Fig. 12.35 CM 3rd-order LP all-pole LC ladder filter

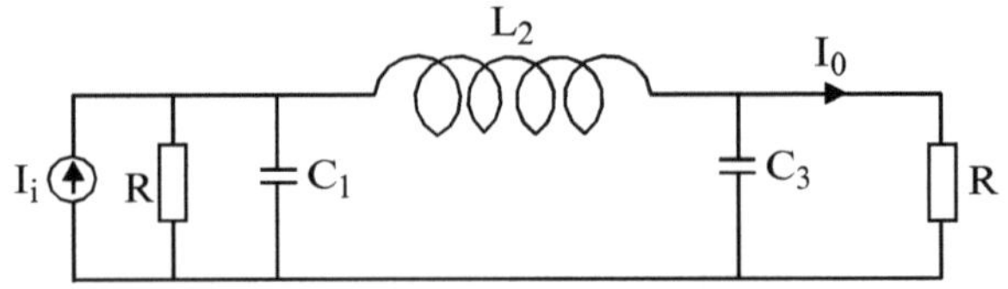

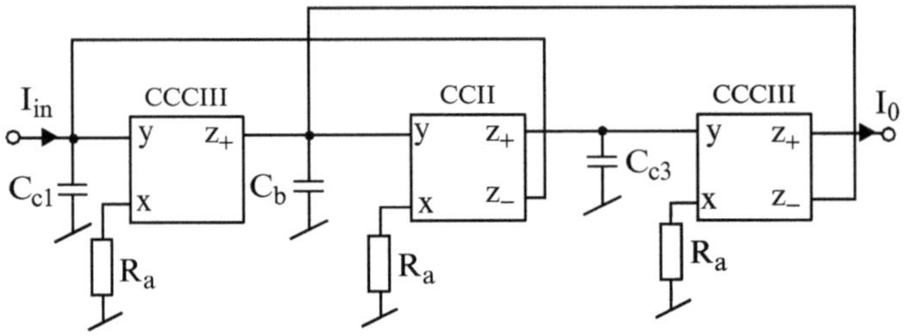

Fig. 12.36 Active realization of all-pole filter derived by Hwang et al. (Adapted from [109] © 2007 Springer)

12.2.6 Filter Design Using CCIII

The third generation CC (CCIII±) which provides $I_y = -I_x$, $V_x = V_y$ and $I_z = \pm I_x$ was introduced by Fabre [103] as a particularly useful element for active current probes and for sensing current in floating elements and making it available from a high input impedance node. However, subsequently, the CCIII found numerous other applications including those of realizing universal biquads. In this section, we outline some of the prominent biquad structures using CCIIIs as active elements.

Hwang-Chen-Li method [102] of higher order filter design Figure 12.35 shows a passive CM 3rd-order LP all-pole LC ladder filter and its active realization is given in Fig. 12.36.

Assuming ideal CCs, a routine circuit analysis yields the following CM transfer function:

$$\frac{I_0}{I_i} = \frac{1}{s^3 R_a^3 (C_{c1} CbC_{c3}) + s^2 R_a^2 C_b (C_{c1} + C_{c3}) + sR_a (C_{c1} + C_b + C_{c3}) + 2} \tag{12.136}$$

The corresponding active realization is shown in Fig. 12.36 which employs three CCs (one DO-CCII, one DO-CCIII and one CCIII), three grounded capacitors and three grounded resistors.

As another example, the circuit of Fig. 12.37 shows a passive realization of CM 3rd-order LP elliptic LC ladder filter and its active realization is shown in Fig. 12.38. Again assuming ideal CCs, through a straight forward circuit analysis, the CM transfer function of Fig. 12.38 can be obtained as:

Fig. 12.37 CM 3rd-order LP elliptic LC ladder filter

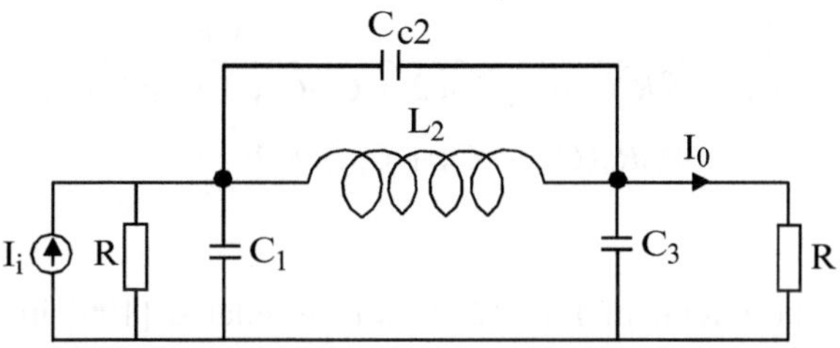

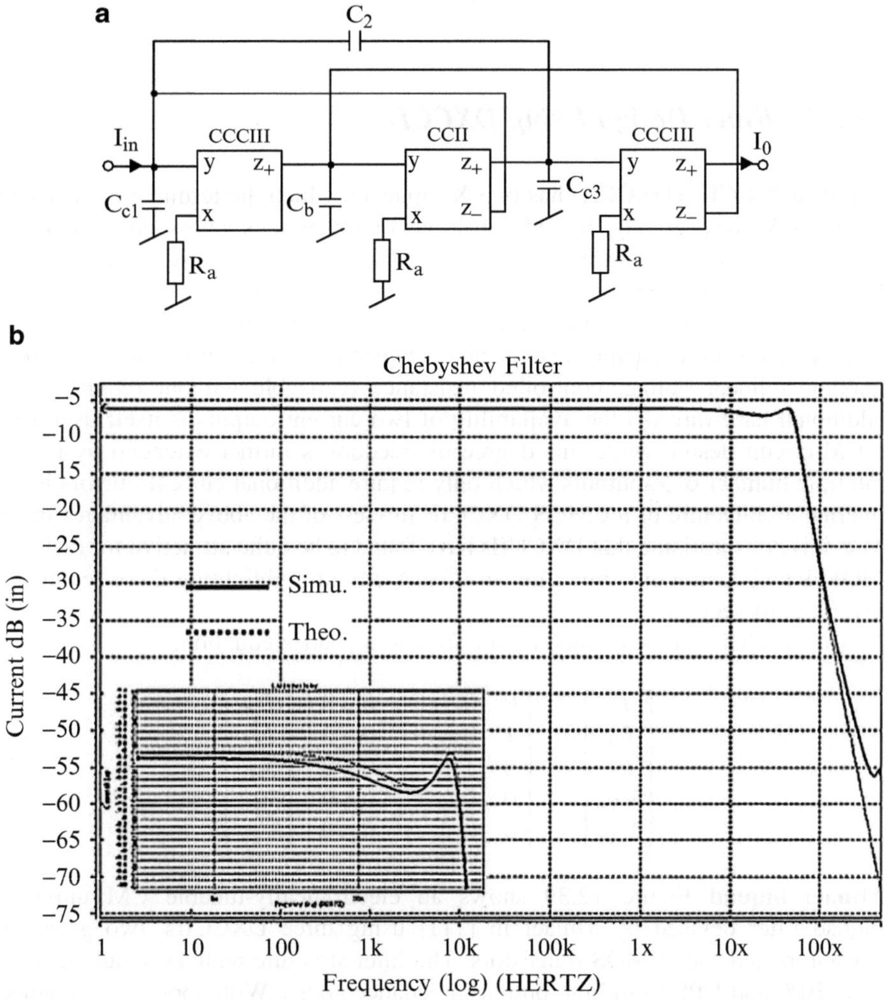

Fig. 12.38 Active realization of CM elliptic filter (Adapted from [109] © 2007 Springer) (**a**) circuit configuration (**b**) The magnitude response of the third order Chebyshev LPF

$$\frac{I_0}{I_{in}} = \frac{s^2R_a^2C_bC_{c2} + 1}{s^3R_a^3C_b(C_{c1}C_{c2}c2 + C_{c2}C_{c3} + C_{c3}C_{c1}c1) + s^2R_a^2C_b(C_{c1} + 2C_{c2} + C_{c3}) + sR_a(C_{c1} + C_b + C_{c3}) + 2} \quad (12.137)$$

The circuit of Fig. 12.38 was simulated [109] in SPICE using TSMC 0.35 μm CMOS process parameters and the circuit designed for 1 dB ripple and 50 MHz bandwidth which is shown here in Fig. 12.38c.

12.2.7 Filter Design Using DXCCII

The dual-X CCIIs (DXCCII) has two X-inputs on which the terminal voltages are V_y and $-V_y$ respectively. In addition it has two Z-outputs whose output currents follow the respective X-terminal input currents, this property is particularly useful from two different viewpoints: (i) availability of complementary signals on the two X-input terminals helps in eliminating the square non-linearity of a MOSFET with its drain and source terminals connected between the two X-terminals and thereby making a linear voltage-controlled resistance realizable without requiring any additional hardware (ii) the availability of two current outputs is itself useful for a CM circuit design while this degree of freedom is further extended by having multiple number of Z-outputs which only require additional current mirrors in the internal architecture of a CMOS DXCCII. In view of the above advantages therefore, it is not surprising that DXCCIIs have found to be quite attractive for realizing MOSFET-C continuous time filter in CM. A circuit highlighting these features is given in this section.

The DXCCII can be defined by the following matrix equation:

$$\begin{bmatrix} I_Y \\ I_{Zp} \\ I_{Zn} \\ V_{Xp} \\ V_{Xn} \end{bmatrix} = \begin{bmatrix} 0 & 0 & 0 & 0 & 0 \\ 1 & 0 & 0 & 0 & 0 \\ 0 & 1 & 0 & 0 & 0 \\ 0 & 0 & 1 & 0 & 0 \\ 0 & 0 & -1 & 0 & 0 \end{bmatrix} \begin{bmatrix} I_{Xp} \\ I_{Xn} \\ V_Y \\ V_{Zp} \\ V_{Zn} \end{bmatrix} \quad (12.138)$$

Minaei biquad Figure 12.39 shows an electronically-tunable CM universal biquad filter devised by Minaei in [111] using three DXCCIIs, two grounded capacitors and four NMOS transistors. The filter structure realizes simultaneously LPF, BPF and HPF from high output impedance nodes. With appropriate connections of output currents a notch and APF can also be obtained. The filter parameters ω_0 and bandwidth can be controlled electronically.

Assuming ideal DXCCIIs and let all MOSFETs M_i ($i = 1-4$) are operating in triode region, then the resistance of the NMOS transistors in Fig. 12.39 can be

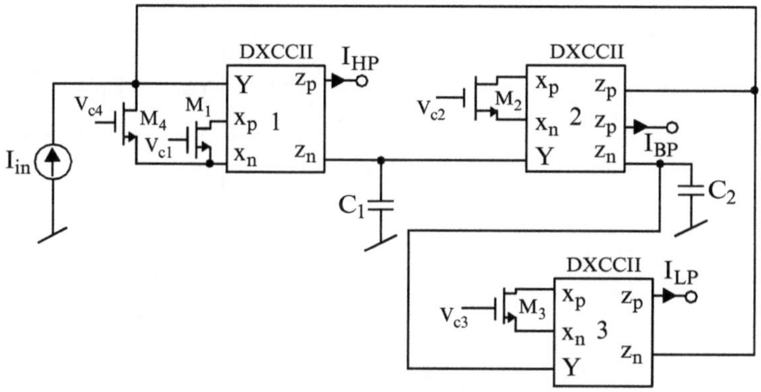

Fig. 12.39 CM universal biquad filter proposed by Minaei [111]

obtained as: $R_i = \frac{1}{K_{ni}(V_{Ci}-V_{Tn})}$, where V_{Tn} is the threshold voltage, $K_{ni} = \mu_n C_{ox}\left(\frac{W}{L}\right)_i$ is the transconductance parameter of the ith NMOS transistor and V_{Ci} is the D.C. control voltage through which resistance value of transistors can be controlled. The circuit of Fig. 12.39 contains two integrators and a summer/subtractor hence the circuit is an equivalent of KHN-biquad.

A straight forward circuit analysis of Fig. 12.39 yields the following CM filter responses:

$$\frac{I_{LP}}{I_{in}} = \frac{\frac{4(R_1+R_4)}{C_1C_2R_1R_2R_3}}{D(s)};\frac{I_{BP}}{I_{in}} = \frac{-\frac{2(R_1+R_4)}{C_1R_1R_2}s}{D(s)} \text{ and } \frac{I_{HP}}{I_{in}} = \frac{\left(\frac{R_4}{R_1}\right)s^2}{D(s)} \tag{12.139}$$

where

$$\mathrm{D(s)} = s^2 + s\frac{2(R_1+R_4)}{C_1R_1R_2} + \frac{4(R_1+R_4)}{C_1C_2R_1R_2R_3} \tag{12.140}$$

Notch and APF can also be realized by adding relevant output currents for the condition $R_1 = R_4$. The filter parameters ω_0, Q_0 and bandwidth (ω_0/Q_0) can be obtained as:

$$\omega_0 = 2\sqrt{\frac{R_1+R_4}{C_1C_2R_1R_2R_3}}, Q_0 = \sqrt{\frac{C_1R_1R_2}{C_2R_3(R_1+R_4)}}, \frac{\omega_0}{Q_0} = \frac{2(R_1+R_4)}{C_1R_1R_2}, \tag{12.141}$$

From the above equation (12.141), ω_0 can be controlled by R_3 without disturbing bandwidth, also ω_0 can be tuned without changing Q_0 through R_2 x R_3 keeping the ratio of R_2/R_3 constant and Q_0 can be varied through R_2/R3 without keeping R_2xR_3

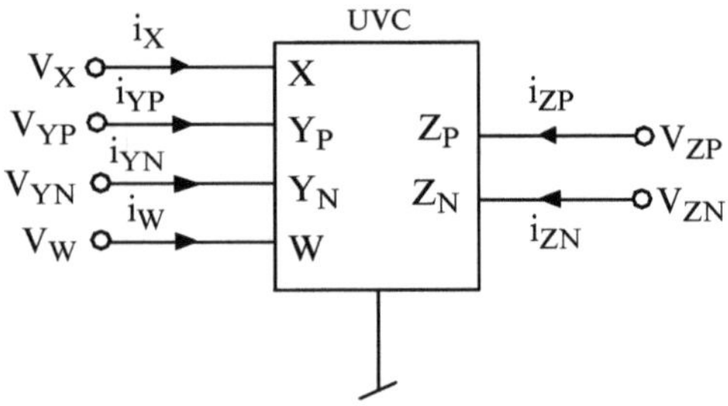

Fig. 12.40 Symbolic notation of the UVC

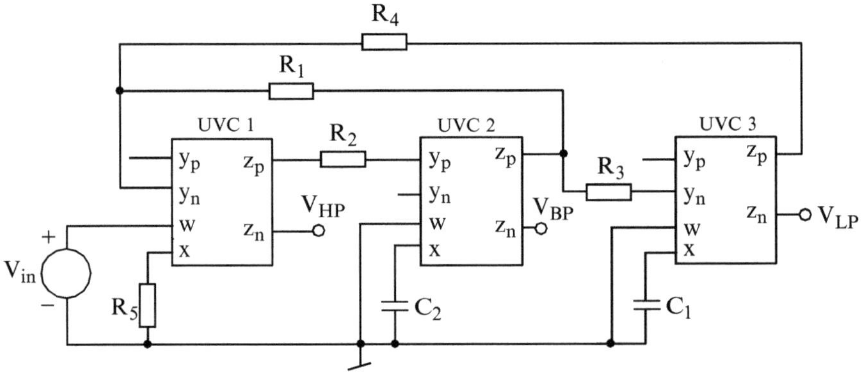

Fig. 12.41 KHN equivalent VM biquad proposed by Koton et al. [200]

constant without altering the value of ω_0. The filter circuit offers low input impedance and high output impedance.

12.2.8 Filter Design Using UVC

The so-called voltage conveyor was presented as dual to the CC as numerous generalizations of CCs were being developed in the literature. Minarcik and Vrba [197] introduced the universal voltage conveyors (UVC) as a six port element characterized by the following hybrid matrix whereas its symbolic notation is shown in Fig. 12.40. Several authors have described a number of filter configurations using UVCs.

Koton, Herencsar and Vrba biquad In the following, we present a VM KHN equivalent biquad using UVCs. This circuit is shown in Fig. 12.41.

$$\begin{bmatrix} i_X \\ v_{YP} \\ v_{YN} \\ v_{ZP} \\ v_{Zn} \\ i_W \end{bmatrix} = \begin{bmatrix} 0 & 1 & -1 & 0 & 0 & 0 \\ 0 & 0 & 0 & 0 & 0 & 1 \\ 0 & 0 & 0 & 0 & 0 & 1 \\ 1 & 0 & 0 & 0 & 0 & 0 \\ -1 & 0 & 0 & 0 & 0 & 0 \\ 0 & 0 & 0 & 0 & 0 & 0 \end{bmatrix} \begin{bmatrix} v_X \\ i_{YP} \\ i_{YN} \\ i_{ZP} \\ i_{ZN} \\ v_W \end{bmatrix} \quad (12.142)$$

The presented configuration in Fig. 12.41 uses three UVCs, two grounded capacitors and five resistors.

Assuming ideal UVCs, a routine circuit analysis of Fig. 12.41 gives the following VM transfer functions:

$$\frac{V_{HP}}{V_{IN}} = \frac{s^2 C_1 C_2 \left(\frac{1}{R_1} + \frac{1}{R_4}\right)}{D(s)} \quad (12.143))$$

$$\frac{V_{BP}}{V_{IN}} = -\frac{sC_1 \left(\frac{1}{R_1} + \frac{1}{R_4}\right)\left(\frac{1}{R_2}\right)}{D(s)} \quad (12.144)$$

$$\frac{V_{LP}}{V_{IN}} = -\frac{\left(\frac{1}{R_1} + \frac{1}{R_4}\right)\left(\frac{1}{R_2 R_3}\right)}{D(s)} \quad (12.145)$$

where

$$\mathrm{D(s)} = s^2 C_1 C_2 \left(\frac{1}{R_5}\right) + sC_1 \left(\frac{1}{R_1 R_2}\right) + \frac{1}{R_2 R_3 R_4} \quad (12.146)$$

The expressions of ω_0 and Q_0 can be given by:

$$\omega_0 = \sqrt{\frac{R_5}{C_1 C_2 R_4 R_2 R_3}} \text{and} \, Q_0 = R_1 \sqrt{\frac{C_2 R_2}{C_1 R_3 R_4 R_5)}} \quad (12.147)$$

From the above equation (12.147), Q_0 can be varied independently by R_1, if $R_2 = R_3$ and ω_0 can be controlled without altering Q_0 by varying R_2 and R_3 simultaneously and equally. The workability of this circuit has been experimentally verified using the UVC-NIC 0520 IC which has been produced in 0.35 μm technology by Brno University of technology in cooperation with the AMI semiconductor, Czech, Ltd. now known as ON semiconductor, Czech Republic, Ltd).

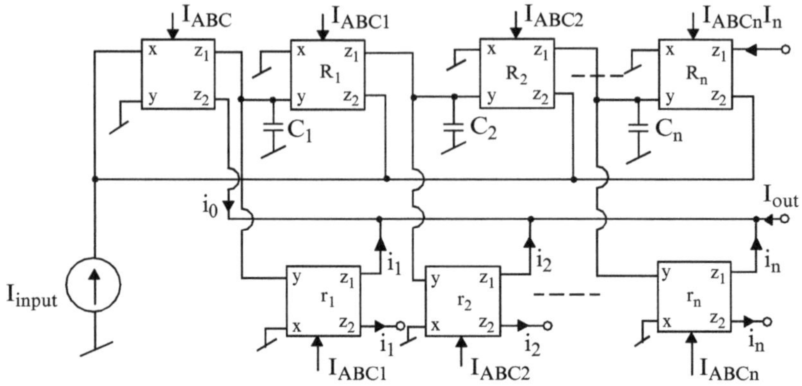

Fig. 12.42 Realization of the nth-order CM transfer function proposed by Abuelma'atti and Tasadduq [119]

12.2.9 Filter Design Using CFCCII

It has been conceptualized by some authors that if an additional current follower is included in the CCII+ thereby making it a current follower CCII (CFCCII) then such a modification becomes not only an useful artifice for synthesizing CM higher order filters but can also lead to interesting CM universal biquads . A plus-type current follower second generation CC (CFCCII$_p$) is characterized by the following hybrid matrix:

$$\begin{bmatrix} I_y \\ V_x \\ I_z \\ I_0 \end{bmatrix} = \begin{bmatrix} 0 & 0 & 0 & 0 \\ 1 & 0 & 0 & 0 \\ 0 & 1 & 0 & 0 \\ 0 & 1 & 0 & 0 \end{bmatrix} \begin{bmatrix} V_y \\ I_x \\ V_z \\ V_0 \end{bmatrix} \tag{12.148}$$

Figure 12.42 shows a translinear CFCCII$_p$-based realization of nth-order current transfer function using only grounded capacitors. The presented circuit configuration is capable of supporting (n + 2) high impedance output currents simultaneously. By appropriate connection of output currents, the circuit has the ability of supporting a wide range of current transfer functions.

Abuelma'atti-Tasadduq nth order transfer function realization using CFCCII The method can be explained by considering a general form of nth order transfer function as:

$$\frac{I_{out}}{I_{input}} = \frac{a_0 + a_1 s + a_2 s^2 + \ldots . + a_{n-1} s^{n-1} + a_n s^n}{b_0 + b_1 s + b_2 s^2 + \ldots . + b_{n-1} s^{n-1} + b_n s_n} \tag{12.149}$$

As per the methodology of [119], the active realization of this function is shown in Fig. 12.42.

The various design equations are given by:

$$a_0 = \frac{R_n}{r_n}; \frac{R_{n-1}}{r_{n-1}} = \frac{a_1}{b_1}; \frac{R_{n-2}}{r_{n-2}} = \frac{a_2}{b_2}; \ldots..; \frac{a_{n-1}}{b_{n-1}} = \frac{R_1}{r_1}; \frac{a_n}{b_n} = 1 \quad (12.150)$$

$$b_0 = 1; b_1 = C_n R_n \frac{b_2}{b_1} = C_{n-1}R_{n-1}; \ldots.; \frac{b_{n-1}}{b_{n-2}} = C_2R_2; \frac{b_n}{b_{n-1}} = C_1R_1 \quad (12.151)$$

where $R_n = \frac{2I_{ABCn}}{V_T}, r_n = \frac{2I_{abcn}}{V_T}$ are the resistances on port-X of nth translinear $CFCCII_p$, I_{ABCn} and I_{abcn} are the auxiliary bias currents of the nth translinear $CFCCII_p$ and V_T is the thermal voltage. An interesting feature of this structure is that the coefficients of numerator and denominator of nth order transfer function are independently controllable of each other.

12.2.10 Filter Design Using OFCC

Like a CCII and a CFOA, an operational floating CC (OFCC) is yet another building block which was proposed by Toumazou, Payne and Lidgey with the objective of creating alternative building blocks which overcome the well-known bandwidth conflict of the traditional VM op-amp. It has been demonstrated in an earlier chapter of this monograph that OFCCs indeed make it possible to realize all the basic controlled sources with a minimum number of active and passive elements without requiring any component matching and within the assured property of maintaining same constant bandwidth even under the variability of the gain.

An ideal OFCC can be characterized by the following matrix:

$$\begin{bmatrix} i_y \\ v_x \\ v_w \\ i_{z+} \\ i_{z-} \end{bmatrix} = \begin{bmatrix} 0 & 0 & 0 & 0 & 0 \\ 1 & 0 & 0 & 0 & 0 \\ 0 & Z_t & 0 & 0 & 0 \\ 0 & 0 & 1 & 0 & 0 \\ 0 & 0 & -1 & 0 & 0 \end{bmatrix} \begin{bmatrix} v_y \\ i_x \\ i_w \\ v_{z^+} \\ v_{z^-} \end{bmatrix} \quad (12.152)$$

where Z_t is the impedance between X and W ports.

In this section, we outline another application of the OFCCs in realizing a universal biquad employing OFCCs.

Ghallab, Badawy, El-Ela and El-Said biquad Figure 12.43 shows a universal MISO-type VM fitter configuration proposed by Ghallab and Badawy in [122] using two OFCCs, two capacitors and five resistors.

Assuming ideal OFCCs, a straight forward circuit analysis of Fig. 12.43 yields the following relation between the inputs and the output of the circuit:

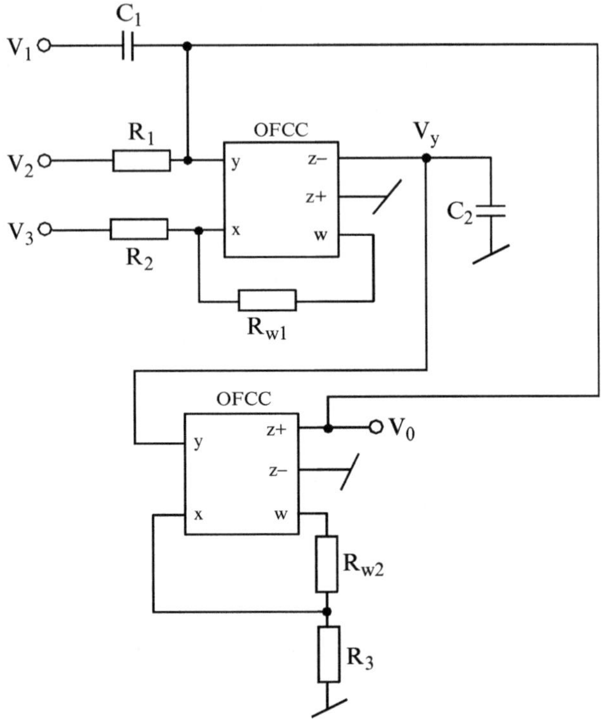

Fig. 12.43 VM universal biquad proposed by Ghallab et al. [122]

$$V_0 = \frac{s^2 V_1 + s\left(\frac{1}{C_1 R_1}\right) V_2 + \left(\frac{1}{C_1 C_2 R_2 R_3}\right) V_3}{s^2 + s\left(\frac{1}{C_1 R_1}\right) + \frac{1}{C_1 C_2 R_2 R_3}} \tag{12.153}$$

From the above equation, all the second-order standard filter responses can be obtained subject to the following conditions:

HPF: if $V_2 = V_3 = 0$ (grounded) and $V_1 = V_{in}$ (input signal); BPF: if $V_1 = V_3 = 0$ and $V_2 = V_{in}$; LPF: if $V_1 = V_2 = 0$ and $V_3 = V_{in}$; Notch: if $V_2 = 0$ and $V_1 = V_3 = V_{in}$ and APF: if $V_1 = -V_2 = V_3 = V_{in}$

The filter parameters ω_0 and Q_0 can be obtained as:

$$\omega_0 = \sqrt{\frac{1}{C_1 C_2 R_2 R_3}} \text{and} Q_0 = R_1 \sqrt{\frac{C_1}{C_2 R_3 R_2)}} \tag{12.154}$$

From the above equation (12.154), it is clear that ω_0 can be varied by R_2 or R_3 and Q_0 can be tuned by the ratio R_1/R_2, R_1/R_3 or C_1/C_2. Hence, both ω_0 and Q_0 are orthogonally controllable.

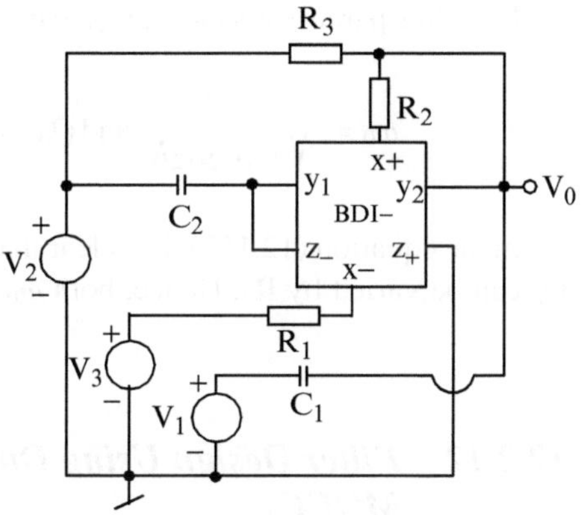

Fig. 12.44 Universal biquad filter configuration proposed by Chen [205]

12.2.11 *Filter Design Using Balanced-dual-input Dual-output-CC (BDI-DOCC)*

In this section, we present an interesting biquad filter configuration using yet another variety of CCs called balanced dual input-dual output CC (BDI-DOCC) which is shown here in Fig. 12.44.

A BDI-DOCC is characterized by the following terminal equations:

$$V_{X^+} = V_{Y1}, V_{X^-} = V_{Y2}, I_{Z^+} = I_{X^+} \text{ and } I_{Z^-} = -I_{X^-} \tag{12.155}$$

Chen biquad using BDI-DOCC The VM MISO-type filter configuration shown in Fig. 12.44 employs one BDI-DOCC, two capacitors and three resistors. This circuit which was proposed by Chen in [205] offers (i) realization of all standard second-order filters from the same topology (ii) no requirement of matching condition(s) of components (iii) low active and passive sensitivities and (iv) orthogonal control of ω_0 and Q_0.

Assuming ideal BDI-DOCC, a routine circuit analysis yields the following VM transfer function:

$$V_0 = \frac{s^2 V_1 + \frac{s}{C_1}\left(\frac{1}{R_2} + \frac{1}{R_3}\right) V_2 + \left(\frac{1}{C_1 C_2 R_1 R_2}\right) V_3}{s^2 + \frac{s}{C_1}\left(\frac{1}{R_2} + \frac{1}{R_3}\right) + \frac{1}{C_1 C_2 R_2 R_1}} \tag{12.156}$$

Using the equation (12.156) all standard second-order filter responses can be found as follows:

HPF: if $V_2 = V_3 = 0$ (grounded) and $V_1 = V_{in}$ (input signal); BPF: if $V_1 = V_3 = 0$ and $V_2 = V_{in}$; LPF: if $V_1 = V_2 = 0$ and $V_3 = V_{in}$; Notch: if $V_2 = 0$ and $V_1 = V_3 = V_{in}$ and APF: if $V_1 = -V_2 = V_3 = V_{in}$.

The filter parameters ω_0 and Q_0 can be determined as:

$$\omega_0 = \sqrt{\frac{1}{C_1 C_2 R_2 R_1}} \text{and} Q_0 = \left(\frac{R_3}{R_2 + R_3}\right)\sqrt{\frac{C_1 R_2}{C_2 R_1)}} \tag{12.157}$$

From the equation (12.157), it is clear that ω_0 can be controlled by R_1 and /or R_2 and Q_0 can be varied by R_3. Hence, both ω_0 and Q_0 are orthogonally controllable.

12.2.12 Filter Design Using Dual/Multi Output CCs (DOCC/MOCC)

In this section of this chapter we include some universal filter configurations using dual output CCs (DOCC) and multiple-output CCs (MOCCs) from amongst a large number of circuits available in literature [124–194]. In fact, DOCCs/MOCCs have been two of the most prominent varieties of CCs which have been more popularly employed than others in the analog circuits' literature.

In the following therefore, we highlight a variety of universal biquads employing DOCCs/MOCCs.

Senani-Singh-Singh-Bhaskar universal CM biquad Figure 12.45 shows a set of tunable universal CM biquads realizable with only three multi-output-current-conveyors (MOCCs) and all grounded passive elements proposed by Senani et al. [130].

The realized structures provide tunability of all the three filter parameters namely, bandwidth, angular frequency and gain (H_0). When MOCCs are implemented as current-controlled-conveyors with multiple outputs, one of the new circuits becomes a novel electronically controllable universal CM biquad.

Assuming MOCCs to be characterized by $i_y = 0$, $v_x = v_y$, $i_z+ = i_x$ and $i_z^- = i_x^-$, it can be shown that the current transfer functions of all the configurations remain same and can be given by:

$$\frac{I_{01}}{I_{in}} = \frac{\left(\frac{r_2}{r_1}\right)}{D(s)}; \frac{I_{02}}{I_{in}} = \frac{-\left(\frac{r_2}{r_1}\right)(sC_2R_2)}{D(s)} \text{and} \frac{I_{03}}{I_{in}} = \frac{\left(\frac{r_2}{r_1}\right)s^2C_1C_2R_1R_2}{D(s)} \tag{12.158}$$

where

$$D(s) = s^2C_1C_2R_1R_2 + sC_2R_2 + 1 \tag{12.159}$$

The different filter parameters of the realized functions can be given by:

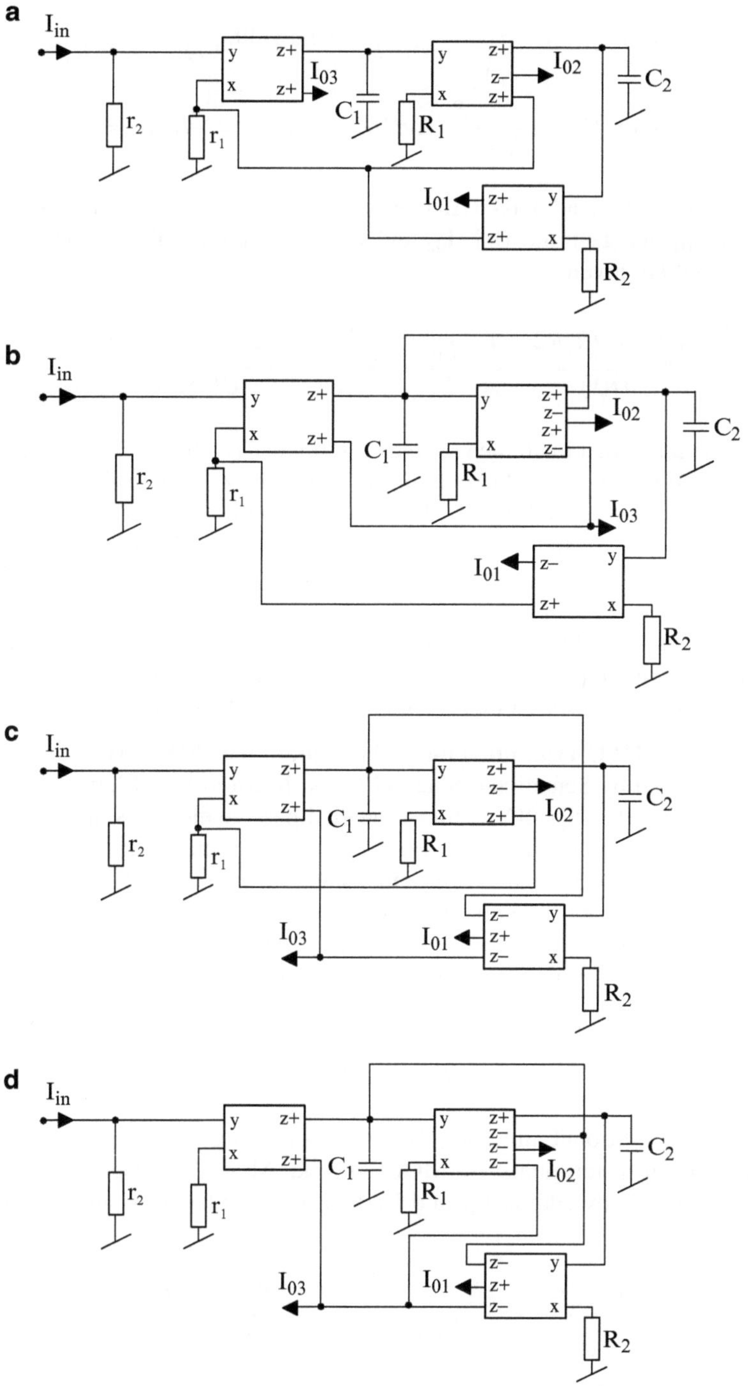

Fig. 12.45 SIMO type Current mode biquads proposed by Senani et al. [130]

$$\text{angular frequency}\,(\omega_0) = \sqrt{\frac{1}{C_1 C2 R_1 R_2}};\ \text{bandwidth}\left(\frac{\omega_0}{Q_0}\right) = \frac{1}{C_1 R_1}; \qquad (12.160)$$

$$\text{quality factor}\,(Q_0) = \sqrt{\frac{C_1 R_1}{C_2 R_2}}\ \text{and}\, H_{0LP} = H_{0HP} = -H_{0BP} = \frac{r_2}{r_1} \qquad (12.161)$$

Band stop (BS) and all pass (AP) responses can be realized (in all cases) by summing (I_{01} and I_{03}) and (I_{01}, I_{02} and I_{03}) respectively. The transfer functions thus obtained are given by:

$$\frac{I_{04}}{I_{in}} = \frac{\left(\frac{r_2}{r_1}\right)\left(s^2 C_1 C2 R_1 R2 + 1\right)}{D(s)};\ \frac{I_{05}}{I_{in}} = \frac{\left(\frac{r_2}{r_1}\right)\left[s^2 C_1 C2 R_1 R2 - sC_2 R_2 + 1\right]}{D(s)} \qquad (12.162)$$

This is to be noted that in all the four biquads, no realization constraints are required in BS and AP realizations. Also, in all the responses Z_{in} is resistive (equal to r_2), however, by using an additional MOCC as a current follower (CF) with virtual ground at its input, all four filter configurations can be modified to yield ideal zero input resistance. The configuration of Fig. 6.25d is convertible into an electronically controllable biquad by connecting the x-terminal of all the three MOCCs to ground thereby eliminating (due to being short circuited) r_1, R_1 and R_2 and using in their places the parasitic input resistances of the three MOCCs namely, R_{x1}, R_{x2} and R_{x3}. The details of the resulting circuit would be presented subsequently in Chap. 10.

Senani-Singh SIMO-type universal CM biquad The SIMO-type universal CM biquad presented by Senani and Singh [125] is shown in Fig. 12.46. The proposed circuit uses only three MOOCs, four grounded passive elements and is capable of providing all five standard filter functions without requiring any component-matching conditions or change of terminals/active elements. The filter parameters ω_0 and bandwidth are orthogonally tunable.

Assuming MOCCs are to be defined by $i_y = 0$, $v_x = v_y$, $i_z+ = {}^{i}{}_x{}^{+}$ and $i_z{}^{-} = i_x{}^{-}$, a routine analysis of the circuit yields the following transfer functions:

$$\frac{I_{LP}}{I_{in}} = \frac{1}{D(s)};\ \frac{I_{BP}}{I_{in}} = \frac{-s(C_1 R_2)}{D(s)};\ \frac{I_{HP}}{I_{in}} = \frac{s^2 C_1 R_1 C_2 R_2}{D(s)} \qquad (12.163)$$

Furthermore, additional inverting HPF and inverting LPF current outputs are realizable from Z-terminals of MOCC-1 and MOCC-3 respectively. Finally, notch is realizable by adding I_{HP} and I_{LP} and AP is realized by summing I_{BP} also, thereby yielding

$$\frac{I_{notch}}{I_{in}} = \frac{s^2 C_1 R_1 C_2 R_2 + 1}{D(s)};\ \frac{I_{AP}}{I_{in}} = \frac{s^2 C_1 R_1 C_2 R_2 - sC_1 R_2 + 1}{D(s)} \qquad (12.164)$$

where

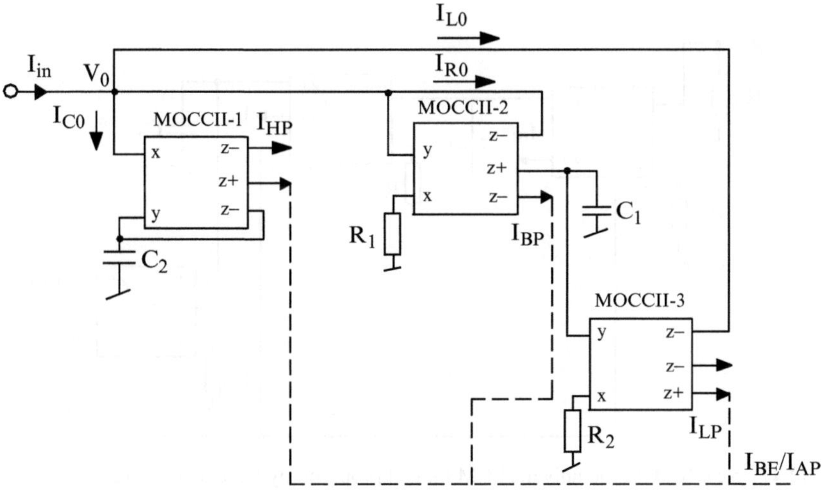

Fig. 12.46 The new universal SIMO-type current-mode biquad proposed by Senani and Singh [125]

$$D(s) = s^2C_1R_1C_2R_2 + sC_1R_2 + 1 \tag{12.165}$$

Senani-Singh-Singh biquad A minimal realization of SIMO-type universal CM biquad employing three MOCCs and six grounded passive elements which offers tunability of all the three filter parameters namely, H_0, ω_0 and bandwidth while retaining all other advantages of its predecessor has been presented by Senani et al. [126] as shown in Fig. 12.47.

Assuming MOCCs are to be characterized by by $i_y = 0$, $v_x = v_y$, $i_z + = i_x$ and $i_z^- = i_x^-$ the three current transfer functions realized by the circuit are given by:

$$\frac{I_{01}}{I_{in}} = \frac{\left(\frac{R_3}{R_4}\right)s^2C_1R_1C_2R_2}{D(s)};\frac{I_{02}}{I_{in}} = -\frac{\left(\frac{R_3}{R_4}\right)sC_2R_2}{D(s)};\frac{I_{03}}{I_{in}} = \frac{\left(\frac{R_3}{R_4}\right)}{D(s)} \tag{12.166}$$

A notch response is realized by summing I_{01} and I_{03} ($I_{04} = I_{01} + I_{03}$) and an APF can be realized by summing I_{02} also ($I_{05} = I_{01} + I_{02} + I_{03}$) thereby yielding:

$$\frac{I_{04}}{I_{in}} = \frac{\left(\frac{R_3}{R_4}\right)(s^2C_1R_1C_2R_2 + 1)}{D(s)} \text{ and } \frac{I_{05}}{I_{in}} = \frac{\left(\frac{R_3}{R_4}\right)\{(s^2C_1R_1C_2R_2 - sC_2R_2 + 1)\}}{D(s)} \tag{12.167}$$

where

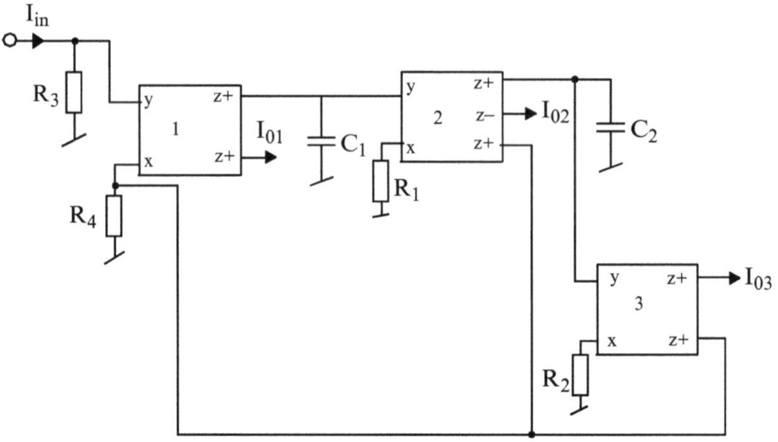

Fig. 12.47 New SIMO-type universal CM biquad proposed by Senani et al. [126]

$$D(s) = s^2C_1R_1C_2R_2 + sC_2R_2 + 1 \tag{12.168}$$

The various filter parameters of the realized filter can be given by:

$$\omega_0 = \sqrt{\frac{1}{C_1C_2R_1R_2}}; \mathrm{BW} = \frac{1}{C_1R_1}; \mathrm{H}_0 = \frac{R_3}{R_4} \tag{12.169}$$

Thus, it is seen that all the three filter parameters are tunable through separate grounded resistors. The workability of all filter responses in Fig. 12.47 was confirmed by PSPICE simulations for $f_0 = 1$ MHz, $Q_0 = 0.707$ and $H_0 = 10$. The component values used were $R_1 = R_2 = 7$ kΩ, $R_3 = 10$ kΩ, $R_4 = 1$ kΩ, $C_1 = 0.0159$ nF and $C_2 = 0.0319$ nF.

Abdalla-Bhaskar-Senani biquad A configuration for realising a CM universal filter and dual-mode quadrature single resistor controlled oscillator was proposed by Abdalla et al. in [127] which employs three dual output CCs (DOCCII) with five grounded passive elements is shown in Fig. 12.48. The circuit is capable to realize all the five basic filter functions without requiring any component matching alongwith tunability of filter parameters through grounded resistors.

With appropriate modification, the structure can also be converted into a quadrature oscillator which can give two VM outputs and as well as two CM outputs in quadrature. The two modes of opration for Fig. 12.48 are:

(i) *CM SIMO-type filter*: The various filter functions realized by the configuration are given as:

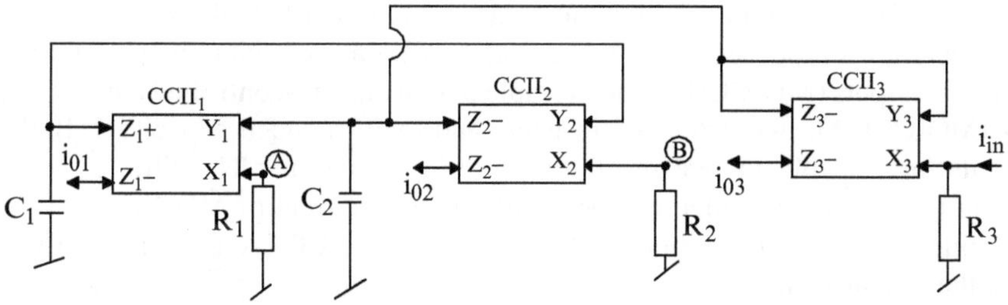

Fig. 12.48 The configuration for realizing CM universal biquad and dual-mode Quadrature oscillator proposed by Abdalla et al. [127]

$$\frac{i_{01}}{i_{in}} = \frac{s\left(\frac{1}{R_1 C\ 2}\right)}{D(s)};\ \frac{i_{02}}{i_{in}} = \frac{\frac{1}{C_1 C_2 R_2 R_1}}{D(s)};\ \frac{i_{03}}{i_{in}} = -\frac{\left(s^2 + \frac{1}{C_1 C_2 R_2 R_1}\right)}{D(s)} \tag{12.170}$$

$$\frac{(i_{02} + i_{03})}{i_{in}} = \frac{-s^2}{D(s)};\ \frac{(i_{01} + i_{03})}{i_{in}} = -\frac{\left\{s^2 - s\left(\frac{1}{C_2 R_1}\right) + \left(\frac{1}{C_1 C_2 R_2 R_1}\right\}\right.}{D(s)} \tag{12.171}$$

where

$$D(s) = s^2 + s\frac{1}{C_2 R_3} + \frac{1}{C_1 C_2 R_2 R_1} \tag{12.172}$$

(ii) *CM/VM quadrature single resistance controlled oscillator (SRCO):*

The circuit shown in Fig. 12.48 can be modified by making = 0 and getting another Z_1+ output terminal from $CCII_1$ and connecting the same to node (C). Thus, the output voltages in VM are considered to be the voltage at node (B) equal to v_{01} and voltage at node (A) equal to V_{02} whereas the output currents in CM are considered to be (i_{01}) and (i_{02}). A routine circuit analysis of this structure yields the following characteristic equation (CE) (for both CM and VM quadrature SRCOs)

$$\text{CE} : s^2 + s\left\{\frac{1}{C_2}\left(\frac{1}{R_3} - \frac{1}{R_1}\right)\right\} + \frac{1}{C_1 C_2 R_2 R_1} = 0 \tag{12.173}$$

From the above CE, the condition of oscillation and frequency of oscillation are found to be:

$$\text{CO} : R_1 \leq R_3 \text{ and FO} : \frac{1}{2\pi}\sqrt{\frac{1}{C_1 R1 C_2 R_2}} \tag{12.174}$$

Therefore, it is seen that both CO and FO are independently controllable by two different resistors the former by R_1 and the latter by R_2.

Chunhua-Haiguang-Yan Universal biquads Figure 12.49 shows two universal filtering circuits proposed by Chunhua et al. [128]. Each circuit uses two CCIIs with multiple outputs (MO-CCIIs), one multiple output current-controlled current amplifier (MO-CCCA), two grounded resistors and two grounded capacitors. Both the circuits are capable to realize all five filtering functions. The filter parameters ω_0 and Q_0 can be controlled independently by bias current of MO-CCCA.

Using the ideal port relations of MO-CCII and MO-CCCA, a routine analysis of both the circuit yields:

$$I_0 = \frac{I_1 - sI_2R_1C_1K - s^2I_3C_1C_2R_1R_2}{1 + sC_1R_1K + s^2C_1C_2R_1R_2} \quad \text{for Fig. 12.49a} \tag{12.175}$$

and

$$I_0 = \frac{K(-I_1 - sI_2R_1C_1 + s^2I_3C_1C_2R_1R_2)}{1 + sC_1R_1K + s^2C_1C_2R_1R_2} \quad \text{for Fig. 12.49b} \tag{12.176}$$

where K is the current gain and is given by

$$K = \left(\frac{I_B}{2I_A}\right)I_i \tag{12.177}$$

Both the circuits of Fig 12.49 were simulated in SPICE using the CMOS MOCC implementation devised by Surakampontorn et al. [206]. The component values were chosen to attain $f_o = 159$ KHz, $Q_o = 1$. The MSOFET model parameters were taken from MOSIS based on 0.5 μm CMOS technology. Figure 12.49c shows the variability of the f_o with respect to the resistors in the circuit, which confirm the workability of the circuit.

Senani's MISO-type CM biquad A universal MISO CM biquad employing all grounded passive elements and only two dual output CCs (DOCCs) was presented by Senani [131]. The proposed configuration is shown in Fig. 12.50 which realizes all the basic generic filter responses providing explicit current outputs. Electronic tunability of the filter parameters can also be accomplished by replacing the resistors R_1, R_2 and R_3 by appropriate CMOS grounded resistors. The port characteristics of DOCC are given by $i_y = 0$, $v_x = v_y$, $i_{z+} = +i_x$ and $i_{z-} = -i_x$, a routine circuit analysis gives an expression for I_0 in terms of input currents:

$$I_0 = \frac{I_2\left(s^2 + \frac{s}{R_1C_2}\right) - s\left(\frac{1}{C_2R_2}\right)I_1 + \left(\frac{1}{C_1R_3C_2R_2}\right)I_3}{s^2 + s\left(\frac{1}{R_1C_2}\right) + \frac{1}{C_1C_2R_3R_2}} \tag{12.178}$$

The various filter functions can be realized from equation (12.183) as follows:

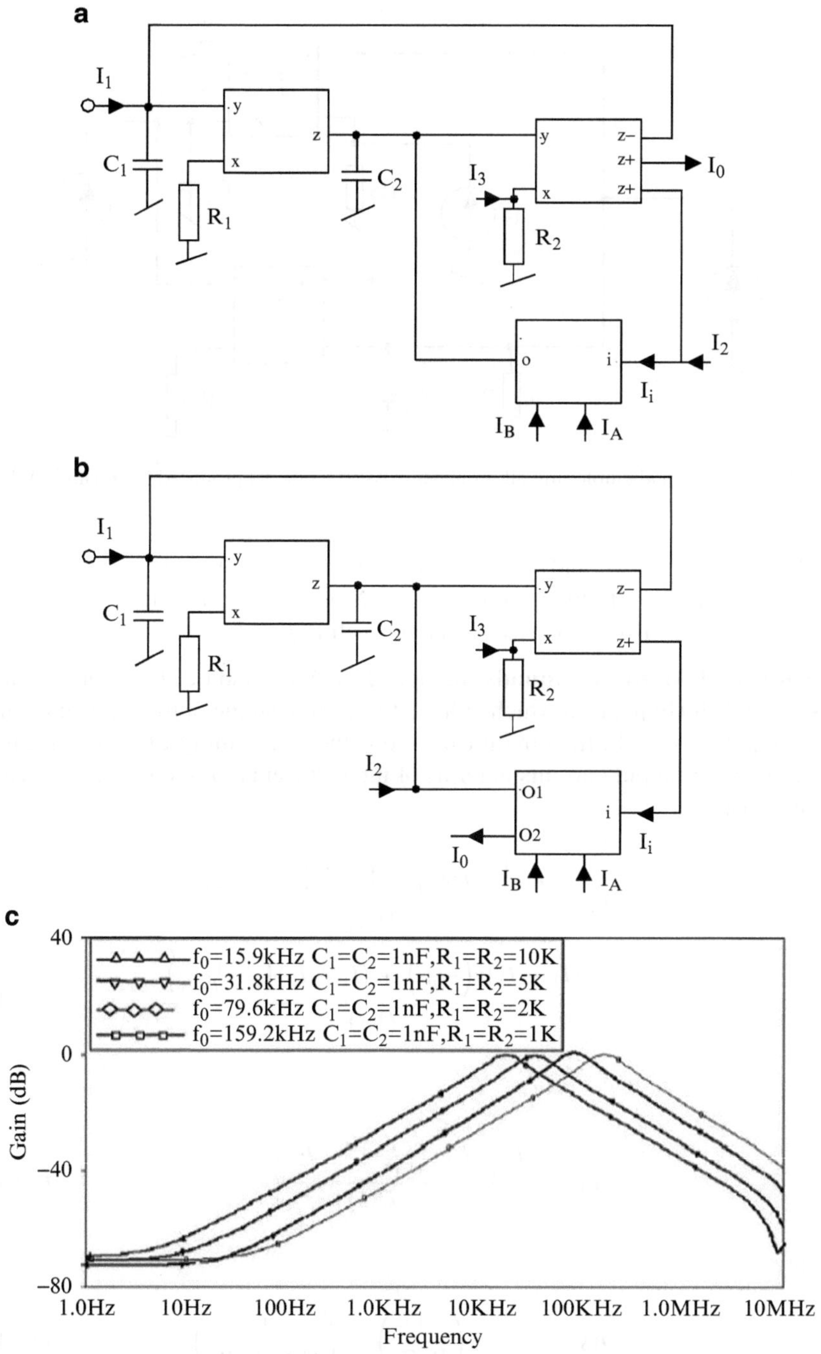

Fig. 12.49 Universal MISO-type CM biquads proposed by Chunhua et al. (Adapted from [128] © 2008 Elsevier GmbH)

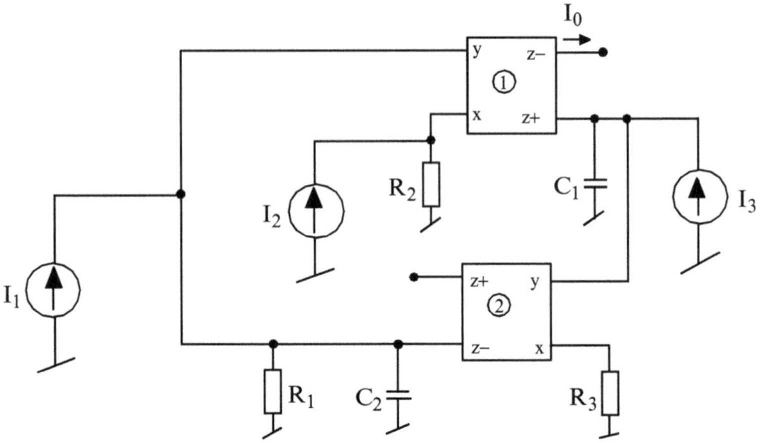

Fig. 12.50 CM biquad employing all grounded passive elements proposed bySenani [131]

LPF: if $I_1 = I_2 = 0$ and $I_3 = I_{in}$; BPF: if $I_3 = I_2 = 0$ and $I_1 = I_{in}$; HPF: if $I_1 = I_2 = I_{in}$ and $I_3 = 0$ along with $R_1 = R_2$; Notch: if $I_1 = I_2 = I_{in} = I_3$ along with $R_1 = R_2$ and APF: if $I_1 = I_2 = I_{in} = I_3$ along with $R_1/2 = R_2$.

Wang- Lee MIMO-type biquads In Fig. 12.51 Wang and Lee [133] proposed two versatile MIMO biquads in which either a three input single output or a single input three output universal filter circuit can be obtained. Assuming CCs to be ideal, the expressions for output currents in terms of input currents for the Fig. 12.51a can be obtained as:

$$I_{01} = \frac{I_3 s^2 - I_2 s\left(\frac{1}{C_2 R_2}\right) + I_1\left(\frac{1}{C_1 C_2 R_1 R_2}\right)}{D(s)} \tag{12.179}$$

$$I_{02} = \frac{(I_1 - I_3)s\left(\frac{1}{C_1 R_1}\right) + (I_1 + I_2)\left(\frac{1}{C_1 C_2 R_1 R_2}\right)}{D(s)} \tag{12.180}$$

and

$$I_{03} = \frac{(I_3 - I_1)s\left(\frac{1}{C_1 R_1}\right) + (I_2 + I_3)s^2}{D(s)} \tag{12.181}$$

where

$$D(s) = s^2 + s\left(\frac{1}{R_2 C_2}\right) + \left(\frac{1}{C_1 C_2 R_1 R_2}\right) \tag{12.182}$$

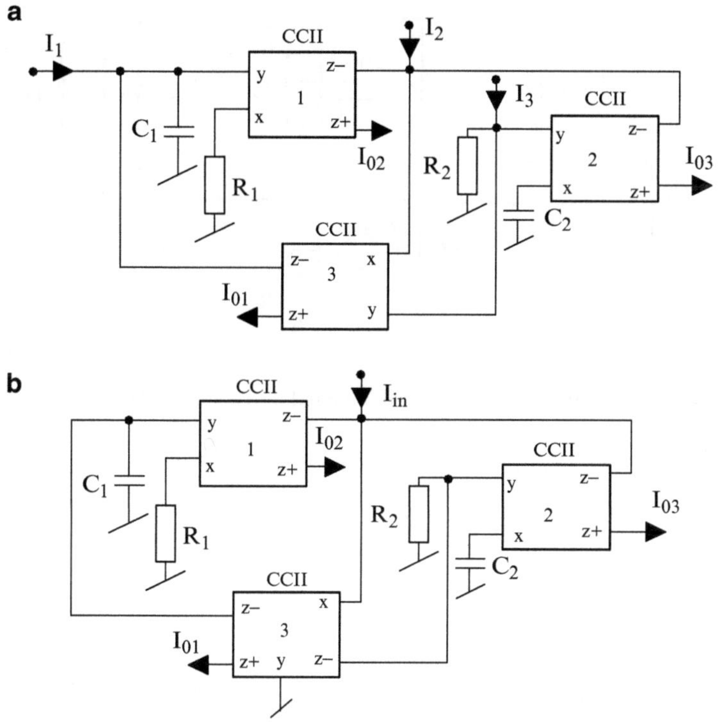

Fig. 12.51 CM Biquads using CCs proposed by Wang and Lee [133] (**a**) MIMO-type (**b**) SIMO-type

Now if I_{01} as the output current, then the various filter responses realized are given by:(i) HPF: $I_1 = I_2 = 0$ and $I_3 = Iin$; BPF: $I_1 = I_3 = 0$ and $I_2 = Iin$ and LPF: $I_3 = I_2 = 0$ and $I_1 = Iin$. Figure 12.51a can also function as a single input three output universal biquad filter if $I_1 = I_3 = 0$ and $I_2 = I_{in}$. The three current transfer functions can be given by:

$$\frac{I_{01}}{I_{in}} = \frac{-s\left(\frac{1}{C_2R_2}\right)}{D(s)};\frac{I_{02}}{I_{in}} = \frac{\left(\frac{1}{C_1R_1C_2R_2}\right)}{D(s)}\text{ and}\frac{I_{03}}{I_{in}} = \frac{s^2}{D(s)} \tag{12.183}$$

Therefore, the configuration yields BPF at I_{01}, LPF at I_{02} and HPF at I_{03} simultaneously. The Notch and APF can easily be obtained by combining appropriate currents.

Figure 12.51b shows the single input three output filter configuration. The structure uses multiple output CCII in place of CCIII. Using standard ideal port relations of the CCs, the three filter transfer functions can be obtained as:

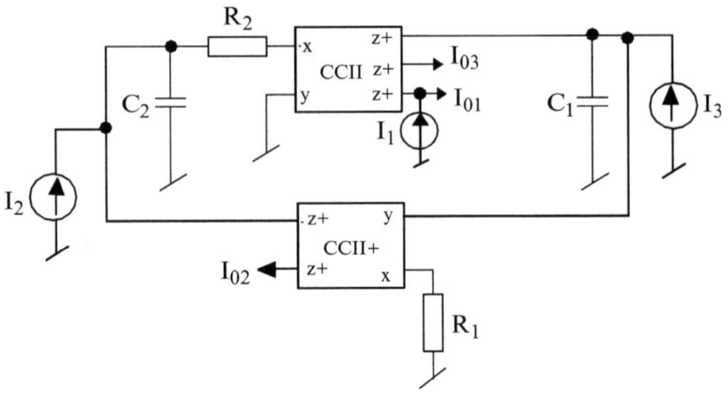

Fig. 12.52 Universal CM CCII-based biquad proposed by Tu et al. [132]

$$\frac{I_{01}}{I_{in}} = \frac{-s\left(\frac{1}{C_2R_2}\right)}{D(s)};\ \frac{I_{02}}{I_{in}} = \frac{\left(\frac{1}{C_1R_1C_2R_2}\right)}{D(s)}\ \text{and}\ \frac{I_{03}}{I_{in}} = \frac{s^2}{D(s)} \tag{12.184}$$

where $D(s) = s^2 + s\left(\frac{1}{R_2C_2}\right) + \left(\frac{1}{C_1C_2R_1R_2}\right)$

remains same as in equation (12.182). Thus, BPF, LPF and HPF responses can be obtained at I_{01}, I_{02} and I_{03} respectively. Notch and APF can also be realized by appropriate connection (s) of the currents I_{01}, I_{02} and I_{03}. Since resistors are grounded the electronic tunability of the filter parameters can be obtained easily by replacing grounded resistor(s) by MOSFET(s).

Tu-Chang-Liao biquad Figure 12.52 shows a versatile insensitive CM three-input and two-output universal biquad proposed by Tu et al. in [132] employing one multiple output CCII+, one dual output CCII+ and four passive elements. The circuit gives simultaneous realization of five generic CM responses without any matching condition(s).

A routine circuit analysis of Fig. 12.52 gives the following CM transfer functions for $I_1 = I_2 = I_{in}$ and $I_3 = 0$

$$\frac{I_{01}}{I_{in}} = \frac{\left(s^2 + \frac{1}{C_1C_2R_1R_2}\right)}{D(s)};\ \frac{I_{02}}{I_{in}} = \frac{-\left(\frac{1}{C_1C_2R_1R_2}\right)}{D(s)};\ \frac{I_{03}}{I_{in}} = \frac{-s\left(\frac{1}{C_2R_2}\right)}{D(s)} \tag{12.185}$$

and a non-inverting HPF can be obtained as $I_{HPF} = (I_{01} + I_{02})$. An APF can be realized by combining I_{01} and I_{03} i.e., $I_{APF} = (I_{01} + I_{03})$. Thus, in realizing all the above functions no cancellation constraints have been used. The expression for D (s) is given by

$$D(s) = s^2 + s\left(\frac{1}{R_2C_2}\right) + \left(\frac{1}{C_1C_2R_1R_2}\right) \tag{12.186}$$

On the other hand, I_{01} and I_{03} in terms of input currents have been obtained as:

$$I_{01} = \frac{I_1\left\{s^2 + s\left(\frac{1}{C_2R_2}\right) + \left(\frac{1}{C_1C_2R_1R_2}\right)\right\} - I_2s\left(\frac{1}{C_2R_2}\right) - I_3\left(\frac{1}{C_1C_2R_1R_2}\right)}{D(s)} \tag{12.187}$$

$$I_{03} = \frac{-I_2s\left(\frac{1}{C_2R_2}\right) - I_3\left(\frac{1}{C_1C_2R_1R_2}\right)}{D(s)} \tag{12.188}$$

From the above equations, the following five generic filter responses can be derived: LPF: $I_1 = I_2 = 0$, $I_3 = I_{in}$ and $I_{out} = I_{01} = I_{03}$; BPF: $I_1 = I_3 = 0$, $I_2 = I_{in}$ and $I_{out} = I_{01} = I_{03}$; $I_1 = I_2 = 0$, HPF: $I_1 = I_2 = I_3 = I_{in}$ and $I_{out} = I_{01}$; Notch: $I_2 = 0$, $I_1 = I_3 = I_{in}$ and $I_{out} = I_{01}$; and APF: $I_1 = I_2 = I_{in}$, $I_3 = 0$ and $I_{out} = I_{01} + I_{03}$. The expression for D(s) remains same as equation (12.190).

Abuelma'atti-Bentrcia mixed-mode biquad The biquads described so far attempt to realize all the five basic functions either in VM (where both input and output variables are voltages) or in CM (where the operating variables are currents). Apart from VM and CM biquads a number of authors have also described universal biquads in which input variable is current but output variable is voltage (the resulting circuits being called transimpedance type biquads) or circuits in which input is voltage but outputs are currents (the resulting circuits are called as trans admittance biquads). There have been several attempts in evolving universal biquads using CCs which realize all the basic five filtering functions in two or more of the four possible modes namely, VM, CM transimpedance mode (TIM), and trans admittance mode(TAM) such circuits are referred to as mixed mode universal biquads. In the following we describe number of prominent circuit configurations of this type.

Figure 12.53 shows a novel mixed-mode filter proposed by Abuelma'atti and Bentrcia [134] employing five CCIIs, seven resistors and two grounded capacitors. All standard second-order filter responses can be obtained in both CM and VM from the same topology. The bandwidth and angular frequency can be controlled electronically independently. Assuming ideal CCIIs, a routine circuit analysis gives:

$$\begin{aligned} V_{out} &= \left(\frac{C_1C_2R_1R_5R_6R_4R_7}{R_2}\right)\left(\frac{N_r(s)}{D_r(s)}\right) \text{ and} \\ I_{out} &= \left(-\frac{C_1C_2R_1R_5R_6R_4}{R_2}\right)\left(\frac{N_r(s)}{D_r(s)}\right) \end{aligned} \tag{12.189}$$

where

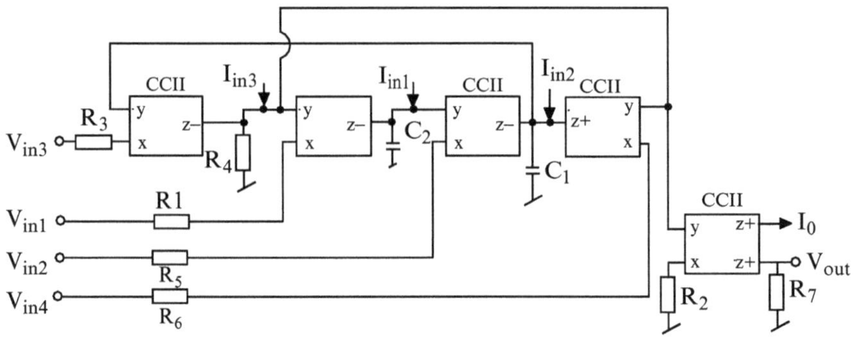

Fig. 12.53 Mixed-mode CCII-based configuration proposed by Abuelma'atti and Bentrcia [134]

$$N_r(s) = s^2 V_{in3} - s\left(\frac{V_{in4}}{C_1 R_6} + \frac{V_{in2}}{C_1 R_5}\right) + \frac{V_{in1}}{C_1 C_2 R_1 R_5} + s^2 R_3 I_{in3} - s\left(\frac{I_{in2}}{C_1}\right) + \frac{I_{in1}}{C_1 C_2 R_5} \tag{12.190}$$

and

$$D_r(s) = s^2 C_1 C_2 R_1 R_3 R_5 R_6 + s C_2 R_1 R_4 R_5 + R_4 R_6 \tag{12.191}$$

From equations (12.189) and (12.190), various inverting and non-inverting mixed-mode responses in both VM and CM can be obtained as:

(a) Realization of VM filter responses ($I_{in1} = I_{in2} = I_{in3} = 0$):

HPF: $V_{in1} = V_{in2} = V_{in4} = 0, V_{in3} = V_{in}$; LPF: $V_{in3} = V_{in2} = V_{in4} = 0, V_{in1} = V_{in}$
BPF: $V_{in1} = V_{in2} = V_{in3} = 0, V_{in4} = V_{in}$ or $V_{in1} = V_{in3} = V_{in4} = 0, V_{in2} = V_{in}$
Notch: $V_{in1} = V_{in3} = V_{in}$ and $V_{in2} = V_{in4} = 0$ with $R_1 = R_4$ and APF: $V_{in1} = V_{in3} = V_{in4} = V_{in}, V_{in2} = 0$ and $R_3 = R_4$; or $V_{in1} = V_{in2} = V_{in3} = V_{in}, V_{in4} = 0$ and $R_3 = R_4$, $R_5 = R_6$.

In the above cases, the polarities are referred to the VM outputs, the mixed mode outputs can be obtained with reverse polarities.

(b) Realization of CM filter responses ($V_{in1} = V_{in2} = V_{in3} = V_{in4} = 0$):

BPF: $I_{in1} = I_{in3} = 0, I_{in2} = I_{in}$; LPF: $I_{in2} = I_{in3} = 0, I_{in1} = I_{in}$ HPF: $I_{in1} = I_{in2} = 0$, $I_{in3} = I_{in}$; Notch: $I_{in2} = 0, I_{in3} = I_{in} = I_{in1}$ and APF: $I_{in2} = I_{in3} = I_{in} = I_{in1}$ and $R_1 = R_4 = R_6$.

By appropriate selection(s) of input(s), the other modes namely, transimpedance and transconductance filter responses can also be realized.

The filter parameters ω_0 and (ω_0/Q_0) can be given by:

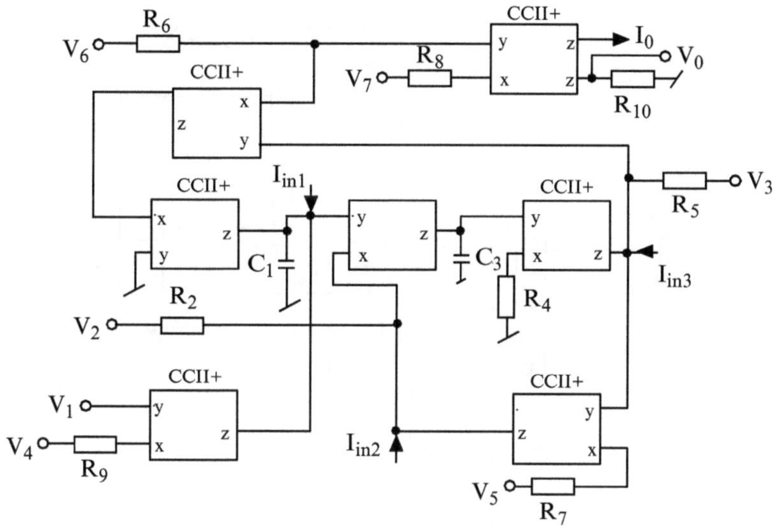

Fig. 12.54 Mixed-mode CCII-based filter proposed by Abuelma'atti et al. [135]

$$\omega_0 = \sqrt{\frac{R_4}{C_1C_2R_1R_3R_5}} \text{ and } \frac{\omega_0}{Q_0} = \frac{R_4}{C_1R_3R_6} \tag{12.192}$$

Hence, ω_0 can be tuned by adjusting the resistances R_1 and /or R_5 independently without disturbing the bandwidth which is controllable by R_6.

Abuelma'atti-Bentrcia-Al-Shahrani mixed-mode biquad Another mixed-mode CCII-based filter configuration was presented by Abuelma'atti et al. [135] as shown in Fig. 12.54.

This filter circuit is capable of realizing all the standard filter functions from the same topology. The filter parameters ω_0 and (ω_0/Q_0) are independently controllable. Assuming ideal CCIIs+, the analysis of the proposed filter yields the following transfer functions:

$$V_0 = \left(\frac{R_{10}}{R_8}\right)\left\{\frac{N_1(s)+N_2(s)}{D_r(s)}\right\} \text{ and } I_0 = \left(\frac{1}{R_8}\right)\left\{\frac{N_1(s)+N_2(s)}{D_r(s)}\right\} \tag{12.193}$$

$$\begin{aligned} N_1(s) = {} & s^2\frac{C_1C_3}{R_5}(V_3-V_7) + s\frac{C_1}{R_4}\left\{\frac{1}{R_7}(V_5-V_7) - \left(\frac{1}{R_2}\right)V_2\right\} \\ & + \frac{1}{R_2R_4}\left\{\frac{1}{R_6}(V_6-V_7) - \frac{1}{R_9}(V_4-V_1)\right\} \end{aligned} \tag{12.194}$$

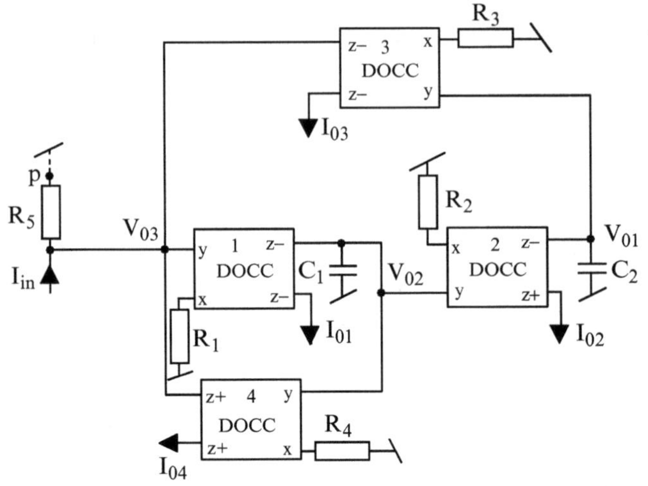

Fig. 12.55 Dual–mode biquad presented by Singh et al. [136]

$$N_2(s) = s^2 C_1 C_3 I_{in3} - s\left(\frac{C_1}{R_4}\right) I_{in2} + \frac{I_{in1}}{R_2 R_4}$$

$$D_r(s) = s^2 \frac{C_1 C_3}{R_5} + s\left(\frac{C_1}{R_4 R_7}\right) + \frac{1}{R_2 R_4 R_6} \tag{12.195}$$

From the above equations, the various inverting and non-inverting mixed-mode filter responses, with input voltage or current and output voltage or current, can be obtained.

Singh- Singh-Senani biquad The filter configuration presented in Fig. 12.55 uses four dual output CCs (DOCCs), five resistors and two grounded capacitors [136]. If the current source I_{in} deleted and V_{in} applied at P, the circuit is essentially the same VM biquad reported in [195], however, if P grounded and I_{in} applied, the same circuit yields the three output currents. Assuming ideal DOCCs, a routine circuit analysis gives the following current transfer functions:

$\frac{I_{01}}{I_{in}} = \left(\frac{R_5}{R_3}\right) \frac{\frac{1}{C_1 C_2 R_1 R_2}}{D(s)}$; $\frac{I_{02}}{I_{in}} = \left(\frac{R_5}{R_2}\right) \frac{\left(\frac{-s}{C_1 R_1}\right)}{D(s)}$ and $\frac{I_{03}}{I_{in}} = \left(\frac{R_5}{R_1}\right) \frac{s^2}{D(s)}$ where

$$D(s) = s^2 + s\left(\frac{R_5}{R_4 R_1 C_1}\right) + \frac{R_5}{R_1 R_2 R_3 C_1 C_2} \tag{12.196}$$

By adding I_{01} and I_{03} a notch filter is realized and adding I_{01}, I_{02} and I_{03} anAPF can be realized without any additional hardware requirement. The same circuit can also function as mixed-mode biquad.

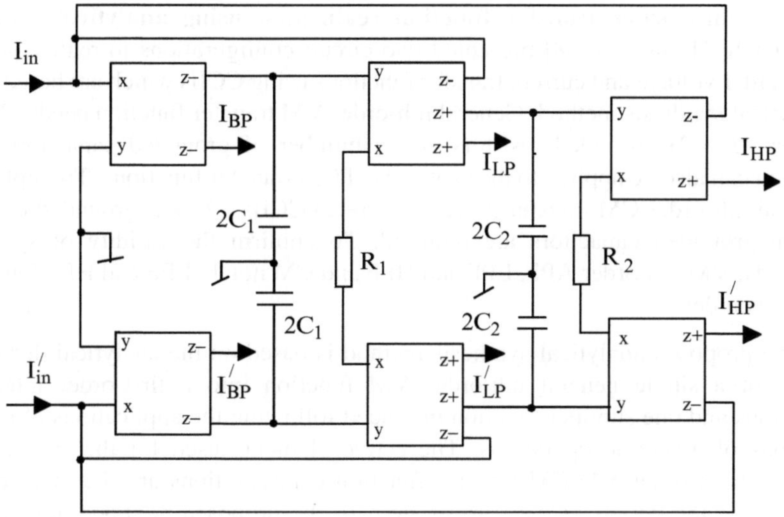

Fig. 12.56 Fully-differential CM filter proposed by Kumar [137]

Fully-differential biquad Fully differential filters offer the advantages of immunity from common mode noise signals, lower harmonic distortion and enhanced dynamic range etc. An exemplary fully-differential CM biquad [137] is shown in Fig. 12.56.

Assuming a multiple output CCII (MOCCII) is to be characterized by $i_y = 0$, $v_x = v_y$, $i_z+ = i_z- = i_x$, a straight forward circuit analysis of the circuit shown in Fig. 12.56 yields:

$$\frac{I_{BP}}{I_{in}} = H_0 \frac{-\left(\frac{\omega_0}{Q_0}\right)s}{D(s)}; \frac{I_{LP}}{I_{in}} = H_0 \frac{{\omega_0}^2}{D(s)}; \frac{I_{HP}}{I_{in}} = H_0 \frac{s^2}{D(s)} \tag{12.197}$$

where $D(s) = s^2 + s\left(\frac{1}{R_1C_1}\right) + \frac{1}{R_1R_2C_1C_2}$; $H_0 = 1$, ${\omega_0}^2 = \frac{1}{R_1R_2C_1C_2}$, and

$$\frac{\omega_0}{Q_0} = \frac{1}{R_1C_1} \tag{12.198}$$

From the above equations, it is clear that a notch filter is realizable by adding I_{HP} and I_{LP} and by combining I_{HP}, I_{BP} and I_{LP} an APF can be obtained without any component matching condition(s) except equal valued capacitors (which is inherent in all the earlier fully differential configurations as well). The filter parameters bandwidth and angular frequency are orthogonally controllable through R_1 and R_2 while the gain is unity in all the responses.

Horng's nth order transfer function realization using analytical synthesis approach Horng in [138] presented two circuit configurations to realize general high-order voltage and current transfer functions using CCIIs which are based on an analytical synthesis method. General nth-order VM transfer function needs (2n + 1) CCIIs, (2n + 2) grounded resistors and n number of grounded capacitors. This method can also be applied to the synthesis of CM transfer functions. To implement general nth-order CM transfer function, (2n + 2) CCIIs, (2n + 2) grounded resistors and n grounded capacitors are required. To confirm the validity of synthesis methods, VM 3rd order APF, LPF and HPF and CM notch, LPF and HPF functions have been derived.

The proposed analytical synthesis method is based on the analytical decomposition of a single general nth-order VM function into n first-order integrator functions and one summer function produced following the applications of manipulations of algebraic operations. The active elements used for the synthesis of general high-order VM/CM transfer function configurations are the multi-output CCIIs. For the details of the various steps of the proposed methodology and the resulting circuit implementations, the reader is referred to [138].

Concluding Remarks

In Chap. 6 we have presented a number of filter configurations using the basic CCs. By contrast, in this chapter we focused on numerous other filter configurations using different variants of CCs which however, are not available commercially as off-the-shelf ICs. From the spectrum of circuits presented in this chapter, it is, however, clear that the configurations using the considered variants of CCs do offer a number of advantageous features over those structures which are realized with conventional CCs.

In view of the above, therefore, these structures although not implementable in discrete designs (except at the cost of employing more number of commercially available IC CCs) but still are potential candidates for integrating analog filters along with digital circuits as part of a complete system on the chip. In addition, the material presented in this chapter together with that to be presented in Chaps. 13, 14 and 15, may also be a good motivation for IC designers to come up with stand-alone ICs of the modified CCs considered here and in the quoted chapters.

References

1. Hassan TM, Mahmoud SA (2010) New CMOS DVCC realization and applications to instrumentation amplifier and active-RC filters. Int J Electron Commun (AEU) 64:47–55
2. Yuce E (2009) Voltage-mode multifunction filters employing a single DVCC and grounded capacitors. IEEE Trans Instrum Meas 58:2216–2221
3. Minaei S, Yuce E (2010) Novel voltage-mode all-pass filter based on using DVCCs. Circ Syst Sig Process 29:391–402

4. Tangsrirat W, Channumsin O (2011) Voltage-mode multifunctional biquadratic filter using single DVCC and minimum number of passive elements. Ind Pure Appl Physics 49:703–707
5. Horng JW, Jhao ZY (2013) Voltage-mode universal biquadratic filter using single DVCC. ISRN Electron: Article ID 125746, 5p
6. Chen HP, Shen SS (2007) A versatile universal capacitor-grounded voltage-mode filter using DVCCs. ETRI J 29:470–476
7. Tsukutani T, Sumi Y, Yabuki N (2007) Novel current-mode biquadratic circuit using only plus type DO-DVCCs and grounded passive components. Int J Electron 94:1137–1146
8. Horng JW (2010) DVCCs based high input impedance voltage-mode first-order all pass, high pass, and low pass filters employing grounded capacitor and resistor. Radioengineering 19:653–656
9. Tsukutani T, Tsunetsugu H, Sumi Y, Yabuki N (2010) Electronically tunable first-order all-pass circuit employing DVCC and OTA. Int J Electron 97:285–293
10. Maheshwari S, Mohan J, Chauhan DS (2010) High input impedance voltage-mode universal filter and quadrature oscillator. J Circ Syst Comput 19:1597–1607
11. Khateb F, Khatib N, Koton J (2011) Novel low-voltage ultra-low-power DVCC based on floating-gate folded cascode OTA. Microelectron J 42:1010–1017
12. Horng JW, Hsu CH, Tseng CY (2012) High input impedance voltage-mode universal biquadratic filters with three inputs using three CCs and grounding capacitors. Radioengineering 21:290–296
13. Chen HP (2012) Tunable versatile current-mode universal filter based on plus-type DVCCs. Int J Electron Commun (AEU) 66:332–339
14. Soliman AM (2009) Generation and classification of Kerwin-Huelsman-Newcomb circuits using the DVCC. Int J Circ Theor Appl 37:835–855
15. Ibrahim MA, Minaei S, Kuntman H (2005) A 22.5 MHz current-mode KHN-biquad using differential voltage current conveyor and grounded passive elements. Int J Electron Commun (AEU) 59:311–318
16. Horng JW, Hou CL, Chang CM, Chou HP, Lin CT (2006) High input impedance voltage-mode universal biquadratic filter with one input and five outputs using current conveyors. Circ Syst Sig Process 25:767–777
17. Minaei S, Ibrahim MA (2009) A mixed-mode KHN-biquad using DVCC and grounded passive elements suitable for direct cascading. Int J Circ Theor Appl 37:793–810
18. Horng JW (2010) Lossless inductance simulation and voltage-mode universal biquadratic filter with one input and five outputs using DVCCs. Analog Integr Circ Sig Process 62:407–413
19. Cajka J, Vrba K (2000) Novel universal RC-active filters using DVCC elements. J Electr Eng 51:207–208
20. Cajka J, Dostal T, Vrba K (2002) High-order low pass filters using DVCC elements. Radioengineering 11:14–17
21. Horng JW, Hou CL, Chang CM, Chung WY (2006) Voltage-mode universal biquadratic filters with one input and five outputs. Analog Integr Circ Sig Process 47:73–83
22. Maheshwari S (2008) Current-mode filters with high output impedance and employing only grounded components. WSEAS Trans Electron 5:238–243
23. Maheshwari S (2008) High performance voltage-mode multifunction filter with minimum component count. WSEAS Trans Electron 5:244–249
24. Mahmoud SA (2008) Low voltage wide range CMOS differential voltage current conveyor and its applications. Contemp Eng Sci 1:105–126
25. Maheshwari S (2009) Analogue signal processing applications using a new circuit topology. IET Circ Devices Syst 3:106–115
26. Horng JW (2012) Voltage-mode multifunction biquadratic filters employing single DVCC. Int J Electron 99:153–162
27. Maheshwari S, Mohan J, Chauhan DS (2013) Novel voltage-mode cascadable all-pass sections employing grounded passive components. J Circ Syst Comput 22:1200065-1–1200065-2
28. Jiang J, Zhou X, Xu W (2013) Nth-order current transfer function synthesis using DVCCs: signal-flow graph approach. Int J Electron 100:482–496

29. Imran A, Arora D, Kumar R (2014) Dual DVCC based voltage-mode digitally programmable biquadratic filter. Circ Syst 5:1–6
30. Ibrahim MA, Minaei S, Kuntman H (2006) DVCC based differential-mode all-pass and notch filters with high CMRR. Int J Electron 93:231–240
31. Chang CM, Chen HP (2003) Universal capacitor-grounded voltage-mode filter with three inputs and a single output. Int J Electron 90:401–406
32. Ibrahim MA, Kuntman H, Cicekoglu O (2003) First-order all-pass filter canonical in the number of resistors and capacitors employing a single DDCC. Circ Syst Sig Process 22:525–536
33. Horng JW, Chiu WY, Wei HY (2004) Voltage-mode high pass, band pass and low pass filters using two DDCCs. Int J Electron 91:461–464
34. Ibrahim MA, Kuntman H (2004) A novel high CMRR high input impedance differential voltage-mode KHN-biquad employing DO-DDCCs. Int J Electron Commun (AEU) 58:429–433
35. Ibrahim MA, Kuntman H, Cicekoglu O (2005) Single DDCC biquads with high input impedance and minimum number of passive elements. Analog Integr Circ Sig Process 43:71–79
36. Chen HP, Wu KH (2006) Single DDCC-based voltage-mode multifunction filter. IEICE Trans Fundam E89-A:1–3
37. Horng JW, Hou CL, Chang CM, Lin YT, Shiu IC, Chiu WY (2006) First-order allpass filter and sinusoidal oscillators using DDCCs. Int J Electron 93:457–466
38. Chang CM, Soliman AM, Swamy MNS (2007) Analytical synthesis of low-sensitivity high-order voltage-mode DDCC and FDCCII-grounded R and C all-pass filter structures. IEEE Trans Circ Syst-I 54:1430–1443
39. Lee CL, Chang CM, Hou CL, Horng JW (2007) Multiple-mode universal biquad filter using two DDCCs. Int J Electr Eng 14:279–285
40. Chen HP (2007) Universal voltage-mode filter using only plus-type DDCCs. Analog Integr Circ Sig Process 50:137–139
41. Chiu WY, Horng JW (2007) High-input and low-output impedance voltage-mode universal biquadratic filter using DDCCs. IEEE Trans Circ Syst-II 54:649–652
42. Chen HP, Shen SS, Wang JP (2008) Electronically tunable versatile voltage-mode universal filter. Int J Electron Commun (AEU) 62:316–319
43. Lee WT, Liao YZ (2008) New voltage-mode high- pass, band-pass and low-pass filter using DDCC and OTAs. Int J Electron Commun (AEU) 62:701–704
44. Chen HP, Wu KH (2008) Novel voltage-mode multifunction filter using only two DDCCs. J Circ Syst Comput 17:1161–1172
45. Chen HP, Chu PL (2008) Versatile voltage-mode multifunction biquadratic filter employing DDCCs. IEICE Electron Express 5:769–775
46. Horng JW (2008) High input impedance voltage-mode universal biquadratic filter with three inputs using DDCCs. Circ Syst Sig Process 27:553–562
47. Chen HP (2009) Versatile universal voltage-mode filter employing DDCCs. Int J Electron Commun (AEU) 63:78–82
48. Yuce E (2010) A novel floating simulation topology composed of only grounded passive components. Int J Electron 97:249–262
49. Chen HP (2010) High-input impedance voltage-mode multifunction filter with four grounded components and only two plus-type DDCCs. Active Passiv Electron Comp 2010: Article ID 362516, 5p
50. Maheshwari S, Gangwar A (2011) Versatile voltage-mode universal filter using differential difference current conveyor. Circ Syst 2:210–216
51. Lee CN (2011) Fully cascadable mixed-mode universal filter biquad using DDCCs and grounded passive components. J Circ Syst Comput 20:607–620
52. Prommee P, Somdunyakanok M (2011) CMOS-based current-controlled DDCC and its applications to capacitance multiplier and universal filter. Int J Electron Commun (AEU) 65:1–8

53. Ibrahim MA, Minaei S, Yuce E (2011) All-pass sections with high gain opportunity. Radioengineering 20:3–9
54. Horng JW, Chiu WY (2011) High input impedance DDCC-based voltage-mode universal biquadratic filter with three inputs and five outputs. Ind J Pure Appl Physics 18:183–190
55. Liao WB, Gu JC (2011) SIMO type universal mixed-mode biquadratic filter. Ind J Pure Appl Physics 18:443–448
56. Chiu WY, Horng JW (2012) Voltage-mode high pass, low pass and notch biquadratic filters using single DDCC. Radioengineering 21:297–303
57. Chiu WY, Horng JW (2012) High input impedance voltage-mode universal biquadratic filter with three inputs and six outputs using three DDCCs. Circ Syst Sig Process 31:19–30
58. Horng JW, Chiu TY, Jhao ZY (2012) Tunable versatile high input impedance voltage-mode universal biquadratic filter based on DDCCs. Radioengineering 21:1260–1268
59. Chaturvedi B, Maheshwari S (2012) An ideal voltage-mode all-pass filter and its application. J Commun Comput 9:613–623
60. Kumngern M, Khateb F, Dejhan K, Phasukkit P, Tungjitkusolmun S (2013) Voltage-mode multifunction biquadratic filters using new ultra-low-power differential difference current conveyors. Radioengineering 22:448–457
61. Hwang YS, Liu A, Wang SF, Yang SC, Chen JJ (2013) A tunable Butterworth low-pass filter with digitally controlled DDCC. Radioengineering 22:511–517
62. Saied AB, Salem SB, Masmoudi DS (2013) A quadrature oscillator based on a new "optimized DDCC" all-pass filter. Circ Syst 4:498–503
63. Stornelli V, Ferri G (2013) A 0.18 μm CMOS DDCCII for portable LV-LP filters. Radioengineering 22:434–439
64. Tangsrirat W, Channumsin O (2011) High-input impedance voltage-mode multifunction filter using a single DDCCTA and grounded passive elements. Radioengineering 20:905–910
65. Tangsrirat W, Channumsin O, Pukkalanun T (2013) Resistorless realization of electronically tunable voltage-mode SIFO-type universal filter. Microelectron J 44:210–215
66. Channumsin O, Pukkalanun T, Tangsrirat W (2012) Voltage-mode universal filter with one input and five outputs using DDCCTAs and all- grounded passive components. Microelectron J 43:555–561
67. Channumsin O, Tangsrirat W (2013) Single-input four-output voltage-mode universal filter using single DDCCTA. Microelectron J 44:1084–1091
68. Tangsrirat W, Channumsin O, Pukkalanun T (2013) Universal voltage-mode SIFO-type biquad with fully MOS-C realization using DDCCTAs. Ind J Pure Appl Phys 51:516–522
69. Jaikla W, Khateb F, Siripongdee S, Supavarasuwat P, Suwanjan P (2013) Electronically tunable current-mode biquad filter employing CCCDTAs and grounded capacitors with low input and high output impedance. Int J Electron Commun (AEU) 67:1005–1009
70. Alzaher H, Elwan H, Ismail M (2000) CMOS baseband filter for WCDMA integrated wireless receivers. Electron Lett 36:1515–1516
71. Chang CM, Al-Hashimi BM, Wang CL, Hung CW (2003) Single fully differential current conveyor biquad filters. IEE Proc Circ Devices Syst 150:394–398
72. Chang CM, Chen HP (2005) Single FDCCII-based tunable universal voltage-mode filter. Circ Syst Sig Process 24:221–227
73. Mahmoud SA (2007) Fully differential CMOS CCII based on differential difference transconductor. Analog Integr Circ Sig process 50:195–203
74. Chen HP (2008) Voltage-mode FDCCII-based universal filters. Int J Electron Commun (AEU) 62:320–323
75. Lee CN, Chang CM (2009) Single FDCCII-based mixed-mode biquad filter with eight outputs. Int J Electron Commun (AEU) 63:736–742
76. Gur F, Anday F (2009) Simulation of a novel current-mode universal filter using FDCCIIs. Analog Integr Circ Sig Process 60:231–236
77. Liao YZ, Chen HP, Lee WT (2009) Versatile universal voltage-mode filter employing minimum components. IEICE Electron Express 6:1246–1252

78. Chen HP (2009) Single FDCCII-based universal voltage-mode filter. Int J Electron Commun (AEU) 63:713–719
79. Kacar F, Yesil A (2010) Voltage mode universal filters employing single FDCCII. Analog Integr Cir Sig Process 63:137–142
80. Kacar F, Metin B, Kuntman H, Cicekoglu O (2010) A new high-performance CMOS fully differential second-generation current conveyor with application example of biquad filter realization. Int J Electron 97:499–510
81. Maheshwari S, Mohan J, Chauhan DS (2009) Voltage-mode cascadable app-pass sections with two grounded passive components and one active element. IET Circ Devices Syst 4:113–122
82. Kacar F, Kuntman H (2010) New current-mode filters using single FDCCII with grounded resistors and capacitors. Online J Electron Electr Eng 2:341–345
83. Chen HP (2010) Versatile multifunction universal voltage-mode biquadratic filter. Int J Electron Commun (AEU) 64:983–987
84. Maheshwari S, Mohan J, Chauhan DS (2011) Novel cascadable all-pass/notch filters using a single FDCCII and grounded capacitors. Circ Syst Sig Process 30:643–654
85. Metin B, Herencsar N, Pal K (2011) Supplementary first-order all-pass filters with two grounded passive elements using FDCCII. Radioengineering 20:433–437
86. Kacar F, Yesil A, Kuntman H (2012) Current-mode biquad filters employing single FDCCII. Radioengineering 21:1269–1278
87. Mohan J, Maheshwari S (2014) Additional high-input low-output impedance voltage-mode all-pass sections. J Circ Syst Comput 23:1450077, 14pages
88. Ozoguz S, Toker A, Cicekoglu O (2000) First-order allpass sections-based current-mode universal filter using ICCIIs. Electron Lett 36:1443–1444
89. Minaei S, Yuce E, Cicekoglu O (2006) ICCII-based voltage-mode filter with single input and six outputs employing grounded capacitors. Circ Syst Sig Process 25:559–566
90. Soliman AM (2007) Generation of grounded capacitor ICCII-based band-pass filters. J Circ Syst Comput 16:553–566
91. Soliman AM (2007) Voltage mode and current mode Tow Thomas bi-quadratic filters using inverting CCII. Int J Circ Theor Appl 35:463–467
92. Toker A, Zeki A (2007) Tunable active network synthesis using ICCIIs. Int J Electron 94:335–351
93. Soliman AM (2008) New grounded capacitor current mode band-pass low-pass filters using two balanced output ICCII. J Active Passiv Electron Devices 3:175–184
94. Soliman AM (2008) Kerwin Huelsman Newcomb filter using inverting CCII. J Active Passiv Electron Devices 3:273–279
95. Soliman AM (2008) Current mode filters using two output inverting CCII. Int J Circ Theor Appl 36:875–881
96. Soliman AM (2009) New current mode low-pass filter using identical single output current conveyors. J Active Passiv Electron Devices 4:21–33
97. Horng JW (2011) Current-mode universal biquadratic filter with five inputs and one output using three ICCIIs. Ind J Pure Appl Physics 49:214–217
98. Chen HP (2012) Versatile current-mode universal biquadratic filter using plus-type dual-output ICCIIs. Ind J Pure Appl Physics 50:188–198
99. Ibrahim MA (2013) Electronically tunable single-input multi-output current-mode biquad filter suitable for easy cascading. Eng Tech J 31:1972–1981
100. Chen HP (2014) Voltage-mode multifunction biquadratic filter with one input and six outputs using two ICCIIs. Sci World J 2014:Article ID 432570, 7p
101. Salama KN, Elwan HO, Soliman AM (2001) Parasitic-capacitance-insensitive voltage-mode MOSFET-C filters using differential current voltage conveyor. Circ Syst Sig Process 20:11–26
102. Soliman AM, Madian AH (2009) MOS-C KHN filter using voltage Op Amp, CFOA, OTRA and DCVC. J Circ Syst Comput 18:733–769

103. Fabre A (1995) Third-generation current conveyor: a new helpful active element. Electron Lett 31:338–339
104. Maheshwari S, Khan IA (2001) Novel first order all-pass sections using a single CCIII. Int J Electron 88:773–778
105. Kuntman H, Cicekoglu O, Ozoguz S (2002) A modified third generation current conveyor, its characterization and applications. Frequenz 56:47–54
106. Maheshwari S, Khan IA (2004) Novel first-order current-mode all-pass sections using CCIII. Active Passive Electron Comp 27:111–117
107. Yuce E, Metin B, Cicekoglu O (2004) Current-mode biquadratic filters using single CCIII and minimum number of passive elements. Frequenz 58:225–228
108. Cam U (2005) A new transadmittance type first-order allpass filter employing single third generation current conveyor. Analog Integr Circ Sig Process 43:97–99
109. Hwang YS, Chen JJ, Li JP (2007) New current-mode all-pole and elliptic filters employing current conveyors. Electr Eng 89:457–459
110. Chunhua W, Keskin AU, Yang L, Qiujing Z, Sichuan D (2010) Minimum configuration insensitive multifunctional current-mode biquad using current conveyors and all-grounded passive components. Radioengineering 19:178–184
111. Minaei S (2009) Electronically tunable current-mode universal biquad filter using dual-X current conveyors. J Circ Syst Comp 18:665–680
112. Minaei S, Yuce E (2010) Unity/variable-gain voltage-mode/current-mode first-order all-pass filters using single dual-X second-generation current conveyor. IETE J Res 56:305–312
113. Maheshwari S, Chaturvedi B (2012) High-input low-output impedance all-pass filters using one active element. IET Circ Devices Syst 6:103–110
114. Maheshwari S, Chaturvedi B (2013) Additional high input low output impedance analog networks. Active Passiv Electron Comp 2013: Article ID 574925, 9p
115. Mohan J (2013) Single active element based current-mode all-pass filter. Int J Comput Appl 82:23–27
116. Beg P (2014) Tunable first-order resistless all-pass filter with low output impedance. Sci World J 2014: Article ID 219453, 6p
117. Gunes EO, Anday F (1995) Realization of current-mode low pass filters using CFCCIIs. Electron Lett 31:2161–2162
118. Gunes EO, Anday F (1996) Realization of current-mode universal filter using $CFCCII_p$s. Electron Lett 32:1081–1082
119. Abuelma'atti MT, Tassaduq NA (1998) High-order current transfer function synthesis using translinear current conveyors. Frequenz 52:76–78
120. Keskin AU, Cam U (2007) Insensitive high-output impedance minimum configuration SITO-type current-mode biquad using dual-output current conveyors and grounded passive components. Int J Electron Commun (AEU) 61:341–344
121. Yucel F, Yuce E (2014) CCII based more tunable voltage-mode all-pass filters and their quadrature oscillator applications. Int J Electron Commun (AEU) 68:1–9
122. Ghallab YH, Badawy W (2006) The operational floating current conveyor and its applications. J Circ Syst Comp 15:351–372
123. Pandey N, Nand D, Khan Z (2013) Single-input four-output current mode filter using operational floating current conveyor. Active Passiv Electron Comp 2013: Article ID 318560, 8p
124. Biolek D, Vrba K, Cajka J, Dostal T (2000) General three-port current conveyor: a useful tool for network design. J Electr Eng 51:36–39
125. Senani R, Singh AK (2002) A new universal current mode biquad filter. Frequenz 56:55–59
126. Senani R, Singh AK, Singh VK (2003) New tunable SIMO type current mode universal biquad using only three MOCCs and all grounded passive elements. Frequenz 57:160–161
127. Abdalla KK, Bhaskar DR, Senani R (2012) Configuration for realizing a current mode universal filter and dual mode quadrature single resistor controlled oscillator. IET Circ Devices Syst 6:159–167

128. Chunhua W, Haiguang L, Yan Z (2009) Universal current mode filter with multiple inputs and one output using MOCCII and MO-CCCA. Int J Electron Commun 63:448–453
129. Jiang J, He Y (2009) Tunable frequency versatile filters implementation using minimum number of passive elements. Analog Integr Circ Sig Process 59:53–64
130. Senani R, Singh VK, Singh AK, Bhaskar DR (2005) Tunable current mode universal biquads employing only three MOCCs and all grounded passive elements: additional new realizations. Frequenz 59:1–5
131. Senani R (2006) New universal current mode biquad employing all grounded passive components but only two DOCCs. J Active Passiv Electron Devices 1:281–288
132. Tu SH, Chang CM, Liao KP (2002) Novel versatile insensitive universal current mode biquad employing two second generation current conveyors. Int J Electron 89:897–903
133. Wang HY, Lee CT (2001) Versatile insensitive current mode universal biquad implementation using current conveyors. IEEE Trans Circ Syst-II 48:409–413
134. Abuelma'atti MT, Bentrcia A (2004) A novel mixed-mode CCII based filter. Active Passiv Electron Comp 27:197–205
135. Abuelma'atti MT, Bentrcia A, Al-Shahrani SM (2004) A novel mixed mode current conveyor based filter. Int J Electron 91:191–197
136. Singh VK, Singh AK, Senani R (2005) Dual function capability of recently proposed four current conveyor based VM biquad. J Circ Syst Comput 14:51–56
137. Kumar P (2006) FTFNs and their applications in circuit synthesis and design. PhD thesis (Supervisor Raj Senani), Faculty of technology, University of Delhi, vol 4, pp 141–172
138. Horng JW (2012) Analytical synthesis of general high-order voltage/current transfer functions using CCIIs. Microelectron J 43:546–554
139. Abuelma'atti MT, Khan MH (1995) Low component current-mode universal filter. Electron Lett 31:2160–2161
140. Elwan HO, Soliman AM (1996) A novel CMOS current conveyor realization with an electronically tunable current mode filter suitable for VLSI. IEEE Trans Circ Syst-II 43:663–670
141. Soliman AM (1996) New inverting-non-inverting band pass and low pass biquad circuit using current conveyors. Int J Electron 81:577–583
142. Horng JW, Weng RM, Lee MH, Chang CW (1997) Universal active current filter using two multiple current output OTAs and one CCIII. Int J Electron 82:241–247
143. Fabre A, Saaid O, Wiest F, Boucheron C (1997) Low power current-mode second order band pass IF filter. IEEE Trans Circ Syst-II 44:436–446
144. Zhang X, Ni X, Iwahashi M, Kambayashi N (1998) Realization of universal active complex filter using CCIIs and CFCIIs. IEICE Trans Fundament 2:244–251
145. Toker A, Ozoguz S (2000) Insensitive current-mode universal filter using dual output current conveyors. Int J Electron 87:667–674
146. Ozoguz S, Acar C, Toker A, Gunes EO (2001) Derivation of low-sensitivity current-mode CCII-based filters. IEE Proc Circ Devices Syst 148:115–120
147. Sagbas M, Fidanboylu K, Bayram MC (2004) A new current-mode multifunction filter with high impedance outputs using minimum number of passive elements. Int J Sig Process 1:1304–4494
148. Tsukutani T, Edasaki S, Sumi Y, Fukui Y (2006) Current-mode universal biquad filter using OTAs and DO-CCII. Frequenz 60:11–12
149. Pandey N, Paul SK, Bhattacharyya A, Jain SB (2007) Insensitive mixed mode biquad using reduced number of active and passive components. J Active Passive Electron Devices 2:117–125
150. Sagbas M, Koksal M (2007) Voltage-mode three-input single-output multifunction filters employing minimum number of components. Frequenz 61:3–4
151. Pandey N, Paul SK (2006) Multi-input single-outputs universal current mode biquad. J Active Passive Electron Devices 1:229–240
152. Pandey N, Paul SK, Jain SB (2008) Voltage mode universal filter using two plus type CCIIs. J Active Passive Electron Devices 3:165–173

153. Pandey N, Paul SK, Jain SB (2008) New high-input impedance voltage-mode universal biquad; multi input multi output. J Active Passive Electron Devices 3:93–100
154. Soliman AM (2008) The CCII+ and the ICCII as basic building blocks in low-pass filter realizations. Int J Circ Theor Appl 36:493–509
155. Lahiri A (2011) New canonic active RC sinusoidal oscillator circuits using second-generation current conveyors with application as a wide-frequency digitally controlled sinusoid generator. Active Passiv Electron Comp Article ID:274394, 8p
156. Vrba K, Cajka J, Zeman V (1998) N-th order filters using balanced-output CCII/-conveyors. Radioengineering 7:16–19
157. Abuelma'atti MT, Alzaher HA (1998) Versatile active biquad based on dual-output second-generation current conveyors. Active Passiv Elec Comp 20:151–155
158. Gunes EO, Toker A, Ozoguz S (1999) Insensitive current-mode universal filter with minimum components using dual-output current conveyors. Electron Lett 35:524–525
159. Zhang X, Ni X, Iwahashi M, Kambayashi N (1999) Realization of universal active complex filter using CCIIs and CFCCIIs. Analog Integr Circ Sig Process 20:129–137
160. Sedef H, Acar C (1999) Simulation of resistively terminated LC ladder filters using a new basic cell involving current conveyors. Microelectron J 30:63–68
161. Toker A, Ozoguz S, Cicekoglu O (2000) A new current-mode multifunction filter with minimum components using dual-output current conveyors. IEICE Trans Fundam 83-A:2382–2384
162. Gunes EO, Ozoguz S, Toker A (2000) Insensitive current-mode universal filter with low component spread using dual-output current conveyors. AEU 54:127–132
163. Toker A, Ozoguz S (2001) Integrable current-mode filter realization using dual output current conveyors for low-frequency operation. Int J Electron Commun 55:145–149
164. Cicekoglu O, Tarim N, Kuntman H (2002) Wide dynamic range high output impedance current mode multifunction filters with dual output current conveyors. Int J Electron Commun 56:55–60
165. Horng JW (2002) Current conveyors based current-mode universal biquadratic filter. J Chin Inst Electr Eng 9:147–150
166. Hwang YS, Hung PT, Chen W, Liu SI (2002) Systematic generation of current mode linear transformation filters based on multiple output CCIIs. Analog Integr Circ Sig Process 32:123–134
167. Tsukutani T, Sumi Y, Higashimura M, Fukui Y (2005) Current mode universal biquad circuit using MO-OTAs and DO-CCII. In: Proceedings of IEEE international symposium on circuits and systems, vol 2, Institute of Electrical and Electronics Engineers, New York, pp 1589–1592
168. Horng JW, Hou CL, Chang CM, Shie JY, Chang CH (2007) Universal current filter with single input and three outputs using MOCCIIs. Int J Electron 94:327–333
169. Tsukutani T, Sumi Y, Fukui Y (2007) Novel current-mode biquad filter using OTAs and DO-CCII. Int J Electron 94:99–105
170. Keskin AU, Cam U (2007) Insensitive high-output impedance minimum configuration SITO-type current mode biquad using dual-output current conveyors and grounded passive components. Int J Electron Commun 61:341–344
171. Chen HP, Chen K, Chen CY, Chen MS (2011) Versatile CCII-based universal current-mode biquadratic filter. PIERS Proc Suzhou China: 488–491
172. Koksal M, Sagbas M (2008) A versatile signal flow graph realization of a general current transfer function. Int J Electron Commun 62:33–40
173. Horng JW (2011) High output impedance current-mode universal biquadratic filters with five inputs using multi-outputs CCIIs. Microelectron J 42:693–700
174. Horng JW, Hou CL, Guo YS, Hsu CH, Yang DY, Ho MJ (2012) Low input and high output impedances current mode first order all pass filter employing grounded passive components. Circ Syst 3:176–179

175. Chen HP (2013) Versatile current-mode universal biquadratic filter using DO-CCIIs. Int J Electron 100:1010–1031
176. Mahmoud SA, Soliman EA (2013) Novel CCII-based field programmable analog array and its application to a sixth order Butterworth LPF. Adv Microelectron Eng 1:9–16
177. Lee CN (2013) Versatile universal current-mode and transresistance mode biquadratic filter using two MOCCIIs and grounded passive components. J Circ Syst Comput 22:1250077-1–1250077-18
178. Maghami MH, Sodagar AM (2013) Low power, low voltage, dual-output, second generation current conveyor and its application in low pass filter design. J Circ Syst Comput 22:1350044-1–1350044-10
179. Alzaher H, Tasadduq N, Al-Ees O (2013) Implementation of reconfigurable nth-order filter based on CCII. Analog Intregr Circ Sig Process 75:539–545
180. Abdalla KK, Bhaskar DR, Senani R (2010) Configuration for realizing a current mode universal filter and dual-mode quadrature single resistor controlled oscillator. IET Circ Devices Syst 6:159–167
181. Eajka J, Novotny V (2003) Two mode high order filters using three port voltage conveyors. ElectronicsLetters.com. 30:1–9
182. Khanday F, Shah N (2013) A low voltage and low-power Sinh-domain universal biquadratic filter for low-frequency applications. Turk J Elec Eng Comp Sci 21:2205–2217
183. Horng JW, Hou CL, Chang CM, Chung WY, Wei HY (2005) Voltage mode universal biquadratic filters with one input and five outputs using MOCCCIIs. Comput Electr Eng 31:190–202
184. Minaei S, Yuce E (2009) All grounded passive elements current mode all-pass filter. J Circ Syst Comput 18:31–43
185. Al-Shahrani SM (2007) CMOS wideband auto-tuning phase shifter circuit. Electron Lett 43:804–805
186. Abdalla KK (2013) Universal current mode biquad employing dual output current conveyors and MO-CCCA with grounded passive elements. Circ Syst 4:83–88
187. Cicekoglu O (2001) High output impedance current mode four function filter with reduced number of active and passive elements using the dual output current conveyor. Analog Integr Circ Sig Process 28:201–204
188. Soliman AM (1995) Current mode universal filter. Electron Lett 31:1420–1421
189. Soliman AM (2008) Current-mode universal filters using current conveyors. Circ Syst Sig Process 27:405–427
190. Sagbas M, Koksal M (2008) Current-mode state-variable filter. Frequenz 62:37–42
191. Wang C, Zhao Y, Zhang Q, Du S (2009) A new current mode SIMO-type universal biquad employing multi-output current conveyors (MOCCIIs). Radioengineering 18:83–88
192. Metin B, Minaei S, Cicekoglu O (2007) Enhanced dynamic range analog filter topologies with a notch/all-pass circuit example. Analog Integr Circ Sig Process 51:181–189
193. Tangsrirat W, Channumsin O (2011) Minimum-component current-mode universal filter. Ind J Pure Appl Phys 49:137–141
194. Horng JW, Chiu TY, Hsiao CP, Huang GT (2013) Three-inputs-one-output current-mode universal biquad using two current conveyors. J Circ Syst Comput 22. doi: 10.1142/S021812661340001X, 12pages
195. Singh AK, Senani R (2002) A new four-CC-based configuration for realizing a voltage mode biquad filter. J Circ Syst Comput 11:213–218
196. Biolek D, Cajka J, Vrba K, Zeman V (2002) Nth-order allpass filters using current conveyors. J Electr Eng 53:50–53
197. Minarcik M, Vrba K (2006) Low-output and high-input impedance frequency filters using universal voltage conveyor for high-speed data communication systems. In: Proceedings of international conference on networking ICN 2006, Mauritius, pp 155–158
198. Minarcik M, Vrba K (2008) Single-input six-output voltage-mode filter using universal voltage conveyors. IEICE Sig Process E91-A:2035–2037

199. Koton J, Vrba K, Herencsar N (2009) Tunable filter using voltage conveyors and current active elements. Int J Electron 96:787–794
200. Koton J, Herencsar N, Vrba K (2010) Single-input three-output variable Q and ω_0 filters using universal voltage conveyors. Int J Electron 97:531–538
201. Jerabek J, Vrba K (2010) SIMO type low-input and high-output impedance current-mode universal filter employing three universal current conveyors. Int J Electron Commun (AEU) 64:588–593
202. Koton J, Herencsar N, Vrba K (2011) KHN-equivalent voltage-mode filters using universal voltage conveyors. Int J Electron Commun (AEU) 65:154–160
203. Akerberg D, Mossberg K (1974) A versatile active RC building block with inherent compensation for the finite bandwidth of the amplifier. IEEE Trans Circ Syst 21:75–78
204. Elwan HO, Soliman AM (1997) Novel CMOS differential voltage current conveyor and its applications. IEE Proc Circ Devices Syst 144:195–200
205. Chen HP (2006) Novel voltage-mode universal biquad filter. Electron World 1842:41
206. Surakampontorn W, Riewruja V, Kumwachara K, Dejhan K (1991) Accurate CMOS-based current conveyors. IEEE Trans Instrum Meas 40:699–702
207. Metin B, Cicekoglu O (2009) Component reduced all-pass filter with a grounded capacitor and high-impedance input. Int J Electron 96:445–455
208. Altun M, Kuntman H, Minaei S, Sayin OK (2009) Realization of nth–order current transfer function employing ECCIIs and application examples. Int J Electron 96:1115–1126
209. Herencsar N, Koton J, Vrba K (2009) Single CCTA-based universal biquadratic filters employing minimum components. Int J Comput Electr Eng 1:309–312
210. Ayten UE, Sagbas M, Sedef H (2010) Current mode leapfrog ladder filters using a new active block. Int J Electron Commun (AEU) 64:503–511
211. Maheshwari S (2013) Current conveyor all-pass sections: brief review and novel solution. Sci World J 2013: Article ID 429391, 6p
212. Yildiz HA, Toker A, Ozoguz S (2013) Biquadratic filter applications using a fully-differential active-only integrator. Radioengineering 22:3–13

Chapter 13
Sinusoidal Oscillator Realizations Using Other Types of Current Conveyors

Abstract A variety of SRCOs realized with numerous types of CCs introduced in the literature (such as OFC, MOCCII, DDCCCII, DVCC, DVCCC, ICCII, DVCCCA and FDCCII) have been discussed. The main focus has been on including some representative circuit configurations (from amongst a large number of oscillator circuits reported in literature) which possess some specific attractive features of practical interest.

13.1 Introduction

In Chap. 7 of this monograph, we presented a number of novel circuits for realizing single-element-controlled oscillators exhibiting a number of desirable features. The circuits proposed therein are all practically realizable using commercially available current conveyors using IC-AD844, using IC op-amps uA741/LF356 in conjunction with transistor arrays CA3096 or using IC OPA660/860. In spite of an abundance of CC-based SRCOs based upon the *basic* types of three-terminal CCs (namely, CCI ± and CCII±), many researchers throughout the world have still toiled with the idea of devising SRCOs using other variants of CCs not commercially available yet, for instance, see references [1–52] and the references cited therein. The basic motivation and inspiration for doing has been a number of interesting features which can be obtained using the newly emerging variants of CCs which somehow are not attainable with oscillators using the normal types of CCs.

The purpose of this chapter is to present some selected sinusoidal oscillator topologies which use these extended or generalized new varieties of current conveyors. It may be noted that although some of the circuits described herein may also be implemented with AD844 (using several of them) however, the real potential of these circuits will be realized only when they are fully integrated in bipolar and/or CMOS technology, because it has been shown elsewhere in this monograph that all extended and generalized forms of current conveyors can be implemented in bipolar or CMOS technology.

© Springer International Publishing Switzerland 2015

R. Senani et al., *Current Conveyors*, DOI 10.1007/978-3-319-08684-2_13

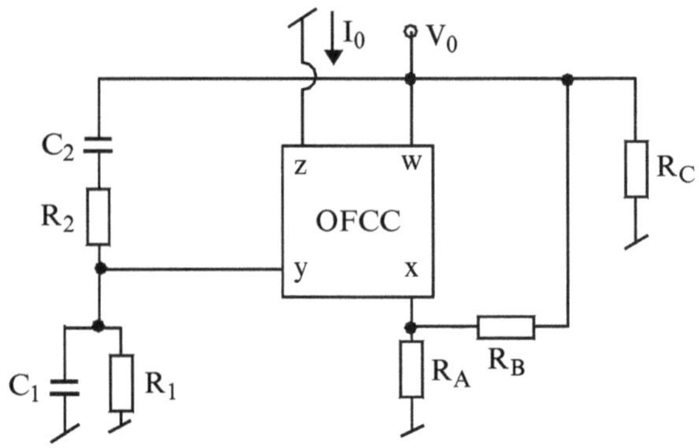

Fig. 13.1 Single OFCC based oscillator with dual outputs proposed by Celma and Martinez [1]

13.2 A Dual-Mode Sinusoidal Oscillator Using a Single Operational Floating Current Conveyor

An operational floating current conveyor (OFCC; as defined in Chap. 10), because of being a four terminal building block, provides the flexibility of having a low output impedance voltage output as well as a high output impedance current output. This is immediately of interest when a circuit is needed for realizing a fixed frequency oscillator which should be able to provide a voltage mode as well as current mode output. One such circuit was advanced by Selma and Martinez in [1] and is shown here in Fig. 13.1.

The characteristic equation (CE) of the circuit of Fig. 13.1 is given by:

$$s^2 + s\left\{\frac{1}{R_1C_1} + \frac{1}{R_2C_2} + (1 - k_0)\frac{1}{R_2C_1}\right\} + \frac{1}{R_1R_2C_1C_2} = 0 \tag{13.1}$$

where $k_0 = (1+\frac{R_B}{R_A})$

Thus, the condition of oscillation can be adjusted by varying R_A or R_B. It is seen that both voltage and current outputs are available from this circuit.

The aspect of determining the limit cycle stability in CM oscillators using OFC has been investigated by Celma and Martinez in [2].

The oscillator of Fig. 13.1 was practically tested and its operation verified by realizing the OFC using two CCIIs in the IC CCII01 from LTP Electronics.

13.3 ICCII-Based Grounded-Capacitor (GC) SRCO

Several authors have described new oscillator topologies using ICCII, for example, [26, 27, 33, 34, 42]. An exemplary ICCII-based oscillator has been shown in Fig. 13.2 which employs a single ICCII, two grounded capacitors and three resistors and was proposed by Horng, Wang and Yang in [42].

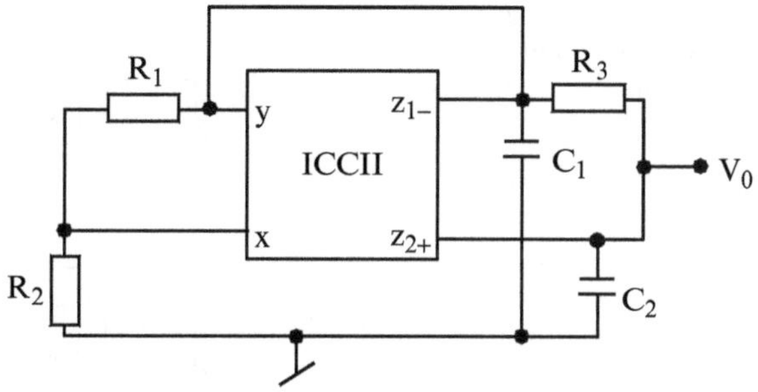

Fig. 13.2 ICCII based sinusoidal oscillator proposed by Horng et al. [42]

The analysis of this oscillator, assuming ideal ICCII, yields the following CE:

$$s^2C_1C_2R_1R_2R_3 + sR_1\{R_2(C_1 + C_2) - C_2R_3\} + 2R_2 = 0 \tag{13.2}$$

From the above CE, CO and FO can be determined as:

$$\text{CO} : \left(R_2 - \frac{C_2R_3}{C_1 + C_2}\right) \leq 0 \tag{13.3}$$

$$\text{FO} : \omega_0 = \sqrt{\frac{2}{C_1C_2R_1R_3}} \tag{13.4}$$

Thus, from equations (13.3) and (13.4), it is clear that CO can be adjusted by the grounded resistor R_2 whereas FO is independently controllable through R_1. A completely CMOS version is also possible by replacing all the resistors by CMOS resistors in which case the circuit can be made to function as a voltage controlled oscillator.

It may be noted that an ICCII can be realized from a DVCC by grounding its non-inverting Y-terminal. Thus, the circuit of Fig. 13.2 was simulated in SPICE by using the DVCC of [53] in the manner described above biased with ± 1.25 V DC power supply. Using TSMC 0.18 μm CMOS technology, the circuit could produce oscillations of frequency 1.73 MHz with % THD being no more than 2.22 %.

13.4 ICCII-Based All Grounded Passive Elements (AGPE) SRCO

Of late, there have many attempts for systematic synthesis of oscillators employing all grounded elements, based upon the use of pathological elements and the concept of nodal admittance matrix (NAM) expansions, for, instance see [35, 38, 39]. However, in the following we present a fully uncoupled oscillator supposedly derived by intuitive method by Toker, Kuntman, Cicekoglu and Discigil in [10].

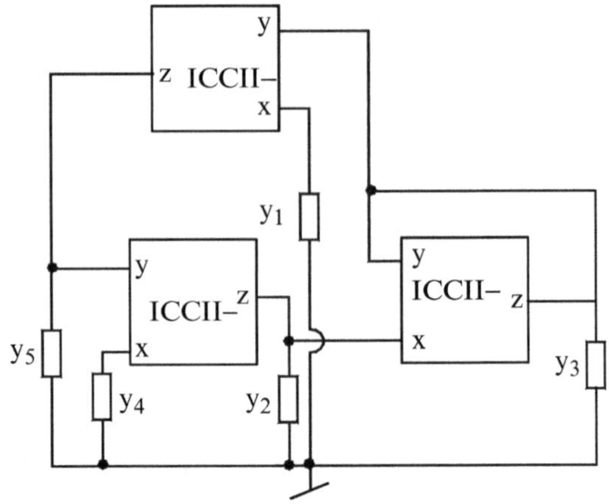

Fig. 13.3 AGPE based SRCO proposed by Toker et al. [10]

This fully-uncoupled oscillator (i.e., having both CO and FO fully-uncoupled) uses three ICCIIs and six grounded passive elements and is shown in Fig. 13.3.

Assuming ideal ICCIIs, the proposed oscillator circuit in Fig. 13.3 has the following CE:

$$y_1y_4 + y_2y_5 - y_3y_5 = 0 \tag{13.5}$$

From the above CE, two different versions of SRCO can be obtained by appropriate choice(s) of elements:

(i) if $y_1 = \frac{1}{R_1}, y_2 = \left(sC_2 + \frac{1}{R_2}\right), y_3 = \frac{1}{R_3}, y_4 = \frac{1}{R_4}$ and $y_5 = sC_5$

$$\text{CO} : (R_3 - R_2) \leq 0 \tag{13.6}$$

$$\text{FO} : \omega_0 = \sqrt{\frac{1}{C_5C_2R_1R_4}} \tag{13.7}$$

(ii) if $y_1 = sC_1, y_2 = \left(sC_2 + \frac{1}{R_2}\right), y_3 = sC_3, y_4 = sC_4$ and $y_5 = \frac{1}{R_5}$

$$\text{CO} : (C_2 - C_3) \leq 0 \tag{13.8}$$

$$\text{FO} : \omega_0 = \sqrt{\frac{1}{C_1C_4R_2R_5}} \tag{13.9}$$

Out of the two SRCO versions, version (i) uses a minimum number of capacitors (two) and four resistors while version (ii) employs four capacitors and two resistors. Thus, version (i) of SRCO is superior as compared to version (ii), although both the oscillators qualify for obtaining fully-uncoupled condition of oscillation and frequency of oscillation.

The workability of he circuits described has been verified by using the CMOS ICCII from [54] using 1.2 μm CMOS process model parameters used in SPICE simulations with component values designed to achieve an oscillation frequency of 39.78 KHz while the CMOS ICCII was biased with ± 2.5 V DC power supplies. The frequency generated by SPICE simulations was found to be 38.76 KHz which is seen to be in excellent agreement with theory.

13.5 Explicit Current Output (ECO) SRCO Using All Grounded Passive Components

A DDCC has already been introduced in an earlier chapter. If we combine DDCC + and DDCC- into a single unit it can be called a differential difference complimentary current conveyor (DDCCC). It was shown by Gupta and Senani in [9] that a DDCCC is a versatile building block to implement an explicit current output SRCO as shown here in Fig. 13.4 which employs two DDCCCs and five passive elements (three resistors and two capacitors) all of which are grounded.

The CO and FO for this circuit are given by:

$$R_1 \geq R_3 \tag{13.10}$$

$$f_0 = \frac{1}{2\pi\sqrt{C_1 C_2 R_2 R_3}} \tag{13.11}$$

Thus, CO is controllable by R_1 while FO is controllable independently by R_2.

The feasibility of the circuit of Fig. 13.4 was confirmed by the DVCC circuit of [53] and modifying it to become a DDCCC by creating an extra complementary Z-output. The variation of oscillation frequency with the resistor R_2 exhibited an excellent correspondence with the theoretical plot thereby establishing the correctness of the formulation.

For a fixed-frequency sinusoidal oscillator with all grounded elements and providing an ECO, see [48] whereas for a number of other ECO oscillators using Transconductance second generation Current Conveyor (TCCII), Generalized Current Conveyor (GCC) and second generation Controlled Current Conveyor (CCCII) can be found in [3, 12, 31] respectively.

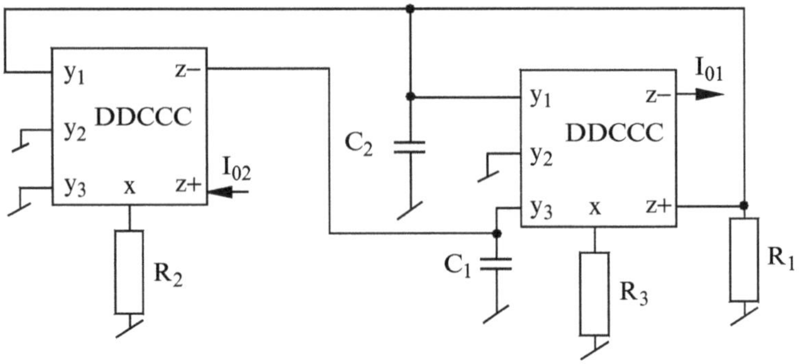

Fig. 13.4 SRCO using two DDCCCs proposed by Gupta and Senani [9]

13.6 Grounded-Capacitor Current-Mode SRCO Using a Single DVCCC

There has been considerable interest in involving SRCO employing both grounded capacitors which is known to be an attractive feature from the view point of IC Implementation as well as ease of eliminating/accommodating various parasitic capacitances of the active elements. With this motivation in mind, GC-based SRCOs have been attempted using almost every kind of active building block. Among these, the circuits which employ only a single active element are of particular interest because of reason of reduced power consumption, low component count and also the theoretical novelty from various considerations. A GC-SRCO which provides an explicit current output while providing the following other advantageous features: (i) use of only a single active building block (ii) employment of only two GCs and no more than three resistors (iii) provision of non-interacting control of frequency of oscillation (FO) as well as condition of oscillation (CO) and (iv) availability of a simple CO and unconstraint tuning law for the FO.

Although there have been several oscillators using DVCCs such as [13, 17], a circuit possessing all the above mentioned desirable properties, while using a single differential voltage complimentary current conveyor (DVCCC), was first proposed by Gupta and Senani in [7] and is shown here in Fig. 13.5.

The CO and FO for this circuit are given by:

$$\mathrm{CO}:\ R_1 \leq R_3\left[\frac{C_2}{C_1+C_2}\right] \tag{13.12}$$

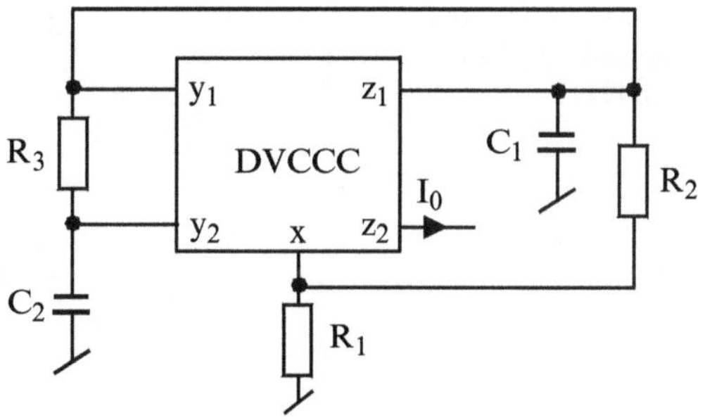

Fig. 13.5 GC-SRCO with explicit-current-output proposed by Gupta and Senani [7]

$$\text{FO}: f_0 = \frac{1}{2\pi}\sqrt{\frac{2}{C_1C_2R_2R_3}} \tag{13.13}$$

It is, therefore, seen that in this circuit, CO can be adjusted through R_1 without affecting the FO, which also can be independently controlled through R_2 and an explicit current output from high output impedance terminal Z_2 is available.

For SPICE simulation of the circuit, the DVCCC was realized using the CMOS DVCC of [53] using 1.2 μm level 3 MOSFET parameters obtained from MOSIS. The DVCCC circuit was biased with ± 3.3 V DC power supply. The oscillator was found to behave satisfactorily over the entire frequency range of observation (70–470 KHz).

13.7 FDCCII-Based SRCOs

Although there have been several studies [15, 23, 46], on evolving oscillators using the so-called fully differential CCII (FDCCII). in this section, we present two circuits which attempt all the properties of the circuit described in Sect. 13.6 but with a difference; the active building block used instead of DVCCC, is as a consequence of which it is possible to have not only both the capacitors grounded but also all the resistor grounded too. The two circuits belonging to this class proposed by Chang, Al-Hashmi, Chen, Tu and Wan [11] are shown in Fig. 13.6.

Both of these circuits have the same characteristic equation from which the resulting CO and FO are found to be

$$\text{CO}: R_1 = R_3 \tag{13.14}$$

$$\text{FO}: f_0 = \frac{1}{2\pi}\sqrt{\frac{1}{C_1C_2R_2R_3}} \tag{13.15}$$

The above equations show that the FO of these oscillators can be controlled by R_2 while CO is controllable independently by R_1.

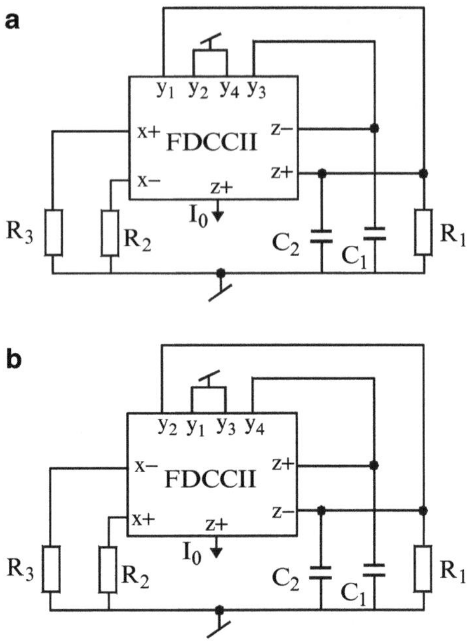

Fig. 13.6 CM SRCOs proposed by Chang et al. [11]

The two oscillators described above were simulated in SPICE by realizing the CMOS FDCCII using the implementation from [55] using 0.5 μm level 3 CMOS process parameters. Additional current mirrors were used to produce additional Z-terminals. The biasing used was ± 5 V DC. The components were chosen to get the theoretical frequency as 2.25 MHz whereas simulations exhibited the frequency as 2.24 MHz thus, establishing the practical viability of the proposed circuits.

13.8 CM Quadrature Oscillator (QO) Using DVCCs

In 2011, Maheshwari and Chaturvedi [45] proposed high output impedance CM QO using three DVCCs and four grounded passive elements. The circuit has the flexibility to generate three more QOs by appropriate connections of input terminals. Assuming ideal DVCCs (DVCC has already been defined in earlier chapter), the CE of the proposed QO, as shown in Fig. 13.7, is given by:

$$s^2 + s\left(\frac{1}{C_1R_1} - \frac{1}{C_2R_3}\right) + \frac{1}{C_1C_2R_3}\left(\frac{1}{R_2} - \frac{1}{R_1}\right) = 0 \tag{13.16}$$

from which the expressions for CO and FO are obtained as:

$$\text{CO}: \ (C_2R_3 - C_1R_1) \le 0 \tag{13.17}$$

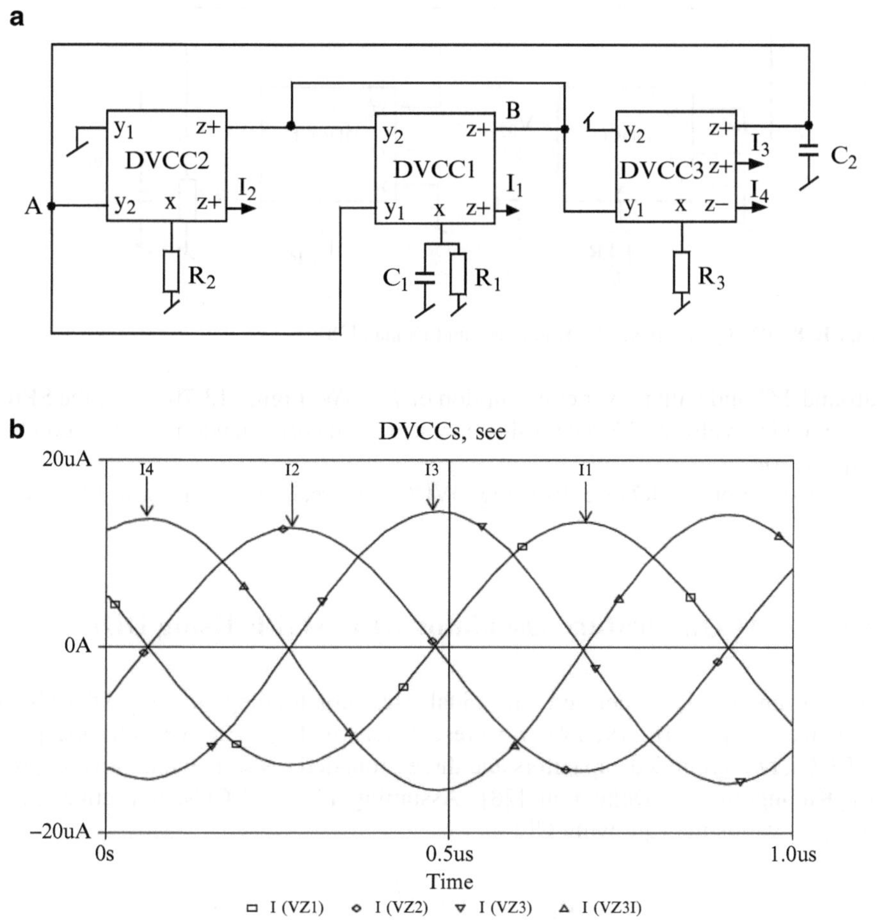

Fig. 13.7 CM QO proposed by Maheshwari and Chaturvedi (Adapted from [45] © 2011 John Wiley & Sons, Ltd.) (**a**) The circuit configuration (**b**) The four quadrature current outputs at 1 MHz

$$\text{FO}: \ \omega_0 = \left[\frac{1}{C_1 C_2 R_3}\left(\frac{1}{R_2} - \frac{1}{R_1}\right)\right]^{1/2} \ \text{provided } R_1 > R_2 \qquad (13.18)$$

Hence, from the above equations, it is clear that FO can be controlled by R_2 independently. The four quadrature output currents are given by $I_4 = -I_3$, $I_2 = -I_1$ and $I_3 = -j\omega R_2 C_2 I_2$, thus for $\omega R_2 C_2 = 1$, all the quadrature outputs have equal amplitude.

The operation of the circuit was verified by SPICE simulations using CMOS DVCCs realizable in 0.5 μm CMOS technology with level 3 MOSFET parameters and with DC biasing supply taken as ±2.5 V DC, component values chosen to achieve an oscillation frequency of 1.02 MHz. The SPICE generated frequency was found to be 1.0 MHz. The error in frequency was merely 1.9 %, with % THD of

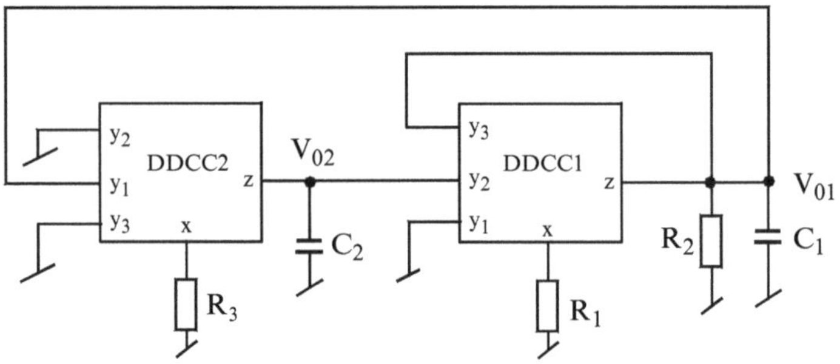

Fig. 13.8 VM QO proposed by Kumngern and Dejhan [28]

around 2 % and with power consumption of 7.5 mW. Figure 13.7b shows the SPICE generated waveforms. These results represent good correspondence with theoretical expectations.

For some other QO circuits using DVCCs, the readers may refer to [22, 50].

13.9 VM Quadrature Oscillator with AGPE Using DDCCs

Several authors have derived sinusoidal oscillator topologies using DDCCs, for instance see [14, 18, 19, 24], we present here in Fig. 13.8 VM QO using two DDCCs, two grounded capacitors and three grounded resistors which was proposed by Kumngern and Dejhan in [28]. Assuming ideal DDCCs, a routine circuit analysis yields the following CE:

$$s^2C_1C_2R_1R_3 + sC_2R_3\left\{\frac{R_1}{R_2} - 1\right\} + \frac{R_2}{R_1} = 0 \tag{13.19}$$

The CO and FO from the above equation are given by:

$$\text{CO}: \ R_1 \leq R_2 \tag{13.20}$$

$$\text{FO}: \ \omega_0 = \frac{1}{R_1}\sqrt{\frac{R_2}{C_1C_2R_3}} \tag{13.21}$$

It can be observed from the above equations that CO can be controlled by R_1 or/ and R_2 and FO can be varied through R_3 without disturbing CO. Hence, CO and FO are orthogonally controllable.

The phase difference φ between V_{01} and V_{02} has been found to be:

$$\varphi = \pi - \tan^{-1}(\omega R_3C_2) \tag{13.22}$$

At $\omega = \omega_0$, from equation (13.22), φ can be determined as $\varphi = \pi/2$ hence, both V_{01} and V_{02} are in quadrature.

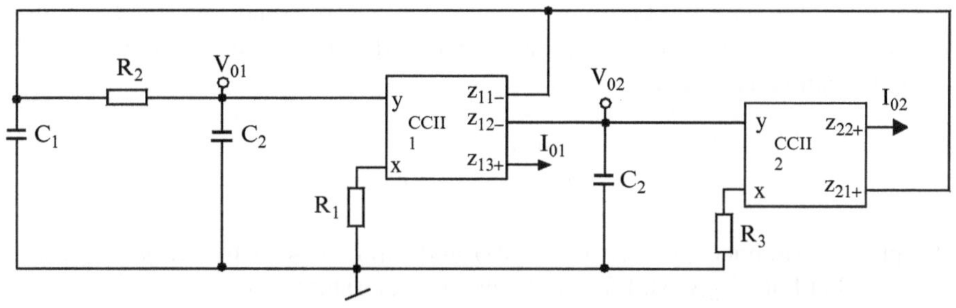

Fig. 13.9 MOCCII- based dual mode quadrature oscillator proposed by Horng [44]

The circuit operation was verified by a simple DVCC implementation using 0.5 μm CMOS process parameters with the circuit biased from ± 2.5 V DC and component values chosen to have oscillation frequency of 649 KHz. The SPICE generated waveform exhibited a frequency of 640 KHz with % THD as 1.02 % only.

13.10 MOCCII-Based VM/CM QO

Horng [44] proposed a 3rd-order MOCCIIs based VM/CM QO with two MOCCIIs, three GCs and three resistors. The presented circuit is shown in Fig. 13.9 which offers controllability of both CO and FO independently.

Assuming ideal MOCCIIs, an analysis of circuit in Fig. 13.9 gives the following CE:

$$s^3C_1C_2C_3R_1R_2R_3 + s^2C_3R_1R_3(C_1 + C_2) + sC_3R_3 + 1 = 0 \tag{13.23}$$

The expressions for CO and FO from the above equation are found to be:

$$\text{CO} : R_3 = \frac{C_1C_2R_2}{C_3(C_1 + C_2)} \tag{13.24}$$

$$\text{FO} : \omega_0 = \sqrt{\frac{1}{C_1C_2R_1R_2}} \tag{13.25}$$

Thus, both CO and FO can be independently controlled the former by R_3 and the latter by R_1.

The output voltages V_{01} and V_{02} can be expressed as

$$\frac{V_{02}(s)}{V_{01}(s)} = -\frac{1}{sR_1C_3} \tag{13.26}$$

The phase difference (φ) between $V_{02}(s)$ and $V_{01}(s)$ is found to be $\varphi = \pi/2$, thus, both $V_{02}(s)$ and $V_{01}(s)$ are in quadrature. Similarly, the two output currents $I_{02}(s)$ and $I_{01}(s)$ can be related as:

$$\frac{I_{02}(s)}{I_{01}(s)} = -\frac{1}{sR_3C_3} \tag{13.27}$$

The phase difference (φ) between $I_{02}(s)$ and $I_{01}(s)$ is seen to be: $\varphi = \pi/2$, thus, ensuring that both $I_{02}(s)$ and $I_{01}(s)$ are also in quadrature.

From the above analysis, it is seen that the oscillator circuit is capable of providing both VM and CM quadrature signals simultaneously.

SPICE simulations were carried out to check the validity of the proposed circuit using MOCC implementations based on those in [56] using 0.18 μm level 3 MOSFET parameters from TSMC. The circuit performance was found to be in good agreement with the theoretical values.

There are several other publications which deal with ideas related to the one discussed here, for instance, for MOCC- based oscillators see [5, 21, 36, 40, 47, 52] and for MOCCC-based oscillators see [4, 6, 16, 25, 29, 30, 37, 49].

13.11 VM/CM QO Using FDCCII

A more flexible and versatile FDCCII-based SRCO was proposed by Horng, Hou, Chang, Chou, Lin in [20] which is shown in Fig. 13.10 which has all the features of the earlier circuits but in addition, this circuit, with one additional Z-terminal incorporated with the FDCCII architecture, is able to provide two explicit current mode outputs which are in quadrature to each other.

The CO and FO for this circuit are given by:

$$\text{CO}: \ \mathrm{R}_1 = \mathrm{R}_2 \tag{13.28}$$

$$\text{FO}: \ \omega_0 = \sqrt{\frac{1}{C_1C_2R_3R_2}} \tag{13.29}$$

Thus, CO can be controlled by R_1 and FO can be varied independently through R_3.

Analyzing this QO taking into account the various non-idealities such as the voltage tracking errors and current tracking errors, the terminal relationships of voltages and currents are found to be governed by the following equations:

$$V_{xa} = \alpha_{a1}v_{y1} - \alpha_{a2}v_{y2} + \alpha_{a3}v_{y3},\ V_{xb} = -\alpha_{b1}v_{y1} + \alpha_{b2}v_{y2} + \alpha_{b4}v_{y4}i_{y1} = i_{y2} = i_{y3} = i_{y4} = 0,\ i_{zai} = \pm\beta_{ai}i_{xa}$$

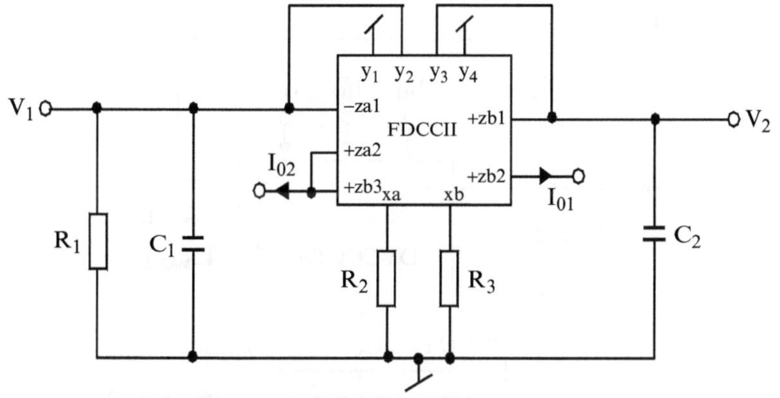

Fig. 13.10 CM QO proposed by Horng et al. [20]

Using the above non-ideal parameters, the non-ideal CO and FO become:

$$R_1 = \frac{R_2}{\alpha_{a2}\beta_{a1}} \tag{13.30}$$

$$\omega_0 = \sqrt{\frac{\alpha_{b2}\alpha_{a3}\beta_{a1}\beta_{b1}}{C_1C_2R_3R_2}} \tag{13.31}$$

From the above equations, it is seen that both CO and FO differ slightly from their ideal expressions but still they are controllable independently.

The quadrature output voltages can be expressed (under steady state) as:

$$V_1 = \omega C_2 R_3 e^{j90^0} V_2 \tag{13.32}$$

and the quadrature output currents can be given by $I_{0ut1} = \omega C_2 R_3 e^{j90^0} I_{0ut2}$ for $R_2 = R_3$

The FDCCII from [55] was used to verify the validity of this circuit. The MOSFET parameters from 0.18 μm CMOS process parameters from TSMC were employed and the FDCCII was biased from ±2.5 V DC power supply. Good agreement between the theoretical and simulation results was observed.

13.12 Electronically-Programmable Dual-Mode QO Using a DVCCCTA and Only Two GCs

The building block differential voltage current conveyor transconductance amplifier (DVCCTA) [32, 51] has been introduced earlier in Chap. 10. The differential voltage current-controlled conveyor transconductance amplifier (DVCCCTA). is an electronically-controllable extension of this. Since it has two parameters which

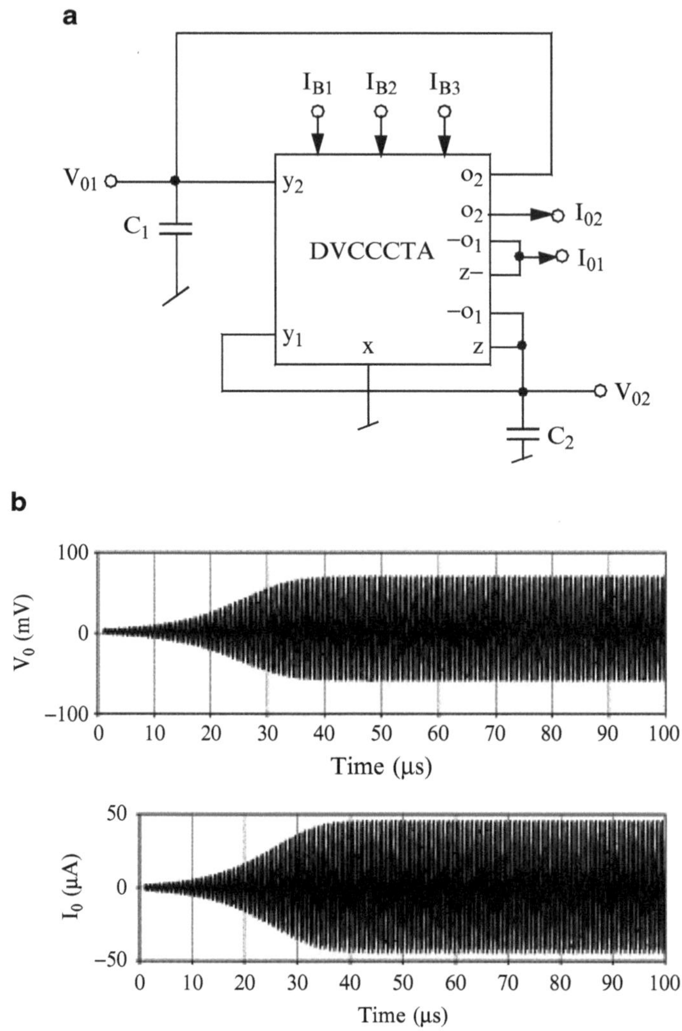

Fig. 13.11 DVCCCTA based QO proposed by Jaikla et al. [43] (**a**) Circuit configuration (**b**) The growing oscillation of quadrature oscillations for VM and CM (**c**) Steady state waveforms for the quadrature VM and CM outputs (**d**) The frequency spectrums for VM and CM outputs

are electronically controllable namely R_x of the front end CC and the transconductance gain g_m, it follows that any circuit built around this building block can, therefore, take advantage of reducing two external resistors and using in their place *implicitly* Rx and $1/g_m$ as electronically-variable resistors. That this can indeed be done was demonstrated by Jaikla, Siripruchyanun and Lahiri [43] who demonstrated that a resistor-less dual-mode QO can indeed be made from only a single DVCCCTA block. Their proposition is shown in Fig. 13.11.

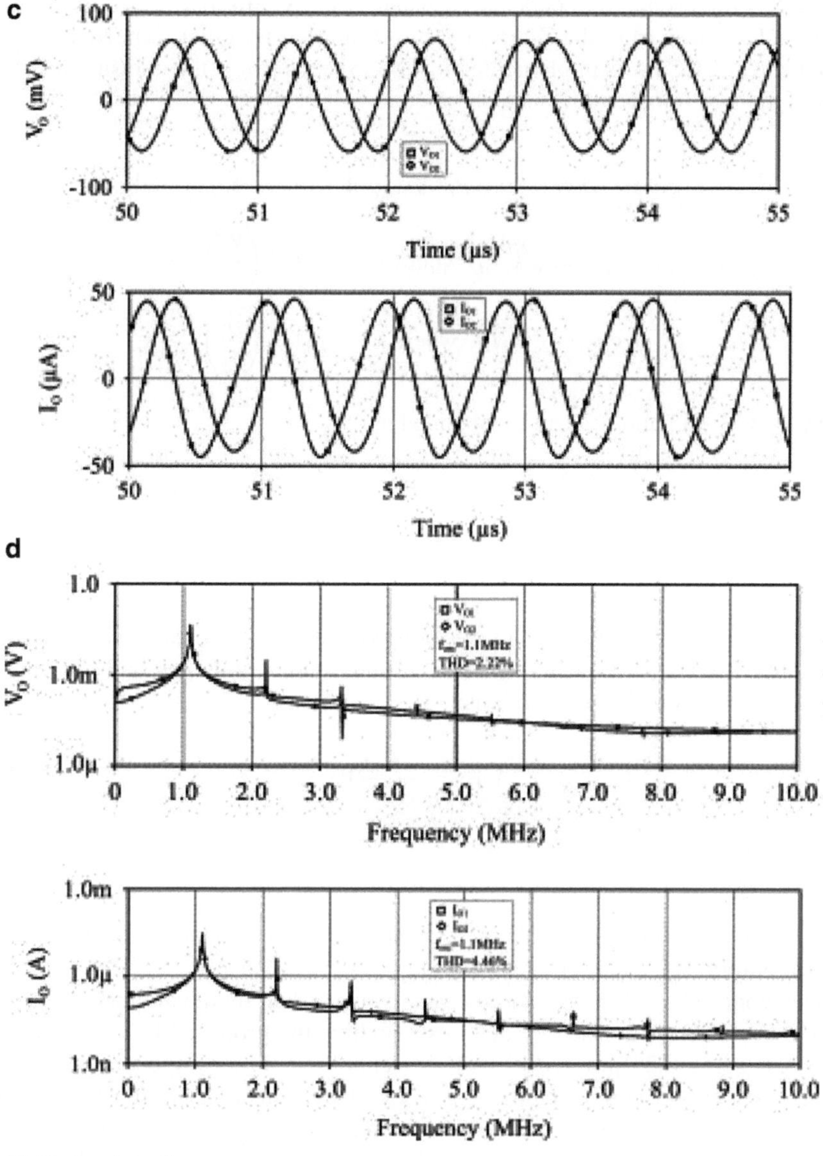

Fig. 13.11 (continued)

Assuming ideal DVCCCTA and performing routine circuit analysis of the circuit shown in Fig. 13.11 gives the following CE:

$$s^2C_1C_2R_x + sC_1(R_xg_{m1} - 1) + g_{m2} = 0 \tag{13.33}$$

where R_x represents the parasitic resistance at X-terminal, g_{m1}, g_{m2} are the transconductances from the Z-terminal to -O_1 and O_2 terminal respectively.

When DVCCCTA is implemented in CMOS technology, the parasitic resistance R_x and the two transconductance parameters will be given by:

$$R_x = \sqrt{\frac{1}{8k_1 I_{B1}}} \text{ where } k_1 = \mu_p C_{ox}\left(\frac{W}{L}\right) = \mu_n C_{ox}\left(\frac{W}{L}\right) \text{ and} \tag{13.34}$$

$$g_{m1} = \sqrt{k_2 I_{B2}} \text{ where } k_2 = \mu_n C_{ox}\left(\frac{W}{L}\right) \tag{13.35}$$

$$g_{m2} = \sqrt{k_3 I_{B3}} \text{ where } k_3 = \mu_n C_{ox}\left(\frac{W}{L}\right) \tag{13.36}$$

From the CE, CO and FO can be obtained as:

$$\text{CO}: g_{m1} R_x \leq 1 \text{ or } k_2 I_{B2} < 8k_1 I_{B1} \tag{13.37}$$

$$\text{FO}: \omega_0 = \sqrt{\frac{g_{m2}}{C_1 C_2 R_x}} = \frac{(8k_1 k_3 I_{B1} I_{B3})^{1/4}}{(C_1 C_2)^{1/2}} \tag{13.38}$$

Hence, it is seen that CO and FO are independently electronically-variable by I_{B2} (g_{m1}) and I_{B3} (g_{m2}) respectively. The various output voltages and currents are related by the following expressions:

$$V_{02} = j\frac{\omega_0 C_1}{g_{m2}} V_{01} \tag{13.39}$$

$$I_{02} = j\frac{\omega_0 C_2}{g_{m2}} I_{01} \tag{13.40}$$

Thus, from the above equations, it is clear that both voltage output signals and current output signals are in quadrature.

The workability of the proposed circuit was verified by the authors of [43] by a CMOS DVCCCTA devised by them using which the proposed QO was simulated in SPICE using 0.25 μm TSMS CMOS technology. The component values were chosen to give oscillation frequency of 1.1 MHz. Figure 13.11b shows the building up of the oscillations, Fig. 13.11c shows the steady state waveforms for the CM and VM outputs while Fig. 13.11d shows the frequency spectra of the CM and VM output waveforms. THE THD was found to be 3 % and 5 % respectively for VM and CM outputs. These results confirm the practical workability of the proposed QO circuit of Fig. 13.11.

Concluding Remarks
This chapter has focused on some selected sinusoidal oscillator configurations using a large variety of extended CC types. For brevity, only a few significant circuits have been included from amongst a large number available as reflected in the list of references [1–52] at the end of this chapter. For further reading, still more additional references are provided for the interested readers in Appendix at the end of this monograph.

The inclusion of various circuits in this chapter was guided by considerations such as employment of grounded capacitors, employment of AGPEs, availability of explicit voltage and current outputs from low output impedance and high output impedance terminals respectively, availability of QOs in VM, CM or dual-mode etc. It must be emphasized again that although none of these circuits can be actually used in practice due to the non-availability of the employed building blocks as off-the-shelf ICs, however, since all the building blocks employed are implementable in bipolar and CMOS technologies, the described circuits which offer very interesting features should provide good enough motivation for the IC designers to think and work in a direction to make these exciting varieties of CCs available as stand-alone ICs. Alternatively, the described topologies can also be employed as an integratable part of the complete system-on-chip which requires a sinusoidal oscillator to be integrated on the same IC chip using bipolar/CMOS technology.

References

1. Celma S, Martinez PA (1995) Transformation of sinusoidal oscillators using universal active elements. IEE Proc Circ Devices Syst 142:353–356
2. Celma S, Martinez PA (1995) On the limit cycle stability in current-mode oscillators. Int J Electron 79:163–169
3. Soliman AM (1998) Generalized voltage and current conveyors: practical realizations using CCII. IEICE Trans Fundam E-81:973–975
4. Abuelma'atti MT, Al-Qahtani MA (1998) A new current-controlled multiphase sinusoidal oscillator using translinear current conveyors. IEEE Trans Circ Syst-II 45:881–885
5. Soliman AM (1998) Current mode CCII oscillators using grounded capacitors and resistors. Int J Circ Theor Appl 26:431–438
6. Abuelma'atti MT, Tasadduq NA (1998) A novel current-controlled oscillator using translinear current conveyors. Frequenz 52:123–124
7. Gupta SS, Senani R (2000) Grounded-capacitor current-mode SRCO-novel application of DVCCC. Electron Lett 36:195–196
8. Papazoglou CA, Karybakas CA (2000) An electronically tunable sinusoidal oscillator suitable for high frequencies operation based on a single dual-output vriable-gain CCII. Analog Integr Circ Sig Process 23:31–44
9. Gupta SS, Senani R (2001) Comment: CMOS differential difference current conveyors and their applications. IEE Proc Circ Devices Syst 148:335–336
10. Toker A, Kuntman H, Cicekoglu O, Discigil M (2002) New oscillator topologies using inverting second-generation current conveyors. Turk J Electr Eng 10:119–129

11. Chang CM, Al-Hashimi BM, Chen HP, Tu SH, Wan JA (2002) current mode single resistance controlled oscillators using only grounded passive components. Electron Lett 38:1071–1072
12. Barthelemy H, Meillere S, Kussener E (2002) CMOS sinusoidal oscillators based on current-controlled current conveyors. Electron Lett 38:1254–1256
13. Hou CL, Chen YT, Huang CC (2003) The oscillators using a single DVCC. Tamkang J Sci Eng 6:183–187
14. Aggarwal V (2004) Novel canonic current mode DDCC based SRCO synthesized using a genetic algorithm. Analog Integr Circ Sig Process 40:83–85
15. Martinez PA, Martinez SBMM (2005) Generation of two integrator loop variable frequency sinusoidal oscillator. Int J Electron 92:619–629
16. Maheshwari S, Khan IA (2005) Current-controlled third order quadrature oscillator. IEE Proc Circ Devices Syst 152:605–607
17. Aggarwal V, Kilinc S, Cam U (2006) Minimum component SRCO and VFO using a single DVCCC. Analog Integr Circ Sig Process 49:181–185
18. Horng JW, Hou CL, Chang CM, Lin YT, Shiu IC, Chiu WY (2006) First order allpass filter and sinusoidal oscillators using DDCCs. Int J Electron 93:457–466
19. Kilinc S, Jain V, Aggarwal V, Cam U (2006) Catalogue of variable frequency and single-resistance-controlled oscillators employing a single differential difference complementary current conveyors. Frequenz 60:142–150
20. Horng JW, Hou CL, Chang CM, Chou HP, Lin CT, Wen YH (2006) Quadrature oscillators with grounded capacitors and resistors using FDCCIIs. ETRI J 28:486–494
21. Khan IA, Hasan S (2006) Current mode four phase quadrature oscillator using CCIIs. J Active Passiv Electron Devices 1:273–279
22. Kumar P, Keskin AU, Pal K (2007) DVCC-based single element controlled oscillators using all-grounded components and simultaneous current–voltage mode outputs. Frequenz 61:141–144
23. Horng JW, Hou CL, Chang CM, Cheng ST, Su HY (2008) Current or/and voltage-mode quadrature oscillators with grounded capacitors and resistors using FDCCIIs. WSEAS Trans Circ Syst 3:129–138
24. Toker A, Ozoguz S (2008) Comment on Fist order all pass filter and sinusoidal oscillators using DDCCs. Int J Electron 95:867
25. Pandey N, Paul SK (2008) A novel electronically tunable sinusoidal oscillator based on CCCII (−IR). J Active Passiv Electron Devices 3:135–141
26. Lahiri A (2009) Additional realizations of single-element-controlled oscillators using single ICCII-. Int J Comput Electr Eng 1:303–306
27. Soliman AM (2009) Generation of oscillators based on grounded capacitor current conveyors with minimum passive components. J Circ Syst Comput 18:857–873
28. Kumngern M, Dejhan K (2009) DDCC-based quadrature oscillator with grounded capacitors and resistors. Active Passiv Electron Comp: Article ID 987304, 4p
29. Maheshwari S (2010) Current-mode third-order quadrature oscillator. IET Circ Devices Syst 4:188–195
30. Kumngern M, Chanwutitum J, Dejhan K (2010) Electronically tunable multiphase sinusoidal oscillator using translinear current conveyors. Analog Integr Circ Sig process 65:327–334
31. Sotner R, Hrubos Z, Slezak J, Dostal T (2010) Simply adjustable sinusoidal oscillator based on negative three port current conveyors. Radioengineering 19:446–454
32. Lahiri A, Jaikla W, Siripruchyanun M (2010) Voltage-mode quadrature sinusoidal oscillator with current tunable properties. Analog Integr Circ Sig Process 65:321–325
33. Soliman AM (2010) On the generation of CCII and ICCII oscillators from three op-Amps oscillator. Microelectron J 41:680–687
34. Soliman AM (2010) Generation of three oscillator families using CCII and ICCII. Int J Electron Commun (AEU) 64:880–887
35. Soliman AM (2010) Generation of current conveyor based oscillators using nodal admittance matrix expansion. Analog Integr Circ Sig Process 65:43–59

36. Lahiri A (2011) Current- mode variable frequency quadrature sinusoidal oscillators using two CCs and four passive components including grounded capacitors: a supplement. Analog Integr Circ Sig Process 68:129–131
37. Lahiri A (2011) Deriving (MO) (I) CCCII based second-order sinusoidal oscillators with non-interactive tuning laws using state variable method. Radioengineering 20:349–353
38. Soliman AM (2011) Current conveyor based or unity gain cells based two integrator loop oscillators. Microelectron J 42:239–246
39. Soliman AM (2011) Transformation of a floating capacitor oscillator to a family of grounded capacitor oscillators. Int J Electron 98:289–300
40. Lahiri A (2011) New canonic active RC sinusoidal oscillator circuits using second-generation current conveyors with application as a wide-frequency digitally controlled sinusoidal generator. Active Passiv Electron comp: Article ID 274394, 8p
41. Beg P, Siddiqi MA, Ansari S (2011) Multi output filter and four phase sinusoidal oscillator using CMOS DX-MOCCII. Int J Electron 98:1185–1198
42. Horng JW, Wang ZR, Yang TY (2011) Single ICCII sinusoidal oscillators employing grounded capacitors. Radioengineering 20:608–613
43. Jaikla W, Siripruchyanun M, Lahiri A (2011) Resistorless dual-mode quadrature sinusoidal oscillator using a single active building blocks. Microelectron J 42:135–140
44. Horng JW (2011) Current/voltage-mode third orders quadrature oscillator employing two multiple outputs CCIIs and grounded capacitors. Ind J Pure Appl Phys 49:494–498
45. Maheshwari S, Chaturvedi B (2011) High output impedance CMQOs using DVCCs and grounded components. Int J Circ Theor Appl 39:427–435
46. Soliman AM (2012) On oscillator circuits using two output CCII, DVCC and FDCCII. J Active Passiv Electron Devices 7:325–343
47. Lahiri A (2012) Current-mode variable frequency quadrature sinusoidal oscillators using two CCs and four passive components including grounded capacitors. Analog Integr Circ Sig Process 71:303–311
48. Chaturvedi B, Maheshwari S (2012) Second order mixed mode quadrature oscillator using DVCCs and grounded components. Int J Comput Appl 58:42–45
49. Maheshwari S, Verma R (2012) electronically tunable sinusoidal oscillator circuit. Active Passiv Electron Comp: Article ID719376, 6p
50. Maheshwari S (2013) Voltage-mode four-phase sinusoidal generator and its useful extensions. Active Passiv Electron Comp: Article ID 685939, 8p
51. Chen HC, Chen CY (2014) CMOS realization of single-resistance-controlled and variable frequency dual-mode sinusoidal oscillators employing a single DVCCTA with all-grounded passive components. Microelectron J 45:226–238
52. Yucel F, Yuce E (2014) CCII based more tunable voltage-mode all-pass filters and their quadrature oscillator applications. Int J Electron Commun (AEU) 68:1–9
53. Elwan HO, Soliman AM (1997) A novel CMOS differential voltage current conveyor and its applications. IEE Proc Circ Devices Syst 144:195–200
54. Awad A, Soliman AM (1999) Inverting second generation current conveyors: the missing building blocks. Int J Electron 86:413–432
55. Al-Adawy AA, Soliman AM, Elwan HO (2000) A novel fully differential current conveyor and its applications for analog VLSI. IEEE Trans Circ Syst-II 47:306–313
56. Surakampontorn W, Riewruja V, Kumwachara K, Dejhan K (1991) Accurate CMOS based Current conveyors. IEEE Trans Instrum Meas 40:699

Chapter 14
Second Generation Applications of Other Types of Current Conveyors in Realizing Synthetic Impedances

Abstract Chosen from a vast amount of literature in the area of impedance simulation using CCs, a number of novel synthetic impedance circuits have been described using the new variants of CCs (such as DOCCII, DVCC, CCIII, DXCCII, MICCII, DDCC and FDCCII etc.) for realizing both grounded and floating forms of inductors and other related elements, which possess a number of interesting features.

14.1 Introduction

In Chap. 5 of this monograph, we have outlined key contributions made in the area of immittance simulation using the normal type of 3-port current conveyors. It was highlighted therein that one of the most prominent feature of employing current conveyors for this purpose is the remarkable property that a floating impedance (FI) can be realized using CCs *without requiring any component matching conditions*- a feat which is impossible to be achieved by the circuits based upon op-amps. Another notable feature of the CC-based circuits is that it is invariably feasible to realize a variety of simulated impedances using CCs with a bare minimum number of external passive components needed for the intended type of immittance function, for instance see [1–5]. It may also be mentioned that a majority of the circuits described in Chap. 5 can be practically implemented with commercially available ICs such as AD844 (with CCII+realized with one AD844 and CCII- realized by two such devices) or op-amps uA741/LF356 in conjunction with transistor arrays CA3096 or OPA660/860.

In this chapter, we re-consider the problem of realizing synthetic grounded and floating immittances (FI) but our main focus here is on the circuits which have been proposed in the literature (during the past two decades or so) using the extended versions and/or new varieties of CCs introduced by numerous researchers see [1–53]. Although these new varieties of CCs are not yet available as of-the-shelf ICs numerous bipolar and/or CMOS implementations of all the variants of CCs have been suggested by the various researchers from time to time. The circuits presented here are superior to those presented in Chap. 5 in that they offer one or more of the

© Springer International Publishing Switzerland 2015
R. Senani et al., *Current Conveyors*, DOI 10.1007/978-3-319-08684-2_14

following advantageous features: reduction in total number of active and passive components used, facilitation of electronic-tunability of the realized impedance value, lesser influence of the various parasitic impedances or non-ideal parameters of the active elements employed, use of grounded capacitor(s) or employment of all grounded passive elements, as desirable from the view point of IC implementation. It is believed that these advantages might also stimulate the circuit designers and IC manufacturers to come up with stand-alone ICs of any of the different variants of the CCs or the generalized current conveyors described in Chap. 10, which may eventually bring the described circuits to a more practical footing.

14.2 Simulated Lossless Floating Inductance Using Only Two CCs and Three Passive Components

In Chap. 5 it was shown that lossless floating inductance and other elements such as floating FDNR require either two CCs and five passive elements or three passive elements and three CCs. Thus, using the conventional type of CCs, any circuit which can realize floating inductor/FDNR using only three passive elements and only two CCs, was not possible. However, if the CCII can be so modified that it can have two complimentary Z- output terminals then it has been found [7] that it does become possible to realize a lossless FI with only two dual output CCIIs (DOCCII) and a bare minimum of only three passive components!

One such circuit was presented by Ananda Mohan in [7] and is shown here in Fig. 14.1.

It is straight forward to determine, using the characterizing equations of the DOCCII, that the floating inductance simulated between ports 1 and 2 is given by $L_{eq} = C_o R_1 R_2$. This circuit clearly has the advantage of employing a grounded capacitor as compared to the two-CC- based and three-CC- based FI simulation circuits described in Chap. 5, all of which employ a floating capacitor.

14.3 DVCC-Based Floating Inductance/FDNR with *All Grounded Passive Elements*

Sedef and Acar [8] demonstrated that using two differential voltage current conveyors (DVCCs)[1] with dual complimentary outputs (characterized by $i_{y1} = 0$, $i_{y2} = 0, v_y = v_{y1} - v_{y2}, i_{z1} = +i_x$ and $i_{z2} = -i_x$), it becomes possible to realize floating impedance using only two DVCCs[2] along with all the three impedances being

[1] The differential voltage Current Conveyor was first introduced by K. Pal in [43].

[2] For a class of floating inductors realizable with DVCCs along with other types of CCs, using only three passive elements, see [35].

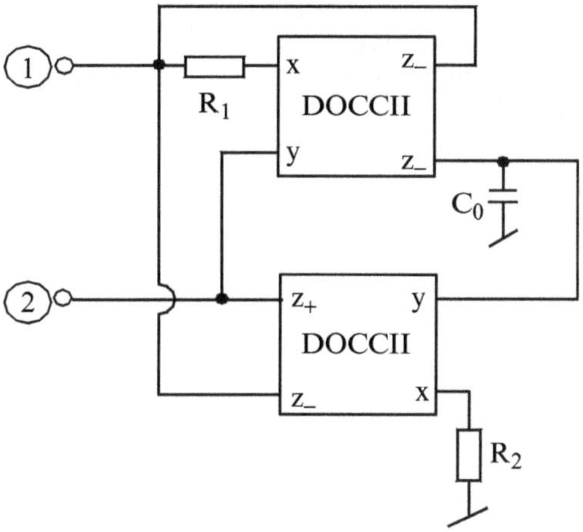

Fig. 14.1 Floating inductance simulation using DOCCII proposed by Ananda Mohan [7]

grounded. Their proposition is shown here in Fig. 14.2 and is characterized by the Y-matrix:

$$[Y] = \left(\frac{Z_3}{Z_1 Z_2}\right)\begin{bmatrix} 1 & -1 \\ -1 & 1 \end{bmatrix} \tag{14.1}$$

It may be noted that from the same circuit, it is possible to realize a lossless floating inductance by choosing Z_1 and Z_2 as resistors and Z_3 as a capacitor and an ideal FDNR by choosing Z_3 as a resistor and Z_1 and Z_2 as capacitors. In both the cases, it is possible to realize variable elements through a single variable resistance. Furthermore, with the resistor implemented by MOSFET-based voltage-controlled grounded resistor such as the one in [54], the circuit can also realize electronically-variable floating inductance or floating FDNR.

The workability of this circuit has been verified by using a CMOS DVCC (see Fig. 1b of [8]) based on 1.2 μm CMOS model parameters obtained through MOSIS. The FDNR was used to design a 4th order Butterworth LC band pass filter with $f_0 = 100$ KHz and BW = 20 KHz. Very good correspondence between theoretical and simulation results was obtained which is indicative of the practical viability of the circuit of Fig. 14.2.

14.4 Simulated Inductors Employing CCIII

Several authors have employed third generation current conveyors (CCIIIs) to realize a variety of simulated immittances for example see [6, 9, 22, 28]. CMOS realisations of CCIII are already known in literature, for example see [11, 55]. Wang and Lee [10] presented a systematic approach to synthesize R-L and C-D immittances employing a single CCIII along with a minimum number of passive

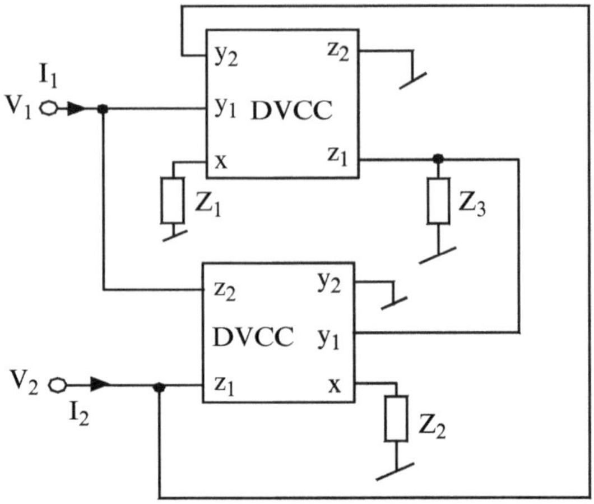

Fig. 14.2 Floating FDNR employing two DVCCs proposed by Sedef and Acar [8]

components by formulating a generalized four-node structure (including one node as a ground) in which a CCIII was embedded with its three terminals terminated into the remaining three nodes of the structure. This generalized configuration gave birth to a number of special cases out of which here we have included four specific circuits which have the desirable feature of employing a grounded capacitor and the use of a minimum of only two resistors. These circuits are shown in Fig. 14.3. The circuits of Fig. 14.3a and b use non-inverting CCIII while the circuits of Fig. 14.3c and d employ inverting CCIII. Assuming ideal CCIII±, an analysis of these circuits gives:

$$Y_{ina} = \frac{1}{R_1} + \frac{1}{R_2} + \frac{1}{sC_0R_1R_2} = Y_{inc} \tag{14.2}$$

and

$$Z_{inb} = R_1 + R_2 + sC_0R_1R_2 = Z_{ind} \tag{14.3}$$

From the above expressions, it is clear that the circuits of Fig. 14.3a and c simulate a parallel R-L immittance while Fig. 14.3b and d simulate a series R-L immittance without the requirements of any component-matching or cancellation condition. Furthermore, the application of the RC: CR transformation on these four circuits would yield circuits capable of simulating parallel C-D and series C-D immittances respectively.

For verifying the workability of the circuits, a CMOS CCIII proposed by Piovaccari [55] with level 28 MOS model parameters provided by TSMC 0.6 μm CMOS technology were employed. The biasing supply was ± 2.5 V and the circuits were used to realize simple filters whose SPICE simulation results were found to be in good agreement with theory.

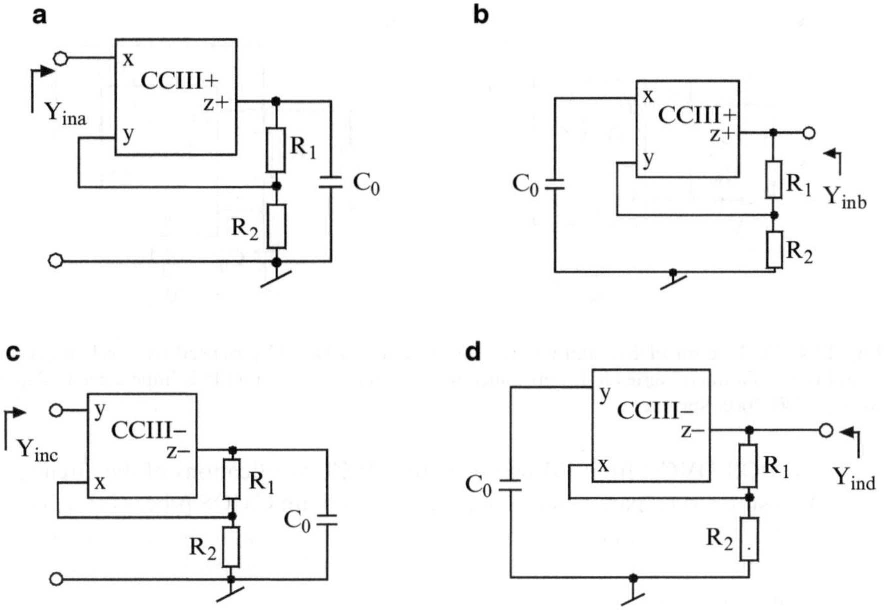

Fig. 14.3 Some Minimal realizations of R-L immittances using a single CCIII as proposed by Wang and Lee [10]

14.5 Grounded R-L and C-D Immittances Using a Single DVCC

Circuits for the simulation of grounded series R-L and parallel R-L immittances employing a single DVCC, a grounded capacitor and two resistors, without any component- matching-constraints, were proposed by Incekaraoglu and Cam in [14] and are shown here in Fig. 14.4. Series C-D and parallel C-D immittances can be obtained from the same circuits by applying RC: CR transformation. A routine circuit analysis, assuming ideal DVCC, yields the following expressions:

$$Z_{ina} = \frac{V_{in}}{I_{in}} = (R_1 + R_2) + sC_0R_1R_2 \tag{14.4}$$

and

$$Y_{inb} = \frac{I_{in}}{V_{in}} = \left(\frac{1}{R_1} + \frac{1}{R_2}\right) + \frac{1}{sC_0R_1R_2} \tag{14.5}$$

Hence, the circuit of Fig. 14.4a simulates series R-L impedance while the circuit of Fig. 14.4b simulates a parallel R-L admittance.

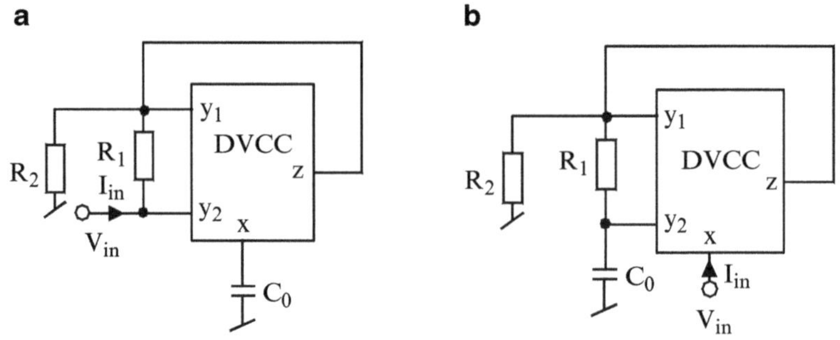

Fig. 14.4 Realization of R-L immittances using a single DVCC proposed by Incekaraglu and Cam (**a**) Simulation of series R-L impedance (**b**) Simulation of parallel R-L impedance (Adapted from [14] © 2005 Springer

The CMOS DVCC from [9] was used for SPICE verifications of the circuits of Fig. 14.4 using model parameters from MIETEC 0.5 μm CMOS process and DVCC biased with ± 2.5 V supply. The circuits were used to design simple second order filters. The filter characteristics obtained from SPICE simulations exhibited good agreement with theory.

Another grounded positive and negative parallel R-L immittance simulator circuit was by Yuce [17] is shown here in Fig. 14.5. An analysis of this circuit gives the following admittance:

$$Y_{in} = \frac{I_{in}}{V_{in}} = \pm\left(\frac{1}{R_1} + \frac{1}{R_2}\right) \pm \frac{1}{sCR_1R_2} \tag{14.6}$$

Thus, the circuit simulates positive or negative parallel R-L impedance. In the equation (14.6), plus sign implies the use of a DVCC+ while negative sign implies a DVCC-.

14.6 Electronically-Controllable Gyrator and Grounded Inductor Using DXCCII

The dual-X CCII (DXCCII) essentially combines the significant features of both CCII and ICCII and has been shown to be particularly interesting building block from the view point of MOSFET-C filter design. The dual-X nature of the X-terminals facilitates easy cancellation of the square non-linearity of the MOSFETs because if the drain and source of the MOSFET are connected between two X-terminals, the imposition of complementary signals on the drain and source terminals of the MOSFET ensures the cancellation of the square nonlinearity. In view of this, it therefore, looks feasible that this basic idea may also be useful in realizing MOSFET-based gyrator and simulated inductors. This was first

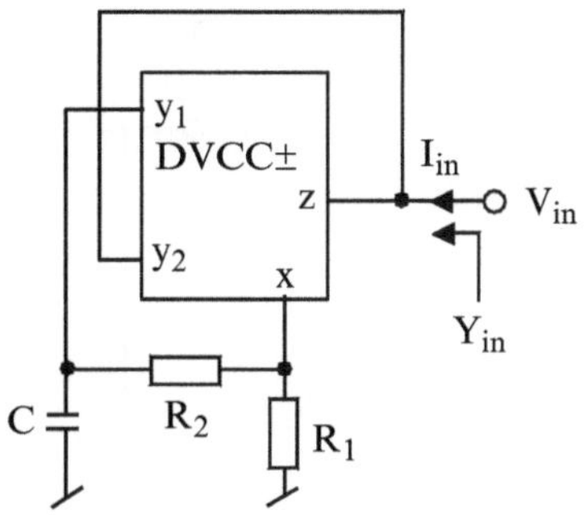

Fig. 14.5 The configuration proposed by Yuce for the realization of positive/negative parallel R-L impedances (Adapted from [17] © 2006 Springer)

demonstrated by Zeki and Toker in [16] by realizing a gyrator and a simulated inductor with the feature of electronic-controllability through external gate control voltages applied to the MOSFETs.

The gyrator circuit proposed by Zeki and Toker in [16] is shown here in Fig. 14.6a. By a straight forward analysis, the y-matrix of this circuit is found to be:

$$[Y] = \begin{bmatrix} 0 & 2\mu C_{ox}\left(\frac{W}{L}\right)_2 (V_{G2} - V_{th}) \\ -2\mu C_{ox}\left(\frac{W}{L}\right)_1 (V_{G1} - V_{th}) & 0 \end{bmatrix} \quad (14.7)$$

In view of the above, it is obvious that this gyrator can be used as a simulated inductor by terminating either port 2 or port 1 into a grounded capacitor. For instance, with a capacitor C_1 connected at port 2, the input impedance looking into the port 1 is given by:

$$Z_{in} = \frac{sC_1}{4\mu^2 C_{ox}^2 \left(\frac{W}{L}\right)_1 \left(\frac{W}{L}\right)_2 (V_{G1} - V_{th})(V_{G2} - V_{th})} \quad (14.8)$$

Hence, grounded inductor value (L_{eq}) is given by:

$$L_{eq} = \frac{C_1}{4\mu^2 C_{ox}^2 \left(\frac{W}{L}\right)_1 \left(\frac{W}{L}\right)_2 (V_{G1} - V_{th})(V_{G2} - V_{th})} \quad (14.9)$$

which can be controlled by V_{G1} or V_{G2} or by both.

The validity of the DXCCII-based gyrator of Fig. 14.6a was verified by realizing a CM biquad filter (see Fig. 14.2b of [16]) wherein the CMOS DXCCII from [56] was used to simulate the inductor with ±2.5 V DC power supplies and biasing current of $I_B = 100$ μA using BSIM3v3 model parameters of AMS 0.35 μm CMOS process. The capacitors of values $C_1 = 7$ pF and $C_2 = 10$ pF were used. The channel length (L) and channel width (W) for MOSFETs M_1 and M_2 was taken as 1.5 μm. The variation of cutoff frequency with gate voltage is shown in Fig. 14.6b which confirms the validity of the simulated inductor.

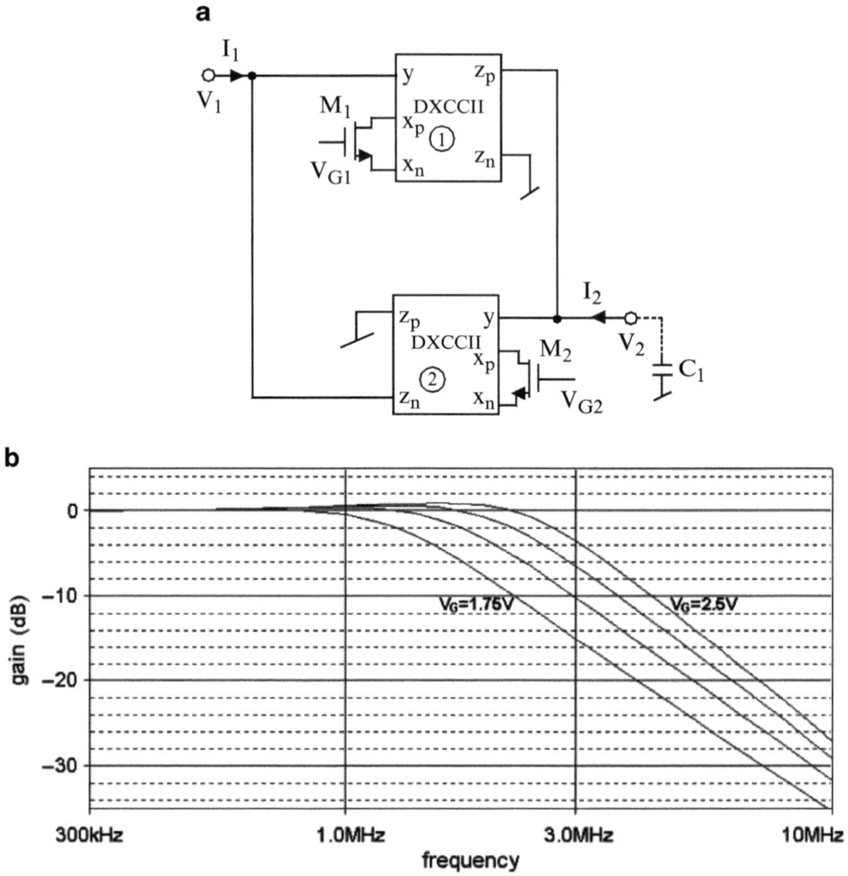

Fig. 14.6 DXCCII-based electronically controllable gyrator proposed by Zeki and Toker (Adapted from [16] © 2005 Elsevier) (**a**) Circuit configuration (**b**) Variation of cutoff frequency with gate voltage

14.7 Grounded Inductor Simulation Using the Modified Inverting CCII (MICCII)

A modified inverting second generation CC (MICCII) [19] is characterized by the following hybrid matrix:

$$\begin{bmatrix} V_x \\ I_y \\ I_z \end{bmatrix} = \begin{bmatrix} -\beta & 0 & 0 \\ 0 & 0 & 0 \\ 0 & \pm 2\alpha & 0 \end{bmatrix} \begin{bmatrix} V_y \\ I_x \\ V_z \end{bmatrix} \tag{14.10}$$

where α represents the non-ideal voltage gain and β represents the non-ideal current gain of the MICCII (ideally equal to unity). The positive and negative signs of α

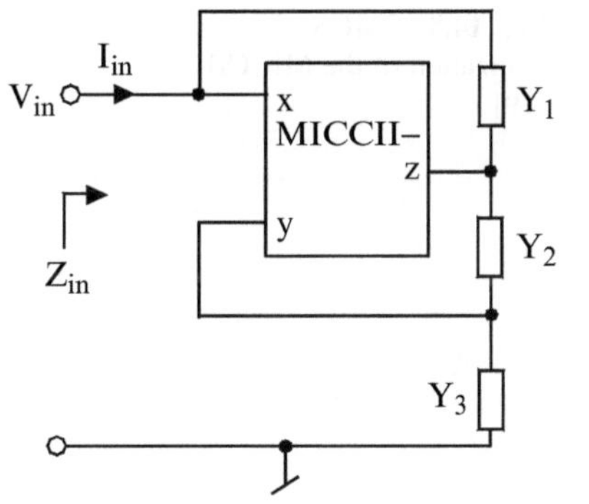

Fig. 14.7 Realization of the grounded inductor employing MICCII- [19]

denote the positive (MICCII+) and negative (MICCII-) respectively. The expression for the input admittance of the circuit of Fig. 14.7 (assuming ideal MICCII-) is given by:

$$Y_{in} = \frac{I_{in}}{V_{in}} = \frac{y_1 y_3}{2y_2} + y_1 - \frac{y_3}{2} \tag{14.11}$$

In the above equation, if we select $y_1 = 1/R$, $y_3 = 2/R$ and $y_2 = sC$, the circuit of Fig. 14.7 would simulate a grounded inductance (L_{eq}) of value CR^2. Hence, a lossless grounded inductor employing a minimum number of passive and active elements is achieved.

Taking into account the non-idealities of the MICCII-, the non-ideal expression for the input admittance is derived as:

$$Y_{in} = \frac{(2\alpha - 1)}{sCR^2\alpha\beta} + \frac{2\alpha + 2\alpha\beta - \beta - 3}{2R\alpha\beta} \tag{14.12}$$

Thus, the circuit realizes an inductor (L_{eq}) in parallel with a resistor (R_{eq}) whose values are given by the following expressions:

$$L_{eq} = \frac{CR^2\alpha\beta}{(2\alpha - 1)} \text{ and } R_{eq} = \frac{2R\alpha\beta}{(2\alpha + 2\alpha\beta - \beta - 3)} \tag{14.13}$$

It may be pointed out that the parameter α of the MICCII- can be changed by varying the bias voltages of the MICCII-(as shown in Fig. 14.8). Thus, assuming $\beta = 1$ and α as a variable, five different categories of inductors can be realized: (i) for $\alpha > 1$, a parallel combination of lossless inductor with positive resistance (ii) for $\alpha = 1$, a lossless inductor (iii) for $0.5 < \alpha < 1$, a parallel combination of lossless inductor with negative resistance (iv) for $0 < \alpha < 0.5$, a parallel combination of negative inductor with a negative resistance and (v) for $\alpha < 0$, a parallel combination of lossless inductor with a positive resistance employing MICCII+.

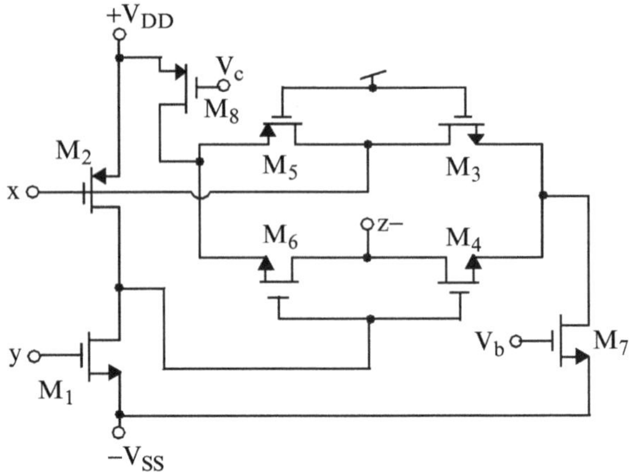

Fig. 14.8 CMOS realization of the MICCII- [19]

The circuit was tested using the CMOS MICCII of Fig. 14.8 with ± 2.5 bias voltage, with $V_B = -0.604$ and $V_c = -0.25$ V. TSMC 0.35 μm CMOS model parameters were chosen. The simulated inductor was used to design a third order Butterworth high pass filter designed for a cut off frequency of 1.59 MHz. Close correspondence between simulation generated frequency response and the theoretical one confirmed the viability of the proposed circuit.

14.8 DO-CCII-Based Synthetic Floating Immittances

Minaei et al. [21] presented a structure as shown in Fig. 14.9 which uses two DO-CCIIs and three passive components to realize general floating impedance. The configuration can simulate an FI, floating capacitance (FC), floating FDNR and floating admittance converter (FAC) without any passive component matching condition, subject to the appropriate choice of various circuit admittances. Assuming ideal DO-CCIIs, an analysis of this circuit yields the following short circuit admittance matrix:

$$[Y] = \frac{y_1 y_2}{y_3}\begin{bmatrix} 1 & -1 \\ -1 & 1 \end{bmatrix} \tag{14.14}$$

Now depending upon the selection of the various passive components, the following floating elements can be obtained: If one selects y_1 and y_2 as resistors with y_3 as capacitor, an FI is realized; if $y_1$and y_3 are chosen as resistors with y_2 as a capacitor, an FC is realized; if y_1 and y_2 are taken as capacitors with y_3 as a resistor, an FDNR is simulated and finally, if y_1 and y_3 are chosen as resistors, with $y_2 = y(s)$ the circuit functions as a FAC. In the realizations of FC, FDNR and FAC only resistors are

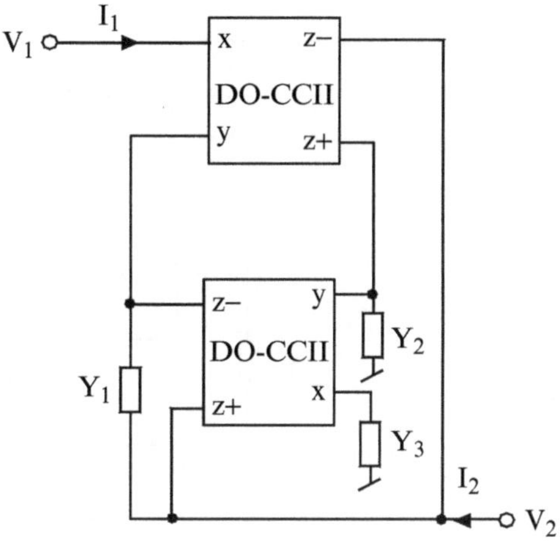

Fig. 14.9 Simulation of FI, FC, FDNR and admittance converter proposed by Minaei et al. (Adapted from [21] © 2006 Springer)

connected at the terminal –X; no capacitors are ever connected at the X-terminals of the DO-CCIIs, which make the circuit highly suitable for high frequency operation.

14.9 A General Circuit for Converting a Grounded Immittance into Floating Immittance

A general structure for converting a grounded admittance into a floating admittance was proposed by Yuce et al. [20] employing one DO-CCII, one CCII-, one CCII + and two passive elements which is shown in Fig. 14.10. In general, the input admittance function of a grounded circuit can be expressed as:

$$y_g = \frac{a_m s^m + a_{m-1} s^{m-1} + \ldots a_1 s + a_0}{b_n s^n + b_{n-1} s^{n-1} + \ldots b_1 s + b_0} \tag{14.15}$$

where **m** and **n** are arbitrary integer numbers, m$\geq$0 and n$\geq$0. Assuming ideal CCIIs, a routine circuit analysis of this circuit gives the following short circuit admittance matrix:

$$[Y] = \left(\frac{y_2}{y_1}\right) y_g \begin{bmatrix} 1 & -1 \\ -1 & 1 \end{bmatrix} = \mathrm{k}\ y_g \begin{bmatrix} 1 & -1 \\ -1 & 1 \end{bmatrix} \tag{14.16}$$

where $\mathrm{k} = \frac{R_1}{R_2}$ is a multiplier constant for floating immittance converter. In the circuit of Fig. 14.10, if CCII- and CCII + are interchanged, the value of k will be negative.

A synthetic FI can be realized if we select $y_1 = sC_1$, $y_2 = 1/R_2$ and $y_g = 1/R_g$

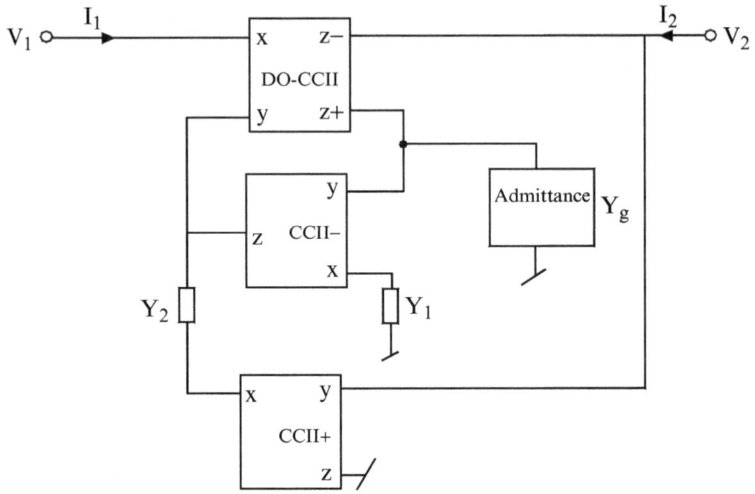

Fig. 14.10 General Configuration for converting grounded admittance into the floating admittance proposed by Yuce et al. (Adapted from [20] © 2006 Springer)

Taking into account the non-idealities of the CCs, namely, $I_{(z+)j} = \alpha_j I_{xj} (j = 1, 3)$, $I_{(z-)n} = -\gamma_n I_{xn} (n = 1, 2)$, $I_{ym} = 0$ and $V_{xm} = \beta_m V_{ym} (m = 1, 2, 3)$, the short circuit admittance matrix becomes

$$[Y]_{nonideal} = \left(\frac{y_2}{y_1 \alpha_1 \beta_1 \beta_2 \gamma_2}\right) y_g \begin{bmatrix} 1 & -\beta_1 \beta_3 \\ -\gamma_1 & \gamma_1 \beta_1 \beta_3 \end{bmatrix} \tag{14.17}$$

It is clear from the above equation that no lossy term appears in this short circuit admittance matrix even under non-ideal conditions.

The workability of the proposed configuration was tested using the CMOS DO-CCII presented by the authors themselves (see Fig. 3 of [20]), biased with ± 1.5 V DC supply simulations were performed in SPICE using 0.35 μm TSMC CMOS technology parameters. Simulation results of a 3rd order Butterworth filter designed by using the proposed simulator, for a cut-off frequency of 159.15 KHz confirmed the validity of proposed circuit.

14.10 Compensated Negative Impedance Converter

Yuce [25] proposed two negative impedance converters (NIC) using two CCs and a resistor. The configuration shown in Fig. 14.11a is comprised of two DO-CCIIs and is capable of reducing both non-ideal gain and parasitic impedance effects simultaneously. The circuit of Fig. 14.11b employs three CCII + s along with a resistor but can reduce only parasitic impedance effects. Since only plus type CCIIs are used in this structure, the configuration can be realized with commercially available

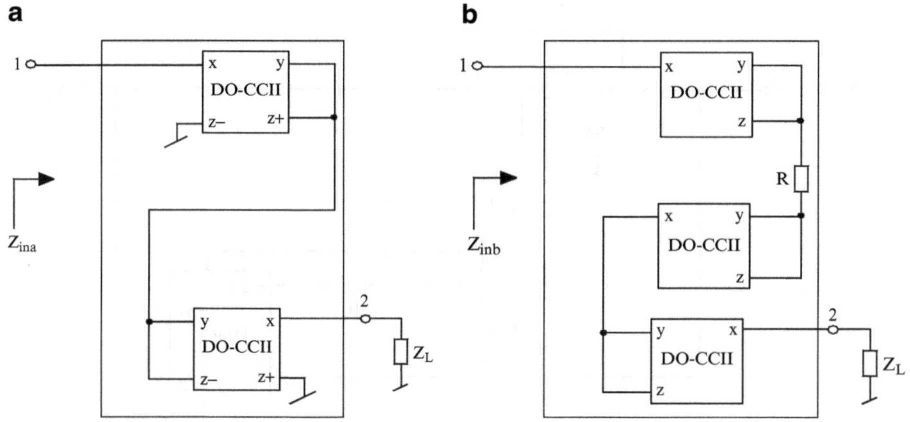

Fig. 14.11 Compensated NICs proposed by Yuce [25]

AD844 (which contains a CCII + and unity gain voltage buffer). If the Y and Z terminal parasitic impedances are ignored, the input impedance looking into the input terminals of the circuit of Fig. 14.11a is obtained as:

$$Z_{ina}(s) = Z_{x1} - \frac{\alpha_{01}\beta_{01}}{\beta_{02}\gamma_{02}} Z_{x2}(s) - \frac{\alpha_{01}\beta_{01}}{\beta_{02}\gamma_{02}} Z_L(s) \quad (14.18)$$

In the circuit of Fig. 14.11b, if only X-terminal parasitic effects of the DO-CCII are considered, the expression for its input impedance can be obtained as:

$$Z_{inb}(s) = R_{x1} + R_{x2} - R_{x3} - R - Z_L(s) \quad (14.19)$$

Thus, if we select $(R_{x1} + R_{x2} - R_{x3}) = R$, the input impedance of the proposed NIC in Fig. 14.11b becomes $Z_{inb}(s) = -Z_L(s)$ and a perfect compensation is achieved. For detailed SPICE simulation results of this circuit, the reader is referred to [25].

14.11 DDCC-Based FI with Improved Low Frequency Performance

Although a lossless FI can be realized by using two DVCCs with only three passive elements (two resistors and a grounded capacitor), Yuce and Minaei [31] presented a lossless FI structure using exactly the same number of active and passive components but with the advantageous features of (i) accommodating the effect of finite input resistance R_x of both the DDCCs and (ii) ensuring that the Z-port parasitics appear directly across port-1 and port-2 respectively. Their circuit arrangement is shown in Fig. 14.12 and has been shown to offer improved low

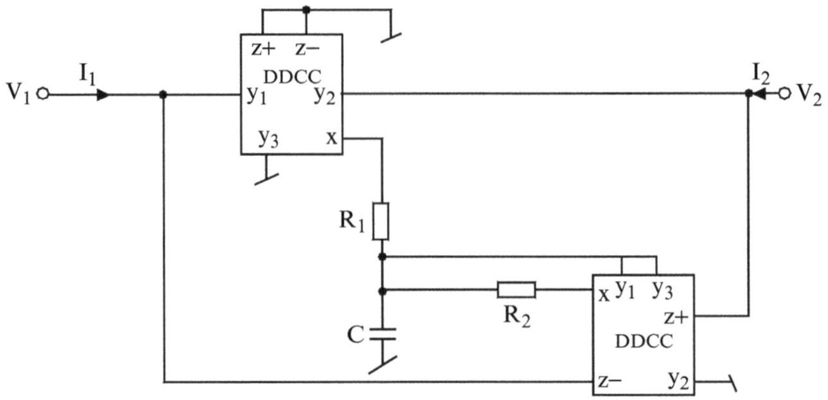

Fig. 14.12 FI with improved low frequency performance proposed by Yuce and Minaei (Adapted from [31] © 2009 Elsevier)

frequency performance. However, the structure has the drawback of requiring a matching condition ($R_1 = R_2$) to simulate a lossless FI.

The equivalent circuit of Fig. 14.12 with parasitic impedances is shown in Fig. 14.13, where L_{eq} and R_{eq} are given by the following expressions:

$$L_{eq} = \left(C + C_{y21} + C_{y23}\right)(R_{x1} + R_1)(R_{x2} + R_2) \tag{14.20}$$

$$R_{eq} = (R_{x1} + R_1)(R_{x2} + R_2)\left(1/(R_1 + R_{x1}) - 1/(R_1 + R_{x1}) + 1/\left(R_{y21}\|R_{y23}\right)\right) \tag{14.21}$$

R_{eq} can be made zero if we select:

$$1/(R_1 + R_{x1}) = 1/(R_2 + R_{x2}) - 1/\left(R_{y21}\|R_{y23}\right) \tag{14.22}$$

On the other hand, if we choose:

$$1/(R_1 + R_{x1}) \geq 1/(R_2 + R_{x2}) - 1/\left(R_{y21}\|R_{y23}\right) \tag{14.23}$$

this will ensure the stability of the circuit operation.

The authors of [31] have proposed as many as eight different topologies of FIs, however, five of them require five CCs of various types while only three configuration therein were implemented with only two CCs and the circuit of Fig. 14.12 is one of these. For a rigorous analysis of all the FI circuits and their detailed SPICE simulation results based upon a DDCC implemented with 0.35 μm TSMC CMOS technology parameters, the readers are referred to [31].

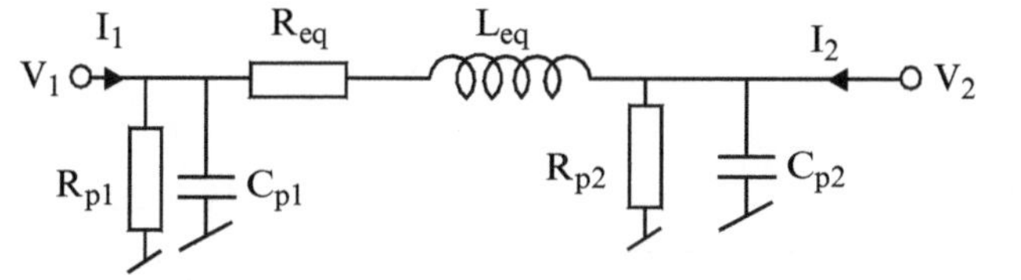

Fig. 14.13 Equivalent circuit of FI in Fig. 14.12 with parasitic impedances (Adapted from [31] © 2009 Elsevier)

14.12 Floating Simulator Employing DO-CCII and OTA

Sagbas et al. presented a FI function simulator [29] employing one DO-CCII, one OTA and two grounded passive components. Their proposed structure can simulate an *electronically-variable* positive/negative floating inductor or floating capacitor or floating resistor by appropriate selection of the two admittances, without any component-matching conditions.

The values of the simulated FI elements are *electronically-controllable* through the biasing current of the OTA or through the grounded resistor or the capacitor.

Assuming ideal devices, a routine circuit analysis of the circuit of Fig. 14.14 yields the following short circuit admittance matrix:

$$[Y] = \left(\frac{y_1}{y_2}\right) g_m \begin{bmatrix} 1 & -1 \\ -1 & 1 \end{bmatrix} \tag{14.24}$$

From the above equation, it is clear that one can simulate a FI, a capacitor or an electronically-variable resistor by appropriate selection of the admittances y_1 and y_2.

The functionality of the simulated inductor of [29] was tested by employing this in the design of a 5th order Chebyshev low pass filter designed for a cut-off frequency of 1 MHz. The FI was realized with $R_1 = 1K\Omega$, $C_2 = 17$ pF and $g_m = 0.1$ ms. The CMOS DOCCII therein (see Fig. 2 of [29]) was based onTSMC 0.35 μm CMOS technology parameters. The simulation results of the filter with a passive inductor and the simulated inductor of Fig. 14.14 demonstrated good agreement.

14.13 DO-CCCII Based Lossless Floating Inductance Simulator Employing a Grounded-Capacitor

A number of authors have employed CCCII to realize a variety of grounded and floating impedances such as [12, 13, 18, 24, 36, 48]. Sagbas et al. in [32] employed one DO-CCCII, one OTA and a grounded capacitor (Fig. 14.15) to realize a lossless grounded-capacitor FI simulator, without the requirement of any

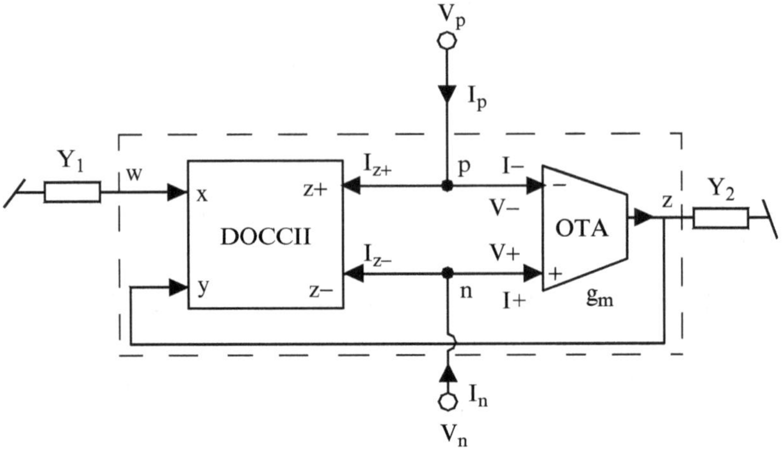

Fig. 14.14 Floating Simulator configuration proposed by Sagbas et al. (Adapted from [29] © 2009 Springer)

component-matching conditions. The simulated FI can be controlled electronically through the bias current of the DO-CCCII or by varying the bias current of the OTA. Assuming ideal DO-CCCII and OTA, the circuit analysis of the circuit of Fig. 14.15 gives the following short circuit admittance matrix:

$$[Y] = \left(\frac{g_m}{sCR_x}\right)\begin{bmatrix} 1 & -1 \\ -1 & 1 \end{bmatrix} \tag{14.25}$$

where R_x is the X-terminal parasitic resistance and is given by $R_x = V_T/2I_0$ and where V_T is the thermal voltage and I_0 is the DC bias current of the DO-CCCII.

Thus, from the above equation, the value of simulated FI is given by:

$L_{eq} = (CV_T^2/I_B I_0)$ where I_B is the bias current of the OTA.

Thus, it is seen that the inductance value can be controlled electronically through either I_B or I_0. By grounding either of the ports, the presented configuration can simulate an electronically adjusted grounded inductor.

The electronic-tunability of the proposed FI has been verified by simulating the DO-CCCII and OTA by bipolar implementations based upon PR100N and NR100N bipolar transistors from the transistor arrays ALA400 from AT&T with circuits biased from ± 2.5 V DC supplies. The simulator was used in a 4th order LC band pass filter whose SPICE simulation results were found to agree with theoretical plots. The simulated inductor exhibited good frequency response from about 1 KHz till about 1 MHz.

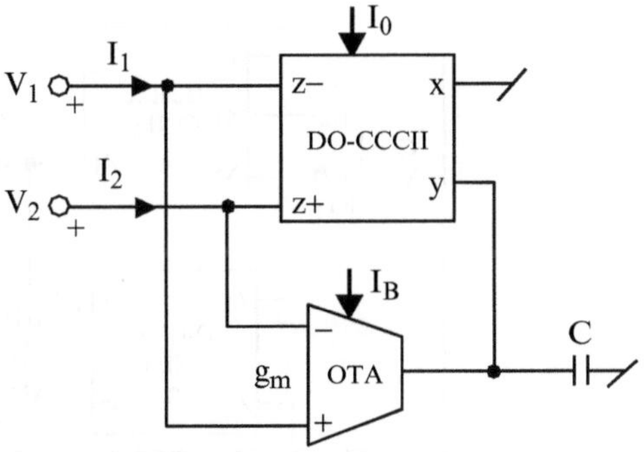

Fig. 14.15 Electronically controllable FI simulator proposed by Sagbas et al. (Adapted from [32] © 2009 Elsevier)

14.14 Resistor-Less Simulated FI Using DXCCII

Figure 14.16 shows a resistor-less simulated FI using three dual –X CCIIs (DXCCII), three NMOS transistors (operating in triode region) and a grounded capacitor propose by Saad and Soliman in [37]. Since drain and source terminals of all the three MOSFETs have complementary signals this enables cancellation of the square non-linearity in the expression for drain current in all the three cases and each MOSFET, therefore, realizes a linear voltage-controlled-resistor whose value is given by

$$R_i = \frac{1}{2\mu_n C_{ox}\left(\frac{W}{L}\right)_i (V_{Gi} - V_{th})} \tag{14.26}$$

Furthermore, since M_2 and M_3 have a common gate voltage V_{G2}, it therefore follows that $R_2 = R_3$.

With the above in mind, the equations for the 2-port can be written as:

$$I_1 = I_{Z2A} = I_{X2A} = I_{D1A} = 2\mu_n C_{ox}\left(\frac{W}{L}\right)_1 (V_{G1} - V_{th})V_3 \tag{14.27}$$

Similarly

$$I_2 = I_{Z1A} = I_{X1A} = -I_{D1A} = -I_1 \tag{14.28}$$

The current in the capacitance (I_C) is given by:

$$\begin{aligned} I_C &= -(I_{Z1C} + I_{Z2B}) = -(I_{X1C} + I_{X2B}) = -(-I_{D2C} + I_{D3B}) \\ &= 2\mu_n C_{ox}\left(\frac{W}{L}\right)(V_{G2} - V_{th})(V_1 - V_2) \text{ for} \left(\frac{W}{L}\right)_2 = \left(\frac{W}{L}\right)_3 \end{aligned} \tag{14.29}$$

Therefore, the voltage across the capacitor (V_3) will be obtained as

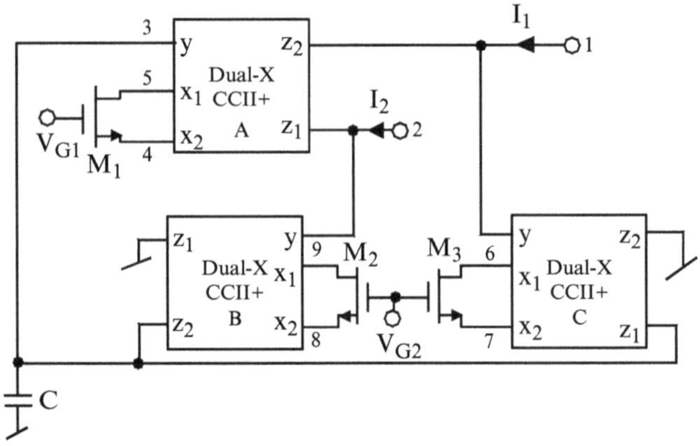

Fig. 14.16 Resistorless simulation of FI using DXCCIIs proposed by Saad and Soliman (Adapted from [37] © 2010 John Wiley & Sons Ltd.)

$$V_3 = \left(\frac{1}{sC}\right) 2\mu_n C_{ox}\left(\frac{W}{L}\right)(V_{G2} - V_{th})(V_1 - V_2) \tag{14.30}$$

Hence, the Y-matrix of the circuit is given by:

$$[Y] = \left(\frac{4}{sC}\right)\left\{\mu_n C_{ox}\left(\frac{W}{L}\right)\right\}^2 (V_{G1} - V_{th})(V_{G2} - V_{th})\begin{bmatrix} 1 & -1 \\ -1 & 1 \end{bmatrix} \tag{14.31}$$

for equal aspect ratios of the MOSFETs.

Thus, the value of the simulated FI is found to be:

$$L_{eq} = \frac{C}{4\left(\mu_n C_{ox}\left(\frac{W}{L}\right)\right)^2 (V_{G1} - V_{th})(V_{G2} - V_{th})} \tag{14.32}$$

Hence, the simulated inductance can be controlled by both V_{G1} and V_{G2}.

14.15 Tunable MOSFET-C FDNR Using a Single DXCCII

Kacar et al. [38] presented a tunable MOSFET-C FDNR configuration employing a single DXCCII, two capacitors and a MOSFET as a resistor as shown in Fig. 14.17. Assuming ideal DXCCII, the input impedance of the circuit is found to be:

$$Z_{in} = \frac{1}{s^2 D} = \frac{2}{s^2 C^2 R} \quad \text{where } R = \frac{1}{2\mu_n C_{ox}\left(\frac{W}{L}\right)(V_c - V_{th})} \tag{14.33}$$

Thus, the value of the simulated FDNR (D) is given by:

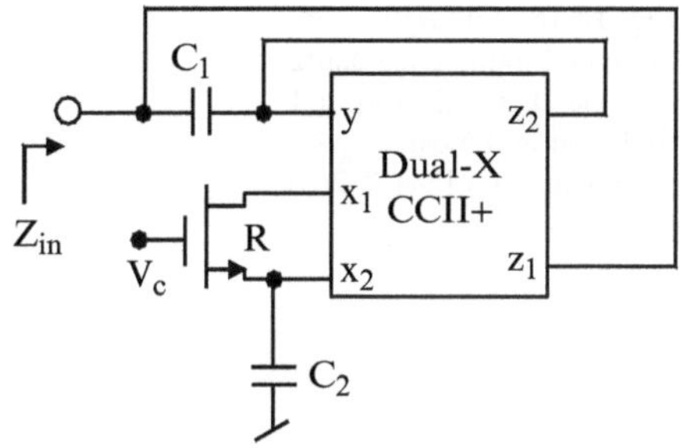

Fig. 14.17 Tunable MOSFET-C FDNR realization proposed by Kacar et al. employing a single DXCCII (Adapted from [38] © 2010 Elsevier)

$D = C^2\mu_n C_{ox}\left(\frac{W}{L}\right)(V_c - V_{th})$ which is tunable through the control voltage V_c.

The circuit was used to design a fifth order elliptic filter design requiring two grounded FDNRs which were simulated from the circuit of Fig. 14.17 using TSMC CMOS 0.35 μm process model parameters with DXCCII biased with ±1.5 V DC. The filter was designed for a cut-off frequency of 1.59 MHz and worked well.

14.16 DXCCII-Based Grounded Inductance Simulation

Another single DXCCII-based grounded inductance simulator[3] was proposed by Kacar and Yesil [34] using two resistors and a capacitor. Assuming ideal characteristics of the DXCCII, a routine circuit analysis of the circuit shown in Fig. 14.18 yields the following expression for the input admittance:

$$Y_{in} = \frac{2}{sC_1R_1R_2} - \frac{(R_1 - 4R_2)}{R_1R_2} \tag{14.34}$$

Now if we select $R_1 = 4R_2$, the input impedance becomes $Z_{in} = (sCR_1R_2/2)$, thus, an inductance of value $L_{eq} = (C_1R_1R_2/2)$ has been simulated.

The circuit was tested [39] in SPICE using the CMOS DXCCII from [57] in the design of a 3rd order Butterworth high pass filter and a band pass filter, both requiring one grounded inductor. The CMOS DXCCII was biased with ± 1.5 V DC supply with MOS transistors simulated using TSMC CMOS 0.35 process. The circuit was designed for $f_0 = 1.59$ MHz. The simulation results agreed well with the expectations.

[3] For an interesting work demonstrating the use of DXCCII in realizing a number of positive/negative, lossy/loss-less inductance simulation circuits, see [34].

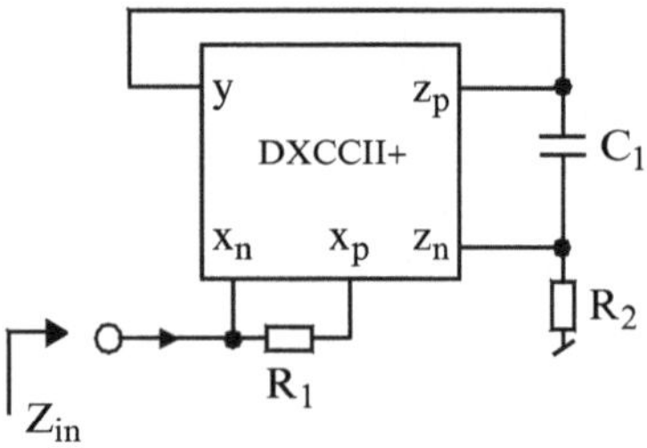

Fig. 14.18 Simulation of grounded inductance using a single DXCCII proposed by Kacar and Yesil (Adapted from [34] © 2010 Elsevier)

14.17 FI Simulators with Only Two DVCCs

In 2010, Horng [39] presented a FI configuration employing two DVCCs, two resistors and a grounded capacitor which did not require any component matching conditions. Assuming ideal DVCCs, the short circuit admittance matrix of the circuit, through a routine analysis, is found to be:

$$[Y] = \left(\frac{1}{sC_1R_1R_2}\right)\begin{bmatrix} 1 & -1 \\ -1 & 1 \end{bmatrix} \tag{14.35}$$

Thus, from above equation, a lossless FI is realized with its inductance value $L_{eq} = C_1R_1R_2$.

The circuit shown in Fig. 14.19, although simulates a FI without any component matching condition, but employs one floating resistor. Yuce in [40] presented another configuration, shown here in Fig. 14.20, which employs exactly the same number of active and passive elements and exhibits the same properties but uses all grounded components. This circuit topology can simulate FI, FC and FDNR elements by selecting appropriate passive elements (resistor (s) or capacitor (s)).

Assuming ideal DVCCs, a straight forward analysis of this simulator yields the following short circuit admittance matrix:

$$[Y] = \left(\frac{y_1y_2}{y_3}\right)\begin{bmatrix} 1 & -1 \\ -1 & 1 \end{bmatrix} \tag{14.36}$$

From the equation (14.36) the various floating elements can be realized as follows: an FI is realized by choosing y_3 as a capacitors with remaining admittances as resistors; a floating capacitor is realized by choosing y_1 as a capacitor with y_2, y_3 being resistors; and FDNR is realizable by having C_1, C_2 as capacitor with y_3 as a resistor.

To verify the workability of the proposed floating simulator of Fig. 14.20, it was simulated [40] in SPICE using 0.35 μm TSMC CMOS technology parameters and the DVCC structure based upon that in [58], with DC bias voltage of $\pm$ 1.5 V. Component values were chosen to realize an inductance value of 0.1 mH and the magnitude and phase plots were obtained. Very good correspondence between theoretical and simulated plots has been observed up to a frequency of about 20 MHz.

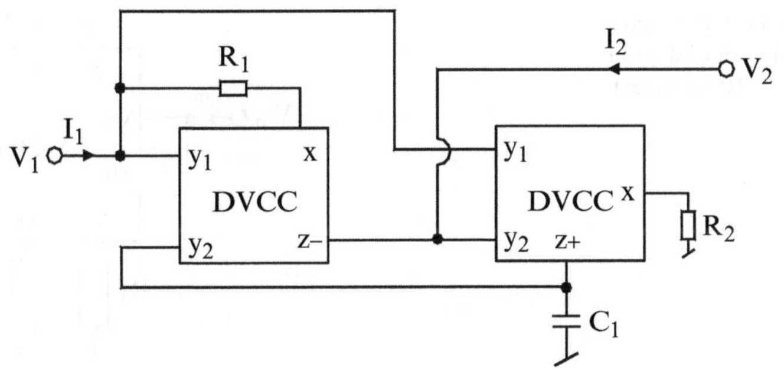

Fig. 14.19 Simulation of FI using DVCCs proposed by Horng (Adapted from [39] © 2010 Springer)

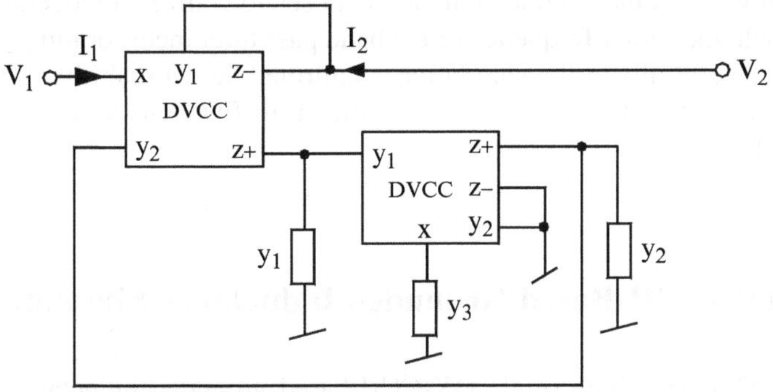

Fig. 14.20 Floating simulator proposed by Yuce [40]

14.18 Lossless Grounded Inductor Using a Single FDCCII and Three Grounded Passive Elements

FDCCII has already been defined and characterized in an earlier chapter of this monograph. In this section, we present an application of the FDCCII in the realization of a grounded inductance employing a minimum number of grounded passive elements without any component-matching condition. Figure 14.21 shows a circuit configuration using a single FDCCII and three grounded passive components proposed by Kacar [41].

A routine circuit analysis (assuming ideal FDCCII) of this circuit gives the following expression of its input impedance: $Z_{in} = sCR_1R_2$. Thus, an inductor of value $L_{eq} = CR_1R_2$ has been simulated.

The circuit has been tested in SPICE using the CMOS FDCC of [59] with MOS transistors simulated using TSMC 0.35 μm process model parameters and

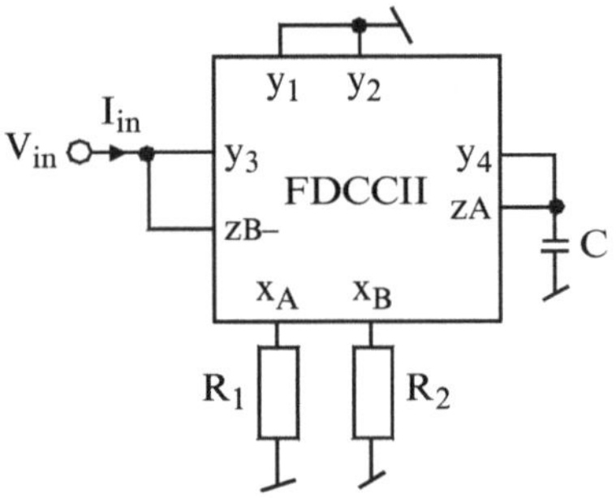

Fig. 14.21 FDCCII-based grounded inductor (Adapted from [41] © 2010 Elsevier)

component value of the simulated inductor chosen to realize an inductance of value 2.53 mH with the center frequency of the band pass filter incorporating this inductor as 1 MHz. The results of the simulations confirmed the workability of the circuit.

For a single FDCCII-based circuits simulating FDNR-type grounded impedances, see [42].

14.19 DX-CCII-Based Grounded Inductance Simulators

Figure 14.22 shows three single DX-CCII-based grounded inductance simulators proposed by Metin [5]. The presented structures employ three resistors and one capacitor and do not require any passive component matching condition. Assuming ideal DX-CCII, the input impedance of these circuits can be obtained as:

Figure 14.22a: This circuit simulates parallel-RL admittance of the form

$$Y_{in} = \frac{1}{R_{eq}} + \frac{1}{sL_{eq}} \tag{14.37}$$

$$R_{eq} = \frac{R_1R_2R_3}{R_1R_2 + R_2R_3 - R_1R_3}\ ;\ L_{eq} = CR_1R_3 \tag{14.38}$$

Figure 14.22b: This circuit also simulates a parallel RL admittance with equivalent parameters given by

$$R_{eq} = \frac{R_1R_2}{4R_2 - R1}\ ;\ L_{eq} = \left(\frac{1}{4}\right)CR_1R_3 \tag{14.39}$$

Figure 14.22c: This circuit simulates a series RL impedance of the form

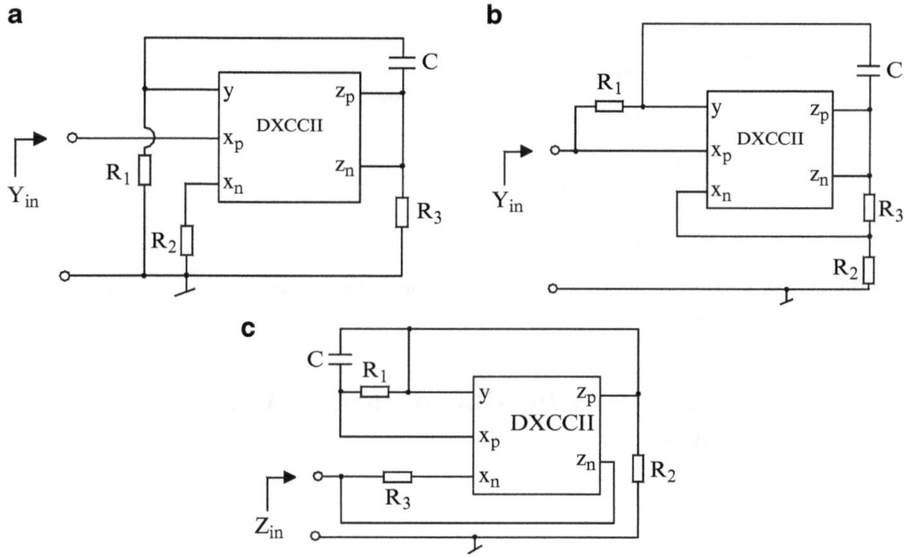

Fig. 14.22 Single DX-CCII-based inductance simulators proposed by Metin [5]

$$Z_{in} = R_{eq} + sL_{eq} \tag{14.40}$$

$$R_{eq} = \left(R_2 + \frac{2R_2R_3}{R_1} - R_3\right);\ L_{eq} = 2CR_2R_3 \tag{14.41}$$

The circuits were tested using a CMOS DXCCII (Fig. 14.3 of [5]) with the simulated inductors used to design a fifth order band pass filter with two simulated inductors of values 50 μH and 100 μH. MOS transistors of DXCCII were simulated using TSMC 0.35 μm CMOS process parameters. The DC bias voltage was taken as ±2.5 V. Simulations demonstrated the usable frequency range of the circuits as 10 MHz.

Another grounded inductance simulator circuit has been presented by Myderrizi et al. [45] employing single DX-CCII and three passive elements which can be operated in 30 kHz–30 MHz frequency range. This circuit is shown in Fig. 14.23. Assuming ideal DX-CCII a routine circuit analysis of this circuit gives:

$$Z_{in}(s) = \frac{V_{in}}{I_{in}} = sCR^2 \text{ if } R_1 = R_2 = R \tag{14.42}$$

Thus, in the circuit of Fig. 14.23 the value of inductance can be controlled by grounded resistances R_1 and R_2.

Through SPICE simulations, it has been demonstrated [43] that the inductance simulation circuit of Fig. 14.23 operates pretty well between 30 KHz and 10 MHz.

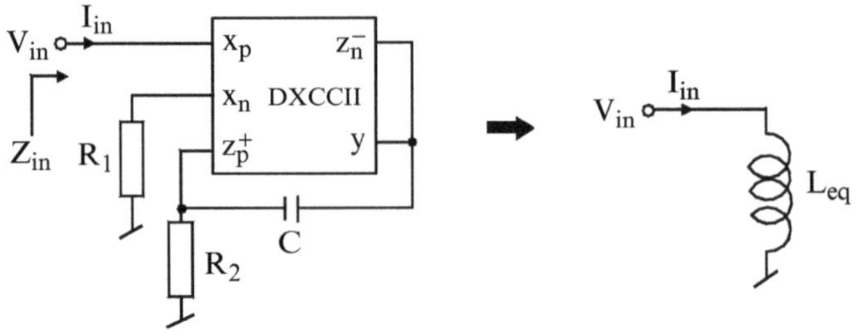

Fig. 14.23 DX-CCII-based grounded inductance simulator proposed by Miderizzi et al. (Adapted from [45] © 2011 Elsevier)

14.20 Grounded-Capacitor-Based Floating Capacitance Multiplier

Prommee and Somdunyakanok [46] proposed a grounded capacitor-based floating capacitance multiplier circuit using three current controlled differential difference current conveyors (CCDDCC) without using any external resistor. Out of the three CCDDCCs, two CCDDCCs are single output and one CCDDCC is multi-output type. The CCDDCC is very similar to DVCC except that DVCC has only two Y-input terminals and CCDDCC has three Y- input terminals. Also, the intrinsic resistance of X- input terminal of CCDDCC has finite resistance of value Rx which can be tuned by the bias current. A CCDDCC can be characterized by the following matrix equation:

$$\begin{bmatrix} V_X \\ I_{Y1} \\ I_{Y2} \\ I_{Y3} \\ I_Z \end{bmatrix} = \begin{bmatrix} R_X & 1 & -1 & 1 & 0 \\ 0 & 0 & 0 & 0 & 0 \\ 0 & 0 & 0 & 0 & 0 \\ 0 & 0 & 0 & 0 & 0 \\ \pm 1 & 0 & 0 & 0 & 0 \end{bmatrix} \begin{bmatrix} I_x \\ V_{Y1} \\ V_{Y2} \\ V_{Y3} \\ V_Z \end{bmatrix} \tag{14.43}$$

Assuming ideal CCDDCCs, a routine circuit analysis of Fig. 14.24 yields the following short circuit admittance matrix:

$$[Y] = sC\left(\frac{R_{X2}}{R_{X1}}\right)\begin{bmatrix} 1 & -1 \\ -1 & 1 \end{bmatrix} \tag{14.44}$$

Thus, from above equation, a lossless FC is realized with its capacitance value

$$C_{eq} = C\left(\frac{R_{X2}}{R_{X1}}\right) = C\sqrt{\frac{I_{B1}}{I_{B2}}} \tag{14.45}$$

Therefore, it is clear that C_{eq} can be controlled by the bias currents of CCDDCC1 and CCDDCC2.

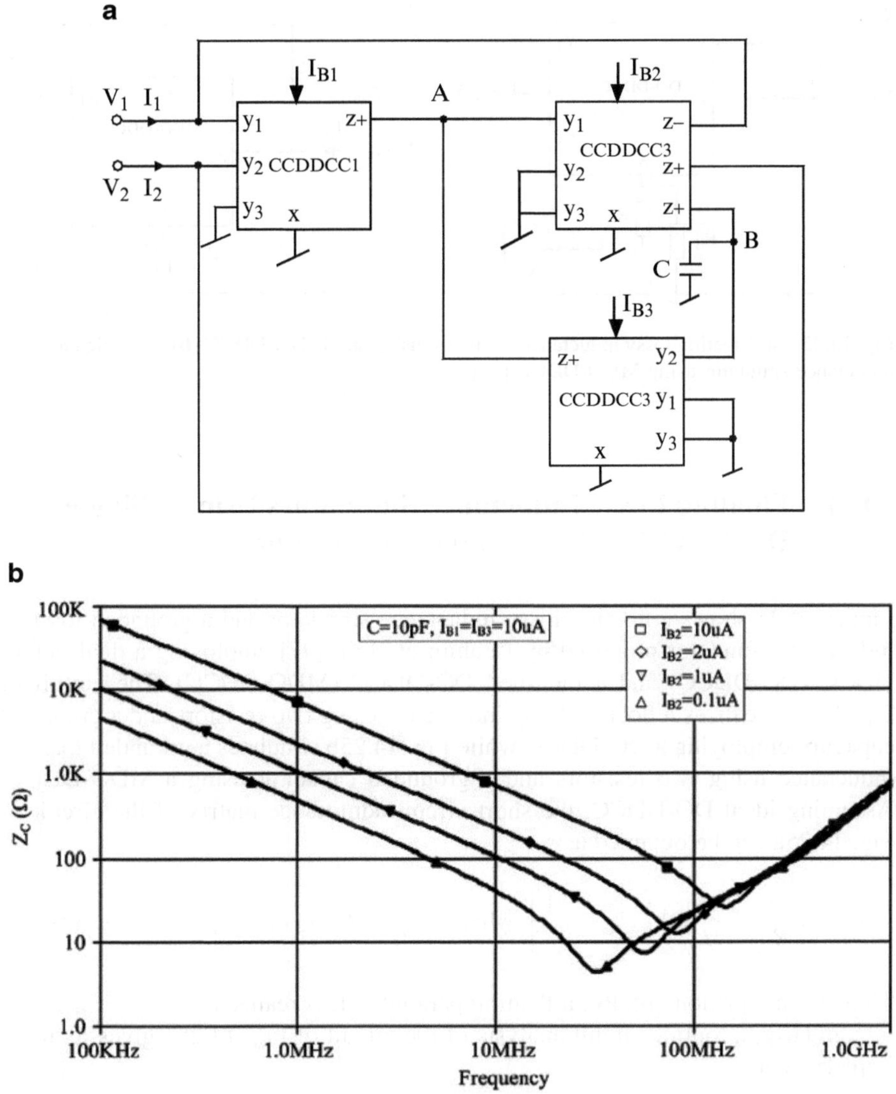

Fig. 14.24 Grounded capacitor floating capacitance multiplier (Adapted from [46] © 2011 Elsevier) (**a**) The circuit configuration (**b**) Variation of the magnitude of the realized equivalent capacitance

The workability of the capacitance multiplier was confirmed by realizing the circuit using the CCDDCCs by the CMOS implementation of [60] (reproduced also in Fig. 4 of [46]). The model parameters of TSMC 0.25 μm CMOS technology were used with the circuit biased with ±1.25 V DC. The magnitude function of the simulated capacitance shown in Fig. 14.24 confirms the variability of the realized capacitance value through the bias current I_{B2}.

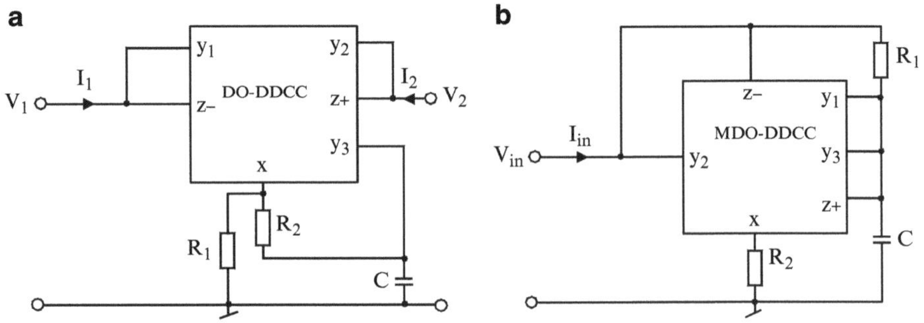

Fig. 14.25 (**a**) Floating lossy inductance simulator using single DO-DDCC (**b**) Grounded lossless inductance simulator using MDO-DDCC [47]

14.21 Floating Lossy Inductance Simulators Using a Single DO-DDCC and a Grounded Capacitor

Figure 14.25 shows a floating lossy inductance simulator and a grounded lossless inductance simulator proposed by Ibrahim et al. in [47] employing a dual-output DDCC (DO-DDCC) and a modified DO-DDCC (MDO-DDCC). The circuit of Fig. 14.25a realizes a floating lossy inductance using two resistors and a grounded capacitor employing a DO-DDCC while Fig. 14.25b simulates a grounded lossless inductance using two resistors and a grounded capacitor using a MDO-DDCC. Assuming ideal DO-DDCC, the short circuit admittance matrix of the circuit of Fig. 14.25a can be obtained as:

$$[Y] = \left(\frac{1}{R_{eq}} + \frac{1}{sL_{eq}}\right)\begin{bmatrix} 1 & -1 \\ -1 & 1 \end{bmatrix} \text{ where } R_{eq} = R_1 \| R_2 \text{ and } L_{eq} = CR_1R_2 \quad (14.46)$$

Thus, from equation (14.46), a floating parallel R-L is realized.

Similarly, a routine circuit analysis of the circuit of Fig. 14.25b gives its input impedance as:

$$Z_{in} = \frac{V_{in}}{I_{in}} = 2sCR^2 \text{ for } R_1 = R_2 = R \quad (14.47)$$

Therefore, from equation (14.47), it is seen that a grounded lossless inductance of value $L_{eq} = 2CR^2$ has been simulated.

The SPICE simulations have indicated that using DO-DDCC and MIDO-DDCC obtained from modifications of the CMOS DDCC architecture of [61] with bias voltage of ± 1.5 V DC and model parameters of 0.35 μm TSMC CMOS technology, the operational frequency range of the FI has been found to be 10–400 KHz and that of the GI as 5–700 KHz.

14.22 Grounded Inductance Simulator Using DCCII

A second-generation CC with current differencing capability is known as differential CC (DCCII). The DCCII is a five–port active analog building block and can ideally be characterized by the following matrix equation:

$$\begin{bmatrix} V_{XN} \\ V_{XP} \\ I_{Z+} \\ I_{Z-} \\ I_Y \end{bmatrix} = \begin{bmatrix} 0 & 0 & 1 \\ 0 & 0 & 1 \\ 1 & -1 & 0 \\ 1 & -1 & 0 \\ 0 & 0 & 0 \end{bmatrix} \begin{bmatrix} I_{XP} \\ I_{XN} \\ V_Y \end{bmatrix} \tag{14.48}$$

Figure 14.26 shows two grounded inductance simulators employing single DCCII with three passive components proposed by Metin [50].

The input admittance of circuits in Fig. 14.26a and b can be obtained using straight forward circuit analysis assuming ideal DCCII as:

$$Y_{ina} = \frac{1}{(R_2 - R_1) + sCR_1R_2} \tag{14.49}$$

$$Y_{inb} = \frac{1}{sCR_1R_2} + \frac{R_2 - R_1}{R_1R_2} \tag{14.50}$$

Thus, circuit of Fig. 14.26a simulates series R-L impedance while the circuit of Fig. 14.26b realizes parallel R-L impedance.

To verify the workability of the simulated inductors a Bipolar DCCII realizable with transistor arrays ALA 400 from AT & T were used [50]. The DCCII was biased with ± 2.5 V DC power supply and the simulated inductors were used to design a second-order BPF. The simulation results were found to be in good agreement with the theory. By realizing the DCCII with commercially available IC 844 the circuit was similarly tested in hardware also which confirmed the workability of the circuit

DCCII based other circuits for inductance simulation have also been presented in [27], however, the circuits therein do not employ a grounded capacitor.

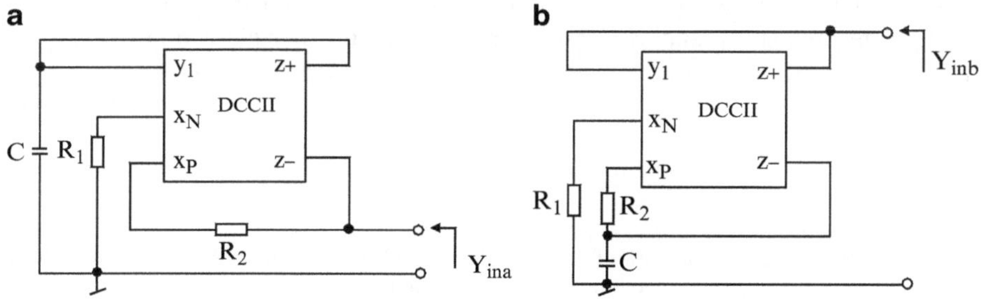

Fig. 14.26 Grounded-capacitor inductance simulation circuits proposed by Metin [50]

Concluding Remarks

In this chapter we presented grounded and floating immittance realization methods and circuits using different varieties of CCs such as DO-CCIIs, DVCCs, CCIIIs, DXCCIIs, MICCIIs, DDCCs, DO-CCCIIs and FDCCIIs. It must be recalled that using a single CCII although one can realize a grounded lossless inductor but it requires as many as five passive elements namely four resistors and a capacitor with a constraint in terms of a specified relation between the four resistors to enable the realization of lossless inductor.

By contrast, in this chapter it has been established that using various modified versions of the CCs only three passive elements along with a single active element are enough to perform the same function. Although a cancellation condition is still needed in some cases, however, this is compensated by the advantage of employing a grounded capacitor in several cases. A particularly novel circuit results when an FDCCII is employed in the arrangement proposed by Kacar [41] in which a lossless inductor results from a single FDCCII and only three passive elements- *all of which are grounded.*

In case of realizing a floating inductor, particularly the lossless one, it was implied by several authors that if this is needed to be accomplished by only three passive elements then at least three CCs are needed while if the realization is to be attained with only two CCs then at least five passive elements are needed.[4]

By contrast, in this chapter, a variety of circuits have been included which demonstrate that if modified forms of CCs offering increased capability and more flexibility are employed then there are numerous ways in which floating impedances of specific kind or those of general nature can be indeed realized with only two active elements and no more than three passive elements. In most of such circuits, at least one grounded capacitor is assured however, in some cases, such as in DVCC- based floating FDNR of Sedef and Acar all the three passive elements can be grounded.

It is also interesting to note that in mixed-source arrangement such as those employing DO-CCCII and OTA the only grounded element used is grounded capacitor along with the advantage of electronic tunability of the realized inductance value. On the other hand, to realize electronically-tunable inductors in both grounded and floating forms, the DXCCII appears to be a particularly interesting element since a MOSFET connecting between its input terminals X_P and X_N results in cancellation of square non-linearity

(continued)

[4] It is interesting to note that in view of the equivalence of CCII- and nullor, these conjectures have their basis in some important theorems existing in literature in the context of gyrator realization using nullors [62].

(continued)

thereby enabling the MOSFET to realize linear VCRs which result in electronic control of simulated inductance.

In view of the novelty of the FDCCII as outlined above, it appears that its use in FI synthesis is an interesting problem in research which has not been attempted so far.

It may be mentioned that other than the variety of CCs considered here, a number of other building blocks closely related to CCs, such as Electronically-tunable CCII (ECCII) [26], dual output-Controlled Current Conveyor (DO-CCCII) [23, 33], Z-copy Current Controlled Current inverting transconductance amplifier (ZC-CCCITA) [49], Current controlled Current Conveyor transconductance amplifier (CCCCTA) [51] and voltage differencing Current Conveyor (VDCC) [52, 53], have also been employed by several researchers in grounded/floating impedance simulation.

Lastly, it must be emphasized that several researchers/authors have proposed a variety of CMOS implementation of the extended types of CCs which along with the novelty of their applications may provide the needed motivation for IC industry to come up with off- the-shelf ICs of any of the varieties of CCs considered in this and other chapters of this monograph.

References

1. Fabre A, Saaid O (1993) Novel translinear impedance convertor and band pass filter applications. Electron Lett 29:746–747
2. Abuelma'atti MT (1998) Comment: floating inductance simulation based on current conveyors. Electron Lett 34:1037
3. Ananda Mohan PV (2005) Floating capacitance simulation using current conveyors. J Circ Syst Comput 14:123–128
4. Soliman AM (2010) New CCII and ICCII based realizations of L-C and L-R mutators. Circ Syst Sig Process 29:1181–1191
5. Metin B (2011) Supplementary inductance simulator topologies employing single DXCCII. Radioengineering 20:614–618
6. Liu SI, Yang CY (1996) Higher-order immittance function synthesis using CCIIIs. Electron Lett 32:2295–2296
7. Ananda Mohan PV (1998) Grounded capacitor based grounded and floating inductance simulation using current conveyors. Electron Lett 34:1037–1038
8. Sedef H, Acar C (2000) A new floating FDNR circuit using differential voltage current conveyors. Int J Electron Commun (AEU) 54:297–301
9. Kuntman H, Gulsoy M, Cicekoglu O (2000) Actively-simulated grounded lossy inductors using third generation current conveyors. Microelectron J 31:245–250
10. Wang HY, Lee CT (2000) Systematic synthesis of R-L and C-D immittances using single CCIII. Int J Electron 87:293–301
11. Minaei S (2003) A new high performance CMOS third generation current conveyor (CCIII) and its application. Electr. Engineering J 85:147–153
12. Minaei S, Cicekoglu O, Kuntman H, Turkoz S (2002) Electronically tunable, active only floating inductance simulation. Int J Electron 89:905–912

13. Khan IA, Zaidi MH (2003) A novel ideal floating inductor using translinear conveyors. Active Passiv Electron Comp 26:87–89
14. Incekaraoglu M, Cam U (2005) Realization of series and parallel R-L and C-d Impedances using single differential voltage current conveyor. Analog Integr Circ Sig Process 43:101–104
15. Gift SJG (2004) New simulated inductor using operational conveyors. Int J Electron 91:477–483
16. Zeki A, Toker A (2005) DXCCII-based tunable gyrator. Int J Electron Commun (AEU) 59:59–62
17. Yuce E (2006) Comment on Realization of series and parallel R-L and C-D impedances using single differential voltage current conveyor. Analog Integr Circ Sig Process 49:91–92
18. Yuce E (2006) On the realization of the floating simulators using only grounded passive components. Analog Integr Circ Sig Process 49:161–166
19. Yuce E, Minaei S, Cicekoglu O (2005) A novel grounded inductor realization using a minimum number of active and passive components. ETRI J 27:427–432
20. Yuce E, Cicekoglu O, Minaei S (2006) CCII-based grounded to floating immittance converter and a floating inductance simulator. Analog Integr Circ Sig Process 46:287–291
21. Minaei S, Yuce E, Cicekoglu O (2006) A versatile active circuit for realizing floating inductance, capacitance, FDNR and admittance converter. Analog Integr Circ Sig Process 47:199–202
22. Yuce E, Minaei S, Cicekoglu O (2006) Limitations of the simulated inductors based on a single current conveyors. IEEE Trans Circ Syst-I 53:2860–2867
23. Yuce E, Minaei S (2007) A new active network suitable for realizing ladder filters and transformer simulator. J Circ Syst Comput 16:29–41
24. Minaei S, Yuce E (2008) Realization of tunable active floating inductance simulators. Int J Electron 95:27–37
25. Yuce E (2008) Negative impedance converter with reduced nonideal gain and parasitic impedance effects. IEEE Trans Circ Syst-I 55:276–283
26. Riewruja V, Petchmaneelumka W (2008) Floating current-controlled resistance converters using OTAs. Int J Electron Commun 62:725–731
27. Metin B, Herencsar N, Horng JW (2014) DCCII-based novel lossless grounded inductance simulators with no element matching constrains. Radioeng J 23:532–4538
28. Horng JW, Hou CL, Chang CM, Yang H, Shyu WT (2009) Higher-order immittance functions using current conveyors. Analog Integr Circ Sig Process 61:205–209
29. Sagbas M, Ayten UE, Sedef H, Koksal M (2009) Floating immittance function simulator and its applications. Circ Syst Sig Process 28:55–63
30. Yuce E, Minaei S (2009) On the realization of simulated inductors with reduced parasitic impedance effects. Circ Syst Sig Process 28:451–465
31. Yuce E, Minaei S (2009) Novel floating simulated inductors with wider operating-frequency ranges. Microelectron J 40:928–938
32. Sagbas M, Ayten UE, Sedef H, Koksal M (2009) Electronically tunable floating inductance simulator. Int J Electron Commun (AEU) 63:423–427
33. Lahiri A (2009) Comment on ‘Electronically tunable floating inductance simulator’. Int J Electron Commun (AEU) 63:878
34. Kacar F, Yesil A (2010) Novel grounded parallel inductances simulators realization using a minimum number of active and passive components. Microelectron J 41:632–638
35. Soliman AM (2010) On the realization of floating inductors. Nat Sci 8(5):167–180
36. Metin B, Minaei S (2010) Parasitic compensation in CC-I based circuits for reduced power consumption. Analog Integ Circ Sig Process 65:157–162
37. Saad RA, Soliman AM (2010) On the systematic synthesis of CCII-based floating simulators. Int J Circ Theor Appl 38:935–967
38. Kacar F, Metin B, Kuntman H (2010) A new CMOS dual-X second generation current conveyor (DXCCII) with an FDNR circuit application. Int J Electron Commun (AEU) 64:774–778

39. Horng JW (2010) Lossless inductance simulation and voltage-mode universal biquadratic filter with one input and five outputs using DVCCs. Analog Integ Circ Sig Process 62:407–413
40. Yuce E (2010) A novel floating simulation topology composed of only grounded passive components. Int J Electron 97:249–262
41. Kacar F (2010) New lossless inductance simulators realization using a minimum active and passive components. Microelectron J 41:109–113
42. Kacar F, Yesil A (2011) FDCCII-based FDNR simulator topologies. Int J Electron 99:285–293
43. Pal K (1989) Modified current conveyors and their applications. Microelectron J 20:37–40
44. Swamy MNS (2011) Mutators, generalized Impedance converters and inverters, and their realization using generalized current conveyors. Circ Syst Sig Process 30:209–232
45. Myderrizi I, Minaei S, Yuce E (2011) DXCCII-based grounded inductances simulators and filter applications. Microelectron J 42:1074–1081
46. Prommee P, Somdunyakanok M (2011) CMOS-based current-controlled DDCC and its applications to capacitance multiplier and universal filter. Int J Electron Commun (AEU) 65:1–8
47. Ibrahim MA, Minaei S, Yuce E, Herencsar N, Koton J (2012) Lossy/lossless floating/grounded inductance simulation using one DDCC. Radioengineering 21:3–10
48. Sagbas M, Ayten UE, Sedef H, Koksal M (2012) Reply to comment on 'Electronically tunable floating inductance simulator'. Int J Electron Commun (AEU) 66:86–88
49. Herencsar N, Lahiri A, Koton J, Vrba K, Metin B (2012) Realization of resistorless lossless positive and negative grounded inductor simulators using single ZC-CCCITA. Radioengineering 21:264–272
50. Metin B (2012) Canonical inductor simulators with grounded capacitors using DCCII. Int J Electron 99:1027–1035
51. Sagbas M (2014) Electronically tunable mutually coupled circuit using only two active components. Int J Electron 101:364–374
52. Prasad D, Ahmad J (2014) New electronically-controllable lossless synthetic floating inductance circuit using single VDCC. Circ Syst 5:13–15
53. Kacar F, Yesil A, Minaei S, Kuntman H (2014) Positive/negative lossy/lossless grounded inductance simulators employing single VDCC and only two passive elements. Int J Electron Commun (AEU) 68:73–78
54. Wang Z (1990) Novel voltage-controlled grounded resistor. Electron Lett 26:1711–1712
55. Piovaccari A (1995) CMOS integrated third generation current conveyor. Electron Lett 31:1228–1229
56. Zeki A, Toker A, Cicekoglu O (2002) The dual-X current conveyor (DXCCII): a new active device for tunable continuous-time filters. Int J Electron 89:913–923
57. Kacar F, Metin B, Kuntman H (2010) A new dual-X CMOS second generation current conveyor (DXCCII) with a FDNR circuit application. AEU Int J Electron Commun 64:774–778
58. Elvan HO, Soliman AM (1997) Novel CMOS differential voltage current conveyor and its applications. IEE Proc Circ Devices Syst 144:195–200
59. Kacar F, Metin B, Kuntman H, Cicekoglu O (2009) A new high-performance CMOS fully differential second-generation current conveyor with application example of biquad filter realization. Int J Electron 97:499–510
60. Prommee P, Angkeaw K, Somdunyakanok M, Dejhan K (2009) CMOS-based near zero-offset multiple input max-min circuits and its applications. Analog Integ Circ Sig Process 61:93–105
61. Chiu W, Liu SI, Tsao HW, Chen JJ (1996) CMOS differential difference current conveyor and their applications. IEE Proc Circ Devices Syst 143:91–96
62. Adams KM, Deprettere E (1974) On the realization of gyrators by nullors and resistors. Int J Circ Theor Appl 2:287–290

Chapter 15
Second Generation Miscellaneous Linear/Nonlinear Applications of Various Types of Current Conveyors

Abstract Miscellaneous linear and non-linear applications of varieties of CCs have been elaborated in this chapter. The functional circuits described include instrumentation amplifier, PID controllers, single ended to differential converters, precision rectifiers, astable and monostable multivibrators, square/triangular wave generator and chaos generator, analog multiplier/dividers, squares and square rooters, mutators and memristive and memcapacitive emulators.

15.1 Introduction

A number of nonlinear applications of normal CCIIs have been described in Chap. 8. In this chapter, we discuss a number of other nonlinear applications but the current conveyors employed are newly introduced variants of CCs which have been described earlier in Chap. 10. The newer variants of CCs due to their different characterization are shown to offer some flexibility in the design as well as a number of advantages such as reduced component count, electronic controllability or multifunction capability. The resulting circuits are usually attractive for bipolar and/or CMOS implementation in IC technology.

15.2 PID Controllers

The proportional plus integral plus derivative (PID) controllers are very useful building blocks of control systems employed in process control industry. PID controllers contain three terms: proportional, integral and derivative and it is desirable that each of them are controllable separately for manual or adaptive tuning. PID controllers are normally realized with voltage mode op-amps and RC components and typically require up to four op-amps.

However, if CCs are employed, the number of active and passive components can be reduced to a bare minimum, particularly if the dual output version of CCIIs is employed. Furthermore, using CCs, PID controllers can be realized for both

© Springer International Publishing Switzerland 2015
R. Senani et al., *Current Conveyors*, DOI 10.1007/978-3-319-08684-2_15

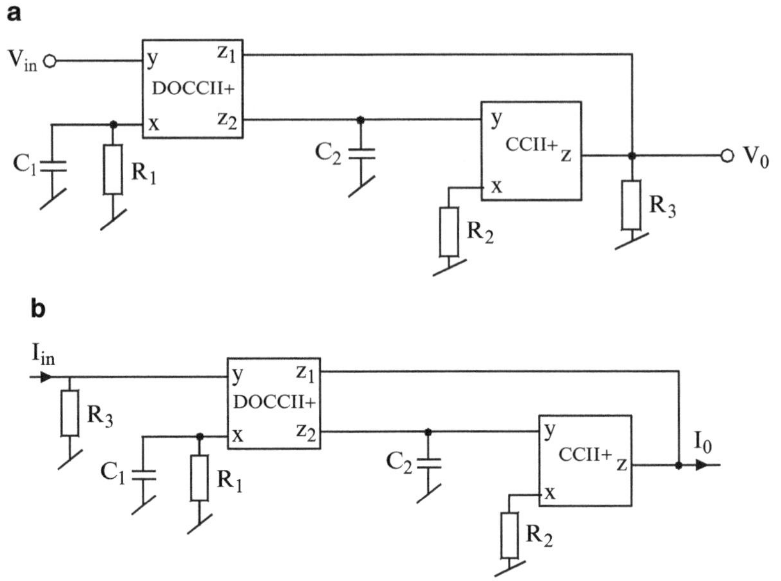

Fig. 15.1 CCII-based PID controllers (**a**) voltage-mode (**b**) current-mode (Adapted from [1] © 2005 Elsevier)

voltage mode as well as current mode signals. Two such exemplary circuits proposed by Yuce et al. [1] are shown here in Fig. 15.1a and b. In these circuits, a DO-CCII + has been employed which can be obtained by adding one extra terminal to CCII + and its defining equations are $I_y = 0$, $I_{z\ 1} = \alpha_1 I_x$, $I_{z2} =$, $\gamma_1 I_x$, $V_x = \beta_1 V_y$ whereas CCII + is assumed to be characterized by $I_y = 0$, $I_z = \alpha_2 I_x$, $V_x = \beta_2 V_y$ and finally, negative CCII is assumed to have same characterization as that of CCII + with only modification being $I_y = 0$, $I_z = -\ \alpha I_x$, $V_x = \beta V_y$. In the characterization of the three current conveyors, the parameters α, α_1, α_2 and γ_1 are the current gains between ports Z and X while parameters β, β_1, β_2 are the voltage gains between ports X and Y; under ideal conditions, all of them are equal to unity.

The voltage mode PID controller of Fig. 15.1a is characterized by:

$$V_o(s) = \left(K_{pv} + \frac{1}{ST_{iv}} + sT_{dv}\right)V_{in}(s) \tag{15.1}$$

whereas the current mode PID controller Fig. 15.1b is characterized by:

$$I_o(s) = \left(K_{pi} + \frac{1}{ST_{ii}} + sT_{di}\right)I_{in}(s) \tag{15.2}$$

where the proportional gain, the integral time constant and the derivative time constant parameters for both the PID controllers are given by:

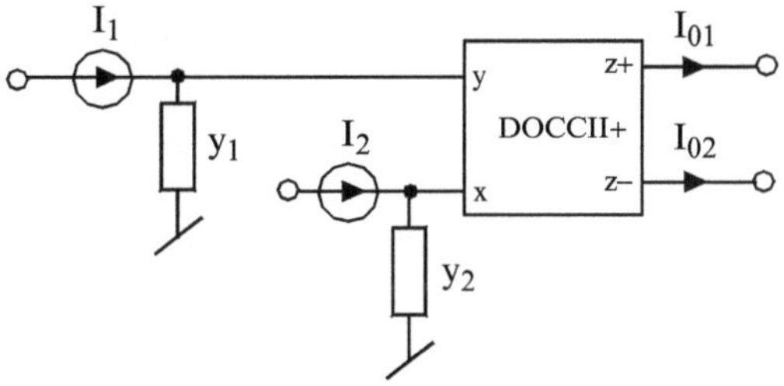

Fig. 15.2 Current-mode PID controller and instrumentation amplifier (Adapted from [2] © 2010 Elsevier Ltd.)

$$K_p = \alpha_1\beta_1\frac{R_3}{R_7} + \alpha_2\beta_1\beta_2\gamma_1\frac{C_1R_3}{C_2R_2} \tag{15.3}$$

$$T_i = \frac{C_2R_1R_2}{\alpha_2\beta_1\beta_2\gamma_1R_3} \tag{15.4}$$

$$T_d = \alpha_1\beta_1C_1R_3 \tag{15.5}$$

It is interesting to note that all the three relevant parameters can be tuned or adjusted arbitrarily or independently by various resistance values or alternatively by gain variable current conveyors through the parameters α_1, α_2, β_1, β_2 and γ_1. A novel feature of both the current mode and voltage mode PID controllers is that they employ five passive elements only all of which are grounded and do not require any kind of component matching. These properties make these circuits good candidates for realizing PID controllers in integrated circuit form.

Using DO-CCII, a very versatile configuration was reported by Yuce and Minaei in [2] which employs only one active element and only two admittances and can realize not only a PID controller but also an instrumentation amplifier in current mode depending upon appropriate choice of the two admittances.

If admittance Y_1 is selected as a parallel combination of a resistor R_1 and a capacitor C_1 with admittance Y_2 has chosen as a series combination of a resistor R_2 and capacitor C_2, the circuit realizes PID controller with the following transfer function:

$$H_k(s) = K_{pk} + \frac{1}{sT_{ik}} + sT_{dk} \tag{15.6}$$

In equation (15.6), the proportional gain (K_p), the integral time constant (T_i) and the derivative time constant (T_d) parameters are given by:

$$K_{p1} = \alpha\beta\left(\frac{R_1}{R_2} + \frac{C_2}{C_1}\right) - a \tag{15.7}$$

$$K_{p2} = -\beta\gamma\left(\frac{R_1}{R_2} + \frac{C_2}{C_1}\right) + a \tag{15.8}$$

$$T_{i1} = \frac{C_1 R_2}{\alpha\beta} \tag{15.9}$$

$$T_{i2} = -\frac{C_1 R_2}{\beta\gamma} \tag{15.10}$$

$$T_{d1} = \alpha\beta C_2 R_1 \tag{15.11}$$

$$T_{d2} = -\beta\gamma C_2 R_1 \tag{15.12}$$

In this case also, it may be noted that the parameters T_d and T_i can be controlled through R_1 and R_2 respectively whereas the constant K_p can be adjusted by the multiplier constant, **a**, which can be controlled if the DOCCII is realized as an electronically-controllable current conveyor. Finally, if a CMOS DOCCII is employed, the resistors R_1 and R_2 could be both CMOS voltage-controlled-resistors thereby permitting the change of their values through external control voltages.

When both the admittances are chosen as resistors, then the same circuit realizes a current mode instrumentation amplifier with the two outputs being given by:

$$I_{01} = \alpha\left(\frac{\beta R_1}{R_2} I_1 - I_2\right) \tag{15.13}$$

$$I_{02} = -\left(\frac{\beta R_1}{R_2} I_1 - I_2\right)\gamma \tag{15.14}$$

The expression for CMRR is given by:

$$\text{CMRR} = \frac{(\beta R_1/2R_2) + (1/2)}{(\beta R_1/R_2) - 1} \tag{15.15}$$

From equation (15.15), it is seen that for the case of unity gain i.e. $R_1 = R_2$, CMRR approaches infinity.

It has been demonstrated by Yuce and Minaei [2] that using a CMOS DOCCII implemented in 0.13 μm TSMC CMOS technology with DC power supply voltages = ± 1 V, the circuit has performed as stipulated by theoretical analysis.

15.3 Wide-Band Controllable Low Noise Amplifiers

A new application of current conveyors namely, the design of wide band controllable low noise amplifiers (LNAs), was proposed by Godara and Fabre in [3]. The proposed structure is shown symbolically in Fig. 15.3a whereas 15.3b shows a transistor level implementation of the same basic idea. The new LNAs have been shown to provide the following important advantages (i) complete elimination of the requirement of any passive components (ii) wide band performance, with stable

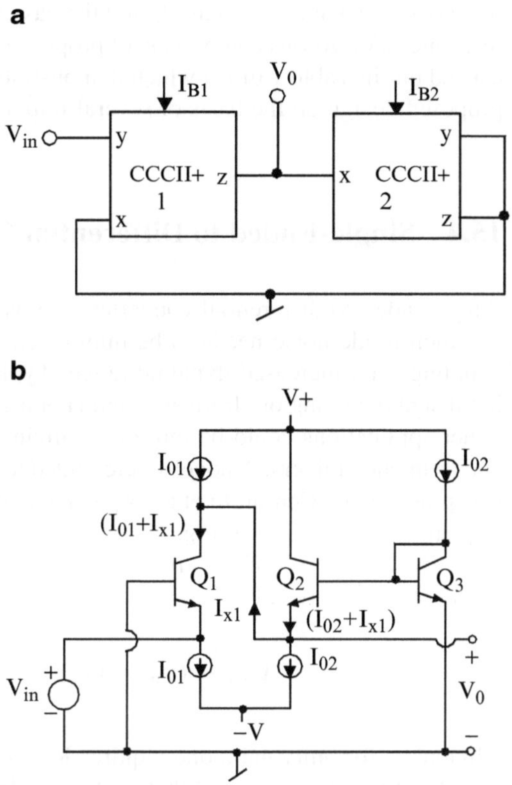

Fig. 15.3 CCII-based LNAs [3] (**a**) symbolic representation (**b**) transistor level implementation

frequency responses from zero to 3GHz and (iii) easy control of the gain by the ratio of two DC bias currents over a wide range (0 dB to 20 dB).

A straight forward analysis of the circuit of Fig. 15.3 shows that its voltage gain is given by

$$\frac{V_o}{V_{in}} = \frac{r_{x2}}{r_{x1}} \tag{15.16}$$

Since $r_{xi} = \frac{V_T}{2I_{Bi}}$, $i = 1, 2$, it follows that, $\frac{V_o}{V_{in}} = \frac{I_{B1}}{I_{B2}}$ where I_{B1} and I_{B2} are respectively the DC bias currents of CCCII1 and CCCII2.

Likewise, a detailed analysis of the transistorized version (Fig. 15.3b) also demonstrates that its voltage gain is given by the ratio of the control current I_{02} and the bias current I_{01} i.e.

$$\frac{V_o}{V_{in}} = \frac{I_{01}}{I_{02}} \tag{15.17}$$

It has been suggested that since the noise of this LNA is demonstrated by the base resistance of transistor Q_1 and the collector current I_c, the emitter length was increased to 10 μm to increase I_c and the base was cut into five parts to reduce base resistance. A rigorous comparison of proposed LNA with the existing ones has been carried out in Table 1 of [3] which demonstrates the comparative superiority of the proposed design on the basis of several important performance criteria.

15.4 Single-Ended to Differential Converters

Single-ended to differential converters are usually required in applications where common mode noise needs to be minimized along with the rejection to parasitic couplings and increased dynamic range. Typical examples are the cases when an input signal coming out from an antenna or a sensor is normally referred to ground. Other applications could be those of realizing fully differential filters, multipliers and frequency mixers. A novel single-ended to differential converter using DOCCII was proposed by Centurelli et al. [4] which is shown in Fig. 15.4a with the DOCCII implementation shown in Fig. 15.4b.

A routine analysis of the structure of Fig. 15.4a shows that the differential output of the circuit is given by

$$V_{diff} = (V_{01} - V_{02}) = -\left(\frac{R_2 + R_3}{R_1}\right)V_{in} \tag{15.18}$$

Obviously, for unity gain, one requires $R_2 = R_3 = R_1/2$

It has been demonstrated that in this single-ended to differential converter, the gain is fixed by only the external resistances and using rail-to-rail current conveyors, a rail-to rail converter may be obtained.

15.5 Precision Rectifiers Revisited

In this section we highlight some interesting precision full wave rectifiers employing different variants of CCs from amongst the various configurations developed by a number of researchers [5–10].

15.5.1 Precision Full Wave Rectifier Proposed by Koton, Herencsar and Vrba

Koton et al. [5] introduced new precision full-wave rectifiers using current conveyors and universal voltage conveyors. In the following, we present two specific topologies from their prepositions which are particularly attractive because each

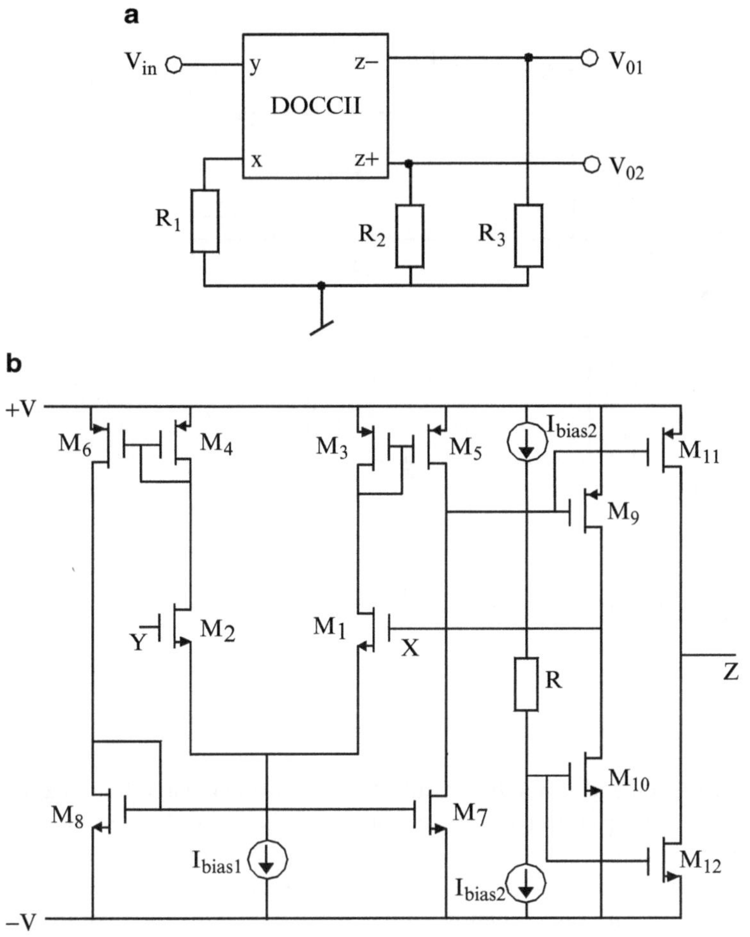

Fig. 15.4 (**a**) DOCCII-based single-ended to differential converter (**b**) CMOS implementation (Adapted from [4] © 2005 Springer Science + Business Media, Inc.)

one of them requires only two current conveyors with multiple outputs and no more than two diodes. The first circuit, which implements a current mode precision full-wave rectifier, is shown in Fig. 15.5a. The circuit has two current outputs, each one of them being given by the following equation:

$$i_{O+}(t) = |i_{in}(t)|, i_{O-}(t) = -|i_{IN}(t)| \tag{15.19}$$

It may be noted that this rectifier has very low input impedance and a very high output impedance at both the output terminals.

For the voltage mode precision rectifier of Fig. 15.5b, a straight forward analysis reveals that

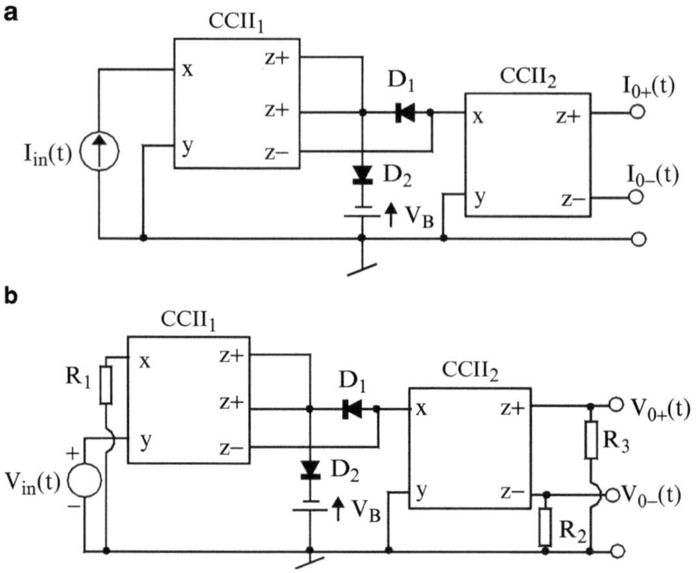

Fig. 15.5 Precision full-wave rectifiers [5] (**a**) current-mode (**b**) voltage-mode

$$v_{O+}(t) = \frac{R_3}{R_1}\left|\mathrm{v_{in}(t)}\right|,.\ v_{O-}(t) = -\frac{R_2}{R_1}\left|\mathrm{v_{in}(t)}\right|, \tag{15.20}$$

The experimental and simulation results have confirmed that although as compared to the traditional op-amp based precision rectifiers, the CC-based precision rectifiers presented in [5] have somewhat larger DC error but on the other hand, they have superior transient response. These circuits are obviously superior to their op-amp based counterpart when the input signals to be rectified are in the vicinity of 1 MHz or so.

15.5.2 *Kumngern's Full Wave Rectifier*

Kumngern [6] demonstrated that a novel full wave rectifier can be made from only two differential difference current conveyors (DDCC) without requiring any diodes. His proportion is shown in Fig. 15.6.

The operation of this circuit can be explained as follows. When V_{in} is positive, the DDCC1is ON, the voltage V_{in} is followed by the DDCC1 and appears at the X-terminal and hence, at the output of the circuit. On the other hand, when V_{in} is negative, this negative voltage is followed by DDCC2 and is conveyed to its X terminal. The relation between input voltage V_{in} and output voltage V_{out} can be expressed as:

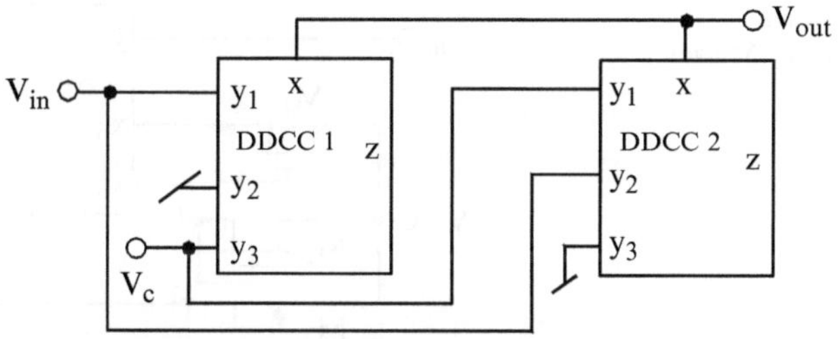

Fig. 15.6 Full-wave rectifier employing DDCCs proposed by Kumngern [6]

$$\begin{array}{l} V_{in} > 0;\, V_{out} = V_{in} \ : \text{DDCC1 is ON,} \\ V_{in} < 0;\, V_{out} = -V_{in} \ : \text{DDCC2 is ON} \end{array} \tag{15.21}$$

From the above, it therefore follows that the output of the rectifier circuit is given by $V_{out} = |V_{in}|$ and the circuit therefore, implements a full-wave rectifier.

It is interesting to note that this circuit has the required high input impedance and low output impedance and because of using exclusively MOSFETs only to implement DDCCs, the circuit is highly suitable for IC implementation in CMOS technology. SPICE simulation results with 0.5 μm MIETEC CMOS process parameters with DDCCs biased from a ±2.5 V DC power supply have demonstrated the workability of the proposed circuit as well as its applicability to various non-linear analog signal processing circuits.

15.5.3 Precision Rectifier Proposed by Minaei and Yuce

Minaei and Yuce [7] also presented a precision full-wave rectifier fully implementable in CMOS technology employing a DXCCII (Fig. 15.7). Assuming the normal characterization of a DXCCII, a straight forward analysis of this circuit reveals its output voltage to be given by:

$$V_o(t) = |V_{in}(t)| = \left(V_{in}(t)_+ - V_{in}(t)_-\right) \tag{15.22}$$

SPICE simulation results based on CMOS DXCCII implemented in 0.25 μ CMOS technology with DC power supply voltages of ±1.25 V have demonstrated the successful operation of this circuit for sinusoidal input signal of the order of 150 mV at 1 MHz. However, a limitation of this circuit is that it does not operate well for input signals with amplitude less than 20 mV due to the Z-terminal parasitic capacitance and resistance of the DXCCII. In conclusion, this active only rectifier circuit possesses high input impedance which makes it easily cascadable and highly suitable for integration in CMOS technology.

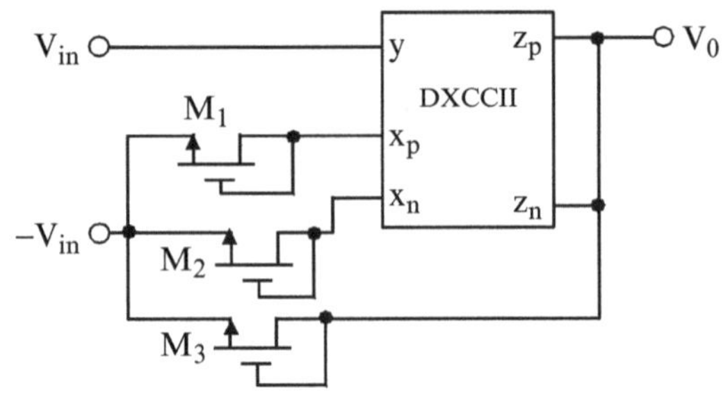

Fig. 15.7 Full-wave rectifier proposed by Minaei and Yuce [7]

15.6 Multivibrators and Relaxation Oscillators

Apart from the relaxation oscillators based upon normal type of CC such as the one in [11], there have been a number of investigations on implementing square/triangular wave generators using one specific variety of current conveyors namely the differential voltage current conveyor (DVCC). In this section, we give a brief account of a variety of relaxation oscillators using DVCC.

15.6.1 Chien's Square/Triangular Wave Generator

Chien [12] proposed a DVCC-based voltage-controlled dual slope operation square/triangular wave generator which is shown in Fig. 15.8. The basic principle of this circuit can be explained as follows:

The connection Y_1-Z provides a positive feedback path in the circuit of Fig. 15.8a which leads to the saturation of DVCC-2 due to which its output voltage would be either in the positive saturation level V_0^+ or negative saturation level V_0^-. The relevant wave forms associated with this circuit are shown in Fig. 15.8b where V_{TH} and V_{TL} represent the upper and lower threshold levels. Assuming that $V_{01} = V_0^+$ and $R_2 > R_1$ it follows that DVCC-I makes it possible to force a constant current flowing through the capacitor C so that the output voltage V_{02} is positive going ramp whose slope is given by

$$\frac{dV_{02}(t)}{dt} = \frac{V_{TH} - V_{TL}}{T_1} = \frac{V_{01} - V_B}{R_3 C} = \frac{V_0^+ - V_B}{R_3 C} \tag{15.23}$$

The positive ramp continues until V_{02} reaches the upper threshold level V_{TH} at which point the output voltage V_{01} of DVCC-II switches to negative saturation levelV_0^-.

The circuit then enters off- duty cycle operation and the expressions for I_{x2} and I_{z2} are given by:

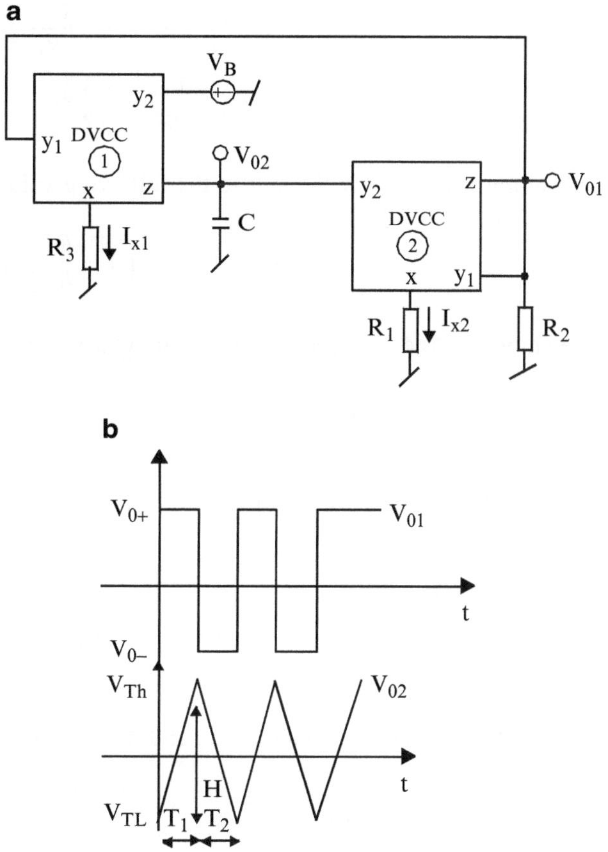

Fig. 15.8 DVCC based voltage-controlled dual slope square/triangular wave generator (Adapted from [12] © 2012 Elsevier Ltd.)

$$I_{x2} = \frac{V_{01} - V_{02}}{R_1} = \frac{V_0^+ - V_{02}}{R_1} \tag{15.24}$$

$$I_{z2} = \frac{V_{01}}{R_2} = \frac{V_0^+}{R_2} \tag{15.25}$$

Because of V_{01} being at V_0^-, this reverses the output current of DVCC-I hence, V_{02} is a negative going ramp. As soon as V_{02} reaches the lower threshold level V_{TL}, V_{01} switches back to positive saturation level V_0^+ and this cycle goes on repeating itself. V_{TH} and V_{TL} can be calculated by using the following equations:

$$\frac{dV_{02}(t)}{dt} = \frac{V_{TL} - V_{TH}}{T_2} = \frac{V_{01} - V_B}{R_3 C} = \frac{-V_0^+ - V_B}{R_3 C} \tag{15.26}$$

$$I_{x2} = \frac{V_{01} - V_{02}}{R_1} = \frac{-V_0^+ - V_{02}}{R_1} \tag{15.27}$$

$$I_{z2} = \frac{V_{01}}{R_2} = \frac{-V_0^+}{R_2} \tag{15.28}$$

and setting $I_{x2} = I_{z2}$, which leads to the following values of V_{TL} and V_{TH}

$$V_{TH} = -V_{TL} = \left(1 - \frac{R_1}{R_2}\right) V_0^+ \tag{15.29}$$

The amplitude (peak-to-peak) of the triangular wave form is given by:

$$V_{02} = (V_{TH} - V_{TL}) = 2\left(1 - \frac{R_1}{R_2}\right) V_0^+ \tag{15.30}$$

The periods of the square and triangular waves are given by:

$$T_1 = \frac{2R_3C(1 - (R_1/R_2))}{V_c} \tag{15.31}$$

$$T_2 = \frac{2R_3C(1 - (R_1/R_2))}{2 - V_c} \tag{15.32}$$

Finally, the oscillation frequency (f_0) and duty cycle (D_a) are given by:

$$f_0 = \frac{1}{T_1 + T_2} = \frac{V_C(2 - V_C)}{4R_3C(1 - (R_1/R_2))} \tag{15.33}$$

$$D_a = \frac{T_1}{T_1 + T_2} = \left(1 - \frac{V_C}{2}\right) \tag{15.34}$$

where

$$V_C = \left(1 - \frac{V_B}{V_0^+}\right) \tag{15.35}$$

It is interesting to note that f_0 can be independently controlled by R_3 thereby resulting in single resistance controlled oscillator. Also, the amplitude of the triangular wave is proportional to the ratio (R_1/R_2) whereas the duty cycle is controllable by the control voltage V_c. Furthermore, if one takes $V_c << 1$ the interval T_2 is negligibly small and can be ignored and the circuit can then be used as saw-tooth waveform generator for which the frequency of oscillation is given by:

$$f_0 = \frac{1}{T_1} = \frac{V_C}{2R_3C(1 - (R_1/R_2))} \tag{15.36}$$

15.6.2 Switch-Controllable Bi-stable Multivibrator

Chien in [13] demonstrated that a simple switch-controllable bistable multivibrator with dual hysteresis modes can be made from only a single DVCC and two external resistors along with a double-pole double-throw (DPDT) switch. This circuit is shown in Fig. 15.9a. The operation of this circuit can be explained as follows. It is assumed that this circuit will be operated with an external voltage signal V_{in} which can be an existing square/triangular wave generator. The shorting of the terminals Y_1 and Z along with the resistor R_2 provides a positive feedback in the circuit. As a consequence, the DVCC is saturated and the output V_0 would be either at positive saturation level V_0^+ or negative saturation level V_0^-. The transfer characteristics of the circuit are shown in Fig. 15.9b. By proper connection of the DPDT switch, this circuit exhibits hysteresis characteristics of two different kinds which have been termed as clockwise (CW) mode or counter-CW (CCW) mode which are shown respectively in Fig. 15.9b and c.

For CW mode, the DPDT switch is made to connect node 1 to node 3 and node 2 to node 4. It is easy to see that the terminal currents I_X and I_Z are respectively given by:

$$I_X = \frac{V_0 - V_{in}}{R_1}\text{; and } I_Z = \frac{V_0}{R_2} \tag{15.37}$$

The hysteresis characteristics for CW mode can be explained as follows. Assuming that V_0 is at positive saturation level V_0^+ and also let initial input $V_{in} = 0$. Further, let $R_2 > R_1$, then from equation (15.37) it follows that I_x would be larger than I_z. Thus, it is ensured that V_0 will be in the level V_0^+ initially. If a positive V_{in} is applied now, the current I_x will be decreasing gradually. Now because I_z is larger than I_x, V_0 switches from the state V_0^+ to V_0^-. This explains the path 1 in Fig. 15.9b. Similarly, in path 2 the output is initially at the level V_0^- and hence, current I_z is greater than current I_x. Now if input signal V_{in} decreases from zero and is negative the current I_x is increases until it becomes larger than I_z, due to which V_0 switches from V_0^- to V_0^+, this explains path 2 of the transfer characteristics of Fig. 15.7b. The upper and lower threshold voltages V_{TH} and V_{TL} can be determined from the instant of time when I_x and I_z are equal. The mathematical expression from V_{TH} and V_{TL} in CW mode are given by

$$V_{TH} = \left(1 - \frac{R_1}{R_2}\right)V_0^+ \text{ and } V_{TL} = \left(1 - \frac{R_1}{R_2}\right)V_0^- \tag{15.38}$$

For CCW mode, the DPDT switch enables in connections of terminals 1 to 4 and

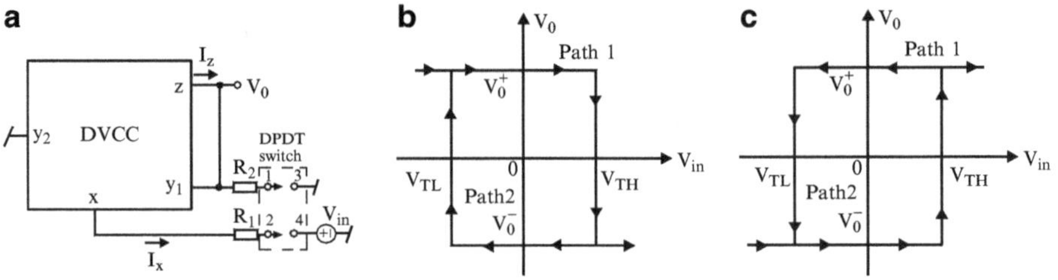

Fig. 15.9 (**a**) Switch-controllable DVCC-based bistable multivibrator (**b**) transfer characteristics in CW mode (**c**) transfer characteristics in CCW mode (Adapted from [13] © 2011 Elsevier Ltd.)

2 to 3. Obviously, in such a case, the expressions for the currents I_x and I_z change as follows:

$$I_x = \frac{V_0}{R_1} \text{ and } I_z = \frac{V_0 - V_{in}}{R_2} \tag{15.39}$$

The transfer characteristics for the CCW mode can be explained similar to the case of CW mode [14] which is omitted to conserve the space. The expressions of two threshold voltages in this case are given by:

$$V_{TH} = \left(\frac{R_2}{R_1} - 1\right)V_0^+ ; \text{ and } V_{TL} = \left(\frac{R_2}{R_1} - 1\right)V_0^- \tag{15.40}$$

A voltage controlled bistable multivibrator is obtained by connecting an external voltage source V_B at the Y_2 terminal of the DVCC as shown in Fig. 15.10. With the introduction of additional controlled voltage it is obvious that the threshold voltages in both CW and CCW modes will become a function of this control voltage also. By routine analysis, the modified expressions for the threshold voltages determined by the following expressions

In CW mode:

$$V_{TH} = \left(1 - \frac{R_1}{R_2}\right)V_0^+ - V_B ; \text{ and } V_{TL} = \left(1 - \frac{R_1}{R_2}\right)V_0^- - V_B \tag{15.41}$$

and in CCW mode:

$$V_{TH} = \left(\frac{R_2}{R_1} - 1\right)V_0^+ + \left(\frac{R_2}{R_1}\right)V_B ; \text{ and } V_{TL} = \left(\frac{R_2}{R_1} - 1\right)V_0^- + \left(\frac{R_2}{R_1}\right)V_B \tag{15.42}$$

The effectiveness of the above bistable multivibrators have been confirmed by implementing the DVCC by AD844AN ICs and from the breadboard prototypes the operating frequency of the proposed bistable circuits was found to be around 500 KHz.

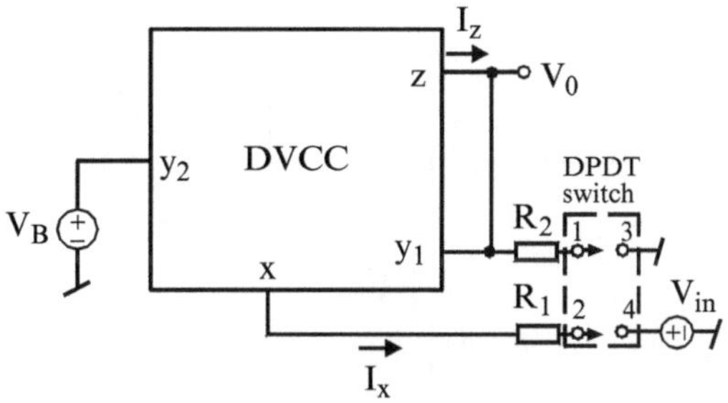

Fig. 15.10 DC voltage controlled DVCC bistable multivibrator (Adapted from [13] © 2011 Elsevier Ltd.)

15.6.3 Single DVCC-Based Monostable Multivibrators

Chien and Lo in [14] presented two novel single DVCC based monostable multivibrators which require fewer components than, for example, a traditional op-amp based multivibrator. These circuits are shown above in Fig. 15.11.

The operation of these circuits can be explained as follows:

Firstly, it may be noted that the connection from Z to Y_1 along with R_2 constitutes the required positive feedback path which facilitates the saturation of the DVCC due to which its output could be only in one of the two saturation levels V_0^+ and V_0^-. The trigger input V_{trg} is a rising edge signal which can be provided by an external function generator. The permanent stable state of the circuit before the application of the trigger pulse can be easily shown to be V_0^+ (Fig. 15.11c) by noting that in the stable state ($0 < t < T_1$), the capacitor C is open circuited and V_c is clamped by the diode D or the analog switch M in the circuit of 15.9b. If R_2 is much greater than R_1 and since $I_x = \frac{V_0}{R_1} = \frac{V_0^+}{R_1}, I_z = \frac{V_0}{R_2} = \frac{V_0^+}{R_2}$, it follows that I_x is more positive than I_z which ensures that output voltage V_0 will be in positive saturation level V_0^+.

When a positive edge triggering signal is applied at $t = T_1$ the circuit enters into a quasi-stable state ($T_1 < t < T_2$). When this happens, I_z becomes more positive than I_x hence, the output voltage V_0 changes abruptly from V_0^+ to V_0^-. This results in the discharge of the capacitor through R_1 from $t = T_1$. In the quasi-stable state, the expressions for I_x, I_z and V_c are given by:

$$V_c(t) = V_0^-\left(1 - e^{-\left(t - T_1/R_1C\right)}\right) \tag{15.43}$$

$$I_X = \frac{V_0 - V_C}{R_1} = \frac{V_0^- - V_C}{R_1} \tag{15.44}$$

$$I_z = \frac{V_0}{R_2} = \frac{V_0^-}{R_2} \tag{15.45}$$

When at a time $t = T_2$ the capacitor voltage drops to the lower threshold voltage V_{TL} the expression for which is determined when $I_x = I_z$ and is given by:

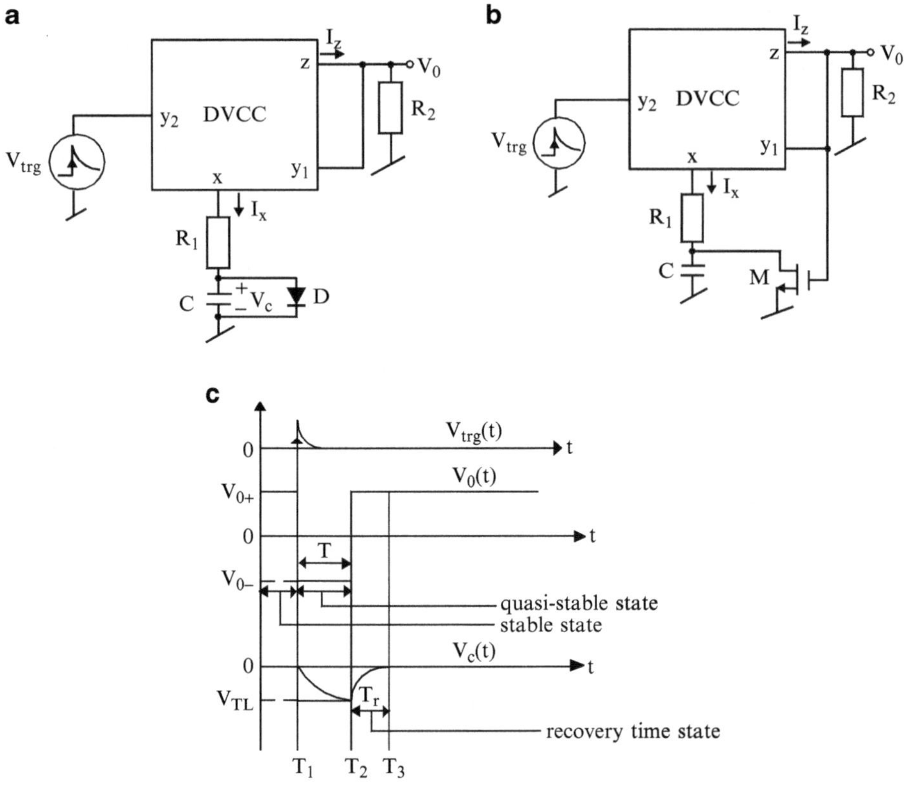

Fig. 15.11 DVCC based monostable multivibrators (**a**) the basic circuit (**b**) the modified circuit shortened recovery time (**c**) corresponding waveforms (Adapted from [14] © 2011 Elsevier Ltd.)

$$V_{TL} = V_c(T_2) = \left(1 - \frac{R_1}{R_2}\right)V_0^- \tag{15.46}$$

By substituting the value of $V_c(T_2)$ in the general expression for the capacitor voltage time period T which is found to be:

$$T = T_2 - T_1 = R_1 Cin\left(1 - \frac{V_{TL}}{V_0^-}\right) = R_1 Cin\left(\frac{R_2}{R_1}\right) \tag{15.47}$$

By a routine analysis, the recovery times T_r for both the circuits are defined as $T_r = (T_3\text{-}T_2)$ and are found to be:

$$T_r = T_3 - T_2 \cong R_1 C\ln\left(2 - \frac{R_1}{R_2}\right) \text{ if } R_{DR} > > R_1 \tag{15.48}$$

where R_{DR} is the turn-off resistance of the diode D.

$$T_r = T_3 - T_2 \cong R_{DS} C \ln\left[1 + \frac{R_1(R_2 - R_1)}{R_2 R_{DS}}\right] \text{ if } R_{DS} < < R_1 \tag{15.49}$$

From the above, it is easy to visualize that, since R_{DS} (turn-on resistance of the analog switch) is usually much larger than R_1, the recovery time of the second monostable circuit will be much shorter than the first one.

The effectiveness of both the circuits has been confirmed through SPICE simulations as well as hardware implementations by realizing DVCC with AD844AN ICs and using CD4066 as the analog switch where from the effectiveness of the second circuit in shortening the recovering time has been confirmed in both the cases.

15.6.4 Chien's Relaxation Oscillators

Two circuits for realizing relaxation oscillators using DVCC were presented by Chien in [15]. The first circuit can generate a symmetrical square wave with 50 % duty cycle while the second circuit can further control the duty cycle by adjusting the tuning factor of the potentiometer. These circuits are shown in Fig. 15.12.

The operation of the circuit of Fig. 15.12 a can be explained as follows: Let us assume that V_0 is in the high state i.e., $V_0 = V_0^+$ if we take $R_1 > R_2$ then, I_X is more positive than I_Z and consequently, V_0 is guaranteed to be in the state V_0^+. The capacitor charges exponentially from low threshold (V_{LT}) to high threshold (V_{HT}). When capacitor voltage becomes slightly greater than V_{TH}, this also implies that now I_X would have become slightly less than I_Z, as a consequence of which V_0 switches from V_0^+ state to V_0^- state. From a routine analysis [15], the two threshold voltages are found to be:

$$V_{TH} = \left(1 - \frac{R_1}{R_2}\right) V_0^+ \text{ and } V_{TL} = -\left(1 - \frac{R_1}{R_2}\right) V_0^+ \tag{15.50}$$

If we consider the total time period in which the output waveform remains high as T_1, from a straight forward analysis, its value is given by:

$$T_1 = R_1 C \ln\left(\frac{2R_2}{R_1} - 1\right) \tag{15.51}$$

Similarly, the time period in which the output wave remains in low state i.e., V_0^- is found to be:

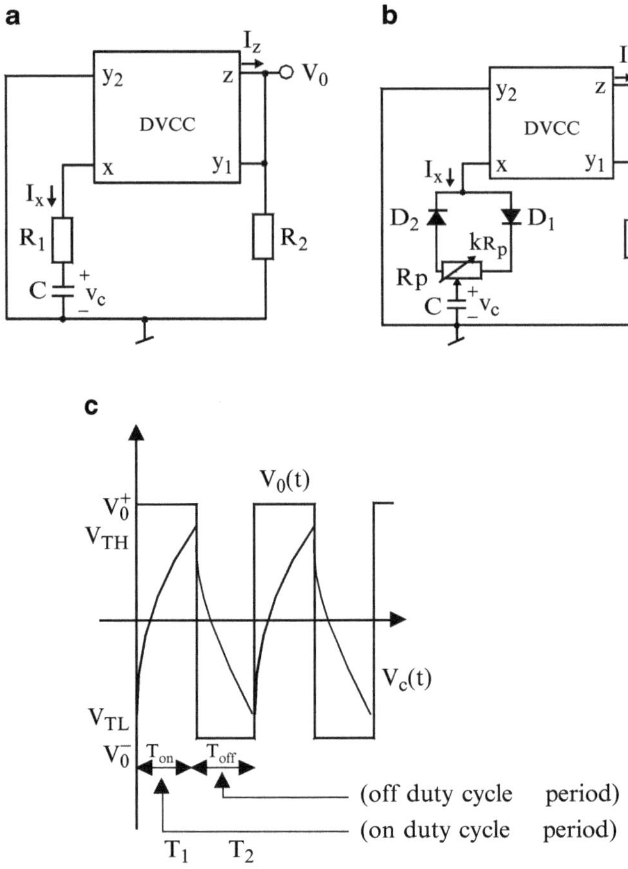

Fig. 15.12 The relaxation oscillators using a single DVCC and their relevant waveforms [15] (**a**) circuit with 50 % duty cycle (**b**) circuit with variable duty cycle (**c**) corresponding waveforms

$$T_2 = R_1 C \ln\left(\frac{2R_2}{R_1} - 1\right) \tag{15.52}$$

Thus, the frequency of the output square wave is found to be:

$$f_0 = \frac{1}{T_1 + T_2} = \frac{1}{2R_1 C \ln\left(\frac{2R_2}{R_1} - 1\right)} \tag{15.53}$$

A slightly modified form of this circuit to facilitate non-50 % duty cycle (Fig. 15.12b) obtained by splitting the resistance R_1in to two parts through a potentiometer (kR_P) and ensuring that during the charging and discharging of the capacitor C only one of these two parts is coming into picture. From the diagram, it can be seen that the diode D_1 is ON and diode D_2 is OFF during the charging of the capacitor (for the duration T_1) whereas on the other hand, during the discharging of the capacitor, the roles of the diodes are reversed i.e., D_2 is ON and D_1 is OFF due to which the lower part of the potentiometer comes into picture. With a routine but detailed analysis, it has been shown that duty cycle (δ) turns out to be:

$$\delta = \frac{1}{1 + \left(\frac{1-k}{k}\right)\left(\frac{\ln\left(\frac{2R_2}{(1-k)R_P} - \frac{k}{1-k}\right)}{\ln\left(\frac{2R_2}{R_P} + k - 1\right)}\right)} \tag{15.54}$$

where k is the tuning factor of the potentiometer. Thus, the value of δ other than 50 % can be achieved by proper selection of the parameter k.

15.6.5 Chien's DO-DVCC-Based Square/Triangular Wave Generator

A novel square/triangular waveform generator using single DO-DVCC and three grounded passive components was advanced by Chien in [16] and is shown here in Fig. 15.13

The operation of this circuit can be explained as follows. The Y_1-Z^+ connection along with the resistor R_2 forms the positive feedback path due to which the DVCC saturates and hence, its output could be either in state V_0^+ or V_0^-. In the on-duty cycle period (T_1), V_{01} is at positive saturation level in the beginning. In order this to happen, the current I_x must have a strong positive charge than I_z. To accomplish this R_2 needs to be larger than R_1. Thus, capacitor C is charged, causing V_{02} to be a positive going ramp. The rate at which the capacitor charges linearly along with the values of I_x and I_z are given by the following equations:

$$\frac{dV_{02}(t)}{dt} = \frac{V_{01}}{R_2C} = \frac{V_0^+}{R_2C} = \frac{V_{TH} - V_{TL}}{T_1} \tag{15.55}$$

$$I_X = \frac{V_{01} - V_{02}}{R_1} = \frac{V_0^+ - V_{02}}{R_1} \tag{15.56}$$

$$I_z = I_{z1} = I_{z2} = \frac{V_{01}}{R_2} = \frac{V_0^+}{R_2} \tag{15.57}$$

The positive ramp continues until V_{02} reaches the upper threshold level (V_{TH}) at which time the output V_{01} switches to low state V_0^-. Now the capacitor voltage V_{02} starts discharging linearly resulting in a negative going ramp at V_{02}. During this period, when the capacitor voltage crosses the lower threshold level (V_{TL}), the circuit resumes the high state and this operation goes on repeating. The equations

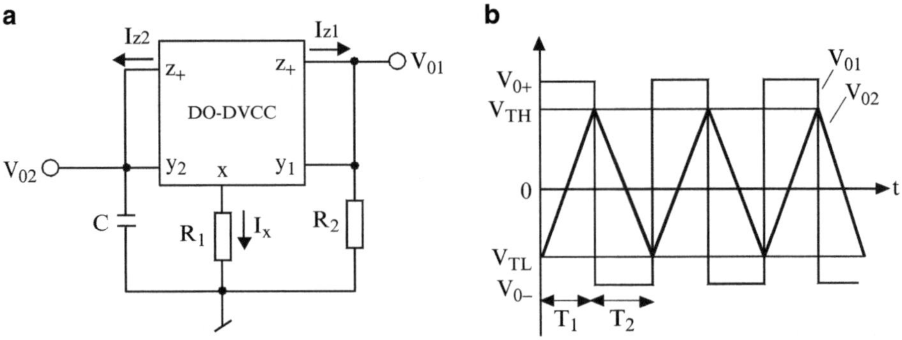

Fig. 15.13 DO-DVCC based square/triangular waveform generator employing grounded passive components [16] (**a**) proposed circuit (**b**) its relevant waveforms

for the capacitor voltage following a negative ramp and the currents I_x and I_z in this state are given by the following equations:

$$\frac{dV_{02}(t)}{dt} = \frac{V_{01}}{R_2 C} = \frac{-V_0^+}{R_2 C} = \frac{V_{TL} - V_{TH}}{T_2} \tag{15.58}$$

$$I_X = \frac{V_{01} - V_{02}}{R_1} = \frac{-V_0^+ - V_{02}}{R_1} \tag{15.59}$$

$$I_z = I_{z1} = I_{z2} = \frac{V_{01}}{R_2} = \frac{-V_0^+}{R_2} \tag{15.60}$$

Using the above equations and setting $Ix = I_Z$, the expressions for the threshold voltages are found to be:

$$V_{TH} = -V_{TL} = \left(1 - \frac{R_1}{R_2}\right) V_0^+ \tag{15.61}$$

The oscillation frequency of the waveforms generated by this circuit is found to be:

$$f_0 = \frac{1}{T_1 + T_2} = \frac{1}{4R_2 C(1 - (R_1/R_2))} \tag{15.62}$$

From the experiments based upon realizing DVCC using commercially available AD844AN ICs, it has been found that in the discrete bread boarded versions of the circuit, the maximum attainable oscillation frequency of the circuit has been found to be several hundred KHz (500–800 KHz) which is still larger than those attainable from square/triangular waveform generators using op-amps or OTAs (which were found to be well below 100 KHz). In view of the simplicity of the structure coupled with its good performance, it appears that the circuit may find applications in instrumentation, communication and signal processing systems.

15.7 Wide-Band Impedance Matching Circuits

Another novel application of CCs was proposed by Godara and Fabre [17, 18] who demonstrated that a CC can be used as a versatile wide band impedance matching device offering a vastly superior performance as compared to the other traditional passive elements solutions. The CC based impedance matching configuration is the first flexible matching circuit to adapt to any impedance to any desired value and is also the smallest matching circuit ever and is a rare example of really wide band matching. The schematic presentation of the new matching impedance circuit is shown in Fig. 15.14 which is, in fact, a CCCII configured as voltage follower. The basic core of the impedance conversion circuit is composed of BJTs Q_1–Q_2–Q_3–Q_4. The DC bias current I_0 is copied to the required branches of this translinear cell through a current repeater M_1–M_2–M_3 and current mirror composed of M_4–M_5. The input V_{in} is applied at port-Y which has a very high intrinsic resistance R_y. The output is taken from port-X where the output impedance consists of resistance $R_x = V_T/2I_0$ (where V_T is the thermal voltage) in series with a small parasitic inductance L_x. It is important to note that different value of R_x can be obtained by changing the DC bias current I_0.

15.8 Sample and Hold Circuits

A simple CC-based sample and hold circuit presented by Hwang et al. [19] is shown in Fig. 15.15. The circuit is operated from the clock signal ϕ_1. When ϕ_1 is HIGH, the circuit is in sample mode and the input signal V_{in} is sampled and stored across the capacitor C_1. When ϕ_1 is LOW the circuit is in hold mode and the output V_{out} will be copied from the voltage sampled across C_1. The output voltage is, therefore, given by $V_{out} = \alpha_v V_{in}$ where α_v is the voltage gain of the current conveyor between ports X and Y; the other characterizing equations being $i_y = 0$ and $i_z = \alpha_i i_x$ where both α_v and α_i are nominally unity.

Another sample and hold circuit proposed by Hwang et al. [20] is shown in Fig. 15.16 which employs a fully balanced CCII (FBCCII). When clock ϕ_1 is HIGH the circuit is in sample mode and the input signal is sampled on the capacitor C_s. When ϕ_1 is LOW the circuit is in hold mode and input signal is held by the capacitor C_s simultaneously. The outputs of the sample and hold circuit namely V_{XP} and V_{XN} trace the input signals V_{YP} and V_{YN} and can be expressed as:

$$V_{out} = V_{XP} - V_{XN} = \frac{1}{1 + \frac{1}{A\beta}}(V_{YP} - V_{YN}) \tag{15.63}$$

where A is the DC gain of the FDCCII and β is the feedback factor of FDCCII which is equal to unity.

Fig. 15.14 Wideband impedance matching circuit employing CCC [17]

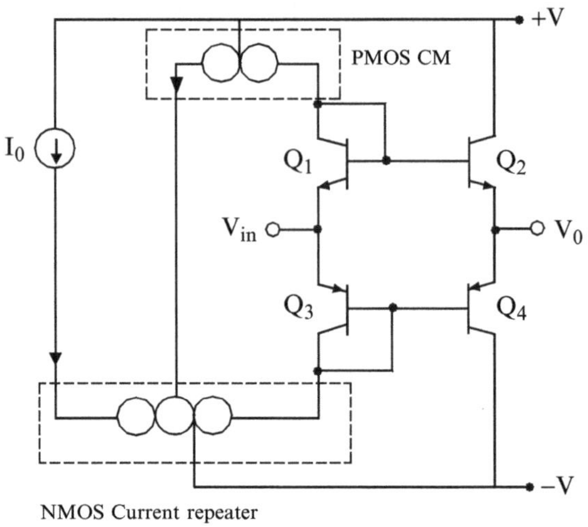

Fig. 15.15 CCII-based sample and hold circuit proposed by Hwang et al. (Adapted from [19] © 2007 Springer Science + Business Media)

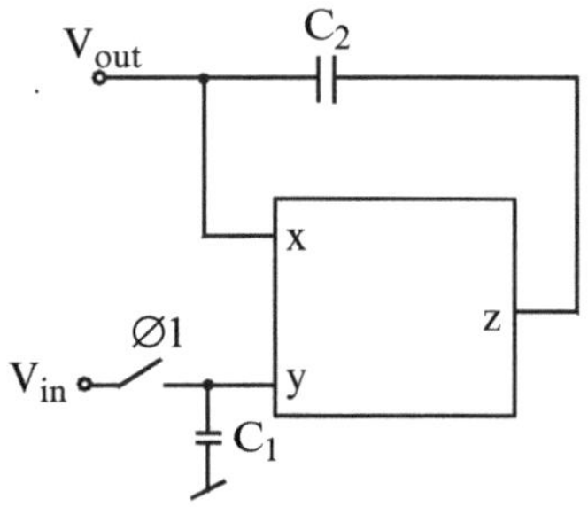

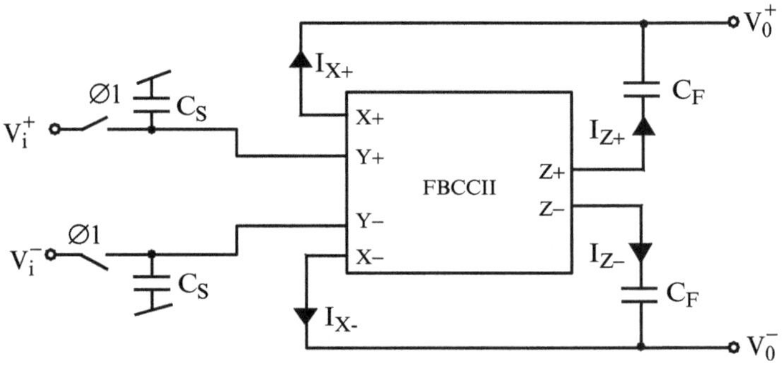

Fig. 15.16 FBCCII-based sample and hold circuit [20]

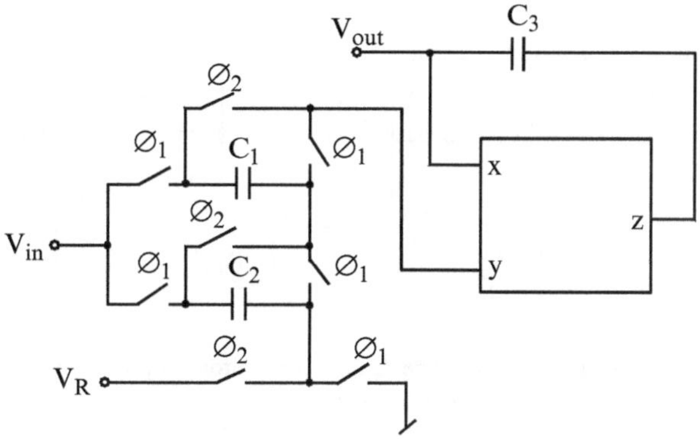

Fig. 15.17 CCII-based MDAC proposed by Hwang et al. (Adapted from [19] © 2007 Springer Science + Business Media)

15.9 CCII-Based Digital-to-Analog Converter

A FBCCII-based multiplying digital-to-analog converter (MDAC) due to Hwang et al. [19] is shown here in Fig. 15.17. Its operation can be explained as follows. When the clock pulse φ_1 is HIGH, both the capacitors C_1 and C_2 sample the input V_{in} and both the capacitors are connected in series and to ground. In the next phase, when φ_2 is HIGH the bottom plate of the capacitor C_2 gets connected to the voltage V_R. The two capacitors are still connected in series but the top plate of the capacitor $C_1$is now connected to Y-terminal of the CCII. The output of the circuit is given by

$$V_{out} = V_R + 2V_{in} \tag{15.64}$$

The output of the circuit is determined by noting that I_z and I_x are equal to make the output voltage settle faster while the feedback capacitor C_3 plays a major role in reducing high frequency and switching noises. For $V_R = 0$, the input/output behavior of the circuit is shown in Fig. 15.17, where V_{in} is the sine wave of 1 V p-p at 2 MHz frequency and the square wave signal (with an amplification by a factor of two) is the output waveform.

15.10 Chaos Generators: Revisited

There have been numerous attempts in making chaotic oscillators using CCs as well as numerous other building blocks as is evident from a large amount of literature in this area, for instance see [21, 22, 23] and the references cited therein. Gandhi [24] presented a DOCCII based implementation which nearly follows the same methodology as adopted in the CFOA-based circuit proposed by Senani and Gupta earlier in [25]. Gandhi's preposition is shown in Fig. 15.18 where the Chua's diode made from two DOCCII is nearly analogist to the two CFOA based implementation proposed in [25] while the simulated inductor with the inductive current

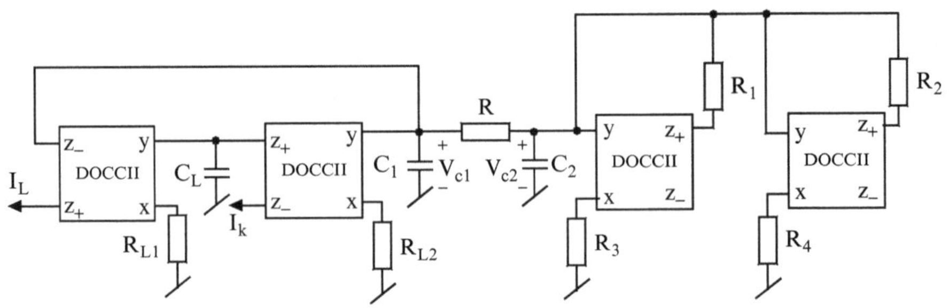

Fig. 15.18 Current mode Chua's oscillator proposed by Gandhi (Adapted from [24] © 2005 Springer Science + Business Media)

available as a current output is made from two DOCCII two grounded resistors and a grounded capacitor is analogous to the three CFOA based inductance simulator employed in Senani, Gupta implementation [25] with inductive current available from the Z output of one of the CFOAs therein.

Gandhi has demonstrated that with DOCCIIs replaced by multiple output CCII, a MOCCII based Chua's oscillator is obtained which can be very easily adopted to create hyper chaotic circuits by appropriate connections and couplings of several such units together with a considerable reduced hardware as compared to other available solutions for generating hyper chaos.

15.11 Realization of Chua Family of Nonlinear Network Elements: Mutators, Rotators, Reflectors and Scalars

Mutators: Chua introduced four new network elements namely, the mutator, the scalor, the reflector and the rotator [26, 27] which have been found to be quite useful in nonlinear network synthesis. Three basic mutators were defined originally: the C-R mutator, the L-C mutator and the L-R mutator. The first one transforms a nonlinear resistance into a nonlinear capacitance, the second converts a nonlinear capacitance into nonlinear inductance and the last one transforms a nonlinear resistance into nonlinear inductance. Some of the mutators can be related to the class of active elements known as generalized impedance converters/inverters (GIC/GII) but all mutators are not equivalent to GIC/GII.

The mutators are normally characterized by transmission matrices and based upon the fact that more than one characterization is possible to achieve an intended *mutation*, the basic types of mutators have been further classified into C-R-mutator type1, C-R-mutator type 2, L-R mutator type 1, L-R-mutator type 2, L-C-mutator type 1 and L-C-mutator type 2. Later on, the concept of mutators had even been extended beyond the territory of R, L, C elements to the domain of FDNR (an element having $Z(s) = 1/Ds^2$) and FDNC (an element having $Z(s) = Ms^2$)

thereby leading to D-R-mutators and M-R-mutators. More recently, mutators have been defined even to convert memresistive elements into memcapacitive and meminductive elements.

A type 1 C-R mutator is characterized by $v_1 = v_2$ and $i_1 = -CR\frac{di_2}{dt}$ while type 2 C-R mutator is characterized by $v_1 = -i_2R$, $i_1 = \frac{Cdv_2}{dt}$. Similarly, a type 1 L-R mutator, is characterized by $v_1 = CR\frac{dv_2}{dt}$, $i_1 = -i_2$ while type 2 L-R mutator is characterized by $v_1 = CR_1R_2\frac{di_2}{dt}$, $i_1 = \frac{V_2}{R_3}$. Lastly, type 1 L-C- mutator is characterized by $v_1 = -i_2R$, $i_1 = \frac{V_2}{R_3}$ while type 2 L-C mutator is characterized by $v_1 = C_2R_2\frac{dv_2}{dt}$, $i_1 = \frac{1}{C_1R_1}\int i_2dt$.

From a careful inspection of the above equations, it is not difficult to infer that it should be possible to realize C-R mutators of both types, L-R mutator type 1 and LC-mutator type 1 by using only two CCIIs while the remaining two mutators namely, L-R mutator type 2 and L-C mutator type 2 may need at least three CCIIs.

Rotators: R−, L−, or C − rotators are active 2-port elements which can rotate, $i - v$, $\varphi - i$ or $q - v$ curve by a prescribed angle of rotation about the origin.

Reflectors: A reflector is an active element which can reflect a given $i - v, \varphi - i$ or $q - v$ curve with respect to a straight line (through the origin) having an angle Θ with the horizontal axis. Corresponding to the three curves under consideration, the reflectors can be classified into three types, namely R−, L − or C − reflector.

Scalars: A scalar is a 2-port element which can scale the current, voltage or power of a device or a nonlinear element. Thus, there are three basic types of scalars namely, a voltage scalar, a current scalar and a power scalar. Scalars can also be used to scale $\varphi - i$ curve of a nonlinear inductor or $q - v$ curve of a nonlinear capacitor.

Since the introduction of the second generation CC by Sedra and Smith in 1970, they were the first to demonstrate in [28] that CCII ± appear to be very interesting building blocks for realizing mutators, reflectors and rotators. Subsequently, the problem of realizing mutators using current conveyors has been widely investigated and a larger number of CC-based realisations of mutators are now known.

Some representative realisations of the four of the above mentioned six basic mutator types are shown in Fig. 15.19.

We have enumerated here only the CC-II based realisations proposed by Sedra and Smith in [28] to give a glimpse of the type of circuits which make mutators. For further studies the readers are referred to references [29–35] and those cited therein.

15.12 Memcapacitance and Meminductance Emulators

With the announcement of the practical realization of a memristive device by Strukov et al. in 2008 [36] from Hevelett-Packard laboratories, the fourth fundamental element *Memristor* predicted by Chua in 1971 [37] became a very

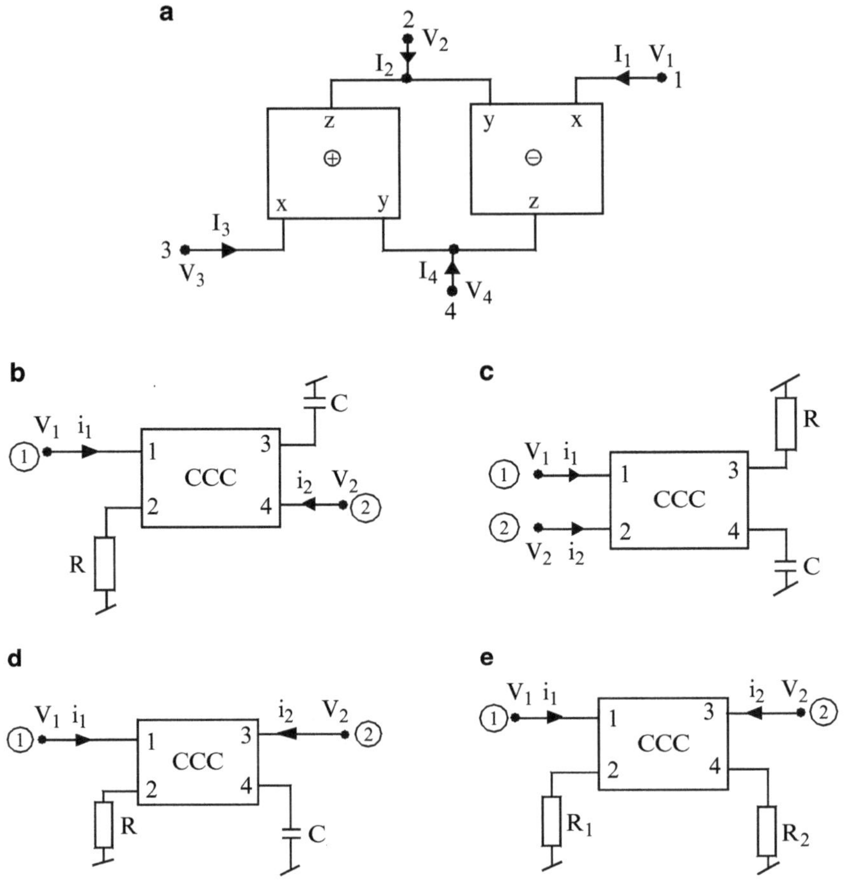

Fig. 15.19 Realization of mutators (**a**) The 4-port Composite Current Conveyor (CCC) characterized by $v_1 = v_2$, $i_2 = i_3$, $v_3 = v_3$ and $i_4 = -\ i_1$. (**b**) L-R mutator Type-1 (**c**) C-R mutator Type-1 (**c**) C-R Mutator Type-1 (**d**) C-R mutator Type-2 (**e**) L-C mutator Type-1

prominent practical element attracting worldwide attention of researchers and soon new applications of Memristors started appearing in literature, for instance see [38–43, 26–31, 33–37]. Soon it was demonstrated that the Memcapacitance and Meminductance are two other fundamental elements capable of possessing memory of past dynamics.

Pershin and Ventra [40] demonstrated that meminductive and memcapacitive elements can be emulated by employing CCs in conjunction with an available memristor. Two circuits which can simulate floating meminductance and memcapacitance while using four single output CCII + devices are shown here in Fig. 15.19. In each of the two circuits, the memristive element is characterized by the following equations:

$$I_M = [R_M(x, V_M, t)]^{-1} V_M \tag{15.65}$$

$$\frac{dx}{dt} = f(x, V_M, t) \tag{15.66}$$

where I_M and V_M are the current through and the voltage across the memristive system respectively, R_M is the memristance (time-dependent resistance), x is a vector of internal state variables, f is a device-specific function and t denotes implicit and explicit time dependence of corresponding elements. It is assumed that a memristor realized by a physical material/system is available from which meminductive and memcapacitive systems are to be created.

The first circuit of Fig. 15.20 (Fig. 15.20a) transforms the time-dependent response of a memristive system into that of a meminductive response while the circuit of Fig. 15.20b converts the same into time response of a memcapacitive system. By a straight forward analysis using CCII + characterization, it turns out that the first circuit, between ports 1 and 2 is characterized by the following equation:

$$I = \frac{1}{CRR_M(x, \phi, /(RC), t)} \phi \equiv [L(x, \phi, t)]^{-1} \phi \tag{15.67}$$

$$\frac{dx}{dt} = f(x, \phi, /(RC), t) \tag{15.68}$$

where flux ϕ is given by $\phi(t) = \int_0^t (V_1(t') - V_2(t'))dt'$. Thus, the above equations describe a *floating* flux-controlled meminductive system.

In a similar way, the circuit of Fig. 15.20b is characterized by the following equations:

$$Q = \frac{L}{RR_M(x, (V_1 - V_2), t)} (V_1 - V_2) \equiv C(x, (V_1 - V_2), t)(V_1 - V_2) \tag{15.69}$$

$$\frac{dx}{dt} = f(x, (V_1 - V_2), t) \tag{15.70}$$

The above equations imply a floating voltage-controlled meminductive system.

If dual output CCIIs are employed the number of active elements can be reduced to two in each case to realize the same elements. Two such circuits are shown in Fig. 15.21a and b respectively. Note that while the circuits of Fig. 15.21 cannot be realized practically those of Fig. 15.20 due to the use of only CCII + s can be practically implemented using AD844AN ICs. The above examples demonstrate that CCII + s have potential applications in realizing circuits involving memristive, meminductive and memcapacitive elements.

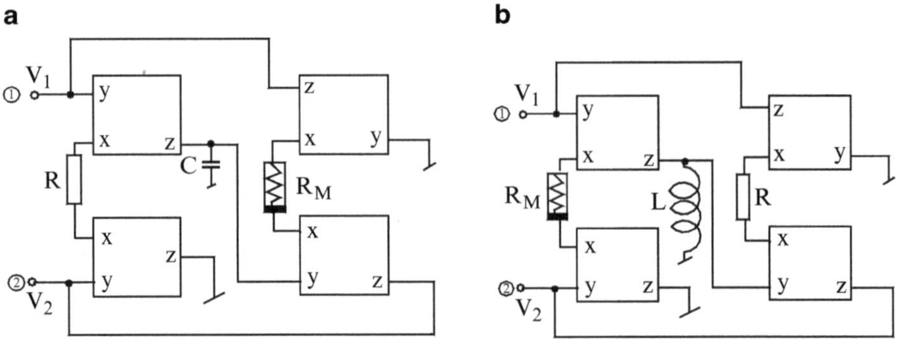

Fig. 15.20 Emulators of meminductive and memcapacitive systems using CCs [40] (**a**) meminductive (**b**) memcapacitive

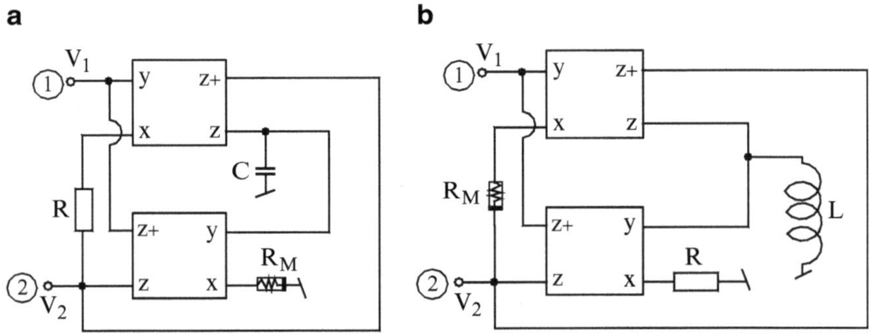

Fig. 15.21 Emulators of meminductive and memcapacitive systems using DOCCs [40] (**a**) meminductive (**b**) memcapacitive

Concluding Remarks

This chapter demonstrated a variety of advanced applications[1] of CCs employing basic CCs as well as the variants of CCs such as MO-CC, DVCC and FDCCII in areas like PID controllers, precision rectifiers, single-ended to differential converters, relaxation oscillators, multivibrators, square/triangular waveform generators, sample and hold, D/A converters, chaos generators, low noise amplifiers mutators and the exciting new area of emulating the memristive, meminductive and memcapacitive systems. It is

(continued)

[1] For some applications not included here, such as synthesis of open circuit impedance matrix of LTI multiport using CCs CCII based high impedance input stage for biomedical applications and for an interesting tutorial on some other applications of CCs, the readers are referred to [44–48], respectively.

(continued)

hoped that the material covered here gives ample proof of the versatility of the CCs, particularly those of the DVCC with dual outputs and the FDCCII. This should serve as a stimulus for IC manufacturers to think of making such varieties of CCs available as off-the-shelf ICs.

References

1. Yuce E, Tokat S, Kizilkaya A, Cicekoglu O (2006) CCII-based PID controllers employing grounded passive components. Int J Electron Commun 60:399–403
2. Yuce E, Minaei S (2010) New CCII-based versatile structure for realizing PID controller and instrumentation amplifier. Microelectron J 41:311–316
3. Godara B, Fabre A (2008) A new application of current conveyors: the design of wideband controllable low-noise amplifiers. Radioengineering 17:91–100
4. Centurelli F, Diqual M, Ferri G, Guerrini NC, Scotti G, Trifiletti A (2005) A novel dual-output CCII-based single ended to differential converter. Analog Integr Circ Sig Process 43:87–90
5. Koton J, Herencsar N, Vrba K (2011) Current and voltage conveyors in current-and voltage-mode precision full-wave rectifiers. Radioengineering 20:19–24
6. Kumngern M (2011) Precision full-wave rectifier using two DDCCs. Circ Syst 2:127–132
7. Minaei S, Yuce E (2008) A new full-wave rectifier circuit employing single dual-X current conveyor. Int J Electron 95:777–784
8. Koton J, Herencsar N, Vrba K (2010) Minimal configuration precision full-wave rectifier using current and voltage conveyors. IEICE Electron Express 7:844–849
9. Monpapassorn A (2013) Low output impedance dual CCII full-wave rectifier. Int J Electron 100:648–654
10. Kumngern M, Knobnob B, Dejhan K (2010) High frequency and high precision CMOS Half-wave rectifier. Circ Syst Sig Process 29:815–836
11. Marcellis AD, Carlo CD, Ferri G, Stornelli V (2013) A CCII-based wide frequency range square waveform generator. Int J Circ Theor Appl 41:1–13
12. Chien H-C (2012) Voltage-controlled dual slope operation square/triangular wave generator and its application as a dual mode operation pulse width modulator employing differential voltage current conveyors. Microelectron J 43:962–974
13. Chien HC (2011) Switch-controllable dual-hysteresis mode bistable multivibrator employing single differential voltage current conveyor. Microelectron J 42:745–753
14. Chien HC, Lo YK (2011) Design and implementation of monostable multivibrators employing differential voltage current conveyors. Microelectron J 42:1107–1115
15. Chien HC (2013) Design and implementation of relaxation generators: new application circuits of the DVCC. Int J Electron 100:227–244
16. Chien HC (2013) Square/triangular wave generator using single DO-DVCC and three grounded passive components. Am J Electr Electron Eng 1:32–36
17. Godara B, Fabre A (2007) Versatile wideband impedance matching circuit based on current conveyors. Electron Lett 43:37–38
18. Godara B, Fabre A (2008) The first active tunable wideband impedance matching circuit. Electroscope Appl Electron 2008:7p
19. Hwang YS, Chen JJ, Wu SY, Liao LP, Tsai CC (2007) A new pipelined analog-to-digital converter using current conveyors. Analog Integr Circ Sig Process 50:213–220
20. Hwang YS, Wang SF, Sheu P-W, Chen JJ (2008) Novel FBCCII-based sample and hold and MDAC circuits. Int J Electron 95:1111–1117

21. Petrzela J, Hrubos Z, Gotthans T (2011) Modeling deterministic chaos using electronic circuits. Radioengineering 20:438–444
22. Ozoguz S, Elwakil AS, Kennedy MP (2002) Experimental verification of the butterfly attractor in a modified Lorenz system. Int J Bifurc Chaos 12:1627–1632
23. Elwakil AS, Kennedy MP (2000) Novel chaotic oscillator configuration using a diode-inductor composite. Int J Electron 87:397–406
24. Gandhi G (2006) An improved Chua's circuit and its use in hyperchaotic circuit. Analog Integr Circ Sig Process 46:173–178
25. Senani R, Gupta SS (1998) Implementation of Chua's chaotic circuit using current feedback op-amp. Electron Lett 34:829–830
26. Chua LO (1968) Synthesis of new nonlinear network elements. Proc IEEE 56:1325–1340
27. Chua LO (1967) The rotator-A new network component. Proc IEEE 55:1566–1577
28. Smith KC, Sedra A (1970) Realization of the Chua family of new nonlinear network elements using the current conveyor. IEEE Trans Circ Theor 17:137–139
29. Soliman AM (1972) Two new LC mutators and their realizations. IEEE Trans Circ Theor 19:171–172
30. Soliman AM (1973) New L-R mutators and their models. Int J Electron 34:753–756
31. Higashimura M, Fukui Y (1988) Type 1 mutator using current conveyor and its application to immittance simulation. Int J Electron 64:377–383
32. Ikeda K, Tomita Y (1976–1990) A realization of third-order CR oscillator using D-R mutator. IEICE Trans E71:51–55
33. Yasushi T, Masahiro A (2001) A multi-port *D-R* mutator using CCIIs with current followers. Int J Electron 88:31–39
34. Soliman AM (2010) New CCII and ICCII based realizations of L-C and L-R mutators. Circ Syst Sig Process 29:1181–1191
35. Swamy MNS (2011) Mutators, Generalized impedance converters and inverters, and their realization using generalized current conveyors. Circ Syst Sig Process 30:209–232
36. Strukov DB, Snider GS, Stewart DR, Williams RS (2008) The missing memristor found. Nature 453:80–83
37. Chua LN (1971) Memristor-the missing circuit element. IEEE Trans Circ Theor 18:507–519
38. Ventra MD, Pershin YV, Chua LO (2009) Circuit elements with memory: memristors, memcapacitors, and meminductors. Proc IEEE 97:1717–1724
39. Pershin YV, Ventra MD (2010) Memristive circuits simulate memcapacitors and meminductors. Electron Lett 46:517–518
40. Pershin YV, Ventra MD (2011) Emulation of floating memcapacitors and meminductors using current conveyors. Electron Lett 47:243–244
41. Esch J (2009) Circuit elements with memory: memristors, memcapacitors, and meminductors. Proc IEEE 97:1715–1716
42. Biolek D, Biolkova V (2010) Mutator for transforming memristor into memcapacitor. Electron Lett 46:1428–1429
43. Yu DS, Liang Y, Chen H, Iu HHC (2013) Design of a practical memcapacitor emulator without grounded restriction. IEEE Trans Circ Syst-II 60:207–211
44. Hou CL, Yeh SN, Huang CC (2004) A synthetic network using current conveyors of the open-circuit impedance matrix of a linear time-invariant multiport. Int J Electron 91:165–174
45. Ferri G, Stornelli V, Simone AD (2011) A CCII-based high impedance input stage for biomedical applications. J Circ Syst Comput 20:1441–1447
46. Rajput SS, Jamuar SS (2007) Advanced applications of current conveyors: a tutorial. J Active Passiv Electron Devices 2:143–164
47. Ghallab YH, Badawy W, Kaler VIS, Maundy BJ (2005) A novel current-mode instrumentation amplifier based on operational floating current conveyor. IEEE Trans Instrum Meas 54:1941–1949
48. Elwakil AS, Soliman AM (1997) Current mode chaos generator. Electron Lett 33:1661–1662

Part V
Concluding Remarks and References for Further Reading

Chapter 16
Recent Advances and Future Directions of Research

Abstract Some recent advances made in the methods of systematic synthesis of circuits using CCs, hardware implementation of CCs, development of field programmable analog arrays using CCs and the applications of CCs in digital logic circuits have been pointed out and future directions of research on Current Conveyors have been indicated.

16.1 Introduction

In this last chapter of the monograph we would be highlighting some of the ideas related to CCs and their applications which are of relatively more recent origin and on which some useful work has just started appearing in current literature. While outlining the various new innovations, we would also point out some ideas for further research as well as some open problems which appear to have remained unattended so far. It is hoped that this exposure would provide the readers enough motivation for further work in the area.

16.2 Pathological Representations of Various Current Conveyors and Their Use in Systematic Circuit Synthesis

After the introduction of CCs in 1968 and 1970, for about a decade or so, the relation between the characterization of these elements and the well-known pathological elements the nullator, norator and nullor prevalent then was neither well known nor explicitly understood and mentioned in the technical literature. However, subsequently, several researchers not only explicitly stated but also started exploiting the equivalence of a CCII- with a three terminal nullor in the analysis and synthesis of active networks using CCs. It was soon understood that a CCII- is a three terminal nullor in which the junction of nullator and norator serves as the X-terminal of the CCII, the free end of the nullator represents the terminal-Y and

© Springer International Publishing Switzerland 2015

R. Senani et al., *Current Conveyors*, DOI 10.1007/978-3-319-08684-2_16

the free end of the norator represents terminal-Z. On the other hand, it was found that a CCII + could not be represented by only nullators and norators and did require two nullors along with two equal-valued resistors for its nullor representation.

In [1] Awad and Soliman introduced two new pathological elements which they called a *current mirror* (CM) and a *voltage mirror* (VM) and argued that with the introduction of these two new elements, the set of the pathological elements has been completed. This development made it possible to represent a CCII + with a nullator and a CM thereby not requiring any resistor. Subsequently, it has been shown in a number of publications that by a judicious combination of nullators, norators, CMs and VMs, almost all the building blocks can be represented without requiring any redundant resistor. Of late, even the floating versions of CM and VM have been proposed in the literature [2].

In reference [3], Haigh and his co-workers formulated a systematic method of synthesizing a class of linear active-RC networks such as synthetic impedances using nullator- norator formalism as a basis. The introduction of a number of new pathological elements by a number of researchers has, thus, further enriched the literature. These pathological elements are currently being used in evolving systematic derivation procedures for a larger class of functional circuits that include mutators, biquad filters, single resistance controlled oscillators etc.

This field of systematic active network synthesis using pathological elements, thereby ultimately leading to practicable designs of the considered circuits in terms of their realizations using various types of current conveyors, is currently a very popular area of research. Needless to say, many important new results might be waiting to be discovered through this methodology of synthesis. The interested readers are referred to [4, 5–17] for the details of the various significant works done in this direction.

16.3 Recent Advances in the Hardware Implementation of Current Conveyors

In this section, we present some recent innovations made in the hardware implementation of current conveyors which have a bearing on the numerous variants of CCs discussed in this monograph.

16.3.1 New CCII Implementation Based Upon Modified Bipolar Translinear Cell

There has been a large number of CC implementations based upon the four transistor mixed-translinear-cell. Recently Merz et al. [18] proposed a new mixed translinear cell composed of eight bipolar transistors as shown in Fig. 16.1.

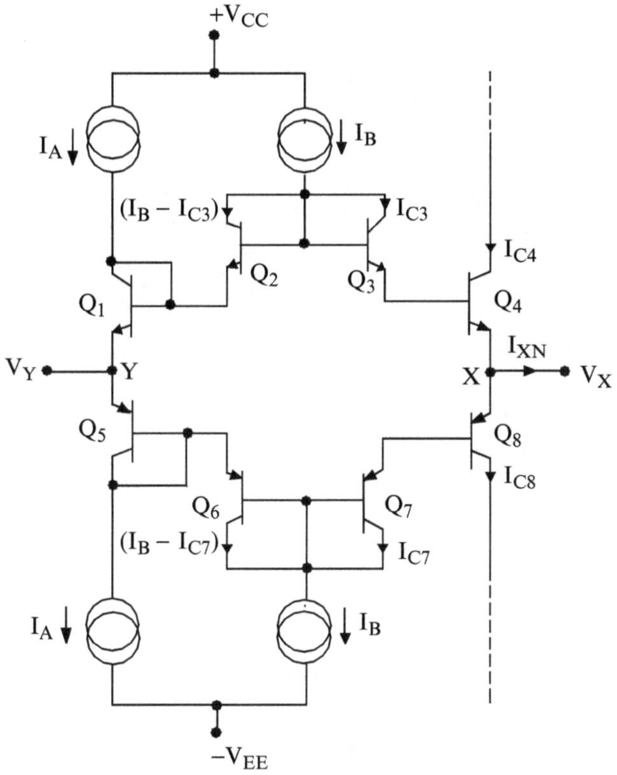

Fig. 16.1 New mixed translinear cell [18]

An analysis of this circuit shows that the V-I characteristic of this modified cell is composed of a positive term and a negative term of hyperbolic sine functions which results in the reduction of the non-linear terms. The same authors have proposed a new schematic of a CCCII [18] using this formulation (shown in Fig. 16.2) which has been verified to offer better linearity as compared to the traditional four transistor mixed- translinear-cell.

In another application involving two-CCCII-based instrumentation amplifier [18] this circuit has performed better (in terms of linearity and increased input voltage range) than the circuit based upon four transistor mixed- translinear-cell.

An interesting problem appears to be the investigation of the use of this new eight transistor cell in the realization of other varieties of CCs.

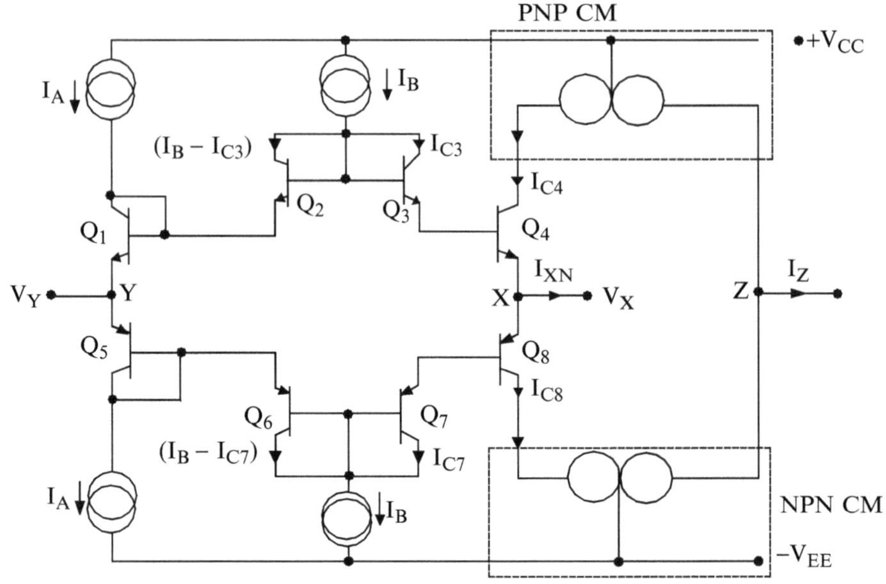

Fig. 16.2 Simplified version of the schematic of a CCCII [18]

16.3.2 Bi-CMOS CCCII Realizations

In view of the potential applications of the CCCII in making available electronically-tunable circuit functions, several authors have tried to improve upon the basic structure of the CCCII itself. An interesting innovation in this regard has been to combine the best properties of the BJTs and the MOSFETs thereby resulting in the proposal of Bi-CMOS CCCII realizations. It may be recalled that in conventional realization of the CCCII, as well as in the Bi-CMOS CC proposed earlier, PNP transistors are also engaged for handling the AC signals and consequently, the maximum frequency of operation of the realized CCCIIs is significantly reduced. Recently, Psychalinos and Souliotis [19] have presented a new CCCII topology for Bi-CMOS implementation (shown in Fig. 16.3) which offers the advantage of low voltage operation as well as improvement in the maximum frequency of operation. It is shown that a slight modification in the topology also enables to obtain negative intrinsic resistance R_X.

In the low voltage CCCII topology under consideration, a two BJT translinear loop with MOS current mirrors for DC bias current supply as well as duplication of X-port current to yield same current at the Z port has been incorporated. This circuit is shown in Fig. 16.3.

According to the translinear principle the currents in transistors Q_2 and Q_1 are related according to the following equation.

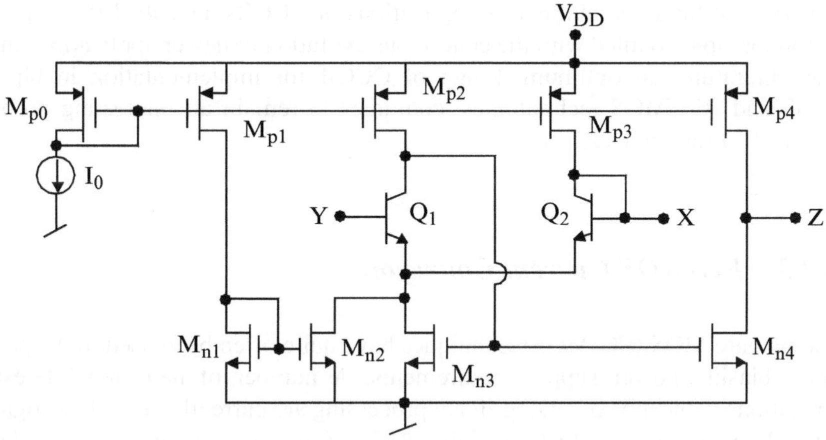

Fig. 16.3 Low-voltage CCCII proposed by Psychalinos and Souliotis (Adapted from [19] © 2010 Springer Science + Business Media)

$$I_0 + i_x = I_0.e^{\frac{vX - vY}{vT}} \quad (16.1)$$

Expending the right hand side of the above equation into Taylor series and considering a first order approximation the X port current is related to the differential input V_x-V_y according to the following formula

$$i_x \cong \frac{I_0}{V_T} \cdot (v_x - v_y) \quad (16.2)$$

From which it is seen that $R_x \cong V_T/I_0$.

This topology has following advantageous features as compared to the traditional bipolar CCCII architecture.

(i) The minimum power supply requirement is equal to $V_{BE} + 2V_{DS,sat}$ instead of 2 ($V_{BE} + V_{CE,\ sat}$). In view of this it is seen that this architecture has a capability for operation in a relatively lower voltage environment.
(ii) Whereas in the conventional translinear structure the translinear loop is formed from two pairs of npn and pnp transistor, in the present case the translinear loop is made in only single pair of npn transistor. Thus, the drawback of limited performance of the integrated pnp transistor (such as a low value of β and high parasitics) are eliminated and therefore this topology should be capable of operating at relatively higher frequency. Furthermore the handling of AC signals by PMOS transistor is avoided which is also helpful in improving maximum frequency of operation.

The new CCCII offers a bandwidth of the current gain in excess of 100 MHz and can operate from a single low voltage DC power supply of 1.5 V.

In view of the large number of applications of CCCIIs described in Chap. 9 of this monograph, coupled with the continuous evolution of newer applications in the recent literature, an optimum design of CCCII for implementation in bipolar, CMOS and Bi-CMOS technologies, continues to remain an interesting problem worthy of further considerations.

16.3.3 FG-MOS Current Conveyors

Floating gate MOS (FG-MOS) techniques have quite often been used in the past to reduce biasing power supply requirements. A number of new and interesting applications in the area of analog signal processing are currently being investigated. A well-known attractive feature of the FG-MOS transistor is that it can be fabricated in all prevalent CMOS technologies; however, the so-called double poly CMOS technology is preferred choice for implementing FG-MOS based circuits. FG-MOS circuits have been employed in the past in realizing a number of analog circuit building blocks such as op-amps, OTAs, transconductors etc.

Recently, there have been a number of efforts in realizing FG-MOS based current mirrors and current conveyors. In [20] Khateb et al. have presented the design of low-voltage, ultra-low power, Class AB CC based on folded-cascode OTA with floating-gate differential pair. It has been shown in [21] that the architecture for an FG-MOS CCII could operate from a power supply voltage as low as ±0.4 V, has rail-to-rail voltage swing capability and consumes a total quiescent power of merely 9.5 μW [21]. More recently, Fani and Farshidi [22] presented a FG-MOS implementation of the fully differential CCCII (FDCCCII) circuit which is reproduced here in Fig. 16.4.

This circuit was implemented in 0.18μm TSMC CMOS technology, operated from ±0.8 V DC power supply and was found to offer a bandwidth more than 1GHz with voltage transfer ratio and current transfer ratio of the order of 0.94 and 0.93 respectively and has been shown to offer significant improvements over earlier designs. Because of the advantages offered by the FG-MOS CC architectures, there appears to be enough scope for further refinements in this area.

16.3.4 Design of CCII Employing Bacterial Foraging Optimization

Recently, Chatterjee et al. [23] have demonstrated the utility of *bacterial foraging optimization* technique in the design of CCs. The technique has been applied on the design of a CMOS positive second generation current conveyors (CCII+). The objective set out was to simultaneously minimize the parasitic X port input resistance R_x and maximize the high end cut off frequency of the current signal.

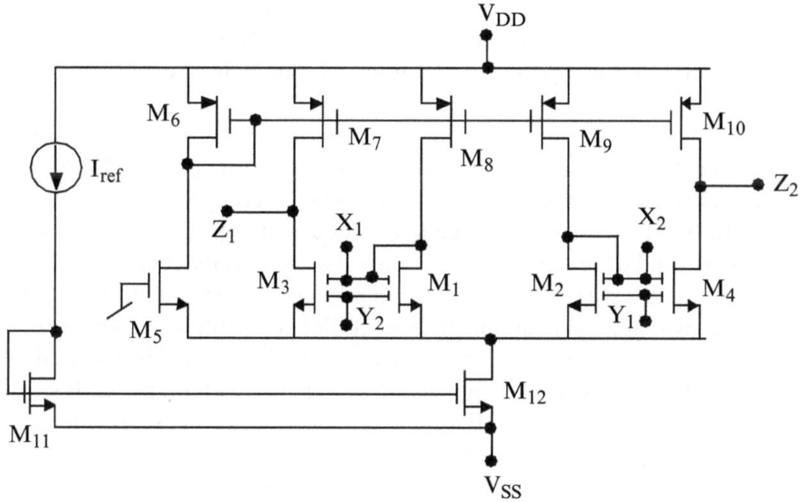

Fig. 16.4 FDCCCII proposed by Fani and Farshidi (Adapted from [22] © 2013 Springer Science + Business Media)

The bacterial foraging optimization solves a bi-objective optimization algorithm to obtain physical dimensions of the PMOS and NMOS transistors of the CMOS CCII +. Through a series of simulation experiments conducted for several bias currents, it has been demonstrated that the proposed algorithm provides satisfactorily better results in comparison to those produced by *particle swarm optimization* and *differential evolution algorithm* employed for the same design problem.

This approach yields an interesting design strategy for evolving improved CMOS hardware not only for CCII + but also to large varieties of CCs and their extended forms discussed in the various chapters of this monograph. This is another area which appears to have lot of potential for further research to improve the design of CCs, considering the fact that although a large number of CMOS CC implementations are known in the literature but not much work appears to have been done on optimizing the design of the developed CMOS CC architectures.

16.4 Current-Conveyor-Based Field Programmable Analog Arrays (FPAA)

Although field programmable gate arrays (FPGA) have been in existence since long, the notion of field programmable analog arrays (FPAA) is relatively new innovation. A FPAA consists of a number of reconfigurable basic analog cells which can be interconnected by a programmable architecture to perform a given function. Although FPAAs are currently available from some manufacturers, they are still not very popular because of the lack of real market and because of a number of performance limitations such as precision, offset, noise etc. An elementary cell of the analog arrays based upon Current Conveyor was first introduced by Premont et al. in 1998 [24] as the first step towards producing, eventually, a fully-integrated FPAA.

The second major step in this direction came after more than a decade when Mahmoud and Soliman [25] presented a low-voltage, low-power FPAA in which the configurable analog block (CAB) design was based upon a three bit digitally controlled fully differential second generation current conveyor (DCFDCCII) and the FPAA consisted of seven such cabs. The cabs are directly connected together in a hexagonal lattice arrangement. The workability of the proposed FPAA has been demonstrated by realizing a variable gain amplifier, a tunable second order low pass filter and a tunable second order band pass filter with the DCFCCII realized by using a fully-differential buffer, power amplifier and a three-bit CMOS current division network. The completed FPAA was realized in 90 nm CMOS technology, with a DC power supply voltages of ± 0.5 V.

In another paper by same authors [26], a similar FPAA, consisting of seven hexagonally arranged cells was shown to realize a sixth order, tunable Butterworth low pass filter for WLAN/WiMAX receivers with cutoff frequency tunable from 5.2 MHz to 16.9 MHz.

From these representative developments on FPAAs using digitally-programmable fully differential current conveyors, it is apparent that this area of research has not been explored much and there appears to be enough scope for refining what has been developed so far or introducing new FPAA structures with CABs based upon other varieties of CCs; for example, Universal Current Conveyors, Universal Voltage Conveyors, fully-balanced Current Conveyors etc.

16.5 Applications of the Current Conveyors in Realizing Logic Functions and Digital Circuits

Undoubtedly, almost all the applications of the CCs have been in the domain of analog circuits and functions. Recently, some attempts have been made in demonstrating that one variety of CCs, namely, CCCII can also be employed to realize XOR and XNOR functions; the latter by a composite connection of two CCCII + s along with a resistor [27]. This opens a new area of research i.e. that of realizing various logic functions and digital circuits using CCCs.

16.6 Newer Varieties of Current Conveyors of More Recent Origin

Although, more than two dozen varieties of CCs have been advanced so far, an account of which was given in Chap. 10 of this monograph, a number of new varieties of CCs have been proposed even while this monograph was under preparation. Two examples of the new varieties of CCs which have been introduced very recently are the *voltage summing CC* and *Z-copy controlled-gain voltage differencing CC* [28, 29]. Therefore, it appears that the search for newer varieties of CCs is still on!

Concluding Remarks
In this chapter, we have outlined some of the ideas related to the hardware implementation and applications of Current Conveyors which are being investigated in recent literature and which appear to offer enough scope for further research. It is hoped that interested readers, with the help of exhaustive list of references [1–27, 30–50] given at the end of this chapter will find enough ideas for further work and investigations.

16.7 Epilogue

This monograph has presented, in a comprehensive manner, the significant works done on CCs, their variants, their numerous applications (spread over a thousand journal papers and conference publications) and their hardware implementations.

It may, however, be noted that in spite a very rich repertoire of circuits employing CCs available in the literature and documented in this monograph, although the CCs have received vide acceptability by the academicians, it appears that the practicing engineers and circuit designers are still hesitant to use CCs and current mode circuits in their designs. The only exception appears to be the AD844 Current feedback operational amplifier, which even though is internally composed of a translinear CCII + followed by a translinear voltage follower.

Another very curious phenomenon to be observed is that there are a number of commercially available ICs which are labelled as high-frequency and/or high-slew -rate op-amps or OTAs, but inside their architecture lies essentially a CC-hardware. Notable examples in this category are AD844, OPA 660, OPA860 and OPA 2662 all of which have translinear CCs inside. Thus, it is ironical that the professional engineers and designers have no hesitation in using current-mode techniques and current conveyors inside the chips but are still hesitant in trying current-mode application circuits employing the very same building blocks explicitly.

It is hoped that this monograph, which contains more than 500 CC-based circuits, the performance of some of which has been experimentally verified by their proposers using commercially available devices (such as AD844/LF356 along with mixed transistor arrays LM3096 or using OPA660/OPA860) or has been confirmed by SPICE simulations of the bipolar/CMOS/BiCMOS implementations of the CCS or their variants, may provide the needed motivation for the circuit designers to try these exotic CC-based circuit topologies as alternatives to the conventional topologies using op-amps or OTAs, to performs the same functions.

In view of the tremendous potential of the CCCIIs for realizing a large variety of electronically-controllable functions as well as significant advantages attainable by numerous application circuits realizable with the new variants of CCs coupled with the fact that almost all varieties of CCs are implementable in CMOS technology, several of the topologies described in this monograph hold the potential of becoming part of a complete system integrated on the same chip along with digital circuits.

Although a large variety of CCs proposed by numerous researchers have been included in this monograph and in spite of the fact that universal CCs and universal voltage conveyors have also been proposed, it would appear that a *truly universal* CC building block from which all the reported varieties of CCs (over two dozen of them) can be derived as special cases does not appear to have been discovered so far and the creation of the same is still open to investigation.

Lastly, we may admit that although we have tried to be as comprehensive as possible in collecting and citing all possible references (journal articles) on CCs, it is quite possible that due to the enormity of the task, we might have omitted some genuinely important references *inadvertently*. Such omissions are obviously not intentional and the authors, therefore, request the aggrieved researchers to point out the omission(s) so that they can be appropriately cited and acknowledged in the subsequent edition of this monograph.

References

1. Awad IA, Soliman AM (2002) On the voltage mirrors and the current mirrors. Analog Integr Circ Sig Process 32:79–81
2. Wang HY, Chang SH, Chiang NH, Nguyen QM (2014) Symbolic analysis using floating pathological elements. Adv Intell Syst Comput 238:379–387
3. Haigh DG, Tan FQ, Papavassiliou C (2005) Systematic synthesis of active-RC circuit building-blocks. Analog Integr Circ Sig Process 43:297–315
4. Wang HY, Chang SH, Jeang YL, Huang CY (2006) Rearrangement of mirror elements. Analog Integr Circ Sig Process 49:87–90
5. Wang HY, Liu CY, Chang SH (2010) New nullor-mirror equivalences. Int J Electron Commun (AEU) 64:828–832
6. Soliman AM (2010) Generation of CCII and ICCII based Wien oscillators using nodal admittance matrix expansion. Int J Electron Commun (AEU) 64:971–977
7. Soliman AM (2011) Pathological realizations of the DCVC (CDBA) and applications to oscillators and filters. Int J Electron Commun (AEU) 65:985–992
8. Soliman AM (2011) Generation and classification of CCII and ICCII based negative impedance converter circuits using NAM expansion. Int J Circ Theor Appl 39:835–847
9. Soliman AM (2012) A note on the generation of generalized impedance converter circuits using NAM expansion. Circ Syst Sig Process 31:1147–1157

10. Soliman AM (2012) Three -port gyrator circuits using transconductance amplifier or generalized conveyors. Int J Electron Commun (AEU) 66:286–293
11. Huang WC, Wang HY, Cheng PS, Lin YC (2012) Nullor equivalents of active devices for symbolic circuit analysis. Circ Syst Sig Process 31:865–875
12. Soliman AM (2012) Classification and pathological realizations of transconductance amplifiers. J Circ Syst Comput 21:17. doi:10.1142/S0218126612500107
13. Soliman AM (2012) Pathological realizations of BOTA and FDDTA using grounded resistors. J Circ Syst Comput 21:23. doi:10.1142/S0218126612500259
14. Soliman AM (2013) Generation of third-order quadrature oscillator circuits using NAM expansion. J Circ Syst Comput 22:13. doi:10.1142/S0218126613500606
15. Lin WC, Wang HY, Liu CY, Lee TF (2013) Symbolic analysis of active device containing differencing voltage or current characteristics. Microelectron J 44:354–358
16. Sanchez-Lopez C, Ruiz-Pastor A, Ochoa-Montiel O, Carrasco-Aguilar MA (2013) Symbolic nodal analysis of analog circuits with modern multiport functional blocks. Radioengineering 22:518–525
17. Soliman AM (2011) Synthesis of oscillators using limit variables and NAM expansion. Active Passiv Electron Comp 2011:13. doi:10.1155/2011/131546
18. Merz N, Kiranon W, Wongtachathum C, Pawarangkoon NW (2012) A modified bipolar translinear cell with improved linear range and its applications. Radioengineering 21:736–745
19. Psychalinos C, Souliotis G (2010) Low-voltage current controlled current conveyor. Analog Integr Circ Sig Process 64:129–135
20. Khateb F, Khatib N, Kubanek D (2012) Novel ultra-low power class AB CCII + based on floating gate folded cascode OTA. Circ Syst Sig Process 31:447–464
21. Khateb F, Khatib N, Kubanek D (2012) Low-voltage ultra- low current conveyor based on Quasi-floating gate transistors. Radioengineering 21:725–735F
22. Fani R, Farshidi E (2013) A FG-MOS based fully differential current controlled conveyor and its applications. Circ Syst Sig Process 32:993–1011
23. Chatterjee A, Fakhfakh M, Siarry P (2010) Design of second-generation current conveyors employing bacterial foraging optimization. Microelectron J 41:616–626
24. Premont C, Grisel R, Abouchi N, Chante JP (1998) A current conveyor based field programmable analog array. Analog Integr Circ Sig Process 17:105–124
25. Mahmoud SA, Soliman EA (2011) Low voltage current conveyor-based field programmable analog array. J Circ Syst Comput 20:1677–1701
26. Mahmoud S, Soliman E (2013) Novel CCII-based filed programmable analog array and its application to a sixth-order Butterworth LPF. Radioengineering 22:440–447
27. Yasin MY (2013) A novel bipolar XOR/XNOR realization using translinear type 2nd generation current controlled current conveyor designed in 45nm CMOS technology. ACTA Electrotehnica 54:3–6
28. Tekin SA (2014) Voltage summing current conveyor (VSCC) for oscillator and summing amplifier applications. J Microelectron Electron Compon Mater 44:159–167
29. Sotner R, Herencsar N, Jerabek J, Prokop R, Kartci A, Dostal T, Vrva K (2014) Z-copy controlled-gain voltage differencing current conveyor: advanced possibilities in direct electronic control of first-order filtek. Elektronika IR Elektrotechnika 20:77–83
30. Andreou AG (1999) Exploiting device physics in circuit design for efficient computational functions in analog VLSI: book chapter 4 in Low- voltage/low-power integrated circuits and systems IEEE Press, Edgar Sanchez, Wiley-IEEE Press, pp 85–132
31. Ananda Mohan PV (2000) Comments on A transformation of obtaining CCII-based adjoint of Op-Amp-based circuits. IEEE Trans Circ Syst-II 47:233
32. Aronhime P, Wang K, Qian T (2001) Generalization of theorem for replacing CCIIs by CFOAs in current-mode circuits. Analog Integr Circ Sig Process 28:27–33
33. Sato T, Wada K, Takagi S, Fujii N (2002) Extension of current conveyor concept and its applications. IEICE Trans Fundam E85-A:414–421

34. Centurelli F, Diqual M, Ferri G, Guerrini NC, Scotti G, Trifiletti A (2005) A novel dual-output CCII based single-ended to differential converter. Analog Integr Circ Sig Process 43:87–90
35. Gianni C, Pennisi S, Scotti G, Trifiletti A (2007) The universal circuit simulator: a mixed signal approach to n-port network and impedance synthesis. IEEE Trans Circ Syst-I 54:2178–2183
36. Godara B, Fabre A (2008) A new application of current conveyors: the design of wideband controllable low-noise amplifiers. Radioengineering 17:91–100
37. Cuautle ET, Frias DM, Lopez CS, Villasenor MAD (2008) Synthesis of CCII-s by superimposing VFs and CFs through genetic operations. IEICE Electron Express 5:411–415
38. Raikos G, Vlassis S, Psychalinos C (2012) 0.5 V bulk-driven analog building blocks. Int J Electron Commun (AEU) 66:920–927
39. Horng JW (2012) Analytical synthesis of general high-order voltage/current transfer functions using CCIIs. Microelectron J 43:546–554
40. Metin B, Pal K, Cicekoglu O (2012) A new approach for high-input impedance in voltage mode filters using first-generation current conveyor in place of second-generation current conveyor. Int J Electron 99:131–139
41. Khateb F, Jaikla W, Kumngern M, Prommee P (2013) Comparative study of sub-volt differential difference current conveyor. Microelectron J 44:1278–1284
42. Mahmoud SA, Soliman EA (2013) Multi-standard receiver baseband chain using digitally programmable OTA based on CCII and current division networks. J Circ Syst Comput 22:20. doi:10.1142/S0218126613500199
43. Fakhfakh M, Pierzchala M (2013) Synthesis of active inductors using SFG stamps. Microelectronics J 44:1107–1122
44. Ercan H, Alci M (2013) A new design for a BI-CMOS controlled current conveyor. Elecktronika IR Elektrotechnika 19:56–60
45. Khateb F, Kumngern M, Spyridon V, Psychalinos C (2014) Differential difference current conveyor using Bulk-driven technique for ultra-low-voltage applications. Circ Syst Sig Process 33:159–176
46. Pushkar KL, Bhaskar DR, Prasad D (2014) Voltage-mode new universal biquad filter configuration using a single VDIBA. Circ Syst Sig Process 33:275–285
47. Thankachan S, Pattanaik M, Rajput SS (2008) A ±0.5V BiCMOS class-A current conveyor. World Acad Sci Eng Tech 2:473–476
48. Abbas Z, Olivieri M, Yakupov M, Ripp A (2013) Design centering/yield optimization of power aware band pass filter based on CMOS current conveyor (CCCII+). Microelectron J 44:321–331
49. Seguin F, Godara B, Alicalapa F, Fabre A (2004) A gain-controllable wide-band low-noise amplifier in low-cost 0.8μm Si BiCMOS technology. IEEE Trans Microw Tech 52:154–160
50. Erdal C, Kuntman H, Kafali S (2004) A current controlled conveyor based proportional-integral-derivative (PID) controller. J Electr Electron Eng 4:1243–1248

Appendix 1: Additional References for Further Reading

1. Palmisano G, Palumbo G, Pennisi S (1999) CMOS current amplifiers. The Springer International Series in Engineering and Computer Science, vol 499, p 159. doi:10.1007/978-14615-5135-5
2. Vrba K, Cajka J (2000) Filter and oscillator design using general current conveyors. Electron Lett.com, Brno University of Technology, F/O Electr Eng Informatics, vol 118, Purkynova, Brno, p 61200
3. Biolek D, Cajka J, Vrba K (2002) Current-mode universal biquad. In: Fourth IEEE international Caracas conference on devices, circuits and systems, Seaport Conference Center, Aruba, Dutch Caribbean, 17–19 Apr 2002
4. Cajka J, Dostal T, Vrba K (2002) Nth-order all pass filter design. In: 12th international Czech-Slovak scientific conference, Radioelektronika, Bratislava, Slovak Republic: 60/1-60/4, 14–16 May 2002
5. Vrba K, Cajka J (2002) High-order low pass filters operating in two modes. In: International conference research in telecommunication technology RTT'2002, Zilina-Slovak Republic, pp 3–6, 17–19 Sept 2002
6. Novotny V, Vrba K (2004) LC Ladder filter emulation by structures with current conveyors. In: Proceedings of the 4th WSEAS international conference on signal processing, computational geometry & artificial vision, ISCGAV' 2004, Wisconsin, USA
7. Silapan P, Jaikla W, Siripruchyanun M (2008) A high-performance current-mode precision full-wave rectifier based on BiCMOS-CCCDBAs. In: IEEE Asia Pacific conference on circuits and systems, APCCAS 2008, Macao, pp 1486–1489
8. Herencsar N, Vrba K (2008) Tunable current-mode multifunction filter using universal current conveyors. In: Proceedings of the third international conference on system, pp 1–6. doi:10.1109/ICONS.2008.29
9. Ghallab YH, Badawy W, Kaler Karan VIS, El-Ela MA, El-Said MH (2002) A new second-order active universal filter with single input and three outputs using operational floating current conveyors. In: The 14th international conference on microelectronics, ICM 2002, Beirut, Lebanon, pp 42–45, 11–13 Dec 2002
10. Sotner R, Slezak J, Dostal T (2010) Influence of mirroring of current output responses through grounded passive elements. In: Proceedings of the 20th international conference radioelektronika, Brno, pp 1–4, 19–21 Apr 2010
11. Chang CM, Chen FC, Hou CL, Horng JW (2011) Universal current-mode biquad filter using single inverting second-generation current conveyors. In: Proceedings of the 10th WSEAS international conference instrument measure circuits systems (IMCAS 11), Benice, pp 35–41, 8–10 March 2011
12. Kuntman H, Cicekoglu O, Ozoguz S, Karacivi B (2000) Universal current-mode filter implemented with the modified third generation current conveyor. In: Proceedings of the

© Springer International Publishing Switzerland 2015
R. Senani et al., *Current Conveyors*, DOI 10.1007/978-3-319-08684-2

IEEE nordic signal process symposium (NORSIG'OO) Kolmarden, Sweden, pp 165–169, 13–15 June 2000
13. Sotner R, Hrubos Z, Herencsar N, Jerabek J, Dostal T, Vrba K (2014) Precise electronically adjustable oscillator suitable for quadrature signal generation employing active elements with current and voltage gain control. Circ Syst Sig Process 33:1–35
14. Minaei S, Yildiz M, Metin B, Cicekoglu O (2004) New active realizations of floating lossless inductance and R-L impedance simulators. In: The 2nd annual IEEE Northeast workshop circuits on systems, Montreal, Que, Canada, pp 313–316, 20–23 June 2004
15. Maundy B, Gift S, Aronhime P (2007) Realization of a GIC using hybrid current conveyor/operational amplifier circuits, 50th Midwest Symp on Circ and Syst.: pp. 163–166. doi:10.1109/MWSCAS.2007.4488562
16. Ferri G, Guerrini NC, Romanato R, Scotti G, Trifiletti A (2007) CCII-based high-valued inductance simulators with minimum number of active elements. In: 18th European conference on circuits theory and design, Seville, pp 440–443, 27–30 Aug 2007
17. Khateb F, Vavra J, Biolek D (2010) A novel current-mode full-wave rectifier based on one CDTA and two diodes. Radioengineering 19:437–445
18. Vavra J, Bajer J (2011) Current-mode multiphase sinusoidal oscillator based on current differencing units. In: ELECO 2011 7th international conference electrical and electronics engineering, Bursa, pp 332–336, 1–4 Dec 2011
19. Herencsar N, Koton J, Vrba K, Lahiri A (2012) Single GCFDITA and grounded passive elements based general topology for analog signal processing applications. In: ICN 2012: the eleventh international conference networks, Saint Gilles, pp 59–62, Feb 29-Mar 5 2012
20. Biolkova V, Kolka Z, Biolek D (2009) Fully balanced voltage differencing buffered amplifier and its applications. In: 52nd IEEE international midwest symposium circuits and systems, MWSCAS, Cancun, pp 45–48, 2–5 Aug 2009
21. Gaudet VC, Gulak PG (1998) CMOS implementation of a current conveyors-based field-programmable analog array. In: 31st asilomar conference of signal systems and computers, Pacific Grove, pp 1156–1159, 2–5 Nov 1998
22. Wadsworth DC (1990) The performance and applications of a new current conveyor integrated circuit. In: IEEE international symposium on circuits systems, 1, New Orleans, pp 287–290, 1-3 May 2009
23. Fabre A, Alami M (1990) A versatile translinear cell-library to implement high performance analog ASICS. In: Euro, ASIC, 90, Paris, pp 89–94, 29 May-1 June 1990
24. Chan KM, Duncan RA, Sedra AS (1991) High-performance current amplifier and current conveyor. In: IEEE international symposium circuits and systems, Singapore, pp 1825–1828, 11–14 June 1991
25. Karybaka CA, Horopanitis EE (1998) V-I scalar circuits based on CCIIs. Electron Lett 34 (4):317–318
26. Pandey N, Paul SK, Bhattacharyya A (2005) Sinusoidal oscillator-a new configuration based on current conveyor. In: Proceedings of XXVIIIth general assembly, New Delhi, 27, 23–29 Oct 2005
27. Discigil M, Toker A, Cicekoglu O, Kuntman H (2000) New oscillator topologies based on a single inverting second generation current conveyor. In: Proceedings of the IEEE nordic signal processing symposium (NORSIG'OO) Kolmorden, pp 469–472, 13–15 June 2000
28. Loescharantaramdee C, Kiranon W, Sangpisit W, Yadum W (2001) Multiphase sinusoidal oscillators using translinear current conveyors and only grounded passive components. In: Proceedings of the 33rd southeastern symposium on systems theory, Athens, pp 59–63, 18–20 Mar 2001
29. Salama K, Soliman A (1999) Novel MOS-C quadrature oscillator using the differential current voltage conveyor. In: Proceedings of the 42nd midwest symposium on circuits and systems, vol 1, Las Cruces, pp 279–282, 8–11 Aug 1999

30. Barthelemy H, Fabre A (1996) 20–90 MHZ current –controlled sinusoidal oscillator. In: Proceedings of the 22nd European solid state circuits conference, vol 96, Neuchatel, pp 56–59, 17–19 Sep 1996
31. Anand VK, Sagar V, Gupta AN (2013) High performance temperature insensitive current mode rectifier without diode. IOSR JECE 6:76–89
32. Mahmoud Soliman A (2005) Low voltage high current gain CMOS digitally controlled fully differential CCII. In: IEEE ISCAS 2005, Kobe, vol 2, pp 1000–1003, 23–26 May 2005
33. Torteanchai U, Sarasitthithum K, Kumngern M (2011) CMOS versatile precision rectifier using current conveyor. In: 8th international conference electrical engineering/electronics, computer, telecommunication information technology, Khon Kaen, pp 78–81, 17–19 May 2011
34. Kumngern M (2010) CMOS precision full-wave rectifier using current conveyor. In: IEEE international conference of electron devices and solid-stage circuits, Hong Kong, pp 1–4, 15–17 Dec 2010
35. Langhammer L, Jerabek J (2013) Precision full-wave rectifiers with current active elements and current biasing. Int J Adv Telecommun Electrotechn Sig Syst 2:79–83
36. Koli K, Halonen K (1992) Current–conveyor subscriber line interface circuits. ISCAS, Sam Diego
37. Maundy BJ, Aronhime PB (1996) Switched-capacitor current conveyor building blocks. In: IEEE international symposium on circuits system, Atlanta, vol 1, pp 345–348
38. Soliman EA, Mahmoud SA (2009) New CMOS fully differential current conveyor and its application in realizing sixth order complex filter. In: IEEE ISCAS, Taipei, Japan, pp 57–60
39. Giir F, Anday F (2007) FDCCII based voltage mode filter design using signal flow-graphs. In: IEEE 15th conference signal processing and communication, ESIKISEHIR, pp 1–3
40. Temizyurek C, Myderrizi I (2004) A novel three-input one-output voltage-mode universal filer using differential difference current conveyor (DDCC). In: IEEE MELECON, Dubrovnik, Croatia, pp 103–106
41. Somdunyakanok M, Angkeaw K, Prommee P (2011) Floating-capacitance multiplier based on CCDDCCs and its application. In: TENCON, Bali, pp 1367–1370
42. Al-Shahrani SM (2004) Fully differential second-order filter. In: 47th IEEE international midwest symposium on circuits systems, pp 299–301
43. Kumar M, Srivastava MC, Kumar U (2010) Current conveyor based multifunction filter. IJCSIS 7:104–107
44. Abuelma'atti MT, Al-Daghreer H (1997) New single–resistor controlled sinusoidal oscillator circuit using unity-gain current followers. Active Passive Electron Comp 20:105–109
45. Fongsamut C, Fujii N, Surakampontorn W (2005) Two new RC oscillators using CCIIs. In: Proceedings of ISCIT, Beijing, vol 2, pp 1138–1141
46. Keskin AU (2004) Design of minimum component oscillators using negative impedance approach based on different single active elements. In: IEEE MELECON, Dubrovnik, vol 1, pp 83–85
47. Popovic J, Nikolic B, Current KW, Pavasovic A, Vasiljevic (2000) Low-power CMOS RC oscillators based on current conveyors. In: Proceedings of 22nd international conference on microelectronics, NIS, vol 2, pp 691–694
48. Toker A, Cicekoglu O (1998) Oscillator design techniques with second generation current conveyors. In: Proceedings of the 6th international conference OPTIM, Brasov, Romania, vol 3, pp 659–662
49. Popovic J, Nikolic B, Current KW, Pavasovic A, Vasiljevic D (1998) CMOS implementation of low-power oscillator based on the modified Fabre-Normand current conveyors. In: IEEE international conference electron circuits and systems, Lisboa, vol 3, pp 349–352
50. Martinez PA, Aldea C, Sabadell J, Celma S (1998) Approach to the realization of state variable based oscillators. In: IEEE international conference on electron, circuits and systems, Lisboa, vol 3, pp 139–142

51. Filanovsky IM (2001) Current conveyor, Voltage conveyor, Gyrator. In: Proceedings of 44th IEEE midwest symposium on circuits and systems, Dayton, OH, vol 2, pp 314–317
52. Cataldo GD, Ferri G, Pennisi S (1998) Active capacitance multipliers using current conveyors. In: Proceedings of 1998 IEEE international symposium on circuits and systems, Monterey, CA, vol 2, pp 343–346
53. Vrba K, Cajka J (1997) High-order one-port elements for low pass filter realization. J Electr Eng 48:31–34
54. Higashimura M, Fukui Y (1991) RC active realization of mutually coupled circuit. IEEE Int Symp Circ Syst (ISCAS) 3:1343–1346
55. Ishida M, Higashimura M, Fukui Y, Ebisutani K (1988) Synthesis of immittance function using current conveyors. IEEE Int Symp Circ Syst (ISCAS) 3:2681–2684
56. Higashimura M, Fukui Y (1988) Novel method for realizing higher-order immittance function using current conveyors. IEEE Int Symp Circ Syst (ISCAS) 3:2677–2680
57. Dostal T, Pospisil J (1982) Current and voltage conveyors-A family of three-port immittance converters. In: Proceedings of ISCAS, Roma, pp 419–422
58. Salawu RI (1980) Realization of frequency dependent negative resistance. Microelectron Reliab 20:853–857
59. Svoboda JA (1989) Analyzing networks containing current conveyors. Int J Electron 67 (6):899–906
60. Horng JW, Lee MH, Hou CL (1995) Universal active filter using four OTAs and one CCII. Int J Electron 78:903–906
61. Soliman AM (1995) Theorem relating a class of op-amp and current conveyor circuits. Int J Electron 79:53–61
62. Celma S, Martinez PA, Sabadell J (1996) A Transformation method for equivalent infinite-gain Op-amp to unity-gain CCII networks. IEEE Trans Circ Syst-I 43:61–63
63. Wang Ke Chang, Aronhime P, Desai M (1996) New network theorems for current-mode circuit design. In: ISCAS'96 1, Atlanta, GA, pp 365–368
64. Sinsky JH, Westgate CR (1996) A new approach to designing active MMIC tuning elements using second-generation current conveyors. IEEE Microw Guid Wave Lett 6:326–328
65. Papazoglou CA, Karybakas CA (1998) A transformation to obtain CCII-based adjoint of op-amp-based circuits. IEEE Trans Circ Syst-II 45:894–898
66. Rathore TS, Khot UP (2012) Current conveyor equivalent circuits. Int J Eng Technol 4:1–7
67. Keskin AU, Toker A (2004) A NIC with impedance scaling properties using unity gain cells. Analog Integr Circ Sig Process 41:85–87
68. Hou AS (2004) A wide bandwidth isolation amplifier design using current conveyors. Analog Integr Circ Sig Process 40:31–38
69. Hua WG, Fukui Y, Kubota K, Watanabe K (2011) Voltage-mode to current-mode conversion by an extended dual transformation. IEEE Int Symp Circ Syst (ISCAS) 3:1833–1836
70. Natarajan S (1991) High frequency active compensation of integrators using CCII and their stability analysis. IEEE Int Symp Circ Syst (ISCAS), Singapore, 3:1432–1435
71. Pospisil J (1983) Resistive multiport network synthesis using arbitrary network matrix decomposition. ECCTD, Stuttgart, Germany, pp 1–6
72. Farag Fathi A (2009) Digitally programmable CMOS current-conveyor circuit for filter design. In: International symposium on signal circuits and systems, Lasi, pp 9–10
73. Tokmakci M, Alci M (2008) A CMOS membership function circuit employing single current differencing buffered amplifier. In: Informacije MIDEM, vol 38, Ljubljana, pp 140–143
74. Alarcon E, Romer IA, Poveda A, Porta S, Salamero LM (2004) CCII-based analog integrated circuit for sliding-mode control of switching power converters. Analog Integr Circ Sig Process 38:203–213
75. Singh D, Afzal N, Choudekar P, Yadav SK (2014) CMOS digitally programmable grounded inductor. In: International conference signal processing and integrated networks (SPIN), Noida, pp 492–496

76. Vrba K, Cajka J (2002) RC oscillators. In: Electron devices system Brno, Czech Republic, pp 243–245
77. Petchakit W, Petchakit S (2005) New floating capacitance multipliers. In: Proceedings of the 28th electrical engineering conference (EECON-28), Phuket, Thailand, pp 1233–1236
78. Michal V, Premont C, Pillonnet G, Abouchi N (2010) Single active element PID controllers. In: 20th international conference radioelektronika (Radioelektronika), Brno, pp 1–4, 19–21 Apr 2010
79. Abuelma'atti MT, Alsuhaibani ES, Obadi ASB, Khalifa ZJ (2013) Independent control of the frequency and condition of oscillation: a caution. Int J Electron 100:384–392
80. Herencsar N, Lahiri A, Vrba K, Koton J (2012) An electronically tunable current-mode quadrature oscillator using PCAs. Int J Electron 99:609–621
81. Metin B, Pal K, Cicekoglu O (2012) A new approach for high-input impedance in voltage mode filters using first-generation current conveyor in place of second-generation current conveyor. Int J Electron 99:131–139
82. Jaikla W, Siripruchyanun M, Biolek D, Biolkova V (2010) High-output-impedance current mode multiphase sinusoidal oscillator employing current differencing transconductance amplifier-based all pass filters. Int J Electron 97:811–826
83. Pal K, Psychalinos C (2010) All-pass filters realized using the current-controlled CCII with intrinsic negative resistance. Int J Electron 97:491–498
84. Al-Absi MA (2009) A novel highly accurate current mirror. Int J Electron 96:781–786
85. Herencsar N, Koton J, Vrba K, Lattenberg I (2010) New voltage-mode universal filter and sinusoidal oscillator using only single DBTA. Int J Electron 97:365–379
86. Sharma S, Gupta R (2012) Voltage controlled current conveyor and its application. In: Proceedings of the international conferences on advances in computer science and electronics engineering, Delhi, India, pp 389–393
87. Chun-Hua C, Zhi-Xing S, Chang-yong Y, Yan Z (2008) Analytical synthesis method of the digitally programmable voltage mode universal biquad based on current conveyors. J Commun 29:129–134
88. Vrba K, Cajka J, Zeman V (1998) N-th order filters using balanced-output CCII+/−conveyors. Radioengineering 7:16–19
89. Salema SB, Fakhfakh A, Loulou M, Loumeau P, Masmoudi N (2004) A 2.5 V 0.35 μm CMOS current conveyor and high frequency high-Q band pass filter. In: Proceedings of the 16th international conference on microelectronics, Tunis, Tunisia, pp 328–333
90. Biolek D, Vrba K, Cajka J, Dostal T (2000) General three-port current conveyor: a useful tool for network design. J Electron Eng 51:36–39
91. Vrba K, Cajka J, Zeman V (1996) Floating RC networks using current conveyors. Radioengineering 5:8–11
92. Minaei S, Kaymak D, Ibrahim MA, Kuntman H (2002) New CMOS configuration for current-controlled conveyors (CCCIIs). In: Proceedings of IEEE ICCSC, St Petersburg, Russia, pp 62–65
93. Jongchanachavawat W, Kumngern M, Dejhan K (2007) High-value floating controlled resistor. In: Proceedings of the 2007 ECTI international conference, Thailand, pp 165–167
94. Jaikla W, Siripruchyanun M (2007) Low-component electronically controllable dual-mode universal biquad filter using DO-CCCIIs. In: Proceedings of Asia-Pacific conference on communications, Bangkok, pp 331–334
95. Jaikla W, Siripruchyanun M (2007) A current-mode alternative topology for realization of universal biquad filter. In: IEEE international symposium on integrated circuits (ISIC-2007), Singapore, pp 616–619
96. Minaei S, Yuce E (2011) T0-CCII based voltage-mode universal biquadratic filter. In: IEEE CCECE, Niagara Falls, Canada, pp 000520–000523
97. Srisakultiew S, Siripruchyanun M (2013) A synthesis of electronically controllable current mode PI, PD and PID controllers employing CCCDBAs. Circ Syst 4:287–292

98. Ayten UE, Sagbas M, Sedef H (2012) Electronically tunable sinusoidal oscillator circuit with current and voltage outputs. Int J Electron 99:1133–1144
99. Yesil A, Kacar F, Gurkan K (2014) Lossless grounded inductance simulator employing single VDBA and its experimental band-pass filter application. Int J Electron Commun (AEU) 68:143–150
100. Soliman AM (2005) Low voltage high current gain CMOS digitally controlled fully differential CCII variable gain amplifier application example. ISCAS 2:1000–1003
101. Loizos DN, Sotiriadis PP, Cauwenberghs G (2008) A translinear SiGe BiCMOS current controlled oscillator with 80 Hz-800 MHz tuning range. Analog Integr Circ Sig Process 57:107–115
102. Sotner R, Jerabek J, Herencsar N, Hrubos Z, Dostal T, Vrba K (2012) Study of adjustable of gains for control of oscillation frequency and oscillation condition 3R-2C oscillator. Radioengineering 21:392–402
103. Siripruchyanun M, Phattanasak M, Jaikla W (2007) Temperature-insensitive current conveyor based floating simulator topology. In: International symposium on integrated circuits (ISIC 07), Singapore, pp 65–68
104. Tangsrirat W, Dumawipata T, Unhavanich S (2004) Simulation of electronically tunable lossless floating inductor using current controlled differential current voltage conveyors. In: International symposium on communications and information technology (ISCIT 2004), Sapporo, Japan, pp 39–42
105. Yakout MA, Abdelfattah AI, Elbazz AS (2000) BiCMOS current conveyors: design and application. IEEE Int Symp Circ Syst 3:463–466
106. Popovic J, Pavasovic A, Vasiljevic D (1997) Low-power high bandwidth CMOS current conveyor. Proceedings of the international conference on microelectronics, Yugoslavia, vol. 2, pp 693–696
107. Lu CW, Shen YC, Sheu ML (2007) A high-driving class –AB buffer amplifier with a new Pseudo source follower. In: IFIP International conference very large scale integration. (VLSI-SoC2007), Atlanta, GA, Canada, pp 105–109
108. Nishio M, Sato H, Suzuki (1985) A gyrator constructed by CCII with variable current transfer ratio. In: Proceedings of ISCAS, Kyoto, Japan, pp 93–96
109. Wilson B, Lidgey FJ, Toumazou C (1988) Current mode signal processing circuits. In: ISCAS, Helsinki, Finland, pp 2665–2667
110. Wilson B (1989) Performance analysis of current conveyors. Electron Lett 25:1596–1598
111. Guo-Hua W, Fukui Y, Kubota K, Watanabe K (2011) Voltage-mode to current-mode conversion by an extended dual transformation. In: ISCAS, Singapore, pp 1833–1836
112. Payne A, Toumazou C (1992) High frequency self-compensation of current-feedback devices. IEEE Int Symp Circ Syst 3:1376–1379
113. Bruun E, Olesen OH (1992) Conveyor implementations of generic current mode circuits. Int J Electron 73(1):129–140
114. Carlosena A, Moschytz GS (1994) Design of variable-gain current conveyors. IEEE Trans Circ Syst-I 41:79–81
115. Fabre A, Barthelemy H (1994) Composite second-generation current conveyor with reduced parasitic resistance. Electron Lett 30:377–378
116. Cabeza R, Carlosena A, Serrano L (1994) Unified approach to the implementations of universal active devices. Electron Lett 30:618–620
117. Toumazou C, Lidgey J (1994) Current-feedback Op-amps: a blessing in Disguise? IEEE Circ Dev Mag 10:34–37
118. Moschytz GS, Carlosena A (1994) A classification of current-mode single-amplifier biquads based on a voltage-to-current transformation. IEEE Trans Circ Syst-II 41:151–156
119. Fabre A (1994) New formulations to describe translinear mixed cells accurately. IEE Proc Circ Dev Syst 141:167–173
120. Hou CL, Shen B, Hwang J (1995) Novel applications of operational mirrored amplifiers with buffers. Chung Yuan J 23:133–139

121. Celma S, Sabadell J, Martinez P (1995) Universal filter using unity-gain cells. Electron Lett 31:1817–1819
122. Acar C (1996) Nth-order voltage transfer function synthesis using a commercially available active component: signal-flow graph approach. Electron Lett 32:1933–1934
123. Bruun E (1996) Noise properties of CMOS current conveyors. IEEE Int Symp Circ Syst 1:144–147
124. Zhang X, Kambayashi N, Shinada Y (1997) A realization of active current-mode resonator with complex coefficients using CCIIs. IEICE Trans Fundam E-80(2):413–415
125. Hyeon Lee C, Cornish J, McClellan K, Choma J (1997) Current-mode approach for wide gain bandwidth product architecture. IEEE Trans Circ Syst-II 45: 626–631
126. Fabre A, Alami M (1997) A precise macromodel for second generation current conveyors. IEEE Trans Circ Syst-I 44:639–642
127. Horng JW, Weng RM, Lee MH, Chang CW (1997) Universal active current filter using two multiple current output OTAs and one CCIII. Int J Electron 82:241–247
128. Wang HY, Lee CT (1998) Realization of R-L and C-D immittances using single FTFN. Electron Lett 34:502–503
129. Dostal T (1998) High frequency state-variable biquadratic active filters. Radioengineering 7:1–6
130. Abuelma'atti M, Alzaher HA (1998) Multi-function active-only current-mode filter with three inputs and one output. Int J Electron 85:431–435
131. Ismail AM, Soliman AM (2000) Novel CMOS current feedback Op-Amp realization suitable for high frequency applications. IEEE Trans Circ Syst-I 47:918–921
132. Cam U, Cicekoglu O, Kuntman H (2000) Universal series and parallel immittances simulators using four terminal floating nullors. Analog Integr Circ Sig Process 25:59–66
133. Emami S, Wada K, Tagaki S, Fujii N (2000) A novel class A CMOS current conveyor. IEEE Int Symp Circ Syst (ISCAS) 4:453–456
134. Ozcan S, Kuntman H, Cicekoglu O (2000) A novel multi-input single-output filter with reduced number of passive elements using single current conveyor. In: Proceedings of 43rd IEEE midwest symposium on circuits and systems, Lansing MI, vol 3, pp 1030–1032
135. Aronhime P, Wang K, Qian T (2001) Generalization of Theorem for replacing CCIIs by CFOAs in current-mode circuits. Analog Integr Circ Sig Process 28:27–33
136. Barthelemy H, Koudobine I, Landeghem DV (2001) Bipolar-low power operational transresistance amplifier based on first generation current conveyor. IEEE Trans Circ Syst-II 48:620–625
137. Rybin AI, Dostal T (2001) Nodal analysis of circuits containing current conveyors. Radioengineering 10:1–4
138. Cam U, Cicekoglu O, Kuntman H (2001) Novel lossless floating immittance simulator employing only two FTFNs. Analog Integr Circ Sig Process 29:233–235
139. Gift SJG (2001) Hybrid current conveyor-operational amplifier circuit. Int J Electron 88:1225–1235
140. Ozoguz S, Elwakil AS, Kennedy MP (2002) Experimental verification of the butterfly attractor in a modified Lorenz system. Int J Bifurcat Chaos 12:1627–1632
141. Elwakil AS, Ozoguz S, Kennedy MP (2002) Creation of a complex butterfly attractor using a novel Lorenz-type system. IEEE Trans Circ Syst-I 49:527–530
142. Mohsen A (2003) F-TRIA01: a fast bipolar transimpedance amplifier. Analog Integr Circ Sig Process 36:199–206
143. Abuelma'atti MT (2003) Active-only immittance simulators. Frequenz 57:8–11
144. Toker A, Ozoguz S (2003) Tunable all pass filter for low voltage operation. Electron Lett 39:175–176
145. Gupta SS, Senani R (2003) Realization of current-mode SRCOs using all grounded passive elements. Frequenz 57:26–37

146. Tangsrirat W, Unhavanich S, Dumawipata T, Surakampontorn W (2003) Single-input and three-output current-mode biquadratic filters using multiple-output OMAs. J KMITNB 13:12–16
147. Cakir C, Cam U, Cicekoglu O (2005) Novel all pass filter configuration employing single OTRA. IEEE Trans Circ Syst-II 52:122–125
148. Kaabi H, Motlagh MRJ, Ayatollahi A (2005) A novel current-conveyor-based switched-capacitor integrator. IEEE Int Symp Circ Syst 2:1406–1408
149. Keskin AU (2005) Voltage-mode notch filters using single CDBA. Frequenz 59:225–228
150. Wisetphanichkij S, Binthawihok A, Dejhan K (2005) A new adjustable grounded/floating resistance circuit with enhanced dynamic range based on mixed-translinear loops. IEEE Int Symp Commun Inform Technol (ISCIT) 1:267–270
151. Klinc S, Jain V, Aggarwal V, Cam U (2006) Catalogue of variable frequency and single-resistance-controlled oscillators employing a single differential difference complementary current conveyor. Frequenz 60:142–146
152. Tsukutani T, Edasaki S, Sumi Y, Fukul Y (2006) Current-mode universal biquad filter using OTAs and DO-CCII. Frequenz 60:237–240
153. Fongsamut C, Anuntahirunrat K, Kumwachara K, Surakampontorn W (2006) Current-conveyor-based single-element-controlled and current-controlled sinusoidal oscillators. Int J Electron 93:467–478
154. Godara B, Fabre A (2007) Versatile wideband impedance matching circuit based on current conveyors. Electron Lett 43:37–38
155. Herencsar N, Vrba K (2007) Current conveyors-based circuits using novel transformation method. IEICE Electron Express 4:650–656
156. Yuce E (2007) On the implementation of the floating simulators employing a single active device. Int J Electron Commun (AEU) 61:453–458
157. Jamuar SS, Sharma S, Rajput SS (2008) Analog signal processing using FGMOS based structures: a tutorial. J Active Passive Electron Dev 3:109–124
158. Prasad D, Bhaskar DR, Singh AK (2008) Realization of single-resistance-controlled sinusoidal oscillator: a new application of the CDTA. WSEAS Trans Electron 5:257–259
159. Chauhan W, Haiguang L, Yan Z (2009) Universal current-mode filter with multiple inputs and one output using MOCCCII and MO-CCCA. Int J Electron Commun (AEU) 63:448–453
160. Soliman AM (2009) Bode-type equalizers using current conveyors. J Circ Syst Comput 3:433–442
161. Tsukutani T, Sumi Y, Higashimura M, Fukui Y (2005) Current-mode universal biquad circuit using MO-OTAs and DO-CCII. IEEE Int Symp Circ Syst 2:1589–1592
162. Gupta SS, Senani R (2005) Grounded-capacitor SRCOs using a single differential difference complimentary current feedback amplifier. IEE Proc Circ Dev Syst 152:38–48
163. Shah NA, Rather MF, Iqbal SZ (2006) SIFO electronically tunable current-mode cascadable active only universal filter. J Active Passive Electron Dev 1:327–334
164. Khan AA, Bimal S, Dey KK, Roy SS (2010) A simple methodology for sinusoidal oscillator design based on simulation of differential equation using AD844 configured as second generation current conveyor. Ind J Sci Tech 3:684–686
165. Yasin MY, Gopal B (2011) High frequency oscillator design using a single 45 nm CMOS current controlled current conveyor (CCCII+) with minimum passive components. Circ Syst 2:53–59
166. Biolek D, Bajer J, Biolkova V, Kolka Z, Kubicek M (2011) Z copy controlled Gain current differencing buffered amplifier and its applications. Int J Circ Theor Appl 39:257–274
167. Khan AA, Turaigi MA, El-Ela MA (1994) Operational floating current Conveyor: characteristics, modeling and applications. In: The eleventh conference on microelectron, Hamamatsu, pp 59–62
168. Hasan S, Khan IA (2009) Multi-scroll chaos generator using current conveyors. IMPACT 2009:229–232

169. Schmid H, Moschytz GS (1997) Tunable CCII-MOSFET-C filter biquads for video frequencies. In: Proceedings of ECCTD 97, Budapest, Hungary, vol 1, pp 82–87
170. Ibrahim MA (2009) The inverting second-generation current conveyor in implementation of transimpedance-mode Tow-Thomas filter. In: International conference on communication computer and power (ICCCP'09), Oman, pp 179–182
171. Tangsrirat W (2007) Current differencing transconductance amplifier-based current-mode four-phase quadrature oscillator. Indian J Eng Mater Sci 14:289–294
172. Alpaslan H, Yuce E (2013) New grounded inductor simulator using unity gain cells. Indian J Pure Appl Phys 51:651–656
173. Tangsrirat W, Mongkolwai P, Pukkalanun T (2012) Current-mode high-Q band pass filter and mixed-mode quadrature oscillator using ZC-CFTAs and grounded capacitors. Indian J Pure Appl Phys 50:600–607
174. Tangsrirat W, Bunruang K (2012) Current-mode universal filter topology with electronic tuning property using single current follower. Indian J Pure Appl Phys 50:133–137
175. Lahiri A (2011) Low-frequency quadrature sinusoidal oscillates using differencing buffered amplifiers. Indian J Pure Appl Phys 49:423–428
176. Satansup J, Tangsrirat W (2011) Realization of current-mode KHN-equivalent biquad filter using ZC-CFTAs and grounded capacitors. Indian J Pure Appl Phys 49:841–846
177. Sharma S, Rana S, Pal K (2010) Realization of variable Q band pass filter using low-voltage CCII. Indian J Pure Appl Phys 48:600–602
178. Sharma S, Rana S, Pal K (2010) Realization of high Q notch/all pass filter using low-voltage current conveyor. Indian J Pure Appl Phys 48:435–437
179. Tangsrirat W, Tanjaroen W (2010) Current-mode sinusoidal quadrature oscillator with independent control of oscillation frequency and condition using CDTAs. Indian J Pure Appl Phys 48:363–366
180. Sharma S, Pal K, Rajput SS, Mangotra LK, Jamuar SS (2009) Low-voltage variable current gain CCII based all-pass/notch filter. Indian J Pure Appl Phys 47:149–152
181. Tangsrirat W (2009) Novel minimum-component universal filter and quadrature oscillator with electronic tuning property based on CCCDBAs. Indian J Pure Appl Phys 47:815–822
182. Shah NA, Quadri M, Iqbal SZ (2008) High output impedance current-mode all pass inverse filters using CDTA. Indian J Pure Appl Phys 46:893–896
183. Shah NA, Rather MF (2006) Realization of voltage-mode CCII-based all pass filter and its inverse version. Indian J Pure Appl Phys 44:269–271
184. Sharma S, Rajput SS, Pal K, Mangotra LK, Jamuar SS (2006) Low-voltage CCII based all-pass/notch filter. Indian J Pure Appl Phys 44:871–874
185. Soliman AM (2009) Adjoint network theorem and floating element in the NAM. J Circ Syst Comput 18:597–616
186. Gift SJG (2004) The operational conveyor and its application in a novel band-pass filter. West Indian J Eng 26:35–44
187. Alaybeyoglu E, Guney A, Altun M, Kuntman H (2014) Design of positive feedback driven current-mode amplifiers Z-copy CDBA and CDTA, and filter applications. Analog Integr Circ Sig Process (in press, June 2014)
188. Dostal T, Biolek D, Vrba K (2002) Adjoint voltage–current mode transformation for circuits based on modern current conveyors. In: Fourth IEEE international Caracas conference on devices circuits and systems, Aruba, pp T034-1–T034-4
189. Ercan H, Tekin SA, Alci M (2012) Voltage-and current-controlled high CMRR instrumentation amplifier using CMOS current conveyors. Turk J Elec Eng Comp Sci 20:547–555
190. Mohan J, Maheshwari S (2013) Two active elements based all-pass section suited for current –mode cascading. Int J Electr Robot Electron Commun Eng 7:52–55
191. Un M, Kacar F (2008) Third generation current conveyor based current-mode first order allpass filter and quadrature oscillator. J Electr Electron Eng 8:529–535
192. Khan IA, Afzal N, Khan MR (2009) Digitally programmable impedance multiplier using CCIIs with high resolution capability. J Active Passive Electron Dev 8:247–257

193. Khan AA, Bimal S, Dey KK, Roy SS (2002) Current conveyor based R-and C-multiplier circuits. Int J Electron 56:312–316
194. Azhari SJ, Kaabi H (2000) AZKA cell, the current-mode alternative of Wheatstone bridge. IEEE Trans Circ Syst-I 47:1277–1284
195. Nandi R (1977) Comment on 'Precise noninverting operator realization with high-resistive input impedance. Proc IEEE 65:1206–1207
196. Ozoguz S, Toker A, Cicekoglu O (1999) New current mode universal filter using only four (CCII+)s. Microelectron J 30:255–258
197. Higashimura M, Fukui Y (1998) Universal cascadable current mode filter using four CCIIs with grounded passive elements. In: Proceedings of the IEEE Asia pacific conference on circuits and systems, Chiangmai, pp 259–262
198. Celma S, Sabadell J, Martinez P (1995) Universal filter using both current and voltage follower. In: Proceedings of 20th international conference on microelectronics, NIS, vol 2, pp 781–784
199. Abuelma'atti MT (1992) New current mode active filters employing first generation current conveyor. In: Proceedings of 35th midwest symposium on circuits and systems, vol 1, pp 310–313
200. Biolek D, Siripruchyanun M, Jaikla W (2008) CCII and OTA based current mode universal biquadratic filter. In: The 6th PSU engineering conference, Songkla, Thailand, pp 238–241
201. Tsukutani T, Higashimura M, Sumi Y, Fukui Y (2004) Current mode universal biquadratic circuit using OTAs and DO-CCII. In: Proceedings of the 47th IEEE international midwest symposium on circuits and systems, Hiroshima, Japan, pp I-45–I-48
202. Summart S, Thongsopa C, Jaikla W (2014) CCCIIs-based sinusoidal quadrature oscillators with non-interactive control of condition and frequency. Indian J Pure Appl Phys 52:277–283
203. Tangsrirat W (2013) Floating simulator with a single DVCCTA. Ind J Eng Mater Sci 20:79–86
204. Dong-Sheng Yu, Yan L, Iu HHC, Yi-Hua H (2014) Mutator for transferring memresistor emulator into meminductive and memcapacitive circuits. Chin Phys B 23:070702-1-070702
205. Yesil A, Kacar F (2013) Electronically tunable resistorless mixed mode biquad filters. Radioengineering 22(4):1016–1025

Index

A
Active filter, 13, 137, 138, 145, 146, 156, 172, 173, 177–181, 277
Active gyrator, 101–103, 193
All pass (AP), 10, 138–142, 144, 146, 150, 151, 156, 157, 166, 167, 181, 204, 275, 276, 280, 282, 388, 389, 421, 422
All pole filters, 285, 409, 410
Amplifier, 3, 7–12, 20, 26, 27, 29, 39, 49–53, 72, 83–90, 100, 113, 128, 151, 154, 182, 207, 209, 221, 222, 229, 247, 248, 254, 262–265, 274, 288, 298, 316–318, 320, 332, 347, 368, 369, 499–502, 524, 531, 536, 537
Analog arrays, 536
Analog circuits, 3, 4, 13, 59, 83–129, 234, 288, 345, 348, 358, 419, 536
Analog switch, 232–234, 511, 513
AP. *See* All pass (AP)
Attractor, 245, 247

B
Bacterial foraging optimization, 534–535
Band elimination, 85
Band pass (BP), 71, 144, 145, 148, 170, 280, 402, 467, 480, 483, 486, 536
Band stop (BS), 421
BiCMOS, 13, 77, 254–258, 301, 302, 346, 367, 538
Biquad, 9, 12, 92, 141–173, 181, 253, 254, 272–282, 319, 320, 328, 334, 336, 353–355, 367–387, 389–409, 412–415, 417–435
Biquadratic filter, 24, 150, 151, 154, 162, 370–372, 375, 382, 390, 395, 401
Bi-stable multivibrator, 349, 509–511
BP. *See* Band pass (BP)
Bruton's transformation, 283
BS. *See* Band stop (BS)
Building block, 3, 5–13, 47, 53, 55, 83–129, 177, 225, 230, 243, 253, 263, 274, 280, 281, 283, 290, 299, 304, 312, 313, 320, 323, 324, 326, 328, 329, 334, 335, 338, 341, 345–358, 367, 395, 406, 416, 417, 448, 451–453, 459, 461, 470, 491, 493, 497, 519, 521, 530, 534, 537, 538
Bulk-driven MOSFET, 74
Butterworth, 102, 169, 173, 174, 176, 286, 467, 474, 476, 483, 536

C
Canonical, 103, 191, 389
Capacitance multiplier, 105, 270, 271, 318, 487–490
Capacitance simulator, 270
Center frequency, 148, 486
CF. *See* Current follower (CF)
CFOA. *See* Current feedback operational amplifier (CFOA)
Channel length modulation, 70
Chaos, 247, 520
Chaos generator, 243, 519–520, 524
Chaotic oscillator, 83, 243–247, 519
Characteristic equation, 197, 425, 448–450, 453, 454, 456, 457, 460
Chebyshev filter, 411, 479

© Springer International Publishing Switzerland 2015
R. Senani et al., *Current Conveyors*, DOI 10.1007/978-3-319-08684-2

Chua diode, 245, 519
Chua's circuit, 245
Circuit analysis, 64, 151, 152, 154–156, 164, 167, 172, 369, 370, 372, 374, 376, 378, 380, 382, 384, 388, 389, 394, 397, 402, 403, 407, 409, 410, 412, 414, 417, 418, 425, 427, 430, 431, 434, 460, 469, 475, 479, 480, 483, 485, 487, 488, 490, 491
CM. *See* Current mirror (CM); Current-mode (CM)
CM filter. *See* Current-mode (CM) filter
CMOS
  architecture, 10, 12, 63, 254
  implementation, 10, 13, 59–77, 232, 315, 316, 319, 321, 323–325, 327, 333, 334, 336, 338, 339, 465, 488, 493, 497, 503, 532, 534, 538
  invertors, 60, 67–68
  realization, 10, 59, 61, 66, 192, 256, 320–323, 328, 336, 383, 474
  resistors, 449
  technology, 59, 66, 73, 227, 234, 265, 303, 341, 367, 369, 374, 375, 377, 402, 427, 447, 449, 454, 460, 461, 468, 476, 479, 484, 488, 490, 500, 505, 534, 536, 538
Common mode rejection ratio (CMRR), 24, 25, 87, 265, 391, 500
Compensated NIC, 476–477
Complementary outputs, 319
Component matching, 3, 9, 84, 100, 105–107, 118, 129, 155, 161, 166, 171–173, 243, 276, 278, 279, 376, 378, 383, 390, 393, 400, 405, 417, 422, 424, 435, 465, 468, 469, 474, 479–480, 484–486, 499
Constant bandwidth, 84–90, 417
Constant current source, 4, 5, 22, 24, 69
CR. *See* Current repeater (CR)
Current amplifiers, 7, 9, 12, 39, 88, 89, 262, 288, 316, 332
Current controlled resistance, 98, 258, 271, 297
Current conveyor, 3–13, 17, 33, 59–77, 83–182, 192, 218, 253–304, 311–341, 345, 367, 447–431, 465–493, 497–525, 529–530
Current differencing, 339, 356, 491
Current feedback operational amplifier (CFOA), 53, 205, 243, 263, 290, 416, 519, 520, 537
Current follower (CF), 18, 73, 76, 232, 281, 356, 415, 421
Current mirror (CM), 10, 18, 19, 22, 25, 26, 28, 29, 33–36, 38–41, 43–47, 49, 50, 59–61, 65, 72–74, 89, 139, 146, 166, 182, 200, 219, 237, 254, 255, 287, 288, 299, 301, 313, 317–319, 321, 334, 338, 339, 346, 412, 453, 517, 530, 532, 534
Current mode (CM), 3, 9, 10, 12, 13, 24, 52, 64, 83, 88–94, 99, 111, 137–139, 144, 161, 209, 210, 230, 253, 274–280, 282, 283, 285, 292–295, 318, 319, 347, 350, 352, 353, 357, 358, 367, 420, 422, 448, 498–500, 503, 504, 520, 537
  filter, 161, 163, 168, 169, 173, 181, 209, 274, 320, 347, 374, 403, 412, 432, 435
  multiplier, 99, 294–295, 347, 352, 353, 357
  SRCO, 452–453
Current repeater (CR), 18, 33, 44, 47, 59, 61, 122, 144, 169, 196, 254, 255, 257, 299, 313, 334, 468, 469, 473, 517
Current source, 4, 5, 24, 39–41, 47, 61, 64, 65, 69, 72, 100, 219, 221, 248, 281, 316, 329, 433
Cut off frequency, 24, 176, 285, 474, 476, 479, 483, 534

D

DC biasing, 50, 73, 75, 206, 246, 257, 292, 455
Design constraints, 140, 141, 197
Differential current amplifier, 39
Differential difference amplifier, 11, 316
Differential mirror,
Differential voltage controlled current source (DVCCS), 21, 109, 120
Differentiators, 7, 83, 90–92, 128, 241, 248
Digital to analog convertor, 3, 519
Divider, 3, 12, 99, 225–230, 238, 240, 254, 294–295, 335, 347, 350, 352–354, 357, 358
Double-pole double-throw (DPDT) switch, 509
Dual mode, 280, 368, 424, 425, 434, 448, 457, 459–461
DVCCS. *See* Differential voltage controlled current source (DVCCS)

E

ECO. *See* Explicit current output (ECO)
Electronically controllable, 20, 24, 40, 97, 121, 125, 253, 262, 264, 270–274, 285, 292, 304, 323, 337, 347, 355, 409, 421, 459, 470–472, 479, 481, 500, 538
Electronically tunable, 39, 97, 120, 254, 259, 261, 265–272, 328, 393, 405, 412, 493, 532

Electronic control, 44, 55, 125, 148, 200, 253, 322, 493
Electronic resistor, 381, 382
Equivalent resistance, 34, 76, 103, 106, 259, 264, 328
Explicit current output (ECO), 209–210, 213, 427, 451–453

F
Fabrication, 195, 205, 236
FAC. *See* Floating admittance convertor (FAC)
FDNC. *See* Frequency dependent negative conductance (FDNC)
FDNR. *See* Frequency dependent negative resistance (FDNR)
Feedback, 12, 19, 20, 25, 53, 64, 66, 69, 76, 86, 192, 201, 238, 315, 345, 349, 506, 509, 511, 515, 517, 519, 537
FET, 148, 163, 194, 200
FI. *See* Floating immittances (FI)
Floating admittance convertor (FAC), 474–475
Floating capacitor, 102, 110, 139, 285, 406, 466, 479, 484
Floating gate-MOS (FG-MOS), 534
Floating immittances (FI), 8, 118, 313, 465, 474–476, 491
Floating impedance, 96, 97, 100, 105–111, 113, 118, 120, 122, 128, 129, 313, 348, 465, 474, 479, 493
Floating inductor, 8, 100, 107, 108, 110–112, 114–120, 267, 269, 320, 350, 466, 479, 492
Floating negative impedance, 95–98, 320, 321
Floating negative inductance, 270
Floating simulator, 479, 480, 484, 485
Four terminal floating nullor (FTFN), 96, 347–349
Frequency dependent negative conductance (FDNC), 9, 53, 83, 104, 111, 123, 520–521
Frequency dependent negative resistance (FDNR), 9, 83, 99–128, 177, 193, 194, 247, 269–271, 283, 284, 320, 466–468, 474, 475, 482–484, 486, 492, 520
Frequency doubler, 223–225, 254, 292, 293
Frequency response, 21, 72, 167, 174, 285, 401, 402, 474, 480, 501
Frequency-shift keying (FSK) generator, 298
Frequency stability, 200, 201
FTFN. *See* Four terminal floating nullor (FTFN)
Full wave rectifier, 217–219, 221–225, 292, 293, 502–506
Fully differential, 10, 12, 323–324, 329, 395, 434–435, 453, 502, 534, 536
Fully-uncoupled, 210–213, 289, 449–451
Function generator, 7, 83, 98–99, 511
Fuzzy function, 230–232

G
Gain bandwidth conflict, 84–86, 263
Grounded capacitors (GC), 9, 99, 100, 103, 106, 110, 111, 120, 139, 140, 148, 153, 165, 166, 169, 173, 177, 180, 195–200, 203, 210, 213, 245, 266, 268, 271, 279, 283–289, 318, 350, 355, 382, 384, 386, 390, 392, 393, 395, 397, 400, 402, 405, 409–410, 415, 425, 431, 433, 435, 448–449, 452–453, 456, 461, 466, 468, 469, 471, 477, 479–481, 484, 487–492, 520
Grounded inductance, 100–104, 268, 269, 347, 356, 473, 483–488, 491

H
Hardware, 59, 158, 388, 410, 434, 491, 520, 535, 537
Hardware implementation, 4, 13, 17–30, 179, 311, 367, 513, 530–535, 537
Higher order filters, 137–182, 283, 385, 406, 409–410
High pass (HP), 102, 144, 148, 170, 280, 402, 474, 483
Hybrid matrix, 6, 35, 109, 312, 315, 324, 326, 368, 388, 393, 413, 415, 472
Hyperbolic, 531
Hysteresis, 240, 243, 509

I
IC implementation, 49, 59, 148, 155, 164, 169, 196, 257, 296, 345, 374, 466, 505
Immittance, 8, 113, 115, 118, 123, 137, 177, 312–314, 349, 465, 467–470, 475–476
Impedance converter, 7, 83, 93–98, 107–114, 128, 312, 320, 321, 476–477, 520
Impedance inverter, 7, 10, 320, 321
Impedance matching, 138, 517, 518
Impedance synthesis, 83–129
Input impedance, 19, 21, 25, 38, 43–46, 49, 55, 64, 73, 91, 92, 101, 102, 111, 125, 127, 138, 140, 145, 258, 263, 268, 279, 281, 296, 299, 300, 313, 316, 337, 349, 368, 376, 377, 382, 384, 388, 391, 393, 395, 406, 409, 413, 471, 477, 482, 483, 485, 486, 503, 505

Instrumentation amplifiers, 83, 87–89, 128, 209, 254, 263–265, 320, 368, 369, 499, 500, 531
Integrators, 12, 83, 90–92, 128, 175–176, 178, 205–207, 210, 242, 248, 322, 336, 400, 412, 435
Inverting amplifier, 50, 51, 85, 86, 88, 151, 206, 207, 229, 274
Inverting integrator, 90, 206

K

Kerwin–Huelsman–Newcomb (KHN) biquad, 146–148, 274, 275, 380, 384, 385, 391, 392, 407, 412

L

Ladder filter, 175, 177, 283–285, 319, 409–411
LNA. *See* Low noise amplifier (LNA)
Logic circuits, 536
Low component count, 103–104, 154, 452
Low noise amplifier (LNA), 500–502, 524
Low pass (LP), 103, 144, 148, 156, 170, 175, 280, 285, 402, 409–411, 479, 536

M

Matrix equation, 11, 124, 271, 323, 326, 335, 356, 400, 412, 488, 491
Memcapacitance, 521–524
Meminductance, 521–524
Memristor, 521–523
Mixed mode, 10, 137, 144, 236, 280–283, 324, 368, 384, 385, 392, 393, 398, 431–433
Mixed-mode biquad, 254, 282, 392–393, 397–399, 431–434
Mixed translinear cell (MTC), 34, 36, 44, 53, 87, 253, 254, 258, 296, 299, 338, 345–358, 530, 531
Model parameters, 43, 375, 427, 451, 467, 468, 470, 471, 483, 486, 488, 490
Monolithic integrated circuit, 39, 153–154
Monostable multivibrator, 511–513
MOSFET, 9, 11, 61, 64–66, 68–70, 72–74, 76, 125, 127, 128, 148, 175–176, 200, 225–227, 229, 231, 232, 234, 264, 265, 290–291, 296, 317, 318, 328, 329, 331, 336, 382, 386, 399–400, 405–407, 410, 412, 429, 453–455, 458, 459, 470, 471, 481, 482, 493, 505, 532
MOSFET-based gyrator, 125, 470
MOSFET-C FDNR, 482–483
MOSFET-C filter, 175, 400, 470
MOS resistive circuit (MRC), 175, 225, 227
MOS transistor, 65, 231, 290, 483, 486
MTC. *See* Mixed translinear cell (MTC)
Multifunction, 138, 272–274, 280–283, 369, 370, 376, 377, 497
Multifunction biquad, 144–173
Multi-phase, 192, 203–209
Multiplication, 12, 294, 353
Multiplier, 3, 9, 10, 12, 99, 105, 225, 226, 254, 270, 271, 294, 318, 324, 335, 347, 350, 352–354, 357, 475, 487–490, 500, 502
Multivibrator, 349, 506–516, 524
Mutator, 7, 8, 39, 105, 520–522, 524, 530
Mutually coupled circuit, 124–125

N

NAM. *See* Nodal admittance matrix (NAM)
Negative impedance converter (NIC), 7, 8, 83, 93–97, 118, 125, 126, 128, 312, 320, 321, 476–477
Negative inductors, 266, 352, 473
Negative resistance, 83, 97, 258–262, 473
Network synthesis, 520, 530
NIC. *See* Negative impedance converter (NIC)
Nodal admittance matrix (NAM), 449
Non interacting control, 163, 289, 452
Non-inverting amplifier, 50, 52, 84–88, 368
Non-inverting integrator, 90, 175, 205, 206, 242
Non-linear network, 8, 520–521
Norator, 97, 121, 320, 321, 529, 530
Notch, 143–146, 149–152, 154, 155, 157–164, 166, 167, 170, 171, 173, 275, 276, 280, 282, 370, 372–374, 377–383, 385, 387, 391, 400, 402–407, 412, 417, 419, 422, 423, 427, 429, 430, 432, 434, 435
npn mirror, 19, 34
Nullator, 97, 121, 320, 321, 529, 530
Nullor, 96–98, 114–116, 118, 121, 122, 347–349, 492, 529, 530

O

Offset cancelling, 61, 68–70
Offset voltage, 36, 66, 196, 219, 221, 223
OMA. *See* Operational mirrored amplifier (OMA)
Op-amp, 3, 7, 8, 11, 12, 17–22, 24–29, 53, 83–86, 88, 89, 96, 99, 100, 103, 105, 114, 120–121, 128, 129, 139, 145–147, 166, 167, 177, 182, 191, 192, 196, 197, 200, 213, 217, 227, 237, 243, 258, 263, 290, 296, 332, 335, 348, 349, 367, 407, 417, 447, 465, 497, 504, 511, 516, 534, 537, 538

Operational floating conveyor, 9, 89, 313–315, 416
Operational mirrored amplifier (OMA), 25–30, 49, 113, 145, 167, 182, 236
Operational transconductance amplifier (OTA), 8, 17, 20–25, 29, 41, 53, 55, 74–75, 120–121, 177, 192, 303, 312, 348, 353, 354, 357, 479, 480, 492, 516, 534, 537, 538
Orthogonal tunability, 155, 167
Oscillator, 3, 71, 83, 137, 191–213, 217, 253, 318, 347, 424, 447–461, 506–516, 530
OTA. *See* Operational transconductance amplifier (OTA)
Output impedance, 19, 25, 38, 44, 49, 52, 66, 92, 138, 140, 182, 209, 222, 223, 262, 263, 276, 279, 281, 299, 315, 353, 381, 385, 390, 405, 412, 413, 448, 453, 454, 461, 503, 505, 517

P
Parasitic capacitance, 26, 87, 88, 195, 196, 210, 349, 406, 452, 505
Parasitic impedance, 71, 76, 87, 93, 111, 203, 210, 213, 466, 476–479
Parasitic resistance, 12, 42–44, 68, 88, 239, 266, 302, 377, 460, 480
Particle swarm optimization, 535
Pathological element, 10, 320, 321, 449, 529, 530
Phase difference, 203–206, 456–458
PID control, 290
PID controller, 290, 291, 497–500, 524
Piece-wise linear, 231, 245, 247
pnp mirror, 18, 27, 34, 47
Positive feedback, 19, 506, 509, 511, 515
Power supply current sensing technique, 25, 30, 49, 139, 227, 337
Pseudo exponential, 234–236
Pulse, 239, 240, 248, 511, 519

Q
Quadrature, 192, 203–209, 213, 288–290, 298, 323, 424, 425, 454–460
Quadrature signals, 458

R
Rail-to-rail, 60, 65, 72–74, 502, 534
Ramp, 506, 507, 515, 516
Rectifiers, 3, 209, 217–225, 254, 290–294, 502–506, 524
Reflector, 7, 8, 520–521
Relaxation oscillator, 83, 243, 248, 253, 320, 506–516, 524
Rotator, 7, 8, 520–521

S
Sample and hold, 517–519, 524
Saturation level, 506, 507, 509, 511, 515
Scalar, 8, 105, 520–521
Schmitt trigger, 175, 237–243, 349
Simulated inductor, 83, 101, 162, 163, 177, 243, 245, 352, 467–471, 474, 479, 480, 482, 486, 491, 493, 519
Single resistance controlled oscillator (SRCO), 191–203, 209–210, 213, 348, 349, 424–425, 447–454, 458, 508, 530
SPICE, 43, 64, 70, 76, 139, 219, 232, 234, 246, 265, 278, 281, 286, 288, 289, 298, 302, 369, 401, 402, 405, 407, 410, 424, 427, 449, 451, 453–456, 458, 461, 468, 470, 476–478, 480, 483, 484, 486, 487, 490, 505, 513, 538
Squarer, 9, 12, 125, 128, 225–230, 254, 295, 296
Square rooting, 228, 229, 295–299
Square wave, 237, 239–243, 253, 506–509, 515–516, 519, 524
SRCO. *See* Single resistance controlled oscillator (SRCO)
Steady state, 459–461
Summers, 7, 12, 83, 92–94, 254, 335, 412, 435
Synthetic transformer, 271, 272

T
TAM. *See* Transadmittance mode (TAM)
Temperature compensation, 5, 41–43
Temperature-sensitive, 4
Threshold voltage (V_T), 4, 41, 74, 87, 239, 244, 258, 259, 262, 264, 353, 405, 412, 416, 480, 509–511, 517
TIM. *See* Transimpedance mode (TIM)
Topology, 33, 49, 64, 66, 67, 70, 72, 87, 89, 137, 144, 146, 151, 191, 204, 221, 241, 274, 289, 346, 349, 368, 369, 371, 372, 392, 395, 418, 431, 432, 447, 448, 456, 461, 478, 484, 502, 532, 533, 538
Tracking errors, 101, 169, 458
Transadmittance mode (TAM), 144, 385, 393, 431
Transconductance, 24, 41, 53, 66, 67, 72, 83, 120, 226, 231, 303, 337, 394, 397, 412, 432, 451, 460
Transconductance amplifier, 316, 345, 353–358, 393

Transconductors, 10, 60, 65–66, 68, 125–129, 303, 320–322, 324, 337, 534
Transfer characteristics, 61, 65, 221, 232, 238–240, 369, 509, 510
Transfer function, 85, 90, 138, 140, 141, 143–145, 147, 148, 150, 154, 161–164, 167, 170, 173, 179–181, 207, 275, 277, 279, 332, 369, 372, 376, 380, 381, 384–386, 388, 389, 394–400, 403–405, 407, 410, 414–416, 418, 421–423, 429, 430, 432, 433, 435
Transformation, 8, 122, 144, 169, 175–178, 196, 283, 468, 469
Transimpedance mode (TIM), 144, 386, 393, 431
Transistor array, 8, 9, 17, 26, 28, 29, 40, 44, 46, 47, 61, 86, 89, 145, 166, 167, 200, 213, 227, 229, 299, 367, 447, 465, 480, 491, 538
Translinear, 7, 9, 34–36, 41, 43, 46, 47, 50, 53, 71, 87, 89, 221, 223, 237, 254–256, 258, 276, 287, 296, 299, 300, 303, 313, 315, 415, 416, 517, 530–533, 537, 799
  principle, 46, 47, 532
  voltage follower, 54, 537
Triangular wave, 239, 242, 506–509, 515–516
Triangular wave form, 234, 240, 242, 508, 515, 516, 524
Trigger, 240, 511
Triode region, 127, 175, 194, 225, 227, 328, 407, 412, 481
Tunability, 24, 60, 118, 121, 138, 148, 169, 182, 265, 277, 279, 288, 345, 421, 423, 424, 427, 429, 466, 480, 492

U

Universal active filter, 145, 167, 173, 277
Universal biquad, 12, 144, 150–152, 154–156, 158, 159, 161–163, 166, 169, 181, 274–281, 319, 334, 355, 367, 368, 371, 377, 387, 390, 392, 393, 395, 399, 402, 406, 408, 409, 413, 415, 417–419, 425, 429, 431

V

VCR. *See* Voltage-controlled-resistance (VCR)
Vector summation, 229, 230
VF. *See* Voltage follower (VF)
VLSI, 3, 61, 341, 367
VM. *See* Voltage mirror (VM)
VM filter, 146, 154–157
Voltage buffer, 9, 50, 52, 53, 145, 175, 200, 268, 346, 350, 351, 431, 477
Voltage controlled impedance, 128, 129
Voltage controlled oscillator, 200, 335
Voltage-controlled-resistance (VCR), 125–129, 148, 200, 201, 493, 500
Voltage controlled resistor, 125, 163–164, 395, 481, 500
Voltage conveyors (VC), 8, 20, 311, 312, 339, 413, 502, 536, 538
Voltage follower (VF), 7, 27, 49, 50, 54, 55, 61, 66, 76, 90, 103, 128, 156, 157, 173, 175, 199, 200, 209, 221, 232, 337, 516, 517, 537
Voltage gain, 19, 63, 65, 68, 70, 72, 76, 86, 87, 145, 197, 199, 212, 254, 257, 262–265, 303, 337, 472, 501, 517
Voltage mirror (VM), 10, 12, 320, 321, 530
Voltage to current convertor, 20, 218, 225, 227, 240, 336
V_T. *See* Threshold voltage (V_T)

W

Waveform, 90, 169, 170, 175, 205, 207, 208, 218, 222, 237–243, 247, 298, 320, 455, 456, 460, 461, 506, 508, 512–516, 519, 524
Wave form generator, 237–243, 320, 508, 515, 516, 524
Wien bridge oscillator, 191, 192, 243, 244
Wilson current mirror, 19, 25, 28, 35, 36, 43, 47, 299